Initial Report of Continental Scientific Drilling Project of the Cretaceous Songliao Basin (SK-1)

Initial Report of Continental Scientific Drilling Project of the Cretaceous Songliao Basin (SK-1)

Wang Chengshan Feng Zhiqiang Wang Pujun
Stephan A. Graham *et al*.

Science Press
Beijing

AMSTERDAM • BOSTON • HEIDELBERG • LONDON
NEW YORK • OXFORD • PARIS • SAN DIEGO
SAN FRANCISCO • SINGAPORE • SYDNEY • TOKYO

Responsible Editors: Wei Qin, Han Peng

Copyright © 2017 Science Press. Jointly published with Elsevier Inc. All rights reserved.

No part of this publication may be reproduced or transmitted in any form or by any means, electronic or mechanical, including photocopying, recording, or any information storage and retrieval system, without permission in writing from the publisher. Details on how to seek permission, further information about the Publisher's permissions policies and our arrangements with organizations such as the Copyright Clearance Center and the Copyright Licensing Agency, can be found at our website: www.elsevier.com/permissions.

This book and the individual contributions contained in it are protected under copyright by the Publisher (other than as may be noted herein).

Notices
Knowledge and best practice in this field are constantly changing. As new research and experience broaden our understanding, changes in research methods, professional practices, or medical treatment may become necessary.

Practitioners and researchers must always rely on their own experience and knowledge in evaluating and using any information, methods, compounds, or experiments described herein. In using such information or methods they should be mindful of their own safety and the safety of others, including parties for whom they have a professional responsibility.

To the fullest extent of the law, neither the Publisher nor the authors, contributors, or editors, assume any liability for any injury and/or damage to persons or property as a matter of products liability, negligence or otherwise, or from any use or operation of any methods, products, instructions, or ideas contained in the material herein.

ISBN: 978-7-03-053047-9 Science Press, Beijing

British Library Cataloguing-in-Publication Data
A catalogue record for this book is available from the British Library

Library of Congress Cataloging-in-Publication Data
A catalog record for this book is available from the Library of Congress

ISBN: 978-0-12-812928-9

For information all Elsevier publications visit
our website at https://www.elsevier.com/

Publisher: Candice Janco
Acquisition Editor: Amy Shapiro
Editorial Project Manager: Hilary Carr
Production Project Manager: Mohanapriyan Rajendran

Typeset by Science Press, Beijing

Not for sale outside the Mainland of China (*Not for sale in Hong Kong SAR, Macau SAR, and Taiwan, and all countries except the Mainland of China*).

Foreword 1

The exploration of the universe, of the earth's interior and of the oceans are three great endeavors of humanity challenging nature in order to improve living conditions for humans. Scientific drilling is a significant exploration tools in current earth science research. We can directly observe the lithosphere which we know little about and obtain the rock record from its deeper parts through scientific drilling programs. Drilling provides important information about natural resources, disaster prediction and a variety of environmental problems currently facing by modern human society. Therefore, scientific drilling can be thought of as a telescope for us to look into the depths of the earth and far back into the past, in order to better understand earth evolution. Plate tectonics theory that developed in the middle of the 20th century is the result of a revolution in earth science based on ocean scientific drilling. Ocean scientific drilling currently has entered the third stage, Integrated Ocean Drilling Program (IODP), the previous stages being the Deep Sea Drilling Project (DSDP) and Ocean Drilling Program (ODP), with the former initiated in the early 1960s.

Humans aspire to better understand the evolution history of the earth's continents which have complex interactions with the oceans, climate and life on earth, by employing continental scientific drilling as one of the principal tools. To achieve such an objective, Germany, the United States and China set up jointly in February 1996 the International Continental Scientific Drilling Program (ICDP). Presently, about twenty states and member organizations participate in the ICDP. The Ministry of Science and Technology of China has sent delegates to represent the Chinese government as councilors of the Assembly of Governors (AOG) in the ICDP, to assist in policy formulation and important decision making.

Since the implementation of the ICDP, many continental scientific drilling projects have been completed. As a result, discoveries have been made in five key research themes: the physical and chemical processes responsible for earthquakes and volcanic eruptions; the manner in which the earth's climate has changed in the recent and deep past and the reasons for such changes; the effects of major impacts on climate and mass extinctions; how sedimentary basins and hydrocarbon resources originate and evolve; how ore deposits are formed in diverse geologic settings. Two ICDP scientific drilling projects have been completed in China, including the Chinese Continental Scientific Drilling (CCSD) project in Donghai County, Jiangsu Province, and the Lake Qinghai Scientific Drilling Project.

Another ambitious scientific drilling project currently in progress is the Continental Scientific Drilling Project of the Cretaceous Songliao Basin (SK-1) in the Daqing oil field, which represents the largest known oil field in China. The SK-1 drilling process was completed in 2007. This project is supported by the National Basic Research Program of China (973 Program) "Integrated Study on the Cretaceous Major Geological Events and Greenhouse Climate Change", jointly sponsored by the National Ministry of Science and Technology of China and the Daqing Oilfield Company Ltd., and organized by China University of Geosciences, Beijing (CUGB), and the Daqing Oilfield Company Ltd. The Continental Scientific Drilling Project of the Cretaceous Songliao Basin has two major objectives: one is to correlate the oceanic and continental records and determine the principal drivers of climate change in order to assist in future climate change predictions and its influence on the earth's environment; the other is to further test the theory of terrestrial genesis of hydrocarbons, and to explore the formation of intra-continental sedimentary basins, as many are major hydrocarbon provinces. In addition, this project provides the scientific basis for exploration of the Daqing Oilfields, with a yearly production of forty million tons of oil.

In order to share the original data and preliminary scientific results of the SK-1 project with international geoscientists, similarly to the success of the ocean drilling programs, the SK-1 preliminary results are presented here in "Initial Reports of Continental Scientific Drilling Project of the Cretaceous Songliao Basin (SK-1)".

During the writing and publishing of this book, we were delighted to learn that the "Continental Scientific Drilling Project of the Cretaceous Songliao Basin: Continuous High-resolution Terrestrial Archive and Greenhouse Climate Change" was approved by the International Continental Scientific Drilling Program (ICDP) in September 2009. The project led by the CUGB and the Daqing Oilfield Company Ltd. includes scientists from more than ten countries including the United States and Austria, and is spearheaded by Chinese scientists. This project consists of two stages: the first stage named SK-1 which has been successfully completed, includes penetration of strata from the K/T boundary to Upper Cretaceous, and the second stage, named SK-2 which will begin soon, includes penetration of strata from the Lower Cretaceous to J/K boundary. The Ministry of Science and Technology of China will continue to support this important scientific endeavor that represents an important leap forward in Chinese earth sciences and welcomes international scientists to share and participate in the scientific research.

The world is entering a new exciting era to explore the earth in which scientific drilling will provide a quantum leap in our understanding of the deep earth and its deep history. It is expected that the publication of this book will play an important role in promoting scientific drilling and earth system science research. Meanwhile, the Ministry of Science and Technology of China will continuously support the Chinese earth science community to assume its important responsibilities and to play a greater role in this human endeavour.

Wan Gang

万 钢

Minister of Science and Technology of China
January 8, 2010

Foreword 2

Sediment basins around the world host the most important resources for mankind: hydrocarbons and groundwater. Yet, the evolution of many of these basins over geological times is still often poorly studied despite the fact that the continuous deposition of sediments provides often very valuable records of the past.

Depositional basins formed in continental environments are one of the least understood major sediment sinks. Their detailed stratigraphy and structure is the key to shed fresh light on orogenic and on sediment supply processes but especially the environmental and climate evolution encapsuled therein.

Therefore, the International Continental Scientific Drilling program, ICDP endorses continental basin research and is accordingly for example supporting recently two major study programs, namely in the Colorado Plateau in the southwestern United Sates and foremost in the Songliao Basin in northeastern China.

This volume summarizes in great detail the current status of knowledge derived from surface and borehole investigations at the Songliao Basin. Especially the two drilled core holes, SK-1 and SK-2 provide insight into the turbulent greenhouse climate world of the Cretaceous. Research highlights encompass the complex interplay of terrestrial and marine influences, the resulting biotic response and carbon cycling as well as the terrestrial reaction to major oceanic events; all of them studied with most modern methods and instruments.

The close cooperation of a Chinese-led international scientific team, national funding agencies and researchers and engineers of the hydrocarbon industry makes this project outstanding. The mutual benefit for science in getting deep access to a major continental basin and for industry in getting novel interpretations for sedimentary basin and hydrocarbon origins is obvious. In documenting such fruitful work, this volume is setting the stage for the even more pre-eminent deep drilling program that will be executed soon in the Songliao Basin.

Rolf Emmermann

Chair of the ICDP Executive Committee
October 25, 2011

Preface

Scientific drilling began with the "Deep Sea Drilling Project" in 1960s and led early to a major scientific breakthrough and geosciences revolution which has changed our views of the Earth (Wang, 2007). With ocean studies currently highly advanced, there is a particular need for earth system studies to obtain relevant information from the continents. This prompted establishment of the International Continental Drilling Programe (ICDP) (Harms et al., 2007).

Since foundation of the ICDP in 1996, initially launched by Germany, the US and China, interest in the program has expanded to its current participation of over twenty member countries. More than 100 continental scientific drilling boreholes of different depths have already been completed in more than 13 countries. The implementation of the continental scientific drilling has contributed to an increase in our understanding of continental plate motion, crustal stresses and earthquakes, volcanic processes, deep resources, the origin of life, earthquake hazards, global climate change, etc. (Truman, 2000; Detlev et al., 2004). In the spring of 1999, under the direction of Chinese scientists, ODP Leg 184 has been implemented in the South China Sea and has obtained sedimentary records for the past 30 million years. This project not only resolved the history of the South China Sea's climate and environment change, but also elucidated the long-term ocean carbon pool (Wang, 2007). The Chinese Continental Scientific Drilling (CCSD) project in Donghai County, Jiangsu Province, completed in March 2005, was located south of the Sulu ultra-high pressure (UHP) metamorphic belt, which is one of the largest in the world. This was the first time that scientific drilling in an UHP metamorphic belt had been attempted and significant advances were made in the precise dating of deep subduction of voluminous continental material, UHP deep subduction metamorphism and exhumation, protolith formation, upper mantle rheology, discovery of new mantle minerals, subsurface fluid anomalies and subsurface microorganisms (Liu et al., 2001; Hartmut, 2002; Xu et al., 2005; Zhang et al., 2005, 2006). Qinghai Lake is located at the northeastern margin of the Tibetan Plateau, boarding the Huangtu Plateau in the east, and represents the transition between the east-Asia monsoonal humid climate region and the inland arid region. The area is therefore very sensitive to climate and global environment change and was selected as the preferential location for environmental studies in western China, and for determining the role of global climatic change during uplift of the Tibetan Plateau uplift (An, 2006; Colman et al., 2007). The Lake Qinghai Scientific Drilling Project began in July 2005 with a purpose of understanding east-Asia monsoon climate and inland arid climate change, evolution of Qinghai Lake basin structure and the history of Qinghai Lake surface fluctuation (An, 2006; Colman et al., 2007).

In the past hundred years, the global climate has experienced a great change represented by significant warming. Human civilization is faced with an increasing urgency to understand global climate change and its potential effects on ecosystems, resources and habitability. This research has so far been focused on the modern and Quaternary climate record. The way to understanding the earth's climate system demands a fully integrated exploration of the earth and its geological history. Based on this premise, the "Deep Time" research project, which incorporates information of the earth's ancient climate change prior to the Quarternary, studies climate changes and main geological events using sedimentary records in pre-Quaternary rocks, in order to better predict future climate change (Sun and Wang, 2009). A new era in the Ocean Drilling Project (IODP) began in 2003 when it adopted "Environmental change, processes and effects" as one of the three most important scientific areas of ocean research. Over 30 International Continental Scientific Drilling Projects have been completed, which include climate change and global environment, impact structures, geobiosphere and early life, volcanic systems and thermal regimes, mantle plumes and rifting, active faulting, collision zones and convergent margins, natural resources, etc. Half of these projects were targeted to research related to "climate change and global environment". The ability to predict future global climate change has become an important objective for the scientific community with an indirect impact on the well being of the whole humanity.

Recent global climate is characterized by oscillating glacial-interglacial periods under "icehouse" conditions controlled by Milankovich cycles (Miller *et al.*, 1991; Wang, 2000). This is because since the Cenozoic the concentration of greenhouse gases in the atmosphere has been lower than the threshold amount (about 560 ppmV; DeConto and Pollard 2003). If the concentration of CO_2 in the atmosphere exceeds this critical concentration of about 560 ppmV, it will result in melting of the ice cover at the North and South poles, and it is widely held that the earth may be entering a greenhouse climate state with a global climate very different from the present. Asa result of human activity, the CO_2 content in the atmosphere is continuously increasing and it may exceed the critical concentration in the near future. Concentration levels of CO_2 may approach those modeled for the Cretaceous (Daniel *et al.*, 2001; Berner and Kothavala, 2001). Hence, it is particularly important to understand how and why the climate changed in the Cretaceous.

During the past forty years, oceanic scientific drilling projects have targeted the Cretaceous period and the climatic changes that occurred during that time. Research of the ocean response to Cretaceous climate changes is well advanced, but what is little known is the response of the continental environment, and no scientific drilling project has been done to obtain evidence from the Cretaceous terrestrial sedimentary record.

The reason for this is that the Cretaceous continental record is fragmentary and less extensive when compared to the oceanic basins. Sea level in the middle Cretaceous was the highest during the past 250 Ma (Haq *et al.*, 1987). At that time, the global continental areas shrank and the largest continent was that of the East-Asia area (inside cover). One of the areas with an extensive continental Cretaceous record is in China. Oil exploration in the Songliao Basin of northern China has demonstrated that this intercontinental basin, which was mostly a fresh water lake for the duration of the Cretaceous to Early Tertiary time, contains an almost continuous sedimentary record. Thus, the Songliao Basin was selected by Chinese scientists to conduct research into Cretaceous continental climate. Further studies on marine-terrestrial correlation, paleoclimate and paleoenvironment reconstruction, terrestrial response to major geological events, large scale hydrocarbon formations, terrestrial biotic evolution and rapid climate changes under greenhouse climate conditions, will provide important information of the nature and mechanisms of Cretaceous climate change, allowing predictions of future climate change, and provide exploration targets in the Daqing Oilfield.

The "Continental Scientific Drilling Project of Cretaceous Songliao Basin: Continuous High-resolution Terrestrial Archives and Greenhouse Climate Change" is designed as a two stage drilling program. The first stage, SK-1, includes drillingof early Tertiary to Lower-Upper Cretaceous strata. This stage has been completed. The next stage, SK-2, will drill Middle and Lower Cretaceous to the Jurassic—Cretaceous boundary rocks and will begin soon. When completed, the entire Cretaceous continental sedimentary record will be available for paleoclimate studies for the first time globally. In order to share this unique primary data with the international scientific community, this book *Initial Report of Continental Scientific Drilling Project of the Cretaceous Songliao Basin* was compiled and published with Chinese and English language editions.

During drilling of SK-1, to ensure core recovery ratio, we integrated a series of drilling and coring techniques, invented long term preservation techniques, and established ten complete, continuous, high-resolution (centimeter) geological profiles. The Late Cretaceous chronological framework of Chinese terrestrial strata was revised according to the results of this drilling project. SK-1 is the first scientific drilling project to continuously core Cretaceous strata in China and obtained continuous, high-resolution, little disturbed, continental sedimentary records of Middle-Late Cretaceous age that may become a standard for global terrestrial Cretaceous research. SK-1 has made significant advances in areas such as drilling engineering, logging engineering, core repository, etc., and relevant technologies have been applied in these areas to yield important social and economic benefits.

The following researchers have contributed to SK-1 studies: Wang Chengshan (Co-Chief scientist), Feng Zhiqiang (Co-Chief scientist), Wang Pujun (Geological director), Feng Zihui (Geochemical research), Yang Gansheng (Drilling engineering director), Wu Heyong (Site selection), Wan Xiaoqiao (Palaeontology research), Ren Yanguang (Site selection), Huang Yongjian (Sedimentary geochemical research), Chi Yuanlin (Logging engineering), Li Yule (Drilling engineering of SK-1s), Zhu Yongyi (Drilling engineering of SK-1n), Wang Zhongxing (Core preservation), Deng Chenglong (Paleomagnetic research), He Mingyue (Core transportation and preservation).

This initial report is a crystallization of the collective wisdom and many institutes and scholars that have contributed to this book. The writers discussed the content and outline of this book in detail at meetings in Beijing, Changzhou and Guangzhou. Wang Chengshan and Feng Zhiqiang wrote the Preface, Wang Pujun and Wang Chengshan wrote Sections 1 and 2, Feng Zhiqiang, Yang Gansheng, Wang Pujun and Wu Xinsong wrote Section 3, Wang Chengshan, Feng Zhiqiang, Wang Pujun, Deng Chenglong, Wu Xinsong, Huang Yongjian, He Huaiyu, Dong Hailiang, Song Zhiguang,

Wan Xiaoqiao, Cheng Rihui and Wu Huaichun wrote Section 4, Wang Pujun wrotes Section 5, Wang Pujun, Wang Chengshan and Yang Gansheng wrote Appendix.

This book was compiled by Wang Chengshan, Feng Zhiqiang and Wang Pujun, assisted by Gao Youfeng and Gao Yuan. We also thank Dr. Lubomir Jansa, Canadian Geological Survey for his revision.

Special thanks to Professor Wan Gang, Minister of Science and Technology of China, and Dr. Rolf Emmermann, Chair of the ICDP Executive Committee, for contributing Forwords for this book.

Deep continental drilling which can be viewed as a telescope for humans to observe the deep earth and to increase our understanding of the earth evolution, has been carried out all over the world. From the Fennoscandian Arctic, Russia—Drilling Early Earth Project, to the Colorado Plateau Coring Project targeting Early Triassic—Early Jurassic strata in the USA, and to the the recent Lake El'gygytgyn drilling project in Siberia, these projects all prove that the global earth science community is entering a new stage of exploring the Earth's evolution. One consequence of this new exploration stage is the Continental Scientific Drilling Project of the Cretaceous Songliao Basin, the results of which are detailed in this Intial Report.

Wang Chengshan and Feng Zhiqiang

王成善　冯志强

May 2010

Participant

1. Chief Scientists

Wang Chengshan
    Sedimentologist
    China University of Geosciences, Beijing

Feng Zhiqiang
    Petroleum Geologist
    Daqing Oilfield Company Ltd.

2. Advisory Group

Wang Yupu
    Petroleum Engineer
    Daqing Oilfield Company Ltd.

Wang Yuhua
    Petroleum Geologist
    Daqing Oilfield Company Ltd.

Chi Yuanlin
    Petroleum Geologist
    Daqing Oilfield Company Ltd.

Wu Junhui
    Petroleum Engineer
    Daqing Petroleum Administration Bureau

Sun Shu
    Sedimentologist, Academician of the Chinese Academy of Sciences
    Institute of Geology and Geophysics, Chinese Academy of Sciences

Ma Zongjin
    Sedimentologist, Academician of the Chinese Academy of Sciences
    China Earthquake Administration

Zhu Rixiang
    Geophysist, Academician of the Chinese Academy of Sciences Institute of Geology and Geophysics, Chinese Academy of Sciences

Chen Jun
    Geochemist
    Nanjing University

Wan Xiaoqiao
    Stratigraphic Paleontologist
    China University of Geosciences, Beijing

Peng Pingan
    Geochemist
    Guangzhou Institute of Geochemistry, Chinese Academy of Sciences

Ji Qiang
    Paleontologist Institute of Geology, Chinese Academy of Geological Sciences

3. Scientific Team

Wu Heyong
    Petroleum Geologist
    Daqing Oilfield Company Ltd.

Kong Fanjun
    Chief Engineer in Charge of Drilling Procedure of the SK-1
    Daqing Oilfield Company Ltd.

Wang Pujun
    Chief Geologist in Charge of Geology of the SK-1
    Jilin University

Yang Gansheng
    Chief Engineer in Surveillance of Drilling Procedure of the SK-1
    China University of Geosciences, Beijing

Zhang Shihong
    Scientific Director of the SK-1
    China University of Geosciences, Beijing

Ren Yanguang
    Petroleum Geologist
    Daqing Oilfield Company Ltd.

Li Yule
    Petroleum Geologist

Daqing Oil Field Company Ltd.
Feng Zihui
Petroleum Geologist
Daqing Oil Field Company Ltd.
Wang Guomin
Petroleum Geologist
Daqing Oil Field Company Ltd.
Lang Dongsheng
Petroleum Geologist
Daqing Oil Field Company Ltd.
Zhang Ye
Petroleum Geologist
Daqing Oil Field Company Ltd.
Jiang Daohua
Petroleum Geologist
Daqing Oil Field Company Ltd.
Jiang Lijun
Petroleum Geologist
Daqing Oil Field Company Ltd.
Huang qinghua
Stratigraphic Paleontologist
Daqing Oil Field Company Ltd.
Zhang Shun
Petroleum Geologist
Daqing Qil Field Company Ltd.
Song Zhiguang
Organic Geochemist
Guangzhou Institute of Geochemistry,
Chinese Academy of Sciences
Cheng Rihui
Sedimentologist
Jilin University
Li Xianghui
Sedimentologist
Chengdu University of Technology
Wang Yongdong
Stratigraphic Paleontologist
Nanjing Institute of Geology and
Palacontology,
Chinese Academy of Sciences
Huang Yongjian
Sedimentary Geochemist
China University of Geosciences, Beijing
Li Gang
Stratigraphic Paleontologist
Nanjing Institute of Geology and
Palacontology,Chinese Academy of Sciences
Lu Hong
Organic Geochemist
Nanjing Institute of Geology and
Palacontology,
Chinese Academy of Sciences

4.Technical Team

Zhang Weidong
Petroleum Geologist
Daqing Oil Field Company Ltd.
Zhang Shizhong
On Site Drilling Supervisor of the SK-1s
Daqing Petroleum Administration Bureau
Zhu Yongyi
Manager of Drilling Project of SK-1n
Institute of Exploration Techniques,
Chinese Academy of Geological Sciences
Wang Shuxue
On Site Geological Supervisor of the SK-1s
Daqing Oil Field Company Ltd.
Zhang Jian
On Site Drilling Supervisor of the SK-1s
Institute of Exploration Techniques,Chinese
Academy of Geological Sciences
Li Xudong
On Site Drilling Supervisor of the SK-1n
Sixth Geological Brigade, Jiangsu Bureau of
Geology and Mineral
Resources
Wang Guodong
Core Description Technition of the SK-1s
Jilin University
Gao Youfeng
Core Description Technition of he SK-1n
Jilin University
Lin Zhiqiang
On Site Drilling Supervisor of the SK-1s
China University of Geosciences, Beijing
Wang Xiaopeng
On Site Drilling Supervisor of the SK-1n
China University of Geosciences, Beijing
Li Jinshan
Drilling Group Captain of SK-1s
Daqing Petroleum Administration Bureau
Bai Gang
Logging Group Captain of SK-1s
Daqing Oil Field Company Ltd.
Zhang Wei
Drilling Group Party Secretary of SK-1s
Daqing Petroleum Administration Bureau
Bai Zhixi
Coring Engineer
Zhang Yuquan
Coring Engineer
Yuan Fuxiang
Coring Engineer

Li Ziyuan
    Coring Engineer
Sun Shaoliang
    Coring Engineer
Xin Mingfeng
    Logging Engineer
Ban Yudong
    Logging Geologic Technician
Wang Bingyang
    Logging Technician
Yang Bin
    Drilling Engineer
Li Tongrun
    Drilling Technician
Hong Weidong
    Drilling Platform Monitor
Tian Lianyi
    Drilling Machine Room Monitor
Zhang Xiangchen
    Electrician
Yu Kailin
    Drilling Mud Operator
Li Yonggang
    Drilling Mud Operator
Li Jinghua
    Material Supervisor
Bi Naihua
    Driller
Wang Shaofeng
    Driller
Su Yukun
    Driller
Jia Hongqing
    Driller
Ren Jingshi
    Drilling Assistant
Chen Huaimin
    Drilling Assistant
Zheng Zhansheng
    Drilling Assistant
Fu Dongqi
    Drilling Assistant
Zhang Guangming
    Derrick Floor Worker
Xu Guohui
    Derrick Floor Worker
Yu Qingjun
    Derrick Floor Worker
Xu Zhenjun
    Derrick Floor Worker
Zhang Guoping
    Locksmith
Li Jianguo
    Locksmith
Wang Jinchun
    Locksmith
Zhao Hongtao
    Locksmith
Cai Xiang
    Locksmith
Zhang Xilong
    Locksmith
Cai Jingxue
    Locksmith
Gao Tao
    Locksmith
Li Bo
    Well Field Worker
Song Xiaowei
    Well Field Worker
Bi Zhiquan
    Drilling Mud Worker
Meng Xianjin
    Drilling Mud Worker
Zhou Bo
    Drilling Mud Worker
Song Zhenqing
    Drilling Mud Worker
Yao Jun
    Driver
Li Shuanghe
    Driver
Wang Yutao
    Driver
Liu Lang
    Driver
Li Jing
    Electricity Generator Technition
Feng Zhanlong
    Technician
Dai Hongwei
    Logging Group Gathering Worker
Zhou Runhua
    Logging Group Gathering Worker
Wang Fuping
    Logging Group Gathering Worker
Li Wei
    Logging Operator
Cong Xijie
    Logging Operator
Chen Xuan
    Logging Operator
Zhang Qiudong
    Deputy Manager of SK-1n Drilling Team

Henan Provincial Bureau of Geo-exploration and Mineral Development

Yang Liwei
Logging Group Captain of SK-1n
Daqing Oilfield Company Ltd.

Bi Jianguo
Chief of SK-1n No. 1 Drilling Rig
Henan provincial Bureau of Geo-exploration and Mineral Development

Zhang Chuanbo
Logging Group Deputy Captain of SK-1n
Daqing Oilfield Company Ltd.

Zou Dingyuan
External Drilling Technician of SK-1n
Bureau of Geology and Mineral Exploration of Auhui Province

Wu Xiaoming
General Chief of Drilling Mud Technique of SK-1n
China University of Geosciences (Wuhan)

Cai Jihua
Chief of Drilling Mud Technique of SK-1n
China University of Geosciences (Wuhan)

Wang Wenshi
Assistant Engineer of SK-1n Drilling Operation
Institute of Exploration Techniques, Chinese Academy of Geological Sciences

Zhang Wenlong
Deputy Chief of SK-1n No. 1 Drilling Rig
Henan Provincial Bureau of Geo-exploration and Mineral Development

Dong Zhenbo
Deputy Chief of SK-1n No. 1 Drilling Rig
Henan Provincial Bureau of Geo-exploration and Mineral Development

Zhang Zhiming
Assistant Technician of SK-1n Drilling Operation
Institute of Exploration Techniques Chinese Academy of Geological Sciences

Bai Ling'an
Geologic Technician

Guo Yu
Geologic Technician

Zhao Dechao
Instrument Technician

Liang Zhiyong
Data Collectorr

Li Mingyu
Logging Operator

Liu Xiaolei
Logging Operator

Tong Yugang
Logging Operator

Wu Nan
Logging Operator

Wang Yuchun
Gathering Worker

Cui Guang
Gathering Worker

Wang Zhi
Gathering Worker

Li Xiaofen
Drilling Mud Worker

Gu Sui
Drilling Mud Worker

Zhang Xiaojing
Drilling Mud Worker

Liu Yunliang
Drilling Mud Worker

Sun Pinghe
Drilling Mud Worker

Jin Xindong
Drilling Technician

Jia Shaopeng
Drilling Technician

Du Yantong
Drilling Technician

Li Qiang
Drilling Technician

Xu Quanjun
Drilling Technician

Li Hongwei
Drilling Technician

Wu Fuzheng
Material Supervisor

Wang Yaoyao
Data Recorder

Chen Fei
Data Recorder

Li Ziyi
Data Recorder

Zhang Futao
Drilling Mud Worker

Mao Kehua
Drilling Mud Worker

Zou Jianhua
Drilling Mud Worker

Lü Qingsong
Derrick Floor Worker

Lou Hongyuan
Derrick Floor Worker

Li Zhiwei
Driller

Contents

Foreword 1
Foreword 2
Preface

Participant
  1. Chief Scientists
  2. Advisory Group
  3. Scientific Team
  4. Technical Team

Section 1   Geological Background of the Songliao Basin ………… 1
  1.1 **Geologic and Tectonic Evolution of the Songliao Basin** ………… 5
  1.2 **Tectono-stratigraphic Framework of the Songliao Basin** ………… 6
    1.2.1 Pre-rift and Syn-rift Tectonostratigraphic Units ………… 6
    1.2.2 Stratigraphic Sequences of the Early Thermal Subsidence Stage ………… 7
    1.2.3 Stratigraphic Sequences of the Late Thermal Subsidence Stage ………… 9
  1.3 **Palaeoenviromental and Palaeoclimatic Studies of the Songliao Basin** ………… 12
  1.4 **Scientific Objectives of the Continental Scientific Drilling Project of the Cretaceous Songliao Basin** ………… 16
    1.4.1 Stratigraphic Boundaries and Marine-terrestrial Correlation of Stratigraphy 17
    1.4.2 Biotic Response to Terrestrial Environmental Change and the Deep Biosphere (Fossil DNA) ………… 18
    1.4.3 Terrestrial Response to Oceanic Anoxic Events and Formation of Massive Terrestrial Hydrocarbon Source Rocks ………… 19
    1.4.4 Cretaceous Normal Superchron from Terrestrial Records ………… 20

Section 2   Selection of Drilling Site ………… 23
  2.1 **Process of site selection** ………… 24
  2.2 **Site Location** ………… 26
  2.3 **Simulated Lithological Column for Well SK-1 before Drilling** ………… 28

Section 3   Implementation of SK-1 Drilling and Preservation of Cores ………… 33
  3.1 **Drilling Design** ………… 34
    3.1.1 Technical Goals ………… 34
    3.1.2 Drilling and Coring Requirements ………… 34
    3.1.3 "One Well-two Holes" Design Solution ………… 35
    3.1.4 Borehole Structure and Casing Procedure ………… 36
    3.1.5 Drilling Devices ………… 38
  3.2 **Intergrated Coring Technique for the Extremely hong and High-recovery Drilling** ………… 40
    3.2.1 Coring Tools ………… 40
    3.2.2 Coring Techniques ………… 41
    3.2.3 Directional Coring Technique ………… 43
    3.2.4 Fluorescent carboxylation Microsphere Tracing and Sealed Coring

		Technique ⋯⋯⋯⋯ 47
	3.2.5	Hydraulic Core Extraction ⋯⋯⋯ 49
	3.2.6	Log Core Identification Techniques and Core Description ⋯⋯⋯ 50
3.3	Construction Organization and Management of SK-1 ⋯⋯⋯ 51	
3.4	Implementation and Technique of Well Logging ⋯⋯⋯ 52	
	3.4.1	Logging Design of SK-1 ⋯⋯⋯ 53
	3.4.2	Implementation of the SK-1 Well Logging Project ⋯⋯⋯ 59
	3.4.3	Result of the SK-1 Well Logging Project ⋯⋯⋯ 59
3.5	Comparison and Evaluation of the Designed and Implemented one Well-two Holes Quantities ⋯⋯⋯ 59	
	3.5.1	SK-1 North Borehole (SK-1n) ⋯⋯ 59
	3.5.2	SK-1 South Borehole (SK-1s) ⋯⋯ 61
	3.5.3	Quality Evaluation of the SK-1n and SK-1s Engineering ⋯⋯⋯ 62
	3.5.4	Core Depth Location of SK-1 ⋯⋯ 65
3.6	Core Handling and Storage ⋯⋯⋯ 71	
	3.6.1	Core Handling and Storage at the Drilling Site ⋯⋯⋯ 72
	3.6.2	Core Scanning and Storage ⋯⋯⋯ 72
	3.6.3	Cutting, Casting and Long-term Preservation of SK-1 Cores ⋯⋯⋯ 72
	3.6.4	Core Sampling of SK-1 ⋯⋯⋯ 75

Section 4 Preliminary Scientific Results of SK-1 ⋯⋯⋯ 79

4.1	Lithostratigraphy ⋯⋯⋯ 80	
	4.1.1	Methods and Summary of Core Description ⋯⋯⋯ 80
	4.1.2	SK-1 Core Lithology ⋯⋯⋯ 80
	4.1.3	SK-1 Special Deposits and Their Geological Significance ⋯⋯⋯ 85
4.2	Paleomagnetism ⋯⋯⋯ 86	
	4.2.1	Aims and Significance ⋯⋯⋯ 86
	4.2.2	Sampling and Magnetic Data Analysis ⋯⋯⋯ 86
	4.2.3	Preliminary Results ⋯⋯⋯ 87
4.3	Logging Results ⋯⋯⋯ 90	
	4.3.1	Aims and Significance ⋯⋯⋯ 90
	4.3.2	Sampling and Analysis ⋯⋯⋯ 91
	4.3.3	Preliminary Results ⋯⋯⋯ 91
4.4	Inorganic Geochemistry and Mineralogy ⋯⋯⋯ 92	
	4.4.1	Research Objectives and Significance ⋯⋯⋯ 92
	4.4.2	Sampling and Analytical Methods ⋯⋯⋯ 92
	4.4.3	Preliminary Results ⋯⋯⋯ 93
4.5	Chronostratigraphy ⋯⋯⋯ 101	
	4.5.1	Aims and Significance ⋯⋯⋯ 101
	4.5.2	Sampling and Analytical Methods ⋯⋯⋯ 101
	4.5.3	Preliminary Results ⋯⋯⋯ 102
4.6	Geomicrobiology ⋯⋯⋯ 107	
	4.6.1	Research Objectives and Significance ⋯⋯⋯ 107
	4.6.2	Materials and Methods ⋯⋯⋯ 107
	4.6.3	Preliminary Data ⋯⋯⋯ 108
4.7	Organic Geochemistry ⋯⋯⋯ 112	
	4.7.1	Objectives and Significances of Organic Geochemistry Research ⋯⋯⋯ 112
	4.7.2	Sampling and Experimental Methods of Organic Matter Analysis ⋯⋯⋯ 113
	4.7.3	Preliminary Results ⋯⋯⋯ 114
4.8	Stable Isotopes ⋯⋯⋯ 117	
	4.8.1	Aims and Significance of This Study ⋯⋯⋯ 117
	4.8.2	Methods and Sampling ⋯⋯⋯ 118
	4.8.3	Preliminary Results ⋯⋯⋯ 119
4.9	Cyclostratigraphy ⋯⋯⋯ 120	
	4.9.1	Cyclostratigraphy of Sedimentary Strata in SK-1 ⋯⋯⋯ 121
	4.9.2	Floating Astronomical Time Scale for Qingshankou Formation, SK-1s

		122
4.10	**Biostratigraphy**	**126**
4.10.1	Aims and Significance	126
4.10.2	Materials and Methods	126
4.10.3	Preliminary Results	126

Section 5 SK-1 Core Description and Core Photographs **133**

5.1	**Stratigraphic Chart**	**134**
5.1.1	Quantou Formation Members 3 and 4	134
5.1.2	Qingshankou Formation Member 1	149
5.1.3	Qingshankou Formation Members 2 and 3	155
5.1.4	Yaojia Formation	189
5.1.5	Nenjiang Formation Members 1 and 2	203
5.1.6	Nenjiang Formation Members 3, 4 and 5	227
5.1.7	Sifangtai Formation	281
5.1.8	Mingshui Formation	307
5.1.9	Taikang Formation	363

Reference	656
Appendix	667
Acknowledgements	691

Section 1
Geological Background of the Songliao Basin

The Songliao Basin in northeastern China is 750 km long, 330—370 km wide and has a total area of 26×10^4 km². The basin is divided into six structural units: the northern plunge, central downwarp, northeastern uplift, southeastern uplift, southwestern uplift, and western slope (Figure 1.1). The main oil and gas-producing province is in the central downwarp, which includes the Daqing anticline, Qijia-Gulong, Sanzhao and Changling sags and Chaoyanggou terrace.

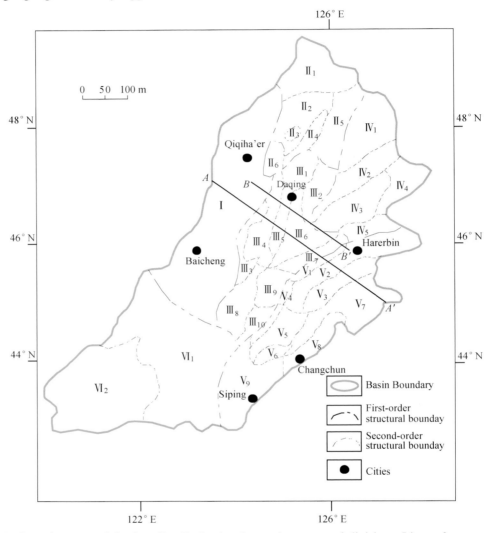

Figure 1.1 Location map of the Songliao Basin showing major structural divisions. Lines of cross section and seismic sections show locations of Figure 1.2 (A-A') and Figure 1.8 (B-B'). I . western slope; II . northern plunge; II$_1$. Nenjiang terrace; II$_2$.Yian sag; II$_3$. Sanxing anticline; II$_4$. Keshan-Yilong anticlinal belt; II$_5$.Qianyuan anticlinal belt; II$_6$.Wuyuer sag; III$_1$. central downwarp; III$_1$, Heiyupao sag; III$_2$. Mingshui terrace; III$_3$. Longhupao-Da'an terrace; III$_4$. Qijia-Gulong sag; III$_5$. Daqing Changyuan anticline; III$_6$. Sanzhao sag; III$_7$. Chaoyanggou terrace; III$_8$. Changling sag; III$_9$. Fuyu uplift; III$_{10}$. Shuangtuozi terrace; IV. northeastern uplift; IV$_1$. Hailun uplift; IV$_2$. Suiling anticline; IV$_3$. Suihua sag; IV$_4$. Qing'an uplift; IV$_5$. Hulan uplift; V. southeastern uplift; V$_1$. Changchunling anticlinal belt; V$_2$. Binxian-Wangfu sag; V$_3$. Qingshankou uplift; V$_4$. Denglouku anticlinal belt; V$_5$. Diaoyutai uplift; V$_6$. Yangdachengzi anticlinal belt; V$_7$.Yushu-Dehuisag; V$_8$. Jiutai terrace; V$_9$. Dehui-Lishu sag; VI. southwestern uplift zone; VI$_1$. Gamatu uplift; VI$_2$. Kailu sag.

The basin contains Jurassic, Cretaceous, Palaeogene and Neogene clastic deposits that are about 10 km thick in the basin centre. Sediment fill thins toward the basin margins resulting in a cross-sectional steer's-head geometry (Figure 1.2). These sediments are underlain by Palaeozoic metamorphic, plutonic and volcanic rocks (Tian and Han, 1993; Gao and Cai, 1997). The basin was formed and filled in four tectonic stages: mantle upwelling, rifting, postrift thermal subsidence and structural inversion. The first stage occurred during the Middle and Late Jurassic, and is characterized by doming, extension and widespread volcanism. Upper Jurassic to Lower Cretaceous pre-rift and syn-rift deposits (pre-rift and syn-rift tectonostratigraphic units) occur only within isolated fault blocks with a maximum thickness of 7000 m in the Shiwu and Dehui sags (Xie et al., 2003). Lower to Upper Cretaceous post-rift deposits (post-rift tectonostratigraphic unit), with a thickness of 3000—4000 m (maximum 6000 m), unconformably overlie syn-rift strata and extend beyond the fault blocks to cover the whole basin. Palaeocene and Eocene deposits (structural inversion tectonostratigraphic unit) with a thickness of 0—510 m, occur only locally in the western part of the basin. A generalized stratigraphic column of the Songliao Basin is shown in Figure 1.3.

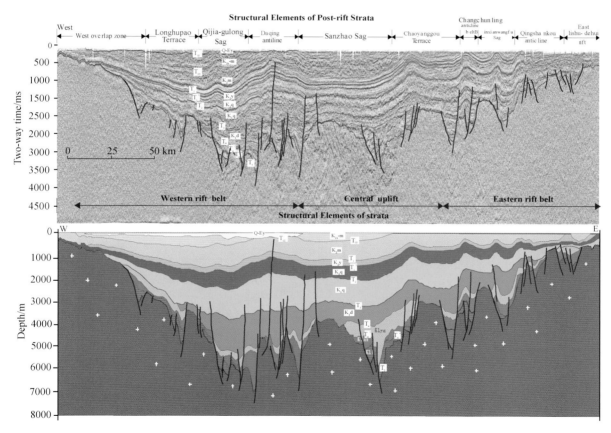

Figure 1.2 Regional seismic section and structural cross section across the central part of the Songliao Basin. Lithologic and formation information derived from geophysical logs and cores are tied to seismic sections. Location of the section is shown in Figure 1.1 (A-A').

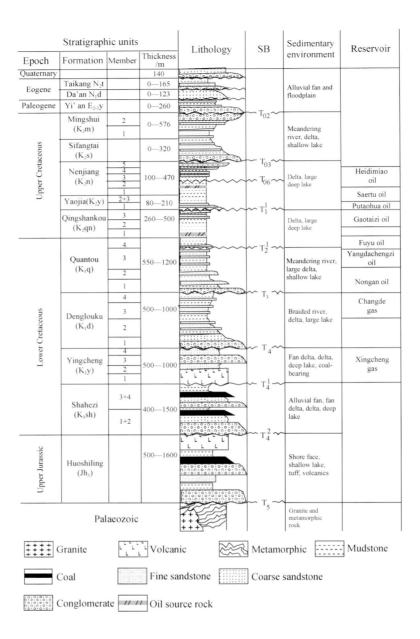

Figure 1.3 Generalized stratigraphic column of the Songliao Basin.

Large-scale geological investigation of Songliao Basin for petroleum started in 1956, and on September 26, 1959, the first highly productive oil well (known as the Songji-3 Gusher) was successfully drilled, initiating the beginning of the Daqing Oilfield. Since then, extensive exploration work has been conducted in the Songliao Basin. By the end of 2000, over 250,000 km of seismic reflection profiles had been completed. The area of the seismic network is 0.5 km × 0.5 km to 2 km × 4 km, and some 3D seismic reflection data have also been acquired. About 50,000 wells have been drilled, with a cumulative drill penetration of about 40 million meters. In recent years, more wells have been drilled to the level of the Lower Cretaceous, that provides reference points for the proposed deeper drilling program, and which form the basis for detailed stratigraphic correlation and derivative reconstructions of Cretaceous paleoenvironments, and permit identification of an optimum drilling location.

1.1 Geologic and Tectonic Evolution of the Songliao Basin

The evolution of the Songliao Basin can be separated into four stages: pre-rift doming, extensional faulting, downwarping (driven by thermal subsidence) and structural inversion.

Pre-rift doming stage: at the end of the Paleozoic, at the northern edge of the North China Plate, the paleo-Asian sea closed along the Chifeng-Kaiyuan to Xi Lamulun River fault zone, which become the basement of the future Songliao Basin. After regional uplift and erosion during early and mid-Indosiniantectonism, and influenced by closing of the Okhotsk belt (between the Siberian and North China-Mongolia Plates) to the north, tensional faulting, differential block movement and strong magmatic activity occurred in the Songliao Basin. During the middle and mate Jurassic, upper mantle perturbation caused crustal extension and thinning leading to the development of a series of fault-controlled depressions in the basin. This was accompanied by magmatism along the faults. Magmatic activity was stronger in the western part of the Songliao Basin, while in the eastern part subsidence along crustal-scale faults was associated with the volcanic-sedimentary deposition with little magmatic activity (Wang *et al.*, 2006).

Extensional faulting stage: During the Early Cretaceous, mantle doming caused further crustal extension. The Sunwu-Shuangliao fracture became active, with uplift of the central horst. Sediments deposited during this time were mainly coarse-grained flysch. The final shape of the Songliao Basin was formed at this time. During deposition of the Shahezi Formation, the Songliao Basin was undergoing extension accompanied by a widespread volcanism. A series of NNE or NE-trending faults were formed. Deposition of the Yingcheng Formation was influenced by westward compression from the Pacific Plate subduction to the east, which terminated the rifting processes. As a result, previously formed depressions began to contract and the rate of tectonic subsidence rate decreased. Areas surrounding the Songliao Basin began to rise.

Downwarping stage: This stage was driven by thermal subsidence. During the Early Cretaceous, lithosphere underlying the Songliao Basin gradually cooled and contracted resulting in large scale subsidence. The rate of subsidence accelerated to form a central deep depression in the area that had been the locus of maximum doming of the mantle. Two large transgressions resulted in deposition of the Qingshankou and Nenjiang formations when up to 3000 m of sediments were deposited during 35 Ma (100—65 Ma). The sediments comprised fluvial, lacustrine, and deltaic lithofacies that contain hydrocarbon. Because of lateral tectonic variability during the early stages of the Songliao Basin development two deposition centers developed: one in the eastern and the other one in the central part of the basin. However, during the middle and late stages of basin development, the eastern depocenter disappeared, which caused faulted depressions to develop in eastern part, sags in the middle part, and a long-lived slope zone to develop in the west of the basin. In the late Early-Cretaceous, as a result of the closing of paleo Mongolia-Okhotsk sea, the adjacent plates in northeast China (that included the Songliao Basin) collided with the Siberian Plate which caused strong regional compression. At the same time, the Sea of Japan began to open which caused in compression of the Songliao Basin (Nenjiang Movement), resulting in uplift, especially in the eastern part of the basin to end the downwarping stage in its evolution.

Structural inversion stage: During the Late Cretaceous, following the Nenjiang tectonism, the geological structure in deep basin was established, the sub-basin crust was slowly rising and the paleolake contracted to a quarter of its former size. Previously deposited sediments were folded. The eastern part of the basin underwent continuous uplift, causing the westward migration of the depocenter. The rate of subsidence was slow. The structural movement in the eastern and central parts of the basin affected deeper strata, while shallow deformation structures were formed in the west. Compression led to the development of reverse faulting, e.g., Hudian, Da'an, Lindian, and Renminzhen faults, and folding, to produce structures such as the the Daqing Changyuan fault structure. During the Tertiary and Quaternary, the deposition of alluvial fans and small-scale lakebeds imply the cessation of major tectonic activity.

1.2 Tectono-stratigraphic Framework of the Songliao Basin

1.2.1 Pre-rift and Syn-rift Tectonostratigraphic Units

The pre-rift and syn-rift tectonostratigraphic units comprise the Upper Jurassic Huoshiling (J_3h), and the Early Cretaceous Shahezi (K_1sh) and Yingcheng (K_1y) formations. The Upper Jurassic Huoshiling Formation, with a thickness of 500—1600 m, was deposited during the doming and initial rifting stage of the basin. It occurs in most fault blocks, and thickens to the south in the Gulong depression. The sequence consists of volcanic flows (mainly andesitic) and volcaniclastics intercalated with clastic fluvial, floodplain and swamp facies. Seismic profiles show that the formation is not confined to the extensional fault blocks, but was deposited before significant extension occurred. This suggests that volcanism was not restricted to the extensional fault blocks.

The Lower Cretaceous Shahezi Formation is restricted to fault blocks in the eastern and central rift belts. The formation is usually 400—1500 m thick, and is thickest in the Xujiaweizi and Changling fault blocks, where it is 2500 m thick in the centre part. Strata consist of grey to black lacustrine and floodplain mudstone and siltstone interbedded with grey sandstone and conglomerate. Seismic profiles and wells reveal that the strata thicken toward the faults in most fault blocks, and thus the faults were active during deposition. A thin layer of felsic tuff and tuff breccia occurs at the base of the formation. Above the volcanic is a succession of fining-upward fluvial and lacustrine deposits followed by coarsening-upward lacustrine deltaic and shoreface sediments. Subsequently, tectonic subsidence accelerated and a large lake was formed in the central part of the rift belt. Black lacustrine mudstone hundreds of meters thick, with an average total organic carbon (TOC) of 2.1%, forms the second and third members of the Shahezi Formation. These mudstone units are highly mature because of deep burial, and are the most important hydrocarbon source rocks in the Songliao Basin.

The Lower Cretaceous Yingcheng Formation was deposited in down-faulted depressions. The formation is typically 500—1000 m thick, but in the Xujiaweizi Fault block, it has a maximum thickness of 2900 m. The formation comprises mainly felsic volcanic, which are the main hydrocarbon reservoirs, interbedded with clastic sediments and several discontinuous coal beds. During deposition of the Yingcheng Formation, volcanism was widespread. Alluvial fans and fan deltas occurred near the main marginal faults, and shallow lacustrine mudstone and sandstone were deposited away from the faults. Conglomerate at the top of the Yingcheng Formation contains volcanic, metamorphic and granitic clasts.

1.2.2 Stratigraphic Sequences of the Early Thermal Subsidence Stage

The tectonostratigraphic unit deposited during the post-rift thermal subsidence period comprises the Denglouku (K_1d), Quantou (K_1q), Qingshankou (K_2qn), Yaojia (K_2y), Nenjiang (K_2n), Sifangtai (K_2s) and Mingshui (K_2m) formations. This unit is bounded at the base by a unconformity, equivalent to seismic horizon T_4, and at the top by another unconformity, equivalent to seismic horizon T_{02}. The upper unconformity is between the Upper Cretaceous Mingshui (K_2m) and Palaeocene Yi'an (Ey) formations. Early post-rift strata are divided into three sequences ($Ⅱ_1$, $Ⅱ_2$ and $Ⅱ_3$) separated by unconformities. The sediments comprise alluvial fan, fluvial, floodplain, shallow lacustrine and deltaic facies.

Sequence $Ⅱ_1$, which corresponds to the Denglouku Formation (K_1d), was deposited during the transition from rift-induced fault-block basins to cooling-induced regional subsidence. During this period, the basin had multiple depositional and subsidence centres. Strata comprise interbedded grey to white structureless sandstone, dark sandy mudstone, various coloured sandstone and mudstone, conglomerate and a thick argillite. The thickness of sequence $Ⅱ_1$ is usually 500—1000 m, with a maximum of 2000 m in the Gulong depression. Depositional limits expanded from the rifting phase and extended to Mingshui in the north and Harbin in the east. Major sediment sources were from the northeast, north, northwest and southeast. Facies tracts are concentric within the basin, with alluvial fans along the basin margin grading inward through alluvial plain, lacustrine delta and shallow lacustrine facies. Deep lacustrine facies were deposited along the two sides of the ancient central uplift, in the Da'an and Xujiaweizi-Sanzhan regions.

Sequence II_2, which corresponds to the first, second and third members of the Quantou Formation, is characterized by red-brown, purple and purple-brown mudstone, coarse greyish-white sandstone and conglomerate of fluvial and floodplain origins deposited in arid or semi-arid conditions. Strata of Sequence II_2 progressively overlap the basin margins. The maximum thickness of the Quantou Formation occurs in the Gulong and Sanzhao depressions around the central uplift zone. N-S trending fluvial channel belt sandstones are common in the first and second members. The third member contains fewer thin fluvial channel sandstones within fine floodplain deposits. Fluvial systems change from braided to meandering channels through the Quantou Formation. Shallow lacustrine and delta deposits occur in the central area of the third member (Figure 1.4). The thickness of sequence II_2 is usually 550—1200 m, with a maximum of 1650 m in the Gulong depression.

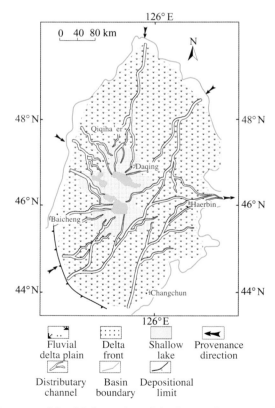

Figure 1.4 Depositional facies map of the third member of the Lower Cretaceous Quantou Formation.

Sequence II_3 consists of the fourth member of the Quantou Formation (K_1q_4) and the Qingshankou Formation (K_2qn), and is 260—500 m thick with a maximum thickness of 635 m. The Quantou Formation is characterized by greyish-green or black mudstone, siltstone and sandstone of fluvial and lacustrine origin

(Figure 1.5). The Qingshankou Formation consists of grey, dark-grey and black mudstone interbedded with oil shale and grey sandstone and siltstone. Deltaic and shallow lacustrine facies dominate both formations. Strata near the base of the sequence contain transitional fluvial to lacustrine facies environments. During early Qingshankou deposition, the lake reached its maximum depth and extent, covering an area of 87,000 km^2. Deep lacustrine black mudstone with a thickness of 60—100 m was deposited across the entire central downwarp, forming the most important petroleum source rocks in the Songliao Basin. The second and third members of the Qingshankou Formation contain large deltaic sediment systems that prograded from the north and west. The areal extent of lacustrine strata diminishes to about 41,000 km^2 in the upper part of the Qingshankou Formation (Yang, 1985). Deep-water lacustrine shale in the second and third members has a thickness of 200—300 m and was deposited across most of the downwarp.

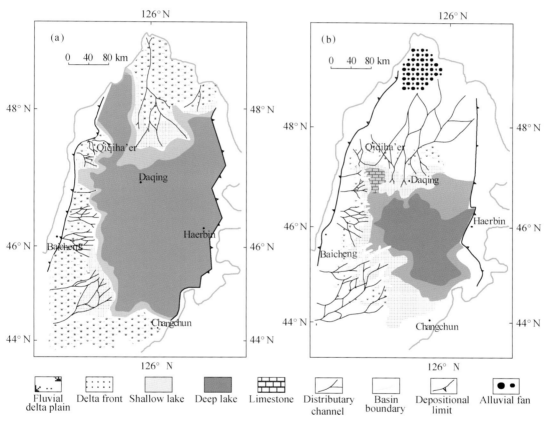

Figure 1.5 Depositional facies map of the first member (a), second and third members (b) of the Upper Cretaceous Qingshankou Formation (K$_2$qn).

1.2.3 Stratigraphic Sequences of the Late Thermal Subsidence Stage

Strata of the late thermal subsidence stage are divided into three sequences separated by unconformities, sequences II$_4$, II$_5$ and II$_6$. The fundamental attribute of these sequences is that they contain facies that fine and deepen upward and are capped by a subaerial unconformity. Late post-rift strata are generally fine grained.

Sequence II$_4$, which corresponds to the Yaojia Formation (K$_2$y) and the first and second members of the Nenjiang Formation (K$_2$n$_{1-2}$), comprises red, grey, greyish-green and black mudstone, siltstone and sandstone of lacustrine, fluvial and deltaic origins. Fluvial sandstone and red floodplain mudstone at the base of the Yaojia Formation onlap the unconformity surface on top of the grey to black shales of the underlying Qingshankou Formation. Mudstone rip-up clasts, mud cracks, calcrete, Rhizo concretions and bioturbation on the unconformity surface in the central part of the basin indicate a period of subaerial exposure and weathering. The remainder of the 80—210-m-thick Yaojia Formation contains a fining-upward succession of fluvial to deltaic to lacustrine facies [Figure. 1.6 (a) and (b)].

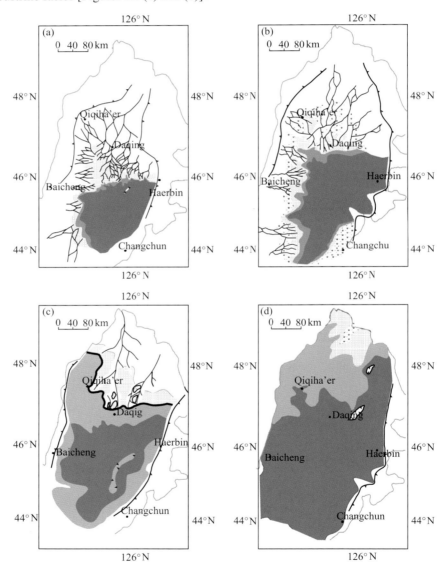

Figure 1.6 Depositional facies map of the first (a), second and third members (b) of the Upper Cretaceous Yaojia Formation (K$_2$y), and the first (c) and second members (d) of the Upper Cretaceous Nenjiang Formation (K$_2$n). Symbols are the same as those in Figure 1.5.

The 100—470-m-thick Nenjiang Formation is dominated by deep-water lacustrine grey to black mudstone, marl, shelly limestone and oil shale interbedded with grey siltstone and fine sandstone. During deposition of the first member of the Nenjiang Formation (K_2n_1), the lake expanded rapidly and reached its maximum extent of $20 \times 10^4 \, km^2$, covering almost the entire basin (Chi *et al.*, 2002) [Figure 1.6 (c) and (d)]. Lake expansion was accompanied by a significant increase in subsidence rate. This is the second phase of large-scale lake expansion, following that of the first member of the Qingshankou Formation. In the first member of the Nenjiang Formation, deep-water lacustrine facies were deposited in the basin centre, turbidite deposits occur in the northern part of the basin and shallow lake-margin facies in the south. In the second member, deep-water lacustrine, organic-rich mudstones with a thickness of 200 m covered nearly the entire basin and form a regional barrier to fluid flow. Shallow lacustrine deposits occur only in the north, west and southeast parts of the basin.

Sequence II$_5$, which corresponds to the third, fourth and fifth members of the Nenjiang Formation, also comprises grey, grey-green and black mudstone, siltstone and sandstone of lacustrine and deltaic origins. During deposition of the third member, deltas shifted rapidly toward the basin center as the lake level dropped. Deltas occur in the northern and southern parts of the basin and in the Zhaodong area. Lacustrine facies sediments occur in the Gulong-Datong area in the central part of the basin (Figure 1.7), and turbidites occur in the Taikang area. Both the lake and the area of sediment accumulation contracted after deposition of the third member, from 110,000 to 10,000 km^2 and from 151,000 to 40,000 km^2, respectively (Ye *et al.*, 2002).

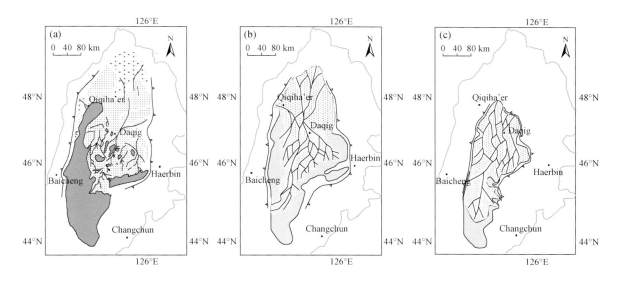

Figure 1.7　Depositional facies map of the third (a), fourth (b) and fifth members (c) of the Upper Cretaceous Nenjiang Formation (K_2n). Symbols are the same as those in Figure 1.5.

Sequence II$_6$ consists of the Sifangtai (K_2s) and Mingshui (K_2m) formations. Differential subsidence across the basin during Sifangtai deposition created thickness variations from 0 to 320 m [Figure 1.8(a)]. Sifangtai strata occur in the central and western parts of the basin, but are absent in the southeastern part. The lower part of the Sifangtai Formation consists of brick-red pebbly sandstone and shale interbedded with brown, grey and grey-green sandstone and muddy siltstone. The middle part of the formation consists of fine grey sandstone and siltstone interbedded with brick-red and mauve shale. Fine clastics dominate the upper part of the formation and comprise red and mauve shale interbedded with grey-green mudstone. These facies represent meandering river, delta and shallow lacustrine environments. Sediment supply was from the north. The Mingshui Formation occurs in the central and western parts of the basin, but is absent in the eastern part. Thickness varies from 0 to 576 m [Figure 1.8(b)].The formation is composed of grey-green, grey, black and brown-red shale and grey-green sandstone. Sediment was supplied from the south, north and east, and this created NNE-aligned facies tracts of braided river, delta, and shallow lacustrine environments from east to west.

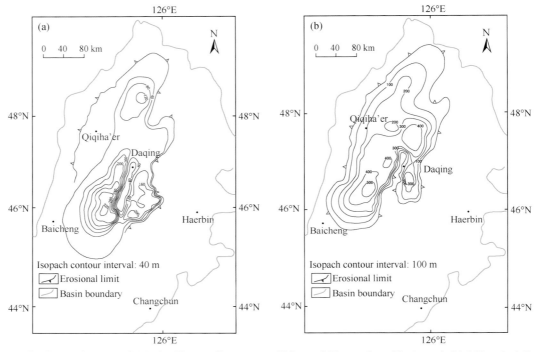

Figure 1.8　Isopach map of (a) the Upper Cretaceous Sifangtai Formation (K_2s), and (b) Mingshui Formation (K_2m).

1.3　Palaeoenviromental and Palaeoclimatic Studies of the Songliao Basin

The Cretaceous provides significant rock records of global climate changes under conditions of greenhouse-type climate (Skelton *et al.*, 2003; Bice *et al.*, 2006). The Songliao Basin offers a unique opportunity to understand the Cretaceous paleoclimate of terrestrial settings, because it contains a nearly complete record of Cretaceous lacustrine sedimentation (Chen, 1987; Chen and Chang, 1994). The Cretaceous paleoclimate of the Songliao Basin is reconstructed from four data sets: spore/ pollen and plant fossils, oxygen isotope data, paleoecology and climatically sensitive deposits.

Researchers have long employed percent ratios of pollen and spore taxa, to estimate temperature (Liu and Leopold, 1994; White et al., 1997; Liu et al., 2002), humidity (van der Zwan et al., 1985; Barron et al., 2006), as well as ecological environment (Hubbard and Boulter, 1983; Kalkreuth et al., 1993; Larsson et al., 2010). In this way, Gao et al. (1999) reconstructed the climate history of the Songliao Basin on the basis of more than 20,000 samples from more than 500 cores. The vegetation landscape of the basin in Cretaceous time was mainly a conifer forest and steppe (Figure 1.9). The Cretaceous atmospheric temperature changed relatively frequently in the Songliao Basin, but was mainly humid to semi-humid and subtropical. Four Early and Late Cretaceous cooling events are recorded by: ① the Huoshiling Fomation and the 1st and 2nd members of the Shahezi Formation; ② the 4th member of the Denglouku Formation; ③ the Nenjiang Formation; ④ the 2nd member of the Mingshui Formation. Three warming events are: ① the 1st and 2nd members of the Denglouku Formation; ② the Qingshankou and Yaojia formations; ③ the Sifangtai Formation. Three semi-arid events are: ① the 3rd and 4th members of the Shahezi Formation; ② the 4th member of the Denglouku Formation; ③ the Sifangtai Formation.

Chamberlain et al. (2013) provide the first reported oxygen isotopic results from the SK-1 of the Songliao Basin (Figure 1.9). The oxygen isotopic data are from ostracods collected that cover the interval from the Qingshankou Formation through to the Mingshui Formation. These data record distinct oxygen isotope shifts. There is a negative oxygen isotope shift in the Qingshankou Formation, with $\delta^{18}O$ values changing from −10‰ to as low as −18‰ (Figure 1.9). Following this negative shift, oxygen isotopes generally tend to increase from the Qingshankou through the Nenjiang formations. In the Sifangtai Formation, $\delta^{18}O$ values decrease and for the most part are more positive thereafter. We also collected marine oxygen isotope data from the Far East (Zakharov et al., 1999, 2009, 2011; Figure 1.9) to compare with other Far East areas at similar latitudes to the Songliao Basin. In general, the trends in the oxygen temperature records from the Far East sites are similar to the oxygen data analyzed in the Songliao Basin by Chamberlain et al. (2013). In the Far East marine record, Cretaceous climate changed relatively frequently, and mainly in temperate, when the temperature was commonly above 5℃ with the highest of >25℃. The temperature range between warming events and cooling events was between 5—10℃, sometimes ranging up to 15—20℃. However, the magnitude of the isotopic shift is much larger in the Songliao Basin than that in the marine sections.

Plant fossils and spore/pollen are abundant and coal is widely distributed (Figure 1.9) in Lower Cretaceous stratigraphic units of the Songliao Basin, indicating a relatively well-developed plant ecosystem (Wang et al., 1995; Gao et al., 1999). Floras reflect a warm and humid temperate climate, as does lacustrine Jehol biota (Huang et al., 1999). As in the oldest Cretaceous stratigraphic unit, a large number of plant and spore/

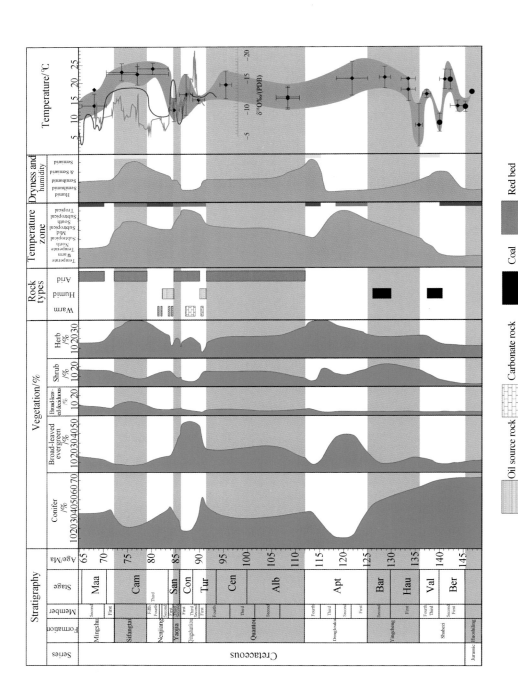

Figure 1.9 Cretaceous paleoclimate evolution of the Songliao Basin. Spore/pollen relative abundances, paleotemperature zones and paleohumidity are derived from Gao et al. (1999), climate-sensitive rock types are from Wang et al. (1994). Temperatures (black curve, black dots, diamonds) are from Zakharov et al. (1999, 2009, 2011). Oxygen isotope data (red curve) are from Chamberlain et al. (2013). Classification of temperature zones is based on the species of spore/pollen spectra that define tropical, tropical-subtropical, subtropical, tropical-temperate, and temperate conditions. Dryness and humidity are based on species of parent plants of spore/pollen fossils subdivided into xerophyte, mesophyte, hygrophyte, helophyte, and hydrophyte types, which correspond to arid, semi-arid, semi-humid, and semi-arid, semi-humid, humid. The red bar = warming event; blue bar = cooling event; yellow bar = semi-arid event.

pollen fossils occur in the overlying stratigraphic units (Wang *et al.*, 1995; Gao *et al.*, 1999). Diverse ecosystems were best developed in the Middle Cretaceous stratigraphic units of the Songliao Basin. At that time, the diversity of genera and species reached a maximum, reflecting well-developed plant and aquatic ecosystems. Many species of foraminifera, chlorophyta and fish first appeared during this period. During deposition of the uppermost Cretaceous stratigraphic units, the fossils are mainly chlorophyta and ostracoda (Wang *et al.*, 1985; Ye *et al.*, 2002). Compared with the Middle Cretaceous mernber, the fossil diversity of the Mingshui biota decreased markedly in numbers of classes and species, which implies an associated shrinkage of the basin and a suppression of the entire ecosystem.

The distribution of coal indicates a range of humid climates under different temperature conditions (Meyerhoff and Teichert, 1971; Boucot *et al.*, 2009). Coal beds occur widely in the Lower Cretaceous Shahezi and Yingcheng Formations and indicate Early Cretaceous humid climates in the Songliao Basin (Figure 1.9). Coal beds are not present in the Denglouku to Mingshui formations, indicating that the paleoenvironment of Songliao Basin changed to a dryer climate with deposition of fluvial to lacustrine facies sediments. Terrestrial red beds are considered to be indicative of a climate that is warm and dry or seasonal with respect to rainfall (Parrish, 1998; Du *et al.*, 2011). In the Songliao Basin four large-scale red bed depositional events are evident in the Quantou Formation, 2nd to 3rd members of the Qingshankou and Yaojia formations (Du *et al.*, 2011), the Sifangtai Formation, and 2nd member of the Mingshui Formation.

Based on all the above information, we conclude that the Songliao Basin was mainly a humid to semi-humid subtropical environment. General trends of the spore/pollen and oxygen isotope data from the Songliao Basin and Far East are similar. Temperature was commonly above 5℃ and reached >25℃; the temperature range between warming events and cooling events was typically 5—10℃, sometimes ranging up to 15—20℃ (Figure 1.9). Global paleovegetation simulations (Upchurch *et al.*, 1998; Sewall *et al.*, 2007) indicate the same conclusion. The major vegetation type of the Songliao Basin was a broad-leaved evergreen forest during the Cretaceous, reflecting a temperate and humid climate; the southern part of the basin was adjacent to arid climate areas and had a slightly arid climate during some periods of the Cretaceous. The temperature was above 0℃ throughout the year, and rainfall was relatively abundant.

The conclusion is also supported by the global paleoclimate reconstruction of climatically sensitive deposits (Boucot *et al.*, 2009). East Asia, including the Songliao Basin, was in the warm temperate zone during most of the Cretaceous and did not experience large changes (Boucot *et al.*, 2009). In a greenhouse world without ice caps, ancient ocean currents became the primary controls of continental climate changes (DeConto *et al.*, 1999; Hay, 2008, 2011). The Songliao Basin was influenced by warm currents flowing northward from the Pacific equatorial regions and cold currents flowing southward from the Arctic region (Puceat *et al.*, 2005; Haggart *et al.*, 2006); these currents approximately correspond to the modern Kuroshio and Oyashio currents, because of Pacific oceanic circulation patterns offshore of the Songliao Basin were broadly similar to those of today, keeping a relative stable state (Gordon, 1973; Klinger *et al.*, 1984).

The Huoshiling Formation represents a humid temperate environment. Overlying the Huoshiling Formation is the Shahezi Formation which indicates more arid conditions and a low temperature, with a decrease in conifers and an increase in shrubs and herbs. However, the presence of coal beds indicates a humid conditions prevailed at times during this period. The Yingcheng Formation indicates a semi-humid to humid subtropical environment, in which the temperature (mainly 15—20℃) decreased and conditions became more humid (Gao *et al*., 1999; Zarkharov *et al*., 2009, 2011). Coal beds occur in this formation and there was a decrease in conifers and an increase in shrubs and herbs. The climate of the Denglouku Formation changed rapidly and sharply. There was a rapid decrease in conifers and a sharp increase in broad-leaved evergreens and herbs. Above the Denglouku Formation is the Quantou Formation, that indicates relatively stable temperature and humidity conditions (Gao *et al*., 1999; Zarkharov *et al*., 2011), and a semi-humid subtropical environment. The presence of red beds and absence of coal beds suggests a more arid environment than the lowest stratigraphical unit and a transition from fluvial to lacustrine facies. The Qingshankou Formation was a humid subtropical-tropical environment; there was a sharp decrease in conifers and a sharp increase in broad-leaved evergreens and herbs (Gao *et al*., 1999). The occurrence of red beds suggests that was abrupt changes in the climate. The climate during deposition of the Yaojia Formation was cooler and more arid than the Qingshankou Formation. Temperatures during the period of deposition of the Nenjiang Formation were about 5—10℃ (Zarkharov *et al*., 1999). The Sifangtai Formation represents a semi-humid subtropical environment, with the occurrence of red beds. The temperature was around 20℃ and there was a decrease in conifers and an increase in broad-leaved deciduous trees and herbs. The Mingshui Formation represents a semi-humid temperate environment (Gao *et al*., 1999; Zarkharov *et al*., 1999, 2011; Chamberlain *et al*., 2013).

1.4 Scientific Objectives of the Continental Scientific Drilling Project of the Cretaceous Songliao Basin

The Continental Scientific Drilling Project of the Cretaceous Songliao Basin, northeastern China, is aimed to recover a nearly complete Cretaceous terrestrial sedimentary record. The recovered cores will provide unique opportunities for the geosciences community to advance the understanding of climate change in the Cretaceous "greenhouse world", and provide documentation of geological events relevant to the carbon cycle during this period. The drilling project will address significant geological questions, such as the identification of important stratigraphic boundaries and marine-terrestrial stratigraphic correlations, reasons for biotic response to terrestrial environmental changes, terrestrial response to Cretaceous oceanic anoxic events, formation of terrestrial petroleum source rocks, and mechanisms for the Cretaceous magnetic Normal Superchron (CNS).

1.4.1 Stratigraphic Boundaries and Marine-terrestrial Correlation of Stratigraphy

The stratigraphy of Cretaceous Songliao Basin has been constructed mainly from domestic biostratigraphic data (Ye *et al.*, 1990, 2002; Gao *et al.*, 1994; Wang *et al.*, 1996; Chen *et al.*, 2003; Zhang *et al.*, 2003; Huang *et al.*, 2007), but any attempt to study terrestrial paleoclimate and paleoenvironment change in the basin has to first accurately date the non-marine strata and correlate this with marine deposits, because marine strata form the basis of the international chronostratigraphic system. The terrestrial-marine correlation of stratigraphy in Cretaceous Songliao Basin will include identification of terrestrial stratigraphic boundaries (J/K and K/T), and temporal correlation of the fluvial-lacustrine succession with the coeval marine succession, and to pinpoint the stage boundaries (Wang *et al.*, 2012; Wan *et al.*, 2012).

The J/K boundary is one of the most problematical Phanerozoic system boundaries, and has long been debated, but is not yet resolved on a global scale. Identification of the J/K boundary in northern European successions is extremely difficult and complicated due to the absence of stratigraphically useful fossil assemblages (Batten, 1996). Understanding of the extinction event at the Jurassic-Cretaceous (J/K) boundary is hampered by problems of faunal isolation and lack of comparative magneto- and biostratigraphy between the Boreal and Tethyan faunal realms. Also, the questions of whether this extinction was a sudden or-in geological terms-prolonged event, and whether regional rather than global agents were important factors, remain open questions. A number of different processes have been invoked as triggers for the J/K extinction, including volcanism, regional tectonic processes, and multiple impact events (e.g., review in McDonald *et al.*, 2006). Dating of the Jurassic-Cretaceous boundary in China has similar difficulties. Fortunately, abundant igneous intrusions and eruptions around the J/K boundary are preserved in the Songliao Basin, and can be dated precisely to aid in the determination of the chronology of the terrestrial J/K boundary (Wang *et al.*, 2002).

Another intriguing aspect of the Songliao Basin is the possibility that the K/T boundary is preserved there. The K/T boundary is defined world-wide by a thin layer that contains shocked minerals and traces of extraterrestrial material, which were derived from a large-scale impact event in present-day Mexico, the 200-km-diameter Chicxulub impact structure (e.g., Schulte *et al.*, 2010). Although there are abundant and well-preserved K/T boundary locations in many parts of the world, it is poorly documented in China or elsewhere in eastern Asia (e.g., Kiessling and Claeys, 2001). Therefore, finding a good K/T boundary in eastern Asia would allow better quantification of ejecta distribution, deposition mechanisms, and environmental effects than has been possible to date. Northeast China is a potential place to find this boundary, because recently some of the project proponents studied several outcrops in this area and found a promising section in the northern part of Songliao Basin, which indicates the presence of at least the uppermost beds of the Maastrichtian from spore/pollen fossils and U-Pb zircondating. According to palynology, charophytes, and magnetostratigraphy studies, we anticipate that the K/T boundary is preserved in the SK-1 core (Wan *et al.*, 2013).

1.4.2 Biotic Response to Terrestrial Environmental Change and the Deep Biosphere (Fossil DNA)

A group of fossil-bearing sedimentary inter-layers within the basal volcanic-sedimentary deposits occur southwest of Songliao Basin, which include fossils of early birds, angiosperms and feathered dinosaurs, as well as bivalves, gastropods, conchostracans, ostracods, shrimp, insects, fish, amphibians, primitive mammals, and various reptiles (the well-known "Jehol Biota"; Ji *et al.*, 1998, 1999, 2001, 2002; Chen *et al.*, 1999). Rather than being endemic to eastern Asia, the Jehol Biota has a large geographic distribution, covering more than half of Europe. Many species of the biota have close relationships with Late Jurassic European counterparts, e.g., both *Confuciusornis* and *Archaeopteryx* are primitive birds, *Sinosauropteryx* has a skeleton construction similar to *Compsognathus*, the ostracod assemblage is similar to that from the Purbeck Bed in England, and *Aeschnidium* occurs in Solnhofen Formation of Germany. In addition, abundant diversified fossil biota are well-preserved in the middle and upper strata of Songliao Basin. Ye *et al.* (1990, 2002) termed the biota from the Denglouku Formation to the Nenjiang Formation as the "Songhuajiang Biota", and the biota from the Sifangtai and Mingshui Formation as the "Mingshui Biota". These biota are mainly composed of micro-fauna and-flora, including abundant ostracoda, conchostraca, charophyta, dinoflagellata, spore/pollen fossils, on the basis of accumulation of the data since the work of 1940s. Abundant of macrofossils, including fish, bivalves, and plants, can also be found in the basin and surrounding areas (Zhang *et al.*, 1976, 1977; Deng *et al.*, 1998; Gu *et al.*, 1999). The evolutionary patterns of the biota will yield much information about the paleoclimate and paleoenvironment during different stages of basin evolution, and provide one of the bases for the stratigraphy of the basin-filling sediments.

Traditionally, reconstruction of past ecosystems has focused on multi-cellular organisms that leave visually discernible fossil remains. However, developments in the characterization of lipid biomarkers over the past twenty years and more recent developments in the analysis of ancient DNA now allow probing of the microbial and algal community in great detail. In thermally mature sediments, the presence of hopanes (bacteria), steranes (algae), highly branched isoprenoids (diatoms), dinosterane (dinoflagellates), methylhopanes (cyanobacteria), irregular isoprenoids (archaea) and isorenieratene derivatives (green sulfur bacteria; Pancost *et al.*, 2004), allows first-order characterization of these communities, i.e., microbial functional groups. In less mature sediments, an even wider range of biomarkers is present, allowing more specific profiling of the microbial community. For example, biomarkers for specific organisms, such as nitrifying pelagic Crenarchaeota, sulfate reducing bacteria, methanotrophs, haptophyte algae and eustigmatophyte algae (Volkman *et al.*, 1998), have all been identified. Such lipid biomarker analyses provide a robust platform for the characterization of ancient microorganisms.

Beyond traditional paleontology, recent developments show that fossil DNA, when preserved, could allow significantly more specific characterization of past microorganisms than is currently possible using lipid biomarkes. Studies have shown great promise in reconstructing paleo- microbial communities and paleoenvironmental conditions through investigations of fossil DNA and molecular biomarkers preserved in sediments and sedimentary rocks (Coolen and Overmann, 2007). Inagaki *et al.* (2005) successfully extracted and amplified DNA from a core sample of black shale of the Cretaceous age collected from Serre des Castets, near Marseilles, France, and concluded that the cores were free of contamination, and the microorganisms observed were indigenous. Based on this work, they proposed the term "Paleome" to describe "the use of preserved DNA and/or microbes to interpret the past". Of course, that term also applies to over 20 years of similar work based on the aforementioned lipid biomarkers. The Continental Scientific Drilling Project of the Cretaceous Songliao Basin provides an unprecedented opportunity for us to further test the "Paleome" concept and to possibly resolve the current controversy (Coolen and Overmann, 2007).

1.4.3 Terrestrial Response to Oceanic Anoxic Events and Formation of Massive Terrestrial Hydrocarbon Source Rocks

The occurrence of Cretaceous OAEs is often accompanied by high-amplitude fluctuations of the ^{13}C record in both the ocean and land-derived carbonates and organic matter, which are interpreted to reflect prominent changes in global carbon cycling in the course of these events (Gröcke *et al.*, 1999; Hesselbo *et al.* 2007). Various lines of evidence also indicate prominent climatic fluctuations during OAE episodes (Kuypers *et al.*, 1999; Wilson and Norris, 2001; Heimhofer *et al.*, 2005), which would have strongly affected continental environments. Some of these effects have already been identified by studying terrestrial signals recorded in marine deposits. This includes drastic changes in weathering and erosion as indicated by Os-isotope patterns during the Toarcian OAE (Cohen *et al.*, 2004). A major shift from C3- to C4-dominated vegetation due to lowering of P_{CO_2} has been proposed by Kuypers *et al.* (1999) for OAE2. In addition, prominent fluctuations in precipitation patterns and continental freshwater run-off were reported for OAE1b (Herrle *et al.*, 2003), OAE1d (Bornemann *et al.*, 2005) and OAE3 (Beckmann *et al.*, 2005). For a better understanding of the interactions between the different regimes of the global ocean-atmosphere system, the role of continental environments during the Cretaceous OAEs needs to be investigated in more detail. For example, ostracods were collected from the SK-1 cores for carbon, oxygen and strontium isotopic analysis and show numerous carbon and oxygen isotope shifts that are both rapid and long-term (Chamberlain *et al.*, 2012, this volume). Also, numerous of red bed horizons are well developed in the Songliao Basin and exposed in the SK-1 cores (Wang *et al.*, 2013). The relationship between their occurrences and Cretaceous Oceanic Red Beds (CORBs) is still unclear (Wang *et al.*, 2005, 2009; Hu *et al.*, 2005).

Another intriguing problem is the relationship between the formation of terrestrial petroleum source rocks to the occurrences of OAEs, as the source rock of the Daqing Oilfield, Member 1 of the Qingshankou Formation and Members 1+2 of the Nenjiang Formation, can be roughly correlated to coeval marine OAEs (Huang and Huang, 1998; Huang *et al.*, 1999) based on current stratigraphic data (Huang *et al.*, 2007). One of the lines of evidence is that the terrestrial source rocks in the Songliao Basin approximately correspond to positive organic carbon spikes, which might be related to the occurrences of OAEs in the ocean (Wang *et al.*, 2005; Huang *et al.*, 2007). Although the Songliao paleo-lake was fresh during most of its life cycle, the source rocks in the Songliao Basin were deposited under saline water conditions, as suggested by paleontological, geochemical and mineralogical data (Wang *et al.*, 1994; Hou *et al.*, 2000; Wang *et al.*, 2001). Some authors argue that salinization of the lake was due to marine incursion during transgressions, leading to stratification of the water column and resulting anoxia favorable to the formation of terrestrial source rocks, and the such transgressions occurred during certain OAEs (Wang *et al.*, 1992; Hou *et al.*, 2000; Wang *et al.*, 2001). However, this hypothesis needs further testing.

A combined approach integrating isotopic, elemental and organic geochemical, as well as palynological techniques, will provide new insights into the terrestrial and lacustrine environmental dynamics during the mid-Cretaceous, a time of major perturbations of climate and the carbon cycle, and for the formation of terrestrial petroleum source rocks in the Songliao Basin. The following questions will be addressed: ① Do the organic carbon burial events in the marine realm have continental counterparts? ② How does a giant Cretaceous lacustrine setting respond to episodes of global OAE? ③ Do we observe significant changes in continental climates during OAE episodes? ④ What is the response of mid-latitude flora to the proposed climate perturbations? ⑤ Did salinity conditions play a role in the formation of terrestrial petroleum source rocks; and what was the cause of the saline conditions?

1.4.4 Cretaceous Normal Superchron from Terrestrial Records

A key piece in the mid-Cretaceous geologic puzzle is the Cretaceous Normal Superchron (CNS), a period in which Earth's magnetic field was so uncharacteristically steady that it did not switch from normal to reversed polarity for about 40 million years (Aptian to Santonian, 124—84 Ma). There are two views on the reversal rates preceding the CNS, which have quite different implications for Earth history and associated events and processes. In one, McFadden and Merrill (1984, 2000) envisioned the CNS as a natural product of the decrease in the reversal rate of polarity due to a change of mantle convection during that time. The other view, supported by Gallet and Hulot (1997) and Hulot and Gallet (2003), holds that the CNS could represent a sudden non-linear transition between a reversing and a non-reversing state of the geodynamo. Presently there is still no final answer for the origin of the CNS.

The recovery of a complete Cretaceous section from the Songliao Basin is essential for answering the above questions, if we are to: ① understand how the CNS evolved over time and how the associated magmatic and tectonic processes are linked; ② determine high-resolution magnetostratigraphy; ③ determine the paleomagnetic polar wandering curve for the Cretaceous in China; ④ provide new knowledge about the geodynamo and Earth's interior. With answers, or partial answers, to these questions we can move on to other important questions: ① Are reversal rate and average paleo-intensity correlated? ② Does the directional dispersion of new data support the earlier finding of McFadden *et al.* (1991) that the secondary harmonic family was low during the CNS compared to times when the reversal rate was high? ③ Do the data support the suggestion of Hulot and Gallet (2003) that the CNS might have occurred as a sudden flip of a highly nonlinear system to a different (non-reversing) state, with essentially no evolutionary change in field behavior leading up to or leading out of the superchron? Recent work on SK-1 cores has shown that the age of the termination of the Cretaceous Normal Superchron was 83.4 Ma and intervals of reverse geomagnetic field polarity of early Santonian age could be present during the late CNS (He *et al.*, 2012). Continuous efforts on the SK-2 cores may give some answers to the above hypothesis.

Section 2
Selection of Drilling Site

The basis of selecting a drilling site for continental scientific drilling is to help resolve key scientific questions in earth science. Three steps are necessary for continental scientific drilling site selection: first, pre-site selection is conducted to find geological sites that are both globally significant and engineeringly feasible; second, the best area to site the drilling in order to fully satisfy the scientific objectives is selected through discussions in international workshops; third, the exact drilling site is determined after precise geological and geophysical observations and a 3D geological and geophysical model is completed (Wang *et al*., 2007). To fulfill these requirements, based on surveys and exploration in the Songliao Basin for more than 50 years, geologists and engineers have determined the drilling site of the SK-1 Drilling Project.

2.1 Process of Site Selection

In March, 2005, when the National Key Basic Research Program (973) "Cretaceous Major Geological Events in Earth Surface System and Greenhouse Climate Change" was proposed, the "Cretaceous Songliao Basin Scientific Drilling" was planned. Key strata (Qingshankou, Yaojia, Nenjiang, Sifangtai formations with a combined thickness of about 1000 m) were targeted to be cored in the Sanzhao area, east of Changyuan, Daqing City. This drilling was termed "Terrestrial Cretaceous Drilling", in which biostratigraphy, magnetostratigraphy, lithostratigraphy, radioactive dating, stable isotope stratigraphy, sequence stratigraphy, and chemostratigraphy were to be studied.

During September 27—28th 2005, at a meeting at the Institute of Oil-Gas Exploration and Development, Daqing Oilfield, Geological delegate Feng Zihui suggested four potential drilling sites (SK-1-1, SK-1-2, SK-1-3, SK-1-4), where the six principle aims of the site selection could be fullfilled.

During a working conference held at China University of Geosciences, Beijing, in January, 2006, based on the stratal characteristics in the Songliao Basin, the SK-1-4 site was selected as the drilling location for SK-1, and located in the Ao'nan nose-like structure of the central part of the Gulong sag, north of the Songliao Basin. Because of tectonic disturbance and a resultant shift of the basin depocenter in late Cretaceous, the Mingshui Formation is not present at the drilling site. Therefore, it was proposed that cores of the Sifangtai and Mingshui formations could be obtained from a current exploitation well located on the western slope of the basin ("SK-1a" well). These two holes were designated as "SK-1" and "SK-1a". This proposal was incorporated into the draft plan of the "China Cretaceous Continental Drilling Project-SK Overall Design".

On February 15, 2006, "SK Engineering Conference" was held in Daqing and a composite profile based on "one well-two holes" was suggested by Wang Da and some other engineers to resolve the problem that depocenter of the Songliao Basin migrated in the basin evolution process. As a consequence of the conference, the original plan was changed, and the proposed "SK-1" site on the Ao'nan nose-like structure in the Gulong sag was renamed "SK-1s". The coring program was changed from total coring to partial coring from the base of Nenjiang Formation Member 2 to the top of Quantou Formation Member 3. The core-supplementary well originally designed for the western slope of the basin was also changed to a new scientific exploration well in the Tahala area within the Qijia-Gulong Sag and designated as "SK-Main Hole or SK-1n", with cores to be taken from the Taikang Formation to the top of Nenjiang Formation Member 1. The two drillholes could be correlated using a basin-wide oil-shale marker bed in Nenjiang Formation Member 2 (Figure 2.1). After studying three-dimensional seismic profiles, the location of drill site was repositioned.

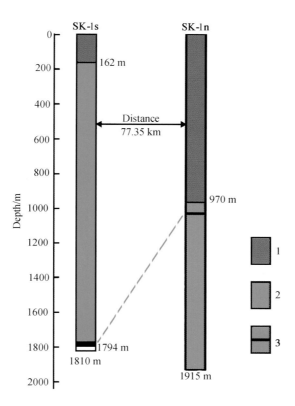

Figure 2.1 Correlation of SK-1s and SK-1n drilling sites. 1. segment not cored; 2. segment cored; 3. oil shale at the base of the second member of the Nenjiang Formation used as a marker bed for stratigraphic correlation within the basin.

On February 20, 2006, scientists and engineers undertook field reconnaissance to examine the drill site. Taking into consideration the ground conditions, the location of the sites was slightly changed. On March 12, during additional field reconnaissance, potential engineering difficulties were recognized and the site was once again slightly adjusted. After examination of 3D seismic profiles across the new site, it was agreed that the new site satisfied drilling requirements. Precise location of the "SK-1s" site was determined by the satellite positioning system by the Logging Company, Daqing Oilfield. The design for "SK-1s" was finalized by the Daqing Oilfield Company late in March 2006.

At a meeting in the field on May 2006, the precise drilling site for SK-1n is discussed. Geologic Director Wang Pujun demonstrated a correlation of the proposed drilling site with the adjacent well Ta21 that confirmed stratigraphic continuity, especially the K/T boundary, and concluded that the designed site was ideal for meeting the scientific objectives of the project. After discussion, the plan was adopted. Location of site SK-1n would not be changed, but drilling depth would decrease from 1810 m to 1780 m, with the coring depth 1764 m. After drilling through the base of Nenjiang Formation Member 2 (1762 m) 2 m core should be obtained, and an additional 16 m was to be drilled without coring. The final design for the SK-1n was completed by the Daqing Oilfield Company in late July 2006.

In July 2006, the two boreholes were designated SK-1s (south borehole) and SK-1n (north borehole).

2.2 Site Location

Six main scientific principles governing the location of the drill site are: ① most continuous strata with almost no hiatuses; ② no important geological events to be missing; ③ minimal drilling thickness; ④ recovery of cores from predominantly deep lake facies; ⑤ recovery of boundaries of the uppermost and lowermost formations in both cores; ⑥ minimal engineering difficulty.

The long-term deepwater sag of the Songliao Basin, where there was a low rate of mudstone deposition resulting in thinner strata thickness and better strata preservation than in adjacent areas, made the sag an ideal area for the SK drilling site. Although the depocenter of the basin shifted several times during the Cretaceous, it primarily remained in the area of the Gulong sag. To ensure stratal continuity and minimal engineering difficulties, a further step to avoid faults and select the area of minimum strata thickness was taken after analysis of 3D seismic data from the top of Quantou Formation Member 3 to the base of Nenjiang Formation Member 2 [Figure 2.2(a),(b)]. As a result, the SK-1s drilling site was sited 2.5 km east of Zhaojiawopeng, Xingfu village, Maoxing town, Zhaoyuan county, Daqing city, Heilongjiang Province (Figure 2.3, Table 2.1). Tectonically, this area lies on the flank of the Ao'nan nose-like structure, Gulong sag, Central Depression

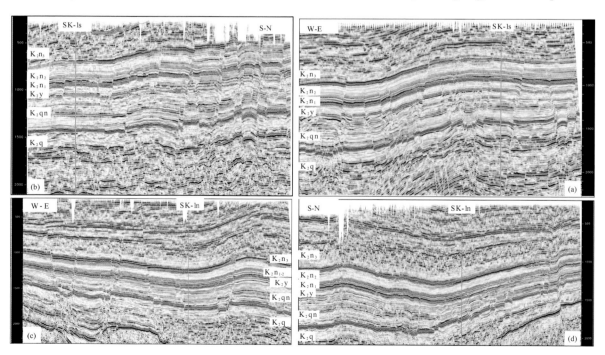

Figure 2.2 Seismic reflection data across the SK-1 drilling site. (a) Seismic section across SK-1s drilling site (west-east transect); (b) Seismic section across SK-1s drilling site (south-north transect); (c) Seismic section across SK-1n drilling site (west-east transect); (d) Seismic section across SK-1n drilling site (south-north transect). K_2q. Quantou Fm.; K_2qn. Qingshankou Fm.; K_2y. Yaojia Fm.; K_2n_1—K_2n_3. Nenjiang Fm. Members 1—3.

Zone of the northern Songliao Basin. Seismic data from adjacent wells show that stratal development from the Nenjiang to Quantou formations is complete in this area. Except for the Quantou Formation, the Nenjiang, Yaojia and Qingshankou formations are dominated by mudstone with minor exogenic debris, and is useful for understanding the Cretaceous paleoclimate record.

According to the stratigraphy from the top of Nenjiang Formation Member 1 to the base of the Taikang Formation, the SK-1n drilling site was finally sited 150 km east to Xiaomiaotunzi, Datong district, Daqing city, Heilongjiang Province (Figure 2.3, Table 2.1). Tectonically, the drill site lies in the Tahala syncline, Gulong sag, Central Depression Zone of the northern Songliao Basin. Seismic data from this site and adjacent wells show that stratal development from Nenjiang Formation Member 1 to the Mingshui Formation is complete, and of small thickness [Figure 2.2(c),(d)]. The target strata are dominated by mudstone, and therefore useful for Cretaceous paleoclimate research.

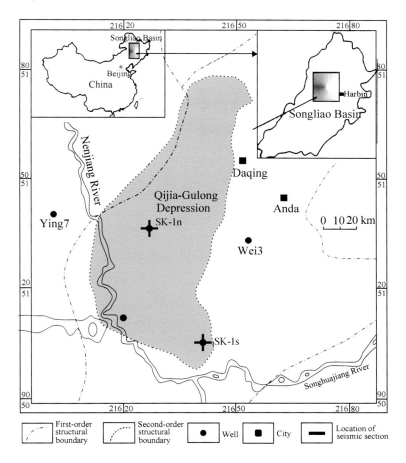

Figure 2.3 Geographic location of SK-1 drilling sites.

Table 2.1　Drilling site selection data of SK-1

Well name	SK-1s	Well type		Scientific drilling	Well trajectory	Straight well
Geographic location	150 m east of Xiaomiaozi Village, Datong District, Daqing City, Heilongjiang Province, China					
Structural location	Tahala Syncline, Gulong Sag, Central Depression Zone, Northern Songliao Basin					
Seismic survey	3D	Inline 782 / Crossline1063				
Earth coordinates	x	5120461 m		Longitude	124°15'56.78"E	
	y	21597675 m		Latitude	46°12'44.22"N	
Designed depth	1810 m			Deepest strata	K_2n_2	
Geographic location	2.5 km east of Zhaojiawopeng, Xingfu Village, Maoxing County, Zhaoyuan Town, Daqing City, Heilongjiang Province					
Structural location	Ao'nan nose-like structure, Gulong Sag, Central Depression Zone, Northern Songliao Basin					
Seismic survey	3D	Inline480/Xline666				
Earth coordinates	x	5049726 m		Longitude	124°40'15.59"E	
	y	21630438 m		Latitude	45°34'14.42"N	
Designed depth		1915 m		Deepest strata	K_2q_3	

2.3　Simulated Lithological Column for Well SK-1 before Drilling

During the scheme demonstration process of SK-1 "one well-two holes", analysis of data from adjacent wells, e.g., stratigraphic units, thickness, top and bottom interface properties of strata, and sedimentary facies, together with seismic profiles, were used to predict the lateral extent of the strata, and determine the geological design of the south and north wellbore. Using logging and core data of nearly 10 wells adjacent to SK-1 (South Hole 4 and North hole 6), a simulated lithologic column predicting geological layers and changes in thickness in the two holes, was made prior to drilling (Figure 2.4, Figure 2.5). This was finally proven to be very important for the SK-1 engineering.

The simulated lithologic column predicted the lithologic characteristics and sequence in SK-1, that included lithology, color, volcanic ash horizons and rock strength, in addition to temperature and pressure gradients, and underground fluid. This data provided a detailed and reliable basis for core description, adjustments to the drilling program and selection of drilling techniques for the different layers encountered during drilling of SK-1.

The actual drilling results indicate a >90% match between the simulated and actual lithological columns and consistency, temperature, pressure and fluid properties. This type of simulated lithological column plays an important role in drilling technology and mud selection. The experimental results in SK-1 indicates that simulated lithological columns should be made before similar scientific drilling is carried out.

Section 2 Selection of Drilling Site | 29

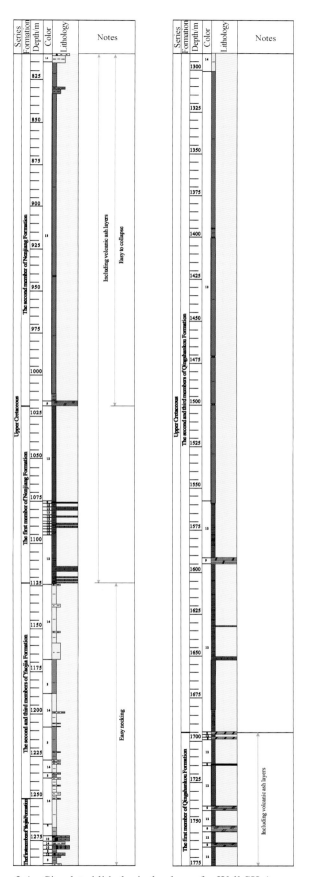
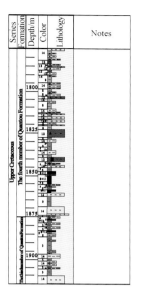
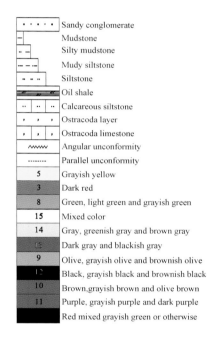

Figure 2.4 Simulated lithological column for Well SK-1s.

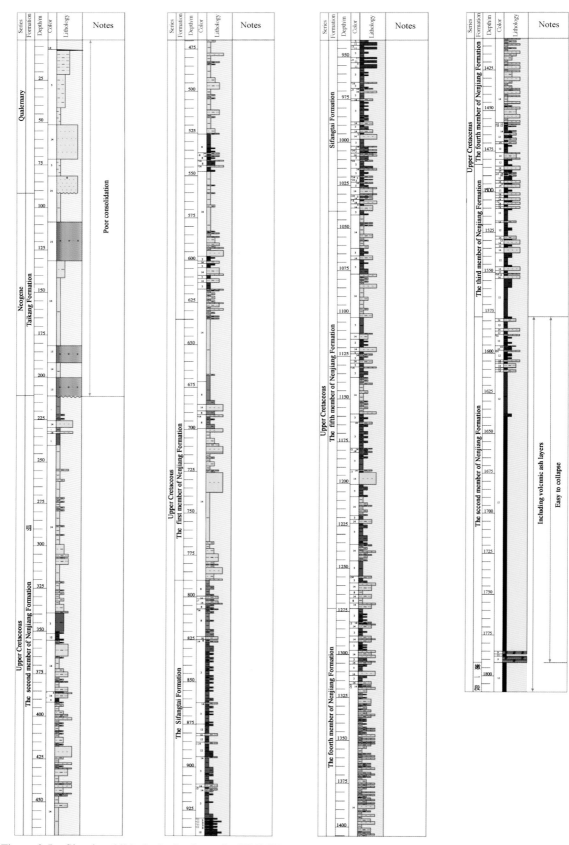

Figure 2.5 Simulated lithological column for Well SK-1n.

Section 3 Implementation of SK-1 Drilling and Preservation of Cores

SK-1 is the first fully cored scientific exploratory well in China aimed at drilling Cretaceous strata and the first terrestrial Cretaceous scientific exploratory well to be drilled under the framework of the International Continental Scientific Drilling Program (ICDP). According to stratigraphy and engineering requirements, and based on extensive experience in scientific drilling initiatives, a complete design concept was applied to SK-1 and this included the four systems/techniques of "economical and flexible one well-two holes design", "advancement by combined drilling technique", "ultra-long high-recovery integrated coring technique and management system," and "core preservation in crystal resin plastic seals-organic glass brackets", organically integrated into the drilling plan of Cretaceous terrestrial strata.

In the borehole design, a one well-two holes design was used for the drilling work, with the deepest hole of 2000 m depth. This not only ensured flexible construction, complete and scientific core recovery, but also enabled implementation in a more economical way to achieve the scientific goals of the project at a lower cost.

Drilling was conducted by high-resolution long-section continuous coring using the "nanometer fluorescent carboxylation microsphere and sealed coring technique", "systematic directional coring technique" and "long-section confined coring technique", including many other techniques first used in China.

In core management and preservation, the "core preservation in crystal resin plastic seals-organic glass brackets" was used. A core repository and a scientific core management system were established for core preservation which laid a solid foundation for future core utilization and subsequent research.

3.1 Drilling Design

3.1.1 Technical Goals

To achieve the objective of obtaining a complete and accurate stratigraphic record within the shortest borehole trajectory, minimizing the engineering difficulty and implementation cost, a special two-hole technique was used for SK-1, i.e., to drill an SK-1 north borehole (SK-1n) in the northern part of the Songliao Basin where strata are well preserved, and an SK-1 south borehole (SK-1s) in the southern part of the basin where the buried depth is shallower, so as to obtain a complete and accurate record of the Cretaceous sediments in the basin.

The goal of the "one well-two holes" technique was to drill a 1780 m deep hole (SK-1n) at Xiaomiaotunzitun in the Datong District, Daqing, Heilongjiang Province, core full length from the surface down to the bore termination and log in-situ, and a 1915 m deep hole (SK-1s) 2.5 km east of Zhaojiawopengtun of Xingfu village, Maoxing town, Zhaoyuan County, Daqing, Heilongjiang Province, a core length from 968.17 m to 1915 m, sample and log in-situ.

Technical requirements for the two SK-1 boreholes were :

(1) Well depth. 1780 m for SK-1n; 1915 m for SK-1s.

(2) Final hole diameter. ϕ140 mm for SK-1n; ϕ215.9 mm for SK-1s.

(3) Well deflexion. SK-1n: 3.0° max for the 0—1000 m section and 3.0°—3.8° for the 1000—1780 m section.

SK-1s: 3.0° max for the 0—1000 m section and 4.2° max for the 1000—1915 m section.

(4) Coring. Continuously cored with full hole recovery of 90% min. For SK-1n, the total core recovery should be 90% min for the section above 1000 m (selected with circlips) and 95% for the section below 1000 m. For SK-1s, the core recovery should be 98% min and the coring footage per barrel should be within 9 m. An additional directional coring was to be made at around 50 m intervals with footage more than 2 m.

(5) Provide assistance in logging, cuttings log sampling and a number of in-well experimental operations.

(6) Completion method. SK-1n: casing cementation for the 0—245 m section and open hole completion for the 245—1811.18 m section; SK-1s: casing cementation for the whole borehole (0—1912 m).

3.1.2 Drilling and Coring Requirements

Drilling and coring was one of the most important scientific objectives of the drilling project and its success would be the basis that decides whether further efforts would be made.

Section 3 Implementation of SK-1 Drilling and Preservation of Cores

(1) For SK-1n, strata were to be cored along a 1602 m section from 50 m above the base of the Taikang Formation (162 m deep) to 2 m below Nenjiang Formation Member 2 (1764 m deep) with a core recovery of 90% min. For SK-1s, strata were to be drilled and cored continuously from 50 m above the base of Nenjiang Formation Member 2 to the bottom of the well (Quantou Formation Member 3) to determine the stratigraphy, with a core recovery of 98% min.

(2) Special coring techniques were to be applied to ensure that the cores recovered remained what they had been underground; the directional error of the cores was to be maintained within the ±15° limit.

(3) The coring footage per barrel should be no more than 9 m. Additional directional coring should be made at around 50 m intervals with footage of more than 2 m.

(4) To allow for core scanning and directional location, the core surface should be unbent, smooth and flat.

(5) Confined coring was to be made when drilling into fragile mudstone or shale sections.

(6) Well cemented strata cores were to be washed with clear water; cores recovered by PVC tubes did not need washing.

(7) Cores recovered were to be horizontally cut off in the conventional way and kept in a core room at the well site specially designed for keeping cores.

(8) A frozen sample was to be recovered every 50 m for special research, following such requirements as to be decided at the site by the project team.

3.1.3 "One Well-two Holes" Design Solution

As the ICDP involves scientific exploratory wells exposed to many uncertainties during the drilling process, more flexible drilling approaches are designed for the drilling construction after technical and economic argumentation. To better secure the scientific objectives, minimize construction risks and secure successful construction, a "one well-two holes" approach was proposed for the drilling project. This approach has been applied in other ICDP projects and has reported very satisfactory results.

KTB of Germany once used a "two-hole" solution, which consists of a large-diameter pilot hole drilled before the formal drilling operation to determine the stratigraphy, and another north borehole that was drilled afterward. For the 9000 m-plus well of KTB, the pilot hole was more than 4000 m deep, which meant additional cost for the drilling operation.

The Chinese Continental Scientific Drilling Project (CCSD) has developed a "flexible two-hole solution" based on the "two-hole" solution applied by KTB in its scientific drilling projects. Essentially, it is an integrated approach consisting of two construction procedures that are applied flexibly according to different construction performances. The first procedure is to drill a 2000 m pilot hole and core along the full depth, and then a 5000 m north borehole around 100 m from the pilot hole, and acceptably leave the first 2000 m of the upper part of the borehole uncored. The second procedure is to directly expand the pilot hole, if the well is of high quality construction and the well structure is uncomplicated, lay the casing, and core and drill down the north borehole, so that the two holes are realized in one. This is called a "flexible two-hole solution", because the "two-hole" solution of KTB remains unchanged while that of CCSD is adaptive to the specific construction conditions encountered. This solution is obviously more economical.

At a "Project Argumentation of SK-1" held in Daqing in February 2006, mainly attended by technical specialists, engineering experts advanced the "one well-two holes" solution for the first time based on the past SE-NW shift event during stratigraphic development and basin evolution of the Songliao Basin. This solution is comprises the drilling project of "SK-1 south borehole" (SK-1s) and "SK-1 north borehole" (SK-1n) that are stratigraphically not repetitive and spatially apart from each other. This solution not only greatly reduced the drilling duration as the two holes are drilled simultaneously, but also saved a large amount of money as the reduced drilling depth more than offset the increased total drilling footage.

3.1.4 Borehole Structure and Casing Procedure

Based on experiences of many countries and after argumentation and investigation, a "one well-two holes" solution was decided for SK-1, under which a $\phi 244.5$ mm (95/8 in) surface casing would be laid when the well had penetrated through the Tertiary into the stable Mingshui Formation mudstone. To prevent hole reduction or collapse due to extended exposure of the Sifangtai Formation strata, a primary backup hole caliper was added between the surface casing and the final hole caliper, and a $\phi 177.8$ mm (7 in) movable casing was inserted into the backup hole caliper according to the practice of the CCSD SDI project. The movable casing had two functions: ① To reduce the annular gap between the drill stem and the surface casing so that small pump duty circulation would be possible for coring and drilling to prevent high-speed slurry flow from eroding the cores and wall, and to avoid well diameter expansion that may impair core recovery. ② If drilling into the lower part proved successful, the coring advancement would be made at a $\phi 140$ mm well diameter (or a $\phi 152$ mm well diameter) until completion; if drilling proved difficult due to complicated lower strata, the movable casing would be pulled out, the hole would be expanded through the complicated horizon and cemented and isolated by a $\phi 177.8$ mm (7 in) technical casing, and then cored and drilled at a $\phi 140$ mm well diameter (or a $\phi 152$ mm well diameter) until completion(Tables 3.1 and 3.7). The well structure is indicated in Figures 3.1 and 3.2. The designed casing approach allowed for coring and drilling at both $\phi 140$ mm and $\phi 152$ mm. By comparison, the $\phi 140$ mm diameter would need lower cost and provide higher drilling efficiency, while the $\phi 152$ mm diameter would result in rougher cores but provide higher core recovery.

Section 3 Implementation of SK-1 Drilling and Preservation of Cores

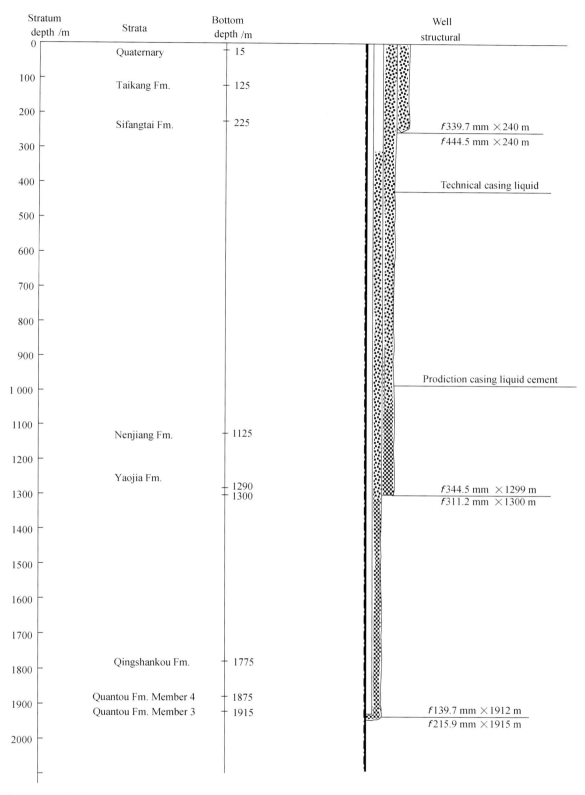

Figure 3.1 Well sketch of SK-1s.

Table 3.1 Bore structure and casing procedure of SK-1n

Spud-in sequence	Bit size/mm × well depth/m	Casing depth/mm × laid depth/m	Stratum unit into which the casing is laid	Annulus liquid cement return depth/m	Remarks
1st spud-in	φ311 × 240	φ244.5 × 240	Mingshui Fm.	Surface	Conventional well cementation
2nd spud-in	φ215.9 × 1020	φ177.8 × 1020	Sifangtai Fm.	Surface	Conventional well cementation. There is possibility that casing is not laid
3rd spud-in	φ140 × 1780				

Table 3.2 Bore structure and casing procedure of SK-1s

Spud-in sequence	Bit size/mm × well depth/m	Casing depth/mm × laid depth/m	Stratum unit into which the casing is laid	Annulus liquid cement return depth/m	Remarks
1st spud-in	φ444.5 × 240	φ339.7 × 240	Nenjiang Fm. Member 5	Surface	Plug-in well cementation
2nd spud-in	φ311.2 × 1300	φ244.5 × 1299	Qingshankou Fm. Members 2+3	Surface	Conventional well cementation
3rd spud-in	φ215.9 × 1915	φ139.7 × 1912	Quantou Fm. Member 3	300	Conventional well cementation

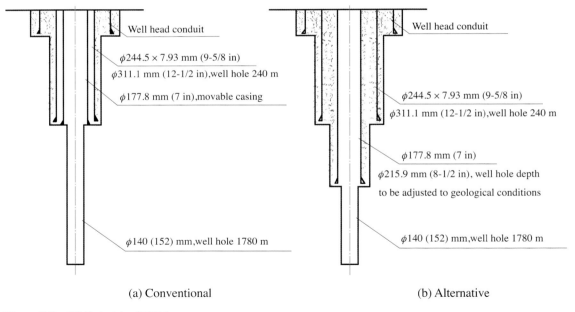

Figure 3.2 Well sketch of SK-1n.

3.1.5 Drilling Devices

To meet the drilling goal of the project, a TSJ-2000 heavy-duty water-feed rig was used for SK-1n, while a ZJ-30DJ rig was used for SK-1s.

1. TSJ-2000 heavy-duty water-feed rig

This machine, completed with its associated auxiliaries, is made by the Shijiazhuang Coal Machine Plant. It is supplied with a 6135AG diesel engine, a 25 M four-leg drill tower and a 110 kW (150 Hp) TBW-1200/7 (1000/8) mud pump. The machine is a modification from a TSJ-1000 water well rig with a unit rope lifting load of 80 kN, a turnplate through diameter of 435 mm, turnplate speed (forward, backward) of 46, 69, 110 and 190 rpms, turnplate torque of 18 kN · m, and turnplate thread tightening torque of 53 kN · m and mainframe weight of 6t. Auxiliaries include a dual hydro-brake, a cylinder thread tightener and a gypsy wheel. ϕ73 mm (2-7/8″) petroleum drill stems are used, with a drilling depth of 2000 m for a ϕ190.5 mm (7-1/2″) bit and 1350 m for a ϕ89 mm (3-1/2″) bit.

The rated load of the typical traveling block and hook for such machines is 640 kN and that of the crown block is 1000 kN. A joint modification with the manufacturer resulted in a 800 kN rated load for the traveling block and hook (the hydro-brake was modified at the same time) and 1200 kN for the crown block.

2. ZJ-30DJ rig

The ZJ30/1700TB rig is a new petroleum drilling device specifically designed for hydrocarbon areas in and around north Shaanxi and Inner Mongolia. It is a combination of the world-leading K-shaped derrick and base assembly suitable for working on 3000 m straight wells and 2500 m directional wells, with parameters in line with GB1806-86, "The Type and Basic Parameters of Petroleum Drilling Rigs". Technical parameters are presented in Table 3.3.

Table 3.3   Technical parameters of ZJ-30DJ rig

Item	Parameters		Item	Parameters	
Nominal drilling scope/m	Stem diameter	ϕ	Derrick	Model	JJ170/41-K
		ϕ		Usable height/m	41
Max hook load/kN	200		Base	Model	DZ170/3.5-T
Winch input power/kW	600			Height/m	Front 2.8 Base 1.4
Nos of traveling block ropes			Diesel engine	Model	PZ12V190B
Wire rope diameter/mm	ϕ28			Nos	2
Turnplate	Model	ZP520		Total installed power/kW	1200×2
	Opening diameter/mm	ϕ520	Transmission mode		

Features include:

(1) It contains the basic equipment configuration of Daqing 130-II drilling rig and characterizes good interchangeability, dependable performance and high operability.

(2) Interlocked operation of two machines, good power property, flexibility for diverse power configura-

tions.

(3) Structural features of the most advanced drilling rigs so far, quick derrick lifting, high stability, provision for integral hauling.

(4) The base of the rig is raised to 5 m, a Cat1 well head device is mounted for drilling operations on high-pressure oil/gas wells and coalbed methane reservoirs.

3.2 Intergrated Coring Technique for the Extremely hong and High-recovery Drilling

3.2.1 Coring Tools

1. Bit selection

The coring bit is the key tool for drilling cores. Whereas the formation level of cores directly influences the recovery level of the cores. Therefore, proper selection of the coring bit will directly decide the quality and performance of coring operations. For confined coring of the hole, a PDC bit with low teeth density and large size cutting teeth that directly penetrate into the water hole was used. This type of bit is good for deep penetration, fast cutting speed and is not easily plugged. There is an annular U-shape trough in the inner chamber of the coring bit to prevent the drilling fluid from eroding the core, and 6 pairs of 10 mm-diameter water return holes in the lower part of the bit to enable the washing fluid to effectively lubricate the core and improve drilling efficiency and core recovery.

2. Transparent composite PVC tubes as confining liners

Transparent compound material PVC was chosen for the inner core liner to protect the loose core structure from damage, reduce the loss of core components, facilitate core extraction and allow the cores to be cut, frozen and transported together with the inner barrel. What's more, the small gap between the core and the PVC tube (around 3 mm) also helps to prevent the core from deformation and damage when enters the barrel. As the PVC tube has sound rigidity and elasticity, and small friction resistance (the coefficient of friction resistance of the PVC tube is only 1/10 that of a steel inner core barrel), this both improves the core quality and favors frozen sampling. Besides, a transparent PVC tube also allows the geological logger to outline the lithology of the cores inside without having to cut the inner barrel. The field application of the PVC tube is shown in Figure 3.3.

3. Core catcher shrinkage

The core catcher is a key part that locks the core inside the core barrel in core cutting and drill lifting at the end of the coring and drilling operation. In our drilling operation, this drilling tool used hydraulic pressure to cut the core, and built the pressure so that the core catcher could shrink. The core catcher was a concealed fully-sealed assembly meshed with the inner surface slope of the bit. It was not in direct contact with the core when recovering it, so did not prevent the core from entering the inner barrel. At the end of coring, the fully-sealed core catcher was closed to protect loose core from dropping to the bottom of the hole when it is extracted. The shrunk core catcher at the end of use is shown in Figure 3.4.

Figure 3.3 Cores by confined coring.

Figure 3.4 Core catcher after coring.

3.2.2 Coring Techniques

1. Technical parameters

The model of the coring tool is BX101; the length of the coring barrel is 5.68 m; the outside diameter of the coring bit is 215.9 mm and the inside diameter is 101.6 mm; the recoverable core diameter is 100 mm, the recoverable core length is 5 m; the outside diameter of the PRV tube is 115 mm, the inside diameter is 108 mm, the length is 5.10 m; the diameter of the pressure steel ball is 50.8 mm, the diameter of the circulation steel ball is 30 mm; the diameter of the suspension pin is 11.5 mm, the material is $45^{\#}$ steel; the minimum caliper of the fully sealed core catcher is 6 mm.

2. Drill combination and work principle

(a) Drill combination: Confined PDC bit + coring barrel + 9178 mm drill collars + 3159 mm drill collars + 127 mm drill stem + square drill stem.

(b) Work principle: The structure of the coring barrel assembly is shown in Figure 3.5. Install the suspension pin 6, connect the pressure device and lay the drill. Before the rig has reached the bottom of the hole, flush the hole at high pump duty first, instead of throwing in steel balls, to remove residues at the bottom of the hole and ensure the coring quality. After washing the inner barrel, throw in a 300 mm steel ball 10 so that the drilling fluid flows annularly from the inner and outer barrels to the bit to cool and carry away cuttings on the bit and help protect the core. When laying down the rig and checking the kelly-in, start the turnplate and begin coring and drilling. At the end of the coring and drilling process, throw a 50.8 mm pressure steel ball 3 into the drill column, build the pressure to 6—8 MPa, snip the suspension pin 6, the inner and outer core barrels will detach from each other, and the inner barrel will travel down. When the pressure device has traveled down a certain distance, the bypass hole of the pressure piston 2 will open, the pump pressure will reduce, and core catcher 15 will deform and shrink under the gravity of the inner barrel (including the core and the PVC tube) and the pressure device so that the core is cut off. The hydraulic pressure tool is especially suitable for loose strata as it breaks away from the disadvantages of traditional mechanical pressure coring tools that contain long pressure devices, limited shear forces and advance invalidation of pin suspension, and allows for washing sediments in the inner core barrel at high discharge levels before drilling.

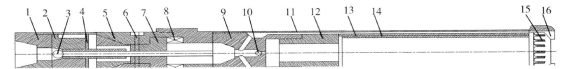

Figure 3.5 Structural sketch of a confined core barrel. 1. upper joint; 2. pressure piston; 3. pressure steel ball; 4. pressure rod; 5. landing nipple; 6. suspension pin; 7. suspension joint; 8. suspension bearing; 9. outer casing reducing joint; 10. circular steel ball; 11. outer core barrel; 12. inner casing reducing joint; 13. inner core barrel; 14. PVC tube; 15. core catcher; 16. coring bit.

3. Coring performance

The designed well sections for confined coring were the 1060—1100 m, 1645—1675 m and 1715—1755 m sections, whose horizons were in Nenjiang Formation Member 1, the upper part of Qingshankou Formation Members 2 and 3, and the upper part of Quantou Formation Member 4. However, from the coring effect of the 1646.01—1647.10 m section in Qingshankou Formation Members 2 and 3, and the 1714.47—1715.27 m section in Quantou Formation Member 4, the good column-forming property, high hardness and long drilling duration (averagely 0.29 m/h) of the cores made it hard to cut the cores and prevented the core catcher from shrinking completely. What was worse, it was impossible for the core catcher to clip off the core. So, to ensure the core recovery and the safety underground, the confined coring of the 1645—1675 m and 1715—1755 m sections was replaced by conventional coring. The performance of confined coring operations is presented in Table 3.4.

Table 3.4 Confined coring operations of SK-1

Barrel	Time	Cored section/m	Footage/m	Core length/m	Recovery/%	Member
1	08:29	971.76—972.26	0.50	4.09	818.00	Nenjiang Fm. Member 1
2	09:04	1060.25—1065.19	4.94	4.94	100.00	Nenjiang Fm. Member 1

Continued

Barrel	Time	Cored section/m	Footage/m	Core length/m	Recovery/%	Member
3	09:04	1065.19—1069.21	4.02	4.02	100.00	Nenjiang Fm. Member 1
4	09:05	1069.21—1074.32	5.11	5.11	100.00	Nenjiang Fm. Member 1
5	09:05	1074.32—1078.82	4.50	4.50	100.00	Nenjiang Fm. Member 1
6	09:05	1078.82—1083.50	4.68	4.68	100.00	Nenjiang Fm. Member 1
7	09:06	1083.50—1087.96	4.46	4.46	100.00	Nenjiang Fm. Member 1
8	09:06	1087.96—1092.54	4.58	4.58	100.00	Nenjiang Fm. Member 1
9	09:06	1092.54—1097.12	4.58	4.58	100.00	Nenjiang Fm. Member 1
10	09:07	1097.12—1100.57	3.45	3.45	100.00	Nenjiang Fm. Member 1
11	09:08	1105.14—1105.24	0.10	0.67	670.00	Nenjiang Fm. Member 1
12	10:20	1646.01—1647.10	1.09	1.09	100.00	Qingshankou Fm. Members 2+3
13	10:24	1714.47—1715.27	0.80	0.80	100.00	Qingshankou Fm. Member 1

3.2.3 Directional Coring Technique

The directional coring technique refers to an etched mark along the axis of the core at the time of coring, with directional data of the core measured and recorded before its original orientation is changed. It is a coring technology that identifies the inclination and dip angle of the stratal crack where the core is located and is essentially intended to return the recovered core to its true state in the strata. The equipment is composed of an directional coring tool, a non-magnetic drill collar, a multi-point electronic inclinometer and a core resetting device.

1. Structure

The directional coring tool is a development from the conventional self-locked coring tools, by replacing the steel ball base at the lower end of the suspension shaft of the conventional self-locked coring tool with a directional base, and replacing the core catcher seat at the lower joint of the inner barrel with a core catcher seat attached with a graver. The structure of the tool is described below:

(a) The multi-point electronic inclinometer is composed of a magnetometer sensor assembly, a data memory and a battery pack, together with some auxiliary sub-assemblies such as a shock absorber, an aligning device, plugs and extension rods, as shown in Figure 3.6. The instrument holder on the multi-point electronic inclinometer is connected to that on the coring tool on a plug-in basis to facilitate instrument installation. During service, it is placed at a suitable point of the non-magnetic drill collar and attached with a high-temperature insulation cylinder to avoid magnetic interference and improve the temperature resistance. At the end of coring, the measurements will be displayed simply by taking the multi-point electronic inclinometer out of the well, connecting it to the surface data processing device and starting the data readout program in the software.

Figure 3.6 Multi-point electronic inclinometer.

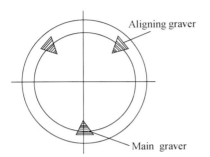

Figure 3.7 Etching knives on core catcher seat.

(b) Three small grooves of unequal round angles are cut in the core inlet at the lower part of the inner cone face of the core catcher seat, and a 55°-vertex angle hard alloy strip of isosceles triangle section is inserted into the groove. Etching knives are distributed at a fixed angle at the downmost end of the core catcher seat, with proper edge exposure height and core feed-in depth to reduce the intrusion resistance of the core and ensure the resolution and accuracy of the directional core mark groove. To guarantee a sound concentricity level between the outer and the inner tubes, the three etching knives are designed to be distributed on the core catcher seat, as shown in Figure 3.7, so that the core orientation is successfully maintained.

(c) The inner surface of the core catcher is composed of a square bosses patch welded with carbon tungsten. In its free state, the inside diameter of the catcher is 2—3 mm smaller than that of the core. During coring and drilling, the core can spread the core catcher and make it move upward along the cone face of the core catcher seat. When cutting the core, the core catcher wraps up the core and slips down the cone face of the core catcher seat under elastic forces. As the carbon tungsten particles have high hardness levels, they can press into the core surface when the core catcher is contracting and thereby increase friction with the core.

2. Technical principle

An azimuth mark groove is designed along the axial direction of the outer circumference of the inner core barrel and the core catcher seat. A key groove is cut in the orientation joint. The centerline of the key groove is on the same line as that of the azimuth mark groove. To ensure the azimuth is fixed, the inner core barrel is directly welded to the landing nipple, and the main graver on the core catcher seat is also on the same line as the azimuth mark groove, so that the measured tool face orientation is the azimuth angle of the core etch mark. During coring and drilling, before the original state of the core changes, mark grooves are continuously cut on the core surface by gravers fixed on the core catcher seat and the azimuth angle, well deflexion angle and well deflexion azimuth angle of the main graver etch mark are measured at fixed intervals by the multi-point electronic inclinometer along with the drilling operation. The coring technician may determine the well depth corresponding to each measuring point with the kelly-in corresponding to the synchronization time. The multi-point electronic inclinometer has a battery supply system that is pre-programmable and can extend the activation of measurement to as long as 36 h. The frequency of data acquisition is also programmable and allows for at least 6 measurement records per 0.3 m core. The multi-point electronic inclinometer has an electronic memory to store measurements made by it and can store as many as 1023 measurements. On the surface, electronic memory system of the inclinometer is preset with the measurement interval, activation time and the state allowing for the most data acquisition and then gently placed into the orienting sub at the beginning of the coring operation. When the electronic inclinometer is pulled out of the well at the end of the measurement, the measurements are uploaded into the surface computer and can be processed on site and outputted into a measurement report. According to the directional parameters measured (the azimuth angle, well deflexion angle and well deflexion azimuth angle of the main graver etch mark), the inclination and dip angle of the rock bed and fracture will be obtained by reset actual measurement or formula calculation.

3. Basic technical parameters and drill combination

1) Technical parameters

Model of the tool: DX101; length of the core barrel: 8.63 m; length of the bit: 0.29 m; outside diameter of the bit: 215.9 mm, inside diameter: 101.26 mm; outside diameter of the multi-point electronic inclinometer: 34.9 mm; vertex angle of the graver: 55°; radius of the main graver: 49.2 mm; radius of the two aligning gravers: 49.5 mm; recoverable core length: 7.0 m; core diameter: 99—100 mm; applicable strata: well cemented medium hard or hard strata.

2) Drill combination

Directional PDC bit + directional core barrel + 178 mm non-magnetic drill collar + 9178 mm bit collars + 2159 mm drill collars + 127 mm drill stem + square drill stem. The structure of the directional coring tool assembly is shown in Figure 3.8.

4. Field application and performance

According to the design, the first barrel should be done in directional coring. After that, directional coring should be conducted at 50 m intervals or every 6 barrels, with a coring length of more than 2 m. According to this rule, satisfactory technical indicators were achieved by directional coring: the diameter of most of the cores was more than 95 mm, which was in line with the requirements of scientific drilling projects, since only cores that are coarse enough would meet the geological design requirements as the re-

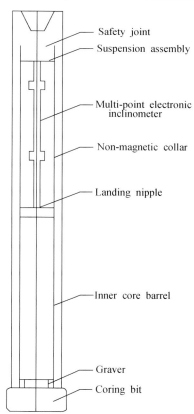

Figure 3.8 Structural sketch of directional coring tool.

search involved element geochemistry, heavy minerals, paleogeomagnetism and micropaleobiology; the inclination and dip angle of the strata were successfully obtained and the geological design requirements were fully met. The directional coring technique was well applied and proved particularly effective on complete rock beds. Figure 3.9 shows cores that were obtained with etch marks. Table 3.5 is the directional coring list at the construction site of SK-1s, where the two empty first barrels resulted from the smaller resistance of the upward/downward movement of the core on the inside surface of the core catcher than the upward/downward movement of the core catcher on the inner cone face of the core catcher seat. When the core barrel was lifted and the core moved down relatively in the inner barrel, the core catcher was not able to move upward along with it. Later, a small modification to the coring tool resulted in satisfactory performance and the fulfillment of the coring objective.

Figure 3.9　Etched core by directional coring.

Table 3.5　Directional coring operations of SK-1s

Barrel	Cored section/m	Footage/m	Core length/m	Recovery/%	Member
1	968.17—968.27	0.10	0.00	0	Nenjiang Fm. Member 2
2	968.27—971.76	3.49	0.00	0	Nenjiang Fm. Member 2
3	1022.50—1025.13	2.63	2.63	100	Nenjiang Fm. Member 1
4	1100.57—1105.14	4.57	3.35	73.30	Nenjiang Fm. Member 1
5	1150.54—1153.42	2.88	2.70	93.75	Nenjiang Fm. Members 2+3
6	1200.72—1204.76	4.04	4.22	104.46	Nenjiang Fm. Members 2+3
7	1249.95—1256.53	6.58	6.58	100	Nenjiang Fm. Member 1
8	1310.00—1315.00	5.00	5.00	100	Qingshankou Fm. Members 2+3
9	1356.21—1361.84	5.63	6.60	117.23	Qingshankou Fm. Members 2+3
10	1410.42—1416.93	6.51	6.51	100	Qingshankou Fm. Members 2+3
11	1464.48—1471.24	6.76	6.76	100	Qingshankou Fm. Members 2+3

Continued

Barrel	Cored section/m	Footage/m	Core length/m	Recovery/%	Member
12	1516.39—1523.23	6.84	6.84	100	Qingshankou Fm. Members 2+3
13	1569.34—1576.07	6.73	6.73	100	Qingshankou Fm. Members 2+3
14	1620.97—1627.69	6.72	6.72	100	Qingshankou Fm. Members 2+3
15	1671.87—1678.73	6.86	6.86	100	Qingshankou Fm. Members 2+3
16	1751.63—1758.46	6.83	6.83	100	Qingshankou Fm. Member 1
17	1806.32—1813.20	6.88	6.88	100	Quantou Fm. Member 4
18	1857.61—1864.54	6.93	6.93	100	Quantou Fm. Member 4
19	1911.70—1915.00	3.30	3.30	100	Quantou Fm. Member 3

3.2.4 Fluorescent Carboxylation Microsphere Tracing and Sealed Coring Technique

This technique is designed to check the extent of the contamination influence of mud on the top surface of the core column throughout the coring process by circulating fluorescent carboxylation microspheres in the well.

1. Target units of fluorescent carboxylation microsphere tracing

The designed horizon was the siltstone bed in the 1815.00—1822.00 m section of Quantou Formation Member 4. As there are certain errors in geological predications, the well section designated could only serve as a reference section. After observing the lithology of the two barrels of cores recovered over the target horizon, resident project team members and well loggers assumed that it was unnecessary to recover a barrel of less than 2 m core to reach to the top of the designed horizon. According to the length of the existing core barrel, it would ultimately be possible to penetrate through the key units of the designed horizon and achieve the tracing objective by fluorescent carboxylation tracing in advance from 1813.20 m. On this basis, we adjusted the original designed horizon and decided to begin the tracing from 1813.20 m.

2. Determination of the tracing method

The difficulties of making a tracing plan include: ① how to ensure substantial and the longest time of contact between the tracer and the core using a limited amount of tracer; ② the plan adopted should not affect core recovery; ③ technical feasibility; ④ dilution of the tracer.

As required for SK-1s, spud-in circulation was to be started to clean up the well body when the coring rig was laid down on the well bottom. After the circulation, steel balls were to be thrown in to seal the top of the core barrel. In the course of drilling, the drilling fluid should only circulate at the point of the bit so as to protect the cores in the core barrel from being affected by the circulating drilling fluid.

(1) The diluted tracer is put directly into the emptied drilling fluid pipe for circulation. The problem is, however, that the concentration of the tracer will be hardly controllable as it will be further diluted in the course of circulation. Besides, steel ball throwing will cause the drilling fluid to circulate between the inner and the outer tubes of the coring barrel, preventing substantial contact with the core; and the circulation duration will be too limited to achieve the purpose of the experiment.

(2) At the end of coring, the tracer is put into a plastic bag and lifted into the well from the center of the drill stem. When it has reached the top point of the core barrel, the plastic bag will be punctured so that the tracer will flow into the core barrel. With the drilling fluid, the tracer will undergo a cycle of circulation in the inner core barrel. The problem is, however, no steel ball is incorporated in this solution during the coring process, causing the core to be injected by mud, and very possibly the core will be too fine or break into fragments. It will be impossible to lift up the core, and a severe coring accident will take place. Besides, the effective circulation duration of the tracer is only 2 or 3 minutes, which is too short.

To solve these problems, after discussion with experienced first-line technical workers, it was decided that sealed coring would be the best choice. Different from conventional seal coring techniques, the sealing fluid was replaced by a mud-diluted fluorescent carboxylation microsphere mixture. This allowed long and extensive contact between the tracer and the core, and made it possible to completely control the concentration of the tracer. Besides, as the drilling team was very familiar with sealed coring techniques, technical feasibility was ensured, and core recovery was not compromised.

After the tracing technique was determined, 100 mL fluorescent carboxylation microsphere solids-latex was diluted with mud to approximately 1090 folds to 109 L.

3. Sealed coring technique

(1) Three courses of seals are included in the sealed coring assembly. After assembly, lay the device into the well steadily and levelly at a controlled rate, to prevent invalid sealing when the pin is snipped under impact.

(2) Resistance encountered in the course of rig laying should not be more than 30 kN. If this happens, pull out the rig, and use a roller bit to dredge the well.

(3) Start the pump for circulation when the rig is laid 1m above the well bottom. After sufficient circulation, lay the rig slowly onto the well bottom.

(4) After the rig is laid on the well bottom, apply 50—110 kN pressure to snip the pin and then drill. Before drilling, first apply 20—30 kN pressure, drill down to 0.3 m, and then normally apply 30—60 kN pressure.

(5) Drilling parameters. Pressure: 30—60 kN; Speed: 60—120 rpm; Discharge: 25—40 L/s.

(6) Drill the core at 1/3 higher pressure than normal for the first 0.5—1 m before core cutting, in order to facilitate thickening of the core so that it is easily grabbed.

(7) Stop the turnplate and the pump before cutting the core. Lift the rig slowly and pull out the core, normally at 15 kN max.

4. Tracing process

At 15:15 p.m., Oct 29, 2006, the inner core barrel was injected with 100 mL mud sealing fluid previously diluted with fluorescent carboxylation microsphere (Fluoresbrite TM Carboxylate Microspheres 0.50 μm) tracer. At 17:00, the inner barrel was full and the sealed coring rig was laid. When the rig reached the bottom of the well, the pump was started for circulation and for cleaning the well body. The pump-start and circulation time was 19:20 to 20:15, and drilling time was 20:15, 29th to 1:15 a.m., 30th, lasting 300 minutes. At 6:00 a.m., 30th, cores were extracted from the barrel. The state of the extracted cores was: the pure siltstone was about 2.5 m, with a 1.7 m continuous section, which was in line with research requirements; the remaining cores were sandy mudstone, argillaceous siltstone and mudstone. Cores from key portions were directly placed into a core box after they were extracted without washing so that the tracing result was not compromised. The entire tracing process was documented, photographed and video recorded.

The actual drilling parameters are presented in Table 3.6.

Table 3.6 Fluorescent carboxylation microsphere tracing log parameters of SK-1s

Bit Size & Model	Well Section /m	Footage /m	Core length /m	Recovery/%	Duration (h: min)	Speed /(m/h)	Drilling Parameters				Performance of drilling fluid		
							Pressure /kN	Speed /rpm	Discharge /(L/s)	Pump pressure /MPa	Density /(g/cm^3)	Viscosity/s	Water loss /mL
MB-FG5G6	1813.20 — 1821.88	8.68	8.68	100	5:00	1.74	35—40	60	28	10.0	1.32	65	3.0

3.2.5 Hydraulic Core Extraction

Conventionally, cores were extracted by knocking on the inner barrel after it is uninstalled and suspended. For large-caliber drill rigs, this not only involves high labor intensity, long extraction duration and dangerous operation, but also subjects the plastic or fragile core samples to deformation and the brittle core samples to damage due to blockage inside the tube, mechanical vibration or free fall of the cores, causing man-made destruction of the primary information of the strata, and adding to the difficulty of core description, scanning, cutting and analysis.

With the hydraulic core extraction device we produced, cores are integrally and evenly pushed out of the inner barrel under the pressure of the mud pump, simply by coupling a mud pipe on the upper joint of the rig, and sealing the annular gap at the lower end of the inner and outer tubes. This device eliminates all aspects of the conventional extraction approach, meaning higher operational safety and significantly lower labor intensity and shorter supporting work time. More importantly, this method creatively ensures damage-free extraction of cores. The successful application of this technical output has provided high-quality physical data of the strata. The principle and field operation of the device are shown in Figure 3.10.

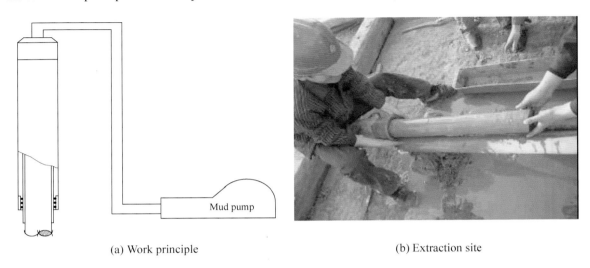

(a) Work principle (b) Extraction site

Figure 3.10 Principle and performance of hydraulic core extraction.

3.2.6 Log Core Ldentification Techniques and Core Description

In order that the top/bottom of the SK-1 cores was still clearly distinguishable after they were extracted and continuously sampled, the direction line of the cores consisted of a black line and a red line, instead of the single line + arrow design applied in previous mud log practice. The marking and direction judgment was made in the following way.

Washed cores were interconnected in the closest attachment according to their broken stubble and wear relation, and a direction line was drawn on the core surface from the shallower point to the deeper point in such a way that the direction line was present on each naturally broken core. The direction line consisted of two parallel straight lines: a red line and a black line. The cores were placed shallow on the left and deep on the right, with the black line on the outside and the red line on the inside (or red at the bottom, black at the top). The two lines were 1 cm apart (Figure 3.11). A marker pen was used to draw the direction line.

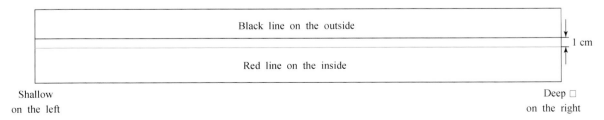

Figure 3.11 Marking of the core directional line.

Core description was conducted in compliance with the "Corporate Standard of PetroChina Co., Ltd" (Q/SY 128-2005), as adjusted and modified to the objectives and requirements for scientific drilling projects, and emphatic description was made on special event strata. Sediment color description was conducted in compliance with the "Geological Society of America: Rock-Color Chart". The stratification criterion was: for general lithologies, the minimum stratification thickness was 5 cm; for thicknesses of less than 5 cm, the term interlayer was used. For special lithologies (e.g., marlite, dolostone, recrystallized limestone, volcanic rocks), the minimum stratification thickness was 2cm; if less than 2 cm, interlayer was used. Strata of variable color, lithology, texture, structure, inclusions or oil/gas content were described layer by layer.

3.3 Construction Organization and Management of SK-1

For a scientific drilling project, economy, efficiency and safety are the foundation for realizing the drilling and scientific objectives. Sound, scientific construction organization and management are indispensable. As a Continental Scientific Drilling project and one of the "National Basic Research Program of China (973 program)" projects, SK-1 drilling is different from ordinary drilling projects. It has its own particularities that have to be investigated to establish a scientific construction organization and management plan, so as to ensure successful implementation of the project and the realization of scientific objectives.

The construction organization of SK-1 drilling fully utilized the experience of Continental Scientific Drilling projects in other countries as well a Chinese Continental Scientific Drilling (CCSD) projects, which combined with the particularities of the project itself, resulted in low consumption and high efficiency of construction without compromising the level of safety.

Management of SK-1n was undertaken by the engineering contractor, Institute of Exploration Techniques, CAGS, as entrusted by China University of Geosciences (Beijing) under contractual authority and responsibility, and construction quality and progress were supervised by a drilling supervisor and a logging supervisor from China University of Geosciences (Beijing). This was carried out in the following manner: the engineering contractor was responsible for the construction design and the daily management of construction; the engineering supervisor was responsible for reviewing and approving the construction design of the construction contractor; the drilling supervisor was responsible for daily supervision of production on behalf of the engineering supervisor; the geological supervisor was responsible for monitoring the whole coring process, describing and studying the results of the drilling operations, and providing adjustment plans as required by the dynamic geological and drilling conditions.

Direct management was implemented in the drilling of SK-1s. Daily construction management was the responsibility of the drilling supervisor as authorized by the engineering supervisor. The task of the geological supervisor was the same as for SK-1n. Construction design of SK-1s was performed by the Drilling Institute of the Daqing Oilfield Administration Bureau as authorized by China University of Geosciences (Beijing). In the course of the construction, the construction contractor was to work in strict accordance with the drilling construction design and the coring construction instructions issued by the 973 program team. A drilling supervisor was appointed from China University of Geosciences (Beijing) and from Daqing Oilfield Exploration Subcompany to control the daily production, and to supervise the construction quality and progress at the site. The drilling supervisor issued daily written construction instructions. Any underground complexity encountered was discussed by the drilling supervisor and engineering and management officers of the construction contractor to resolve the problem. Drilling work of both boreholes was reported on both a daily and weekly basis. With respect to daily reports, the drilling team would report to the drilling subcompany and the drilling group; the mud log team would report to the mud log subcompany; resident project members would report to the chief scientist, geological supervisor, engineering supervisor and Daqing Oilfield Institute of Exploration and Development. In addition to daily reports, weekly reports were also made to the relevant departments and the program team outlining the coring and engineering progress, and geological conditions experienced over the past week.

3.4 Implementation and Technique of Well Logging

Well log is the integral information of the rock type, rock physics and geochemical characteristics of the strata, and an important means for exploiting fossil energies like solid mineral resources, oil and natural gas, and coal. Both lithologic changes and sedimentary cycles of a strata are, either directly or indirectly, reflected in the configuration and amplitude of the logging curve, and are thereby important references for identifying the paleogeography, paleo-environment and paleoclimate in which sedimentary rocks were formed. Logging data, with their good continuity, high vertical resolution and information, offer valuable data for the scientific objectives of scientific drilling projects. Logging provides us with information unobtainable from core testing: it can yield well temperature, pressure and water pressure parameters that can not be obtained by surface core testing. Drilling operations are also dependent on the support of logging. For example, well deflexion, well diameter monitoring, quality inspection of well cementation, and determination of the sticking point and fallen objects in the well. Breakthroughs with respect to geophysical logging interpretation of SK-1n and SK-

1s will provide us with alternative geophysical indicators of paleoclimate changes and thereby serve our scientific objectives.

As unbiased in-situ information at depth under elevated temperature and pressure is provided by the logging, the scope of resolution and exploration of the logging lie between those of surface geophysical and core testing. When rebuilding the core profile of no-core sections, stratal depth in the borehole is generally given by the logging data. Geophysical logging of SK-1n and SK-1s would provide an important geophysical basis for constructing the corresponding relationship between cores and logging curves, determining core depth location, looking for alternative geophysical indicators of paleoclimatic change, establishing a complete organic geochemical profile, and realizing other scientific research objectives of scientific drilling projects.

3.4.1 Logging Design of SK-1

Strata encountered by SK-1 drilling mainly included sandstone and mudstone, gray siltstone, argillaceous siltstone and mudstone, with the mudstone occurring as black, dark gray, purplish red, versicolor and grayish green varieties. For this type of lithological assemblage, a world-leading ECLIPS-5700 logging instrument is used to obtain data including logging data reflecting the lithology (spontaneous gamma-ray GR, spontaneous potential SP, and neutron capture logging PE), three lateral resistivity logging data reflecting the nature of reservoirs and fluids (deep lateral R_{DLL}, shallow lateral R_{SLL}, and micro-lateral R_{MLL}), porosity logging data reflecting rock physics (including acoustic AC, neutron CNL and density DEN), and spontaneous gamma-ray spectral logging (SGR) data reflecting the type of clay minerals present. With these data, reliable parameters are available for core depth location, lithology identification and cyclic strata studies.

Resistivity, acoustic and radioactivity logging were also incorporated in the conventional logging of the project. These included spontaneous gamma-ray spectral logging, spontaneous potential logging and caliper logging that reflect lithologic variability of the strata, resistivity logging that reflects the electric conductivity, and neutron and acoustic time difference logging that reflect porosity variability. No density logging was designed for SK-1n as borehole conditions were unsuitable. Well logging items of the two SK-1 boreholes are presented in Tables 3.7 and 3.8. In addition, logging items derived from new techniques were also designed and implemented for SK-1s, as presented in Table 3.9.

Table 3.7 Well logging items of SK-1s

No	Logged section/m	Stratum	Open hole logging item(incl. standard and combined logging)	Proportion	Well core quantity	Special Requirements (e.g. for drilling fluid and logging time)	Remarks
1	240—1300	Whole well	Dual lateral, spontaneous gamma, caliper, well temperature, fluid resistivity, well deflection and orientation	1:500			Drilled well logging (ECLIPS-5700)

Continued

No	Logged section/m	Stratum	Open hole logging item(incl. standard and combined logging)	Proportion	Well core quantity	Special Requirements (e.g. for drilling fluid and logging time)	Remarks
2	445—970 section with hydrocarbon indication	H	Dual lateral, spontaneous gamma, caliper, well temperature, fluid resistivity, well deflexion and orientation, micro lateral, array induction, spontaneous potential, digital acoustic, compensated neutron, Z density	1:200			
3	970—1300	P	Dual lateral, spontaneous gamma, caliper, well temperature, fluid resistivity, well deflexion and orientation, micro lateral, array induction, spontaneous potential, digital acoustic, compensated neutron, Z density, spontaneous gamma-ray spectral	1:200			
4	1300—1915	F, Y	Dual lateral, spontaneous gamma, caliper, well temperature, fluid resistivity, well deflexion and orientation	1:500			
5	1300—1915	F, Y	Dual lateral, spontaneous gamma, caliper, well temperature, fluid resistivity, well deflexion and orientation, micro lateral, array induction, spontaneous potential, digital acoustic, compensated neutron, Z density, spontaneous gamma-ray spectral	1:200			

Table 3.8 Well logging items of SK-1n

No	Logged section/m	Strata	Open hole logging item (incl. standard and combined logging)	Proportion	Wall core quantity	Special Requirements (e.g. for drilling fluid and logging time)	Remarks
1	228—1798	Whole well	Dual lateral, spontaneous gamma, fluid resistivity, caliper	1:500			Drilled well logging (ECLIPS-5700)
2	228—1798	Whole well	Dual lateral, spontaneous gamma, caliper, micro lateral, spontaneous potential, neutron capture, digital acoustic, compensated neutron, spontaneous gamma-ray spectral	1:200			

Table 3.9 Well logging items by new techniques

No	Logged Section/m	Stratum	Logging Item by New Techniques	Dynamic Stratigraphic Testing	Special Requirements(e.g. for drilling fluid and logging time)	Remarks
1	970—1300	P	FMI/XRMS			MAXIS-500
2	970—1300	P	XMAC			ECLIPS-5700
3	1300—1915	F, Y	FMI/XRMS			MAXIS-500
4	1300—1915	F, Y	XMAC			ECLIPS-5700

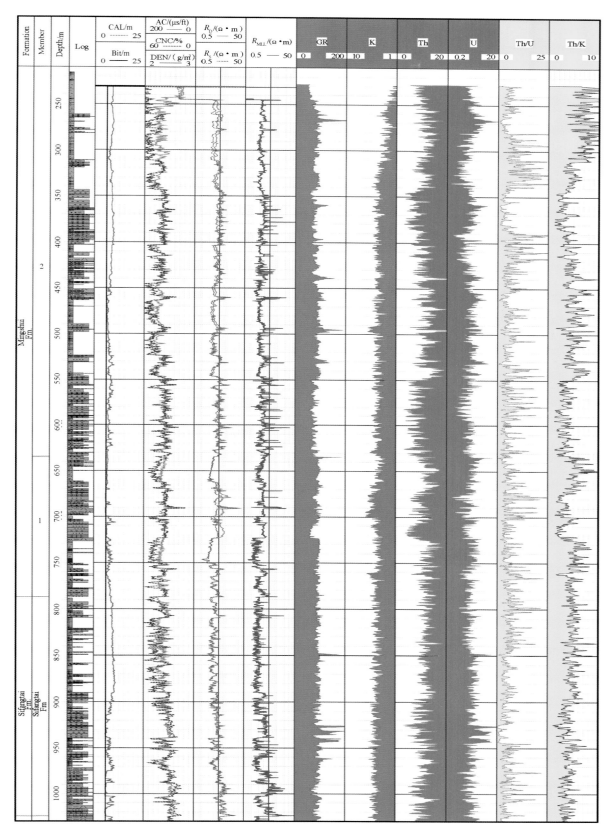

Figure 3.12　Synthetic well log-mud log profile of SK-1n.

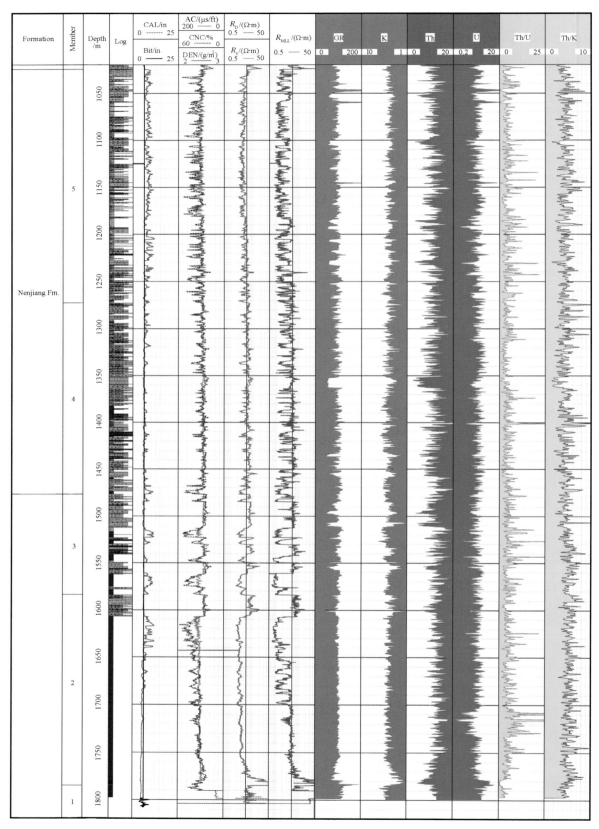

Figure 3.12　Synthetic well log-mud log profile of SK-1n (continued).

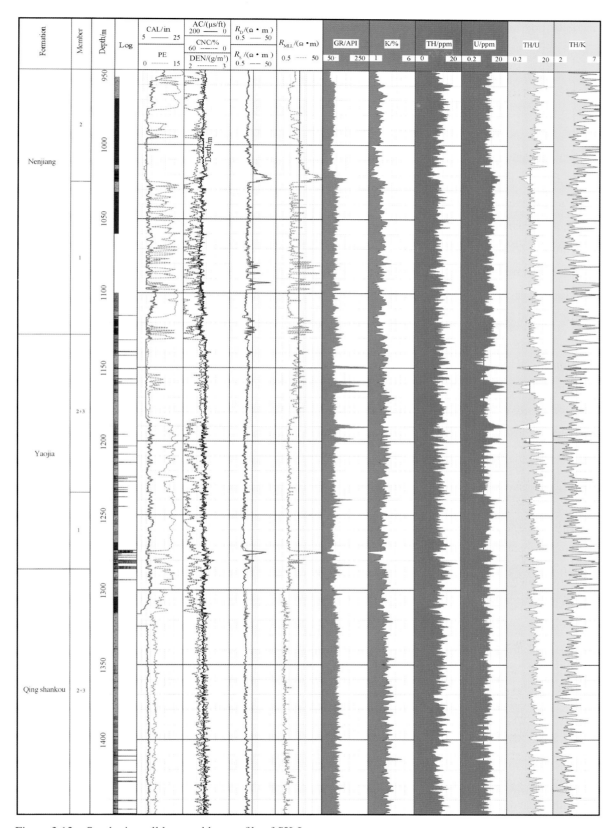

Figure 3.13　Synthetic well log-mud log profile of SK-Is.

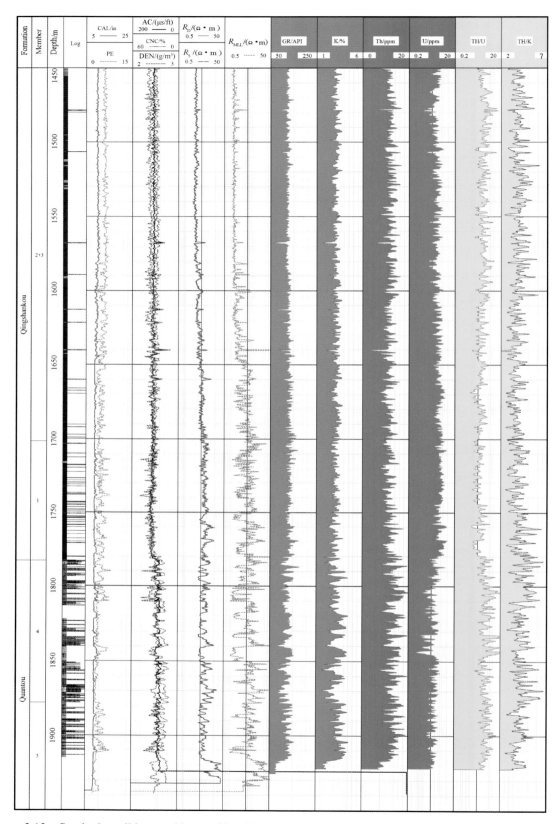

Figure 3.13 Synthetic well log-mud log profile of SK-1s (continued).

3.4.2 Implementation of the SK-1 Well Logging Project

Both the SK-1 boreholes were logged as designed. In addition to conventional well logging methods, additional array induction logging reflecting resistivity variability at different strata depths was conducted in the SK-1s boreholes

On October 23, 2008, 5700-3 logging team of the Liaohe Logging Company, came to the SK-1n site. They used an Atlas 5700 logging instrument and started the logging project on the SK-1n 245—1805 m section, after investigating the well site and the well bore structure. Measurements included digital spectral well logging and synthetic well logging where the latter involved spontaneous gamma logging, caliper logging, spontaneous potential logging, neutron capture section logging, three lateral resistivity logging, acoustic time difference logging and neutron logging. Restricted by the construction conditions, no lithologic density logging was conducted.

On September 21, 2006, the Dagang Oilfield well logging team used an Atlas 5700 logging instrument and began the first logging operation at the SK-1s drilling site after investigating the well site and well bore structure, and completed the operation on September 25. The second logging operation of SK-1s started on November 5, and by 9:00 a.m. November 9, all logging measurements of the SK-1s had been completed.

3.4.3 Result of the SK-1 Well Logging Project

Digital spectral logging of both SK-1 boreholes mainly included spontaneous gamma-ray spectral logging, from which the spontaneous gamma-ray curve, U, Th and K curves were derived after digital data processing. The synthetic well log-mud log profiles of the two SK-1 boreholes are presented in Figures 3.12 and 3.13.

3.5 Comparison and Evaluation of the Designed and Implemented one Well-two Holes Quantities

3.5.1 SK-1 North Borehole (SK-1n)

The drilling and coring operation was suspended during winter after the surface casing was laid, as restricted by the drilling conditions. The second spud-in was started on April 10, 2007 and completed on October 22, 2007, by which time all the designed coring tasks had been completed. The final drilled depth was

1811.18 m. The first spud-in included 90 coring operations totaling a footage of 80.23 m, a core length of 66.71 m, and a recovery of 83.15%; the second spud-in included 285 coring operations, totaling a footage of 1550.18 m, a core length of 1474.94 m, and a recovery of 95.15%. The two spud-ins totaled a coring footage of 1630.41 m, a core length of 1541.65 m, and a core recovery of 94.56%. The drilling and coring project was implemented by the Langfang Institute of Exploration Techniques, Hebei Province. The drilling team was the 2nd hydrogeological engineering geological team of Henan Bureau of Geology and Mineral Exploration and Development. The mud log team was the No. 34 team of the 1st data acquisition brigade, Daqing Oilfield Geological Mud Log Subcompany. The mud project was undertaken by China University of Geosciences (Wuhan). The drilling technicians were appointed from the No. 321 geological team, Anhui Bureau of Geology and Mineral Resources. The 973 project team employed a geological supervisor and an engineering supervisor, and two undergraduate students for core logging and engineering recording at the site.

The designed and implemented quantities of SK-1n are presented in Table 3.10.

Table 3.10 Designed v.s. implemented quantities of drilling, coring and relevant activities of SK-1n

Item		Designed	Implemented	Remarks
Coring		Design well depth: 1810 m	Actual well depth: 1811.18 m	Confined cored horizon concentrated on the bottom of Taikang Formation, consisting of sandy conglomerate and pebbly sandstone, immature, loose.
		Cored section: 162—1794 m	Cored section: 164.77—1795.18 m	
		Total core length: 1632m	Total core length: 1541.65 m	
		Core recovery: ≥ 90%	Core recovery: 94.56%	
		Confined coring technique was used for loose and fragile units	Confined coring: 70 barrels totaling 52.96 m	
Mud log		Record interval: sampled every 10m from well head to coring horizon; sampled every 2 m from starting point of coring till the well bottom	Cuttings were sampled at designed intervals and total 786	Per the "Geological Design of SK-1 North Borehole"
		Drilling duration, gas logging, drilling parameters, gas parameters, drilling fluid parameters	Real-time uninterrupted monitoring and recording with an synthetic well log instrument	
		A microbial sample and the corresponding mud sample at 50 m intervals	33 core samples together with 33 corresponding mud samples were recovered	Per the "Notice on Geomicrobial Sampling of SK-1 North Borehole"
Well log	Conventional log suites	Spontaneous potential, spontaneous gamma, caliper, dual lateral, microsphere focus, spontaneous gamma-ray spectral, compensated acoustic, lithologic density, compensated neutron	Spontaneous potential, spontaneous gamma, caliper, dual lateral, microsphere focus, spontaneous gamma-ray spectral, compensated acoustic, lithologic density, compensated neutron	Logged to well bottom
		Logged section: 245—1810 m	Logged section: 245—1810 m	
Others	Technical requirements	Core surface roughness 0.5 mm max; Well deflexion 4° max; Core diameter 90 mm min;	Most of the cores met the 0.5 mm max surface roughness standard; a few sections failed; Max well deflexion was not more than 1 degree; Core diameter 90—110 mm	
Duration			4958 hours 21 minites	

3.5.2 SK-1 South Borehole (SK-1s)

The drilling and coring project of SK-1s was completed well ahead of the designed schedule, and successfully fulfilled, and even overfulfilled, the design requirements. More than 900m of coring and a design core recovery of 98% min has rarely been achieved in scientific drilling and coring projects in China, and even globally. Core recovery of the SK-1s was as high as 99.73% which is exceptional. The Daqing Oilfield has mainly focused on finding oil, and coring has been mostly targeted at oil reservoir sandstone with very little drilling of mudstone, whereas the designed target unit for the SK-1s was mudstone. This was a significant challenge for the Daqing Oilfield drilling team. The designed and the implemented quantities of SK-1s are presented in Table 3.11.

Table 3.11 Designed v.s. implemented quantities of SK-1s

Item	Designed		Implemented	Remarks
Coring	Design well depth: 1915 m; Cored section: 970—1915 m; Total core length: 945 m; Core recovery: ≥ 98%; Confined coring: ≥ 110 m; Directional coring: 18 barrels min, with total length ≥ 36 m		Actual well depth: 1935 m; Cored section: 968.17—1915 m; Total core length: 944.23 m; Core recovery: 99.73%; Confined coring: 13 barrels, 46.97 m; Directional coring: 19 barrels, 95.44 m; Closed coring: 1 barrel, 8.68 m	Coring at the 1645—1675 m and 1715—1755 m sections was cancelled with the consent of the project team for the sake of the recovery ratio and the safety underground, since the cores in Qingshankou Fm. were too hard, the drilling duration too long (2.5—3 h/m), the core cutting too difficult and the core catcher could hardly shrink
Mud log	Record interval: 0—220 m, sample/10 m; 220—1220 m, sample/5 m; 1220—1320 m, sample/2 m; 1320—1720 m, sample/5 m; 1720—1915 m, sample/2 m		Cuttings were sampled at designed intervals, 464 cuttings samples were recovered	Per the "Logging Design of SK-1 South Borehole"
	Drilling duration, gas logging, drilling parameters, gas parameters, drilling fluid parameters		Real-time uninterrupted monitoring and recording with an synthetic well log instrument	
	A microbial sample and the corresponding mud sample at 50 m intervals		20 core samples and 20 corresponding mud samples	Per the "Notice on Geomicrobial Sampling of SK-1 South Borehole"
Well log	Conventional log suites	Conventional 5700 series log (12 items in total); 6 items at 245.0—1915.0 m section, 1:500; 12 items at 445.0—1915.0 m section, 1:200	Conventional 5700 series log (12 items in total); 6 items at the 245.0—1935.0 m section, 1:500; 12 items at the 445.0—1935.0 m section, 1:200;	20 m Logged to the well bottom, 20 m deeper than designed

Item		Designed	Implemented	Remarks
Well log	Unconventional log suites	Spontaneous gamma-ray spectral, 950—1915 m, 1:100 amplified	Spontaneous gamma-ray spectral, 950—1915 m, 1:100 amplified	Per the "Logging Design of SK-1 South Borehole"
		X-MAC, 950—1915 m, 1:100 amplified	X-MAC, 950—1915 m, 1:100 amplified	
		FMI, 970—1915 m 1:100 amplified	FMI, 970—1915 m 1:100 amplified	
	Cementation quality inspection log suites	Acoustic variable density, magnetic location, spontaneous gamma 0—1300m, 1:200, 250—1915m, 1:200	Acoustic variable density, magnetic location, spontaneous gamma 0—1300m, 1:200, 250—1915m, 1:200	Per the "Logging Design of SK-1 South Borehole"
Others	Carboxylation microsphere tracing	A fluorescent carboxylation microsphere tracing experiment was conducted at 1815.00—1822.00 m, section of Quantou Fm. Member 4 using the "closed coring technique"	The actual well section was 1813.20—1821.88 m. Of the 8.6m cores recovered, siltstone was about 2.5 m and included a 1.7 m continuous section, which was in line with the research requirements	From the coring implementation at the site, it was assumed that the key lithologic sections in the designed target units could absolutely be drilled through by drilling 1.80 m less than the designed depth. This could also save the per barrel cost and duration
	Technical requirements	Core surface roughness 0.5 mm max; max well deflexion 4°; core diameter 90 mm;	Most of the cores met the 0.5 mm max surface roughness standard; a few sections failed; max well deflexion was not more than 1°; core diameter 90—110 mm	
Duration		180 days	100 days (actual drilling duration 78 days 0.43 hours)	

3.5.3 Quality Evaluation of the SK-1n and Sk-1s Engineering

1. Use of logging instruments

SK-1 was drilled on a "one well-two holes" basis, i.e., it consists of a south borehole (SK-1s) and a north borehole (SK-1n). An ECLIPS-5700 system was used for both boreholes. SK-1s was measured between September 21—23, 2006 and between November 4—6, 2006; SK-1n was measured once on October 23, 2007.

2. Evaluation of logging quality

1) Acceptance and assessment of log data

A rigid operation procedure was used in the construction of the two boreholes. The logging graduation was appropriate and conformed to operational rules. The two construction teams employed for the boreholes were well experienced as they had worked extensively in the Songliao Basin and therefore guaranteed the construction quality (Tables 3.12 and 3.13).

Table 3.12　Construction quality evaluation of SK-1s

No.	Curve designation	Logged section/m	Quality assessment	Depth proportion	Remarks
1	DLL-MLL-GR-SP	445.0—1315.0 1295.68—1935.00	Pass	1:200	
2	DAL	445.0—1315.0 1295.68—1935.00	Pass	1:200	
3	HDIL	445.0—1315.0 1295.68—1935.00	Pass	1:200	
4	ZDEN-CN	445.0—1315.0 1295.68—1935.00	Pass	1:200	
5	SL	955.0—1315.0 1295.68—1935.00	Pass	1:200	
6	4CAL	445.0—1315.0 1295.68—1935.00	Pass	1:200	
7	XMAC-II	955.0—1315.0 1295.68—1915.00	Pass	1:200	
8	DLL-MLL-GR-SP	950.0—1315.0 1295.68—1918.00	Pass	1:100	
9	DLL-MLL-GR-SP	245.0—1315.0 1295.68—1935.00	Pass	1:500	
10	4CAL	245.0—1315.0 1295.68—1935.00	Pass	1:500	
11	STAR	968.5—1308.5 1766.00—1885.10	Pass	1:200	MAXIS-500
		Pass rate: 100%			

Table 3.13　Construction quality evaluation of SK-In

No.	Curve designation	Logged section/m	Quality assessment	Depth proportion	Remarks
1	GR	229.972	Pass	1:200	
2	SP	229.972	Pass	1:200	
3	CAL	231.267—1803.730	Pass	1:200	
4	RD	231.572—1803.730	Pass	1:200	
5	RS	231.572—1803.730	Pass	1:200	
6	RMSL	230.657—1803.730	Pass	1:200	
7	DT24	229.972—1803.730	Pass	1:200	
8	CNCF	229.972—1803.730	Pass	1:100	
9	GRSL-U-TH-K	227.990—1790.852	Pass	1:500	
		Pass rate: 100%			

2) Data comparison of relogged sections

SK-1s was measured between September 21—23, 2006 and between November 4—6, 2006, at the 445.0—1315 m and 1295.68—1935.0 m sections. The repeated section measurement was 1295.68—1315 m. From the data obtained, repeatability was high, implying the instrument had normal performance indexes (Figure 3.14).

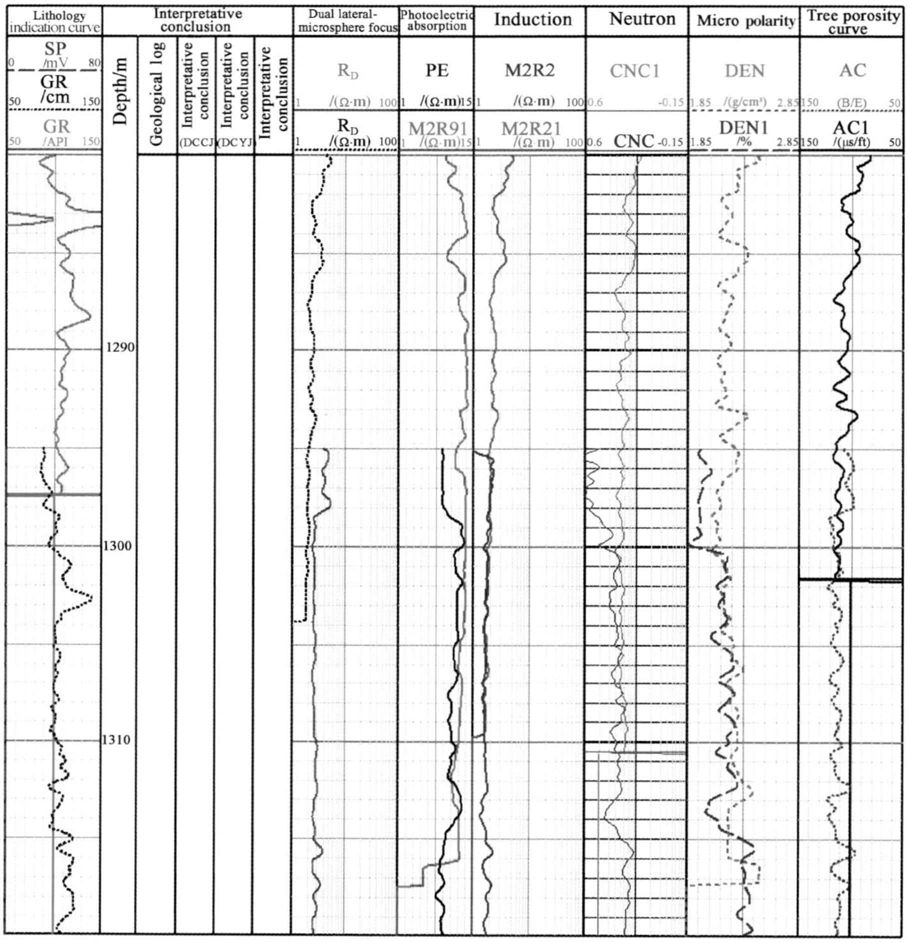

Figure 3.14 Data comparison of relogged SK-1 sections.

3) Data comparison between two boreholes

The "one well-two holes" design of SK-1 takes the blackish brown oil shale present at the bottom of Nenjiang Formation Member 2, Songliao Basin, as the main comparative marker bed, and Nenjiang Formation Member 2 is also a unit having the same lithology in the strata drilled in both boreholes. The log curves best reflecting the lithology are the spontaneous gamma-ray (GR) and the thorium (Th) curves. Theoretically, the GR curve is subject to more adsorption of organic matter than U, while the Th curve is more better at reflecting the lithology. Statistical analysis of the GR and Th log curves of the comparative sections of the two boreholes (SK-1s 958—983 m and SK-1n 1718—1743 m) indicate the same distribution characteristics and roughly consistent characteristic parameters (Figures 3.15 and 3.16).

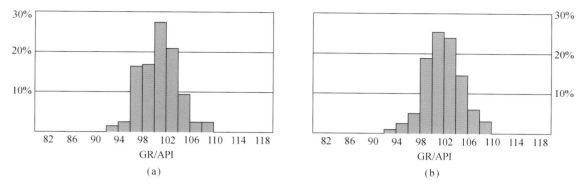

Figure 3.15 GR comparison between SK-1n section 958—983 m (a) and SK-1n section 1718—1743 m (b).

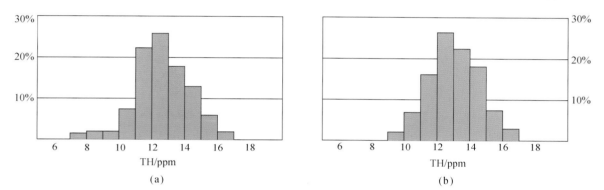

Figure 3.16 Th comparison between SK-1s section 958—983 m (a) and SK-1n section 1718—1743 m (b).

The GR curve values of the SK-1s 958—983m section range from 92 API to 108 API and average 101 API, whereas those of the SK-1n 1718—1743 m section are 87 API, 109 API and 100 API, respectively. Thorium values of the SK-1s 958—983 m section range from 7.7 ppm to 16.4 ppm and average 12.6 ppm; whereas those of the SK-1n 1718—1743 m section are 9.9 ppm, 17.0 ppm API and 13.1 ppm, respectively. The differences between both ranges are less than 5%, suggesting satisfactory consistency between the measurements of the two instruments.

3.5.4 Core Depth Location of SK-1

A total of 2485.89 m of core were recovered from the two SK-1 boreholes, and constitute important geological information for establishing a complete Cretaceous geological-geochemical profile of the Songliao Basin. However, depth deviations are unavoidable between the mud log profile and well log curve of the cores since the cored depth is measured by the length of the drill rig, while the log curve is measured by the length of the cable, and the drill rig has a different contraction coefficient than the cable. The purpose of core depth location is to match the drilled depth with the logged depth and to identify the depth deviation of different cored units on the basis of the logged depth. The critical for investigating the relationship between lithology, physical parameters, oil-bearing properties other geological-geochemical characteristics and logging responses.

1. Method and steps of core location

Core depth location is mainly based on special lithologic marker beds, namely rocks of special lithologies with different electrical properties, such as limestone, marlite, coal, high salinity beds and argillaceous beds in limestone sequences. If a special lithologic marker bed is absent in a sandstone/mud stonesequence, core depth location may still be conducted using the abrupt change in electrical properties displayed at an anomalous lithologic interface.

Core depth location is conducted by the following steps:

(1) Identify a special lithologic marker bed or lithologic section by observing the cores;

(2) Determine the logged depth of the special lithologic marker bed, ideally by identifying the difference between the drilled depth and the logged depth according to the position of the turning point of the micro-electrode curve and the extreme point of the micro-gradient curve;

(3) Locate cores of higher-recovery barrels first, and determine the lifting or lowering value of the core mud log profile of each barrel of core based on the logged depth. Start from the topmost marker bed, extend upward to the top of the cored well section, and then move downward layer by layer so that the lithology coincides with the electricity.

(4) For lower-recovery barrels, core location is achieved by controlling section-to-section control within the top/bottom boundary according to the marker bed and lithologic assemblage.

Pertinent measures are to be taken for core location in special cases.

(1) Broken cores: Measurement errors are generally expected for broken cores, so it is necessary to eliminate this according to the log interpreted thickness, by elongating or compressing, depending on the extent of the breakage.

(2) Worn cores: At the wear point, the core may be moderately pulled apart according to the logged thickness so that the lithology will coincide with the electricity. The pull-apart point is the vacant place.

(3) Where the recovered core is longer than the log interpreted thickness and the core is quite complete: In-proportion location may be done to restore the true length of the core (i.e. the log interpreted thickness of the core), assuming it has expanded when taken up to the earth's surface and its original underground pressure has changed.

(4) Disordered cores: Cores are sometimes disordered when they are extracted. It is therefore necessary to join them carefully, putting the pieces of core back into their original order and document their field condition in every detail. During location, correlate their lithology with their electricity according to the log interpretation and lithologic characteristics.

2. Core location of SK-1n

The coring horizon of SK-1n was from the Tertiary Taikang Formation to the top of the Cretaceous Nenjiang Formation Member 1, at a depth of 164.77—1795.18 m, and a footage of 1630.41 m. The total core length was 1541.66 m and core recovery was 94.56%. Poor rock-forming properties and loose cores in the shallow strata limited core recovery and added to the difficulty of core location.

The coring of SK-1 indicates the composition of the Nenjiang Formation, i.e., dominantly large units of black mudstones and oil shales at the bottom, and sandstone-mudstone interbeds above Nenjiang Formation Member 4 and the Sifangtai and Mingshui formations. Depth location was made using the logging response difference between different lithologies and the logging response, abrupt changes across sandstone-mudstone interfaces, special lithologies of thin layers, oil shales and their special logging response characteristics. The location began with the topmost marker layer, extended upward to the top of the cored well section, and then downward layer by layer so that lithology coincided with electricity readings.

In the SK-1n mudstone profile, marlite strata were not developed, but some thin-layered special lithologies were recovered, such as conglomerate and gritstone. There were two micro-resistivity anomaly highs in the 807—807.4 m and 809.25—809.6 m sections of the Sifangtai Formation. These corresponded to two thin gritstone layers suggesting that the depth of the core recovered at these sections is 2.31 m higher than the logged depth level (Figure 3.17).

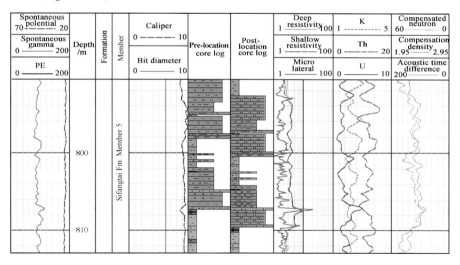

Figure 3.17 Effect of core depth location using thin-layered gritstone.

The type of rocks and their log response are one of the principal basis for depth location of the SK-1n cores. In the case of Figure 3.18, for example, Nenjiang Formation Member 5 is a sandstone-mudstone interbed; on the micro-resistivity curve, the sandstone has significantly higher resistivity than the mudstone: the sandstone is 4 $\Omega \cdot$ m, whereas the mudstone is less than 1.54 $\Omega \cdot$ m. From this, we can deduce the deviation of the core depth from the logged depth, which was 3.96 m at around 1200 m (Figure 3.18).

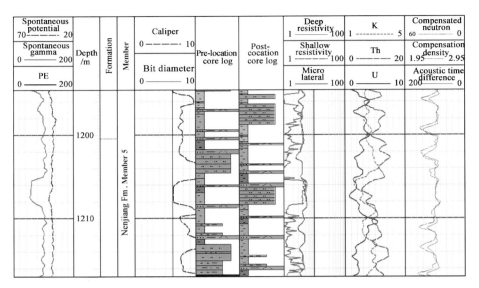

Figure 3.18 Effect of core depth location using thin-layered gritstone.

The extensive oil shale that forms the base of Nenjiang Formation Member 2, is an important marker bed for the regional stratigraphic correlation, as well as the macro marker bed for our core depth location. As the shale is organic-rich and exhibits significant spontaneous gamma-ray spectral log and resistivity log curve anomalies, it is a particularly good marker bed in the SK-1n core (Figure 3.19).

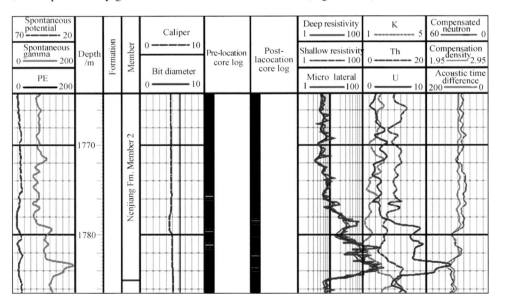

Figure 3.19 Effect of core depth location using oil shale at the bottom of Nenjiang Formation Member 2.

For cores of lower-recovery barrels, "hollow" treatment was undertaken. For example, cores from the thick units of siltstone in the 579.11—592.83 m section of the Mingshui Formation Member 2 were worn and the core length was smaller than the cored footage. Therefore, fine sandstone and pebbly siltstone beds in the siltstone were adopted as the main marker beds for core location (Figure 3.20), and proper "hollow" treatment was made according to the presence of a worn face.

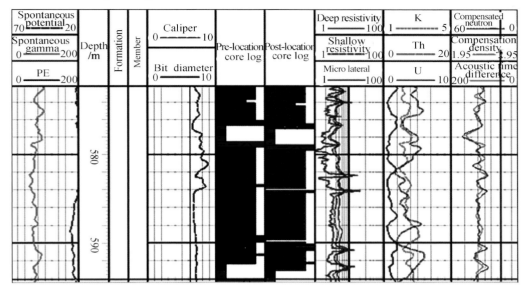

Figure 3.20 "Hollow" treatment in core depth location for worn cores.

For more complete cores whose recovered length was longer than the log interpreted thickness (mainly the mudstone section), in-proportion compressed location was made to restore their true length, i.e., the log interpreted thickness, assuming the cores had expanded when lifted to the surface and that their underground pressure had changed.

3. Core location of SK-1s

The cored sections of SK-1s included Nenjiang Formation Member 2 and Quantou Formation Member 3, at a depth of 968.17—1915 m, footage of 946.83 m, core length of 944.23 m, and core recovery of 99.73%. The high recovery ratio and the world-leading 5700 digital log curve provided a solid basis for correct location of the cores.

Coring results indicate a dominance of mudstone in the SK-1s core profile, except for sandstone-mudstone interbeds in the upper part of Quantou Formation Member 3, Quantou Member 4, and the lower part of Yaojia Formation Member 1. Different considerations were addressed with respect to core depth location of this borehole according to the different lithologic assemblages present. For Quantou Formation Member 3 and Member 4 strata, and Yaojia Formation Member 1 strata, depth location was conducted according to the sandstone-mudstone interbeds and an abrupt change in the logging response when they were intersected. The basal sandstone interface, on the other hand, was usually correlated with an extreme point of the microsphere focus circuit ratio measuring line (Figure 3.21).

For depth location of the thick units of mudstone in the Qingshankou and Nenjiang formations, marlite

(Qingshankou Formation) and dolostone (Nenjiang Formation) within the mudstone were used as the marker beds. Marlite is characterized by high density, low sound wave propagation and low radioactivity on the argillaceous profile, and a noticeable t spike on the resistivity curve (Figure 3.22).

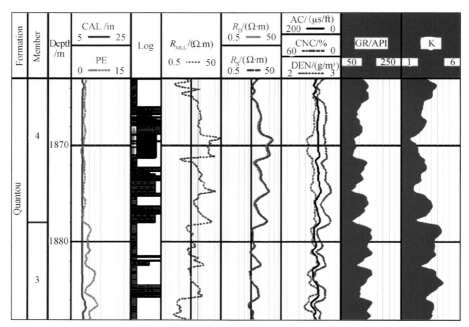

Figure 3.21　Effect of core depth location using lithologic assemblage and abrupt interface.

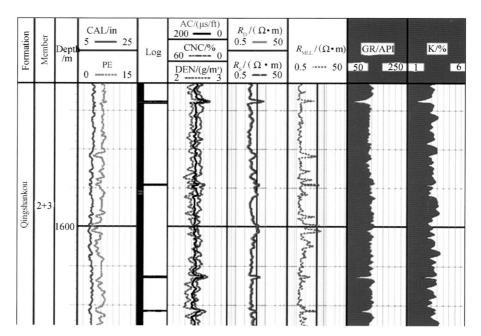

Figure 3.22　Effect of core depth location using thin-layered marlite in large mudstone units.

The lacustrine dolostone horizon in the SK-1 Nenjiang Formation deep lacustrine dark-colored mudstone is also an important marker bed for the location of the Nenjiang Formation cores. The dolostone is indicated by a small positive spike on the resistivity curve and a small negative spike on the spontaneous gamma curve. The unit thickness per layer generally ranges from a few to 12 cm. It is an important reference marker for high-precision core depth location (Figure 3.23).

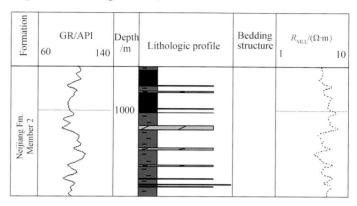

Figure 3.23 Effect of core depth location using thin-layered dolostone in large mudstone units.

The objective of core depth location in high sections of SK-1 was fulfilled by using coarse-to-fine lithologic assemblages, special lithologic marker beds and thin-layered special lithologic sections in sandstone or mudstone, and a depth bridge was constructed for analyzing the relationship between geological, geochemical and geophysical responses.

3.6 Core Handling and Storage

A long time is needed from the core extraction from the barrel to storage, sampling and analysis. Therefore, each part of the process required careful attention. A comprehensive plan for core preservation was made before they were recovered from SK-1, and appropriate arrangements made for temporary preservation at the well site to the final storage at the core repository, to ensure the original state and completeness of the cores. As SK-1 cores are extremely valuable and are recovered in limited quantities, to enable subsequent observation and studies, the reserved part of the cores from the whole well section has to be permanently preserved, while the other part has been used for the present research. It was decided by the project office together with Daqing Oilfield personnel, that a third of the cross section of the cores from the whole well section was to be permanently preserved, and long-term preservation samples of the cores was prepared.

3.6.1　Core Handling and Storage at the Drilling Site

The mud log team was responsible for extracting the SK-1s cores from the barrel, and the cores were released from the upstanding barrel under their own gravity. The field drilling team was responsible for extracting the SK-1n cores from the barrel, with the assistance of the mud log team, using a hydraulic core extraction technique. For this, the core-containing barrel was positioned horizontally. A mud pipe was connected to the top of the barrel. After being pressurized, the core was gradually extruded from the barrel by the mud, at a speed controllable by the pressure applied. This not only kept preserved the extracted cores, but also avoided damage to the core profile that often occurs during conventional extraction. The extracted cores were then interconnected according to their stubble, and mud on the core surface was removed with pure water or, in winter, with water vapor. When the core surface was dry, a direction line consisting of two parallel lines of different colors (black and red) were drawn on it, and the core was measured, labeled, barreled and described. At the well site, the cores were temporarily kept in the core room of the mud log company.

3.6.2　Core Scanning and Storage

When the number of cores at the drilling site had reached a certain quantity, they were transported to the SK-1 core storage room of the Daqing Oilfield core repository for surface image scanning. The scanning work was conducted at two places: SK-1s cores were scanned in the core scanning room of the Daqing Oilfield Geological Mud Log Subcompany; SK-1n cores were scanned in the Daqing Oilfield core repository. The scanning was divided into two types: vertical scanning, i.e., a single side of the core was scanned from top to bottom; outside surface scanning, i.e., rotary scanning of a well preserved core so that the scanning images l reflect the entire outside surface of the core. When surface image scanning was completed, the core was sent to the "SK-1 core storage room" of the Daqing Oilfield core repository that was specifically designed to keep SK-1 cores. The core storage room is furnished with high-power air conditioning and an electric hot plate to regulate humidity and temperature of the core room. Cores were kept on special core shelves arranged for each installation, observation and sampling of the cores.

3.6.3　Cutting, Casting and Long-term Preservation of SK-1 Cores

To ensure the quality of core sample preparation, a "Sample Preparation and Preservation Techniques for SK-1 Cores" project was given to the Material Department of Chengdu University of Technology and the Daqing Oilfield Institute of Exploration & Development, with the aim to investigate casting materials for preparing SK-1 core samples and their application processes, the design and materials for U-shaped core troughs and techniques for preparing core samples.

Studies and comparative experiments on casting performance of 6 different casting materials indicated that crystal resin has a superior overall performance, sound transparency, low viscosity, no contamination to cores, good operability, excellent mechanical properties and high ageing resistance. For this reason, crystal resin was adopted as the casting material for preparing SK-1 core samples. The application process parameters of crystal resin were determined by studies on the influence of accelerants, curing agents, temperature and core moisture on curing time, mechanical properties and application performance of crystal resin. The crystal resin casting material was proven to have excellent ageing resistance through testing and analysis of the color, light transmission and yellow index of UV ageing-resistance samples of the crystal resin, as well as satisfactory infrared absorption spectrum (IR) and X-ray diffraction (XRD) results. The exterior design of the U-shape trough was proposed, and organic glass (PMMA) was taken as the U-shaped trough material for preparing SK-I core samples, as organic glass had suitable sound transparency, superior overall performance, mature manufacture process, and good compatibility with crystal resin. Infrared absorption spectrometry indicated no chemical reaction or characteristic peak of any new material at the interface between the crystal resin and organic glass trough. The interface was well bonded.

The core sample preparation processes were determined according to the above-mentioned research results, and nearly 3000 m of core samples were prepared with excellent engineering application performance. The core sample preparation process includes six steps: core location cutting, core trimming, core casting, cast core cutting, core surface sealing, and core surface polishing.

The cut size of the cutting machine used in the core location cutting process was determined according to the size of the U-shaped trough. The core trimming process is the prerequisite for the core casting process that follows; keeping the core surface free from dust, contamination or water stains was of paramount importance; the core position should be fixed so that it could not be shifted to ensure that the core is as close to its original state underground as possible. The core casting process is central to this series of processes and involves, the preparation of the casting material, casting of the core, gelated and cured shaping of the cores, which are described in more detail below. The cast core cutting process is quite similar to the above-mentioned core location cutting process, and should agree with the size of the organic glass U-shaped trough. The core surface sealing and surface polishing processes are the most important since the surface sealing and polishing effects will directly relate to the visual appearance of the core surface.

Based on the above studies, and in combination with field implementation, the core sample preparation process flowline is given below (Figures 3.24 and 3.25).

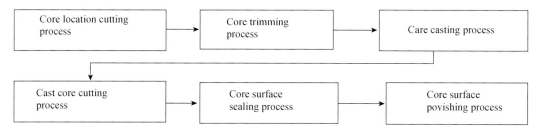

Figure 3.24 Process flow of core sample preparation.

Figure 3.25 Core profiling, casting and polishing of SK-1.

Outline of core sample preparation processes:

(1) Core sample preparation processes include core location cutting, core trimming, core casting, cast core cutting, core surface sealing, and core surface polishing. The core casting process is the core of this series of processes.

(2) The cut size of core samples should agree with the designed and manufactured size of the organic glass or PP U-shaped trough.

(3) As an approximately 3% shrinkage is expected when the casting material in the organic glass trough is cured, a secondary casting will be necessary.

(4) Experiments indicate that the organic glass trough could crack when the baking temperature in the over is more than 80℃.

(5) The cured casting material has good detachment properties from the PP U-shaped trough interface and good bonding properties to the organic glass trough interface.

3.6.4 Core sampling of SK-1

After scanning and dividing surface images of the core, the project team organized a number of unified core sampling operations. The sampling process was: First, each of the research groups made their respective sampling plans, submitted them to the geological supervisor to review, integration and adjustment, and then delivered to the chief scientist for approval. After review and approval, the geological supervisor organized and coordinated operations according to the sampling plan. The "Sampling Principle and Rule for Core Sampling from SK-1" was formulated specific to the project on the basis of the ICDP international standard, and sampling was carried out in accordance with this standard.

Sampling principle: An aggregate of 2485.89 m of cores should be recovered from SK-1, half of which should be left for archiving purposes. This means the cores were very limited, and there were many sampling organizations, so the application of any organization intending to take samples should be carefully discussed and approved by the 973 project office before the operation could be started. The 973 project office would review the application made by the intending organization or individual, coordinate between any two or more organizations that intended to take samples from the same core, so that the limited cores were fully utilized on the maximal research projects. SK-1 cores would be vertically cut in halves. The half marked with the direction line (with a core label on it) would be archived, and samples would be taken from the remaining half. Besides, samples taken from the same position of the core should not be more than 1/4 the cross section of the core or more than 5 cm in length.

Sampling rules: ① The sampling organization or individual should apply with the project office for sampling operations and should sample in accordance with the sampling criteria after approval. ② The sampler should work in accordance with the applied horizon, depth, lithology and size, and fill out a "Core Sampling Record". ③ Any sampling operation should be subject to special supervision, acceptance and registration by the core data room before the samples can be brought of the core repository. ④ Cores with labels or direction lines should not be taken away. ⑤ A sampling device should be used to take samples. Any half or whole core to be taken should be approved by the project office and taken in accordance with the specifications, with a mark at the corresponding place of the core box. ⑥ After use, the remaining core should be immediately put back into the core box, and should be inspected and accepted by the core data room.

When samples had been taken in accordance with the above-mentioned rules, the sampling position, the size of the sample and purpose of the sample were submitted to the project team before the sampling operation is closed. Samples of SK-1 mainly include: lithologic samples (taken at an interval of 1 m/sample), paleogeomagnetic samples (taken at an interval of 20 cm/sample), environmental magnetic samples (taken at an interval of 5 cm/sample), lithogeochemical samples (taken at an interval of 20 cm to 1 m/sample), ostracod fossil samples (taken at an interval of 1—2 m/sample), ostracod valve isotopic samples (taken at an interval of 1 m/sample), organic geochemical samples (taken at an interval of 1m/sample), dinoflagellate fossil samples (taken at an interval of 1 m/sample), sporo-pollen samples (taken at an interval of 1 m/sample), with aggregate samples amounting to more than 30,000. Samples taken are presented in Table 3.14.

Table 3.14 Sampling of SK-1

Type	Size/g	Density/(sample/m)		Qty		Total
		SK-1s	SK-1n	SK-1s	SK-1n	
lithologic	50	0.7	1.5	702	2245	2947
Paleomagnetic	20	4.8	4.1	4533	6302	10835
Enviromagnetic	10	13.7	0	12892	0	12892
Lithogeochemical	100	0.9	0	808	0	808
Ostracod fossil	100	0.5	0.4	435	655	1090
Ostracod shell isotopic	150	1.1	0	1003	0	1003
Organic geochemical	150	0.9	1.0	860	1527	2387
Dinoflagellate fossil	150	0.5	0.5	472	771	1243
Sporopollen	100	0.6	0.4	579	688	1267
Total number of samples				22284	12188	34472

Section 4
Preliminary Scientific Results of SK-1

Scientific drilling is an indispensable tool of modern Earth science research, as it provides the only means of obtaining direct information on processes operating at depth (Harms *et al.*, 2007). Drilling allows for the determination of in situ properties of solid materials and fluids and permits testing of hypotheses and models derived from surface observations. The International Continental Scientific Drilling Program, ICDP, founded in 1996, has demonstrated that the principal goals of understanding composition, structure and geological processes underground can be achieved (Harms *et al.*, 2007). The Continental Scientific Drilling Project of the Cretaceous Songliao Basin has obtained a 2500 m long continuous sedimentary record which geophysical and geochemical data have been obtained and provide the basis for further scientific research.

Based on completion of SK-1 drilling and scientific data obtained, ten continuous geological profiles have been established including lithostratigraphy, biostratigraphy, paleomagnetics, chronostratigraphy, inorganic geochemistry, organic geochemistry, geomicrobiology, logging results, cyclostratigraphy and stable isotopes. This section summarizes preliminary scientific results of the drilling project.

4.1 Lithostratigraphy

4.1.1 Methods and Summary of Core Description

Detailed core description from SK-1 provides the basic information for scientific research including important information about drilling and logging operations. The core description allows the detailed construction of the lithostratigraphic section. Cores from the SK-1 well were described at the drilling site so as to obtain first-hand information about the geologic characteristics of sedimentary sequence. Continuous coring recovered a complete sedimentary sequence from the top of the Cretaceous Quantou Formation to the base of Tertiary Taikang Formation. Methods and principles of core description generally followed the procedures established by the PetroChina Company (Criterion Q/SY 128-2005). According to the specific requirements of scientific drilling, some adaptations were made to the above procedures, e.g., emphasis was placed on the recognition of special events and on rock description of color.

During oil exploration in the Songliao Basin, the focus has been on clastic rocks, with minor attention paid to mudstone. SK-1s drilling site was selected in an area with a complete sedimentary record where there is thick mudstone, and where distinctive beds such as volcanic ash, ostracod-bearing limestone and dolomite were expected. Such lithologies are important for interpreting the geologic history of the basin, and to establish whether coeval anoxic and oxic events, which are known from marine deposits, had also occurred in a continental setting. These events provide important evidence of global climate change and its causes.

During geologic operations in the oil fields of the Songliao Basin, description of core color mostly depends on the experience of the geologist, and no special colorcoding criterion has been used. For example, during previous core descriptions, only a single dark grey color has been used. For the SK-1 core, the Rock Color Chart of the Geological Society of America was used to provide a more precise color description, e.g., dark grey, medium-dark grey, olive-grey, etc.

A bed is described as a sedimentary unit with a minimum thickness of 5 cm; 2—5 cm thick unit are termed a layer and is used to describe a specific lithology, e.g., argillaceous limestone, dolomite, ostracod limestone. Sedimentary units that are <2 cm in thickness, are termed laminae. During core description, 5 major features were recorded: lithology, color, texture, sedimentary structure and contents.

4.1.2 SK-1 Core Lithology

In the SK-1 borehole, continuous coring recovered strata from the top of Quantou Formation (Late Cretaceous) to the base of Taikang Formation (Paleogene). Lithologic characteristics of each formation are summarized below and the formations are described in stratigraphically descending order.

Taikang Formation: Only the basal part of the formation was cored. It is represented by light grey sandy conglomerate and pebbly sandstone.

Mingshui Formation: Strata are dominated by siltstone and mudstone with subordinate argillaceous siltstone and silty mudstone, a few thin-layers of sandy conglomerate, fine-grained and medium-grained sandstone. Color of the sandstone and some of the mudstones is mainly grey. Mudstones are grey, grayish-black, black, mauve, greenish-gray and or grayish-green. Sedimentary structures in fine and medium-grained sandstones are parallel bedding, wave-ripple lamination, cross bedding, minor small-scale trough bedding, and graded bedding. Slump features are common. Mudstones are massively bedded. In grey, grayish-black and black mudstone the bedding is horizontal and ripple laminations occur. Sandy conglomerates have massive structure and deformation structures are present. Sandstones enclose abundant charcoal and plant debris. Sparse clam and spiro fossils have been found. This formation is interpreted to have been deposited in a shallow lake and near-shore lake environment.

Sifangtai Formation: Middle to lower part of the formation is mainly siltstone. The upper part is dominated by mudstone with subordinate argillaceous siltstone and silty mudstone, with a few thin-layers of calcareous siltstone, sandy conglomerate, and mud gravels. Fine- and medium-grained sandstone occurs in the middle to lower part and at the top of the formation. Color of sandstone, conglomerate and some mudstones is gray. Mudstone colors are predominantly mauve, greenish-grey, grayish-green and mottled mauve to grayish-green. Less common are grey, black and grayish-black colors. In siltstone and fine-grained sandstone, sedimentary structures are parallel bedding, wave ripple bedding, wave cross bedding, minor small-scale trough bedding and wedge-shaped bedding. Slump features are common. The mauve, greenish-gray, grayish-green, mottled mauve and grayish-green mudstone are massively bedded, with occasional rippled laminae. In grey, grayish-black, black mudstone horizontal bedding and ripple lamination are present. Sandstones contain abundant charcoal debris, which are more common on some bedding planes. A few clam and spiro fossils have been found. Sediments were deposited in alternating shallow lake and near-shore lake environments.

Nenjiang Formation Member 5: Mauve mudstone-mottled mauve mudstone and grayish-green mudstone dominate, with subordinate grey siltstone, fine-grained sandstone, argillaceous siltstone, some grey mudstone and grayish-green mudstone. Ripple lamination is present in sandstone and some charcoal debris occur in the sandstone. Mudstone is massively bedded. At sandstone mudstone contacts deformation features occasionally occur. The sediments of this formation were deposited in a shallow lake environment.

Nenjiang Formation Member 4: Member 4 is mainly comprised of grey-colored fine-grained sandstone, siltstone, argillaceous siltstone and silty mudstone. Grey mudstone and greenish-grey mudstoneare subordinate lihtologies. In sandstone, foreset cross-bedding, wavy bedding are present, with some trough bedding in siltstone and fine-grained sandstone. Local slumps and erosion surfaces occur. Sandstones enclose sparse charcoal debris. Sediments show several cycles with reversed grain-size gradation. The sedimentary environment was deltaic, specifically delta front.

Nenjiang Formation Member 3: In the middle to lower part of Member 3, black and grey mudstone dominate, while in the upper part grey fine-grained sandstone, siltstone, argillaceous siltstone and silty mudstone are dominant. In the middle to lower part of the formation, sedimentary structures present in sandstone are trough bedding, wave bedding, climbing ripples and deformation bedding. In the upper part, ripple lamination and wavy bedding occur in the sandstone. Dark mudstone is horizontally bedded. The deposition environment changed from semi-deep lake in the lower part to shallow lake in the upper part of Member 3.

Nenjiang Formation Member 2: This is mainly black, horizontally bedded grayish-black mudstone interbeded with sandstone. At the base of Member 2 is inferior oil shale, which is laminated and has an oily smell. At the top of the unit are three reversely graded cycles, including grey, dark-grey silty mudstone, argillaceous siltstone and siltstone. In sandstone, ripple cross bedding and foreset bedding are present. In silty mudstone and argillaceous siltstone, ripple lamination is present, with some beds showing deformation. In black mudstone, ostracoda, conchonstracan and other fossil fragments are present. In sandstone, charcoal debris is occasionally seen. Member 2 is interpreted as being deposited in a semi-deep to deep lake environment.

Nenjiang Formation Member 1: The dominant lithology is dark-grey, medium-dark grey, olive-grey, dark greenish-grey mudstone, interbeded with thin-layered oil shale and ostracod-bearing mudstone. The depositional environment was deep lake.

Yaojia Formation Members 2+3: These members are dominated by greenish-grey, dark greenish-grey, grayish-brown mudstones with interbeded with thin layers of dark greenish-grey, greenish-grey argillaceous siltstone and silty mudstone. The depositional environment was semi-deep lake to deep lake.

Yaojia Formation Member 1: The upper part of Member 1 is comprised of greenish-grey and dark greenish-grey mudstone. Mudstone in the middle to lower part is mainly grayish-brown interbedded with dark greenish-grey argillaceous siltstone, and grayish-brown oil-bearing sandstone. Deposition occurred in a shallow lake with delta-front subfacies. The contact with underlying strata is disconformable.

Qingshankou Formation Members 2+3: The dominant lithology of these members is a dark grey, medium-dark grey, greenish-grey and dark greenish-grey mudstone, interbedded with thin-layers of olive-grey oil shale and ostracod-bearing mudstone. Deposition occurred in a deep lake environment.

Qingshankou Formation Member 1: This member is mainly comprised of dark grey, olive-grey and olive-black mudstone. In the lower part are thin layers of dark grey, medium dark-grey silty mudstone and argillaceous siltstone. The sediments were deposited in a deep lake environment. A disconformity at the base of Member 1 separates the Qingshankou and Quantou formations.

Quantou Formation Member 4: This member is mainly comprised by greenish-grey, dark greenish-grey, brownish-grey and grayish-brown mudstone interbedded with argillaceous siltstone, and some oil-bearing brownish-grey siltstone and fine-grained sandstone. Sediments were deposited on a flood plain with meandering river subfacies.

Quantou Formation Member 3 (not fully penetrated): This member is dominated by greenish-grey, dark greenish-grey, brownish-grey, grayish-brown mudstone and silty mudstone, with thin layers of greenish-grey fine-grained sandstone. The depositional environment was a meandering river flood plain.

Figures 4.1 and 4.2 are stratigraphic columns of SK-1n and SK-1s boreholes.

Section 4 Preliminary Scientific Results of SK-1

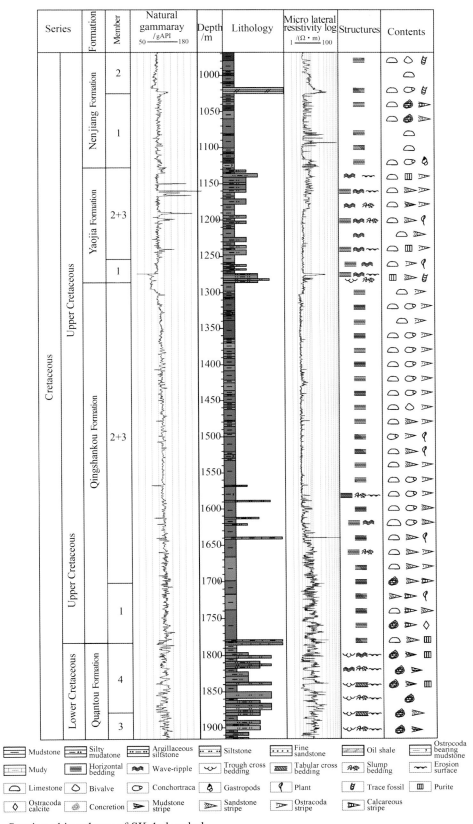

Figure 4.1 Stratigraphic column of SK-1s borehole.

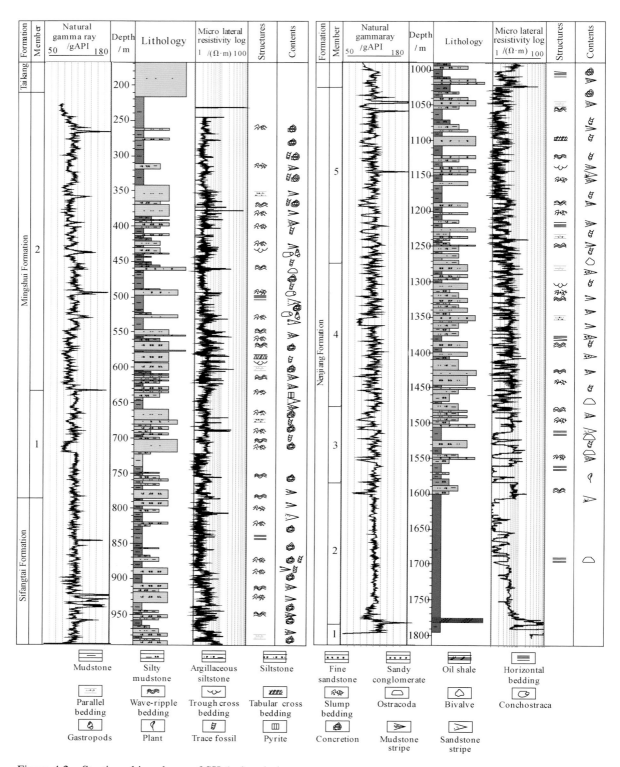

Figure 4.2 Stratigraphic column of SK-1n borehole.

4.1.3 SK-1 Special Deposits and Their Geological Significance

(1) Volcanic ash layers: SK-1 comprises 14 thin (5—40 mm thick) volcanic ash layers in the first and second members and dark shale of the Qingshankou Formation. Loose, gray, pale gray or greenish-gray, the ash layers are composed of fine-grained fragmented quartz and feldspar in a tuffaceous matrix. High-precision U-Pb zircon SIMS geochronology was carried out on five samples (see Chapter 4.5). The volcanic ash layers and their geochronology in the continental strata of the Songliao Basin provide critical marker horizons throughout the entire basin and allow correlation with global Cretaceous marine strata.

(2) Lacustrine subaqueous channel system: Based on analysis of high-resolution 3D seismic data, a large-scale lacustrine subaqueous channel system is identified in the lacustrine mudstone of the first member of the Nenjiang Formation in the central basin. This system is composed of 3 trunk channels and 4 tributary channels, and extended from north to south along the Daqing Placanticline with a maximum length of about 71 km and a maximum width of about 600 m. This study indicates that this subqueous channel system indicate a major river draining into a lake. Interpretation of logging and cores indicates that sand lenses in the channels have high organic content and provide a new target for hydrocarbon reservoir exploration within lacustrine mudstone in the central depression of the Songliao Basin.

(3) Seismites: From the drilling core, four groups of seismites are recognised in the second and third members of Qinshankou Formation. The seismites occur between the lower part of the second member to the lower part of the third member. Structures induced by seismic shaking are preserved in thin layers of sandstone interlayered with thick dark grey mudstone deposited in a deep lake environment, and include microstep faulting, fissured beds, autoclastic breccias structures, load structures, ball and pillow structures (pseudo-nodules), liquefaction structures, enterolithic structures,sand dykes, and pseudo-mud cracks. The thicknesses of the seismites ranges from several millimeters to several centimeters. Seismites in the Qingshankou Formation in SK-1(s) are interpreted as in-situ structures from their soft-sediment deformation and absence of seismo-turbidites. Based on analysis of the temporal-spatial relationship between underwater volcanic eruptions during the deposition period of the Qingshankou Formation revealed by drilling, and displacement on the Sunwu-Shuangliao crustal fault, the seismites are considered to be the result of both normal faulting and volcanic eruption during rapid whole basin subsidence.

(4) Dolomite: Sixty-two layers of bedded dolostone and ellipsoidal dolostone are identified in the Nenjiang Formation. The vertical section, dolostone nodules are convex lenses, and laminations at the edges of some nodules converge toward their long axes. Some nodules have zonal structures in which the core is marl and the rim is dolomite. There are also some "leopard spot" dolomite within marl. All these nodule forms suggest that the dolostone of Nenjiang formation is the product of penecontemporaneous replacement of marl formed during Late Cretaceous marine transgression.

4.2 Paleomagnetism

4.2.1 Aims and Significance

Previous magnetostratigraphic studies were conducted by Fang *et al.* (1990) with relatively poor resolution. We conducted a detailed paleomagnetic study of the SK-1 cores and established high resolution geomagnetic stratigraphy which is compared with the geomagnetic polarity timescale of Cande and Kent (1995). Geomagnetic polarity reversals in SK-1 cores can be correlated with the global chronostratigraphic framework and thus for the first time establish a correlation between the continental Cretaceous stratigraphy of the Songliao Basin and global marine sediments.

4.2.2 Sampling and Magnetic Data Analysis

1. Sampling for paleomagnetics

6302 individual palaeomagnetic samples were collected from SK-1s and 4565 samples from SK-1n, with a total number of 10,867. The average sampling interval was approximately 24 cm.

2. Thermal demagnetization and experimental method

Experimental work was undertaken in the Paleomagnetism and Geochronology Lab in Beijing at the Institute of Geology and Geophysics, Chinese Academy of Sciences. 4459 samples (2546 samples in SK-1n and 1913 samples in SK-1s) were subjected to progressive thermal or thermal and alternating field demagnetization (2323 samples from SK-1n and 2136 samples from SK-1s), which were measured at each step of the treatment. The average interval between analyzed samples was 58 cm.

The equipment used for thermal demagnetization was a MMTD60 or MMTD80 demagnetization furnace manufactured in the UK, and a 2G cryogenic magnetometer demagnetization system for alternating field (AF) demagnetization. Thermal demagnetization was measured at each step of the treatment using either a 2G 760 or 2G 755 cryogenic magnetometer. All of the experimental work was undertaken in a magnetically shielded room (<300 nT).

A small, soft component of magnetization was easily removed during the initial thermal demagnetization, and characteristic remnant magnetization (ChRM) was successfully isolated. Magnetization directions were determined by principal-component analysis using PaleoMag software compiled by Craig H. Jones and Joya Tetreault (declination is random direction REF). In a total of 4459 samples, 2494 of the samples measured (56%) provided a reliable remnant magnetization signal (2043 samples in a magnetite carrier, 451 samples in a hematite carrier). Declinations of samples are random and we constructed the magnetic polarity sequence in SK-1 based on recognition of reversed and normal polarity components (Figure 4.3).

4.2.3 Preliminary Results

1. The magnetic polarity sequence of SK-1

From the data obtained a magnetostratigraphic profile which contains 11 magnetozones in SK-1n and 8 magnetozones in SK-1s can be constructed (Figure 4.3). The major polarity zones in SK-1n are: mixed magnetic zone M (175—317.03 m), reversal magnetic zones R1 (317.03—342.1 m), R2 (530.78—700.88 m), R3 (852.6—887.8 m), R4 (895.8—910.2 m), R5 (1020.4—1739.3 m); normal magnetic zones N1 (342.1—530.78 m), N2 (700.88—852.6 m), N3 (887.8—895.8 m), N4 (910.2—1020.4 m), N5 (1739.3—1800 m). Major polarity zones established in SK-1s are: reversal magnetic zones R5 (955.45—987.95 m), R6 (1144.4—1149.8 m), R7 (1163.85—1175.9 m), R8 (1193.15—1239.9 m); normal magnetic zones N5 (987.95—1144.4 m), N6 (1149.8—1163.85 m), N7 (1175.9—1193.15 m), N8 (1239.9—1915 m).

2. Correlation of polarity reversals in SK-1 with established global geomagnetic polarity timescale

For this correlation, we adopted the geomagnetic polarity timescale of Cande and Kent (1995). Based on the Cretaceous geomagnetic field and isotope chronology data obtained from the Songliao Basin (see Chapter 4.5 and Figure 4.3), N5-R6-N6-R7-N7-R8-N8 in SK-1s polarity column can be correlated with the Cretaceous normal polarity superchron (C34n). SK-1n and SK-1s boreholes can be intercorrelated using the well established basin-wide oil-shale horizon in Nenjiang Formation Member 2, which occurs in SK-1n at 1780 m and in SK-1s at 1020m. Therefore, the polarity zone R5 in SK-1n (North Hole) and the polarity zone R5 in SK-1s (South Hole) can be correlated from the geomagnetic polarity time scale of Cande and Kent (1995) to C33r (Figure 4.3).

In brief, SK-1n (North Hole) records the sedimentary succession between C29r reversed polarity and C34n normal polarity, and SK-1s (South Hole) records the sedimentary succession between C33r reversed polarity and C34n normal polarity. For the detailed correlations see Figure 4.3.

3. Age framework of SK-1 and correlation between continental Cretaceous stratigraphy in the Songliao Basin and global marine sediments

Based on the age framework of SK-1 established by this study, correlation between the continental Cretaceous sequence in the Songliao Basin and global marine sediments can be made (Figure 4.4). Principles used to confirm the age of each formation and member are as follows: first, we adopt the age of the reversed and normal polarity boundaries as the age of the strata boundaries; second, we obtain additional ages of the strata boundaries by linear interpolation or extrapolation.

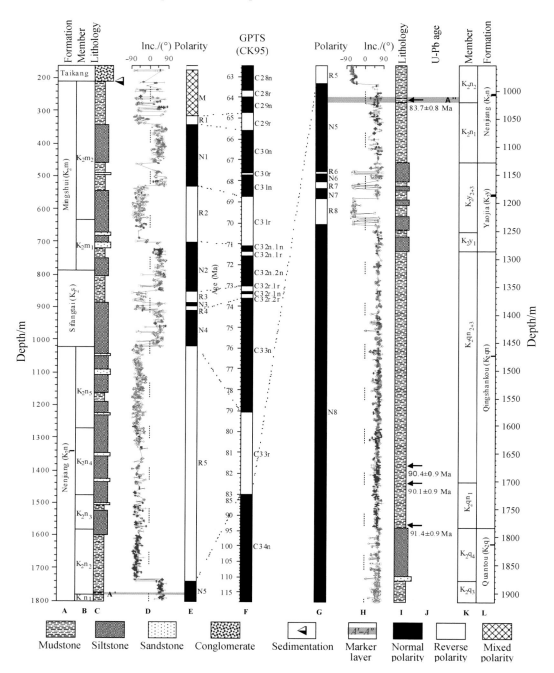

Figure 4.3 Lithostratigraphy, zircon U-Pb dating, paleomagnetic polarities in SK-1, and correlation with global geomagnetic polarity time scale. A—E and G—L are relusts of SK-1n (Deng et al., 2013)and SK-1s (He et al., 2012), respectively, F = paleomagnetic chart CK95 (Cande and Kent, 1995). In D and H, red circles = thermal demagnetization and hematite carrier; purple rhombs = thermal demagnetization and magnetite carrier; blue crosses = mixing demagnetization and magnetite carrier.

In SK-1n (North Hole), the boundary of the Sifangtai and Nenjiang formations is close to the boundary of C33n and C33r. Therefore, the age of the Sifangtai-Nenjiang Formation boundary is 79.075 Ma. In SK-1s (South Hole), the boundary NenJiang Formation Members 1 and 2, the boundary of Qingshankou Formation Members 1 and 2+3, and the boundary of the Quantou and Qingshankou formations are close to zircon U-Pb age dated ash horizons, and designated as 83.7 Ma, 90.1 Ma and 91.4 Ma, respectively.

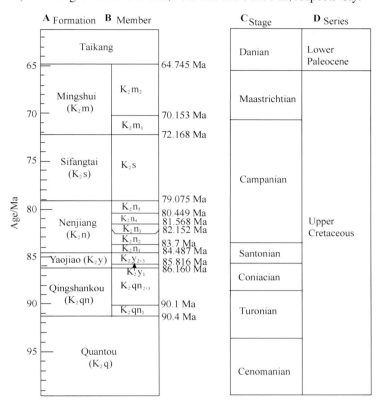

Figure 4.4 Age framework of SK-1 (Deng *et al.*, 2013; He *et al.*, 2012) and correlation with global marine stratigraphy (Skelton *et al.*, 2003; Gradstein *et al.*, 2004).

In SK-1n (North Hole), linear interpolation is used in R2 (matched with C31r) and the boundary age of Mingshui Formation Members 1 and 2 is 70.153 Ma. Using the same method, the boundary age of the Sifangtai and Mingshui formations is 72.168 Ma, the boundary age of Nenjiang Formation Member 5 and 4 is 80.449 Ma, the boundary age of Nenjiang Formation Member 4 and 3 is 81.568 Ma, and the boundary age of Nenjiang Formation Members 3 and 2 is 82.152 Ma (Figure 4.3).

In SK-1s (South Hole), linear interpolation is used according to the 91.4 Ma age at 1780 m and the upper boundary of C34n (83 Ma, terminal of CNS, at 987.95 m), the boundary age of the Nenjiang and Yaojia formations is 84.487 Ma, the boundary age of Yaojia Formation Member 1 and Members 2+3 is 85.816 Ma, and the boundary age of the Qingshankou and Yaojia formations is 86.160 Ma (Figure 4.3).

Based on these data, a preliminary age framework of SK-1 and correlation to global marine stratigraphy was made (Figure 4.4). The integrated chronostratigraphy shows that the age of sediments of the

Qingshankou to Mingshui formations in SK-1 ranges from middle Turonian to Maastrichtian. According to this correlation, the Cretaceous / Paleogene (K/Pg) boundary lies in upper part of the Mingshui Formation.

4.3 Logging Results

4.3.1 Aims and Significance

Geophysical logging technology is continually progressing and using special instruments to measure and analyze various parameters in drill holes in order to analyze in more detail. The main logging techniques include electric, acoustic and radioactive logging.

Aims of obtaining logging results are as follows:

(1) Logging data has the advantage of obtaining continuous, consistent data from different logging techniques and with little of error in depth determination. Core correlation with log data can be established by using various marker beds, such as marl, dolomite, coal, igneous rock such as ash beds or lava flows, high organic carbon beds, which provide uniform depths for sampling and information obtained by logging tools such as lithology, acoustic properties, density, porosity and permeability, electric properties, etc.

(2) Logging data can be used to determine more precise lithology (grain size, argillaceous content), physical properties (density, porosity and permeability, etc.), and particular features of source rocks such as organic carbon content.

(3) Logging data has a high resolution. Depending on the logging instruments used, logging data can have high precision with about 20 cm vertical resolution, especially in mudstone where is little of change in composition and variation in logging parameters. Using signal processing techniques, geophysical data can be useful in paleoclimate research. In 1920, Milankovitch proved the existence of orbital cycles which influence climate (eccentricity, precession, obliquity cycles) and in turn. The environment of sediment deposition as recorded by sedimentary facies, bed thickness and physical properties of these diments. The cyclicity of sedimentary sequences can be recorded by logging data analysed by Fourier or wavelet transform techniques. Using spectrum analysis and rhythm identification methods, sedimentary cycles can be identified, and the cycles then correlated with changes in climate. Such cyclic signals can be extracted from logging data using a special data processing system.

Logging results are also used in source rock identification and distribution. Two aspects of the research are: climate control on source rock deposition and distribution that is related to climate change; determination of the organic carbon content and other geochemical information from both the logging and geophysical data.

Using high resolution logging data, "hidden" cyclicity in the strata can be revealed and correlated to orbital Milankovitch cycles, and the cyclostratigraphy established which can be correlated with climate change.

4.3.2 Sampling and Analysis

(1) Cyclic analysis of recovered cores at SK-1 was conducted, and meter-scale (sixth order), fifth order and fourth order cycles were identified.

(2) Through Fourier transform or wavelet transform, cyclic analysis based on logging data in SK-1 was carried out.

(3) According to organic carbon content in the cores, quantitative relations between logging data and organic carbon was established. This is useful for further source rock interpretation from the logging data.

4.3.3 Preliminary Results

(1) Meter-scale cycle identification was conducted on the Quantou Formation in SK-1s. Sedimentary cycles identified were then compared with cycles determined by deep laterolog (RD) (Figure 4.5).

(2) Relations between sedimentary petrology and geophysical response were analyzed. Natural gamma spectrum logging was considered to better respond to the determination of climate change, and was chosen as an important parameter for paleoclimate studies.

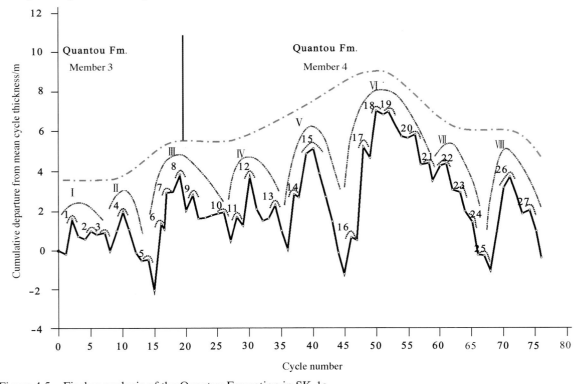

Figure 4.5 Fischer analysis of the Quantou Formation in SK-1s.

(3) According to coherent analysis, logging parameters such as natural gamma spectrum U, Th/U, acoustic travel time AC, and deep resistivity R_D, are all related to source rock organic content (Figure 4.6). Through multiple stepwise regression, a logging interpretation model was used to determine source rock organic content, which may more effectively predict source rock distribution.

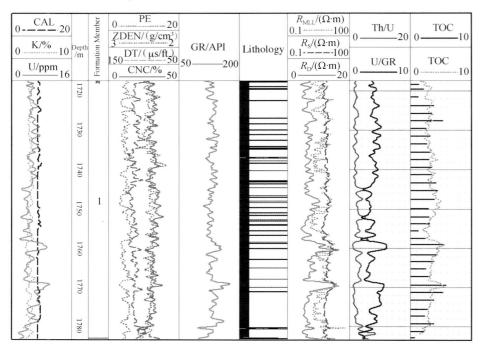

Figure 4.6　Correlation of organic content between calculated data derived from logging and actual measurement.

4.4　Inorganic Geochemistry and Mineralogy

4.4.1　Research Objectives and Significance

Element concentrations, mineralogy and stable isotopes of sediments can sensitively reflect geological processes such as weathering, transport, sedimentation and diagenesis. Inorganic geochemistry and clay minerals provide an essential data base for the reconstruction of paleoclimatic conditions during sediment deposition in the Songliao Basin.

4.4.2　Sampling and Analytical Methods

Samples for elemental and clay mineral analysis were taken from SK-1n, SK-1s and Chao 73—87 wells.

Samples for elemental analysis were collected at intervals of 2—10 m; in some key beds such as Qingshankou Formation Member 1, samples were taken from smaller intervals. Major elements were analyzed by a Philips PW2404 X-ray fluorescence (XRF) spectrometer at Daqing Petroleum Institute, following GB/T14506.28-93 *Silicate Rock Geochemistry Analytical Procedures*. Trace elements were analyzed by Atomic Emission Spectral Analysis, using an Atomic Emission Spectrophotometer ICPS-7510 at Daqing Petroleum Institute.

Samples for clay mineral analysis were undertaken on samples collected at 10—20 m intervals. The clay minerals were analyzed using a D/max-2200-X X-ray diffractometer (XRD) at the Institute of Exploration and Development, Daqing Oilfield and China University of Geosciences, Beijing, following SY/T5163-1995 Clay Mineral X-ray diffraction analytical procedures.

A study of pyrite morphology was focused on Qingshankou Formation Member 1, using reflection microscopy and a Scanning Electron Microscope (SEM) at the China University of Geosciences, Beijing and Chinese Academy of Geological Sciences. Fifty-one samples for pyrite S-isotope analysis were collected from the top 10 m of the Quantou Formation and Qingshankou Formation Member 1, with average sampling interval of 1.8 m. S-isotopes were analyzed using a Mass Spectrometer Delta-S at the Institute of Geology and Geophysics, Chinese Academy of Sciences.

4.4.3 Preliminary Results

1. Major elements

One hundred and seventy samples from SK-1n, SK-1s and Chao 73—87 wells were analysed for major elements (Figure 4.7). Vertical changes of major oxide concentrations from the center to shallow part of the Songliao Basin are summarized below.

(1) CaO: In Quantou Formation Members 3, 4 from SK-1s, CaO is usually <5.0%. At the boundary of the Quantou and Qingshankou formations, CaO increases to a maximum of 19.9% (1780.72 m). From the Qingshankou to Yaojia Formation, CaO is highly variable between 0.6%—10.0%. In Nenjiang Formation Member 1, CaO in the upper and lower parts is variable, while in the middle part is relatively constant. In the upper part of the formation, CaO rapid increases from 1.7% at 1065.00 m to 17.0% at 1045.00 m, and then rapidly decreases to 2.7% at 1029.00 m. At the base of Nenjiang Formation Member 2, CaO increases rapidly to 16.9% at 1021.00 m, and then decreases to 0.3% at 1017.00 m. Above Nenjiang Formation Member 2, CaO is low. In Chao 73—87 well, CaO variation from Quantou Formation Member 4 to Qingshankou Formation Member 1 is similar to that in SK-1s. At the boundary of these two formations, there is a rapid decrease in CaO from 5.0% at 844.02 m to 0.7% at 843.32 m, and then an increase to 6.4% at 840.42 m, followed by a fall to 0.7% at 836.87 m. In the middle to lower part of Qingshankou Formation Member 1, CaO is relatively constant, but within the middle to upper part of Qingshankou Formation Member 1, several spikes occur, such as, 7.1% (797.28 m), 8.8% (786.46 m) and 6.9% (771.90 m).

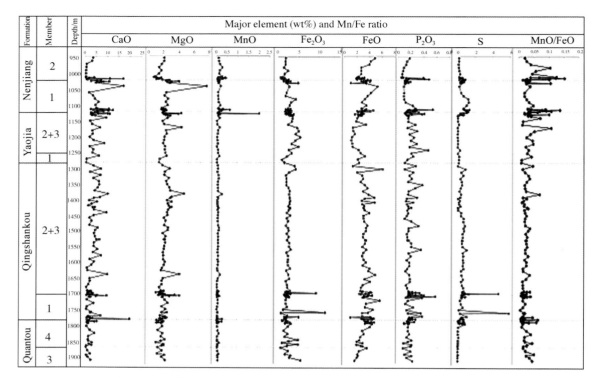

Figure 4.7 Major element (wt%) variation from the Quantou Formation to the Nenjiang Formation in SK-1s. The left column from the bottom up : Quantou Formation (Member 3 and Member 4), Qingshankou Formation (Member 1 and Members 2+3), Yaojia Formation (Member 1 and Members 2+3), Nenjiang Formation (Member 1 and Member 2).

(2) MgO: In Quantou Formation Members 3, 4, in SK-1s, MgO typically varies between 1.0%~2.2%. In Qiangshankou Formation Member 1, MgO is generally similar to that in the Quantou Formation, but at the top increases from 1.8% at 1711.27 m to 4.0% at 1707.27 m, and then rapidly decreases to 1.3% at 1706.27 m. From Qiangshankou Formation Members 2+3 to Nenjiang Formation Member 1, MgO remains comparatively stable (2.0%), but few spikes of 4.0% (1640.11 m), 4.6% (1385.00 m), 4.3% (1175.00 m), and 4.2% (1131.00 m) occur. In the upper part of Nenjiang Formation Member 1, there is a rapid increase in MgO from 2.3% at 1065.00 m to 7.5% at 1045.00 m, followed by a rapid decrease to 2.7% at 1035.00 m. In Nenjiang Formation Member 2, MgO is generally low (usually >2.0%). In Chao 73—87 well, the variation of MgO from Quantou Formation Member 4 to Qingshankou Formation Member 1 is similar to SK-1s. Only one anomaly is shown as a rapid increase from 1.1% at 763.20 m to 2.3% at 761.40 m, and then a slight fall to 1.2%.

(3) MnO: In SK-1s, MnO shows only minor variation in Quantou Formation Member 3 to Nenjiang Formation Member 2, with values usually < 0.1%. Two Mn anomalies occur, one at the top of the Yaojia Formation (2.0% at 1131.00 m) and the other at the base of Nenjiang Formation Member 1 (0.63% at 1120.00 m). Near the boundaries of Quantou-Qingshankou formations and Nenjiang Formation Member 1-Member 2, small increases in MnO from 0.06% at 1793.93 m to 0.3% at 1780.72 m occur and there is an increase from 0.08% at 1028.00 m to 0.4% at 1019.00 m. In Chao 73—87 well, the average MnO from Quantou Formation Member 4 to Qing-

shankou Formation Member 1 is < 0.05%. An Mn anomaly near the boundary is indicated by a sharp increase from 0.02% at 842.92 m to 0.13% at 840.77 m, and followed by a rapid decrease to 0.03% at 837.42 m.

(4) Fe_2O_3: In SK-1s, there are only small-scale variations in Fe_2O_3 from Quantou Formation Member 3 to lower Nenjiang Formation Member 2. In the Quantou Formation, Fe_2O_3 ranges from 0.6%—4.9% and in Qingshankou Formation it is about 2.0%, with two spikes of 11.4% (1759.96 m) and 9.2% (1697.27 m). At the top of Qingshankou Formation, Fe_2O_3 increases slightly and then decreases at the upper boundary. In the Yaojia Formation, Fe_2O_3 is steadily increases to a maximum of 5.4%. In the Nenjiang Formation, Fe_2O_3 is similar to that in the Qingshankou Formation, and is generally <2.0%.

(5) FeO: In SK-1s, FeO steadily increases from 1.7% (1910.67 m) in Quantou Formation Member 3, to 4.8% (1788.93 m) in Member 4. Near the boundary of the Quantou and Qingshankou formations, FeO is at a minimum of 1.0% at 1774.22 m. In Qingshankou Formation Member 1, FeO is variable, with a maximum of 5.7% at 1720.27 m. In Member 2 and 3, the variation is between 3.0%—4.0%, but with one high of 6.1% at 1305.00 m. In the Yaojia Formation, FeO is generally low, averaging about 1.3%, which is lower than in the Qingshankou Formation. From Yaojia Formation Members 2+3 to Nenjiang Formation Member 1, there is an increase in FeO, with small-scale variation at the boundary between the Yaojia and Nenjiang formations. From the top of Nenjiang Formation Member 1 to the base of the Nenjiang Formation, the FeO content is variable. A sharp decrease in FeO is occurs from 5.3% at 1045.00 m to 1.0% at 1034.00 m, followed by a rapid increase to 4.2% at 1025.00 m, a fall to 1.9% (1020.00 m), and an increase to 4.9% at 955.00 m.

(6) P_2O_5: In SK-1s, P_2O_5 in the Quantou Formation shows only a minor fluctuation. In the lower part of Member 1 of the Qingshankou Formation, P_2O_5 rapidly increases from 0.18% at 1773.22 m to 0.38% at 1770.46 m, followed by rapid decrease to 0.2% at 1760.46 m, and a further increase to 0.36% at 1759.96 m; in the upper part of Member 1, P_2O_5 increases from 0.16% at 1720.27 m to 0.38% at 1711.27 m, a decrease to 0.13% at 1710.27 m, and an increase to 0.58% at 1707.27 m. In Qingshankou Formation Members 2 and 3, the P_2O_5 content is low, at between 0.16%—0.19%, with increases of 0.36% (1560.28 m), 0.32% (1484.05 m) and 0.38% (1355.00 m). In the Yaojia Formation, P_2O_5 varies from a minimum of 0.11% (1285.00 m) to maximum of 0.47% (1245.00 m). At the top of the Yaojia Formation, P_2O_5 variability is higher, with a rapid increase from 0.13% (1134.00 m) to 0.41% (1131.00 m), and then a rapid decrease to 0.11% (1129.00 m). At the base of Nenjiang Formation Members 1 and 2, the phosphorous content varies from 0.12% (1128.00 m) to 0.54% (1117.00 m) and 0.11% (1025.00 m) to 0.48% (1022.00 m). In the middle to upper Nenjiang Formation Member 1, P_2O_5 is comparatively stable at between 0.09%—0.12%, but in lower Nenjiang Formation Member 2 the content gradually increases from a minimum of 0.05% (1017.00 m) to 0.18% at 955.00 m. In the Chao 73—87 well, P_2O_5 content is typically less than 0.10% from Quantou Formation Member 4 to Qingshankou Formation Member 1, with spikes of 0.37% at 826.37 m and 0.23% at 818.92 m.

(7) Sulfur: In SK-1s, the sulfur content is usually <0.10%. in Quantou Formation Member 3 and in the middle to lower part of Quantou Formation Member 4. In the upper part of the Quantou Formation, S variability is slightly higher. In Qingshankou Formation Member 1, S is between 0.15%—1.0%, with one promi-

nent spike of 5.8% (1759.96 m). In Qingshankou Formation Members 2+3, S is generally higher than in the Quantou Formation, where it is usually between 0.4%—0.8%, with a spike of 4.6% (1697.27 m) at the base. In the Yaojia Formation, S is lower than in the Qingshankou Formation, but is similar to the Quantou Formation, where is mostly < 0.15%. From the top of the Yaojia Formation to the Nenjiang Formation, the sulfur content shows small-scale variation from 0.04% (1138.00 m) to 1.3% (1118.00 m). In the lower part of the Nenjiang Formation, S is mostly higher than 0.7%, while in the upper part, S gradually decreases to a minimum of 0.19% (1045.00 m). At the base of Nenjiang Formation Member 2, the content of S is low and increases to a maximum of 0.87% (1023.00 m), and then decreases to < 0.20%. In the Chao 73—87 well, S in Quantou Formation Member 4 and Qingshankou Formation Member 1 differs from that in SK-1s. In Quantou Formation Member 4, S is usually < 0.20%. In Qingshankou Formation Member 1, S-variation can be subdivided into three parts; in the lower part the concentration curve is "U" shaped with a minimum of 0.24% and a maximum 0.99%, the middle part has a "W" form with a minimum of 0.10% and maximum of 1.04%, the upper part has comparatively high S with occasionally peaks, a minimum of 0.44% and a maximum 1.71%.

(8) Mn/Fe: In SK-1s, changes in the Mn/Fe ratio are small and the overall ratio is < 0.02. Two spikes occur between the Yaojia and Nenjiang formations, and between Nenjiang Formation Member 1 and Member 2. At the boundary of Yaojia and Nenjiang formations, the Mn/Fe ratio is 0.38 (1131.00 m), while between Nenjiang Formation Member 1 and Member 2 the maximum ratio is 0.98 (1019.00 m).

2. Mineralogy

1) Clay minerals

Clay minerals are important constitutes of sedimentary rocks and are significant for oil exploration in sedimentary basins. Wang *et al*. (1985) studied clay minerals in Cretaceous mudstone from the Songliao Basin, and determined the diagenetic evolution of clay minerals and their relationship to oil and gas distribution. Gao *et al*. (1994) discussed stratigraphic correlation based on clay mineralogy.

Clay minerals were analyzed in 223 samples from SK-1n Taikang Formation to Nenjiang Formation, and 179 samples were taken from SK-1s Quantou Formation Member 3 to Nenjiang Formation Member 2. According to these analyses, the vertical distribution of clay minerals in the middle to shallow parts of the Songliao Basin is as follows (Figure 4.8).

Illite (I): The variation of illite content is minor. From Quantou Formation Member 4 to Qingshankou Formation Member 1, modal illite decreases from 10.1% at 1785.93 m to 3.3% at 1780.72 m, with a minimum of 0.92% at 1770.46 m in Qingshankou Formation Member 1. At the base of Qingshankou Formation Members 2+3, illite reaches 14.4% at 1700.27 m. Near the boundary of the Yaojia and Nenjiang formations, illite decreases from 10.8% at 1130.00 m to 2.9%, and then increases to 10.8% (1105.00 m). There is little illite in Nenjiang Formation Members 1+2. In Mingshui Formation Member 1, the illite content is 17.3% at 667.92 m. Towards the upper part of Mingshui Formation Member 2 (above 420 m), illite rapidly decreases to <2.5%.

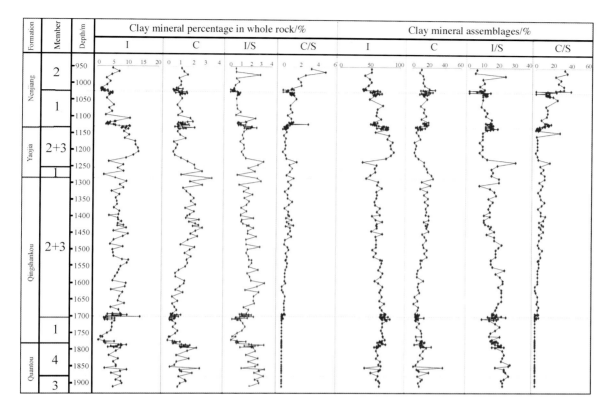

Figure 4.8 Clay mineral distribution from Quantou Formation to Nenjiang Formation in SK-1s. Formations and members in left column are from the base up: Quantou Formation (Member 3 and Member 4), Qingshankou Formation (Member 1 and Members 2+3), Yaojia Formation (Member 1 and Members 2+3), Nenjiang Formation (Member 1 and Member 2).

Chlorite (C): In Quantou Formation Members 3+4, modal chlorite oscillates, with a minimum of 0.3% and a maximum 2.6%. In Qingshankou Formation Member 1 and the lower parts of Members 2+3, chlorite is generally < 1.4%, and gradually increases from the upper to lower part to a maximum of 3.5% (1285.00 m) at the base of the Yaojia Formation. In Yaojia Formation Members 2+3, chlorite decreases from 2.2% (1255.00 m) to 0.43% (1205.00 m). From the upper part of Yaojia Formation Members 2+3 to the lower part of Nenjiang Formation Member 2, chlorite generally shows a "W"-shaped variation, with a peak of 2.1% near the boundary of Nenjiang Formation Member 1 and the Yaojia Formation. From Nenjiang Formation Member 1 to Member 3, the chlorite content is <1.7%. In Member 4, the amount of chlorite increases slightly. In Member 5, the the chlorite content again low at <2%. In the Sifangtai Formation, chlorite content is similar to Nenjiang Formation Member 5. In Mingshui Formation Member 1, chlorite shows a greater variation. In Mingshui Formation Member 2, the amount of chlorite is generally low with one spike of 8.4% (450 m).

Smectite (S): Beneath Nenjiang Formation Member 5, the smectite content is almost zero. In middle to upper part of Nenjiang Formation Member 5, the amount of smectite increases up to a maximum of 6.2%. In the Sifangtai Formation, the smectite content is very low. In the Mingshui Formation, there are two modal spikes of 17.3% (699.27 m) and 25.3% (634.52 m). In Mingshui Formation Member 2, smectite content is again lower than ~8.6%.

Kaolinite (K): Beneath Nenjiang Formation Member 2, modal kaolinite is <1%. In Nenjiang Formation Members 3+4, kaolinite content is higher with modal peaks reaching a maximum of 8.1% (1360.00 m). In Nenjiang Formation Member 5, the kaolinite content decreases to less than 1.5%. In the Sifangtai Formation, the amount of kaolinite is minor with a maximum of 2.8% at 807.40 m. In the Mingshui Formation, kaolinite is generally <1.8%.

Illite/Smectite mixed layer clays (I/S): Illite/Smectite mixed layer clays occur from the Quantou to the Sifangtai Formation. In Quantou-Nenjiang Formation, the I/S content is generally low especially within two lower intervals in Qingshankou Formation Member 1 and between Yaojia Formation Members 2+3 and Nenjiang Formation Member 3. From Nenjiang Formation Member 4 to the Sifangtai Formation, the I/S content is comparatively high, with a few wide-ranging peaks. Generally, at the boundaries of formations or units, the I/S content changes, e.g., at the boundaries of the Qingshankou-Yaojia and Yaojia-Nenjiang Formations, the I/S content is higher; while at Quantou-Qingshankou Formation boundary and Nenjiang-Sifangtai Formation boundary, the amount of I/S clays decreases.

Chlorite/Smectite mixed layers (C/S): Chlorite/Smectite mixed layer clays occur in rocks between Qingshankou Formation Members 2+3 and the Sifangtai Formation, with the content increasing upwards. From Qingshankou Formation Members 2+3 to the Yaojia Formation, the C/S clay content shows only minor variation. From the Nenjiang Formation to the Sifangtai Formation, the C/S clay content increases with wide variation and a maximum of 4.9% (965.00 m) in Nenjiang Formation Member 2.

2) Pyrite morphology

In lacustrine facies rocks of the Qingshankou Formation Member 1 (K_2qn_1), pyrite content is low, but it's morphology varies from single or clustered euhedral crystals, single or clustered framboids to anhedral crystals. In all K_2qn_1 rock samples, single euhedral pyrite crystals occur as cubic, octahedral and spherical forms, but mainly form cubes. The pyrite is generally concentrated along bedding planes. The diameter of euhedral pyrite has a wide range of between 1—15 μm (Figure 4.9), but in most samples the average diameter is 2—4 μm (Figure 4.9). Single pyrite framboids, mostly occur as spherical aggregates of submicron-size. Most pyrite framboids consists of cubic microcrystals that have uniform diameters; some of the clustered cubic euhedral pyrite crystals are surrounded by submicron-sized spherical euhedral crystals. Microcrystals that compose the outer surface of framboids are spherical. Compared to euhedral crystals, the number of framboids is small and their distribution variable. In samples from the depth interval of 1704—1751 m, the diameter of single pyrite framboids ranges from 4—25 μm, and the average diameter is 7.1—18.5 μm. In samples from the depth interval of

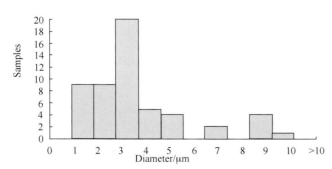

Figure 4.9 Average diameters of euhedral pyrite in K_2qn_1 sediments.

1751—1770 m, the diameter of single pyrite framboids ranges between 3 to 8 μm, with an average diameter of 4.3—6.2μm. In samples from the depth interval of 1770—1783 m, the diameter of single pyrite framboids ranges from 4 to 16 μm, with an average diameter of 6.7—10.4 μm. "Box-and-whisker" plot is a useful measure to describe the change of pyrite framboid size and their distribution in the sedimentary sequence. From the Box-and-whisker structure in Figure 4.10, pyrite framboids occurring in the depth range of 1750—1770 m are on average smaller and less variable in size than those in other depth ranges. The diameter changes of the pyrite framboids in K_2qn_1 suggests that redox conditions of the bottom water of paleolakes were flutulating.

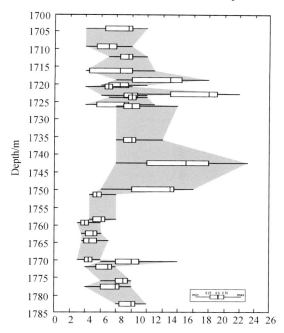

Figure 4.10 Box- and –whisker- Plot show pyrite framboid size distributions from K_2qn_1 sediments. The "boxes" extend from quartile $Q = 0.25$ to $Q = 0.75$. The median value $Q = 0.5$ is marked in each box. The lines extending to the right and left of each box delineate maximum and minimum values.

3. Sulfur isotopes

Of 282 samples from cores of Quantou Formation Member 4, Qingshankou Formation Member 1 in SK-1 of the Songliao Basin, 51 samples, were analyzed for sulfur mass fraction and $\delta^{34}S$ isotope concentration in pyrite and TOC. In Quantou Formation Member 4, the sulfur mass fraction in pyrite ranges between 0.003%—0.96%, and the average is 0.55%; the average values of $\delta^{34}S$ for pyrite ranges between 17.29‰—21.51‰, and the average is 19.47‰; TOC ranges between 0.003%—0.97% and the average is 0.35%. For all samples, the sulfur mass fraction in pyrite ranges between 0.003%—1.74%, and the average is 0.55%; $\delta^{34}S$ ranges between 14.4‰—24.06‰, with an average of 18.61‰; and TOC ranges between 0.003%—8.63%, with an average of 2.77%.

The distributions with depth of sulfur mass fraction of pyrite in Quantou Formation Member 4 and Qingshankou Formation Member 1 core are shown in Figure 4.11. From a depth of 1802 m in the Quantou Formation Member 4, the sulfur mass fraction of pyrite is low throughout the first 10 m, and then increases upward until 1760 m, at the base of Qingshankou Formation Member 1. The sulfur mass fraction maintains at a high level, but declines in the upper 60 m of Qingshankou Formation Member 1. Overall, sulfur mass fraction of pyrite in Qingshankou Formation Member 1 is significantly higher than in the top 10 m of Quantou Formation Member 4.

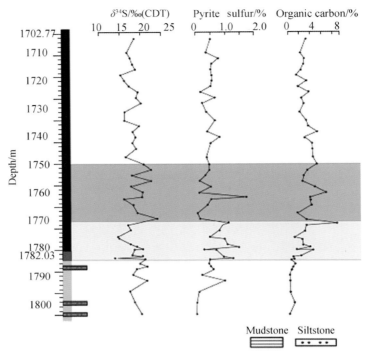

Figure 4.11 Simplified lithology and distribution of sulfur mass fraction of pyrite, $\delta^{34}S$ isotope and TOC content in Quantou Formation Member 4 and Qingshankou Formation Member 1.

The distribution of $\delta^{34}S$ for pyrite in Quantou Formation Member 4 and Qingshankou Formation Member 1 are shown in Figure 4.11. Values of $\delta^{34}S$ of pyrite are maintained at a high level and show some variation at the top of Quantou Formation Member 1 and the throughout the bottom 30 m of Qingshankou Formation Member 4. Values of $\delta^{34}S$ reach the peak of 24.6‰ at 1768 m in Qingshankou Formation Member 1, but slightly decline in the upper 50 m of Qingshankou Formation Member 1. Values of $\delta^{34}S$ in the top 10 m of Quantou Formation Member 4 are higher than those in Qingshankou Formation Member 1. Based on the relationship of sulfate concentration and of $\delta^{34}S$ of pyrite in other environments, and the C-S relationship in samples, pyrite $\delta^{34}S$ values become more positive in Qingshankou Formation Member 1. Together with the C-S relationship, they indicate that during deposition of Qingshankou Formation Member 1 lacustrine sulfate concentration was not high and lay between freshwater and marine environment values. Significantly, the positive increase of pyrite $\delta^{34}S$ in Qingshankou Formation Member 1 is similar to $\delta^{34}S$ of sulfates in Cretaceous marine sediments, that are obviously different from the $\delta^{34}S$ in terrigenous sources. This is because sulfate in lakes are likely be from marine sources. Therefore, during deposition of Qingshankou Formation Member 1 the Songliao Basin was most probably invaded by sea water. Since the sulfate concentration is not high, marine transgressions were periodic and limited.

4.5 Chronostratigraphy

4.5.1 Aims and Significance

Chronostratigraphy studies have been widely undertaken in the Songliao Basin, including paleomagnetism (Fang, 1989), radioactive isotope dating (Wang *et al.*, 1995; Huang *et al.*, 1999), palynostratigraphy (Gao *et al.*, 1999), ostracoda stratigraphy (Li, 2001; Ye *et al.*, 2002; Li and Li, 2005), conchostracan stratigraphy (Li *et al.*, 2004), and carbon-oxygen stable isotope stratigraphy (Wan *et al.*, 2005). From these studies, a chronostratigraphic framework for the basin has been established. However, recent developments in logging and other analytical methods show that new results differ from the older chronostratigraphy framework. One of the differences is radioactive dating. New age data indicate that top of Yingcheng Formation is about 110 Ma (Ding *et al.*, 2007; Jia *et al.*, 2008). This is 9 Ma younger than previously suggested. One of the reasons for this age difference is that previously only a few samples were dated in contrast to coring in SK-1 which has provided a continuous section for chronostratigraphic studies in the Songliao Basin. In this chapter we attempt to use all available data to establish a precise chronostratigraphy framework that provides the basis for intercorrelation of continental and marine Cretaceous strata.

4.5.2 Sampling and Analytical Methods

Three types of samples are used for chronostratigraphy studies: ① volcanic ash; ② rock samples for paleomagnetic studies; ③ ostracoda, pollen and angiosperm fossils for biostratigraphic dating.

Volcanic ash and paleomagnetics samples were analysed at the Institute of Geology and Geophysics, China Academy of Sciences; pollen, fern and angiosperm fossils were studied in the Nanjing Institute of Geology and Palaeontology, Chinese Academy of Sciences; fossil algae were examined at the Institute of Exploration and Development of Daqing Oilfield Company Ltd. Five volcanic ash samples for selected for SIMS zircon U-Pb dating, 4459 rock samples for paleomagnetic measurment, pollen was sampled from each formation in SK-1s and SK-1n, together with ostracoda in SK-1s. Pollen was collected from 20 m intervals on average, and in 5m intervals from some key layers. Densities of paleomagnetic samples and ostracoda-bearing samples are given in Chapters 4.2 and 4.10, respectively. In Well-Jin 6 in the Qijia-Gulong Depression a, volcanic rocks occur in the Qingshankou Formation between 1791.5—1982.5 m. The volcanic rock at 1972.8 m was $^{40}Ar/^{39}Ar$ dated (Figure 4.12).

4.5.3 Preliminary Results

1. SIMS U-Pb zircon dating of volcanic ash in SK-1

Although ash layers are thin and strongly altered, 5 samples were selected according to detailed core description and microscopic analysis for high-resolution SIMS U-Pb dating. The four ages obtained in SK-Is are: (91.4 ± 0.5) Ma (1780 m), (90.1 ± 0.6) Ma (1705 m), (90.4 ± 0.4)Ma (1673 m), (83.7 ± 0.5)Ma (1019 m) (Figure 4.13), which can be interpreted as deposition ages. Although the two ages at 1705 m and 1673 m are reversed, they are within the range of error. Several age peaks are shown in the ash at 1594 m, and the smallest peak at 84.5 Ma is interpreted as the oldest age. These ages provide control for magnetic stratigraphic correlation in SK-1. Moreover, three ash samples in SK-1s are close to formation and/or unit boundaries, and their ages can be used to date these stratigraphic boundaries. The boundaries of Nenjiang Formation Members 1 and 2, Qingshankou Formation Members 1 and 2+3, and the Quantou and Qingshankou formations are thus equated to the 83.7 Ma, 90.1 Ma and 91.4 Ma zircon U-Pb ash ages, respectively (Figures 4.3, 4.4).

2. $^{40}Ar/^{39}Ar$ dating of volcanic rock in the Qingshankou Formation (Well-Jin 6)

A mugearite in Qingshankou Formation Member 2, located 35 m above the base of Qingshankou Formation Member 2 was dated by the $^{40}Ar/^{39}Ar$ method.

$^{40}Ar/^{39}Ar$ plateau age t_p = (88.0 ± 0.3)Ma

A plateau age of (88.0 ± 0.3)Ma for the mugearite is interpreted as the cooling age (Figure 4.12). According to the average sedimentation rate of the Qingshankou Formation, the 35 m depth to the base of Qingshankou Formation Member 2 was estimated to represent about 0.8 Ma. Thus, the boundary of Qingshankou Formation Members 2+3 and Member 1 is considered to approximately correlate to the Coniacian-Turonian boundary.

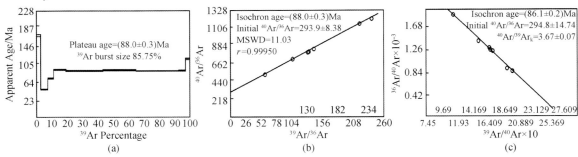

Figure 4.12 Qingshankou Formation mugearite $^{40}Ar/^{39}Ar$ age spectrum (a), correlated isochron diagram (b), and inverse isochron diagram (c).

3. Paleomagnetics stratigraphy

4459 rock samples were analyzed to determined their magnetic polarity to establish a palaeomagnetic polarity profile that could be correlated with GPTS. According to radioactive age dating in the Songliao Basin, the termination of CNS (about 121—83.5 Ma) correlates with the base of Nenjiang Formation Member 2, and magnetic reversal zones in the middle to upper part of Nenjiang Formation Member 2 to Member 5 correlate with C33r. Thus, the age of the Sifangtai-Nenjiang formation boundary is 79.1 Ma. Ages of the Qingshankou-Quantou formation boundary, Qingshanou Formation Member 1 and Members 2+3 boundary, and Nenjiang Formation Member 1 and Member 2 boundary, are obtained by radioactive dating, and the ages of other formation boundaries and/or members are given by interpolation (Figure 4.4, Figure 4.13).

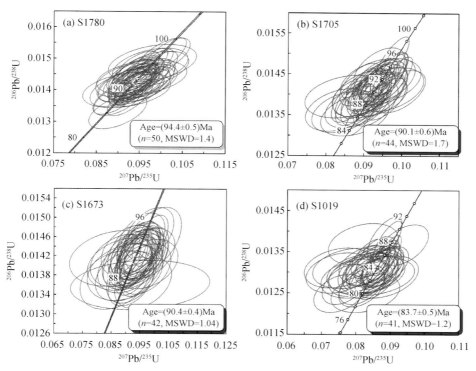

Figure 4.13 SIMS zircon U-Pb ages of volcanic ash in SK-1s.

4. Palynological chronostratigraphy

Pollen is an important age indicator and palynological analysis has been completed for SK-1n and SK-1s. Integrated with palynological morphology and previous studies, the following angiosperm pollens were selected as important age indicators: *Aquilapollenites, Betpakdalina, Borealipollis, Buttinia, Callistopollenites, Complexiopollis, Cranwellia, Integricorpus, Kurtzipites, Lythraites, Ulmipollenites, Ulmoideipites* and *Wodehouseia*. A few gymnosperm pollens and fern spores are also good age indicators, one of the most important being *Quantonenpollenites*. According to fossil pollen and spore assemblages, 8 zones can be established: Zones 1—3 in SK-1s and Zones 2—8 in SK-1n (Figure 4.14). The two boreholes can be correlated from Zones 2 and 3 assemblages.

Many publications have considered the evolution of angiosperm pollen in the Late Cretaceous of North America and Siberia. In Alberta, Canada, Jarzen and Norris (1975), Norris *et al.* (1975) describe the evolution of angiosperm pollen in Late Cretaceous (Table 4.1). After correlation with data in SK-1, the following palynological age framework is established.

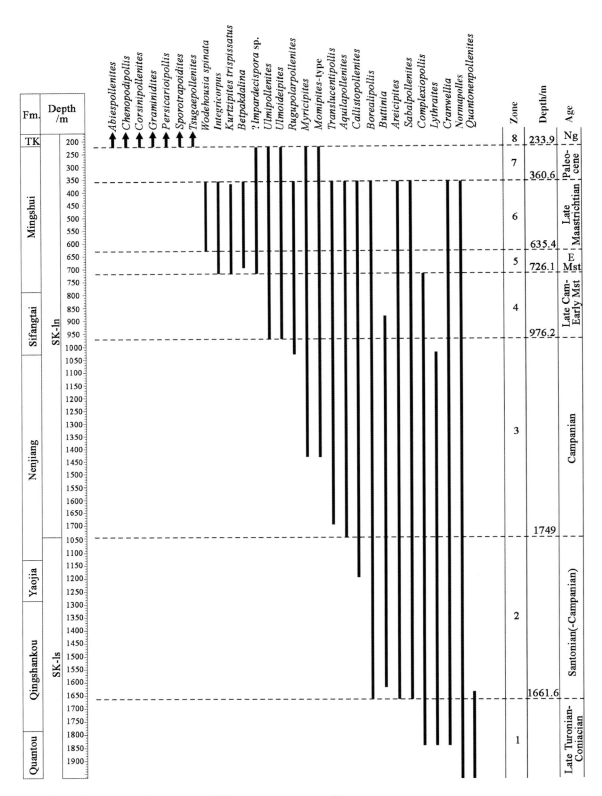

Figure 4.14 Palynological zonation of SK-I showing ranges of key taxa.

Table 4.1 International correlation of the Cretaceous palynological zones of SK-1.

SK-1 well (this study)		Alberta, Canada		W. Interior, Am	Siberia, Russia
Age	Zone	(Jarzen and Norris, 1975; Norris et al., 1975)	(Nichols and Sweet, 1993)	(Nichols and Sweet, 1993)	(Markevitch, 1994)
L. Mst.	6			9—10	XII. *O. lucidus* - *W. avita*
E. Mst.	5				XI. *W. spinata* - *A. subtilis*
L. Cam.-E. Mst.	4			7—8	
Cam.	3	*Cranwellia* Suite	*A.quadrilobus* Zone	4—6	X. *Cranwellia striata* - *Aquilapollenites trialatus*
		Trudopollis Suite			
		Advanced Angiosperm Suite			
		Late Loranthaceous Suite	*A. senonicus* Zone		
		Early Loranthaceous Suite	*Pseudoplicapollis* Zone		
San.—(Cam).	2	Early Triporate Suite	*Proteacidites* Zone	3	IX. *Lobatia involucrata* - *Kuprianipollis santaloides*
L. Tur.—Con.	1	*Nyssapollenites* Suite	*Nyssapollenites* Zone	2	

The age of the Quantou Formation is considered to be much younger than previously thought, with late Early Cretaceous Cenomanian determinations being replaced by Late Turonian—Coniacian. The proposal of an Early Trifoliate Suite by some palynologists (Gao et al., 1999) for the Quantou Formation is, therefore, doubtful. It can instead be correlated with the *Lythraites* Suite of Zhang (1999) and probably with the lower part of the Early Triporate Suite of western Canada (Jarzen and Norris, 1975; Norris et al., 1975), which was re-evaluated by Nichols and Sweet (1993) and considered to be equated with a Proteacidites retusus Biozone of Coniacian—Santonian age.

Consequently, angiosperm pollen in the Qingshankou and Yaojia formations are also comparable with the Early Triporate Suite of western Canada in that both lack *Aquilapollenites*. Age determinations of the Nenjiang and Sifangtai formations are adjusted to Campanian and Late Campanian—Early Maastrichtian, respectively, and the Mingshui Formation is now considered to include the Cretaceous/Paleogene boundary in its upper part whereas most authors have previously regarded it as entirely Maastrichtian.

Our analysis indicates a major palynofloral change at a depth of around 360.6 m in the northern borehole, with many taxa disappearing in the upper part of the Mingshui Formation. Although some of the disappearances do not necessarily indicate that the parent plants no longer existed, a "mass extinction" seems to be indicated from the assemblages examined. However, with nearly 100 m of section (from 360.6 to 263.4 m) being devoid of pollen and spores, analysis of more closely spaced samples around the boundary is required before it will be possible to determine whether this palynological "extinction event" is as significant as it appears at present.

5. Integrated Chronostratigraphy of SK-1

Integrated with radioactive isotopic dating, paleomagnetic stratigraphy, pollen, spore data, the chronostratigraphy of SK-1 has been established and correlated with the global geological timescale (Figure 4.15).

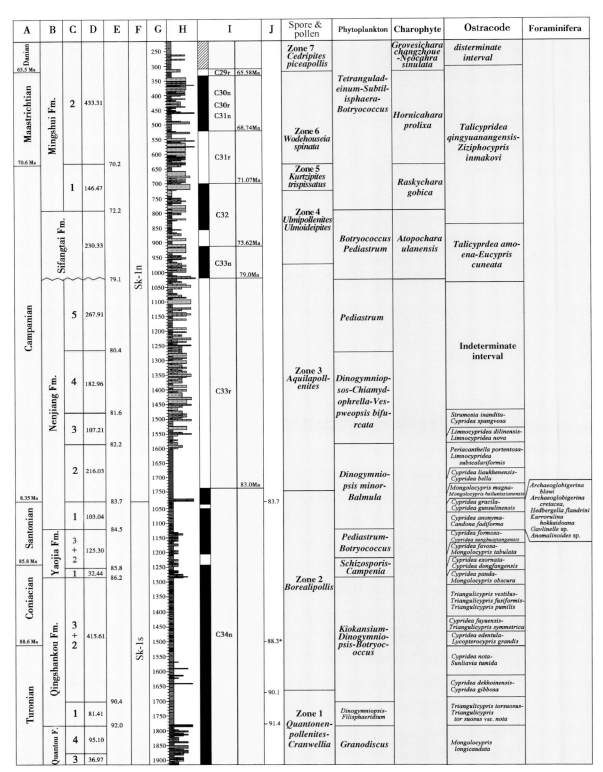

Figure 4.15 Integrated stratigraphic frame of SK-Ⅰ (South and North) based on magnetostratigraphy, SIMS U-Pb age and biostratigraphy. A. Epoch; B. Formation; C. Member; D. Thinkness(m); E. Interpolated boundary age(Ma); F. Core; G. Depth; H. Stratigraphic column; I. Polarity chron; J. SIMS zircou U-Pb and feldspar $^{40}Ar/^{39}Ar$ ages(Ma) (*Ar/Ar age). For legend see Figure 2.3.

4.6 Geomicrobiology

4.6.1 Research Objectives and Significance

Microorganisms are widely distributed in various environments and they actively participate in a number of geochemical processes. Many microbial metabolic processes involve consumption or production of CO_2 and CH_4, two of important green-house gases. Therefore, microorganisms may have significantly affected carbon cycling and global climate change throughout geological history. Modern DNA-based molecular microbiology techniques are capable of identifying microbial species and through comparison with cultured relatives, their physiological characteristics can be inferred and paleoenviromental conditions defined.

Here, we report preliminary results of a pilot study of two shales from SK-1, to determine the microbial community structure. Combined with geochemical and mineralogical characterization, this information will provide important evidence for the interpretation of paleo-depositional environment and paleo-environmental conditions existing in the Songliao Basin during the Cretaceous.

4.6.2 Materials and Methods

1. Sampling

Rock samples were collected at 50 m intervals. Aseptic techniques (sterile gloves, sampling tools, etc.) were used during sample handling to avoid any possible contamination. Rock cores of sufficient size, 120 mm in diameter and 10 cm in length, were hand-picked from the drilling barrel. The selected cores were placed inside sterile plastic bags and stored at −20 ℃ until the time of analysis. The whole sample was typically completed within 10 minutes.

2. Analytical techniques

X-ray diffraction (XRD) and 16S rRNA gene analysis methods were used in this study.

X-ray diffraction (XRD): Shale samples were analyzed by XRD to determine their mineralogical composition. Powder XRD patterns were obtained with a Phillips PW3040/00 X'Pert MPD system, using CuKα radiation with a variable divergent slit and a solid-state detector. The routine powder was 700 W (35 kV, 20 mA). Low background quartz XRD slides (Gem Dugout, Inc., Pittsburg, PA) were used. For analysis, powder samples were tightly packed into a depression (1/4 in diameter) in a glass slide. Mineral identification was performed with automated SearchMatch software and manual inspection.

16S rRNA gene analysis: This gene is conservative across all known prokaryotes. The gene has the following characteristics: ① it is present in all prokaryotic cells and can be used to study the phylogenetic relationships among various microorganisms; ② it is abundant in microbial cells; ③ it is conservative; ④ the molecular weight is appropriate for analysis; ⑤ there is a large database where the 16S rRNA gene of any unknowns can be compared with known, or clone sequences in the dadabase (GenBank). Genomic DNA was directly extracted from the shale samples (Dong et al., 2006) and PCR amplified. The amplified 16 rRNA gene was cloned and sequenced. A phylogenetic tree was constructed using various software (Dong et al., 2006; Jiang et al., 2006).

4.6.3 Preliminary Data

1. Sample description

Sample descriptions are shown in Table 4.2.

Table 4.2 Descriptions for geomicrobiology samples in SK-1s

Sample	Barrel	Depth/m	Lithology
B2-03	156	1117.00	Dark grey/black mudstone with fossils
B2-05	129	1215.00	Dark red silty mudstone with fossils

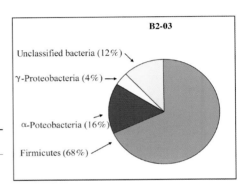

Figure 4.16 Distribution of major phylogenetic groups for B2-05.

2. Phylogenetic analysis

Sixteen clone sequences were obtained for B2-03 and they belong to 4 optional texanomic units (OTUs). The distribution of these OTUs is shown in Figure 4.16 (α-Proteobacteria, 16%; γ-Proteobacteria, 4%; Firmicutes, 68%). Close relatives from the GenBank (based on BLAST results) are given in Table 4.3. The pholygenetic tree is shown in Figure 4.17.

Table 4.3 BLAST results showing similarity of cloned sequences of B2-03 to those from GenBank

Clone number	Relative abundance/%	Group affiliation	Close relatives from GenBank	Similarity /%
1	4.0	Firmicutes	*Alkalibacterium olivapovliticus* (AF143512)	98
27	4.0	Unclassified bacteria	*Alkalibacterium olivapovliticus* (AJ576348)	96
30	4.0	Firmicutes	*Soehngenia saccharolytica* (AY353956)	95
34	12.0	Firmicutes	*Planococcus* sp. ZD22 (DQ177334)	99
31	4.0	γ-proteobacteria	γ-Proteobacterium HTB082 (AB010842)	96
32	4.0	Firmicutes	*Halolactibacillus miurensis* (AB196784)	98
33	16.0	Firmicutes	*Tissierella praeacuta* (X80833)	92
82	28.0	Firmicutes	Uncultured low G+C Gram-positive bacterium (DQ206408)	95
63	16.0	α-proteobacteria	Antarctic bacterium R-9219 (AJ441009)	98
76	8.0	Unclassified bacteria	Uncultured candidate division JS1 bacterium (AJ535219)	97

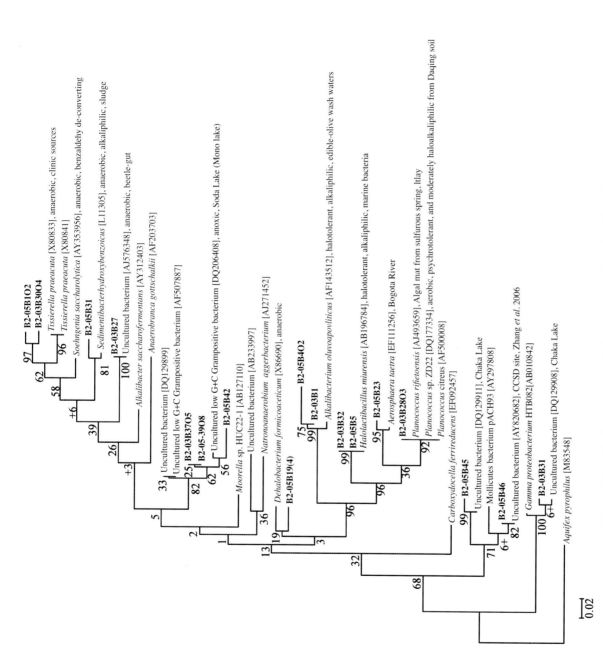

Figure 4.17 Phylogenetic tree of clone sequences from B2-03 and B2-05.

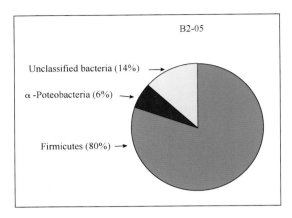

Figure 4.18 Distribution of major phylogenetic groups for B2-05.

Our results indicate that all clone sequences are affiliated with low G + C Gram positives and they are similar to sequences from saline and alkaline environments. The B2-03B37 clone type is the most abundant, accounting for 28% of total number of clones. This clone sequence is most similar to the uncultured environmental sequence (DQ206408) from a soda lake in Inner Mongolia. Clone B2-03B27 is similar to *Alkalibacterium olivapovliticus* (AJ576348) which is anaerobic and was isolated from the midgut and hindgut of humus-feeding larva of *Pachnoda ephippiata*. Clone B2-03B28 is similar to anaerobic and moderately halophilic *Planococcus* sp. ZD22 (DQ177334) (Li *et al.*, 2006) isolated from soils of Daqing, China. Clone B2-03B32 is similar to halophilic and alkaliphilic marine *Halolactibacillus miurensis* (AB196784).

Twenty-one clones were obtained for B2-05 and they belong to 3 OUT (α-Proteobacteria, 6%; Firmicutes, 80%)(Figure 4.18). Close relatives from the GenBank (based on the BLAST results) are listed in Table 4.4. The pholygenetic tree is shown in Figure 4.18.

Table 4.4 BLAST results showing similarity of cloned sequences of B2-03 to those from GenBank

Clone #	Abundance/%	Major group	Close relatives from GenBank	Similarity/%
1	5.9	Firmicutes	*Soehngenia saccharolytica* (AY353956)	95
4	2.0	Firmicutes	*Alkalibacterium olivapovliticus* (AF143512)	94
5	2.0	Firmicutes	*Halolactibacillus miurensis* (AB196784)	97
10	9.8	Unclassified bacteria	Uncultured bacterium (DQ129899)	96
13	2.0	Firmicutes	*Alkalibacterium olivapovliticus* (AF143512)	94
19	2.0	Firmicutes	*Dehalobacterium formicoaceticum* (X86690)	91
23	2.0	Firmicutes	*Aerosphaera taetra* (EF111256)	98
31	3.9	Firmicutes	*Tissierella praeacuta* (X80841)	93
37	3.9	Firmicutes	*Alkalibacter saccharofermentans* (AY312403)	88
44	2.0	Firmicutes	*Dehalobacterium formicoaceticum* (X86690)	89
48	19.6	Firmicutes	Uncultured low G+C Gram-positive bacterium (AF507887)	95
54	23.5	Firmicutes	Mollicutes bacterium pACH93 (AY297808)	92
46	2.0	Unclassified bacteria	Uncultured bacterium (AY820682)	96
52	2.0	α-Proteobacteria	Uncultured α-Proteobacterium (AY921836)	97
60	2.0	Unclassified bacteria	Uncultured bacterium (DQ125843)	98
65	3.9	Firmicutes	*Clostridium* sp. Z-7036 (EF382660)	92
71	3.9	α-Proteobacteria	*Caulobacter leidyia* (AJ227812)	97

Clone #	Abundance/%	Major group	Close relatives from GenBank	Continued Similarity/%
75	2.0	Firmicutes	*Halolactibacillus miurensis* (AB196784)	95
83	2.0	Firmicutes	Uncultured low G+C Gram-positive bacterium (DQ206408)	96
88	2.0	Firmicutes	*Soehngenia saccharolytica* (AY353956)	94
90	2.0	Firmicutes	*Soehngenia saccharolytica* (AY353956)	94

Our results indicate that sample B2-05 is similar to B2-03 in that nearly all clone sequences belong to low G + C Gram-positives, and they are similar to those recovered from saline and alkaline environments. The difference is that sample B2-05 does not have any sequences of γ-Proteobacteria. Clone B2-05B23 is similar to *Aerosphaera taetra* (EF111256) isolated from Bogota River, Columbia. Clone B2-05B5 is similar to halophilic and alkaliphilic marine *Halolactibacillus miurensis* (AB196784). Clone sequences of B2-05 show a low similarity to anaerobic bacteria such as anaerobic and alkaliphilic *Sedimentibacter hydroxybenzoicus* (L11305) and anaerobic *Dehalobacterium formicoaceticum* (X86690).

3. XRD and geochemical results

The XRD analyses were completed at Miami University (Figure 4.19) and the results are given in Table 4.5; geochemical data are listed in Table 4.6.

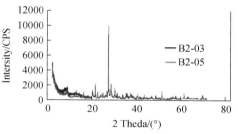

Figure 4.19 XRD patterns of B2-03 and B2-05.

Table 4.5 Mineral composition of B2-03 and B2-05

B2-03	Quartz	Albite	Kaolinite	Calcite	Smectite	Illite	Pyrite
B2-05	Quartz	Calcite	Ca-feldspar	Analcime	Smectite	Illite	

Sample B2-03 contains pyrite (FeS_2), and its occurrence in sedimentary rocks is often related to the decomposition of organic matter (Pan, 1994), indicating an anoxic environment. In sample B2-05, expected hematite is absent, but analcime is present indicating oxic conditions.

Table 4.6 Geochemical results of B2-03 and B2-05

Sample	Total Fe /%	Fe(II)/%	Fe(III)/%	N/%	Total C /%	TOC/%	pH	Soluble salt /(mmhos/cm)	SO_4-S /(μg/g)
B2-03	3.1			0.37	9.23	1.12	10.1	0.95	8
B2-05	3.8	0.453	3.368	0.12	0.73	0.62	9.9	1.05	1

The two samples contain a high amount of total iron. pH is alkaline. The TOC content of B2-03 is much higher than B2-05. High TOC content usually indicates a reducing conditions (Demaison and Moore, 1980; Demaison *et al.*, 1984). This is consistent with the Oceanic Anoxic Event (OAE) represented by black shales in the Cretaceous. The TOC content of B2-05 is low, indicating oxidizing conditions.

4. Conclusion

(1) In this pilot project which is based on only two shale samples from the Songliao Basin, no archaea were detected, which is consistent with previous studies.

(2) Most DNA sequences indicate saline and alkaline environments, consistent with geochemistry of the shale, and probably reflects paleo-environmental conditions of the basin.

(3) Many DNA sequences reflect a lacustrine rather than terrestrial environment, suggesting that the detected DNA was of fossil origin, and therefore should reflect paleo-environmental conditions.

(4) DNA sequences from black and red shales reflect different redox conditions, but more data is needed to confirm this.

4.7 Organic Geochemistry

4.7.1 Objectives and Significances of Organic Geochemistry Research

Ancient lacustrine sediments preserve and carry information about contemporary climate and environmental changes on both regional and global scales. The abundance and composition of sedimentary organic matter, biomarkers and their carbon isotopic composition provide important information for reconstruction of paleo-lacustrine environments and palaeoclimate in the depositional area. The lacustrine sedimentary sequence of the Songliao Basin, which records geological, geochemical and palaeoclimate changes throughout the Cretaceous, is a preferential location for the study of Cretaceous climate and continental environmental response to global geologic events. Application of organic geochemistry to the sedimentary profile of SK-1s in the Songliao Basin, provides a series of high resolution data that identifies abundance of organic matter, specific biomarkers, and their carbon and hydrogen isotopic compositions. Application of terrestrial biomarkers through the vertical profile, allows key lacustrine environmental parameters, such as palaeo-temperature of lake surface water, salinity, stratification and redox conditions of the water column, to be established; reconstruction of the lacustrine environment and palaeoclimate. Furthermore, organic geochemisty should allow us to investigate if here was a response of lacustrine organic matter to OAEs and marine transgressions

This study has both scientific and practical application as it helps provide a better understanding of past global climate change, the response mechanism of a regional lacustrine environment to global climatic change, and its influence on the formation of organic-rich source rocks in the Songliao Basin.

4.7.2 Sampling and Experimental Methods of Organic Matter Analysis

1. Sample collection

Samples were collected at intervals of 0.5—1 m from SK-1s, with attention paid to lithological variation. The SK-1s core represents most of the upper Cretaceous strata in Songliao Basin, including the lower part of the Nenjiang, Yaojia, and Qingshankou Formations, and the upper part of the Quantou Formation.

2. Sample treatment and analyses

Core samples were initially cleaned with ethanol to remove possible surface contamination from their handling, and then ground to 200 mesh size and dried in a vacuum freeze drier. The samples were analyzed for total organic carbon (TOC), carbon isotopic composition of organic matter, abundance of soluble organic matter (SOM), and the fraction component of soluble organic matter (SOM). Gas chromatography and coupled gas chromatography-mass spectrometry (GC-MS) was used for analysis of saturated and aromatic fractions. Some samples were also analyzed for the carbon isotopic composition of individual compounds, such as normal alkanes, branched and cyclic alkanes and aromatic hydrocarbons, and black carbon by the pyrolysis (Rock-Eval) method.

3. Major analytical instruments

(1) LECO CS-400 organic carbon analyzer. It is used to measure organic carbon content, the experiment conditions is the room temperature.

(2) Gas chromatography (GC-FID) was performed using a Thermo Finnigaen Trace Ultra equipped with a split/splitless injector operated in splitless mode. A 30 m JW-DB-5 fused silica capillary column (0.25 mm id, 0.25 μm film thickness) was used with high purity nitrogen as carrier gas. The temperature of injector and detector were 290 ℃ and 300 ℃ , respectively. Samples were injected at 60 ℃ , held for 5 minutes, and the oven temperature was programmed at 3 ℃ /min to 290 ℃ , and then held for 20 min.

(3) GC-MS: GC-MS analysis was performed on a HP 6890 gas chromatography instrument coupled with a Platform Ⅱ mass spectrometry operated in EI mode at 70 eV. A JW-DB-5 capillary column (30 m × 0.25 mm × 0.25μm) was used and the temperature of the injector was 290 ℃ operated in splitless mode. The oven was initially held at 80 ℃ for 5 min, and then programmed at 3 ℃ /min increments to 290 ℃ , at which temperature it was held for 10 min. Helium was the carrier gas. The scanning mass range was 50—600 U.

(4) Carbon isotopic composition of organic matter. The carbon isotopic composition of organic matter was analysed using a LECO CS-400 element analyzer coupled with a Delta Plus XP isotope ratio mass spectrometer, following the procedure of national standard GB/T19145-2003. The basic principle of such analysis is to release all the carbon and hydrogen by burning the kerogen in a prepared device and oxidizing it to CO_2

and H_2O. Subsequently, the CO_2 is sent to the mass spectrometry to analyze the carbon isotopic composition. The standard carbon isotope gas was calibrated using a NS22 crude oil standard provided by IAEA and the isotopic error of parallel analysis is less than 0.5‰.

(5) Pyrolysis chromatography analysis of bulk samples was performed using a Rock-Eval 6 Plus instrument.

All experiments and analysis were performed in the State Key Laboratory of Organic Geochemistry, Geochemistry Institution of Gangzhou, Chinese Academy of Sciences.

4.7.3 Preliminary Results

Figures 4.20 and 4.21 show vertical variations of TOC, carbon isotopic composition of organic matter, and pyrolysis parameters. Both the TOC content and organic carbon isotopic composition show clear variability in different parts of the borehole. Organic abundance varies widely within most of Nenjiang Formation Members 1 and 2, with total thickness 150 m, and K_2q_1 and the lower part of K_2q_1 with a total thickness of more than 153.64 m. These organic-rich strata are excellent source rocks, with high hydrocarbon potential. The organic matter is dominated by type I kerogen, with a hydrocarbon potential index of 11.8—27.0 mg/g and I_H of 355—594 mg/g. The TOC contents of the middle and upper sections of K_2q_{2+3} is lower and organic matter is dominated by type II - III kerogen. The pyrolysis data suggest that the latter sections could be classified as only moderately good source rocks.

The carbon isotopic composition of organic matter in the organic rich section shows is broadly negative, while there is a positive trend with relatively high fluctuations associated with the low-TOC section. Correlation of TOC vs. I_H indicates a general positive relationship in the low-TOC section and a negative relationship between TOC and $\delta^{13}C$ in the high-TOC section. This suggests that the preservation of organic matter during sedimentation may have played an important role in the abundance, type and isotopic composition of the organic matter. When sedimentary conditions were favorable for organic matter preservation, high quality organic matter accumulated and was preserved, and the initial negative carbon isotopic signature has been retained; when conditions were not favorable for organic matter preservation, a large quantity of primary organic matter (mainly represented by Cyanobacteria) was oxidized and biodegraded. Simultaneously, the organic matter originating from plants, which is more resistant to oxidation and degradation, was preserved, which resulted in a low abundance of organic matter of poor quality resulting in an associated positive carbon isotopic signature.

A pair of long-chain alkyl naphthalene series which has not been reported anywhere else has been discovered in the core samples. The occurrence of these compounds is not related to their abundance, type and or burial depth of organic matter. The geochemical significance of these compounds remain unknown, particularly their original source and formation mechanism.

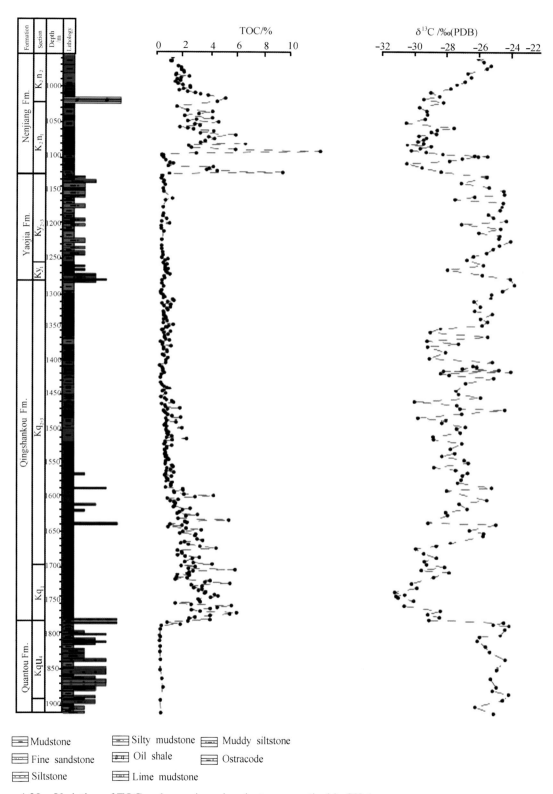

Figure 4.20 Variation of TOC and organic carbon isotopes vertical in SK-1s.

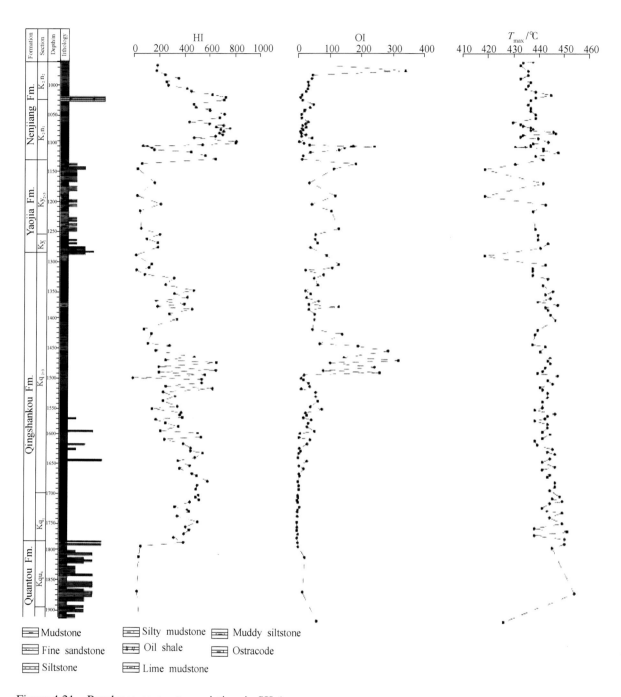

Figure 4.21 Pyrolyses parameter variation in SK-1s.

The isotopic composition of normal alkanes show an unusual positive excursion in the middle of the section which has low TOC. This positive plateau is different from carbon isotopic increases in marine carbonates and marine organic matter of the OAEs. Due to the lack of age data for the strata, the nature and cause of the latter increase remains unknown and is under further investigation.

A wide range of biomarkers was also identified from core samples of SK-1s. The biomarkers are dominated by hopanes including C_{27}—C_{34} hopanes, neohopanes, rearranged hopanes, moretanes and Gammacerane. The abundance of these biomarkers is clearly controlled by burial depth, since the biomarkers occur in much higher abundance with more types of compounds in core samples from the shallower buried Nenjiang Formation. It is worthwhile pointing out that terrestrial source-specific biomarkers such as oleanane, ursane and lupine, etc., have not yet been found in the core samples. In particular, the absence of oleanane indicates that angiosperms were either limited or did not exist, or were not yet developed in the Songliao Basin and its surroundings.

Biomarkers that are indicative of palaeo-salinity, water stratification and possible marine transgressions were also found, such as the Methylated-2-methyltrimethyltride-cylchromans (MTTCs), gammacerane and aryl isoprenoids and 24-n-propylcholestanes etc. Their distribution and environmental significance are currently under investigation.

4.8 Stable Isotopes

4.8.1 Aims and Significance of This Study

The biostratigraphy of the Late Cretaceous Songliao Basin has been well studied, compared with other stratigraphic sequences. The chemical records preserved in sediments are not only an important indicator of paleoenviromental change, but can also provide an alternative stratigraphy proxy. Based on Upper Cretaceous stratigraphic characteristics in the Songliao Basin, oxygen and carbon isotopes are selected as best indicators of environmental changes during sedimentation. Global or regional-scale isotope consistency controlled by geological events could become the basis for regional stratigraphic correlation.

It is difficult to obtain sufficient calcium carbonate from organic-rich lacustrine sediments to carry out stable isotope tests. Ostracods are the only microfossils in considerable abundance that can secrete a calcium carbonate shell in terrestrial freshwater-brackish-salt sediments. Benthic ostracods live in surface sediment, so their chemical composition has a direct relationship with the water conditions. Thus, the main purpose of this study is to use ostracod shells for high-precision stable isotope and trace element analysis.

4.8.2 Methods and Sampling

In 2007, we collected 1003 continuous high-resolution samples at intervals of one meter throughout SK-1s. Ostracod fossil were found in 557 samples. First we classified them (See Section 4.10), and then carried out geochemical analyses.

Oxygen and carbon isotope analysis of the ostracod tests were performed using the phosphoric acid digestion method of McCrea (1950). Between 0.1 and 0.3 mg of ostracods were loaded in sealed reaction vessels, flushed with helium gas, and reacted at 72°C. Carbon dioxide evolved in the vial headspace was sampled using a Finnigan Gas-Bench, and isotope ratios were educed by a Finnigan MAT+XL mass spectrometer at the Stanford University Stable Biogeochemistry Laboratory. Replicate analyses of NBS-19 and MERCK carbonate standards yielded a precision for carbon and oxygen isotopes of ± 2‰ or better. All isotopic results are reported in standard delta notation and corrected to PDB.

Trace element analysis was obtained from selected ostracod samples using inductively coupled plasma-atomic emission spectroscopy (ICP-AES). Samples were digested in HNO_3, diluted with megapure water and filtered. Repeat analyses of prepared blanks and standard solution of varying known concentration indicated detection limits for Sr, Ca and Mg of 0.1 μg/L, 6 μg/L and 2 μg/L, respectively, and precision better than 1 μg/L for Sr and 15 μg/L for Ca and Mg. The analyses represent a precision better than 0.1 mmol/mol for both Sr/Ca and Mg/Ca ratios.

In the Stanford University Stable Biogeochemistry Laboratory, pure ostracod samples were separated and classified. In this study, 252 samples were obtained from the ostracod matrix for stable isotope analysis, and 189 samples were analysed for trace elements. Prior to this, the matrix containing the ostracods and the samples of filling by random. The results show that three group of these samples have no correlation. This also means the the samples of pure ostracod tests are dependable. The final data shows that most of the oxygen isotopic records show a positive correlation with the carbon isotopic record. The oxygen isotopic indicates three general trends, and carbon isotopic record indicates five five trends. Two trace element ratios show a positive correlation, and indicate five general trends for both correlations.

Oxygen isotopic records reveal three general trends:

(1) Between Member 4 of the Quantou Formation and Member 2 of the Qingshankou Formation, there is a general trend of decreasing $\delta^{18}O$. The initial decrease in $\delta^{18}O$ is from −11.5‰ to −18‰. This negative shift of $\delta^{18}O$ ranges approximately from 7‰ to 8‰.

(2) From Member 3 of the Qingshankou Formation to Member 1 of the Nenjiang Formation, $\delta^{18}O$ values of fossil samples gradually increase by 10‰, although the change is not regular. In Member 3 of the Yaojia Formation and Member 2 of the Nenjiang Formation, there are slight decreases in $\delta^{18}O$.

(3) In Member 2 of the Nenjiang Formation, there is a clear decrease in $\delta^{18}O$ from −7‰ to −10‰. This

negative shift is in the range of 3‰.

The carbon isotopic record indicates five general trends:

(1) In the first trend, $\delta^{13}C$ values covary with $\delta^{18}O$ from 4.5‰ to 1‰ between Member 4 of the Quantou Formation and Member 2 of the Qingshankou Formation.

(2) Between 1570 m and 1330 m, $\delta^{13}C$ increases by 3‰, from 1‰ to 4‰, although the change is not regular.

(3) Between 1330 m and 1500 m, there is a sharp decrease in $\delta^{13}C$ by 4.5‰, and where are the lowest $\delta^{13}C$ values are recorded.

(4) From Member 1 of the Yaojia Formation to Member 1 of the Nenjiang Formation, $\delta^{13}C$ increases from –0.5‰ to 4‰.

(5) In Member 2 of the Nenjiang Formation, $\delta^{13}C$ decrease by about 3‰.

The variation in $\delta^{13}C$ is correlated with the evolution of the basin, and further study is in progress.

4.8.3 Preliminary Results

Ostracods secrete shells eight to nine times (in several weeks) before sexual maturity. After deshelling, ostracods rapidly form a new shell on the surface of the mantle by way of biomineralization. Because

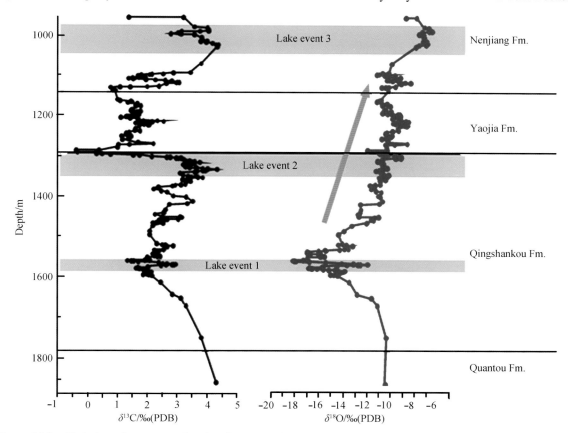

Figure 4.22 Variation of oxygen and carbon isotopes of ostracod shells in SK-1s.

this process is short-lived, we assume that the mineralization process reflects the chemical properties of the water during shell formation, i.e., there is a balance between calcium carbonate and water organisms. Based on analysis and data statistics, preliminary results are as follows. First, carbon, oxygen isotopes and trace elements of ostracod shells from SK-1s preserve a long-term and continuous record of climate history. Second, though oxygen isotope data may not necessarily accurately reflect absolute temperature, it can explain temperature changes during the geological history (Figure 4.22). The major role of the oxygen isotope data is that it reflects basin evolution, compliments local climate information, and contrasts with the marine record over the same period. It is assumed that the beginning of deposition of Member 2 of the Nenjiang Formation may signal global climate cooling. Third, carbon isotopes indicate changes in paleo-productivity. Test results show that ^{13}C enrichment occurred during the formation of hydrocarbon reservoirs, which may have been caused by the preferential removal of ^{12}C by organic matter in a closed basin. This conclusion is consistent with the abundance and diversity of fossil ostracods.

4.9 Cyclostratigraphy

The Cretaceous represents one of the warmest periods of the Phanerozoic and a time when ocean anoxic events (OAEs) developed. Establishing a high-precision chronostratigraphic framework is fundamental for a better understanding of global climate during Cretaceous greenhouse conditions. Owing to the limitations of radiometric age dating and paucity of paleontological data in fresh water lake deposits, the astronomical time scale (ATS), established by tuning astronomical signals recorded in sedimentary strata, plays a critical role in defining and correlating Cretaceous paleoclimatic/paleoceanographic events, including OAEs (e.g., Fiet *et al.*, 2006; Li *et al.*, 2008; Locklair and Sageman, 2008; Mitchell *et al.*, 2008). With the progresses made in the last decade, it is anticipated that an ATS for the entire Cretaceous will be completed in a few years (Hinnov and Ogg, 2007). The current ATS, however, is restricted to marine strata. Therefore, there is need to establish the Cretaceous ATS from terrestrial sedimentary basins so that marine and terrestrial records can be compared and a better understanding of changes in the Cretaceous Earth system can be achieved.

Continental rift basins are unique repositories of long-term palaeoclimate records (Olsen and Kent, 1999). Lacustrine mudstone and shales deposited in continental rift basins, due to their sensitivity to changes in precipitation-evaporation ratios and/or water-level changes, are particularly suitable for high-resolution cyclostratigraphic studies (e.g., Olsen and Kent, 1999; Prokoph and Agterberg, 2000), and for establishing astronomical time scales with potential resolution of 0.02—0.40 Ma (e.g., Hinnov, 2004; Hinnov and Ogg,

2007).

The Songliao Basin in northeastern China is one of the largest Cretaceous continental rift basins in the world. Well-preserved Cretaceous lacustrine deposits in this basin provide a unique opportunity for the construction of a terrestrial Cretaceous ATS. In particular, the Upper Cretaceous Qingshankou Formation (K_2qn) consists of up to 550 m of black mudstone, shale and oil shale that have attracted considerable attention regarding their potential link to Milankovitch climate forcing (e.g., Wu et al., 2007; Cheng et al., 2008). The basal interval of this unit is enriched in oil shale and is considered to represent a lacustrine anoxic event (Huang et al., 1998, 2007). This event, referred to as lacustrine anoxic event 1 (LAE1), has been suspected to be the time equivalent of oceanic anoxic event 2 (OAE2), but the lack of age control has prevented chronostratigraphic correlation with the marine record.

4.9.1 Cyclostratigraphy of Sedimentary Strata in SK-1

A cyclostratigraphic profile is used to evaluate sedimentary cycles in the basin, with the latter being controlled by climate, and in turn controlled by astronomical forcing. Continuous coring of SK-1 from the top of the Quantou Formation to the base of the Nenjiang Formation has provided excellent data for cyclostratigraphic research. According to detailed core descriptions and micro facies division, 862 meter-cycles in 3 groups and 16 subgroups can be recognized in SK-1s, including 76 meter-cycles in Quantou Formation Members 3+4, 503 cycles in the Qingshankou Formation, 151 cycles in the Yaojia Formation, 132 cycles in Nenjiang Formation Members 1+2.

Fischer plots show that the meter-size cycles are well-developed from the base to the top of the strata (Figure 4.23). Five different types of Milankovitch cycles can be identified.

The changes of the five-level cycles are shown by the developed process of the four-level cycles. graphic documentation shows that the five-level cycles are combined in a ratio of 4:1 and 3:1, and the four-level cycles are stacked by five-level cycles in the ratio 2:1—6:1 in the meter cycles developed from the top of the Quantou Formation to the base of the Nenjiang Formation. Through modeling calculation and isotope data, the time-length of each meter cycle is 22.9—29.8 ka, the five-level cycle is 75.0—104.7 ka, and the four-level cycle is 225.0—460.0 ka. Such cycle durations are consistent with Milankovitch orbital cycles, that consist of a precession cycle of 19—24 ka, a short eccentricity cycle of 85—140 ka, and long eccentricity cycle of 350—400 ka. This strongly suggests that the sedimentary cycles in SK-1s are controlled by the Earth's orbit, which has strong control on climate which in turn affects sedimentary processes. Recognition and establishing of the four major Miklankovich cycles in the Songliao Basin lake deposits provides an important tool for intercorrelation of continental and marine strata.

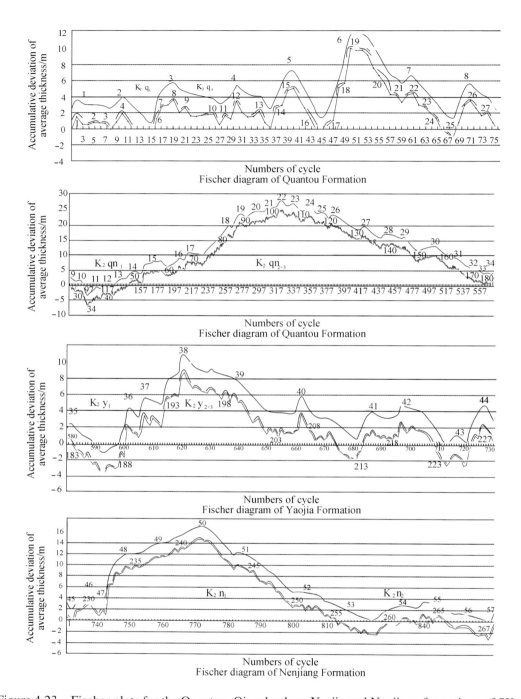

Figure 4.23   Fischer plots for the Quantou, Qingshankou, Yaojia and Nenjiang formations of SK-1s.

4.9.2　Floating Astronomical Time Scale for Qingshankou Formation, SK-1s

1. Data processing

Natural gamma-ray logging records the intensity of gamma rays emitted during the decay of atomic nuclei of radioactive elements in sedimentary rocks. The intensity of gamma rays relates to the amount of ^{40}K,

^{232}Th and ^{238}U in the rocks. Clay and organic particles have a strong capacity to absorb radioactive elements. Gamma ray logging curves can therefore reflect changes in the amount of clay and organic material in sediments, both of which are sensitive to precipitation–evaporation and lake level fluctuations induced by climatic change (e.g., Serra, 1984; Hinnov, 2004). Density logging measures the relative density of the strata in the drill hole and is used to define lithology and porosity.

In the Qingshankou Formation, sampling spacing is 0.125 m for the different logs. The natural gamma logging is between 60—180 API. Low radioactivity correlates with marl and high radioactivity indicates black shale and mudstone. Density logging is between 2.2—2.6 g/cm^3. A density of 1.7—2.3 g/cm^3 at the top of the Qingshankou Formation is probably caused by an increase in porosity. Both the natural gamma log and density log curves show good cyclic variation.

To confirm the relationship between lithologic cycles shown by the logs and orbital cycles, successive wavelet and power spectral analysis was applied to the strata analyses. Before analyzing the stratigraphic cycles, the data are pre-processed by filtering out signals >100 m and <1 m. Power spectral analysis can provide the cycle's frequency and length. The Red fit (Schulz and Mudelsee, 2002) has been used in spectral analysis and Wavelet analysis is employed by the software provided by Torrence and Compo (1998), and the filtering is used in the software of Analyseries 2.0 (Paillard *et al.*, 1996).

2. Preliminary results

Natural gamma logs and density logs were analyzed by continuous wavelet analysis. Cycles of 68 m, 39 m, 9—13.5 m, 3.8—5 m and 1.7—2.5 m are shown in the power spectrum of the two logs (Wu *et al.*, 2008) (Figure 4.24). Generally, the cycle signal in the density logs is less continuous than that of the natural gamma logs, and the power of the spectrum of the former is weaker than the latter. Cycles of 39 m and 9—13.5 m are stable throughout the entire logs, and imply a stable rate of sedimentation. In contrast, cycles of 3.8—5 m and 1.7—2.5 m are unstable throughout the logs, indicating a variable rate of sedimentation. The same cycles in the natural gamma and density logs indicate that they are both forced by the same outer mechanism, and do not represent random noise (Prokoph and Agterberg, 1999).

One of the best ways to identify whether observed cycles in sedimentary strata are controlled by astronomical forcing is to compare the relative ratio of the observed cycles with the relative ratio of the Milankovitch cycles. If they are similar, it would likely indicate that the periods of astronomical forcing controls the periodicities in the strata. The ratios of major periods of 39 m, 9—13.5 m, 3.8—5 m, and 1.7—2.5 m found in the natural gamma-ray and density logging data is approximately 20:5:2:1. Spectral analysis of the full time-series of the gamma-ray logging reveals four distinct dominant cycles of 8.9—14.2 m, 3.8—4.9 m, 2.3—2.6 m, and 1.7—1.9 m (Figure 4.24). The ratios of cycles for five main peaks are 13.3:10.3:4.05:2.43:1.9 (123:95:37.5:22.5:17.6). These ratios are similar to the results of spectral analysis of the insolation curve on June 21, at 65°N from 96.5 to 95 Ma (REF). Therefore, the ratios of the major periods of 39 m, 9—13.5 m,

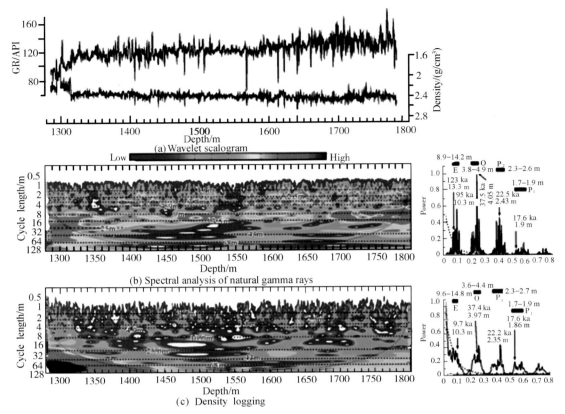

Figure 4.24 Natural gamma and density logging curves of the Qingshankou Formation, SK-1s.

3.8—5 m, and 1.7—2.5 m are considered to be caused by long and short eccentricities, obliquity, and precession, respectively. Guassian band-pass filters were designed for filtering the signals of long and short eccentricity, obliquity and precession (Figure 4.25), and indicates a uniform sedimentation rate 9.55 cm/ka for the Qingshankou Formation over 5.20 Ma.

Because the long eccentricity cycle (405 ka) remains stable at least back to ~250Ma, it was recommended that a Mesozoic astronomically calibrated time scale can be established as a floating time scale using long eccentricity cycles (Laskar et al., 2004; Hinnov and Ogg, 2007). However, the extracted long eccentricity cycles from K_2qn in SK-1s are less stable than the short eccentricity cycles (Figure 4.25), which also show more stable and clear signals than other bands of signals in most wavelet scalograms and spectral analyses. Thus we chose to construct the ATS using short eccentricity cycles.

We chose section M206 in the central depression zone of the basin as a reference section to establish the ATS because this section records stable Milankovitch cycles, has the longest accumulation time, and contains no identifiable stratigraphic discontinuities. The estimated age of ~94Ma for the K_2qn/K_2q boundary (Wang et al., 2007) was used as a reference for curve matching between the short eccentricity cycles from gamma-ray logging and the short eccentricity curve of the La2004 solution. The established ATS from section M206 suggests that the duration of K_2qn is about 5.20 Ma, from 94.27 Ma to 89.07 Ma, i.e., from Late Cenomanian to Early Coniacian (Figure 4.25).

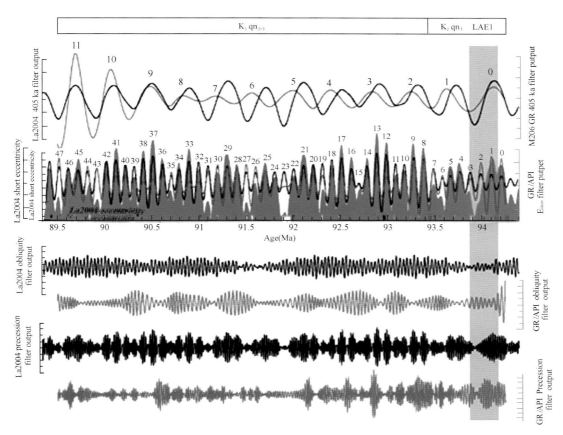

Figure 4.25 Astronomical calibration of short eccentricity cycles from filter output of gamma-ray logging (red) compared with short eccentricity (blue) and full eccentricity (gray) of La2004 solution (Laskar et al., 2004) for the Qingshankou Formation (K_2qn) in M206. Also shown are comparisons between extracted long eccentricity, obliquity and precession cycles from gamma-ray logging of well M206 (red) and those of the La2004 solution (black). Gaussian band pass filters for short (long) eccentricity of La2004 solution are 0.0093 ± 0.0013 (0.002469 ± 0.0003) cycle/ka.

Because available $\delta^{13}C$ org values are sparse and cannot define a positive $\delta^{13}C$ org "excursion", the traditional LAE1 in SLB was defined by the presence of oil shales in the lower K_2qn_1. The thickness of the oil shales ranges from 17.0 m (D501) to 27.0 m (G692 and L2), and records 2—3 short eccentricity cycles and 10—16 precession cycles. Our new ATS indicate the shortest and the longest duration of LAE1 are 210 ka (C503) and 310 ka (M206), respectively.

The LAE1 has been suspected to be the time equivalent of the oceanic anoxic event 2 (OAE2) at the Cenomanian–Turonian boundary (Huang et al., 1998, 2007). The established ATS for the K_2qn confirms this correlation. The onset of the OAE2 has recently been estimated as 94.09 Ma (Sageman et al., 2006) and 94.21/93.72 Ma (Mitchell et al., 2008) in central Colorado and Italy, respectively. These estimations, particularly the onset age of 94.21 Ma, are consistent with the onset age of 94.21—94.18 Ma for the LAE1 in the Songliao Basin. Based on the established ATS, the duration of the lacustrine anoxic event 1 (LAE1) at the base of the Qingshankou Formation is estimated to be 210—310 ka. The onset age of LAE1 (94.21—94.18

Ma) is comparable with that of the oceanic anoxic event 2 (OAE2) at the Cenomanian-Turonian boundary, which has been estimated as 93.72/94.21 Ma or 94.09 Ma (Wu *et al.*, 2009).

4.10　Biostratigraphy

4.10.1　Aims and Significance

Environmental changes influence the occurrence, development, expansion or demise or even extinction of biological species. Usually more than one biological species are present in any environment which can support life. In the Songliao Basin, the biota shows evolutionary changes as represented by the occurrence of the Songhuajiang and Mingshui biota, responding to environmental changes during the basin's depressed period (Quantou to Nenjiang formations) and the "atrophied" period (Sifangtai and Mingshui formations).

The Mesozoic-Cenozoic sedimentary strata in the Songliao Basin are poorly exposed subaerially. Therefore, a continuous mid-Late Cretaceous continental sedimentary record obtained by the SK-1 well provides valuable information for biostratigraphic studies.

4.10.2　Materials and Methods

According to the preservation of fossils in SK-1, we plan to establish the biostratigraphic evolution of ostracods, pollens and spores, and dinoflagellates. Since 2008 we have conducted biostratigraphic research on the SK-1s core. from 2008. Ostracod evolution has been generally established, whereas pollens and spores and dinoflagellate are still being analysed.

Samples for ostracod extraction were taken on average at 1 m intervals throughout the entire core except sandstones of Member 3 of the Quantou Formation. A total of 1000 samples were collected. Samples of 100 g dry weight were dispersed in deionized water for several weeks prior to sieving through a 100 μm sieve. Ostracod shells were picked from the samples under a low-power binocular microscope and then stored in a micropalaeontological specimen box. Identification and taxonomic information of ostracod species is mainly based on the work of Ye *et al.* (2002) and Hou *et al.* (2002).

4.10.3　Preliminary Results

1. Foraminifera

Based on detailed analysis, abundant foraminifera fossils were found in Members 1 and 2 of the Nenjiang Formation. Benthic foraminifera (*Gavlinella* sp., *Anomalinoides* sp., *Pullenia* sp., *Haplophragmoides* sp., *Karrorulina hokkaidoana*, *Clavulinoides* sp.), as well as planktonic foraminifera (*Archaeoglobigerina blowi*, *Archaeoglobigerina cretacea* and *Hedbergella flandrini*), were identified. These foraminifera were widely distributed in the marine Cretaceous and provide direct evidence for marine water incursions into the Songliao Basin during deposition of the Lower Nenjiang Formation.

2. Ostracoda

More than 13 genera and 70 species of ostracoda have been identified in strata extending from Quantou Formation Member 3 to the lower part of Nenjiang Formation Member 2. According to identified changes in the stratigraphic section and referring to previous studies, ostracoda species were grouped into 14 fossil assemblages (Table 4.7).

Table 4.7 Distribution of ostracod fossil assemblages in SK-1s

Epoch	Formation	Member	No.	Assemblages	Main Speices
Late Cretaceous	Nenjiang	2	14	*Cypridea liaukhenensis-Periacanthella magifica*	*Cypridea liaukhenensis, C. stellata, C. spongvosa, Periacanthella magifica, Ilyocypmorpha* sp., *Harbinia hapla*
			13	*Mongolocypris magna-Mongolocypris heiluntszianensis*	*Mongolocypris magnata, M.heiluntszianensis, M.*sp.
		1	12	*Cypridea gracila-Cypridea gunsulinensis*	*Cypridea gunsulinensis, C. gracila, C. acclinia, C. ardua*
			11	*Cypridea anonyma-Candona fabiforma*	*Cypridea anonyma, C. obolonga, C. maculata, Candona fabiforma, C.daqingensis, C. ovta, Advenocypris mundulaformis, Mongolocypris magna, M.heiluntszianensis*
	Yaojia	2-3	10	*Cypridea formasa-Cypridea sunghuajiangensis*	*Cypridea formasa, C.sunghuajiangensis, Mongolocypris tabulate, Lycopterocypris retractilis, Ziziphocypris rugosa*
			9	*Cypridea favosa-Mongolocypris tabulata*	*Cypridea favosa, Mongolocypris tabulata, Lycopterocypris retractilis, Ziziphocypris rugosa*
		1	8	*Cypridea exornata- Cypridea donfangensis*	*Cypridea exornata, C. Donfangensis, Mongolocypris infidelis, M. tera, M. magna, Lycopterocypris retractilis*
	Qingshankou	2+3	7	*Cypridea panda- Lycopterocypris subovatus*	*Cypridea panda, Lycopterocypris subovatus, Mongolocypris obsura, M. tera*
			6	*Triangulicypris vestilus Triangulicypris fusiformis*	*Triangulicypris pumilis, T. fusiformis, T. uniformis, T. vestilus, T. symmetrica, Kaitunia implata, Mongolocypris obsura, M. tera*
			5	*Cypridea edentula-Lycopterocypris grandis*	*Cypridea nota, C. fuyuensis, C. edentula, Lycopterocypris grandis*
			4	*Cypridea nota-Sunliavia tumida*	*Cypridea nota, C. fuyuensis, Triangulicypris torsuosus, Limnocypridea buccerusa, L. succinata, Sunliavia tumida, S. fuyuensis, Kaitunia implata, Ziziphocypris rugosa*
			3	*Cypridea dekhionensis-Cypridea gibbosa*	*Cypridea dekhionensis, C. gibbosa, C. bistyloformis C. tubeaculata, C. unicostata, Triangulicypris torsuosus, T. torsuosus* var. *nota, T. fertilis*
		1	2	*Triangulicypris torsuosus-Triangulicypris torsuosus* var. *nota*	*Triangulicypris torsuosus, T. torsuosus* var. nota
	Quantou	4	1	*Mongolocypris longicaudata*	*Mongolocypris longicaudata, Cypridea* sp.

(1) *Mongolocypris longicaudata* assemblage is mainly present in the upper part of the fourth member of the Quantou Formation. Fossils are rare except for *Mongolocypris longicaudata*, which only occurs in a particular horizon. A small number of *Mongolocypris* sp. and *Cypridea* sp. can also be found. The assemblage is dominated by the *Mongolocypris*.

(2) *Triangulicypris torsuosus-Triangulicypris torsuosus* var. nota assemblage is present in the first member of the Qingshankou Formation. The assemblage is dominated by two species: *Triangulicypris torsuosus* with smooth shell ornamentation and *T. torsuosus* var. nota with nobby shell ornamentation. Samples characteristically have low numbers of genera and species, and there are also a small number of specimens.

(3) *Cypridea dekhoinensis-Cypridea gibbosa* assemblage is located in the lower part of the second member of the Qingshankou Formation. The assemblage is dominated by eight species: *Cypridea dekhoinensis, C. gibbosa, C. bistyloformis, C. tuberculata, C. unicostata, Triangulicypris torsuosus, T. torsuosus* var. nota, and *T. fertilis* with nobby and/or thorn shell ornamentations. The samples characteristically have moderate to high numbers of species, and a greater number of specimens.

(4) *Cypridea nota-Sunliavia tumida* assemblage is located in the middle-upper part of the second member of the Qingshankou Formation. The main ostracoda species in this assemblage are: *Cypridea fuyuensis, C. nota, C. edentula, C. vicina, Sunliavia tumida, Triangulicypris symmertrica, Kaitunia andaensis, Ziziphocypris rugosa* and *Lycopterocypris grandis* mainly with smooth, or reticular shell ornamentations. Large numbers of species and specimens were found in the samples.

(5) *Cypridea edentula-Lycopterocypris grandis* assemblage is mainly located in the upper part of the second Unit of the Qingshankou Formation. The assemblage is dominated by four species: *Cypridea nota, C. fuyuensis, C. edentula, Lycopterocypris grandis*, mainly with smooth or reticular shell ornamentations. Samples characteristically have a moderate number of species and specimens.

(6) *Triangulicypris vestilus-Triangulicypris fusiformis* assemblage is mainly located in the middle part of the third member of the Qingshankou Formation. Significant ostracoda species in this assemblage are: *Triangulicypris pumilis, T. fusiformis, T. uniformis, T. vestilus, T. symmetrica, Kaitunia implata, Mongolocypris obsura, M. tera*. At the base of the formation, a small amount of *Cypridea fuyuensis* appears. The assemblage is dominated by *Triangulicypris*, but progressively more *Mongolocypris* characterized by mixed shell size and smooth ornamentations occurs in stratigraphically younger samples. This assemblage has a moderate number of species and specimens.

(7) *Cypridea panda-Lycopterocypris subovatus* assemblage is mainly present in the upper part of the third member of the Qingshankou Formation. The assemblage is dominated by three species: *Cypridea panda, Lycopterocypris subovatus, Mongolocypris obsura, M. tera,* characterized by mixed shell size, smooth ornamentations and a low numbers of species, but a moderate to high number of specimens.

(8) *Cypridea exornata-Cypridea dongfangensis* assemblage is mainly present in the first member of the Yaojia Formation. The assemblage is dominated by *Cypridea exornata, C. Donfangensis, Mongolocypris infidelis, M. tera, M. magna, Lycopterocypris retractilis*. It is characterized by the mixing of *Cypridea* and *Mon-*

golocypris. Overall, the specimens have a relatively larger, smooth shell, or shallow pits. This assemblage has large number of species, but only a moderate number of specimens.

(9) *Cypridea favosa-Mongolocypris tabulata* assemblage is located in the second Unit of the Yaojia Formation. This assemblage only contains two main ostracoda species: *Cypridea favosa, Mongolocypris tabulata*. Fossil preservation is similar to that in assemblage (8), but with the occurrence of many red fossils stained by iron oxide from surrounding rocks.

(10) *Cypridea formosa-Cypridea sunghuajiangensis* assemblage is present in the third member of the Yaojia Formation. The assemblage is dominated by five species: *Cypridea formosa, Cypridea sunghuajiangensis, Mongolocypris tabulata, Lycopterocypris retractilis, Ziziphocypris rugosa.* Fossil preservation is similar to that in assemblages (8) and (9).

(11) *Cypridea anonyma-Candona fabiforma* assemblage is mainly found in the lower part of the first member of the Nenjiang Formation. The assemblage contains a large variety of ostracoda species, such as *Cypridea anonyma, C. oblonga, C. maculata, Candona fabiforma, C. ovta, Advenocypris mundulaformis, Mongolocypris porrecta. Cypridea anonyma* is dominant with a large valve and occurs in the stratigraphically lower part of the assemblage. In the upper part, *Candona* species with a smaller valve becomes the main constituent. Samples characteristically have a high number of species and specimens.

(12) *Cypridea gracila-Cypridea gunsulinensis* assemblage mainly occurs in the upper part of the first member of the Nenjiang Formation. The assemblage is dominated by four species of ostracoda: *Cypridea gracila, C. gunsulinensis, C. acclinia, C. ardua, Cypridea* characteristic by smooth ornamentation, a swelled ventral valve and a low number of species and specimens.

(13) *Mongolocypris magna-Mongolocypris heiluntszianensis* assemblage is mainly found in black shale near the base of the second member of the Nenjiang Formation. It contains three species: *Mongolocypris magna, M. heiluntszianensis, Cypridea spongvosa. Mongolocypris* is characteristic by a large valveup to 4mm in the length.

(14) *Cypridea liaukenensis-Periacanthella magifica* assemblage is mainly found in the lower part of the second member of the Nenjiang Formation above the black shale of assemblage (13)? with intercalations oil shale. Main fossils are: *Cypridea liaukenensis, C. stellata, C. spongvosa, Periacanthella magifica, Ilyocypmorpha* sp., *Harbinia hapla, Mongolocypris agna*, *M. heiluntszianensis. Cypridea liaukhenensis* with thorn ornamentation is the dominant species. The number of specimens in this zone is low, but it contains a large number of endemic species.

Ecological characteristics of the ostracoda in SK-1s indicate environmental changes in the Songliao Basin during the Cretaceous. In the Quantou Formation, ostracod fauna characterized by *Mongolocypris* is of a low biodiversity and abundance. Ostracod shells are poorly preservedsuggesting a relatively shallow and dynamic water environment. During deposition of Qingshankou Formation Member 1, ostracods were in low abundance and biodiversity. Member 1 is primarily characterized by *Triangulicypris torsuosus* and *Triangulicypris torsuosus* var. nota, which have small shells that are not well preserved and are mostly stained black, or are gray and partly replaced by pyrite. This suggests a deep, anoxic or disoxic lake bottom environment, not conducive for ostracods. During sedimentation of the lower part of Qingshankou Formation Member 2, the appearance of ostracoda with nobby shells as well as increased biodiversity and abundance suggests an improvement of living conditions. Lake-level was gradually decreasing and the water body was mainly calm. In the upper part of the second member of the Qingshankou Formation, ostracods with nobby shells disappeared, while the biodiversity and abundance of ostracoda increased. This may indicate further shallowing of the lake, more transparent water with a higher calcium content. During deposition of the Yaojia Formation biodiversity of ostracoda decreased, but they remained relatively abundant. Ostracoda within this formation are mainly *Cypridea* and *Mongolocypris* which have a thick shell. Some shells are stained slightly red implying deposition in a relatively shallow and oxidized environment.

In the lower part of the Nenjiang Formation, the abundance and biodiversity of ostracoda is high and fossils are well preserved. Ostracoda shells are mainly of a gray color. The above evidence suggests a deep lake environment which provided a favorable niche for ostracoda. In the upper part of Unit 1 of the Nenjiang Formation, ostracods with smooth shells and a swollen posterodorsal area, appear. This may indicate that hydro-dynamic conditions in the basin were not favorable for ostracods, which have low abundance, and their preservation. In the lower part of Nenjiang Formation Member 2, abundant ostracoda with large shells, especially *Mongolocypris*, appear They are found mainly in black shale, which may implies a deep and eutrophic environment in which inter-species competition was low, allowing species such as *Monglocypris* to form larger shells. Ostracoda preserved in strata above the black shale and inter-bedded with the oil shale in the second member of the Nenjiang Formation, are characterized by shells with nodes and spines, as well as large reticulation, suggesting a gradually declining but still relatively high lake-level.

Evolution of the ostracoda throughout the stratigraphic sequence suggests that the lake in the Songliao Basin was very large and provided favorable conditions for aquatic organisms to flourish. Three large lake transgressions occurred during the Cretaceous during deposition of Qingshankou Member 1, and Nenjiang Formation Members 1 and 2. Transgressions resulted in the development of an anoxic environment in the deepest parts of the basin, and this had an influence on production of new species and their expansion. Three episodes of black shale and oil shale deposition in the Songliao Basin provided rich source rocks for hydrocarbon generation. This preliminary study indicates that the appearance of new species coincides with the developments of anoxic conditions in the lake, with biological changes similar to those of marine anoxic events. However, whether such events were controlled by global climatic change remains uncertain.

Based on ten continuous geological profiles, significant progress has been made in the fields of micropaleontology, geochronology, sedimentology and organic geochemistry, etc. For example, foraminifera were first discovered in the Songliao Basin which increases our understanding of the basin evolution; a large-scale lacustrine turbidite channel and fan system identified within the lacustrine mudstone of the the Nenjiang Formation provides a new target for hydrocarbon exploration and development; dolomite in hydrocarbon source rocks is also very useful for hydrocarbon exploration; identification of isoprenoids indicates the presence of green sulfur bacteria, which are significant for interpreting the formation environment of mass terrestrial hydrocarbon source rocks; robust ages of Upper Cretaceous strata are obtained by precise radiometric dating of volcanic ash in the SK-1 drill cores, and combined with paleomagnetic, biostratigraphic and cyclostratigraphic analysis, a chronostratigraphic framework of terrestrial late Cretaceous of China is established. The above scientific results provide a firm basis for a deeper understanding of the Songliao Basin.

Section 5
SK-1 Core Description and Core Photographs

The Continental Scientific Drilling Project of the Cretaceous Songliao Basin (SK-1) obtained cores from Upper Cretaceous Quantou Formation to Tertiary Taikang Formation, with a total footage of 2577.24 m, a total core length of 2485.89 m, and an average core recovery of 96.46%. Core photographic scanning and detailed description were conducted right after coring at the drilling site, to guarantee the accuracy of core photographs and descriptions.

In this section, detailed core description, stratigraphic chart and core scanning photographs are shown in different formations and units.

Note that in the numbers 'x-y-(z)' under core photographs of Chapter 5.2, the first number 'x' correlates to 'Barrel' number in core information table of Appendix. People when reading the core photographs can use the Core Numbers to identify relative information (e.g. depth) from the tables.

5.1 Stratigraphic Chart

In this chapter, the detailed core description and stratigraphic chart are shown from lowermost Quantou Formation Members 3 and 4 to uppermost Taikang Formation in SK-1 cores.

5.1.1 Quantou Formation Members 3 and 4

The overlying formation of the Quantou Formation Memter 4 is the Qingshankou Formation Member 1, the contacting relationship between them is conformable contact; the underlying formation is the Quantou Formation Member 3, the contacting relationship betwecn them is parallel unconformable contact (Figure 5.1).

The detailed description of the Quantou Formation Member 4:

92-1	1782.93—1783.03 m	0.10 m	Calcareous siltstone, medium light gray, thinly bedded, interbeded with medium dark gray mudstone; wavy laminated
92-2	1783.03—1783.09 m	0.06 m	Calcareous siltstone, light gray, thinly bedded; trough cross bedding, few medium-thick laminae
92-3	1783.09—1783.19 m	0.10 m	Calcareous siltstone, medium light gray, thinly bedded, interbedded with medium dark gray mudstone laminae; wavy laminatied
92-4	1783.19—1783.36 m	0.17 m	Calcareous siltstone, light gray, thinly bedded with few medium dark gray mudstone interbeds; trough cross stratification. In the lower part of the section, it is intercalated with very light gray, fine grained sandstone
92-5	1783.36—1783.47 m	0.11 m	Mudstone, medium gray, thinly bedded with few medium light gray siltstone interbeds; horizontal stratification
92-6	1783.47—1783.70 m	0.23 m	Calcareous siltstone, light gray and brownish-gray with a few medium gray laminae; wavy bedding. In the upper part of the section, very light gray, oil-bearing, fine grained sandstone laminae
92-7	1783.70—1784.03 m	0.33 m	Mudstone, interbeded with medium dark gray and greenish-gray argillaceous siltstone; horizontal bedding. Some oil-bearing, calcareous, fine grained sandstone laminae in the lower part of the section
92-8	1784.03—1784.32 m	0.29 m	Sandstone, fine grained, brownish-gray, calcareous, oil bearing; wavy lamination and trough cross bedding
92-9	1784.32—1785.03 m	0.71 m	Sandstone, fine grained, brownish-gray, oil bearing interbeded with medium dark gray mudstone; wavy bedding, erosion surface at the base, few ostracoda shells. A light gray, calcareous, fine grained sandstone laminae in the lower part of the section
92-10	1785.03—1785.18 m	0.15 m	Sandstone, fine grained, calcareous, light gray, medium bedded; trough cross bedding, oil spots, sand balls, and pillows
92-11	1785.18—1785.36 m	0.18 m	Sandstone, calcareous, fine grained, brownish-gray medium bedded; trough cross bedding, parallel bedding in local, sand pillows and balls, some pyrites, erosion surface at the base A mudstone layer with a thickness of 2 cm at the bottom
92-12	1785.36—1785.62 m	0.26 m	Sandstone, calcareous, fine grained, brownish-gray, interbeded with oil-bearing medium. dark gray mudstone; wavy bedding, a few pyrite crystals

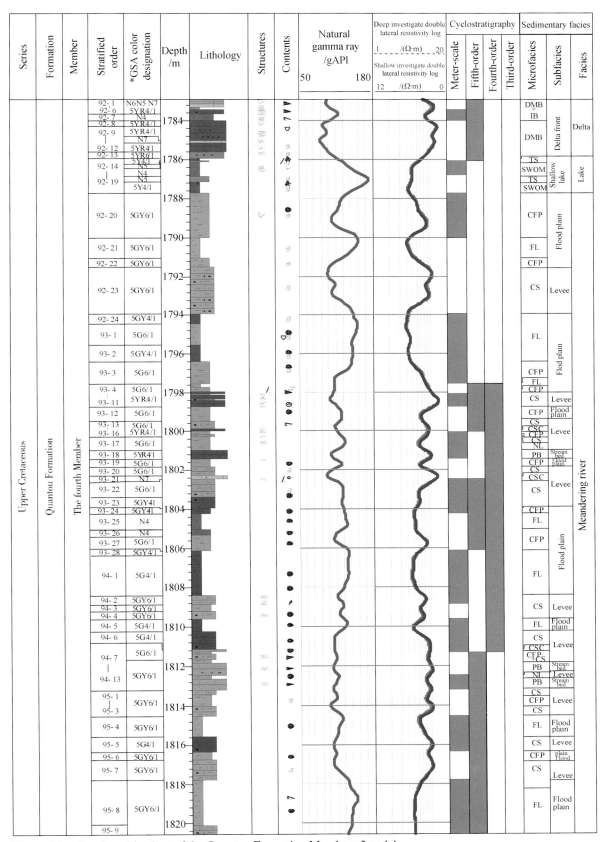

Figure 5.1 Stratigraphic chart of the Quantou Formation Members 3 and 4.

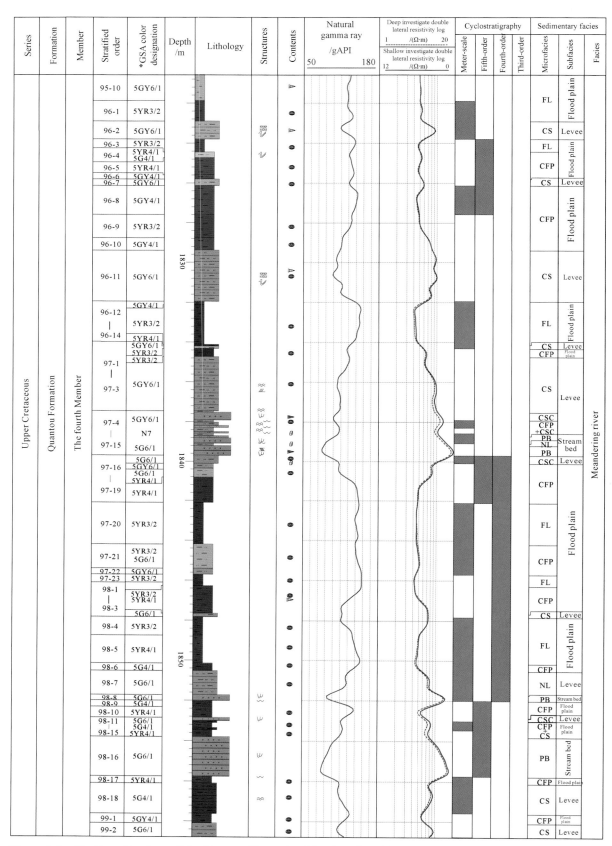

Figure 5.1 Stratigraphic chart of the Quantou Formation Members 3 and 4 (continued).

Section 5 SK-1 Core Description and Core Photographs

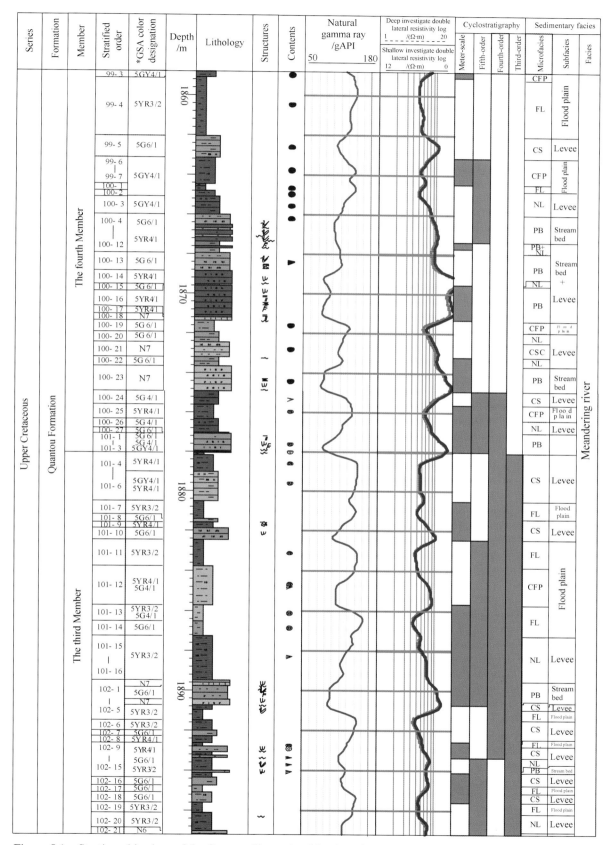

Figure 5.1 Stratigraphic chart of the Quantou Formation Members 3 and 4 (continued).

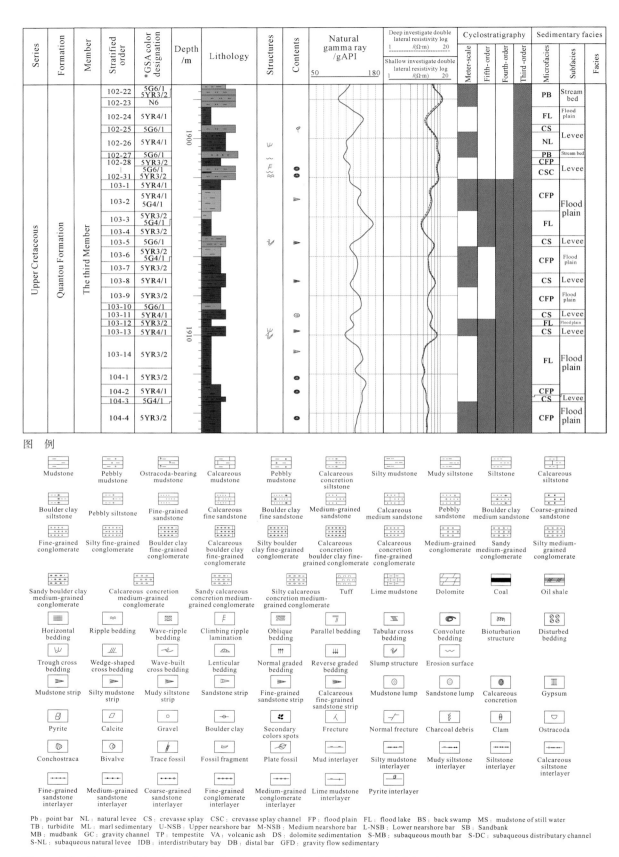

Figure 5.1 Stratigraphic chart of the Quantou Formation Members 3 and 4 (continued).

92-13	1785.62—1785.95 m	0.33 m	Siltstone, oil-bearing, light brownish-gray, parallel bedding with ripple lamination, few pyrites
92-14	1785.95—1786.06 m	0.11 m	Mudstone, calcareous, olive gray thinly bedded some fractures healed by calcites, some are open and form secondary pores
92-15	1786.06—1786.16 m	0.10 m	Mudstone, silty, medium gray, thinly bedded, some pyrites
92-16	1786.16—1786.88 m	0.72 m	Mudstone, medium dark gray, intercalated with light greenish-gray silty mudstone, horizontal bedding, ripple lamination, and lenticular beds fossil fragments on bedding planes
92-17	1786.88—1787.19 m	0.31 m	Mudstone, silty, medium gray, massive some pyrites
92-18	1787.19—1787.28 m	0.09 m	Mudstone, calcareous, olive gray, thinly bedded, fractures filled with calcite, part of the calcite was dissolved forming secondary porosity
92-19	1787.28—1787.73 m	0.45 m	Mudstone, olive gray, massive bedding, some pyrites
92-20	1787.73—1790.03 m	2.30 m	Mudstone, silty, greenish-gray, thickly bedded, deformation structure, some calcareous concretions (maximum diameter is 8 cm), few pyrites
92-21	1790.03—1791.08 m	1.05 m	Mudstone, greenish-gray, thickly bedded, some pyrites
92-22	1791.08—1791.53 m	0.45 m	Mudstone, silty, greenish-gray, massive bedding, some pyrites
92-23	1791.53—1793.91 m	2.38 m	Siltstone, argillaceous, greenish-gray, thickly bedded, some pyrites, a few fine sandstone laminae at the top
92-24	1793.91—1794.44 m	0.53 m	Mudstone, dark greenish-gray, massive bedding, some pyrites
93-1	1794.44—1795.53 m	1.09 m	Mudstone, greenish-gray, massive bedding, calcareous concretions, pyrites and few ostracoda shells
93-2	1795.53—1796.38 m	0.85 m	Mudstone, dark greenish-gray, thickly bedded, calcareous concretions and few pyrites
93-3	1796.38—1797.52 m	1.14 m	Mudstone, silty, greenish-gray, thick bedding, calcareous concretions and few pyrites
93-4	1797.52—1797.71 m	0.19 m	Mudstone, greenish-gray with brownish-gray oil-bearing siltstone intraclasts, few pyrites
93-5	1797.71—1797.94 m	0.23 m	Mudstone, silty, greenish-gray, ripple lamination, some laminae with pyrite
93-6	1797.94—1798.04 m	0.10 m	Siltstone, brownish-gray, oil-bearing, intercalated with silty mudstone laminae, trough cross bedding, trace fossils, erosion surface at the base
93-7	1798.04—1798.11 m	0.07 m	Mudstone, silty, greenish-gray, thinly bedded
93-8	1798.11—1798.29 m	0.18 m	Siltstone, calcareous, brownish-gray, medium bedded fining upward; trough cross bedding few mud pebbles, erosion surface at the base
93-9	1798.29—1798.40 m	0.11 m	Siltstone, argillaceous, greenish-gray, massive bedding
93-10	1798.40—1798.52 m	0.12 m	Siltstone, calcareous, brownish-gray, small trough cross bedding, erosion surface and mud gravels at the base
93-11	1798.52—1798.70 m	0.18 m	Siltstone, calcareous, brownish-gray, oil-bearing intercalated with greenish-gray argillaceous siltstone; small trough cross bedding in the lower section and deformation structures in the middle and upper section, mud gravels

93-12	1798.70—1799.44 m	0.74 m	Mudstone, silty, greenish-gray, massive bedding, calcareous concretions
93-13	1799.44—1799.88 m	0.44 m	Siltstone, argillaceous, greenish-gray with mudstone laminae and few calcareous siltstone laminae
93-14	1799.88—1800.00 m	0.12 m	Siltstone, calcareous, brownish-gray, small trough cross bedding, erosion surface at the base
93-15	1800.00—1800.14 m	0.14 m	Mudstone, silty, greenish-gray, ripple lamination, erosion surface at the base, 1 cm oil-bearing calcareous siltstone above the erosion surface
93-16	1800.14—1800.26 m	0.12 m	Siltstone, argillaceous, greenish-gray with few mudstone laminae, erosion surface at the base
3-17	1800.26—1800.94 m	0.68 m	Mudstone, silty, greenish-gray, thick bedded, ripple lamination; greenish-gray mudstone laminae and oil-bearing calcareous siltstone lumps in the lower part of the core, scour and fill structures at the base
93-18	1800.94—1801.40 m	0.46 m	Sandstone, calcareous, fine grained, brownish-gray, homogenous, erosion surface at the base, few pyrites
93-19	1801.40—1801.78 m	0.38 m	Mudstone, silty, greenish-gray, homogenous, calcareous concretions
93-20	1801.78—1802.25 m	0.47 m	Siltstone, argillaceous, greenish-gray, homogenous, few pyrites
93-21	1802.25—1802.44 m	0.19 m	Sandstone, fine grained, calcareous, light gray, mud gravels, and erosion surface at the base, some vertical fractures
93-22	1802.44—1803.34 m	0.90 m	Siltstone, argillaceous, greenish-gray, thickly bedded, few pyrites and calcareous concretions
93-23	1803.34—1803.94 m	0.60 m	Siltstone, argillaceous, dark greenish-gray, few pyrites and calcareous concretions
93-24	1803.94—1804.19 m	0.25 m	Mudstone, silty, dark greenish-gray, medium bedded, few calcareous concretions
93-25	1804.19—1805.02 m	0.83 m	Mudstone, medium dark gray, with few calcareous concretions, massive bedding
93-26	1805.02—1805.31 m	0.29 m	Mudstone, silty, medium dark gray, massive bedding, calcareous concretions
93-27	1805.31—1806.09 m	0.78 m	Mudstone, silty, greenish-gray with some calcareous concretions and calcareous siltstone lumps
93-28	1806.09—1806.32 m	0.23 m	Mudstone, dark greenish-gray with few calcareous concretions, massive bedding
94-1	1806.32—1808.39 m	2.07 m	Mudstone, dark gray, thickly bedded with calcareous concretions
94-2	1808.39—1808.84 m	0.45 m	Siltstone, argillaceous, greenish-gray, medium bedded, enclosing calcareous concretions, in the upper part of the cores, ripple lamination, in the middle part wedged shaped cross bedding; light gray calcareous siltstone lumps in the middle-lower section, trace fossil at the base
94-3	1808.84—1809.16 m	0.32 m	Mudstone, silty, greenish-gray with calcareous concretions (include size)
94-4	1809.16—1809.59 m	0.43 m	Greenish-gray, medium bedded, argillaceous mudstone with few calcareous concretions, horizontal bedding and ripple lamination
94-5	1809.59—1810.21 m	0.62 m	Mudstone, dark greenish-gray, medium beded with a few calcareous concretions
94-6	1810.21—1811.12 m	0.91 m	Siltstone, argillaceous, dark greenish-gray with calcareous concretions, massive bedding
94-7	1811.12—1811.26 m	0.14 m	Siltstone, greenish-gray with few calcareous concretions

Section 5 SK-1 Core Description and Core Photographs

94-8	1811.26—1811.35 m	0.09 m	Mudstone, silty, greenish-gray with calcareous concretions up to 5 cm in a diameter at the base
94-9	1811.35—1811.62 m	0.27 m	Siltstone, argillaceous, greenish-gray with mudstone laminae and few calcareous concretions, ripple lamination
94-10	1811.62—1811.82 m	0.20 m	Mudstone, silty, greenish-gray with calcareous concretions
94-11	1811.82—1812.42 m	0.60 m	Siltstone, greenish-gray with few mudstone laminae and mud gravels, ripple lamination
94-12	1812.42—1812.61 m	0.19 m	Mudstone, silty, greenish-gray with calcareous concretions
94-13	1812.61—1813.20 m	0.59 m	Siltstone, greenish-gray, with some mudstone laminae, ripple lamination
95-1	1813.20—1813.40 m	0.20 m	Siltstone, argillaceous, greenish-gray with calcareous concretions
95-2	1813.40—1814.00 m	0.60 m	Mudstone, silty, greenish-gray, homogenous, some pyrite
95-3	1814.00—1814.50 m	0.50 m	Siltstone, argillaceous, greenish-gray
95-4	1814.50—1815.55 m	1.05 m	Mudstone, greenish-gray, thickly bedded with calcareous concretions
95-5	1815.55—1816.30 m	0.75 m	Siltstone, argillaceous, greenish-gray, massive bedding
95-6	1816.30—1816.70 m	0.40 m	Mudstone, silty, greenish-gray with calcareous concretions
95-7	1816.70—1817.78 m	1.08 m	Siltstone, argillaceous, greenish-gray with few pyrite crystals
95-8	1817.78—1820.15 m	2.37 m	Mudstone, some greenish-gray sandstone laminae and calcareous concretions
95-9	1820.15—1820.61 m	0.46 m	Siltstone, argillaceous, greenish-gray, massive bedding
95-10	1820.61—1821.88 m	1.27 m	Mudstone, greenish-gray with calcareous concretions and sandstone laminae
96-1	1821.88—1822.90 m	1.02 m	Mudstone, grayish-brown with calcareous concretions
96-2	1822.90—1823.79 m	0.89 m	Siltstone, argillaceous, greenish with light gray calcareous siltstone laminae
96-3	1823.79—1824.43 m	0.64 m	Mudstone, grayish-brown with calcareous concretions, homognous
96-4	1824.43—1824.73 m	0.30 m	Mudstone, silty, brownish-gray and dark greenish-gray with slump structure
96-5	1824.73—1825.59 m	0.86 m	Mudstone, silty, brownish-gray with calcareous concretions
96-6	1825.59—1825.80 m	0.21 m	Mudstone, silty, dark greenish-gray
96-7	1825.80—1826.13 m	0.33 m	Siltstone, argillaceous greenish-gray with calcareous concretions and a few light gray calcareous siltstone laminae and lumps, massive bedding
96-8	1826.13—1827.58 m	1.45 m	Mudstone, silty, dark greenish-gray with grayish-brown spots
96-9	1827.58—1828.76 m	1.18 m	Mudstone, silty, grayish-brow with few calcareous concretions
96-10	1828.76—1829.38 m	0.62 m	Mudstone, silty, dark greenish-gray with few calcareous concretions, some brownish-gray spots
96-11	1829.38—1831.95 m	2.57 m	Siltstone, argillaceous, greenish-gray with few calcareous concretions and calcareous siltstone laminae, ripple lamination and deformation structure
96-12	1831.95—1832.13 m	0.18 m	Mudstone, dark greenish-gray
96-13	1832.13—1834.12 m	1.99 m	Mudstone, grayish-brown with calcareous concretions, massive bedding
96-14	1834.12—1834.20 m	0.08 m	Siltstone, argillaceous, brownish-gray, a few dark greenish-gray spots

97-1	1834.20—1834.32 m	0.12 m	Siltstone, argillaceous, greenish-gray and grayish-brown with few calcareous concretions. Intercalated, 1 cm thick sandstone laminae at the top
97-2	1834.32—1834.73 m	0.41 m	Mudstone, silty, grayish-brown, some greenish-gray spots and calcareous concretions in the middle and lower part of the section
97-3	1834.73—1837.45 m	2.72 m	Siltstone, argillaceous, greenish-gray with some mudstone laminae, a few calcareous concretions, mud gravels and siltstone lumps. Some cross bedding and ripple lamination
97-4	1837.45—1837.52 m	0.07 m	Mudstone, silty, greenish-gray with few siltstone laminae and lumps, ripple lamination
97-5	1837.52—1837.91 m	0.39 m	Sandstone, fine-grained, greenish-gray, trough cross bedding in the lower and upper part of the core and parallel bedding and ripple lamination in the middle, erosion surface at the base, few mudstone laminae in the middle of the section
97-6	1837.91—1838.03 m	0.12 m	Silty mudstone, greenish-gray with calcareous concretions, ripple lamination
97-7	1838.03—1838.19 m	0.16 m	Siltstone, argillaceous, greenish-gray with few mudstone laminae, ripple lamination, erosion surface at the base
97-8	1838.19—1838.28 m	0.09 m	Siltstone, light gray, ripple lamination, erosion surface at the base, mudstone laminae 1 cm thick and mud gravels and pyrites below erosion surface. Core has been most probably turned upside down
97-9	1838.28—1838.50 m	0.22 m	Mudstone, silty, greenish-gray few siltstone lumps, ripple lamination, and small scale trough cross bedding
97-10	1838.50—1838.57 m	0.07 m	Siltstone, light gray with few mud gravels and mudstone laminae, small scale trough cross bedding, erosion surface at the base
97-11	1838.57—1838.71 m	0.14 m	Mudstone, silty, greenish-gray with siltstone lumps, ripple lamination
97-12	1838.71—1838.81 m	0.10 m	Siltstone, greenish-gray, trough cross bedding
97-13	1838.81—1839.08 m	0.27 m	Sandstone, fine-grained, greenish-gray with few mud gravels, trough cross bedding in the lower part of the section and tabular cross bedding in the middle and upper part, erosion surface at the base
97-14	1839.08—1839.20 m	0.12 m	Mudstone, silty, greenish-gray with siltstone laminae and few pyrites, ripple lamination
97-15	1839.20—1839.69 m	0.49 m	Sandstone, fine-grained greenish-gray, trough cross bedding, in the lower part of the core and parallel bedding in the middle part, tabular cross bedding in the upper part of the section; mud gravels and mudstone laminae at the base
97-16	1839.69—1840.10 m	0.41 m	Siltstone, argillaceous, greenish-gray with a few calcareous concretions, thin beds of silty mudstone at top
97-17	1840.10—1840.35 m	0.25 m	Mudstone, silty, greenish-gray with calcareous concretions, deformation structure in some of calcareous concretions
97-18	1840.35—1840.81 m	0.46 m	Mudstone, silty, greenish-gray and brownish-gray, a few calcareous concretions with deformation structure in the lower and middle part of the section
97-19	1840.81—1842.10 m	1.29 m	Mudstone, silty, brownish-gray, greenish-gray spots at the top, few carbonized fossil fragments
97-20	1842.10—1844.20 m	2.10 m	Mudstone, grayish-brown with calcareous concretions, massive bedding

97-21	1844.20—1845.40 m	1.20 m	Mudstone, silty, grayish-brown and greenish-gray with few calcareous concretions
97-22	1845.40—1845.70 m	0.30 m	Mudstone, silty, greenish-gray
97-23	1845.70—1846.11 m	0.41 m	Grayish-brown, medium bedded, mudstone with few calcareous concretions, massive bedding
98-1	1846.10—1846.28 m	0.18 m	Mudstone, grayish-brown with calcareous concretions
98-2	1846.28—1847.65 m	1.37 m	Mudstone, silty, brownish-gray with calcareous concretions, greenish-gray argillaceous siltstone laminae and some lumps
98-3	1847.65—1847.82 m	0.17 m	Siltstone, argillaceous, greenish-gray with some brownish-gray silty mudstone
98-4	1847.82—1848.75 m	0.93 m	Mudstone, grayish-brown with some calcareous
98-5	1848.75—1850.15 m	1.40 m	Mudstone, brownish-gray with calcareous concretions, and few greenish-gray argillaceous siltstone lumps
98-6	1850.15—1850.52 m	0.37 m	Mudstone, silty, dark greenish-gray with few calcareous concretions, and few brownish-gray mudstone laminae at the top
98-7	1850.52—1851.72 m	1.20 m	Siltstone, argillaceous, greenish-gray with few calcareous concretions, some brownish-gray mud gravels
98-8	1851.72—1852.00 m	0.28 m	Sandstone, fine grained, greenish-gray with few greenish-gray mud gravels, parallel bedding in the lower part of the section and small trough cross bedding in the upper part, erosion surface at the base
98-9	1852.00—1852.31 m	0.31 m	Mudstone, silty, dark greenish-gray
98-10	1852.31—1852.85 m	0.54 m	Mudstone, silty, brownish-gray, some calcareous concretions surrounded by greenish-gray silty mudstone layer in the lower part of the section
98-11	1852.85—1853.00 m	0.15 m	Sandstone, fine-grained, greenish-gray with a few argillaceous siltstone lumps, small trough cross bedding
98-12	1853.00—1853.10 m	0.10 m	Siltstone, argillaceous, dark greenish-gray with fine sandstone laminae and lumps, climbing-ripple lamination
98-13	1853.10—1853.35 m	0.25 m	Mudstone, silty, brownish-gray with calcareous concretions (maximum diameter is 3 cm)
98-14	1853.35—1853.45 m	0.10 m	Siltstone, argillaceous, dark greenish with few calcareous concretions
98-15	1853.45—1853.80 m	0.35 m	Mudstone, silty, brownish-gray with calcareous concretions enveloped in greenish-gray silty mudstone
98-16	1853.80—1855.75 m	1.95 m	Sandstone, fine grained, greenish-gray, small trough cross bedding, erosion surface at the base
98-17	1855.75—1856.10 m	0.35 m	Mudstone, silty, brownish-gray with some calcareous concretions
98-18	1856.10—1857.61 m	1.51 m	Siltstone, argillaceous, dark greenish-gray, thick bedded with few calcareous concretions, ripple lamination, greenish-gray fine grained sandstone laminae and lumps
99-1	1857.61—1858.09 m	0.48 m	Mudstone, silty, dark greenish-gray with calcareous concretions
99-2	1858.09—1858.78 m	0.69 m	Siltstone, argillaceous, greenish-gray, some calcareous concretions and calcareous laminae in the upper part of the section, a calcareous siltstone with deformation structure

99-3	1858.78—1859.07 m	0.29 m	Mudstone, silty, dark greenish-gray with few calcareous concretions
99-4	1859.07—1862.01 m	2.94 m	Mudstone, grayish-brown with calcareous concretions, dark greenish-gray silty mudstone laminae at the top and base of the section
99-5	1862.01—1863.11 m	1.10 m	Siltstone, argillaceous, greenish-gray with calcareous concretions, some laminae of calcareous siltstone and fine-grained calcareous sandstone
99-6	1863.11—1864.44 m	1.33 m	Mudstone, silty, dark greenish-gray with calcareous concretions
99-7	1864.44—1864.54 m	0.10 m	Mudstone, dark greenish-gray with some calcareous concretions
100-1	1864.54—1864.77 m	0.23 m	Mudstone, dark greenish-gray with calcareous concretions
100-2	1864.77—1865.04 m	0.27 m	Mudstone, silty, dark greenish-gray with few calcareous concretions
100-3	1865.04—1865.97 m	0.93 m	Siltstone, argillaceous, dark greenish-gray with calcareous concretions
100-4	1865.97—1866.47 m	0.50 m	Siltstone, greenish-gray with some calcareous concretions and a few mud gravels
100-5	1866.47—1866.64 m	0.17 m	Sandstone, fine-grained, brownish-gray with an oil smell, trough cross bedding, slump structures, some greenish-gray siltstone lumps and an erosion surface at the base
100-6	1866.64—1866.77 m	0.13 m	Siltstone, greenish-gray, few micas, massive bedding
100-7	1866.77—1867.00 m	0.23 m	Sandstone, fine-grained, brownish-gray, oil-bearing with greenish-gray siltstone interbeds, slump structures and siltstone lumps in the upper part of the section, through cross bedding at the lower part of the section, erosion surface at the base
100-8	1867.00—1867.10 m	0.10 m	Siltstone, greenish-gray, thinly bedded with a few fine-grained sandstone lumps
100-9	1867.10—1867.33 m	0.23 m	Sandstone, fine-grained, brownish-gray with trace of oil. Enclosed greenish-gray silty mud gravels slump structures in the upper part of the section, erosion surface at the base
100-10	1867.33—1867.57 m	0.24 m	Siltstone, argillaceous, greenish-gray, ripple lamination, deformation structure, load cast at the base, fine grained sandstone laminae and lumps in the lower part of the section
100-11	1867.57—1867.70 m	0.13 m	Sandstone, fine-grained, brownish-gray with argillaceous siltstone laminae, climbing ripple cross lamination, erosion surface at the base
100-12	1867.70—1867.92 m	0.22 m	Siltstone, argillaceous, greenish-gray, ripple lamination, erosion surface at the base, 1 cm thick mudstone laminae below erosion surface
100-13	1867.92—1868.83 m	0.91 m	Siltstone, greenish-gray with oil-bearing fine grained sandstone laminae, and mudstone laminae, a few micas, slump structure and ripple lamination
100-14	1868.83—1869.50 m	0.67 m	Sandstone, fine-grained, brownish-gray with a few mud gravels, micas, oil smell, parallel bedding in the upper part of the section and small scale trough cross bedding in the middle and lower part of the core erosion surface and mud gravels at the base
100-15	1869.50—1869.56 m	0.06 m	Siltstone, argillaceous, greenish-gray with micas and fine sandstone laminae
100-16	1869.56—1871.04 m	1.48 m	Sandstone, fine-grained, brownish-gray, trough cross bedding, tabular cross bedding, climbing cross ripple lamination, and horizontal lamination developed in a sequence from the base to the top of the section

100-17	1871.04—1871.19 m	0.15 m	Sandstone, fine-grained, brownish-gray, with oil smell
100-18	1871.19—1871.33 m	0.14 m	Sandstone, fine-grained calcareous, light gray, erosion surface and a few mud gravels at the base
100-19	1871.33—1871.91 m	0.58 m	Mudstone, silty, greenish-gray with a few carbon dust particles s and calcareous concretions
100-20	1871.91—1872.44 m	0.53 m	Siltstone, argillaceous, greenish-gray with a few carbon dust particles
100-21	1872.44—1873.18 m	0.74 m	Siltstone, light gray with greenish-gray silty mudstone lumps
100-22	1873.18—1873.64 m	0.46 m	Siltstone, argillaceous, greenish-gray with few mud gravels, massive bedding
100-23	1873.64—1874.89 m	1.25 m	Sandstone, fine-grained, light gray, tabular cross bedding in the middle and upper part of the section and trough cross bedding in the lower part erosion surface at the base. A few mud gravels in the lower half of the section
100-24	1874.89—1875.59 m	0.70 m	Siltstone, argillaceous, dark greenish-gray
100-25	1875.59—1876.34 m	0.75 m	Mudstone, silty, brownish-gray with few calcareous concretions, greenish-gray rim around the lumps
100-26	1876.34—1877.01 m	0.67 m	Siltstone, argillaceous, dark greenish-gray with few calcareous concretions and mottled brownish-gray argillaceous siltstone
100-27	1877.01—1877.08 m	0.07 m	Siltstone, greenish-gray, trough cross bedding
101-1	1877.08—1877.90 m	0.82 m	Sandstone, fine-grained, greenish-gray, parallel bedding in the upper part of the section, trough cross bedding and climbing-ripple cross lamination in the lower part, erosion surface at the base; mud gravels and calcareous concretions in the lower part of the section
101-2	1877.90—1877.96 m	0.06 m	Siltstone, argillaceous, dark greenish-gray with mudstone laminae and lumps, ripple bedding
101-3	1877.96—1878.03 m	0.07 m	Sandstone, fine-grained, greenish-gray with mud gravels and calcareous concretions in the lower part of the core, small trough cross bedding, erosion surface at the base

The Overlying formation of the Quantou Formation Member 3 is Quantou Formation Member 4, the contacting relationship between them is parallel unconformable contact: the bottom of Quantou Formation Member 3 is not penetrate (Figure 5.1).

The detailed description of the Quantou Formation Member 3:

101-4	1878.03—1878.16 m	0.13 m	Mudstone, dark greenish-gray, massive
101-5	1878.16—1878.98 m	0.82 m	Mudstone, silty, brownish-gray, some calcareous concretions
101-6	1878.98—1880.47 m	1.49 m	Siltstone argillaceous, dark greenish-gray and brownish-gray, some calcareous concretions in the lower section
101-7	1880.47—1881.38 m	0.91 m	Mudstone, grayish-brown

101-8	1881.38—1881.56 m	0.18 m	Siltstone, argillaceous, greenish-gray, erosion surface at the base
101-9	1881.56—1881.86 m	0.30 m	Mudstone, silty, brownish-gray, ripple lamination, sand pillows and balls in the most upper part of the section
101-10	1881.86—1882.43 m	0.57 m	Siltstone, greenish-gray, small scale trough cross stratification, erosion surface and mud gravels at the base, siltstone bed 2.5 cm thick below erosion surface
101-11	1882.43—1883.78 m	1.35 m	Mudstone, grayish-brown, some irregular calcareous concretions, locally greenish-gray color spots
101-12	1883.78—1885.68 m	1.90 m	Mudstone, brownish-gray and dark greenish-gray, massive bedding, some calcareous concretions
101-13	1883.68—1886.48 m	2.80 m	Mudstone, grayish-brown and dark greenish-gray, massive, few calcareous concretions
101-14	1886.48—1887.23 m	0.75 m	Mudstone, greenish-gray, a few calcareous concretions
101-15	1887.23—1889.49 m	2.26 m	Mudstone, silty, grayish-brown with greenish-gray, siltstone lumps, few calcareous concretions, light gray fine-grained sandstone lumps and laminae in the lower section
101-16	1889.49—1889.59 m	0.10 m	Sandstone, fine-grained, light gray with small trough cross bedding
102-1	1889.59—1889.74 m	0.15 m	Sandstone, calcareous, fine-grained, light gray medium bedded, small trough cross bedding, erosion surface at the base, few mud gravels and trace fossils
102-2	1889.74—1890.44 m	0.70 m	Siltstone, sandy, greenish-gray, thickly bedded slump structures in the upper section and climbing-ripple lamination in the lower section, some brownish-gray mud gravels of irregular shape and size, some micas on the bedding surface, few calcareous fine grained sandstone thin interbeds
102-3	1890.44—1890.74 m	0.30 m	Sandstone, fine-grained, light gray, trough cross bedding, deformation structure in the middle section, intercalated greenish-gray, and brownish-gray mudstone and greenish-gray siltstone
102-4	1890.74—1890.97 m	0.23 m	Mudstone, silty, grayish-brown and greenish-gray, argillaceous siltstone lumps, slump structures and convolute bedding in the upper section, massive bedding in the middle and lower section
102-5	1890.97—1891.46 m	0.49 m	Mudstone, grayish-brown, massive
102-6	1891.46—1891.99 m	0.53 m	Mudstone, silty, greenish-gray, massive bedding, grayish-brown spots
102-7	1891.99—1892.28 m	0.29 m	Siltstone, argillaceous, greenish-gray, massive bedding
102-8	1892.28—1892.62 m	0.34 m	Mudstone silty, brownish-gray, massive bedding, greenish-gray spots
102-9	1892.62—1892.79 m	0.17 m	Mudstone, grayish-brown, massive, calcareous concretions
102-10	1892.79—1893.08 m	0.29 m	Siltstone, greenish-gray, trough cross bedding with brownish-gray silty mudstone lumps and calcareous siltstone lumps, erosion surface at the base

Section 5 SK-1 Core Description and Core Photographs

102-11	1893.08—1893.25 m	0.17 m	Mudstone, silty, brownish-gray, massive bedding, ripple lamination in the lower part of the section, greenish-gray argillaceous siltstone lumps
102-12	1893.25—1893.35 m	0.10 m	Siltstone, greenish-gray, thinly bedded, trough cross bedding, mudstone laminae locally, erosion surface at the base
102-13	1893.35—1893.97 m	0.62 m	Mudstone, silty, grayish-brown, thickly bedded, slump structures, a large number of greenish-gray argillaceous siltstone lumps, siltstone lumps, calcareous concretions and mudstone lamellae
102-14	1893.97—1894.09 m	0.12 m	Sandstone, calcareous, fine-grained, greenish-gray, trough cross stratification, dark greenish-gray mud lamellae
102-15	1894.09—1894.32 m	0.23 m	Mudstone, silty, grayish-brown, massive, few siltstone lumps
102-16	1894.32—1894.74 m	0.42 m	Siltstone, argillaceous, greenish-gray with calcareous siltstone laminae, ripple bedding
102-17	1894.74—1895.09 m	0.35 m	Mudstone, greenish-gray, massive, a few grayish-brown spots
102-18	1895.09—1895.59 m	0.50 m	Mudstone, silty, greenish-gray, massive bedding
102-19	1895.59—1896.12 m	0.53 m	Mudstone, grayish-brown, massive, few calcareous concretions and greenish-gray spots
102-20	1896.12—1897.04 m	0.92 m	Mudstone, silty, grayish-brown, thickly bedded, calcareous concretions and greenish-gray argillaceous siltstone lumps in the lower section
102-21	1897.04—1897.22 m	0.18 m	Siltstone, medium light gray, wavy bedding, a few grayish-brown spots, erosion surface at the base
102-22	1897.22—1897.46 m	0.24 m	Siltstone, argillaceous, greenish-gray and grayish-brown, few calcareous concretions
102-23	1897.46—1898.38 m	0.92 m	Siltstone, medium light gray, massive bedding, few calcareous concretions
102-24	1898.38—1899.30 m	0.92 m	Mudstone, brownish-gray thickly bedded, lumps of greenish-gray silty mudstone in the lower section
102-25	1899.30—1899.69 m	0.39 m	Siltstone, greenish-gray, massive bedding, a few plant fragments, brownish-gray spots in the lower section
102-26	1899.69—1900.64 m	0.95 m	Siltstone, argillaceous, brownish-gray, thickly bedded, some greenish-gray siltstone lumps, a few calcareous concretions
102-27	1900.64—1900.99 m	0.35 m	Sandstone, fine-grained, greenish-gray with brownish-gray spots, small-scale trough cross bedding, erosion surface at the base
102-28	1900.99—1901.39 m	0.40 m	Mudstone, silty, grayish-brown, massive bedding, a few greenish-gray argillaceous siltstone lumps
102-29	1901.39—1901.69 m	0.30 m	Siltstone, greenish-gray, ripple lamination, lumps and lamellae of grayish-brown silty mudstone, calcareous concretions, erosion surface at the base
102-30	1901.69—1901.75 m	0.06 m	Mudstone, grayish-brown, massive bedding
102-31	1901.75—1902.10 m	0.35 m	Siltstone, greenish- ripple lamination, few calcareous concretions

103-1	1902.10—1902.60 m	0.50 m	Mudstone, silty, brownish-gray, massive bedding, few calcareous concretions, locally dark greenish-gray spots
103-2	1902.60—1903.70 m	1.10 m	Mudstone, silty, dark greenish-gray and brownish-gray massive bedding, greenish-gray laminae in the upper section
103-3	1903.70—1904.10 m	0.40 m	Mudstone. grayish-brown and dark greenish-gray, massive bedding
103-4	1904.10—1904.95 m	0.85 m	Mudstone, grayish-brown, massive bedding, a few lumps of greenish-gray argillaceous siltstone in the lower section
103-5	1904.95—1905.50 m	0.55 m	Siltstone, argillaceous, greenish-gray, massive bedding, deformation structure, a few laminae of grayish-brown mudstone
103-6	1905.50—1906.00 m	0.50 m	Mudstone, silty, grayish-brown and dark greenish-gray, massive
103-7	1906.00—1906.85 m	0.85 m	Mudstone, silty, grayish-brown massive bedding, few lumps of greenish-gray silty mudstone
103-8	1906.85—1907.55 m	0.70 m	Siltstone, argillaceous, brownish-gray, massive bedding, few lumps and laminae of grayish-brown mudstone, lumps of greenish-gray argillaceous siltstone in the lower section
103-9	1907.55—1908.35 m	0.80 m	Mudstone, silty, grayish-brown, massive
103-10	1908.35—1908.70 m	0.35 m	Mudstone, silty, greenish-gray, massive bedding, few calcareous concretions
103-11	1908.70—1909.18 m	0.48 m	Siltstone, argillaceous, grayish-brown, massive, grayish-brown mudstone lumps
103-12	1909.18—1909.55 m	0.37 m	Mudstone, grayish-brown, massive, few dark greenish-gray spots and calcareous concretions
103-13	1909.55—1910.05 m	0.50 m	Siltstone, argillaceous brownish-gray mixed with a few irregular-shaped grayish-brown mudstone and greenish-gray silty mudstone, small trough cross bedding, deformation structures in local
103-14	1910.05—1911.70 m	1.65 m	Mudstone, grayish-brown, massive bedding, a few calcareous concretions, few lumps of dark greenish-gray argillaceous siltstone and laminae of greenish-gray siltstone
104-1	1911.70—1912.52 m	0.82 m	Mudstone, grayish-brown, massive bedding, some calcareous concretions.
104-2	1912.52—1913.15 m	0.63 m	Mudstone, silty, brownish-gray, massive some calcareous concretions and dark greenish-gray spots
104-3	1913.15—1913.40 m	0.25 m	Siltstone, argillaceous, dark greenish-gray, massive bedding, some lumps of brownish-gray silty mudstone
104-4	1913.40—1915.00 m	1.60 m	Mudstone, silty, grayish-brown, thickly bedded, calcareous concretions, a few lumps of greenish-gray silty mudstone in the upper section

5.1.2 Qingshankou Formation Member 1

The overlying formation is the Qingshankou Formation Members 2 and 3, the contacting relationship of them is conformable contact; the underlying formation is the Quantou Formation Member 4, the contacting relationship of them is conformable contact (Figure 5.2).

The detailed description of the Qingshankou Formation Member 1:

83-15	1701.52—1701.57 m	0.05 m	Dolomite, olive black, massive bedding
83-16	1701.57—1702.12 m	0.55 m	Mudstone, dark gray, horizontal bedding, dolomite lens at the top, few fossil fragments
83-17	1702.12—1702.15 m	0.03 m	Dolomite, olive gray, massive
83-18	1702.15—1702.77 m	0.62 m	Mudstone, dark gray, horizontal bedding, some calcareous siltstone laminae
84-1	1702.77—1703.13 m	0.36 m	Mudstone, dark gray, horizontal bedding, some fossil fragments
84-2	1703.13—1703.38 m	0.25 m	Dolomite, olive black dolomite, massive bedding
84-3	1703.38—1707.32 m	3.94 m	Mudstone, dark gray, horizontal bedding; dolomite lens and thin-bedded calcareous siltstones in 1703.97 m and 1704.62 m; dissolution pores at 1705.22 m, bitumen filling; greenish gray tuff, 0.5 cm thick at 1705.77 m
84-4	1707.32—1707.49 m	0.17 m	Limestone, white, interbeded with dark gray mudstone in the upper section; oil-bearing crystalline limestone in the lower section
84-5	1707.49—1709.83 m	2.34 m	Mudstone, dark gray, horizontal bedding; crystalline limestone intercalations at 1707.57 m, siltstone laminae at 1708.87 m, 1709.47 m and 1709.77 m; few fossil plate
84-6	1709.83—1709.97 m	0.14 m	Dolomite, olive black, massive bedding
84-7	1709.97—1710.37 m	0.40 m	Mudstone, dark gray, horizontal bedding, fossil fragments
84-8	1710.37—1712.32 m	1.95 m	Mudstone, dark gray horizontal bedding, fossil fragments; vertical fracture filled with calcite
84-9	1712.32—1712.47 m	0.15 m	Dolomite, olive black, massive bedding
84-10	1712.47—1712.77 m	0.30 m	Mudstone, medium dark gray, horizontal bedding, few fossil fragment
84-11	1712.77—1714.47 m	1.70 m	Mudstone, dark gray, horizontal bedding, few fossil fragment; with thin-bedded olive black dolomite
85-1	1714.47—1715.27 m	0.80 m	Mudstone, dark gray, horizontal bedding, few fossil fragment; some thin-bedded olive black dolomite
86-1	1715.27—1717.27 m	2.00 m	Mudstone, dark gray, horizontal bedding, some calcareous siltstone lamellae; ostracoda clast limestone at 1715.82 m and 1715.97 m; thin-bedded crystalline limestone at 1716.12 m and 1716.27 m
86-2	1717.27—1717.42 m	0.15 m	Dolomite, olive black dolomite, massive bedding

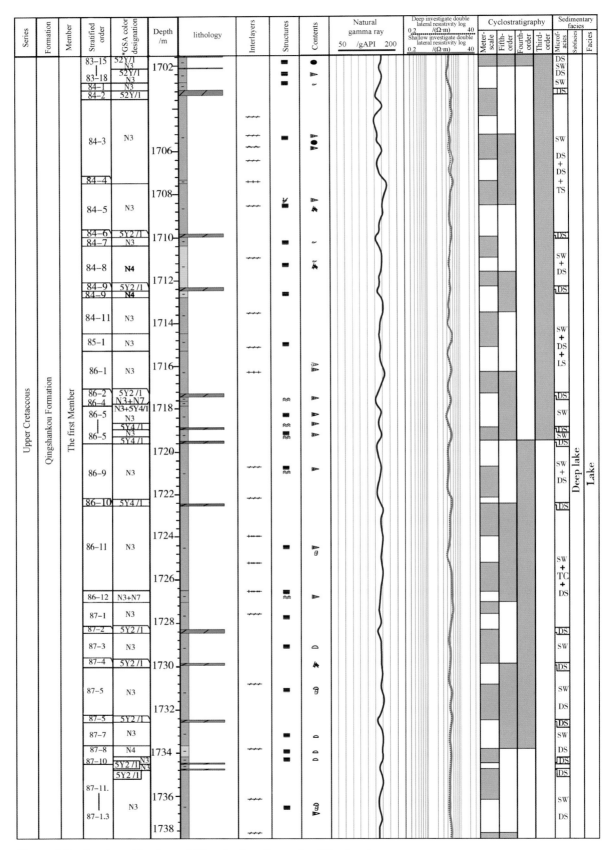

Figure 5.2 Stratigraphic chart of Qingshankou Formation Member 1.

Section 5 SK-1 Core Description and Core Photographs

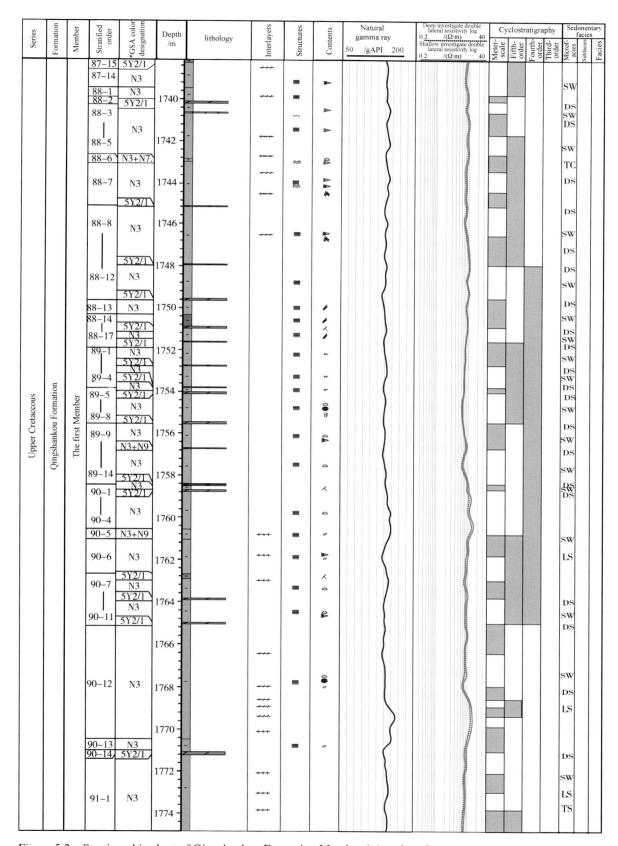

Figure 5.2 Stratigraphic chart of Qingshankou Formation Member 1 (continued).

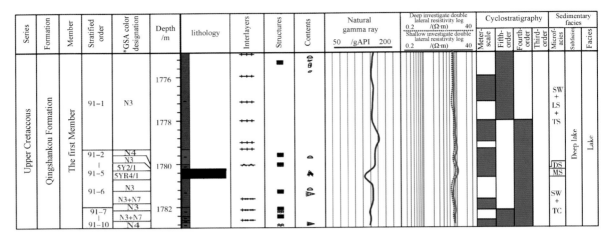

Figure 5.2　Stratigraphic chart of Qingshankou Formation Member 1 (continued). Symbols are the same as those in Figure 5.1.

86-3	1717.42—1717.62 m	0.20 m	Mudstone, dark gray, interbeded with light gray calcareous siltstone, ripple lamination, slight erosion surface at base
86-4	1717.62—1717.84 m	0.22 m	Mudstone, dark gray, interbeded with olive gray dolomite
86-5	1717.84—1718.89 m	1.05 m	Mudstone, dark gray, horizontal bedding, few trace fossil; olive gray dolomite intercalations at 18.27 m; siltstone laminae in the lower section, ripple lamination
86-6	1718.89—1718.96 m	0.07 m	Dolomite, olive, massive bedding
86-7	1718.96—1719.48 m	0.52 m	Mudstone, dark grey, horizontal bedding, few trace fossil
86-8	1719.48—1719.59 m	0.11 m	Dolomite, dark gray, massive bedding
86-9	1719.59—1722.42 m	2.83 m	Mudstone, dark gray, horizontal bedding, gray calcareous siltstone lamellae, and few fossil fragment; thin-bedded dolomite at 1722.27 m
86-10	1722.42—1722.52 m	0.10 m	Dolomite, olive gray, massive bedding
86-11	1722.52—1726.50 m	3.98 m	Mudstone, dark gray, horizontal bedding, few calcareous siltstone laminae, and pyrite
86-12	1726.50—1727.03 m	0.53 m	Mudstone, dark gray intercalated with light gray calcareous siltstone, horizontal bedding, and ripple lamination
87-1	1727.03—1728.28 m	1.25 m	Mudstone, dark gray, horizontal bedding, few thin-bedded dolomite lamellae
87-2	1728.28—1728.46 m	0.18 m	Dolomite, olive black, massive, dissolution pores, filled by bitumen
87-3	1728.46—1729.84 m	1.38 m	Mudstone, dark gray, horizontal bedding, ostracoda fossils
87-4	1729.84—1729.94 m	0.10 m	Dolomite, olive black, massive bedding
87-5	1729.94—1732.46 m	2.52 m	Mudstone, dark gray, horizontal bedding, ostracoda fossils; a pyrite laminae at 1730.83 m; thin-bedded dolomite lamellae locally
87-6	1732.46—1732.56 m	0.10 m	Dolomite, olive black, massive bedding

87-7	1732.56—1733.65 m	1.09 m	Mudstone, dark gray, horizontal bedding, ostracoda fossils
87-8	1733.65—1734.16 m	0.51 m	Mudstone, medium dark gray, horizontal bedding, ostracoda fossils
87-9	1734.16—1734.48 m	0.32 m	Mudstone, dark gray, horizontal bedding, ostracoda fossils; dolomite lens, and laminae
87-10	1734.48—1734.54 m	0.06 m	Dolomite, olive black, massive bedding, dissolution pores, filled by bitumen
87-11	1734.54—1734.73 m	0.19 m	Mudstone, dark gray, horizontal bedding, ostracoda fossils; few dolomite lenses, and lamellae
87-12	1734.73—1734.79 m	0.06 m	Dolomite, olive black, massive bedding, dissolution pores, filled by bitumen
87-13	1734.79—1738.21 m	3.42 m	Mudstone, dark gray, horizontal bedding, ostracoda fossils; dolomite lens, and lamellae
87-14	1738.21—1738.35 m	0.14 m	Mudstone, olive black, massive bedding
87-15	1738.35—1739.54 m	1.19 m	Mudstone, dark gray, horizontal bedding, few ostracoda fossil, some olive black dolomite
88-1	1739.54—1740.19 m	0.65 m	Mudstone, dark gray, horizontal bedding, few trace fossil at some bedding planes, few pyrite laminae
88-2	1740.19—1740.29 m	0.10 m	Dolomite, olive black, massive bedding
88-3	1740.29—1740.70 m	0.41 m	Mudstone, dark gray, horizontal bedding, calcareous siltstone lamellae
88-4	1740.70—1740.74 m	0.04 m	Limestone, ostracoda bearing dark gray, erosion surface at the base
88-5	1740.74—1742.92 m	2.18 m	Mudstone, dark gray, horizontal bedding, calcareous siltstone laminae
88-6	1742.92—1743.06 m	0.14 m	Mudstone, dark gray, interbeded with light gray calcareous siltstone, ripple lamination, erosion surface at the base
88-7	1743.06—1745.15 m	2.09 m	Mudstone, dark gray mudstone, horizontal bedding, two olive black dolomite interlayers at 1743.5 m and 1745.01 m; a calcareous siltstone interbeded at 1744.54 m, erosion surface at base
88-8	1745.15—1745.19 m	0.04 m	Dolomite, olive black, massive bedding
88-9	1745.19—1747.94 m	2.75 m	Mudstone, dark gray, horizontal bedding, intercalated olive black dolomite
88-10	1747.94—1747.98 m	0.04 m	Dolomite, olive black
88-11	1747.98—1749.54 m	1.56 m	Mudstone, dark gray, horizontal bedding, intercalated olive black dolomite at 1743.5 m and 1745.01 m
88-12	1749.54—1749.64 m	0.10 m	Dolomite, olive black
88-13	1749.64—1750.32 m	0.68 m	Mudstone, dark gray, horizontal bedding, few trace fossil
88-14	1750.32—1750.88 m	0.56 m	Mudstone, medium dark gray, horizontal bedding, few trace fossil
88-15	1750.88—1750.98 m	0.10 m	Dolomite, olive black, fracture stained by oil
88-16	1750.98—1751.59 m	0.61 m	Mudstone, dark gray, horizontal bedding, few trace fossils

88-17	1751.59—1751.63 m	0.04 m	Dolomite, olive black
89-1	1751.63—1752.75 m	1.12 m	Mudstone, dark gray, horizontal bedding, fossil fragments on bedding plane
89-2	1752.75—1752.80 m	0.05 m	Dolomite, olive black, massive bedding
89-3	1752.80—1753.80 m	1.00 m	Mudstone, dark gray, horizontal bedding, fossil fragments on bedding plane
89-4	1753.80—1753.85 m	0.05 m	Dolomite, olive black
89-5	1753.85—1754.00 m	0.15 m	Mudstone, dark gray, horizontal bedding, fossil fragments on bedding plane.
89-6	1754.00—1754.10 m	0.10 m	Dolomite, olive black
89-7	1754.10—1755.45 m	1.35 m	Mudstone, dark gray, horizontal bedding, few ostracoda and fossil fragment; two lenses of dolomite at 1754.77 m and 1755.13 m; a pyrite laminae at 1755.28 m
89-8	1755.45—1755.55 m	0.10 m	Dolomite, olive black
89-9	1755.55—1756.71 m	1.16 m	Mudstone, dark gray, horizontal bedding, few ostracoda fossil; four olive gray dolomite laminae at 1756.13 m, 1756.63 m, 1757.13 m and 1758.43 m
89-10	1756.71—1756.76 m	0.05 m	Dolomite, olive black
89-11	1756.76—1758.43 m	1.67 m	Mudstone, dark gray, horizontal bedding, few ostracoda fossil
89-12	1758.43—1758.46 m	0.03 m	Dolomite, olive black
90-1	1758.46—1758.48 m	0.02 m	Dolomite, olive black
90-2	1758.48—1758.71 m	0.23 m	Mudstone, dark gray, horizontal bedding, few ostracoda fossils; an oblique fracture inclined at 30 deg. filled with calcite at 1760.02 m
90-3	1758.71—1758.78 m	0.07 m	Dolomite, olive black
90-4	1758.78—1760.56 m	1.78 m	Mudstone, dark gray, horizontal bedding, few ostracoda fossils
90-5	1760.56—1761.04 m	0.48 m	Mudstone, dark gray, interbeded with white crystalline limestone
90-6	1761.04—1762.68 m	1.64 m	Mudstone, dark gray, horizontal bedding, white crystalline limestone, and some dolomite laminae
90-7	1762.68—1762.91 m	0.23 m	Mudstone, olive black, cut by vertical fracture
90-8	1762.91—1763.91 m	1.00 m	Mudstone, dark gray, horizontal bedding, some ostracoda fossils and fossil fragments
90-9	1763.91—1763.96 m	0.05 m	Olive black dolomite, massive bedding
90-10	1763.96—1765.01 m	1.05 m	Dark gray mudstone, horizontal bedding, ostracoda fossils and fossil fragments in local site
90-11	1765.01—1765.06 m	0.05 m	Dolomite, olive black
90-12	1765.06—1770.46 m	5.40 m	Mudstone, dark gray, horizontal bedding, ostracoda fossils, and fossil fragments dolomite interbeds at 1766.56 m, 1767.96 m and 1769.96 m; a crystalline limestone bed at 1770.16 m

90-13	1770.46—1771.07 m	0.61 m	Mudstone, calcareous, dark gray, horizontal bedding, some crystalline limestone laminae
90-14	1771.07—1771.22 m	0.15 m	Dolomite, olive black
91-1	1771.22—1779.30 m	8.08 m	Mudstone, dark gray, horizontal bedding, fossil fragments on bedding planes, ostracoda fossils, and thin-bedding white crystalline limestone, pyrite laminae at 1775.30 m
91-2	1779.30—1779.99 m	0.69 m	Mudstone, medium dark gray, horizontal bedding, ostracoda fossils
91-3	1779.99—1780.22 m	0.23 m	Mudstone, dark gray, horizontal bedding, greenish gray tuff 1.5 cm thick at 1780.05 m
91-4	1780.22—1780.27 m	0.05 m	Dolomite, olive black
91-5	1780.27—1780.64 m	0.37 m	Marl, brownish-gray, massive bedding, crystalline limestone at the top and bottom
91-6	1780.64—1781.98 m	1.34 m	Mudstone, dark gray, horizontal bedding, ostracoda fossils; few calcareous siltstone laminae at the bottom
91-7	1781.98—1782.40 m	0.42 m	Mudstone, dark gray, interbeded with light gray calcareous siltstone, horizontal bedding, ripple lamination
91-8	1782.40—1782.59 m	0.19 m	Mudstone, dark gray, horizontal bedding, few calcareous siltstone laminae
91-9	1782.59—1782.69 m	0.10 m	Mudstone, dark gray, interbeded with light gray calcareous siltstone, ripple lamination, convolute bedding
91-10	1782.69—1782.93 m	0.24 m	Mudstone, medium dark gray, horizontal bedding, calcareous siltstone laminae

5.1.3 Qingshankou Formation Members 2 and 3

The overlying formation is the Yaojia Formation Member 1, the contacting relationship of them is parallel disconformity; the wnderlying formation is Qingshankou Formation Member 1, the contacting recationship between them is conformity (Figure 5.3).

The detailed description of the Qingshankou Formation Members 2 and 3:

43-14	1285.91—1286.26 m	0.35 m	Mudstone, silty, dark greenish gray massive bedding, few calcareous siltstone lumps
43-15	1286.26—1287.07 m	0.81 m	Mudstone, grayish brown, massive bedding, dark gray spots in the upper and basal parts of the section
43-16	1287.07—1287.87 m	0.80 m	Mudstone, dark greenish gray, intercalated with light gray siltstone, with interrupted horizontal bedding and light gray siltstone laminae
43-17	1287.87—1288.16 m	0.29 m	Mudstone, dark gray, intercalated with light gray siltstone, horizontal laminae, locally lenticular bedding, disturbed bedding, with siltstone, sand balls, and sandy lumps

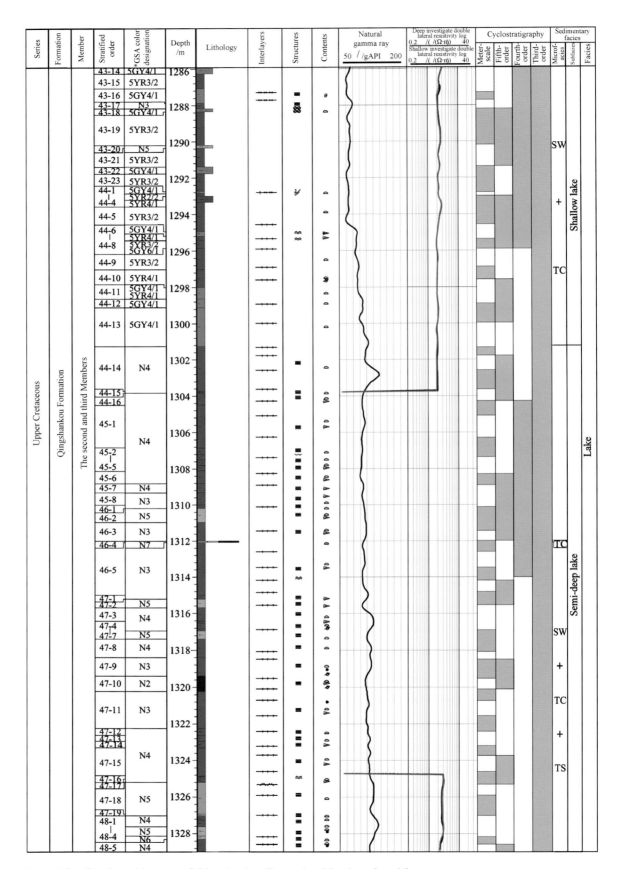

Figure 5.3 Stratigraphic chart of Qingshankou Formation Members 2 and 3.

Section 5　SK-1 Core Description and Core Photographs

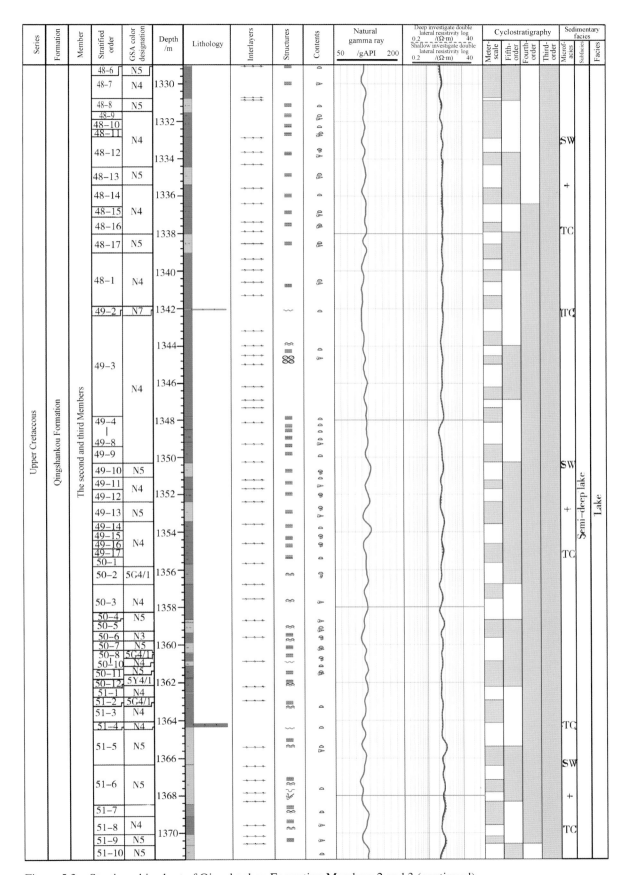

Figure 5.3　Stratigraphic chart of Qingshankou Formation Members 2 and 3 (continued).

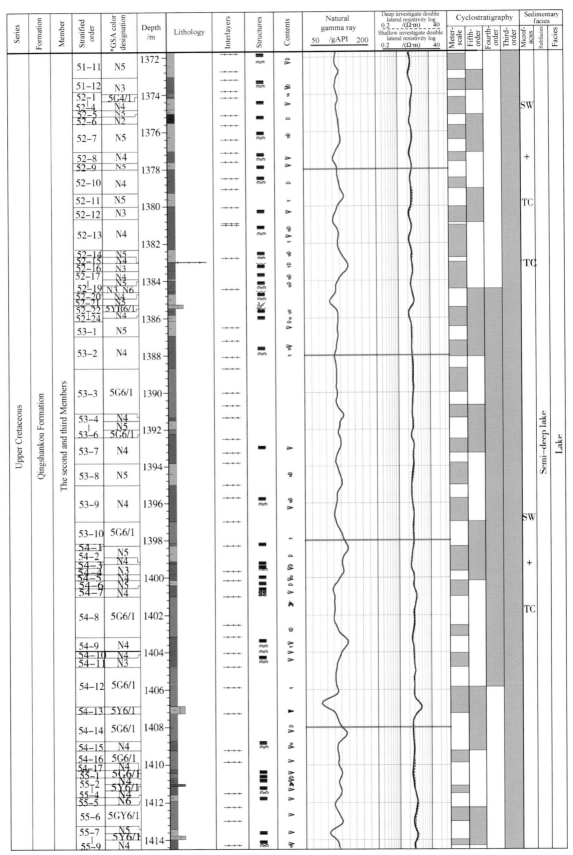

Figure 5.3 Stratigraphic chart of Qingshankou Formation Members 2 and 3 (continued).

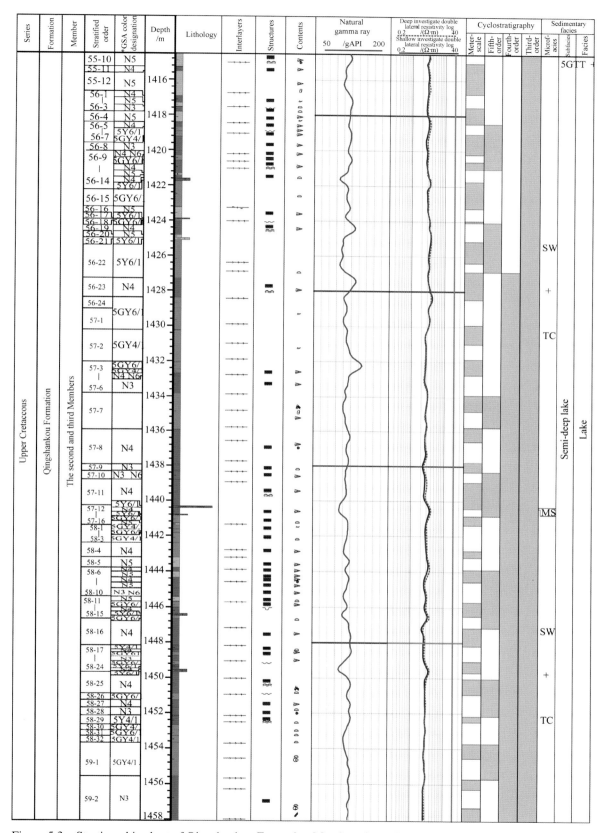

Figure 5.3　Stratigraphic chart of Qingshankou Formation Members 2 and 3 (continued).

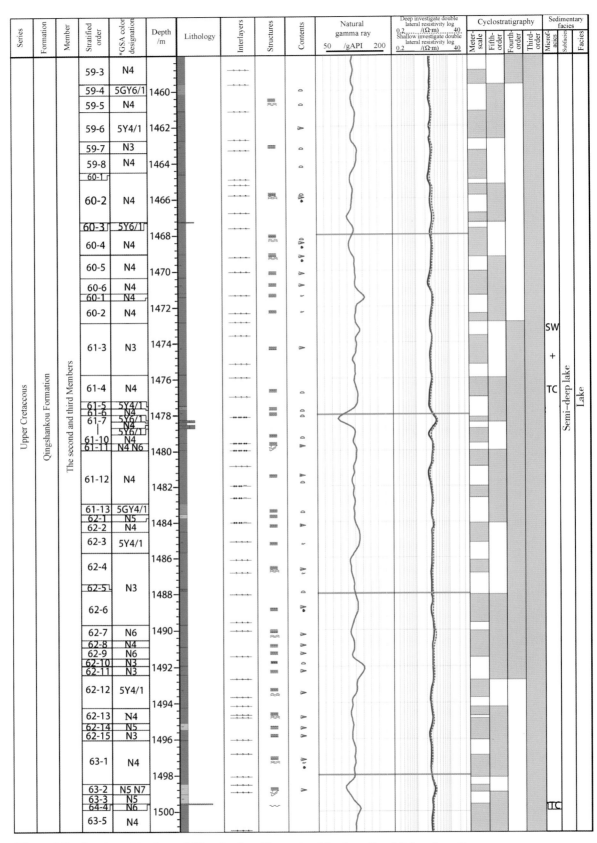

Figure 5.3　Stratigraphic chart of Qingshankou Formation Members 2 and 3 (continued).

Section 5 SK-1 Core Description and Core Photographs

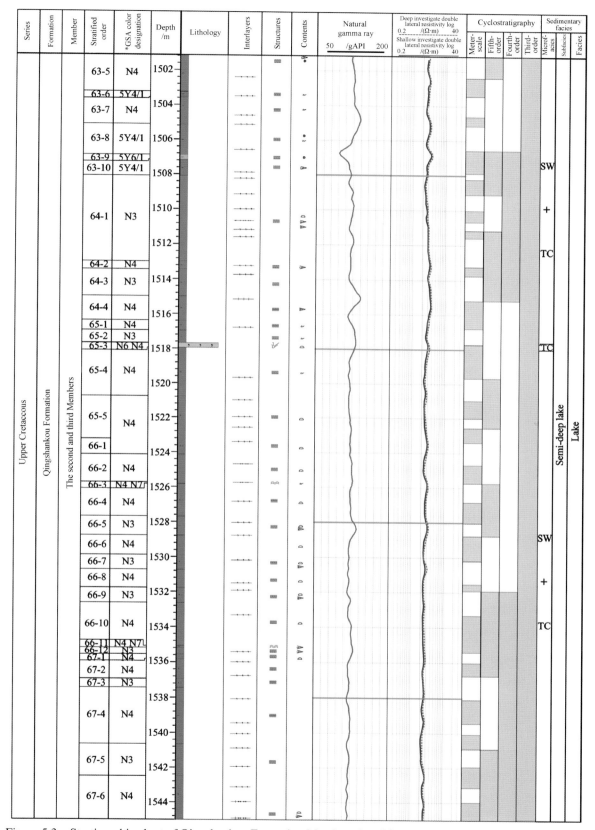

Figure 5.3 Stratigraphic chart of Qingshankou Formation Members 2 and 3 (continued).

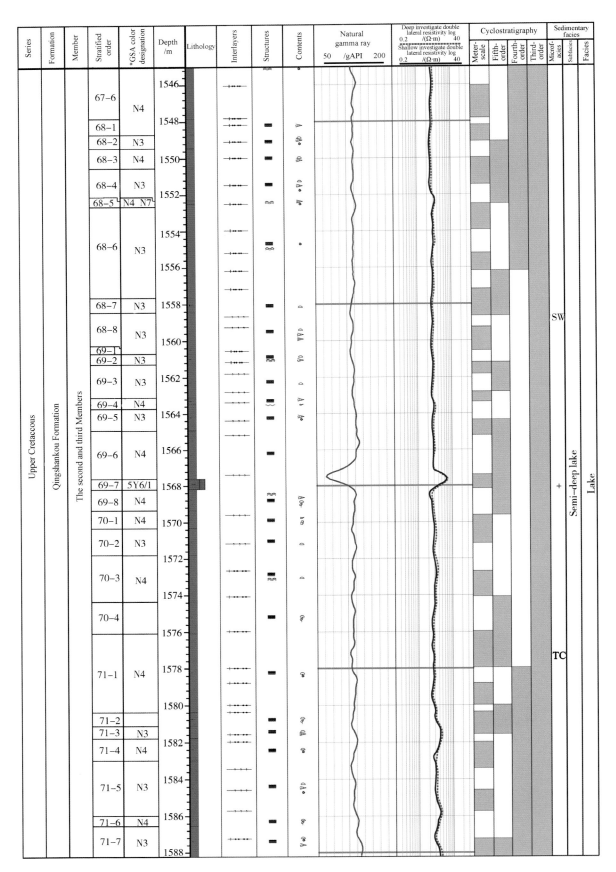

Figure 5.3 Stratigraphic chart of Qingshankou Formation Members 2 and 3 (continued).

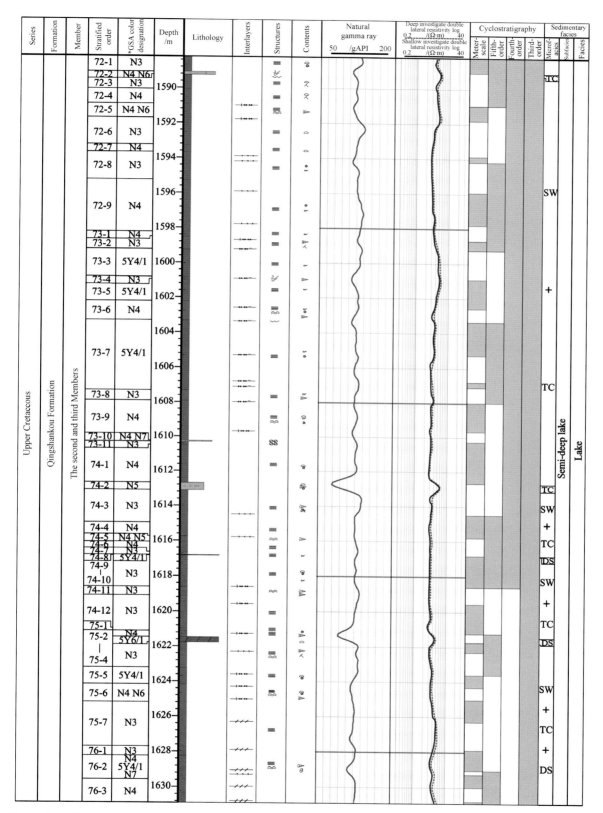

Figure 5.3 Stratigraphic chart of Qingshankou Formation Members 2 and 3 (continued).

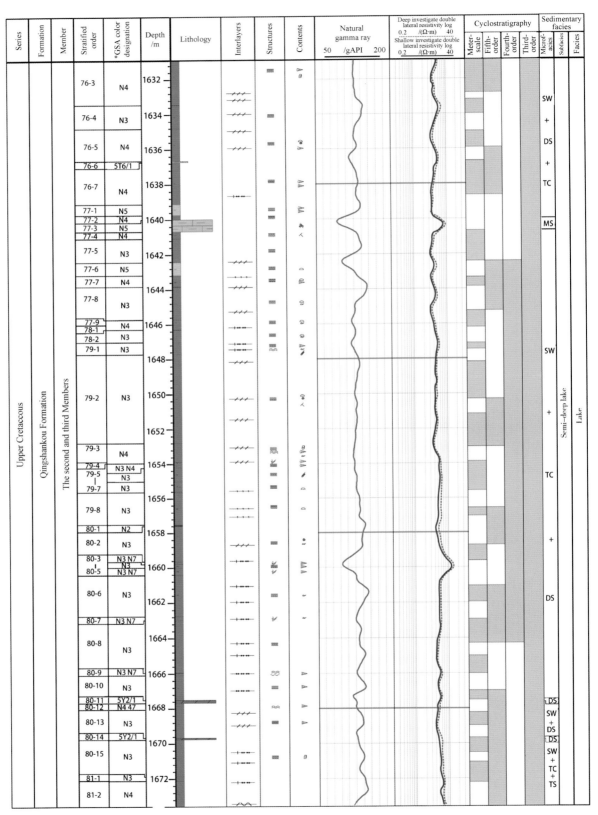

Figure 5.3　Stratigraphic chart of Qingshankou Formation Members 2 and 3 (continued).

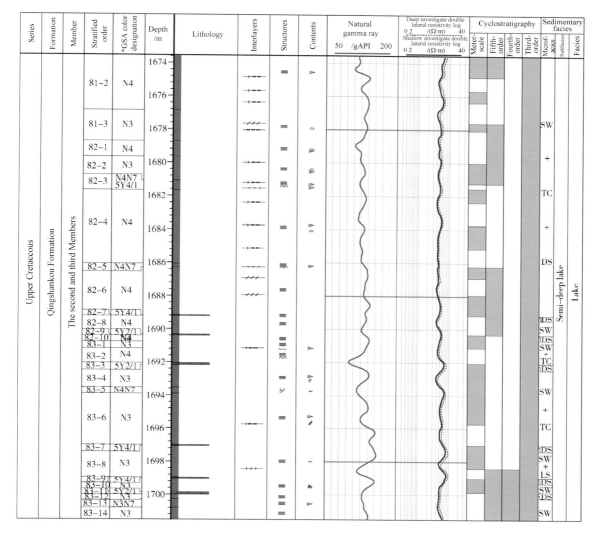

Figure 5.3　Stratigraphic chart of Qingshankou Formation Members 2 and 3 (continued). Symbols are the same as those in Figure 5.1.

43-18	1288.16—1288.34 m	0.18 m	Mudstone, silty, dark greenish, interrupted wavy bedding, deformation bedding, massive bedding in the lower part; light gray calcareous siltstone lens and lumps, few ostracoda fossil
43-19	1288.34—1290.20 m	1.86 m	Mudstone, grayish brown, massive bedding, greenish gray spots locally
43-20	1290.20—1290.34 m	0.14	Mudstone, silty, gray
43-21	1290.34—1291.39 m	1.05 m	Mudstone, grayish brown, massive
43-22	1291.39—1291.77 m	0.38 m	Mudstone, silty, dark greenish gray, massive bedding
43-23	1291.77—1292.63 m	0.86 m	Mudstone, grayish brown, massive bedding
44-1	1292.63—1292.74 m	0.11 m	Mudstone, grayish brown, massive bedding, dark greenish gray muddy siltstone lumps with deformation structures

44-2	1292.74—1292.86 m	0.12 m	Mudstone, dark greenish gray intercalated with very light gray calcareous siltstone, lenticular bedding, and deformation structures in the mudstone; ripple lamination in siltstone, calcareous siltstone laminae, and lumps with few ostracoda fossil
44-3	1292.86—1293.00 m	0.14 m	Mudstone, dark brown and dark greenish gray mudstone lumps, ostracoda fossils, calcareous concretions
44-4	1293.00—1293.33 m	0.33 m	Mudstone, silty, calcareous brownish gray, massive, few ostracoda fossil and calcareous concretion
44-5	1293.33—1294.94 m	1.61 m	Mudstone, grayish brown, massive bedding, few ostracoda fossil and calcareous concretion
44-6	1294.94—1295.18 m	0.24 m	Mudstone, dark greenish gray, horizontal bedding, light gray siltstone laminae
44-7	1295.18—1295.42 m	0.24 m	Mudstone, brownish gray, horizontal bedding with a light gray ostracoda limestone laminae
44-8	1295.42—1295.85 m	0.43 m	Mudstone, grayish brown and greenish gray, massive bedding
44-9	1295.85—1297.01 m	1.16 m	Mudstone, grayish brown, massive bedding, few ostracoda shells, and ostracoda clast limestone lamella
44-10	1297.01—1298.00 m	0.99 m	Mudstone, brownish gray massive bedding, few ostracoda fossil, and ostracoda clast limestone lamella
44-11	1298.00—1298.63 m	0.63 m	Mudstone, brownish gray and dark greenish gray, massive bedding, few ostracoda fossil
44-12	1298.63—1299.11 m	0.48 m	Mudstone, dark greenish gray, massive bedding, grayish black mud gravels, few ostracoda clast limestone lamella
44-13	1299.11—1301.25 m	2.14 m	Mudstone, dark greenish gray, massive bedding, few ostracoda fossil
44-14	1301.25—1303.72 m	2.47 m	Mudstone, medium dark gray, horizontal bedding, ostracoda fossils, some ostracoda clast limestone lamellae
44-15	1303.72—1303.82 m	0.10 m	Mudstone, medium dark gray, ostracoda-bearing, horizontal bedding, ostracoda fossils
44-16	1303.82—1304.50 m	0.68 m	Mudstone, dark gray, intercalated with medium light gray ostracoda clast limestone, horizontal bedding, ripple lamination, ostracoda fossils, and ostracoda clast limestone lamellae
45-1	1304.50—1306.85 m	2.35 m	Mudstone, dark gray, horizontal bedding, ostracoda clast limestone laminae, and ostracoda fossils
45-2	1306.85—1307.21 m	0.36 m	Mudstone, dark gray, ostracoda-bearing, horizontal bedding, with intercalated ostracoda clast limestone lamellae
45-3	1307.21—1307.50 m	0.29 m	Mudstone, dark gray, horizontal bedding, ostracoda clast limestone lamellae, and complete ostracoda fossils
45-4	1307.50—1307.80 m	0.30 m	Mudstone, dark gray, ostracoda-bearing, horizontal bedding, ostracoda clast limestone laminae

45-5	1307.80—1308.10 m	0.30 m	Mudstone, dark gray, horizontal bedding, ostracoda fossils and ostracoda clast limestone laminae
45-6	1308.10—1308.80 m	0.70 m	Mudstone, dark gray, ostracoda-bearing, intercalated with ostracoda clast limestone, horizontal bedding, lenticular bedding locally, ostracoda fossils, and ostracoda clast limestone laminae
45-7	1308.80—1309.33 m	0.53 m	Mudstone, medium dark gray, intercalated with ostracoda clast limestone, horizontal and, lenticular bedding
45-8	1309.33—1310.00 m	0.67 m	Mudstone, dark gray, intercalated with ostracoda limestone, horizontal bedding, ripple lamination
46-1	1310.00—1310.18 m	0.18 m	Mudstone dark gray, horizontal bedding, ostracoda clast limestone lamellae
46-2	1310.18—1310.96 m	0.78 m	Mudstone, gray, horizontal bedding, ostracoda fossils and ostracoda clast limestone laminae
46-3	1310.96—1312.00 m	1.04 m	Mudstone, dark gray, intercalated with ostracoda clast limestone, ripple lamination, ostracoda-clast limestone laminae
46-4	1312.00—1312.07 m	0.07 m	Limestone, light gray ostracoda-bearing
46-5	1312.07—1315.00 m	2.93 m	Mudstone, dark gray, intercalated with ostracoda clast limestone, ripple lamination, ostracoda fossils
47-1	1315.00—1315.22 m	0.22 m	Mudstone, dark gray, horizontal bedding, ostracoda clast limestone laminae
47-2	1315.22—1315.68 m	0.46 m	Mudstone, medium gray, horizontal bedding, ostracoda clast limestone lamellae
47-3	1315.68—1316.41 m	0.73 m	Mudstone, medium dark gray, horizontal bedding, some of ostracoda are of smaller size
47-4	1316.41—1316.53 m	0.12 m	Mudstone, medium dark gray, intercalated with ostracoda clast limestone, horizontal bedding, ripple lamination, some ostracoda clast limestone laminae and ostracoda fossils
47-5	1316.53—1316.82 m	0.29 m	Mudstone, medium dark gray, horizontal bedding, with some ostracoda fossil clasts and relict plant fragments
47-6	1316.82—1316.94 m	0.12 m	Mudstone, medium dark gray, interbedded with ostracoda clast limestone, horizontal bedding, ripple lamination
47-7	1316.94—1317.38 m	0.44 m	Mudstone, medium gray, ostracoda-bearing, horizontal bedding, with calcareous mudstone lamellae in the basal part of the section
47-8	1317.38—1318.40 m	1.02 m	Mudstone, medium dark gray, horizontal bedding, ostracoda clast limestone lamellae and ostracoda fossils
47-9	1318.40—1319.40 m	1.00 m	Mudstone, dark gray, ostracoda-bearing, horizontal bedding, with thin bedded ostracoda clast limestone lamellae some ostracoda, and conchostracan fossils and plant fragments
47-10	1319.40—1320.22 m	0.82 m	Mudstone, grayish black, intercalated with ostracoda clast limestone, horizontal bedding, ripple lamination, ostracoda clast limestone lamellae and silicified fossils
47-11	1320.22—1322.25 m	2.03 m	Mudstone, dark gray, intercalated with ostracoda clast limestone, horizontal bedding, ripple lamination, conchonstracan and ostracoda fossils, ostracoda clast limestone lamellae

47-12	1322.25—1322.62 m	0.37 m	Mudstone, medium dark gray, ostracoda-bearing, horizontal bedding, some ostracoda fossils and ostracoda clast limestone lamellae
47-13	1322.62—1322.95 m	0.33 m	Mudstone, medium dark gray, horizontal bedding, ostracoda fossils
47-14	1322.95—1323.29 m	0.34 m	Mudstone, medium dark gray, interbeded with ostracoda clast limestone, horizontal bedding, ripple lamination, ostracoda clast limestone laminae and lenses
47-15	1323.39—1324.88 m	1.49 m	Mudstone, medium dark gray, horizontal bedding, some ostracoda and conchonstracan fossils, ostracoda clast limestone lamellae
47-16	1324.88—1325.00 m	0.12 m	Mudstone, medium dark gray, interbeded with ostracoda clast limestone, ripple lamination, ostracoda fossils
47-17	1325.00—1325.20 m	0.20 m	Mudstone, medium dark gray, ostracoda-bearing, intercalated with ostracoda clast limestone, horizontal bedding, ripple lamination, ostracoda fossils, fossil fragments on top of the erosion surface
47-18	1325.20—1326.92 m	1.72 m	Mudstone, medium gray, intercalated with ostracoda clast limestone, horizontal bedding, ripple lamination, ostracoda fossils, and an 8 mm thick tuff-like bed at 1325.31 m
47-19	1326.92—1327.02 m	0.10 m	Mudstone, medium gray, interbeded with ostracoda clast limestone, horizontal bedding, ripple lamination, ostracoda fossils and fossil fragments
48-1	1327.02—1327.25 m	0.23 m	Mudstone, medium dark gray, intercalated with ostracoda clast limestone, horizontal bedding, ostracoda fossils
48-2	1327.25—1327.62 m	0.37 m	Mudstone, medium dark gray, horizontal bedding, some ostracoda fossils
48-3	1327.62—1328.12 m	0.50 m	Mudstone, medium gray, ostracoda-bearing, horizontal bedding, ostracoda fossils, few conchonstracan fossil
48-4	1328.12—1328.29 m	0.17 m	Mudstone, medium light gray, horizontal bedding, ostracoda clast limestone lamellae, few conchonstracan fossil
48-5	1328.29—1328.99 m	0.70 m	Mudstone, medium dark gray, intercalated with ostracoda clast limestone, horizontal bedding, ripple lamination, ostracoda fossils
48-6	1328.99—1329.09 m	0.10 m	Mudstone, medium gray, horizontal bedding, ostracoda clast limestone lamella in the basal part of the lower section with ostracoda fossils
48-7	1329.09—1330.78 m	1.69 m	Mudstone, medium dark gray, horizontal bedding, ostracoda clast limestone in the basal part of the section
48-8	1330.78—1331.52 m	0.74 m	Mudstone, medium gray, ostracoda-bearing, horizontal bedding, ostracoda clast limestone lamella
48-9	1331.52—1331.87 m	0.35 m	Mudstone, medium dark gray, horizontal bedding, ostracoda fossils, ostracoda clast limestone lamellae in the basal part of the section, erosion surface
48-10	1331.87—1332.42 m	0.55 m	Mudstone, medium dark gray, ostracoda-bearing, horizontal bedding, few conchonstracan fossils
48-11	1332.42—1332.84 m	0.42 m	Mudstone, medium dark gray, horizontal bedding, some ostracoda and conchonstracan fossils, ostracoda clast limestone lamellae

48-12	1332.84—1334.45 m	1.61 m	Mudstone, medium dark gray, intercalated with ostracoda clast limestone, horizontal bedding, ripple lamination, some conchonstracan fossils, ostracoda clast limestone lamellae
48-13	1334.45—1335.40 m	0.95 m	Mudstone, medium dark gray, horizontal bedding, ostracoda fossils and ostracoda clast limestone laminae
48-14	1335.40—1336.57 m	1.17 m	Mudstone, medium dark gray, horizontal bedding, ostracoda fossils and ostracoda clast limestone lamellae
48-15	1336.57—1337.12 m	0.55 m	Mudstone, medium dark gray, ostracoda-bearing, horizontal bedding
48-16	1337.12—1338.02 m	0.90 m	Mudstone, medium dark gray, horizontal bedding, ostracoda fossils and ostracoda clast limestone lamellae
48-17	1338.02—1339.03 m	1.01 m	Mudstone, medium gray, ostracoda-bearing, intercalated with ostracoda clast limestone, horizontal bedding, ripple lamination, ostracoda clast limestone laminae
49-1	1339.03—1342.01 m	2.98 m	Mudstone, medium dark gray, interbeded with ostracoda clast limestone, horizontal bedding, ripple lamination, disturbed bedding, and ostracoda clast limestone laminae and lumps
49-2	1342.01—1342.06 m	0.05 m	Limestone, light gray with ostracoda, ripple lamination, mudstone laminae in the upper part of the section
49-3	1342.06—1347.80 m		
49-4	1347.80—1347.95 m	0.15 m	Mudstone medium dark gray, horizontal bedding, ostracoda fossils
49-5	1347.95—1348.52 m	0.57 m	Mudstone medium dark gray, interbeded with light gray ostracoda clast limestone, ripple lamination, horizontal bedding, some ostracoda fossils and ostracoda clast limestone lamellae
49-6	1348.52—1348.66 m	0.14 m	Mudstone, medium dark gray, horizontal bedding, with very few ostracoda fossil
49-7	1348.66—1349.20 m	0.54 m	Mudstone, medium dark gray, horizontal bedding, ostracoda fossils of variable size
49-8	1349.20—1349.44 m	0.24 m	Mudstone, medium dark gray, horizontal bedding, some ostracoda clast limestone lamellae
49-9	1349.44—1350.30 m	0.86 m	Mudstone, medium dark gray, intercalated with ostracoda clast limestone, horizontal bedding, ripple lamination, some ostracoda fossils and ostracoda clast limestone lamellae
49-10	1350.3—1351.05 m	0.75 m	Mudstone, medium gray, horizontal bedding, ripple lamination, some ostracoda fossils
49-11	1351.05—1351.69 m	0.64 m	Mudstone, medium dark gray, intercalated with laminated ostracoda clast limestone, horizontal bedding, ripple lamination, some ostracoda and conchonstracan fossils with ostracoda clast limestone laminae
49-12	1351.69—1352.38 m	0.69 m	Mudstone, medium dark gray, ostracoda-bearing, horizontal bedding, some ostracoda and conchonstracan fossils

49-13	1352.38—1353.43 m	1.05 m	Mudstone, medium gray, intercalated with laminated ostracoda clast limestone, horizontal bedding, ripple lamination, with slight scouring surface at the bottom of the ostracoda clast limestone, some ostracoda and conchonstracan fossils with ostracoda clast limestone laminae
49-14	1353.43—1353.91 m	0.48 m	Mudstone, medium dark gray, intercalated with ostracoda clast limestone, horizontal bedding, ripple lamination, some conchonstracan and ostracoda fossils
49-15	1353.91—1354.54 m	0.63 m	Mudstone, medium dark gray, horizontal bedding, some ostracoda and conchonstracan fossils
49-16	1354.54—1354.96 m	0.42 m	Mudstone, medium dark gray, intercalated with laminated ostracoda clast limestone, horizontal bedding, ripple lamination, few conchonstracan and ostracoda fossil, ostracoda clast limestone laminae
49-17	1354.96—1355.23 m	0.27 m	Mudstone, medium dark gray, horizontal bedding, ostracoda fossils
50-1	1355.23—1355.65 m	0.42 m	Mudstone, medium dark gray, horizontal bedding, ostracoda fossils, ostracoda clast limestone lamellae in the basal part of the section
50-2	1355.65—1356.77 m	1.12 m	Mudstone, dark greenish gray, with unclear horizontal ripple lamination, more ostracoda fossils and pyrite-riched layers
50-3	1356.77—1358.65 m	1.88 m	Mudstone, medium dark gray, intercalated with laminated light gray ostracoda clast limestone, ripple lamination
50-4	1358.65—1358.80 m	0.15 m	Mudstone, medium gray, ostracoda clast limestone thinly interbedded with mudstone, ripple lamination, ostracoda fossils
50-5	1358.80—1359.35 m	0.55 m	Mudstone, medium gray, ostracoda-bearing, thinly interbeded with ostracoda clast limestone, ripple lamination, ostracoda fossil, ostracoda clast limestone laminae
50-6	1359.35—1359.88 m	0.53 m	Mudstone, dark gray, thinly interbeded with ostracoda clast limestone, ripple lamination, horizontal bedding, ostracoda fossils and ostracoda clast limestone laminae, few conchonstracan fossils
50-7	1359.88—1360.40 m	0.52 m	Mudstone, medium gray with intercalated with ostracoda clast limestone ripple lamination, horizontal bedding, ostracoda fossils
50-8	1360.40—1360.49 m	0.09 m	Mudstone, dark greenish gray, thinly interbeded with olive gray ostracoda clast limestone, ripple lamination, horizontal bedding, ostracoda fossils, and ostracoda clast limestone laminae
50-9	1360.49—1360.86 m	0.37 m	Mudstone, medium dark gray, intercalated with laminated ostracoda clast limestone, ripple lamination, horizontal bedding, conchonstracan and ostracoda fossils, ostracoda clast limestone laminae
50-10	1360.86—1361.12 m	0.26 m	Mudstone, medium gray, intercalated with laminated ostracoda clast limestone, ripple lamination, horizontal bedding, conchonstracan and ostracoda fossils, ostracoda clast limestone laminae
50-11	1361.12—1361.75 m	0.63 m	Mudstone, olive gray, intercalated with laminated ostracoda clast limestone, ripple lamination, horizontal bedding, ostracoda fossils and ostracoda clast limestone laminae
50-12	1361.75—1361.83 m	0.08 m	Mudstone, medium dark gray, horizontal bedding, few ostracoda fossil

51-1	1361.84—1362.46 m	0.62 m	Mudstone, medium dark gray, horizontal ripple lamination, horizontal bedding, few ostracoda fossil
51-2	1362.46—1362.51 m		
51-3	1362.51—1364.14 m	1.63 m	Mudstone, medium dark gray, intercalated with laminated ostracoda clast limestone, ripple lamination, ostracoda fossils
51-4	1364.14—1364.34 m	0.20 m	Mudstone, medium dark gray, thinly interbeded with ostracoda limestone, horizontal ripple lamination, horizontal bedding, ostracoda fossils, ostracoda clast limestone laminae
51-5	1364.34—1366.34 m	2.00 m	Mudstone, medium gray, horizontal bedding, ostracoda fossils, ostracoda clast limestone laminae
51-6	1366.34—1368.47 m	2.13 m	Mudstone, medium gray, thinly interbeded with ostracoda clast limestone, horizontal ripple lamination, horizontal bedding, deformation bedding, load structure, some ostracoda and conchonstracan fossils, ostracoda clast limestone laminae
51-7	1368.47—1369.10 m	0.63 m	Mudstone, light greenish gray, massive bedding, ostracoda fossil fragments
51-8	1369.10—1370.06 m	0.96 m	Mudstone, medium dark gray horizontal bedding, ostracoda clast limestone laminae
51-9	1370.06—1370.64 m	0.58 m	Mudstone, medium gray, interbeded with ostracoda clast limestone, horizontal bedding and ripple lamination
51-10	1370.64—1371.41 m	0.77 m	Mudstone, medium gray, massive bedding, ostracoda fossils locally more frequent
51-11	1371.41—1373.06 m	1.65 m	Mudstone, medium gray, intercalated with ostracoda clast limestone, horizontal bedding and ripple lamination
51-12	1373.06—1373.77 m	0.71 m	Mudstone, dark gray, intercalated with laminated ostracoda clast limestone, horizontal bedding, ripple lamination, with dark greenish gray spots in the upper part
52-1	1373.77—1373.91 m	0.14 m	Mudstone, dark gray, intercalated with ostracoda clast limestone, horizontal bedding, and ripple lamination
52-2	1373.91—1374.10 m	0.19 m	Mudstone, dark greenish gray, massive bedding, some ostracoda clast limestone lamellae and pyrite grains
52-3	1374.10—1374.17 m	0.07 m	Mudstone, dark gray, horizontal bedding, fragments of ostracoda
52-4	1374.17—1374.80 m	0.63 m	Mudstone, medium dark gray, horizontal bedding, ostracoda clast limestone laminae, with dark greenish gray spots in its basal part
52-5	1374.80—1375.02 m	0.22 m	Mudstone, medium gray, horizontal bedding, with light greenish gray spots, a dark greenish gray ostracoda-bearing mudstone lamella in the upper part
52-6	1375.02—1375.53 m	0.51 m	Mudstone, grayish black, ostracoda-bearing, horizontal bedding
52-7	1375.53—1377.03 m	1.50 m	Mudstone, medium gray, horizontal bedding, some ostracoda and conchonstracan fossils, ostracoda clast limestone lamellae
52-8	1377.03—1377.62 m	0.59 m	Mudstone, medium dark gray, horizontal bedding and ripple lamination, ostracoda clast limestone lamellae, with well preserved ostracoda and several conchonstracan fossil-enriched beds

52-9	1377.62—1377.98 m	0.36 m	Mudstone, medium gray, horizontal bedding, ostracoda clast limestone laminae with ripple lamination
52-10	1377.98—1379.27 m	1.29 m	Mudstone, medium dark gray, ostracoda-bearing, horizontal bedding, ripple lamination, ostracoda fossil fragments and few ostracoda clast limestone lamellae
52-11	1379.27—1379.99 m	0.72 m	Mudstone, medium gray, massive bedding, fossil fragments
52-12	1379.99—1380.64 m	0.65 m	Mudstone, dark gray, horizontal bedding, some ostracoda clast limestone stripes and lamellae
52-13	1380.64—1382.32 m	1.68 m	Mudstone, medium dark gray, horizontal bedding, ostracoda clast limestone laminae, conchonstracan and plant fossils fragments, vertical fractures
52-14	1382.32—1382.99 m	0.67 m	Mudstone, medium gray, horizontal bedding, fossil fragments and conchonstracan fossils, ostracoda clast limestone lamellae
52-15	1382.99—1383.01 m	0.02 m	Mudstone, medium dark gray, ostracoda clast limestone laminae, horizontal ripple lamination
52-16	1383.01—1383.46 m	0.45 m	Mudstone, dark gray, horizontal bedding, few fossil fragments
52-17	1383.46—1383.89 m	0.43 m	Mudstone, medium dark gray, horizontal bedding, vertical fracture filled with ostracoda fossil fragments
52-18	1383.89—1384.00 m	0.11 m	Mudstone, medium dark grey, massive, ostracoda fossil fragments and conchonstracan fossils
52-19	1384.00—1384.58 m	0.58 m	Mudstone, dark gray, intercalated with laminated medium. light gray ostracoda clast limestone, horizontal bedding, ripple lamination, more fossil fragments, conchonstracan fossils and ostracoda clast limestone laminae
52-20	1384.58—1384.70 m	0.12 m	Mudstone, medium dark gray, horizontal bedding, few ostracoda fossil fragments
52-21	1384.70—1385.27 m	0.57 m	Mudstone, medium gray, horizontal bedding, dark greenish gray mudstone laminae in the upper section
52-22	1385.27—1385.47 m	0.20 m	Mudstone, calcareous, light olive gray, thinly interbeded with dark gray mudstone, horizontal bedding, ripple lamination, beds deformed, load structure, ostracoda and conchonstracan fossils in the mudstone
52-23	1385.47—1385.84 m	0.37 m	Mudstone, calcareous, ostracoda-bearing, medium dark gray, interbeded with dark gray mudstone, horizontal bedding, with fossil fragments and conchonstracan fossils in the mudstone
52-24	1385.84—1386.14 m	0.30 m	Mudstone, medium gray, horizontal bedding, some ostracoda fossil fragments and conchonstracan fossils
53-1	1386.14—1386.96 m	0.82 m	Mudstone, medium gray, horizontal bedding, dark greenish gray mudstone in the upper part, some ostracoda clast limestone lamellae, with scouring surface at the bottom
53-2	1386.96—1388.68 m	1.72 m	Mudstone, medium dark gray, intercalated with laminated ostracoda clast limestone, horizontal bedding, ripple lamination, ostracoda clast limestone laminae with, more fossil fragments and conchonstracan fossils

53-3	1388.68—1391.32 m	2.64 m	Mudstone greenish gray, massive bedding, some ostracoda clast limestone lamellae
53-4	1391.32—1391.51 m	0.19 m	Mudstone medium dark gray, massive bedding, ostracoda clast limestone lamella in the upper part of the section
53-5	1391.51—1391.98 m	0.47 m	Mudstone, medium gray, massive bedding
53-6	1391.98—1392.24 m	0.26 m	Mudstone, greenish gray, massive bedding, some medium gray spots
53-7	1392.24—1393.81 m	1.57 m	Mudstone, medium dark gray, horizontal bedding, ostracoda clast limestone lamellae
53-8	1393.81—1395.02 m	1.21 m	Mudstone, medium gray, massive bedding, with conchonstracan fossils enriched locally with few ostracoda fragments and fossils
53-9	1395.02—1396.97 m	1.95 m	Mudstone, medium dark gray, intercalated with ostracoda bearing limestone, horizontal bedding, ripple lamination, ostracoda clast limestone laminae, ostracoda and conchonstracan fossils
53-10	1396.97—1398.16 m	1.19 m	Mudstone, greenish gray, massive bedding, some fossil fragments
54-1	1398.16—1398.28 m	0.12 m	Mudstone, greenish gray, horizontal bedding, ostracoda clast limestone lamellae in the basal part
54-2	1398.28—1399.13 m	0.85 m	Mudstone, medium gray, massive bedding, few ostracoda fossil fragments and green spots
54-3	1399.13—1399.25 m	0.12 m	Mudstone, medium dark gray, horizontal bedding, few ostracoda clasts
54-4	1399.25—1399.82 m	0.57 m	Mudstone, dark gray, horizontal bedding, ostracoda clast limestone laminae, fossil fragments in the basal parts of the section
54-5	1399.82—1400.13 m	0.31 m	Mudstone, medium dark gray, horizontal bedding, some ostracoda and conchonstracan fossils, ostracoda clast limestone laminae
54-6	1400.13—1400.42 m	0.29 m	Mudstone, medium gray, horizontal bedding, ostracoda clast limestone lamellae
54-7	1400.42—1401.01 m	0.59 m	Mudstone, medium dark gray, intercalated with medium. light gray ostracoda clast limestone, horizontal bedding, ripple lamination, disturbed bedding, ostracoda clast limestone laminae
54-8	1401.01—1403.17 m	2.16 m	Mudstone, greenish grey, ostracoda fossils
54-9	1403.17—1403.94 m	0.77 m	Mudstone, medium dark grey, limestone laminae with ostracoda
54-10	1403.94—1404.11 m	0.17 m	Mudstone, medium dark gray, interbeded with ostracoda clast limestone, horizontal bedding, ripple lamination
54-11	1404.11—1404.78 m	0.67 m	Mudstone, dark gray, horizontal ripple lamination, ostracoda clast limestone lamellae in the basal parts more fossil fragments and ostracoda clasts concentrated in the upper part of the section
54-12	1404.78—1406.88 m	2.10 m	Mudstone, greenish gray, massive bedding, fossil fragments in the basal part, ostracoda clast limestone laminae in the middle
54-13	1406.88—1407.27 m	0.39 m	Mudstone, calcareous, light olive gray, massive bedding, greenish gray mudstone lamellae
54-14	1407.27—1408.73 m	1.46 m	Mudstone, greenish gray, massive bedding, ostracoda fossil fragments and ostracoda clast limestone laminae in the basal part

54-15	1408.73—1409.23 m	0.50 m	Mudstone, medium dark gray, horizontal bedding, conchonstracan fossils with few fossil fragments and ostracoda clast limestone laminae in the upper parts, greenish-gray mud gravels in the basal parts
54-16	1409.23—1410.24 m	1.01 m	Mudstone, greenish gray, massive bedding, ostracoda clast limestone lamellae and marl lumps
54-17	1410.24—1410.42 m	0.18 m	Medium dark gray mudstone, horizontal bedding, ostracoda clast limestone stripes with horizontal ripple lamination
55-1	1410.42—1410.71 m	0.29 m	Mudstone, greenish gray, horizontal bedding, with medium gray spots, more ostracoda fossil fragments at its basal section, few complete ostracoda fossil
55-2	1410.71—1411.00 m	0.29 m	Mudstone, medium dark gray, horizontal bedding, ostracoda clast limestone laminae, few green spots in its upper parts
55-3	1411.00—1411.11 m	0.11 m	Mudstone, calcareous, light olive gray massive bedding, a sub horizontal fracture in the middle, filled with calcite
55-4	1411.11—1411.57 m	0.46 m	Mudstone, medium dark gray, interbeded with ostracoda clast limestone, horizontal bedding and ripple lamination
55-5	1411.57—1411.91 m	0.34 m	Mudstone, medium light gray, horizontal bedding, fossil fragments, ostracoda clast limestone laminae in the basal parts
55-6	1411.91—1413.42 m	1.51 m	Mudstone, greenish gray, ostracoda clast limestone laminae more common downwards
55-7	1413.42—1413.77 m	0.35 m	Mudstone, medium gray, horizontal bedding, ostracoda clast limestone laminae
55-8	1413.77—1413.97 m	0.20 m	Mudstone, calcareous, light olive gray, massive bedding
55-9	1413.97—1414.51 m	0.54 m	Mudstone, medium dark gray, horizontal bedding, ostracoda clast limestone lamellae, few conchonstracan fossils
55-10	1414.51—1415.26 m	0.75 m	Mudstone, medium gray, intercalated with ostracoda clast limestone, ripple lamination, horizontal bedding, conchonstracan fossils, more fossil fragments and ostracoda clast limestone lamellae in the upper and middle parts of the section
55-11	1415.26—1415.53 m	0.27 m	Mudstone, medium dark gray, horizontal bedding, ostracoda clast limestone laminae ripple lamination
55-12	1415.53—1416.93 m	1.40 m	Mudstone, medium gray, massive marl lumps in the upper section, some ostracoda clast limestone laminae and locally greenish gray spots, pyrite and ostracoda clast limestone lamellae in the lower section
56-1	1416.93—1417.32 m	0.39 m	Mudstone, medium gray, horizontal bedding, greenish gray mudstone lamella in the upper section, fossil fragments
56-2	1417.32—1417.54 m	0.22 m	Mudstone, medium gray, intercalated with ostracoda clast limestone, horizontal bedding, some conchonstracan fossils
56-3	1417.54—1417.83 m	0.29 m	Mudstone, dark gray, ostracoda-bearing horizontal bedding, few fossil fragments
56-4	1417.83—1418.42 m	0.59 m	Mudstone, medium gray, horizontal bedding, ostracoda clast limestone laminae, fossil fragments
56-5	1418.42—1418.72 m	0.30 m	Mudstone, medium dark gray, horizontal bedding, ostracoda clast limestone laminae

56-6	1418.72—1418.94 m	0.22 m	Mudstone, greenish gray, massive bedding, ostracoda clast limestone lamella
56-7	1418.94—1419.23 m	0.29 m	Mudstone, dark greenish gray, thinly interbeded with light olive gray ostracoda clast limestone, horizontal bedding, ripple lamination
56-8	1419.23—1420.07 m	0.84 m	Mudstone, dark gray, horizontal bedding, locally ostracoda clast limestone laminae
56-9	1420.07—1420.29 m	0.22 m	Mudstone, medium dark gray, intercalated with medium. light gray ostracoda clast limestone, horizontal bedding, with local ripple lamination
56-10	1420.29—1420.60 m	0.31 m	Mudstone, greenish gray and medium dark gray, horizontal bedding, ostracoda fragments and ostracoda clast limestone laminae
56-11	1420.60—1421.31 m	0.71 m	Mudstone, Medium dark gray, mudstone, horizontal bedding, ripple lamination, locally ostracoda clast limestone stripes
56-12	1421.31—1421.44 m	0.13 m	Mudstone, medium gray, ostracoda-bearing horizontal bedding, ripple lamination
56-13	1421.44—1421.60 m	0.16 m	Mudstone, medium dark gray, horizontal bedding
56-14	1421.60—1421.75 m	0.15 m	Mudstone, calcareous, light olive gray, massive bedding
56-15	1421.75—1423.24 m	1.49 m	Mudstone, greenish gray, massive, locally ostracoda fossil fragments
56-16	1423.24—1423.88 m	0.64 m	Mudstone, medium gray, horizontal bedding, locally laminae of calcareous mudstone and ostracoda clast limestone in the basal part
56-17	1423.88—1423.94 m	0.06 m	Mudstone, calcareous, light olive gray
56-18	1423.94—1424.03 m	0.09 m	Mudstone, greenish gray, ostracoda clast limestone lamellae with horizontal ripple lamination in the lower parts
56-19	1424.03—1424.83 m	0.80 m	Mudstone, medium dark gray, horizontal bedding, ostracoda clast limestone laminae with horizontal ripple lamination
56-20	1424.83—1425.00 m	0.17 m	Mudstone, medium gray, massive bedding, ostracoda clast limestone laminae in the basal section
56-21	1425.00—1425.11 m	0.11 m	Mudstone, calcareous, light olive gray with greenish gray mudstone lamella in the upper section
56-22	1425.11—1427.30 m	2.19 m	Mudstone, greenish gray, massive bedding, ostracoda clast limestone lamellae in the lower section
56-23	1427.30—1428.39 m	1.09 m	Mudstone, medium dark gray, horizontal bedding, ostracoda clast limestone laminae and fossil fragments on bedding planes
56-24	1428.39—1429.05 m	0.66 m	Mudstone, greenish gray, massive bedding
57-1	1429.05—1430.25 m	1.20 m	Mudstone, greenish gray, massive bedding, fossil fragments, locally ostracoda clast limestone lamellae
57-2	1430.25—1432.06 m	1.81 m	Mudstone, dark greenish gray massive bedding, fossil fragments, locally ostracoda clast limestone lamellae
57-3	1432.06—1432.49 m	0.43 m	Mudstone, greenish gray, massive, some fossil fragments and plant debris
57-4	1432.49—1432.71 m	0.22 m	Mudstone, dark greenish gray, horizontal bedding, some ostracoda clast limestone laminae with horizontal ripple lamination

57-5	1432.71—1432.87 m	0.16 m	Medium dark gray, mudstone thinly interbeded with light gray ostracoda clast limestone, horizontal bedding, ripple lamination
57-6	1432.87—1433.84 m	0.97 m	Mudstone, dark gray, horizontal bedding, ostracoda clast limestone laminae and plant fossil fragments
57-7	1433.84—1435.93 m	2.09 m	Mudstone, greenish gray, massive bedding, some local irregular fractures, filled with calcite; pyrite and few ostracoda clast limestone lamella in the lower section
57-8	1435.93—1437.94 m	2.01 m	Mudstone, medium dark gray, horizontal bedding, ostracoda clast limestone laminae, locally conchonstracan fossils
57-9	1437.94—1438.32 m	0.38 m	Dark gray mudstone, horizontal bedding, ostracoda fossil fragments
57-10	1438.32—1438.80 m	0.48 m	Mudstone, dark gray, interbeded with light gray ostracoda clast limestone, horizontal bedding, ripple lamination, ostracoda clast limestone laminae, more ostracoda fossils in the middle and lower section
57-11	1438.80—1440.32 m	1.52 m	Mudstone, medium dark gray, intercalated with ostracoda bearing limestone, horizontal bedding, ripple lamination, ostracoda clast limestone laminae, few conchonstracan fossil, locally abundant ostracoda fossil fragments
57-12	1440.32—1440.41 m	0.09 m	Mudstone, calcareous, light olive gray
57-13	1440.41—1440.79 m	0.38 m	Mudstone, medium dark gray, horizontal bedding, Limestone laminae with ostracoda, more fossil fragments in the lower section
57-14	1440.79—1440.86 m	0.07 m	Mudstone, calcareous, light olive gray
57-15	1440.86—1441.32 m	0.46 m	Mudstone, greenish gray, horizontal bedding, some ostracoda fossil fragments
57-16	1441.32—1441.41 m	0.09 m	Mudstone, medium gray, horizontal bedding, ostracoda clast limestone laminae in the basal parts
58-1	1441.41—1441.64 m	0.23 m	Mudstone, dark greenish gray, horizontal bedding, fossil fragments
58-2	1441.64—1441.85 m	0.21 m	Mudstone, greenish gray, horizontal bedding, few ostracoda fossil fragment
58-3	1441.85—1442.40 m	0.55 m	Mudstone, dark greenish gray, horizontal bedding, ostracoda clast limestone laminae, ripple lamination
58-4	1442.40—1443.28 m	0.88 m	Mudstone, dark gray, intercalated with laminated ostracoda clast limestone, horizontal bedding, ripple lamination
58-5	1443.28—1443.79 m	0.51 m	Mudstone, medium gray, horizontal bedding, ostracoda clast limestone laminae with ripple lamination
58-6	1443.79—1444.02 m	0.23 m	Mudstone, medium dark gray, thinly interbeded with ostracoda clast limestone, horizontal bedding, ripple lamination
58-7	1444.02—1444.35 m	0.33 m	Mudstone, medium gray, horizontal bedding, ostracoda clast limestone laminae
58-8	1444.35—1444.64 m	0.29 m	Mudstone, medium dark gray, intercalated with ostracoda clast limestone, horizontal bedding, ripple lamination; a bluish (5B5/1) gray mudstone bed, less than 1 cm, on top of the section, some conchonstracan fossils and ostracoda clast limestone laminae
58-9	1444.64—1444.96 m	0.32 m	Mudstone, medium dark gray, horizontal bedding, few ostracoda clast limestone laminae

58-10	1444.96—1445.42 m	0.46 m	Mudstone, dark gray, intercalated with ostracoda clast limestone, horizontal bedding, ripple lamination
58-11	1445.42—1445.68 m	0.26 m	Mudstone, medium gray, horizontal bedding, ostracoda clast limestone lamellae in the lower section
58-12	1445.68—1446.02 m	0.34 m	Mudstone, greenish gray, horizontal bedding, ostracoda clast limestone laminae
58-13	1446.02—1446.37 m	0.35 m	Mudstone, medium dark gray, massive bedding, ostracoda fossil fragments
58-14	1446.37—1446.50 m	0.13 m	Mudstone, calcareous, light olive gray
58-15	1446.50—1446.66 m	0.16 m	Mudstone, greenish gray, ostracoda fossil fragments
58-16	1446.66—1448.17 m	1.51 m	Mudstone, medium dark gray, horizontal bedding, few ostracoda clast limestone lamellae
58-17	1448.17—1448.40 m	0.23 m	Mudstone, olive gray, horizontal bedding, some ostracoda fossil fragments
58-18	1448.40—1448.55 m	0.15 m	Mudstone, medium dark gray, horizontal bedding, some ostracoda clast limestone laminae
58-19	1448.55—1448.75 m	0.20 m	Mudstone, greenish gray, massive bedding, ostracoda fossil fragments, some irregular fractures in the upper section, filled with dark gray mudstone
58-20	1448.75—1449.07 m	0.32 m	Mudstone, dark gray, intercalated with ostracoda clast limestone, horizontal bedding, ripple lamination
58-21	1449.07—1449.27 m	0.20 m	Mudstone, greenish gray, horizontal bedding
58-22	1449.27—1449.39 m	0.12 m	Mudstone, calcareous, light olive gray
58-23	1449.39—1449.52 m	0.13 m	Mudstone, medium dark gray, horizontal bedding, few limestone lumps with ostracoda, calcite veins and ostracoda-bearing mudstone in the lower section
58-24	1449.52—1449.68 m	0.16 m	Mudstone, calcareous, light olive gray
58-25	1449.68—1450.91 m	1.23 m	Mudstone, medium dark gray, intercalated with ostracoda clast limestone, horizontal bedding, ripple lamination
58-26	1450.91—1451.27 m	0.36 m	Mudstone, greenish gray, horizontal bedding, ostracoda fossil fragments
58-27	1451.27—1451.68 m	0.41 m	Mudstone, Medium dark gray, horizontal bedding, conchonstracan fossils, ostracoda clast limestone laminae
58-28	1451.68—1452.14 m	0.46 m	Mudstone, dark gray, horizontal bedding, conchonstracan fossils, ostracoda clast limestone laminae ripple laminated
58-29	1452.14—1452.67 m	0.53 m	Mudstone, olive gray, horizontal bedding, local ostracoda clast limestone lamellae
58-30	1452.67—1452.99 m	0.32 m	Mudstone, dark greenish gray, locally ostracoda fossil fragments
58-31	1452.99—1453.29 m	0.30 m	Mudstone, greenish gray, massive, locally ostracoda fossil fragments
58-32	1453.29—1453.69 m	0.40 m	Mudstone, dark greenish gray, massive locally ostracoda fossil fragments
59-1	1453.69—1455.57 m	1.88 m	Mudstone, dark greenish gray, massive bedding, some ostracoda fossils, few bivalve fossil

59-2	1455.57—1458.07 m	2.50 m	Mudstone, dark gray, horizontal bedding, ostracoda fossils, few bivalve fossil, locally medium gray, ostracoda clast, limestone laminae and some escape trace ichnofossils (penetrating the fossil fragment beds)
59-3	1458.07—1459.62 m	1.55 m	Mudstone, medium dark gray, horizontal bedding, ostracoda fossils, few bivalve fossil fragments, with greenish gray mudstone bed on the top, locally laminated medium. gray ostracoda clast limestone
59-4	1459.62—1460.24 m	0.62 m	Mudstone, greenish gray, locally enriched ostracoda fossils, with local light greenish gray mudstone and ostracoda clast limestone laminae and lumps
59-5	1460.24—1461.15 m	0.91 m	Mudstone, medium dark gray, horizontal bedding, ostracoda clast limestone lamellae, scouring surface at the bottom
59-6	1461.15—1462.78 m	1.63 m	Mudstone, olive gray, indistinct horizontal bedding, locally dark gray mudstone laminae and lumps; with ostracoda clast limestone lamellae in the lower part of section
59-7	1462.78—1463.46 m	0.68 m	Mudstone, dark gray, horizontal bedding, ostracoda clast limestone lamellae in local site and some ostracoda fossils
59-8	1463.46—1464.49 m	1.03 m	Mudstone, medium dark gray, indistinct horizontal bedding, ostracoda fossils
60-1	1464.48—1464.66 m	0.18 m	Mudstone, medium dark gray, ostracoda clast limestone laminae
60-2	1464.66—1467.28 m	2.62 m	Mudstone, medium dark gray, thinly interbeded with dark gray ostracoda clast limestone, horizontal bedding, ripple lamination, abundant dark gray, light olive gray ostracoda clast limestone lamellae and with few conchonstracan fossil
60-3	1467.28—1467.33 m	0.05 m	Mudstone, calcareous, light olive gray, few ostracoda clast limestone lumps
60-4	1467.33—1469.08 m	1.75 m	Mudstone, medium dark gray, thinly interbeded with dark gray ostracoda clast limestone, horizontal bedding, ripple lamination, more dark gray and light olive gray ostracoda clast limestone lamellae; few conchonatracan fossil
60-5	1469.08—1470.38 m	1.30 m	Mudstone, medium dark gray, intercalated with dark gray ostracoda clast limestone, horizontal bedding, ripple lamination
60-6	1470.38—1471.24 m	0.86 m	Mudstone, medium dark gray, horizontal bedding, few dark gray ostracoda clast limestone lamella
61-1	1471.24—1471.47 m	0.23 m	Mudstone, dark gray, horizontal bedding, few calcareous fossil fragment
61-2	1471.47—1472.89 m	1.42 m	Mudstone, medium dark gray, horizontal bedding, few ostracoda fossil
61-3	1472.89—1475.79 m	2.90 m	Mudstone, dark gray, horizontal bedding, ostracoda and conchonstracan fossils and several laminae of ostracoda clastic limestone
61-4	1475.79—1477.54 m	1.75 m	Mudstone, medium dark gray, horizontal bedding local ostracoda fossils greenish gray mudstone laminae and lumps, on the top of the section, some ostracoda clast limestone lamellae
61-5	1477.54—1477.74 m	0.20 m	Mudstone, dark greenish gray, horizontal bedding, few ostracoda fossil
61-6	1477.74—1478.31 m	0.57 m	Mudstone, medium dark gray, horizontal bedding, few ostracoda fossil and fossil fragments; few light gray muddy siltstone and greenish gray mudstone laminae and lumps

61-7	1478.31—1478.44 m	0.13 m	Mudstone, light olive gray, ostracoda fossil fragments at the bottom, some medium dark gray mudstone lumps with calcite fillings in contraction fissures
61-8	1478.44—1478.56 m	0.12 m	Mudstone, medium dark gray, horizontal bedding, few ostracoda fossil and fossil fragment; locally light gray mud siltstone and greenish gray mudstone laminae
61-9	1478.56—1478.78 m	0.22 m	Mudstone, light olive gray, massive, some ostracoda clast limestone lamellae
61-10	1478.78—1479.58 m	0.80 m	Mudstone, medium dark gray, horizontal bedding, locally light gray muddy siltstone lamellae
61-11	1479.58—1479.98 m	0.40 m	Mudstone, medium dark gray, intercalated with medium. light gray siltstone, horizontal bedding, ripple lamination, deformed and, convolute bedding; medium, light gray siltstone lamellae
61-12	1479.98—1482.98 m	3.00 m	Mudstone, medium dark gray, horizontal bedding, locally with mud siltstone laminae, few ostracoda fossil
61-13	1482.98—1483.55 m	0.57 m	Mudstone, dark greenish gray, horizontal bedding, few ostracoda fossil and other fossil fragments
62-1	1483.55—1483.78 m	0.23 m	Mudstone, medium gray, horizontal bedding, ripple lamination, ostracoda fossils
62-2	1483.78—1484.55 m	0.77 m	Mudstone, medium dark gray, horizontal bedding, calcareous siltstone laminae; some irregular fractures in the upper section, filled with calcite; locally calcareous siltstone lumps
62-3	1484.55—1485.74 m	1.19 m	Mudstone, olive gray, horizontal bedding, ostracoda fossil fragments
62-4	1485.74—1487.77 m	2.03 m	Mudstone, dark gray, horizontal bedding, ostracoda fossil fragments and ostracoda clast limestone laminae
62-5	1487.77—1487.87 m	0.10 m	Mudstone, dark gray, ripple lamination, few complete ostracoda fossils
62-6	1487.87—1489.74 m	1.87 m	Mudstone, dark gray, horizontal bedding, ostracoda clast limestone laminae, conchonstracan fossils
62-7	1489.74—1490.58 m	0.84 m	Mudstone, dark gray, interbeded with medium light gray ostracoda clast limestone, horizontal bedding, ripple lamination
62-8	1490.58—1490.97 m	0.39 m	Mudstone, medium dark gray, horizontal bedding, ripple lamination, ostracoda bearing limestone laminae
62-9	1490.97—1491.58 m	0.61 m	Mudstone, dark gray, intercalated with ostracoda clast limestone, horizontal bedding, ripple lamination
62-10	1491.58—1491.99 m	0.41 m	Mudstone, dark gray, ostracoda-bearing, horizontal bedding, few ostracoda clast limestone laminae
62-11	1491.99—1492.48 m	0.49 m	Mudstone, dark gray, horizontal bedding, few ostracoda clast limestone laminae
62-12	1492.48—1494.30 m	1.82 m	Mudstone, olive gray, horizontal bedding, ostracoda clast limestone laminae, ripple lamination
62-13	1494.30—1495.14 m	0.84 m	Mudstone, medium dark gray, horizontal bedding, ostracoda clast limestone laminae, ripple lamination

62-14	1495.14—1495.51 m	0.37 m	Mudstone, medium gray horizontal bedding, ostracoda clast limestone laminae, ripple lamination
62-15	1495.51—1496.09 m	0.58 m	Mudstone, dark gray, horizontal bedding, ostracoda clast limestone laminae, ripple lamination
63-1	1496.09—1498.51 m	2.42 m	Mudstone, medium dark grey, horizontal bedding, few ostracoda clast laminae in the middle and lower part of the section, fossil fragments on bedding planes, more conchostracan fossils at the bottom
63-2	1498.51—1499.08 m	0.57 m	Mudstone, medium gray, intercalated with light gray ostracoda clast limestone, ripple lamination, locally bedding is deformed, few calcareous siltstone laminae and lumps
63-3	1499.08—1499.59 m	0.51 m	Mudstone, medium gray, horizontal bedding, fossil fragments, few ostracoda clast limestone laminae
63-4	1499.59—1499.64 m	0.05 m	Siltstone, calcareous, medium light gray, horizontal ripple lamination, load structure, medium dark gray mudstone laminae
63-5	1499.64—1503.24 m	3.60 m	Mudstone, medium dark gray, horizontal bedding, ostracoda clast limestone laminae, conchostracan fossils and fossil fragments
63-6	1503.24—1503.57 m	0.33 m	Mudstone, olive gray, horizontal bedding, fossil fragments
63-7	1503.57—1505.12 m	1.55 m	Mudstone, medium dark gray, horizontal bedding, more fossil fragments
63-8	1505.12—1506.89 m	1.77 m	Mudstone, olive gray, horizontal bedding, conchostracan fossils and fossil fragments
63-9	1506.89—1507.16 m	0.27 m	Mudstone, light olive gray, intercalated with olive gray mudstone, massive bedding, few conchonstracan fossil
63-10	1507.16—1508.09 m	0.93 m	Mudstone, olive gray, horizontal bedding, ostracoda clast limestone laminae, fossil fragments
64-1	1508.09—1512.99 m	4.90 m	Mudstone, dark gray, horizontal bedding, few ostracoda fossil, ostracoda clast limestone and ostracoda-bearing siltstone stripes in local site; dark greenish gray mudstone and ostracoda clast limestone in the middle; ostracoda clast limestone and light gray calcareous siltstone stripes and lamellae in the lower section
64-2	1512.99—1513.45 m	0.46 m	Mudstone, medium dark gray, horizontal bedding, few ostracoda fossil, locally ostracoda clast limestone and ostracoda-bearing siltstone stripes.
64-3	1513.45—1514.99 m	1.54 m	Mudstone, dark gray, horizontal bedding, few ostracoda fossil
64-4	1514.99—1516.39 m	1.40 m	Mudstone, medium dark gray, horizontal bedding, few ostracoda fossil, locally ostracoda-bearing siltstone and crystalline limestone laminae in local site
65-1	1516.39—1516.96 m	0.57 m	Mudstone, medium dark gray, horizontal bedding, more ostracoda fossil fragments
65-2	1516.96—1517.69 m	0.73 m	Mudstone, dark gray, horizontal bedding, ostracoda fossil fragments
65-3	1517.69—1517.96 m	0.27 m	Mudstone, medium light gray, ostracoda clast limestone intercalated with laminated medium dark gray mudstone, olistostromic? structure, compressional deformation bedding, ostracoda clast limestone laminae

65-4	1517.96—1520.76 m	2.80 m	Mudstone, medium dark gray, horizontal bedding, more ostracoda clasts and fossil fragments
65-5	1520.76—1523.23 m	2.47 m	Mudstone, medium dark gray, ostracoda-bearing, horizontal bedding, ostracoda fossils and fragments
66-1	1523.23—1524.13 m	0.90 m	Mudstone, medium dark gray, ostracoda-bearing, horizontal bedding, more ostracoda fossils in locally enriched beds, light gray few calcareous siltstone lamellae
66-2	1524.13—1525.73 m	1.60 m	Mudstone, medium dark gray, horizontal bedding, more ostracoda fossils enriched locally light gray siltstone intercalations
66-3	1525.73—1525.83 m	0.10 m	Mudstone, medium dark gray, thinly interbeded with light gray calcareous siltstone, horizontal ripple lamination, few ostracoda fossil
66-4	1525.83—1527.68 m	1.85 m	Mudstone, medium dark gray, ostracoda-bearing, horizontal bedding, more ostracoda fossils locally
66-5	1527.68—1528.78 m	1.10 m	Mudstone, dark gray, horizontal bedding, locally enriched ostracoda fossil in beds, light gray siltstone laminae
66-6	1528.78—1529.88 m	1.10 m	Mudstone, medium dark gray, massive,, ostracoda fossils enriched locally
66-7	1529.88—1530.73 m	0.85 m	Mudstone, dark gray, ostracoda-bearing, horizontal bedding, more ostracoda fossils, locally light gray calcareous siltstone laminae
66-8	1530.73—1531.78 m	1.05 m	Mudstone, medium dark gray, horizontal bedding, ostracoda fossils enriched locally
66-9	1531.78—1532.63 m	0.85 m	Mudstone, dark gray, ostracoda-bearing, horizontal bedding, ostracoda fossils and calcareous siltstone laminae
66-10	1532.63—1535.09 m	2.46 m	Mudstone, medium dark gray, horizontal bedding, ostracoda fossils enriched locally
66-11	1535.09—1535.19 m	0.10 m	Mudstone, medium dark gray, interbeded with light gray ostracoda clast limestone, horizontal ripple lamination
66-12	1535.19—1535.55 m	0.36 m	Mudstone, dark gray, horizontal bedding, ostracoda clast limestone lamellae
67-1	1535.55—1535.81 m	0.26 m	Mudstone, medium dark gray, horizontal bedding, ostracoda fossils
67-2	1535.81—1536.94 m	1.13 m	Mudstone, medium dark gray, ostracoda-bearing, horizontal bedding, few ostracoda clast limestone laminae and conchonstracan fossils in the lower part
67-3	1536.94—1537.44 m	0.50 m	Mudstone, dark gray, ostracoda-bearing, horizontal bedding, few conchonstracan fossil and ostracoda clast limestone laminae
67-4	1537.44—1540.65 m	3.21 m	Mudstone, medium dark gray, horizontal bedding, few conchonstracan fossil and ostracoda clast limestone laminae
67-5	1540.65—1542.50 m	1.85 m	Mudstone, dark gray, horizontal bedding, few ostracoda clast limestone laminae
67-6	1542.50—1547.78 m	5.28 m	Mudstone, medium dark gray, horizontal bedding, ripple lamination, ostracoda clast limestone laminae, ostracoda, and conchonstracan fossils enriched in local beds

68-1	1547.78—1548.68 m	0.90 m	Mudstone, medium dark gray, horizontal bedding, ostracoda fossils, and ostracoda clast limestone laminae locally
68-2	1548.68—1549.43 m	0.75 m	Mudstone, dark gray, horizontal bedding, ostracoda and conchonstracan fossils; ostracoda clast limestone lamellae
68-3	1549.43—1550.51 m	1.08 m	Mudstone, medium dark gray, horizontal bedding, more ostracoda fossils, ostracoda clast limestone laminae locally
68-4	1550.51—1552.28 m	1.77 m	Mudstone, dark gray, horizontal bedding, more ostracoda and conchonstracan fossil fragments; locally ostracoda clast limestone lamellae, conchonstracan fossils enriched
68-5	1552.28—1552.48 m	0.20 m	Mudstone, medium dark gray, intercalated with light gray calcareous siltstone laminae, horizontal ripple lamination, conchonstracan fossils in the mudstones, more light gray calcareous siltstone laminae, with a calcareous siltstone lamella at the bottom
68-6	1552.48—1557.63 m	5.15 m	Mudstone, dark gray, horizontal bedding, complete conchonstracan fossils and fragments, locally enriched in beds
68-7	1557.63—1558.43 m	0.80 m	Mudstone, dark gray, ostracoda-bearing mudstone, horizontal bedding, more calcareous siltstone lamellae set in mudstones
68-8	1558.43—1560.38 m	1.95 m	Mudstone, dark gray, horizontal bedding, more ostracoda fossils, light gray calcareous siltstone, and light olive gray ostracoda clast limestone stripes and lamellae
69-1	1560.38—1560.69 m	0.31 m	Mudstone, dark gray, horizontal bedding, ostracoda fossils, and ostracoda clast limestone stripes and lumps
69-2	1560.69—1561.26 m	0.57 m	Mudstone, dark gray, ostracoda-bearing horizontal bedding, ostracoda fossils
69-3	1561.26—1563.10 m	1.84 m	Mudstone, dark gray, horizontal bedding, ostracoda and conchonstracan fossils; ostracoda clast limestone laminae
69-4	1563.10—1563.46 m	0.36 m	Mudstone, medium dark gray with medium light gray ostracoda clast limestone, horizontal bedding, ripple lamination, ostracoda clast limestone laminae, fossil fragments
69-5	1563.46—1564.91 m	1.45 m	Mudstone, dark gray, horizontal bedding, ostracoda clast limestone laminae, ostracoda and conchonstracan fossil fragments
69-6	1564.91—1567.59 m	2.68 m	Mudstone, medium dark gray, horizontal bedding, few ostracoda clast limestone laminae and lamellae
69-7	1567.59—1568.18 m	0.59 m	Mudstone, calcareous, light olive gray, massive bedding
69-8	1568.18—1569.34 m	1.16 m	Mudstone, medium dark gray, horizontal bedding, ostracoda clast limestone laminae ripple lamination and ostracoda fragments on top of the section, few conchonstracan fossils, and other fossil fragments
70-1	1569.34—1570.30 m	0.96 m	Mudstone, medium dark gray, horizontal bedding, more ostracoda fossil, and other fossil fragments with plant relics and pyrite grains
70-2	1570.30—1571.78 m	1.48 m	Mudstone, dark gray, horizontal bedding, ostracoda clast limestone lamellae unevenly distributed

70-3	1571.78—1574.34 m	2.56 m	Mudstone, medium dark gray, horizontal bedding, more ostracoda clast limestone and calcareous siltstone lamellae
70-4	1574.34—1576.07 m	1.73 m	Mudstone, medium dark gray, horizontal bedding, ostracoda fossils and plant fossil fragments
71-1	1576.07—1580.37 m	4.30 m	Mudstone, medium dark gray, horizontal bedding, ostracoda unevenly distributed and locally enriched, conchonstracan fossils, few calcareous siltstone laminae and lamellae
71-2	1580.37—1581.12 m	0.75 m	Mudstone, medium dark gray, ostracoda-bearing horizontal bedding, ostracoda fossils, plant fossils in the middle of the section
71-3	1581.12—1581.77 m	0.65 m	Mudstone, dark gray, horizontal bedding, locally enriched in ostracoda fossils, light gray calcareous siltstone laminae and lumps with horizontal ripple lamination
71-4	1581.77—1582.97 m	1.20 m	Mudstone, medium dark gray, horizontal bedding, ostracoda and conchonstracan fossils enriched locally
71-5	1582.97—1585.97 m	3.00 m	Mudstone, dark gray, horizontal bedding, ostracoda and conchonstracan complete fossils and fragments, few light olive gray ostracoda clast limestone lamellae
71-6	1585.97—1586.52 m	0.55 m	Mudstone, medium dark gray, horizontal bedding, ostracoda fossils locally, few plant fossil fragments
71-7	1586.52—1588.23 m	1.71 m	Mudstone, dark gray, horizontal bedding, ostracoda and conchonstracan fossils, some light olive gray ostracoda clast limestone lamellae
72-1	1588.23—1589.10 m	0.87 m	Mudstone, dark gray, horizontal bedding, ostracoda and conchonstracan fossils locally
72-2	1589.10—1589.26 m	0.16 m	Siltstone, calcareous, medium light gray, with medium. dark gray mudstone laminae, horizontal ripple lamination, compression deformation locally convolute bedding; climbing-ripple lamination on top of the upper section; ostracoda fossil fragments, sand pillows and balls
72-3	1589.26—1590.09 m	0.83 m	Mudstone, dark gray, horizontal bedding, ostracoda fossils; a vertical fracture cross-cutting the core, fracture healed by calcite, with crystal surface covered by oil film; few conchonstracan fossil
72-4	1590.09—1590.93 m	0.84 m	Mudstone, medium dark gray, horizontal bedding, ostracoda fossils; vertical fracture cross-cutting the core, fracture healed by calcite, covered by oil film; few conchonstracan fossil
72-5	1590.93—1591.73 m	0.80 m	Mudstone, medium dark gray, intercalated with medium light gray laminated calcareous siltstone, horizontal bedding, ripple lamination, locally convolute bedding, calcareous siltstone laminae, sand pillow, sand balls
72-6	1591.73—1593.26 m	1.53 m	Mudstone, dark gray, horizontal bedding, ostracoda fossils, few calcareous siltstone laminae
72-7	1593.26—1593.70 m	0.44 m	Mudstone, medium dark gray, horizontal bedding, ostracoda fossils
72-8	1593.70—1595.26 m	1.56 m	Dark gray mudstone, horizontal bedding, few ostracoda clast limestone stripes, conchonstracan fossils enriched on the bedding plane and more fossil fragments

72-9	1595.26—1598.23 m	2.97 m	Mudstone, medium dark gray, horizontal bedding, few ostracoda clast limestone laminae, conchonstracan fossils enriched on the bedding plane and more fossil fragments
73-1	1598.23—1598.52 m	0.29 m	Mudstone, medium dark gray, horizontal bedding, fossil fragments
73-2	1598.52—1599.25 m	0.73 m	Mudstone, dark gray, horizontal bedding, light gray calcareous siltstone laminae at the top of the upper section
73-3	1599.25—1600.85 m	1.60 m	Mudstone, olive gray, horizontal bedding, fossil fragments enriched on bedding planes
73-4	1600.85—1600.98 m	0.13 m	Mudstone, dark gray, disturbed bedding, ripple lamination, calcareous siltstone lumps, and laminae
73-5	1600.98—1602.23 m	1.25 m	Mudstone, olive gray, horizontal bedding, fossil fragments on bedding planes
73-6	1602.23—1603.33 m	1.10 m	Mudstone, medium dark gray, fossil fragments, conchonstracan and ostracoda fossils on bedding planes, calcareous siltstone laminae and lumps; above scouring surface, calcareous siltstone lamellae with fossil fragments
73-7	1603.33—1607.34 m	4.01 m	Mudstone, olive gray, horizontal bedding, more fossil fragments and conchonstracan fossils on bedding planes, few calcareous siltstone laminae and lumps
73-8	1607.34—1607.92 m	0.58 m	Mudstone, dark gray, horizontal bedding, fossil fragments, calcareous siltstone laminae at the basal part of the section
73-9	1607.92—1610.26 m	2.34 m	Mudstone, medium dark gray, horizontal bedding, fossil fragments, ostracoda fossils, few light gray calcareous siltstone laminae conchonstracan fossils
73-10	1610.26—1610.31 m	0.05 m	Siltstone, calcareous, light gray, thinly interbedded with medium dark gray mudstone, ripple lamination on top of the upper section, disturbed bedding in the middle and lower section, locally light gray calcareous siltstone lumps and laminae
73-11	1610.31—1610.49 m	0.18 m	Mudstone, dark gray, horizontal bedding, few ostracoda fossil
74-1	1610.49—1612.70 m	2.21 m	Mudstone, medium dark gray, horizontal bedding, ostracoda and conchonstracan fossils enriched locally
74-2	1612.70—1613.11 m	0.41 m	Siltstone, medium gray, massive bedding, locally horizontal ripple lamination, more ostracoda and conchonstracan fossils, ostracoda clast limestone laminae
74-3	1613.11—1614.99 m	1.88 m	Mudstone, dark gray, horizontal bedding, ostracoda and conchonstracan fossils, ostracoda clast limestone laminae locally
74-4	1614.99—1615.76 m	0.77 m	Mudstone, medium dark gray, horizontal bedding, few ostracoda fossil and other fossil fragments
74-5	1615.79—1616.09 m	0.30 m	Mudstone, medium dark gray with medium gray ostracoda clast limestone, ripple lamination, ostracoda clast limestone lamellae
74-6	1616.09—1616.64 m	0.55 m	Mudstone, medium dark gray, horizontal bedding, few ostracoda fossil and plant fossil fragment
74-7	1616.64—1616.83 m	0.19 m	Mudstone, dark gray, horizontal bedding, plant fossil fragments, locally enriched

74-8	1616.83—1616.85 m	0.02 m	Oil shale, olive gray, laminated
74-9	1616.85—1617.09 m	0.24 m	Mudstone, dark gray, horizontal bedding, plant fossil fragments
74-10	1617.09—1618.61 m	1.52 m	Mudstone, medium dark gray, horizontal bedding; complete ostracoda, conchonstracan fossils shells and plant fossil fragments
74-11	1618.61—1619.11 m	0.50 m	Mudstone, medium dark gray, intercalated with medium dark gray ostracoda clast limestone and medium light gray calcareous siltstone, horizontal ripple lamination, more ostracoda clast limestone and calcareous ostracoda-bearing siltstone laminae
74-12	1619.11—1620.97 m	1.86 m	Mudstone, dark gray, horizontal bedding, few ostracoda fossils
75-1	1620.97—1621.11 m	0.14 m	Mudstone, dark gray, horizontal bedding
75-2	1621.11—1621.47 m	0.36 m	Mudstone, medium dark gray, horizontal bedding, abundant conchonstracan fossils and calcite-bearing siltstone laminae
75-3	1621.47—1621.77 m	0.30 m	Dolomite, light olive gray, massive bedding, ostracoda and conchonstracan fossils on some bedding planes
75-4	1621.77—1623.22 m	1.45 m	Mudstone, dark gray, horizontal bedding, locally horizontal ripple lamination, few ostracoda and conchonstracan fossil ; light gray calcareous siltstone laminae and a nearly vertical fracture
75-5	1623.22—1624.17 m	0.95 m	Mudstone, olive gray, horizontal bedding, fossil fragments with few conchonstracan fossils, ostracoda fossils
75-6	1624.17—1625.21 m	1.04 m	Mudstone, medium dark gray intercalated with medium light gray muddy laminated siltstone, horizontal bedding, ripple lamination, ostracoda and conchonstracan fossils enriched on some of the bedding planes
75-7	1625.21—1627.69 m	2.48 m	Mudstone, dark gray, horizontal bedding with dolomite lamellae, few ostracoda and conchonstracan fossils
76-1	1627.69—1628.24 m	0.55 m	Mudstone, dark gray mudstone, horizontal bedding, dolomite lamellae, with a few ostracoda and conchonstracan fossils
76-2	1628.24—1629.59 m	1.35 m	Mudstone, dark gray, interbedded with light gray ostracoda clast limestone, ripple lamination, ostracoda clast limestone laminae, oil-bearing locally; calcite and pyrite in ostracoda clast limestone lamellae
76-3	1629.59—1633.44 m	3.85 m	Mudstone, medium dark gray, horizontal bedding, light gray and olive gray dolomite lamellae in middle part of the section a complete animal fossil, pyrite lumps locally
76-4	1633.44—1634.79 m	1.35 m	Mudstone, dark gray, horizontal bedding, few ostracoda fossil, dolomite lamellae
76-5	1634.79—1636.64 m	1.85 m	Medium dark gray mudstone, horizontal bedding, few ostracoda and conchostracan fossils locally enriched, few ostracoda clast limestone stripes
76-6	1636.64—1636.72 m	0.08 m	Dolomite, light olive gray, massive bedding
76-7	1636.72—1639.11 m	2.39 m	Mudstone, medium dark gray, horizontal bedding, ostracoda clast limestone lamellae, few light gray calcareous siltstone lamella

77-1	1639.11—1639.74 m	0.63 m	Mudstone, medium gray, horizontal bedding, more ostracoda clast limestone and calcareous siltstone laminae, fossil fragments
77-2	1639.74—1639.99 m	0.25 m	Mudstone, medium dark gray, horizontal bedding, ostracoda and other fossils fragments
77-3	1639.99—1640.66 m	0.67 m	Mudstone, calcareous medium gray, massive bedding, more fractures, filled with calcite
77-4	1640.66—1641.06 m	0.40 m	Mudstone, medium dark gray, horizontal bedding, few fossil fragment
77-5	1641.06—1642.41 m	1.35 m	Mudstone, dark gray, horizontal bedding, few ostracoda fossil and carbonized fossil fragments
77-6	1642.41—1643.16 m	0.75 m	Mudstone, medium gray, horizontal bedding, ostracoda and other fossils fragments; medium gray and olive gray bioclast limestone lumps
77-7	1643.16—1643.81 m	0.65 m	Mudstone, medium dark gray, horizontal bedding, ostracoda clast limestone laminae in the lower section, and more ostracoda fossils
77-8	1643.81—1645.71 m	1.90 m	Mudstone, dark gray, horizontal bedding, few ostracoda fossil and olive gray dolomite lamella
77-9	1645.71—1646.01 m	0.30 m	Mudstone, medium dark gray, horizontal bedding, more ostracoda fossils and fossil fragments
78-1	1646.01—1646.27 m	0.26 m	Medium dark gray, horizontal bedding, more ostracoda fossils and calcareous siltstone lamellae
78-2	1646.27—1647.01 m	0.74 m	Mudstone, dark gray, horizontal bedding, ostracoda fossils
79-1	1647.10—1647.71 m	0.61 m	Mudstone, dark gray, horizontal bedding, calcareous siltstone lamellae, ichnofossils and fossil fragments
79-2	1647.71—1652.77 m	5.06 m	Mudstone, dark gray, horizontal bedding, ostracoda and conchostracan fossils; fossil fragments enriched on bedding plane; a vertical fracture, filled with calcite
79-3	1652.77—1653.86 m	1.09 m	Mudstone, medium dark gray, horizontal bedding, pyrite at the top, dolomite lamellae
79-4	1653.86—1653.93 m	0.07 m	Mudstone, medium dark gray, ostracoda ripple lamination, deformation bedding
79-5	1653.93—1654.22 m	0.29 m	Mudstone, dark gray, thinly interbeded with dark gray ostracoda-bearing mudstone, horizontal bedding, ripple lamination
79-6	1654.22—1655.01 m	0.79 m	Mudstone, dark gray, horizontal bedding, trace fossils
79-7	1655.01—1655.62 m	0.61 m	Mudstone, dark gray, ostracoda-bearing, horizontal bedding, ostracoda fossils, few trace fossil; with ostracoda clast limestone laminae near the base
79-8	1655.62—1657.50 m	1.88 m	Mudstone, dark gray, horizontal bedding, few ostracoda clast limestone lamella, with ostracoda fossils
80-1	1657.50—1657.58 m	0.08 m	Mudstone, siliceous, grayish black, massive
80-2	1657.58—1659.63 m	2.05 m	Mudstone, dark gray, horizontal bedding, conchostracan fossils, fossils fragments enriched on some of bedding planes, few olive gray dolomite lamellae

80-3	1659.63—1659.77 m	0.14 m	Mudstone, dark gray, thinly interbedded with light gray calcareous siltstone, ripple lamination, deformation bedding and convolute bedding, sand pillows and ball structures, irregular calcareous siltstone laminae and lumps
80-4	1659.77—1659.98 m	0.21 m	Mudstone, dark gray, horizontal bedding, light gray calcareous siltstone laminae horizontal ripple lamination
80-5	1659.98—1660.41 m	0.43 m	Mudstone, dark gray, interbeded with light gray calcareous siltstone, interrupted horizontal ripple lamination, convolute bedding and deformation bedding with sand dykes penetrating into calcareous siltstone laminae
80-6	1660.41—1662.75 m	2.34 m	Mudstone, dark gray, horizontal bedding, few fossil fragments
80-7	1662.75—1662.97 m	0.22 m	Mudstone, dark gray, interbeded with light gray siltstone, horizontal bedding, convolute bedding, interrupted ripple lamination and deformation bedding, fossil fragments in the mudstone
80-8	1662.97—1665.93 m	2.96 m	Mudstone, dark gray, horizontal bedding, few fossil fragment
80-9	1665.93—1666.09 m	0.16 m	Mudstone, dark gray, interbeded with light gray siltstone, horizontal ripple lamination and convolute bedding, deformation structure and "olistostromic"? structure calcareous siltstone lamiane
80-10	1666.09—1667.50 m	1.41 m	Mudstone, dark gray, horizontal bedding, calcareous siltstone laminae and fossils
80-11	1667.50—1667.66 m	0.16 m	Dolomite, olive black
80-12	1667.66—1668.03 m	0.37 m	Mudstone, medium dark gray with light gray calcareous siltstone laminae, horizontal ripple lamination, sand pillow and ball structures, calcareous siltstone laminae and lumps
80-13	1668.03—1669.65 m	1.62 m	Mudstone, dark gray, horizontal bedding, few olive gray dolomite lamella
80-14	1669.65—1669.75 m	0.10 m	Dolomite, olive black
80-15	1669.75—1671.87 m	2.12 m	Mudstone, dark gray, horizontal bedding, light gray calcareous siltstone lamellae and lumps horizontal ripple lamination, ostracoda fossils and fossil fragments
81-1	1671.87—1672.10 m	0.23 m	Mudstone, dark gray, horizontal bedding, pyrite laminae
81-2	1672.10—1676.87 m	4.77 m	Mudstone, medium dark gray, horizontal bedding, few ostracoda fossil and fossil fragment; locally light gray calcareous siltstone laminae, lumps and lamellae; greenish grey tuff (?) bed, 0.5 cm thick, in the upper section
81-3	1676.87—1678.73 m	1.86 m	Mudstone, dark gray, horizontal bedding, ostracoda fossils few olive gray dolomite and light gray siltstone lamella
82-1	1678.73—1679.67 m	0.94 m	Mudstone, medium dark gray, horizontal bedding, ostracoda fossil, few carbonized plant fossil fragment, locally light gray calcareous siltstone laminae and lamellae
82-2	1679.67—1681.10 m	1.43 m	Mudstone, dark gray, horizontal bedding, ostracoda fossils locally enriched, few carbonized plant fossil fragment, calcareous siltstone lamella and lumps
82-3	1681.10—1681.67 m	0.57 m	Mudstone, medium dark gray, interbedded with light gray calcareous siltstone and olive gray ostracoda clast limestone, horizontal bedding, ripple lamination many ostracoda and calcareous siltstone lamellae and lumps

82-4	1681.67—1686.07 m	4.40 m	Mudstone, medium dark gray, horizontal bedding, calcareous siltstone laminae and lamellae, beds enriched in ostracoda fossils, few pyrite laminae
82-5	1686.07—1686.29 m	0.22 m	Mudstone, medium dark gray with ostracoda clast limestone, horizontal bedding, ripple lamination, light gray calcareous siltstone laminae in the upper section, and ostracoda clast limestone lamella in the lower section
82-6	1686.29—1689.17 m	2.88 m	Mudstone, medium dark gray, horizontal bedding, few ostracoda fossil and carbonized plant fossil fragments, thin-bedded olive gray dolomite
82-7	1689.17—1689.23 m	0.06 m	Dolomite, olive black, laminae
82-8	1689.23—1690.33 m		
82-9	1690.33—1690.41 m	0.08 m	Dolomite olive black, convolute deformation bedding near the base
82-10	1690.41—1690.77 m	0.36 m	Mudstone, medium dark gray, horizontal bedding, few ostracoda fossil and calcareous siltstone laminae
83-1	1690.77—1691.22 m	0.45 m	Mudstone, dark gray, horizontal bedding, few fossil fragments, light gray calcareous siltstone laminae which locally penetrate into underlying mudstone
83-2	1691.22—1692.05 m	0.83 m	Mudstone, medium dark grey with light gray calcareous siltstone, lens and fossil fragments
83-3	1692.05—1692.19 m	0.14 m	Dolomite, olive black
83-4	1692.19—1693.54 m	1.35 m	Mudstone, dark gray, horizontal bedding, locally horizontal ripple lamination in the light gray calcareous siltstone laminae; carbonized plant and conchostracan fossils on some bedding planes
83-5	1693.54—1693.90 m	0.36 m	Mudstone, dark gray, interbeded with laminated light gray calcareous siltstone; interrupted ripple lamination, deformation bedding and convolute bedding, with sand pillows and balls; some fossil fragments
83-6	1693.90—1697.00 m	3.10 m	Mudstone, dark gray, horizontal bedding, few trace fossils, thin-bedded calcareous siltstone in the middle
83-7	1697.00—1697.07 m	0.07 m	Dolomite, olive black
83-8	1697.07—1698.98 m	1.91 m	Mudstone, dark gray, horizontal bedding, some fossil fragments, two crystalline limestone interlayers in the middle, few calcareous siltstone laminae
83-9	1698.98—1699.06 m	0.08 m	Dolomite, olive black
83-10	1699.06—1699.86 m	0.80 m	Mudstone, dark gray, horizontal bedding, few fossil fragment; thin-bedded crystalline limestone in the lower section, few light gray calcareous siltstone laminae
83-11	1699.86—1700.01 m	0.15 m	Dolomite, olive black, massive bedding
83-12	1700.01—1700.29 m	0.28 m	Mudstone, dark gray, horizontal bedding, few fossil fragment
83-13	1700.29—1700.80 m	0.51 m	Mudstone, dark gray, thinly interbeded with light gray calcareous siltstone, horizontal bedding, calcareous siltstone laminae
83-14	1700.80—1701.52 m	0.72 m	Dark gray mudstone, horizontal bedding, few calcareous siltstone stripe

5.1.4 Yaojia Formation

The overlying formation of the Yaojia Formation Members 2 and 3 is the Nenjiang Formation Member 1, the contacting relationship between them is conformable contact; the underlying formation is the Yaojia Formation Member 1, the contacting relationship between them is parallel unconfor mable contact(Figure 5.4).

The detailed description of the Yaojia Formation Members 2 and 3:

27-2	1128.17—1129.19 m	1.02 m	Mudstone, dark greenish gray with ostracoda clast interbeds
27-3	1129.19—1129.69 m	0.50 m	Mudstone, dark greenish gray with wave bedded light greenish gray ostracoda clast interbeds
27-4	1129.69—1130.98 m	1.29 m	Mudstone, dark greenish gray with a few ostracoda fossils, horizontal bedding
27-5	1130.98—1131.62 m	0.64 m	Mudstone, silty with few ostracoda fossils
27-6	1131.62—1131.74 m	0.12 m	Mudstone, greenish gray interbeded with light gray calcareous siltstone enclosing a few ostracoda fossils, wave ripple bedding
27-7	1131.74—1131.99 m	0.25 m	Mudstone, silty, greenish gray with a few ostracoda fossils
27-8	1131.99—1133.33 m	1.34 m	Siltstone, argillaceous, greenish gray, massive with a few ostracoda fossils
27-9	1133.33—1133.79 m	0.46 m	Mudstone, silty, greenish gray, massive with few pyrites
27-10	1133.79—1135.94 m	2.15 m	Mudstone greenish gray, massive with pyrites and ostracoda clast interbeds
27-11	1135.94—1136.54 m	0.60 m	Siltstone, argillaceous, greenish gray massive, vertical fractures intersecting at nearly 90°
27-12	1136.54—1137.83 m	1.29 m	Siltstone, argillaceous, dark greenish gray, massive
27-13	1137.83—1138.35 m	0.52 m	Mudstone, silty, calcareous, grayish brown with irregular siltstone lumps
27-14	1138.35—1138.49 m	0.14 m	Mudstone, silty, light greenish gray with interbeded ostracoda clasts in the upper part of the section
27-15	1138.49—1138.94 m	0.45 m	Mudstone, silty, dark greenish gray, massive
27-16	1138.94—1139.19 m	0.25 m	Siltstone, argillaceous, dark greenish gray
27-17	1139.19—1139.54 m	0.35 m	Mudstone, silty, grayish green, massive with ostracoda clast interbeds in the lower part of the section
28-1	1139.54—1139.92 m	0.38 m	Siltstone, argillaceous, greenish gray, some pyrite, ostracoda clast laminae, and oil-bearing fine-grained sandstone
28-2	1139.92—1140.82 m	0.90 m	Siltstone, argillaceous, dark greenish gray with pyrites and oil-bearing fine grained sandstone interbeds
28-3	1140.82—1141.62 m	0.80 m	Siltstone, argillaceous, grayish brown with bivalve fossils and nearly vertical fractures
28-4	1141.62—1142.14 m	0.52 m	Siltstone, calcareous, light greenish gray dark greenish gray mudstone and grayish brown silty mudstone, comprise three fining-upward sedimentary cycles graded bedding in the siltstone and massive bedding and parallel bedding in the mudstone

28-5	1142.14—1144.56 m	2.42 m	Mudstone, silty, grayish brown, nearly vertical fracture with oil film on its surface
28-6	1144.56—1146.44 m	1.88 m	Mudstone, silty, dark yellowish brown, massive with shrinkage fractures and vertical fractures, ostracoda fossils in local
28-7	1146.44—1148.01 m	1.57 m	Mudstone, silty, with vertical fractures and oil film on the fracture surface, few ostracods
28-8	1148.01—1148.30 m	0.29 m	Mudstone, silty, grayish brown, thinly interbeded with grayish green mudstone, ripple lamination in the mudstone, some calcareous siltstone lumps
28-9	1148.30—1148.62 m	0.32 m	Mudstone, grayish gray, massive
28-10	1148.62—1148.90 m	0.28 m	Mudstone, dark greenish gray, with calcareous siltstone lumps and a few ostracoda clast lamella, horizontal bedding
28-11	1148.90—1149.49 m	0.59 m	Mudstone, silty, olive gray, massive, intercalated with oil-bearing, light olive gray ostracoda clast-bearing siltstone lamellae and lumps, erosion surface at the base
28-12	1149.49—1150.54 m	1.05 m	Mudstone, silty, grayish green, massive, irregular fractures
29-1	1150.54—1151.22 m	0.68 m	Siltstone, argillaceous, dusky yellowish green with mud gravels locally, irregular fractures
29-2	1151.22—1153.24 m	2.02 m	Mudstone, silty, grayish brown, argillaceous siltstone in the lower section
30-1	1153.24—1155.29 m	2.05 m	Mudstone, silty, grayish brown, fractures filled with mud
30-2	1155.29—1155.39 m	0.10 m	Mudstone, ostracoda-bearing, dark greenish gray, ripple lamination, thin ostracoda clast limestone at the base
30-3	1155.39—1155.98 m	0.59 m	Mudstone, silty grayish brown, massive
30-4	1155.98—1156.05 m	0.07 m	Limestone, pale brown thinly interbeded with dark greenish gray ostracoda-bearing mudstone, graded bedding in the ostracoda clast limestone and horizontal bedding in the mudstone
30-5	1156.05—1157.36 m	1.31 m	Mudstone, silty, grayish brown with few ostracoda fossils
30-6	1157.36—1157.52 m	0.16 m	Mudstone, grayish brown with few ostracoda fossils
30-7	1157.52—1157.69 m	0.17 m	Mudstone, silty, brownish gray, massive, irregular fractures
30-8	1157.69—1158.24 m	0.55 m	Mudstone, silty, brownish gray, irregular fractures, grayish green argillaceous siltstone laminae and lumps
30-9	1158.24—1158.40 m	0.16 m	Siltstone, argillaceous, greenish gray, intercalated with grayish brown mudstone, ripple bedding, deformed bedding locally, erosion surface at the base
30-10	1158.40—1159.19 m	0.79 m	Mudstone, dark greenish gray with ostracoda clast laminae and lens, calcareous siltstone interbeds in the middle and lower part of the section
30-11	1159.19—1159.75 m	0.56 m	Mudstone, medium dark gray, thinly interbedded with oil-bearing pale yellowish brown ostracoda clast limestone, horizontal bedding in the mudstone
30-12	1159.75—1160.33 m	0.58 m	Mudstone, silty, grayish green, massive
30-13	1160.33—1160.55 m	0.22 m	Mudstone, silty, greenish gray, massive

30-14	1160.55—1160.99 m	0.44 m	Mudstone, silty, dark greenish gray, massive
30-15	1160.99—1161.63 m	0.64 m	Mudstone, dark greenish gray, massive, few bivalve shell fragments
30-16	1161.63—1162.24 m	0.61 m	Mudstone, medium dark gray, horizontal bedding
30-17	1162.24—1162.71 m	0.47 m	Mudstone, medium dark gray, intercalated with calcareous siltstone balls, laminae and lumps, few pyrites, horizontal bedding, bedding locally deformed
30-18	1162.71—1163.74 m	1.03 m	Mudstone, medium gray, interbeded with light gray ostracoda-bearing calcareous siltstone, horizontal bedding in mudstone and graded bedding in the calcareous siltstone, few estheria fossils
30-19	1163.74—1164.82 m	1.08 m	Mudstone, medium dark gray, horizontal bedding, ostracoda fossils
30-20	1164.82—1165.15 m	0.33 m	Mudstone, greenish gray, massive
30-21	1165.15—1165.19 m	0.04 m	Mudstone, dark gray
31-1	1165.25—1165.88 m	0.63 m	Mudstone, dark gray, horizontal bedding, ostracoda clast laminae at the base
31-2	1165.88—1166.99 m	1.11 m	Mudstone, dark greenish gray, some completely preserved ostracoda fossils, pyrites
31-3	1166.99—1169.92 m	2.93 m	Mudstone, grayish brown, locally some dark greenish gray spots, few ostracoda fossils
32-1	1169.92—1170.80 m	0.88 m	Mudstone, grayish brown, light gray mud gravels and laminae at the top, a few ostracoda fossils
32-2	1170.80—1170.84 m	0.04 m	Siltstone, calcareous, light gray, ripple lamination, few ostracoda fossils, one interbed of dark greenish gray mudstone
32-3	1170.84—1172.20 m	1.36 m	Mudstone, silty, brownish gray with ostracoda siltstone thin interbeds in the middle, few ostracoda fossils
32-4	1172.20—1172.77 m	0.57 m	Mudstone, silty, calcareous, dark greenish gray with light gray ostracoda clast laminae and lumps
32-5	1172.77—1172.96 m	0.19 m	Mudstone, grayish brown, ostracoda-bearing intercalated with greenish gray ostracoda clast limestone, ripple bedding
32-6	1172.96—1174.34 m	1.38 m	Mudstone, grayish brown, silty intercalated with greenish gray ostracoda clast siltstone, few ostracoda fossils fully preserved
32-7	1174.34—1174.62 m	0.28 m	Massive bedded pale brownish gray silty mudstone, one group vertical irregular fractures filled with grayish brown silty mudstone
32-8	1174.62—1176.92 m	2.30 m	Mudstone, silty, brownish gray with three greenish gray ostracoda clast interbeds
32-9	1176.92—1177.12 m	0.20 m	Mudstone, silty, pale brown
32-10	1177.12—1178.13 m	1.01 m	Mudstone, silty, greenish gray, ostracoda clast siltstone laminae and lumps in the lower section
32-11	1178.13—1178.23 m	0.10 m	Mudstone, silty, medium gray, horizontal bedding, erosion surface at the base
32-12	1178.23—1181.35 m	3.12 m	Mudstone, grayish brown, irregular fractures filled with grayish brown mudstone in the middle-upper section and the bottom, few ostracoda fossils
33-1	1181.35—1183.83 m	2.48 m	Mudstone, grayish brown, massive

33-2	1183.83—1183.99 m	0.16 m	Mudstone, silty, brownish gray, massive, few ostracoda fossils
33-3	1183.99—1185.22 m	1.23 m	Mudstone, grayish brown, intercalated with light greenish gray calcareous siltstone laminae and lumps, ostracoda fossils
33-4	1185.22—1186.05 m	0.83 m	Mudstone, silty, grayish purple mixed with greenish gray calcareous siltstone, massive bedding, lumps and convolute bedding, greenish gray calcareous siltstone lumps
33-5	1186.05—1186.32 m	0.27 m	Mudstone, silty, greenish black, erosion surface at the base, one ostracoda clasts layer with a thickness of 4 mm above the erosion surface
33-6	1186.32—1186.45 m	0.13 m	Mudstone, silty, greenish gray, shrinkage cracks filled with calcareous siltstone in the upper section
33-7	1186.45—1186.55 m	0.10 m	Mudstone, silty, grayish purple, few ostracoda fossils
33-8	1186.55—1186.86 m	0.31 m	Mudstone, silty, dark greenish gray, ostracoda fossils
33-9	1186.86—1187.20 m	0.34 m	Mudstone, silty, dark greenish gray with light gray calcareous siltstone laminae and lumps
33-10	1187.20—1187.30 m	0.10 m	Mudstone, silty, dark greenish gray, ostracoda fossils
33-11	1187.30—1188.13 m	0.83 m	Mudstone, silty, brownish gray with greenish gray calcareous siltstone thin interbeds, massive bedding in the mudstone and ripple lamination in the siltstone, a layer of ostracoda clasts in the lower section
34-1	1188.13—1188.32 m	0.19 m	Mudstone, silty, dark greenish gray, few ostracoda fossils
34-2	1188.32—1188.95 m	0.63 m	Mudstone, silty, grayish black, horizontal bedding, thin layer of ostracoda clast and erosion surface at the base
34-3	1188.95—1190.27 m	1.32 m	Mudstone, silty, dark greenish gray, ostracoda fossils
34-4	1190.27—1190.41 m	0.14 m	Mudstone, dark gray, ripple lamination, shrinkage cracks
34-5	1190.41—1191.00 m	0.59 m	Mudstone dark gray, interbeded with ostracoda limestone, ripple bedded and, mud gravels, erosion surface at the base
34-6	1191.00—1191.58 m	0.58 m	Mudstone, calcareous, dark gray, ostracoda fossils
34-7	1191.58—1192.13 m	0.55 m	Mudstone, dark grey, interbeded with ostracoda limestone, with ripple bedding, and mud gravels and erosion surface at the base
34-8	1192.13—1193.62 m	1.49 m	Mudstone, dark greenish gray, ostracoda fossils
34-9	1193.62—1194.13 m	0.51 m	Mudstone, silty, dark grey, interbeded with dark greenish gray mudstone in the middle section, few ostracoda fossils
34-10	1194.13—1194.38 m	0.25 m	Mudstone, calcareous, dark greenish gray, local ostracoda fossil concentration and plant fragments
34-11	1194.38—1194.58 m	0.20 m	Mudstone, silty, few ostracoda fossils
34-12	1194.58—1194.93 m	0.35 m	Massive, greenish black, few ostracoda fossils
34-13	1194.93—1195.23 m	0.30 m	Mudstone, brownish gray, intercalated with argillaceous siltstone and ostracoda clast laminae, mud gravels and erosion surface at the base
34-14	1195.23—1195.30 m	0.07 m	Mudstone, silty, grayish green, load cast structure at the base

34-15	1195.30—1195.46 m	0.16 m	Mudstone, grayish brown
34-16	1195.46—1195.82 m	0.36 m	Mudstone, silty, pale brown
34-17	1195.82—1196.63 m	0.81 m	Mudstone, silty, dark gray, local ostracoda fossil concentration
34-18	1196.63—1197.49 m	0.86 m	Mudstone, silty, dark greenish gray, ostracoda fossil concentration locally, charcoal debris at the top
34-19	1197.49—1198.45 m	0.96 m	Mudstone, grayish black with ostracoda clast limestone interbeds, ripple bedding locally bedding deformed, erosion surface at the base
34-20	1198.45—1199.14 m	0.69 m	Mudstone, dark greenish gray, with ostracoda clasts, horizontal bedding
34-21	1199.14—1199.59 m	0.45 m	Mudstone Horizontal bedded dark gray mudstone, ostracoda fossils
34-22	1199.59—1199.69 m	0.10 m	Mudstone, dark greenish gray with an ostracoda clast limestone laminae, mud gravels and erosion surface at the base
34-23	1199.69—1200.02 m	0.33 m	Mudstone grayish black interbeded with dark gray ostracoda clast limestone, ripple lamination, erosion surface at the base
34-24	1200.02—1200.53 m	0.51 m	Mudstone, greenish gray locally with ostracoda clasts
35-1	1200.53—1201.04 m	0.51 m	Mudstone, greenish gray with ostracoda clasts
35-2	1201.04—1201.22 m	0.18 m	Siltstone, argillaceous, calcareous, dark greenish gray
35-3	1201.22—1202.03 m	0.81 m	Mudstone, silty, calcareous, greenish gray
35-4	1202.03—1202.30 m	0.27 m	Mudstone, silty, grayish brown mixed with dark greenish gray calcareous argillaceous siltstone, slump structure
35-5	1202.30—1202.99 m	0.69 m	Mudstone, silty, grayish brown with laminae of dark greenish gray calcareous argillaceous siltstone, one ostracoda clast laminae at the base
35-6	1202.99—1203.39 m	0.40 m	Mudstone, silty, ostracoda fossil locally concentrated
35-7	1203.39—1203.66 m	0.27 m	Siltstone, argillaceous, brownish gray, slump structure, few ostracoda fossils
35-8	1203.66—1204.09 m	0.43 m	Mudstone, grayish green, few ostracoda fossils
35-9	1204.09—1204.34 m	0.25 m	Mudstone, dark greenish gray, intercalated with light gray ostracoda clast limestone
35-10	1204.34—1204.75 m	0.41 m	Mudstone, grayish green, few fossils
36-1	1204.76—1205.76 m	1.00 m	Mudstone, grayish brown, few ostracoda fossils
36-2	1205.76—1206.91 m	1.15 m	Mudstone, greenish black with two ostracoda clast interbeds, erosion surface at the base
36-3	1206.91—1208.87 m	1.96 m	Mudstone, grayish brown, some ostracoda clast laminae and argillaceous siltstone thin interbeds
36-4	1208.87—1209.06 m	0.19 m	Siltstone, argillaceous, ostracoda clast-bearing, greenish gray mixed with grayish brown mudstone, slump structure
36-5	1209.06—1209.36 m	0.30 m	Mudstone, dusky brown, local slump structure, dark greenish gray ostracoda-bearing argillaceous siltstone laminae and lumps at the top

36-6	1209.36—1210.56 m	1.20 m	Mudstone, dark greenish gray with greenish black mudstone laminae, a few well preserved ostracoda fossils
36-7	1210.56—1210.84 m	0.28 m	Mudstone, dark greenish gray, ostracoda-bearing intercalated with ostracoda clast thin interbeds, few plant fossils
36-8	1210.84—1211.47 m	0.63 m	Mudstone, greenish black, mixed with dark greenish gray mudstone, ostracoda clast laminae at the top, erosion surface at the base
36-9	1211.47—1211.61 m	0.14 m	Mudstone, dark greenish gray with some ostracoda clast thin interbeds, trace fossils
36-10	1211.61—1211.76 m	0.15 m	Mudstone, silty, brownish black, few ostracoda fossils
36-11	1211.76—1212.36 m	0.60 m	Mudstone, silty, brownish gray, irregular fractures filled with mudstone, few ostracoda fossils and trace fossils
36-12	1212.36—1213.06 m	0.70 m	Mudstone, grayish brown few ostracoda fossils
36-13	1213.06—1214.00 m	0.94 m	Mudstone, dusky brown, few ostracoda fossils
36-14	1214.00—1214.38 m	0.38 m	Mudstone, few ostracoda fossils
36-15	1214.38—1214.63 m	0.25 m	Mudstone, calcareous, light brownish gray, irregular vertical fractures filled with grayish brown mudstone
36-16	1214.63—1214.73 m	0.10 m	Mudstone, dark greenish gray, irregular fractures and greenish gray siltstone
36-17	1214.73—1215.51 m	0.78 m	Mudstone, dusky brown with three greenish gray calcareous siltstone thin interbeds, erosion surface at the base
37-1	1215.51—1215.63 m	0.12 m	Mudstone, grayish brown
37-2	1215.63—1215.77 m	0.14 m	Mudstone, calcareous, brownish gray few plant fossils and ostracoda fossils in siltstone laminae
37-3	1215.77—1216.06 m	0.29 m	Mudstone, grayish brown
37-4	1216.06—1217.06 m	1.00 m	Mudstone, silty, few ostracoda fossils well preserved
37-5	1217.06—1217.46 m	0.40 m	Mudstone with few plant fossils and ostracoda fossils intercalated with grayish green argillaceous siltstone, some irregular vertical fractures
37-6	1217.46—1217.67 m	0.21 m	Mudstone, silty, brownish gray
37-7	1217.67—1218.29 m	0.62 m	Mudstone, grayish brown, few ostracoda fossils well preserved
37-8	1218.29—1219.03 m	0.74 m	Mudstone, silty, few ostracoda fossils
37-9	1219.03—1220.11 m	1.08 m	Mudstone, grayish brown with grayish green argillaceous siltstone laminae and lumps, few ostracoda fossils
37-10	1220.11—1220.15 m	0.04 m	Limestone, argillaceous, grayish green ostracoda clast, slump structure
37-11	1220.15—1221.79 m	1.64 m	Mudstone, silty with two grayish green ostracoda-bearing calcareous siltstone thin interbeds, few ostracoda fossils
37-12	1221.79—1223.51 m	1.72 m	Mudstone grayish brown, few ostracoda fossils well preserved
37-13	1223.51—1223.81 m	0.30 m	Mudstone, silty, brownish gray, few ostracoda fossils

37-14	1223.81—1224.01 m	0.20 m	Mudstone, greenish gray, mixed with brownish gray ostracoda-bearing silty mudstone, some ostracoda fossils
37-15	1224.01—1224.21 m	0.20 m	Mudstone, silty, greenish black, few ostracoda fossils
37-16	1224.21—1224.38 m	0.17 m	Limestone, ostracoda bearing, light olive gray, graded bedding, mud gravels and erosion surface at the base
37-17	1224.38—1225.07 m	0.69 m	Mudstone, silty, grayish brown, mixed with dark greenish gray, slump structure, ostracoda clasts, erosion surface at the base
37-18	1225.07—1225.37 m	0.30 m	Mudstone, silty, grayish brown, few ostracoda fossils
37-19	1225.37—1225.96 m	0.59 m	Mudstone, calcareous, light brownish gray
37-20	1225.96—1226.86 m	0.90 m	Mudstone, silty, grayish brown, irregular fracture and few ostracoda fossils
37-21	1226.86—1227.04 m	0.18 m	Mudstone, grayish brown with light greenish gray spots of silty mudstone, few ostracoda fossils well preserved
38-1	1227.04—1227.54 m	0.50 m	Siltstone, argillaceous, medium dark gray, ostracoda-bearing, deformation bedding, ostracoda fossils and bivalve fossils
38-2	1227.54—1229.91 m	2.37 m	Mudstone, silty, grayish red, some ostracoda fossils
38-3	1229.91—1230.11 m	0.20 m	Mudstone, dark greenish gray, ostracoda clast laminae at the bottom
38-4	1230.11—1231.63 m	1.52 m	Mudstone, grayish red
38-5	1231.63—1231.80 m	0.17 m	Mudstone, grayish red
38-6	1231.80—1232.34 m	0.54 m	Mudstone, grayish red with grayish green ostracoda-bearing mudstone laminae
38-7	1232.34—1232.54 m	0.20 m	Mudstone, grayish green, intercalated with light gray calcareous siltstone, ripple bedding, ostracoda clast laminae and lumps
38-8	1232.54—1232.73 m	0.1 m	Mudstone, grayish green, few ostracoda fossils
38-9	1232.73—1232.94 m	0.21 m	Mudstone, silty, grayish brown, few ostracoda fossils
38-10	1232.94—1233.11 m	0.17 m	Mudstone, dark greenish gray with ostracoda clast thin interbeds at the bottom, erosion surface at the base
38-11	1233.11—1233.26 m	0.15 m	Marl, light olive gray
38-12	1233.26—1233.39 m	0.13 m	Mudstone, dark gray with light gray ostracoda clast limestone interbeds, discontinuous ripple lamination, erosion surface at the base
38-13	1233.39—1233.66 m	0.27 m	Mudstone, grayish brown mixed with dark gray silty mudstone and pale brown silty mudstone interbeds
38-14	1233.66—1234.94 m	1.28 m	Mudstone, grayish brown with pale brown silty mudstone interbeds
38-15	1234.94—1236.48 m	1.54 m	Mudstone, silty, dark gray, few ostracoda fossils, some grayish brown spots
38-16	1236.48—1237.01 m	0.53 m	Mudstone, silty, grayish brown with grayish green silty mudstone interbeds, ostracoda fossils
38-17	1237.01—1237.87 m	0.86 m	Mudstone, silty, dark grey, thinly interbeded with light gray ostracoda clast limestone, ripple bedding and discontinuous ripple bedding, ostracoda fossils
38-18	1237.87—1237.94 m	0.07 m	Mudstone, silty, brownish gray, few ostracoda fossils

38-19	1237.94—1238.09 m	0.15 m	Marly limestone, olive gray
38-20	1238.09—1238.49 m	0.40 m	Mudstone dark gray with light gray ostracoda clast limestone interbeds, horizontal bedding and ripple bedding, erosion surface at the base
38-21	1238.49—1238.72 m	0.23 m	Mudstone, silty, grayish green, pyrites and ostracoda fossils
38-22	1238.72—1238.84 m	0.12 m	Mudstone grayish red, few ostracoda fossils
39-1	1238.84—1238.95 m	0.11 m	Mudstone grayish red, few ostracoda fossils
39-2	1238.95—1239.40 m	0.45 m	Mudstone, grayish red, ostracoda-bearing, dark greenish gray ostracoda clast laminae in the middle and lower sections
39-3	1239.40—1239.56 m	0.16 m	Mudstone, silty, dark greenish gray, mixed with grayish red silty mudstone, ostracoda fossils concentrated locally in, mud gravels; erosion surface at the base
39-4	1239.56—1240.10 m	0.54 m	Mudstone, dark greenish gray, horizontal bedding and ripple bedding l, calcareous siltstone laminae at the bottom, erosion surface at the base
39-5	1240.10—1240.51 m	0.41 m	Mudstone, dark grey, thinly interbeded with ostracoda limestone, horizontal bedding and ripple lamination, mud gravels and erosion surface at the base
39-6	1240.51—1241.74 m	1.23 m	Mudstone, silty, dark greenish gray, ostracoda fossils
39-7	1241.74—1241.94 m	0.20 m	Mudstone, grayish green, shrinkage fractures, few ostracoda fossils
39-8	1241.94—1242.54 m	0.60 m	Mudstone, silty, dark greenish gray, ostracoda fossils
39-9	1242.54—1242.90 m	0.36 m	Mudstone dark gray, thinly interbeded with light olive gray ostracoda clast limestone, horizontal bedding and ripple lamination, erosion surface at the base
39-10	1242.90—1244.34 m	1.44 m	Mudstone, silty, greenish gray, some ostracoda clast laminae
39-11	1244.34—1244.57 m	0.23 m	Dolostone, light olive gray
39-12	1244.57—1245.14 m	0.57 m	Mudstone, silty, massive, dark greenish gray, ostracoda clasts
39-13	1245.14—1245.62 m	0.48 m	Mudstone, dark gray, thinly interbeded with light gray ostracoda clast limestone, graded bedding, ripple lamination, calcite veins, mud gravel and erosion surface at the base
39-14	1245.62—1247.34 m	1.72 m	Mudstone, silty, dark greenish gray pyrites at the top, some ostracoda fossils
39-15	1247.34—1247.68 m	0.34 m	Mudstone gravel, dark greenish gray with few complete ostracoda fossils and fossil fragments
39-16	1247.68—1248.42 m	0.74 m	Mudstone, greenish gray, ostracoda clasts
39-17	1248.42—1248.79 m	0.37 m	Mudstone, dark gray, thinly interbeded with ostracoda clast limestone, ripple bedding, erosion surface at the base
39-18	1248.79—1249.22 m	0.43 m	Mudstone, dark greenish gray, thinly interbeded with ostracoda clast limestone, ripple lamination, deformation bedding at the bottom, erosion surface at the base
39-19	1249.22—1249.95 m	0.73 m	Mudstone, greenish gray, ripple lamination, ostracoda clast laminae
40-1	1249.95—1250.77 m	0.82 m	Mudstone, greenish gray with some ostracoda clast laminae

40-2	1250.77—1252.05 m	1.28 m	Mudstone, dark greenish gray with ostracoda clast limestone thin interbeds, ripple lamination, erosion surface at the base
40-3	1252.05—1252.50 m	0.45 m	Mudstone, grayish black, shrinkage fractures filled with ostracoda clasts
40-4	1252.50—1253.47 m	0.97 m	Mudstone, grayish black with ostracoda clast limestone thin interbeds, ripple lamination, some ostracoda clasts and few complete ostracoda fossils

The overlying formation of Yaojia Formation Member 1 is the Yaojia Formation Members 2 and 3, the contacting relationship between them is parallel unconformable contact; the underlying formation is the Qingshankou Formation Members 2 and 3, the contacting relationship between them is parallel unconformable contact (Figure 5.4).

The detailed description of the Yaojia Formation Member 1:

40-5	1253.47—1256.53 m	3.06 m	Mudstone, greenish gray, some fossils debris
41-1	1256.53—1256.66 m	0.13 m	Mudstone, dark greenish gray, ostracoda fossils
41-2	1256.66—1257.96 m	1.30 m	Mudstone, dark greenish gray, a few ostracoda fossils
41-3	1257.96—1261.09 m	3.13 m	Mudstone, medium dark gray thinly interbedded with medium. light gray ostracoda clast limestone, horizontal bedding and ripple lamination, graded bedding at the bottom, erosion surface at the base
41-4	1261.09—1263.79 m	2.70 m	Mudstone, silty, dark greenish gray ostracoda clast laminae, few ostracoda fossils fully preserved
41-5	1263.79—1266.52 m	2.73 m	Mudstone, medium dark gray with ostracoda clast limestone, horizontal bedding and ripple lamination, graded bedding within the ostracoda limestone, erosion surface at the base
41-6	1266.52—1268.51 m	1.99 m	Massive bedded dark grayish green silty mudstone, few ostracoda fossils preserved completely
42-1	1268.51—1268.91 m	0.40 m	Mudstone, dark gray, ripple lamination with siltstone laminae
42-2	1268.91—1269.23 m	0.32 m	Mudstone, dark gray with some siltstone laminae
42-3	1269.23—1269.35 m	0.12 m	Siltstone, ripple laminated, light gray, mud gravels at the top, erosion surface at the base
42-4	1269.35—1272.31 m	2.96 m	Mudstone, dark gray with ostracoda clast limestone thin interbeds, horizontal bedding in the mudstone and graded bedding in the ostracoda clast limestone, erosion surface at the base
42-5	1272.31—1273.48 m	1.17 m	Mudstone, dark gray thinly interbeded, with ostracoda clast limestone and light gray argillaceous siltstone laminae in the lower section, pyrites, erosion surface at the base
42-6	1273.51—1273.59 m	0.08 m	Mudstone, medium light gray, oil stains

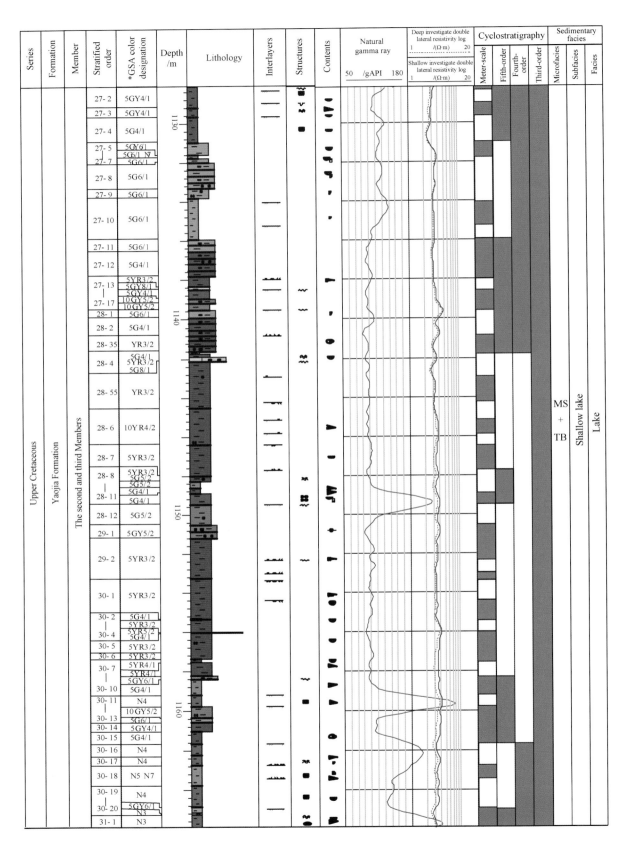

Figure 5.4　Stratigraphic chart of Yaojia Formation.

Section 5 SK-1 Core Description and Core Photographs

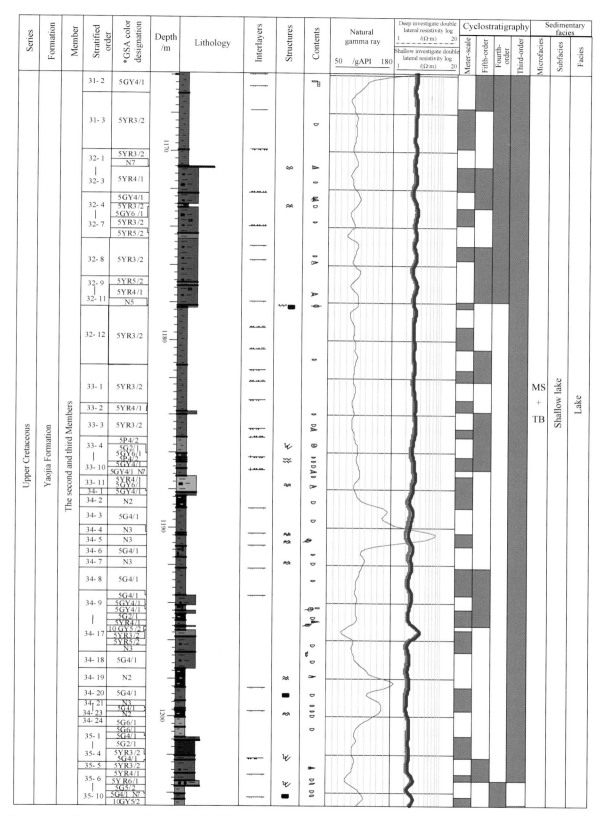

Figure 5.4 Stratigraphic chart of Yaojia Formation (continued).

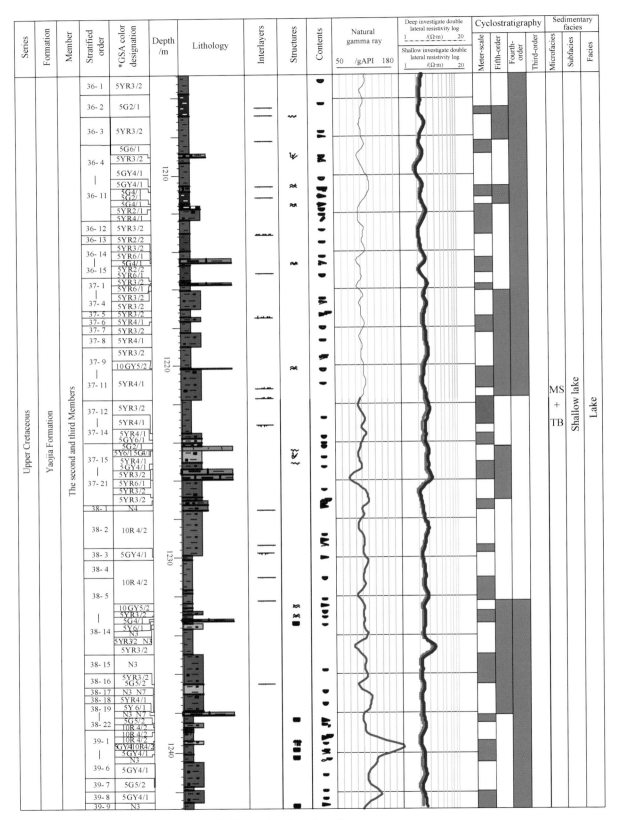

Figure 5.4　Stratigraphic chart of Yaojia Formation (continued).

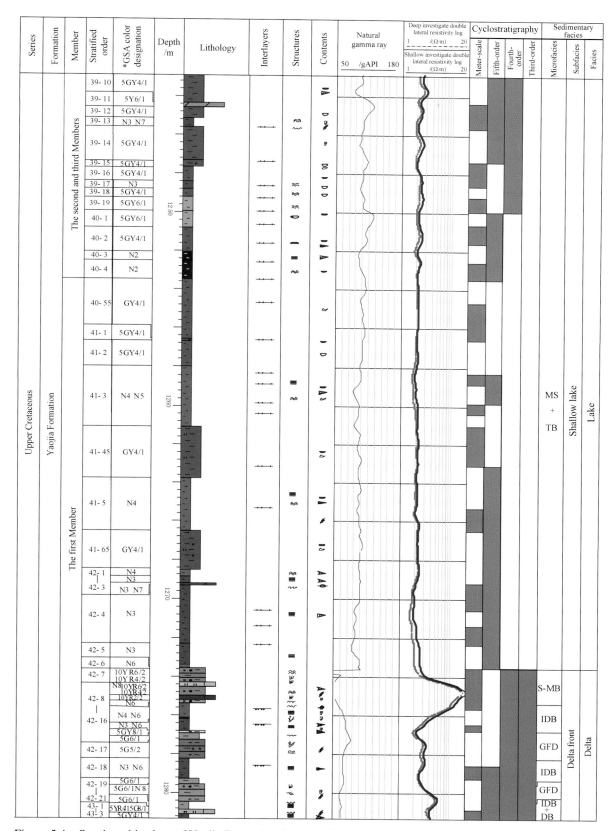

Figure 5.4 Stratigraphic chart of Yaojia Formation (continued).

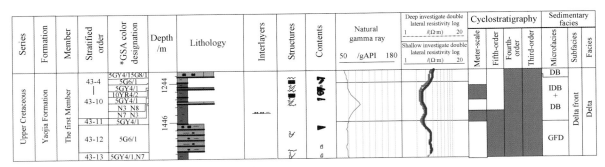

Figure 5.4 Stratigraphic chart of Yaojia Formation (continued). Symbols are the same as those in Figure 5.1.

42-7	1273.59—1274.34 m	0.75 m	Siltstone, pale yellowish brown with dark yellowish brown calcareous siltstone interbeds, small-scale trough cross bedding in the siltstone
42-8	1274.34—1274.57 m	0.23 m	Siltstone calcareous very light gray interbeded with pale yellowish brown argillaceous siltstone, small-scale trough cross bedding, ripple lamination in the argillaceous siltstone, oil stain and oil trace in the siltstone, erosion surface at the base
42-9	1274.57—1275.00 m	0.43 m	Siltstone, argillaceous, pale yellowish brown with dark yellowish brown calcareous siltstone thin interbeds, ripple lamination in the argillaceous siltstone and small-scale trough cross bedding in the calcareous siltstone, erosion surface at the base
42-10	1275.00—1275.21 m	0.21 m	Siltstone, calcareous, dusky yellowish brown, small-scale trough cross bedding in the lower section and ripple lamination in the upper section, erosion surface at the base, pyrites
42-11	1275.21—1275.41 m	0.20 m	Siltstone, argillaceous, medium light gray, ripple lamination oil stain, scour and infill structure
42-12	1275.41—1276.12 m	0.71 m	Mudstone, medium dark gray with medium light gray argillaceous siltstone thin interbeds, horizontal bedding in mudstone and ripple lamination in argillaceous siltstone, plant fossils fragments, pyrites
42-13	1276.12—1276.22 m	0.10 m	Mudstone, medium dark gray with medium light gray argillaceous siltstone thin interbeds, slump structure and convolute bedding, medium light gray argillaceous siltstone lumps and laminae, pyrites
42-14	1276.22—1276.83 m	0.61 m	Mudstone, dark gray with medium. light gray silty mudstone thin interbeds, horizontal bedding and ripple lamination, medium light gray argillaceous siltstone lumps and laminae, pyrites
42-15	1276.83—1276.95 m	0.12 m	Siltstone, argillaceous, light grayish green, deformation bedding, convolute bedding, water-escape structure, mud gravels and erosion surface at the base
42-16	1276.95—1277.30 m	0.35 m	Mudstone, silty, light gray
42-17	1277.30—1278.23 m	0.93 m	Siltstone, argillaceous, greenish gray, deformation bedding, convolute bedding, light greenish gray argillaceous siltstone balls and pillows with inner lamination structures, Fugichnia

42-18	1278.23—1279.25 m	1.02 m	Mudstone, dark gray interbeded with medium. light gray argillaceous siltstone, horizontal bedding and ripple lamination, some oil stain
42-19	1279.25—1279.54 m	0.29 m	Mudstone, silty, greenish gray
42-20	1279.54—1280.49 m	0.95 m	Siltstone, argillaceous, greenish gray interbeded with very light gray calcareous siltstone, horizontal bedding and wave-ripple lamination, pyrites, trace fossils, erosion surface at the base
42-21	1280.49—1280.85 m	0.36 m	Mudstone, Massive bedded greenish gray mudstone
43-1	1280.85—1280.87 m	0.02 m	Mudstone, greenish gray, oil-bearing calcareous siltstone lumps
43-2	1280.87—1281.11 m	0.24 m	Siltstone, calcareous, brownish gray and light greenish gray, discontinuous horizontal bedding, wave-ripple bedding and climbing bedding, locally thin interbeds of mudstone, trace fossils, erosion surface at the base
43-3	1281.11—1281.35 m	0.24 m	Mudstone, dark greenish gray, oil-bearing calcareous siltstone laminae
43-4	1281.35—1281.64 m	0.29 m	Siltstone, calcareous, brownish gray with light greenish gray mudstone thin interbeds
43-5	1281.64—1281.93 m	0.29 m	Mudstone, greenish gray, horizontal bedding, ripple lamination and deformation bedding locally, calcareous siltstone laminae and trace fossils
43-6	1281.93—1282.17 m	0.24 m	Mudstone, dark greenish gray
43-7	1282.17—1282.27 m	0.10 m	Siltstone, calcareous, dark yellowish brown, slump structure, discontinuous horizontal bedding and convolute bedding, sand balls and pillows, erosion surface at the base
43-8	1282.27—1282.69 m	0.42 m	Mudstone, dark greenish gray, few bivalve fossils
43-9	1282.69—1282.92 m	0.23 m	Mudstone, dark gray interbeded with light gray siltstone, horizontal bedding and ripple lamination
43-10	1282.92—1283.05 m	0.13 m	Siltstone, calcareous, light gray interbeded with dark gray mudstone slump structure, convolute bedding, scour and infill structure, calcareous siltstone balls and pillows
43-11	1283.05—1284.07 m	1.02 m	Mudstone, dark greenish gray with, grayish brown spots
43-12	1284.07—1285.50 m	1.43 m	Siltstone, argillaceous, greenish gray, slump structure, deformation structure and convolute structure, mud gravels and erosion surface at the base, calcareous siltstone balls and pillows, pyrites
43-13	1285.50—1285.91 m	0.41 m	Mudstone, dark greenish gray with light gray siltstone thin interbeds, slump structure and convolute bedding, light gray siltstone balls and pillows, pyrites.

5.1.5 Nenjiang Formation Members 1 and 2

The overlying formation of Nenjiang Formation Member 2 is the Nenjiang Formation Member 3, the contacting relationship between them is conformable contact; the underlying formation is the Nenjiang Formation Member 1, the contacting relationship between them is conformable contact (Figure 5.5).

The detailed description of the Nenjiang Formation Member 2:

349-14	1582.93—1583.05 m	0.12 m	Fine-grained sandstone, light gray, wavy bedding, charcoal debris
349-15	1583.05—1583.83 m	0.78 m	Mudstone, dark gray, horizontal bedding, few fine grained sandstone thin interbeds
349-16	1583.83—1583.86 m	0.03 m	Dolomite, olive gray, massive
349-17	1583.86—1584.33 m	0.47 m	Mudstone, dark gray, horizontal bedding, few fine grained sandstone thin interbeds
349-18	1584.33—1584.37 m	0.04 m	Dolomite, olive gray, massive
349-19	1584.37—1586.42 m	2.05 m	Mudstone, dark gray, horizontal bedding, two 5cm-thick fine-grained sandstone laminae at 1584.62 m and 1586.12 m
349-20	1586.42—1586.57 m	0.15 m	Fine-grained sandstone, light gray, wavy bedding, charcoal debris
349-21	1586.57—1587.97 m	1.40 m	Silty mudstone, dark gray, horizontal bedding, few charcoal debris
349-22	1587.97—1588.07 m	0.10 m	Fine-grained sandstone, light gray, wavy cross-bedding, mudstone laminae and charcoal debris
349-23	1588.07—1588.25 m	0.18 m	Mudstone, dark gray, horizontal bedding, few charcoal debris
350-1	1588.25—1588.50 m	0.25 m	Muddy siltstone, gray, massive, with some relict plant fragments
350-2	1588.50—1588.70 m	0.20 m	Siltstone, light gray, wavy bedding, with some mudstone laminae and charcoal debris
350-3	1588.70—1588.85 m	0.15 m	Muddy siltstone, gray massive, with some relict plant fragments
350-4	1588.85—1589.00 m	0.15 m	Siltstone, light gray, wavy bedding, with some mudstone laminae and charcoal debris
350-5	1589.00—1589.25 m	0.25 m	Muddy siltstone, gray, massive, with some relict plant fragments
350-6	1589.25—1589.65 m	0.40 m	Fine-grained sandstone, light gray, wave-ripple bedding
350-7	1589.65—1590.00 m	0.35 m	Muddy siltstone, gray, wavy bedding, with some relict plant fragments and sandstone laminae
350-8	1590.00—1590.50 m	0.50 m	Fine-grained sandstone, light gray, wave-ripple bedding
350-9	1590.50—1590.75 m	0.25 m	Muddy siltstone, gray, wavy bedding, with some relict plant fragments and sandstone laminae
350-10	1590.75—1591.05 m	0.30 m	Fine-grained sandstone, light gray, wave-ripple bedding
350-11	1591.05—1592.60 m	1.55 m	Silty mudstone, dark gray, wavy bedding, with some siltstone laminae and charcoal debris
350-12	1592.60—1592.70 m	0.10 m	Medium sandstone, light gray, wavy bedding
350-13	1592.70—1592.80 m	0.10 m	Mudstone, dark gray, horizontal bedding
350-14	1592.80—1593.37 m	0.57 m	Fine-grained sandstone, light gray, wave-ripple bedding, with some mudstone laminae

Section 5 SK-1 Core Description and Core Photographs

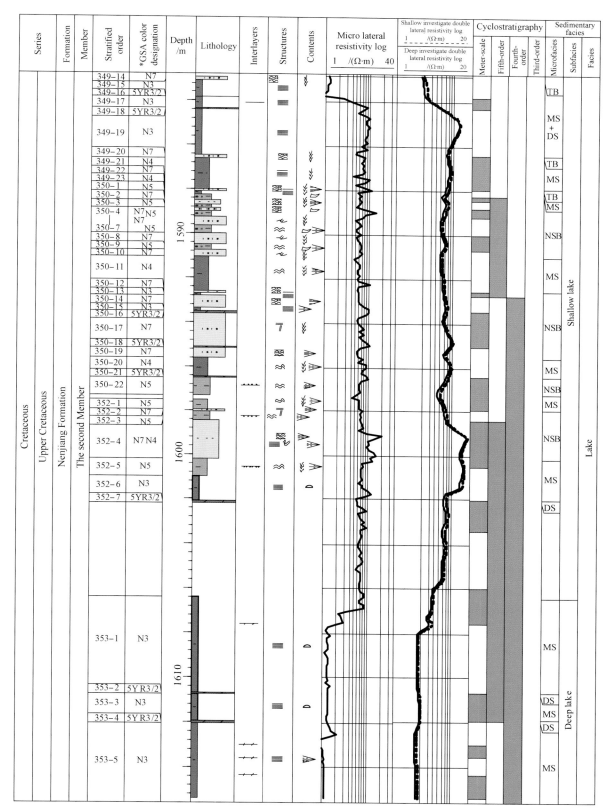

Figure 5.5 Stratigraphic chart of Nenjiang Formation Members 1 and 2.

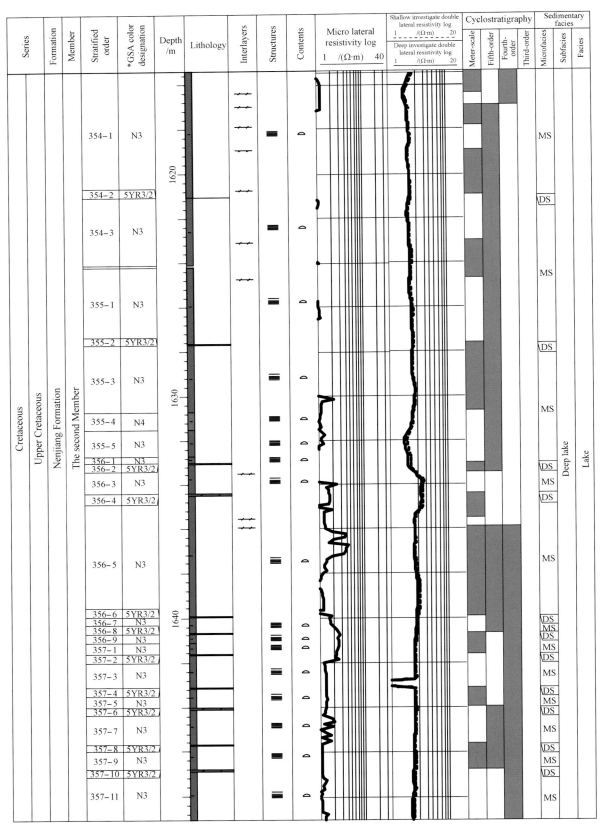

Figure 5.5 Stratigraphic chart of Nenjiang Formation Members 1 and 2 (continued).

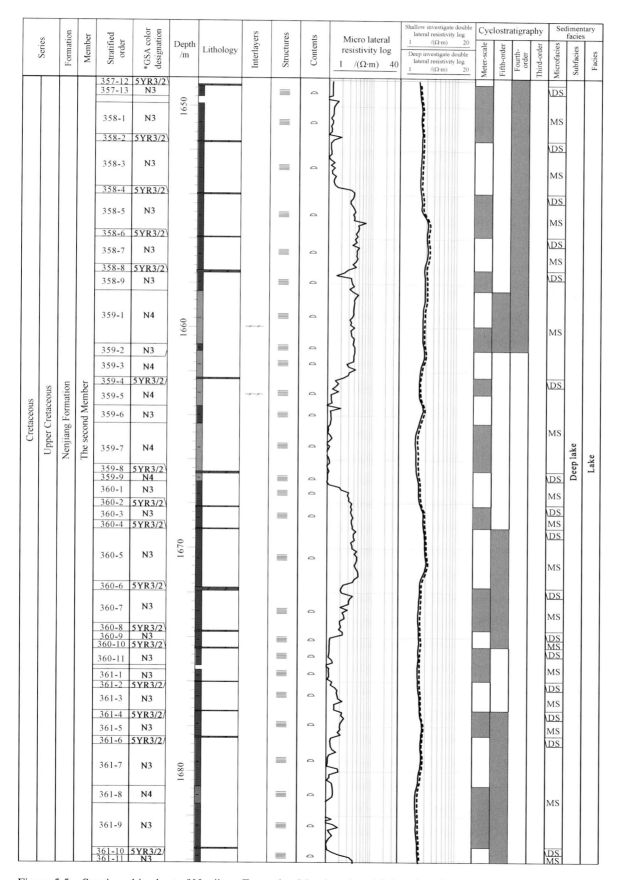

Figure 5.5 Stratigraphic chart of Nenjiang Formation Members 1 and 2 (continued).

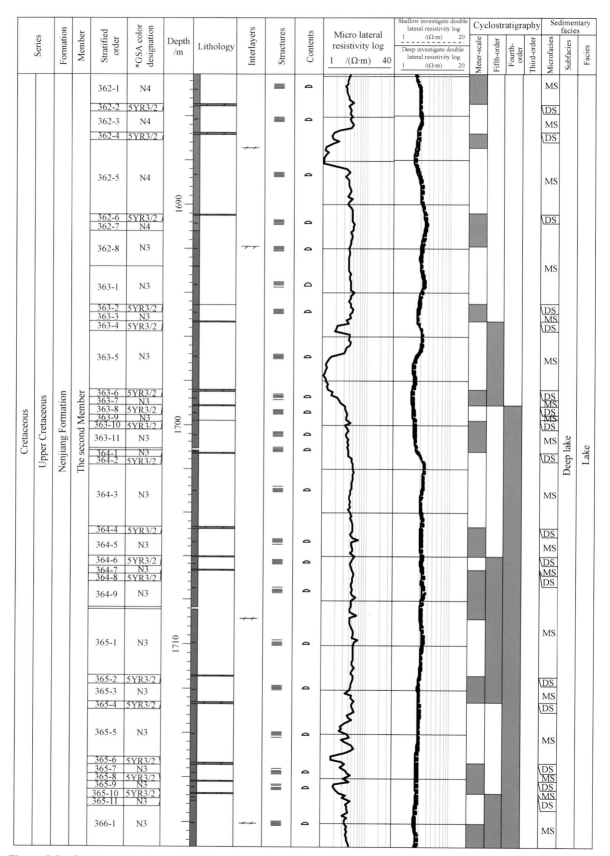

Figure 5.5 Stratigraphic chart of Nenjiang Formation Members 1 and 2 (continued).

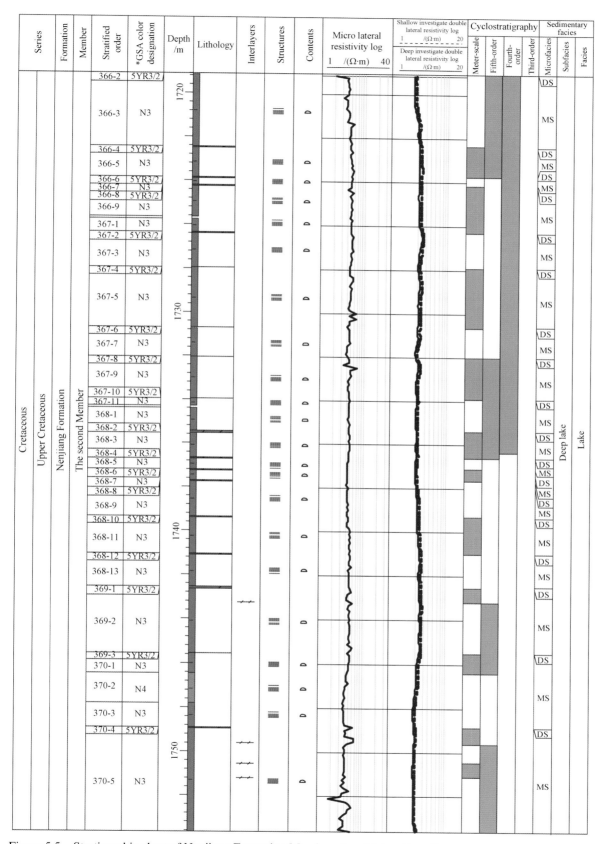

Figure 5.5　Stratigraphic chart of Nenjiang Formation Members 1 and 2 (continued).

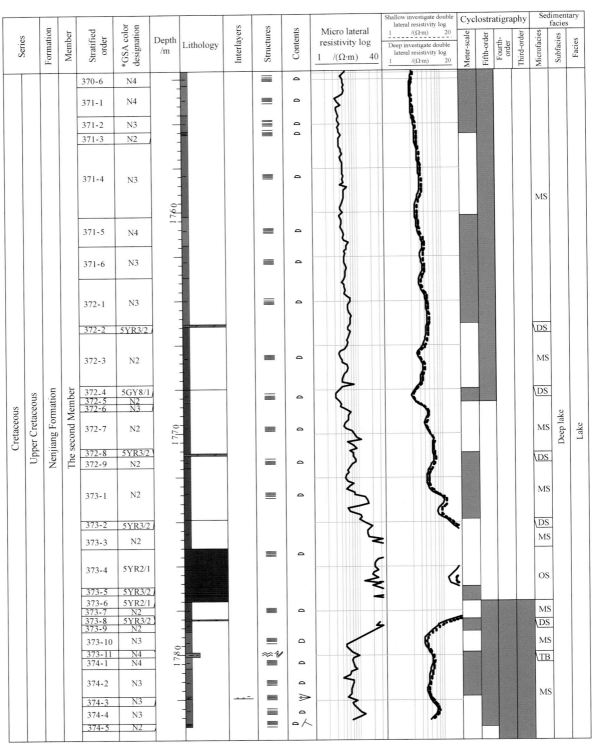

Figure 5.5　Stratigraphic chart of Nenjiang Formation Members 1 and 2 (continued).

Section 5 SK-1 Core Description and Core Photographs

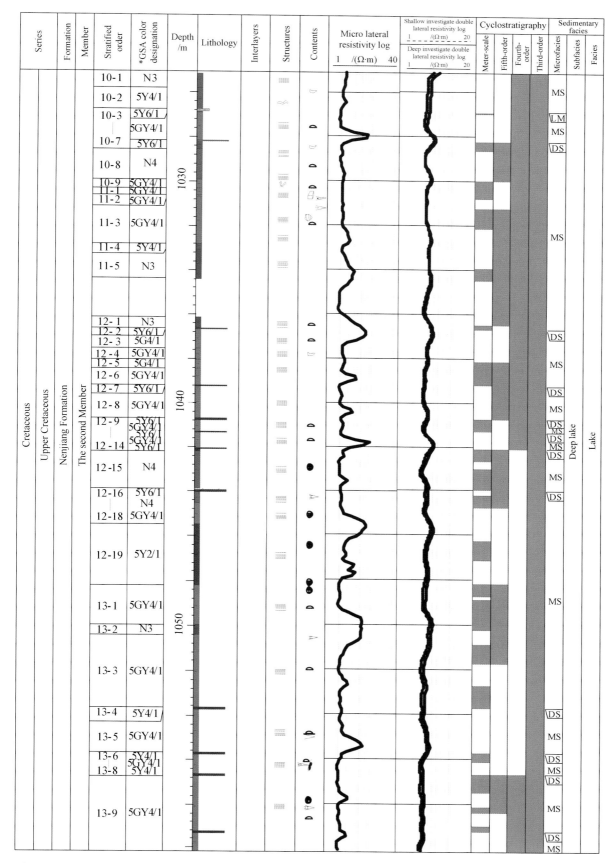

Figure 5.5 Stratigraphic chart of Nenjiang Formation Members 1 and 2 (continued).

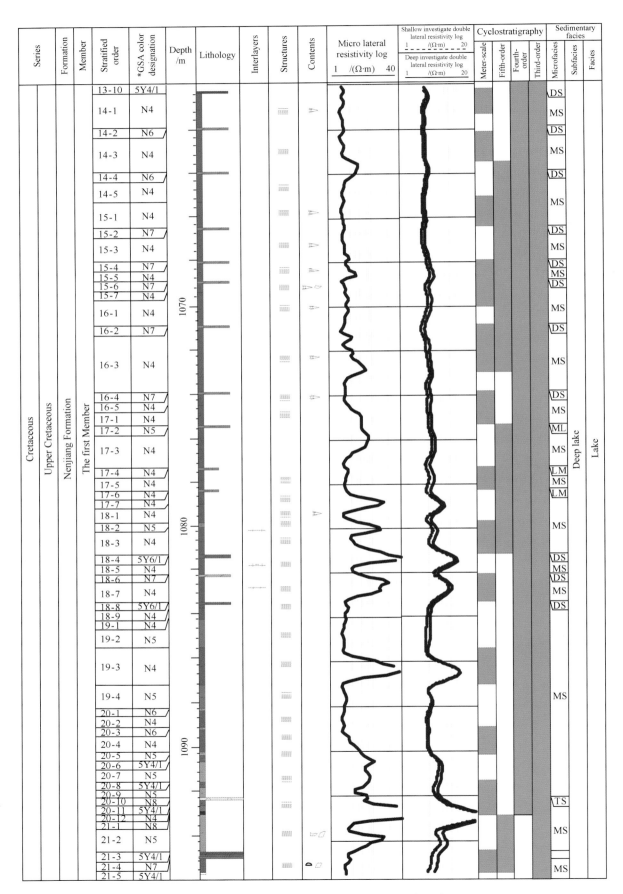

Figure 5.5 Stratigraphic chart of Nenjiang Formation Members 1 and 2 (continued).

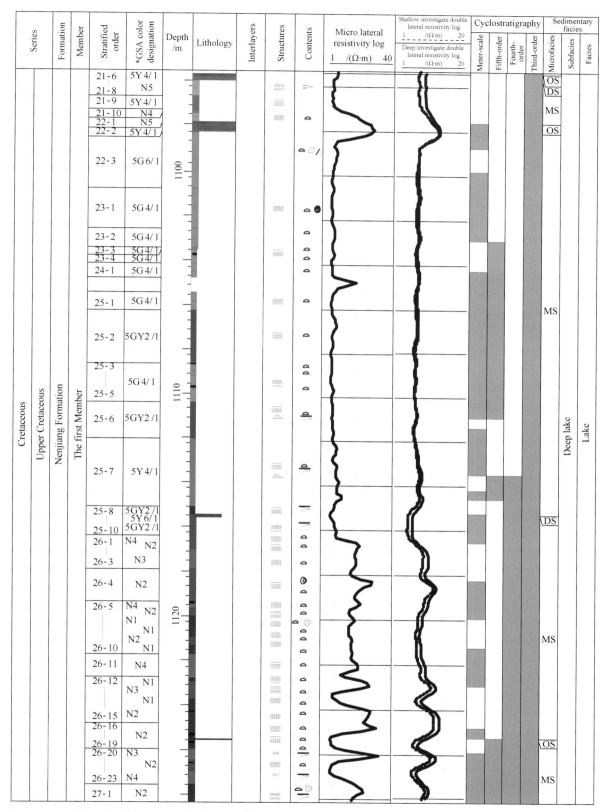

Figure 5.5 Stratigraphic chart of Nenjiang Formation Members 1 and 2 (continued). Symbols are the same as those in Figure 5.1.

350-15	1593.37—1593.50 m	0.13 m	Mudstone, dark gray, horizontal bedding
350-16	1593.50—1593.56 m	0.06 m	Dolomite, olive black, massive
350-17	1593.56—1595.10 m	1.54 m	Medium sandstone, light gray, wavy bedding, parallel bedding, with a lot of charcoal debris
350-18	1595.10—1595.15 m	0.05 m	Dolomite, olive black, massive
350-19	1595.15—1595.60 m	0.45 m	Fine-grained sandstone, light gray, wavy bedding, with some mudstone laminae
350-20	1595.60—1596.45 m	0.85 m	Silty mudstone, dark gray, wavy bedding, with a lot of charcoal debris and siltstone laminae
350-21	1596.45—1596.48 m	0.03 m	Dolomite, olive black, massive
350-22	1596.48—1597.28 m	0.80 m	Muddy siltstone, gray, wavy bedding, with fine-grained laminae and lenticular sandstone beds
352-1	1597.50—1597.95 m	0.45 m	Silty mudstone, gray, wavy bedding, with a lot of charcoal debris and siltstone laminae
352-2	1597.95—1598.05 m	0.10 m	Fine-grained sandstone, light gray, parallel bedding, with mudstone laminae
352-3	1598.05—1598.45 m	0.40 m	Muddy siltstone, wavy bedding, with fine-grained laminae and lenticular sandstone beds
352-4	1598.45—1600.20 m	1.75 m	Siltstone interbedded with mudstone, light gray, wavy bedding, horizontal bedding, wavy cross-bedding and deformed bedding in local
352-5	1600.20—1601.00 m	0.80 m	Silty mudstone, gray, wavy bedding, with a lot of charcoal debris and siltstone laminae
352-6	1601.00—1602.10 m	1.10 m	Mudstone, dark gray, horizontal bedding, few ostracoda shells
352-7	1602.10—1602.18 m	0.08 m	Dolomitic mudstone, olive black, massive
353-1	1606.41—1610.71 m	4.30 m	Mudstone, dark gray, horizontal bedding, few ostracoda shells
353-2	1610.71—1610.75 m	0.04 m	Dolomitic mudstone, olive black, massive
353-3	1610.75—1611.96 m	1.21 m	Mudstone, dark gray, horizontal bedding, few ostracoda shells
353-4	1611.96—1612.04 m	0.08 m	Dolomitic mudstone, olive black, massive
353-5	1612.04—1615.41 m	3.37 m	Mudstone, dark gray, horizontal bedding, fine-grained sandstone laminae
354-1	1615.41—1621.03 m	5.62 m	Mudstone, dark gray, horizontal bedding, few ostracoda shells
354-2	1621.03—1621.06 m	0.03 m	Dolomitic mudstone, olive black, massive
354-3	1621.06—1624.12 m	3.06 m	Mudstone, dark gray, horizontal bedding, few ostracoda shells
355-1	1624.22—1627.67 m	3.45 m	Mudstone, dark gray, horizontal bedding, few ostracoda shells
355-2	1627.67—1627.72 m	0.05 m	Dolomite, olive black, massive
355-3	1627.72—1630.72 m	3.00 m	Mudstone, dark gray, horizontal bedding, few ostracoda shells
355-4	1630.72—1631.53 m	0.81 m	Mudstone, dark gray, horizontal bedding, few ostracoda shells

355-5	1631.53—1632.72 m	1.19 m	Mudstone, dark gray, horizontal bedding, few ostracoda shells
356-1	1632.72—1633.02 m	0.30 m	Mudstone, dark gray, horizontal bedding, few ostracoda shells
356-2	1633.02—1633.05 m	0.03 m	Dolomite, olive black, massive
356-3	1633.05—1634.45 m	1.40 m	Mudstone, dark gray, horizontal bedding, few ostracoda shells
356-4	1634.45—1634.53 m	0.08 m	Dolomite, olive black, massive
356-5	1634.53—1639.95 m	5.42 m	Mudstone, dark gray, horizontal bedding, few ostracoda shells
356-6	1639.95—1639.99 m	0.04 m	Dolomite, olive black, massive
356-7	1639.99—1640.70 m	0.71 m	Mudstone, dark gray, horizontal bedding, few ostracoda shells
356-8	1640.70—1640.73 m	0.03 m	Dolomite, olive black, massive
356-9	1640.73—1641.14 m	0.41 m	Mudstone, dark gray, horizontal bedding, few ostracoda shells
357-1	1641.14—1641.63 m	0.49 m	Mudstone, dark gray, horizontal bedding, few ostracoda shells
357-2	1641.63—1641.67 m	0.04 m	Dolomitic mudstone, olive black, massive
357-3	1641.67—1643.14 m	1.47 m	Mudstone, dark gray, horizontal bedding, few ostracoda shells
357-4	1643.14—1643.17 m	0.03 m	Dolomite, olive black, massive
357-5	1643.17—1644.01 m	0.84 m	Mudstone, dark gray, horizontal bedding, few ostracoda shells
357-6	1644.01—1644.11 m	0.10 m	Dolomite, olive black, massive
357-7	1644.11—1645.67 m	1.56 m	Mudstone, dark gray, horizontal bedding, few ostracoda shells
357-8	1645.67—1645.72 m	0.05 m	Dolomite, olive black, massive
357-9	1645.72—1646.79 m	1.07 m	Mudstone, dark gray, horizontal bedding, few ostracoda shells
357-10	1646.79—1646.87 m	0.08 m	Dolomite, olive black, massive
357-11	1646.87—1649.04 m	2.17 m	Mudstone, dark gray, horizontal bedding, few ostracoda shells
357-12	1649.04—1649.12 m	0.08 m	Dolomite, olive black, massive
357-13	1649.12—1649.59 m	0.47 m	Mudstone, dark gray, horizontal bedding, few ostracoda shells
358-1	1649.90—1651.62 m	1.72 m	Mudstone, dark gray, horizontal bedding, few ostracoda shells
358-2	1651.62—1651.65 m	0.03 m	Dolomite, olive black, massive
358-3	1651.65—1653.95 m	2.30 m	Mudstone, dark gray, horizontal bedding, few ostracoda shells
358-4	1653.95—1654.00 m	0.05 m	Dolomite, olive black, massive
358-5	1654.00—1655.91 m	1.91 m	Mudstone, dark gray, horizontal bedding, few ostracoda shells
358-6	1655.91—1655.96 m	0.05 m	Dolomite, olive black, massive
358-7	1655.96—1657.44 m	1.48 m	Mudstone, dark gray, horizontal bedding, few ostracoda shells
358-8	1657.44—1657.54 m	0.10 m	Dolomite, olive black, massive
358-9	1657.54—1658.38 m	0.84 m	Mudstone, dark gray, horizontal bedding, few ostracoda shells

359-1	1658.38—1660.78 m	2.40 m	Mudstone, dark gray, horizontal bedding, few ostracoda shells
359-2	1660.78—1661.08 m	0.30 m	Mudstone, dark gray, horizontal bedding, few ostracoda shells
359-3	1661.08—1662.28 m	1.20 m	Mudstone, dark gray, horizontal bedding, few ostracoda shells
359-4	1662.28—1662.36 m	0.08 m	Dolomite, olive black, massive
359-5	1662.36—1663.58 m	1.22 m	Mudstone, dark gray, horizontal bedding, few ostracoda shells
359-6	1663.58—1664.38 m	0.80 m	Mudstone, dark gray, horizontal bedding, few ostracoda shells
359-7	1664.38—1666.56 m	2.18 m	Mudstone, dark gray, horizontal bedding, few ostracoda shells
359-8	1666.56—1666.60 m	0.04 m	Dolomite, olive black, massive
359-9	1666.66—1667.02 m	0.36 m	Mudstone, dark gray, horizontal bedding, few ostracoda shells
360-1	1667.02—1668.14 m	1.12 m	Mudstone, dark gray, horizontal bedding, few ostracoda shells
360-2	1668.14—1668.19 m	0.05 m	Dolomite, olive black, massive
360-3	1668.19—1669.14 m	0.95 m	Mudstone, dark gray, horizontal bedding, few ostracoda shells
360-4	1669.14—1669.17 m	0.03 m	Dolomite, olive black, massive
360-5	1669.17—1671.77 m	2.60 m	Mudstone, dark gray, horizontal bedding, few ostracoda shells
360-6	1671.77—1671.92 m	0.15 m	Dolomite, olive black, massive
360-7	1671.92—1673.72 m	1.8 m	Mudstone, dark gray, horizontal bedding, few ostracoda shells
360-8	1673.72—1673.77 m	0.05 m	Dolomite, olive black, massive
360-9	1673.77—1674.48 m	0.71 m	Mudstone, dark gray, horizontal bedding, few ostracoda shells
360-10	1674.48—1674.53 m	0.05 m	Dolomite, olive black, massive
360-11	1674.53—1675.28 m	0.75 m	Mudstone, dark gray, horizontal bedding, few ostracoda shells
361-1	1675.48—1676.00 m	0.52 m	Mudstone, dark gray, horizontal bedding, few ostracoda shells
361-2	1676.00—1676.03 m	0.03 m	Dolomite, olive black, massive
361-3	1676.03—1677.33 m	1.30 m	Mudstone, dark gray, horizontal bedding, few ostracoda shells
361-4	1677.33—1677.35 m	0.02 m	Dolomite, olive black, massive
361-5	1677.35—1678.48 m	1.13 m	Mudstone, dark gray, horizontal bedding, few ostracoda shells
361-6	1678.48—1678.53 m	0.05 m	Dolomite, olive black, massive
361-7	1678.53—1680.68 m	2.15 m	Mudstone, dark gray, horizontal bedding, few ostracoda shells
361-8	1680.68—1681.48 m	0.80 m	Mudstone, dark gray, horizontal bedding, few ostracoda shells
361-9	1681.48—1683.46 m	1.98 m	Mudstone, dark gray, horizontal bedding, few ostracoda shells
361-10	1683.46—1683.51 m	0.05 m	Dolomite, olive black, massive
361-11	1683.51—1684.13 m	0.62 m	Mudstone, dark gray, horizontal bedding, few ostracoda shells
362-1	1684.13—1685.49 m	1.36 m	Mudstone, dark gray, horizontal bedding, few ostracoda shells
362-2	1685.49—1685.53 m	0.04 m	Dolomite, olive black, massive

362-3	1685.53—1686.78 m	1.25 m	Mudstone, dark gray, horizontal bedding, few ostracoda shells
362-4	1686.78—1686.85 m	0.07 m	Dolomite, olive black, massive
362-5	1686.85—1690.45 m	3.60 m	Mudstone, dark gray, horizontal bedding, few ostracoda shells
362-6	1690.45—1690.53 m	0.08 m	Dolomite, olive black, massive
362-7	1690.53—1691.23 m	0.70 m	Mudstone, dark gray, horizontal bedding, few ostracoda shells
362-8	1691.23—1692.80 m	1.57 m	Mudstone, dark gray, horizontal bedding, few ostracoda shells
363-1	1692.80—1694.55 m	1.75 m	Mudstone, dark gray, horizontal bedding, few ostracoda shells
363-2	1694.55—1694.58 m	0.03 m	Dolomite, olive black, massive
363-3	1694.58—1695.30 m	0.72 m	Mudstone, dark gray, horizontal bedding, few ostracoda shells
363-4	1695.30—1695.38 m	0.08 m	Dolomite, olive black, massive
363-5	1695.38—1698.45 m	3.07 m	Mudstone, dark gray, horizontal bedding, few ostracoda shells
363-6	1698.45—1698.50 m	0.05 m	Dolomite, olive black, massive
363-7	1698.50—1699.13 m	0.63 m	Mudstone, dark gray, horizontal bedding, few ostracoda shells
363-8	1699.13—1699.18 m	0.05 m	Dolomite, olive black, massive
363-9	1699.18—1699.80 m	0.62 m	Mudstone, dark gray, horizontal bedding, few ostracoda shells
363-10	1699.80—1699.83 m	0.03 m	Dolomite, olive black, massive
363-11	1699.83—1701.10 m	1.27 m	Mudstone, dark gray, horizontal bedding, few ostracoda shells
364-1	1701.15—1701.25 m	0.10 m	Mudstone, dark gray, horizontal bedding, few ostracoda shells
364-2	1701.25—1701.31 m	0.06 m	Dolomite, olive black, massive
364-3	1701.31—1704.70 m	3.39 m	Mudstone, dark gray, horizontal bedding, few ostracoda shells
364-4	1704.70—1704.75 m	0.05 m	Dolomite, olive black, massive
364-5	1704.75—1706.03 m	1.28 m	Mudstone, dark gray, horizontal bedding, few ostracoda shells
364-6	1706.03—1706.06 m	0.03 m	Dolomite, olive black, massive
364-7	1706.06—1706.65 m	0.59 m	Mudstone, dark gray, horizontal bedding, few ostracoda shells
364-8	1706.65—1706.68 m	0.03 m	Dolomite, olive black, massive
364-9	1706.68—1708.31 m	1.63 m	Mudstone, dark gray, horizontal bedding, few ostracoda shells
365-1	1708.42—1711.40 m	2.98 m	Mudstone, dark gray, horizontal bedding, few ostracoda shells
365-2	1711.40—1711.43 m	0.03 m	Dolomite, olive black, massive
365-3	1711.43—1712.60 m	1.17 m	Mudstone, dark gray, horizontal bedding, few ostracoda shells
365-4	1712.60—1712.70 m	0.10 m	Dolomite, olive black, massive
365-5	1712.70—1715.36 m	2.66 m	Mudstone, dark gray, horizontal bedding, few ostracoda shells
365-6	1715.36—1715.42 m	0.06 m	Dolomite, olive black, massive
365-7	1715.42—1716.12 m	0.70 m	Mudstone, dark gray, horizontal bedding, few ostracoda shells

365-8	1716.12—1716.16 m	0.04 m	Dolomite, olive black, massive
365-9	1716.16—1716.70 m	0.54 m	Mudstone, dark gray, horizontal bedding, few ostracoda shells
365-10	1716.70—1716.73 m	0.03 m	Dolomite, olive black, massive
365-11	1716.73—1717.01 m	0.28 m	Mudstone, dark gray, horizontal bedding, few ostracoda shells
366-1	1717.01—1719.11 m	2.10 m	Mudstone, dark gray, horizontal bedding, few ostracoda shells
366-2	1719.11—1719.14 m	0.03 m	Dolomite, olive black, massive
366-3	1719.14—1722.45 m	3.31 m	Mudstone, dark gray, horizontal bedding, few ostracoda shells
366-4	1722.45—1722.48 m	0.03 m	Dolomite, olive black, massive
366-5	1722.48—1723.81 m	1.33 m	Mudstone, dark gray, horizontal bedding, few ostracoda shells
366-6	1723.81—1723.87 m	0.06 m	Dolomite, olive black, massive
366-7	1723.87—1724.21 m	0.34 m	Mudstone, dark gray, horizontal bedding, few ostracoda shells
366-8	1724.21—1724.25 m	0.04 m	Dolomite, olive black, massive
366-9	1724.25—1725.64 m	1.39 m	Mudstone, dark gray, horizontal bedding, few ostracoda shells
367-1	1725.72—1726.34 m	0.62 m	Mudstone, dark gray, horizontal bedding, few ostracoda shells
367-2	1726.34—1726.39 m	0.05 m	Dolomite, olive black, massive
367-3	1726.39—1727.92 m	1.53 m	Mudstone, dark gray, horizontal bedding, few ostracoda shells
367-4	1727.92—1727.96 m	0.04 m	Dolomite, olive black, massive
367-5	1727.96—1730.69 m	2.73 m	Mudstone, dark gray, horizontal bedding, few ostracoda shells
367-6	1730.69—1730.73 m	0.04 m	Dolomite, olive black, massive
367-7	1730.73—1732.02 m	1.29 m	Mudstone, dark gray, horizontal bedding, few ostracoda shells
367-8	1732.02—1732.06 m	0.04 m	Dolomite, olive black, massive
367-9	1732.06—1733.92 m	1.86 m	Mudstone, dark gray, horizontal bedding, few ostracoda shells
367-10	1733.92—1733.95 m	0.03 m	Dolomite, olive black, massive
367-11	1733.95—1734.32 m	0.37 m	Mudstone, dark gray, horizontal bedding, few ostracoda shells
368-1	1734.39—1735.47 m	1.08 m	Mudstone, dark gray, horizontal bedding, few ostracoda shells
368-2	1735.47—1735.54 m	0.07 m	Dolomite, olive black, massive
368-3	1735.54—1736.64 m	1.10 m	Mudstone, dark gray, horizontal bedding, few ostracoda shells
368-4	1736.64—1736.69 m	0.05 m	Dolomite, olive black, massive
368-5	1736.69—1737.19 m	0.50 m	Mudstone, dark gray, horizontal bedding, few ostracoda shells
368-6	1737.19—1737.24 m	0.05 m	Dolomite, olive black, massive
368-7	1737.24—1737.69 m	0.45 m	Mudstone, dark gray, horizontal bedding, few ostracoda shells
368-8	1737.69—1737.71 m	0.02 m	Dolomite, olive black, massive
368-9	1737.71—1739.36 m	1.65 m	Mudstone, dark gray, horizontal bedding, few ostracoda shells

368-10	1739.36—1739.40 m	0.04 m	Dolomite, olive black, massive
368-11	1739.40—1741.04 m	1.64 m	Mudstone, dark gray, horizontal bedding, few ostracoda shells
368-12	1741.04—1741.09 m	0.05 m	Dolomite, olive black, massive
368-13	1741.09—1742.58 m	1.49 m	Mudstone, dark gray, horizontal bedding, few ostracoda shells
369-1	1742.58—1742.66 m	0.08 m	Dolomite, olive black, massive
369-2	1742.66—1745.53 m	2.87 m	Mudstone, dark gray, horizontal bedding, few ostracoda shells
369-3	1745.53—1745.58 m	0.05 m	Dolomite, olive black, massive
370-1	1745.58—1746.48 m	0.90 m	Mudstone, dark gray, horizontal bedding, few ostracoda shells
370-2	1746.48—1747.83 m	1.35 m	Mudstone, dark gray, horizontal bedding, few ostracoda shells
370-3	1747.83—1748.88 m	1.05 m	Mudstone, dark gray, horizontal bedding, few ostracoda shells
370-4	1748.88—1748.96 m	0.08 m	Dolomite, olive black, massive
370-5	1748.96—1753.68 m	4.72 m	Mudstone, dark gray, horizontal bedding, few ostracoda shells
370-6	1753.68—1754.29 m	0.61 m	Mudstone, dark gray, horizontal bedding, few ostracoda shells
371-1	1754.29—1755.59 m	1.30 m	Mudstone, dark gray, horizontal bedding, few ostracoda shells
371-2	1755.59—1756.34 m	0.75 m	Mudstone, dark gray, horizontal bedding, few ostracoda shells
371-3	1756.34—1756.49 m	0.15 m	Mudstone, grey black, horizontal bedding, few ostracoda shells, with an oil smell
371-4	1756.49—1760.19 m	3.70 m	Mudstone, dark gray, horizontal bedding, few ostracoda shells
371-5	1760.19—1761.49 m	1.30 m	Mudstone, dark gray, horizontal bedding, few ostracoda shells
371-6	1761.49—1762.94 m	1.45 m	Mudstone, dark gray, horizontal bedding, few ostracoda shells
372-1	1762.94—1765.05 m	2.10 m	Mudstone, dark gray, horizontal bedding, few ostracoda shells
372-2	1765.05—1765.15 m	0.10 m	Dolomite, olive black, massive
372-3	1765.15—1767.97 m	2.82 m	Mudstone, grey black, horizontal bedding, few ostracoda shells, with an oil smell
372-4	1767.97—1767.98 m	0.01 m	Tuff, greenish gray
372-5	1767.98—1768.55 m	0.57 m	Mudstone, grey black, horizontal bedding, few ostracoda shells, with an oil smell
372-6	1768.55—1768.56 m	0.01 m	Mudstone, dark gray, horizontal bedding, few ostracoda shells
372-7	1768.56—1770.90 m	2.34 m	Mudstone, grey black, horizontal bedding, few ostracoda shells, with an oil smell
372-8	1770.90—1771.00 m	0.10 m	Dolomite, olive black, massive
372-9	1771.00—1771.53 m	0.53 m	Mudstone, grey black, horizontal bedding, few ostracoda shells, with an oil smell
373-1	1771.53—1773.88 m	2.35 m	Mudstone, grey black, horizontal bedding, few ostracoda shells, with an oil smell

373-2	1773.88—1773.89 m	0.01 m	Dolomite, olive black, massive
373-3	1773.89—1775.26 m	1.37 m	Mudstone, grey black, horizontal bedding, few ostracoda shells, with an oil smell
373-4	1775.26—1776.93 m	1.67 m	Oil shale, olive black, laminated, with an oil smell
373-5	1776.93—1776.94 m	0.01 m	Dolomite, olive black, massive
373-6	1776.94—1777.62 m	0.68 m	Oil shale, olive black, laminated, with an oil smell
373-7	1777.62—1778.43 m	0.81 m	Mudstone, grey black, horizontal bedding, few ostracoda shells, with an oil smell
373-8	1778.43—1778.46 m	0.03 m	Dolomite, olive black, massive
373-9	1778.46—1778.93 m	0.47 m	Mudstone, grey black, horizontal bedding, few ostracoda shells, with an oil smell
373-10	1778.93—1779.88 m	0.95 m	Mudstone, dark gray, horizontal bedding, few ostracoda shells
373-11	1779.88—1780.06 m	0.18 m	Silty mudstone, dark gray, wavy bedding, bedding locally deformed
374-1	1780.06—1780.56 m	0.50 m	Mudstone, dark gray, horizontal bedding, few ostracoda shells
374-2	1780.56—1781.86 m	1.30 m	Mudstone, dark gray, horizontal bedding, few ostracoda shells
374-3	1781.86—1781.96 m	0.10 m	Mudstone interbeded with muddy siltstone, dark gray, horizontal bedding, muddy siltstone laminae
374-4	1781.96—1783.01 m	1.05 m	Mudstone, dark gray, horizontal bedding, few ostracoda shells
374-5	1783.01—1783.16 m	0.15 m	Mudstone, grey black, horizontal bedding, few ostracode shells, with an oil smell

The overlying formation of the Nenjiang Formation Member 1 is the Nenjiang Formation Member 2, the contacting relationship between them is conformable contact (Figure 5.5).

The detailed description of the Nenjiang Formation Member 1:

10-1	1025.13—1025.82 m	0.69 m	Mudstone, dark gray, horizontal bedding, fossil fragments occasionally
10-2	1025.82—1026.77 m	0.95 m	Mudstone, olive-gray, massive bedding, conchoidal fracture, horizontal ripple bedding at the base, fossil fragments occasionally
10-3	1026.77—1026.87 m	0.10 m	Lime mudstone, light olive gray, massive
10-4	1026.87—1026.90 m	0.03 m	Mudstone, light olive gray, intercalated with light olive gray siltstone, wavy bedding
10-5	1026.90—1028.17 m	1.27 m	Dolomite, light olive gray, lenticular bedding at the base, synsedimentary deformation structures (water-escape structure, liquefaction-induced deformation structure, interlayer synsedmentary concretion) in the middle part, interlayer fluidized flow-dragged structure, echelon foreset structure, and fill structure in the upper part
10-7	1028.20—1028.59 m	0.39 m	Mudstone, olive gray, horizontal bedding, fossil fragments

10-8	1028.59—1029.91 m	1.32 m	Mudstone, medium dark gray, horizontal lamination, plentiful ostracoda fossils, some fossil fragments, flat lenticular ostracoda limestone at 1028.82 m
10-9	1029.91—1030.32 m	0.41 m	Mudstone, dark greenish-gray, interbedded with medium light gray calcareous coarse siltstone, discontinuous bioclast layer at the top, slump, deformation corrugation structures in the middle part, sand balls, sand pillow, load structure, flame structure, roll-up structure, deformation bedding and lenticular bedding (in-situ slump, not moved), lenticular and banded sandstone at the base, horizontal ripple bedding inside, no clear erosion surface, discontinuous bioclast laminae at the top
11-1	1030.32—1030.64 m	0.32 m	Mudstone, dark greenish gray, massive, ostracoda fossils and pyrite
11-2	1030.64—1030.75 m	0.11 m	Mudstone, dark greenish-gray, interbedded with medium light gray calcareous coarse siltstone, horizontal lamination, wave lamination, deformation lamination in mudstone, parallel bedding in coarse sandstone, deformation and roll-up structures, calcareous fossil fragments in coarse siltstone
11-3	1030.75—1032.82 m	2.07 m	Mudstone, dark greenish gray, horizontal bedding, with conchostracon and ostracoda fossils
11-4	1032.82—1033.09 m	0.27 m	Mudstone, olive gray, horizontal bedding
11-5	1033.09—1034.41 m	1.32 m	Mudstone, dark gray, horizontal bedding
12-1	1036.13—1036.65 m	0.52 m	Mudstone, dark gray, horizontal lamination, incomplete ostracoda fossils at the base, syngenetic light olive gray dolomite concretion at 1036.54m (occurred along the strata, horizontal lamination inside)
12-2	1036.65—1036.68 m	0.03 m	Dolomite, light olive gray, interbedded with dark gray mudstone, horizontal bedding
12-3	1036.68—1037.53 m	0.85 m	Mudstone, dark greenish gray, horizontal bedding, with ostracoda fossils occasionally
12-4	1037.53—1038.03 m	0.50 m	Mudstone, greenish gray, horizontal bedding
12-5	1038.03—1038.43 m	0.40 m	Mudstone, dark greenish gray, horizontal bedding
12-6	1038.43—1039.18 m	0.75 m	Mudstone, dark greenish gray, horizontal bedding
12-7	1039.18—1039.24 m	0.06 m	Dolomite, olive black, massive
12-8	1039.24—1040.70 m	1.46 m	Mudstone, dark greenish gray, horizontal bedding
12-9	1040.70—1040.79 m	0.09 m	Dolomite, light olive gray, with concretion occurred along the strata, dispersive lamination inside
12-10	1040.79—1041.29 m	0.50 m	Mudstone, dark greenish gray, horizontal bedding, with ostracoda fossil fragments
12-11	1041.29—1041.35 m	0.06 m	Dolomite, light olive gray, horizontal bedding
12-12	1041.35—1042.05 m	0.70 m	Mudstone, dark greenish gray, horizontal bedding, with ostracoda fossil fragments
12-13	1042.05—1042.12 m	0.07 m	Dolomite, light olive gray
12-14	1042.12—1042.21 m	0.09 m	Mudstone, olive gray, horizontal bedding, with dolomite strips
12-15	1042.21—1043.93 m	1.72 m	Mudstone, medium dark gray, horizontal lamination, two light olive gray dolomite concretions at 1042.70 m and 1042.83 m, deepest 7 cm, thinnest 3 cm

12-16	1043.93—1044.07 m	0.14 m	Dolomite, light olive black, massive
12-17	1044.07—1044.57 m	0.50 m	Mudstone, medium dark gray, horizontal lamination, rudite dolomite, sand dolomite strips and lens at the top (due to turbidity current), erosion surface at the base
12-18	1044.57—1045.53 m	0.96 m	Mudstone, dark greenish-gray, horizontal lamination, two light olive gray dolomite concretions at 1044.79 m and 1044.86 m, thickest 5 cm, thinnest 2 cm
12-19	1045.53—1048.24 m	2.71 m	Mudstone, olive black, horizontal lamination, two light olive gray dolomite concretions at 1046.27 m and 1048.24 m, thickest 5 cm, thinnest 2 cm, calcareous bioclast layer at 1047.12 m
13-1	1048.24—1049.98 m	1.74 m	Mudstone, dark greenish-gray, horizontal lamination, continuous bedding in sandstone around and bypass the edge of concretion, ostracoda fossils occasionally, synsedimentary light olive gray dolomite concretion (diameter 12 cm) at 1048.69 m
13-2	1049.98—1050.44 m	0.46 m	Mudstone, dark gray, massive. Mudstone, dark gray, massive structure
13-3	1050.44—1053.70 m	3.26 m	Mudstone, dark greenish-gray, horizontal lamination, ostracoda fossils occasionally, two 1 cm-thick discontinuous light olive gray dolomite laminae at 1050.44 m, 5 cm thick interlayer at 1056.88 m, 1—2.5 cm thick interlayer at 1051.76 m
13-4	1053.70—1053.80 m	0.10 m	Dolomite, olive black, massive
13-5	1053.80—1055.71 m	1.91 m	Mudstone, dark greenish-gray, horizontal lamination, integral ostracoda fossils with size of 1 mm, banded or flaky ostracoda locally
13-6	1055.71—1055.81 m	0.10 m	Dolomite, olive gray, horizontal bedding
13-7	1055.81—1056.64 m	0.83 m	Mudstone, dark greenish-gray, horizontal lamination, ostracoda fossils occasionally, 2 cm olive gray dolomite laminae at 1056.12 m, with 4 biologic fugichnia above (strata penetrated in horizontal mudstones in lower and upper parts of dolomite section), 2nd 0.5—2 cm thick olive gray dolomite layer at 1056.14 m
13-8	1056.64—1056.74 m	0.10 m	Dolomite, olive black, massive
13-9	1056.74—1060.14 m	3.40 m	Mudstone, dark greenish-gray, horizontal lamination, integral ostracoda fossils with size of 1 mm, banded or flaky ostracoda locally, 5 cm thick olive gray dolomite layer at 1057.46 m, a dolomite concretion (up to 6 cm thick) at 1058.16 m, with a calcspar vein, a dolomite concretion (up to 6 cm thick) at 1058.39 m, 5—8 cm thick dolomite layer at 1058.94 m, a dolomite concretion (up to 5 cm thick) at 1059.39 m, 5—8 cm thick dolomite layer at 1059.50 m
13-10	1060.14—1060.25 m	0.11 m	Dolomite, olive black, massive
14-1	1060.25—1061.82 m	1.57 m	Mudstone, medium dark gray, horizontal bedding, discontinuous marlite laminae with thickness of 1mm, ostracoda occasionally
14-2	1061.82—1061.86 m	0.04 m	Dolomite, light gray, massive
14-3	1061.86—1063.85 m	1.99 m	Mudstone, dark brown, horizontal bedding
14-4	1063.85—1063.9 m	0.05 m	Dolomite, light gray, massive

14-5	1063.90—1065.19 m	1.29 m	Mudstone, dark brown, horizontal bedding
15-1	1065.19—1066.34 m	1.15 m	Mudstone, dark brown, horizontal bedding
15-2	1066.34—1066.37 m	0.03 m	Dolomite, light gray, massive
15-3	1066.37—1067.85 m	1.48 m	Mudstone, dark brown, horizontal bedding
15-4	1067.85—1067.89 m	0.04 m	Dolomite, light gray, oil-bearing bedded, unclear bedding, strong oil smell
15-5	1067.89—1068.69 m	0.80 m	Mudstone, dark brown, horizontal bedding
15-6	1068.69—1068.78 m	0.09 m	Dolomite, light gray, horizontal bedding
15-7	1068.78—1069.21 m	0.43 m	Mudstone, medium dark gray, horizontal bedding, discontinuous marlite laminae locally, fossil fragments
16-1	1069.21—1070.71 m	1.50 m	Mudstone, dark brown, horizontal bedding
16-2	1070.71—1070.76 m	0.05 m	Marl, light gray, horizontal wavy ripple bedding
16-3	1070.76—1073.71 m	2.95 m	Mudstone, dark brown, horizontal bedding
16-4	1073.71—1073.74 m	0.03 m	Dolomite, light gray, massive
16-5	1073.74—1074.32 m	0.58 m	Mudstone, dark brown, horizontal bedding
17-1	1074.32—1075.27 m	0.95 m	Mudstone, dark brown, horizontal bedding
17-2	1075.27—1075.30 m	0.03 m	Marl, gray, massive
17-3	1075.30—1077.22 m	1.92 m	Mudstone, dark brown, horizontal bedding
17-4	1077.22—1077.47 m	0.25 m	Mudstone, dark gray, horizontal bedding
17-5	1077.47—1078.17 m	0.70 m	Mudstone, dark brown, horizontal bedding
17-6	1078.17—1078.26 m	0.09 m	Mudstone, dark gray, horizontal bedding
17-7	1078.26—1078.82 m	0.56 m	Mudstone, dark brown, horizontal bedding
18-1	1078.82—1079.62 m	0.80 m	Mudstone, dark brown, horizontal bedding
18-2	1079.62—1079.82 m	0.20 m	Mudstone, gray, horizontal bedding
18-3	1079.82—1081.12 m	1.30 m	Mudstone, medium dark gray, horizontal bedding, light gray marlite interlayer at 1080.92 m
18-4	1081.12—1081.34 m	0.22 m	Dolomite, light olive gray, with horizontal bedding
18-5	1081.34—1081.97 m	0.63 m	Mudstone, dark brown, horizontal bedding
18-6	1081.97—1082.00 m	0.03 m	Dolomite, light gray, horizontal bedding
18-7	1082.00—1083.28 m	1.28 m	Mudstone, medium dark gray, horizontal bedding, 1cm thick dolomite interlayer separately at 1082.47 m and 1082.62 m, oil bearing
18-8	1083.28—1083.36 m	0.08 m	Dolomite, light olive, unclear bedding, poorly zonal (breccia), with fractures, and strong oil smell
18-9	1083.36—1083.50 m	0.14 m	Mudstone, dark brown, horizontal bedding
19-1	1083.50—1084.00 m	0.50 m	Mudstone, dark brown, horizontal bedding

19-2	1084.00—1085.30 m	1.30 m	Mudstone, gray, horizontal bedding
19-3	1085.30—1086.90 m	1.60 m	Mudstone, dark brown, horizontal bedding
19-4	1086.90—1087.96 m	1.06 m	Mudstone, gray, horizontal bedding
20-1	1087.96—1088.06 m	0.10 m	Mudstone, light gray, horizontal bedding
20-2	1088.06—1088.86 m	0.80 m	Mudstone, dark brown, horizontal bedding
20-3	1088.86—1088.96 m	0.10 m	Mudstone, light gray, horizontal bedding
20-4	1088.96—1089.86 m	0.90 m	Mudstone, dark brown, horizontal bedding
20-5	1089.86—1090.06 m	0.20 m	Mudstone, gray, horizontal bedding
20-6	1090.06—1090.36 m	0.30 m	Mudstone, olive-gray, horizontal bedding, with oil smell, possible oil generating
20-7	1090.36—1091.29 m	0.93 m	Mudstone, gray, horizontal bedding
20-8	1091.29—1091.53 m	0.24 m	Mudstone, olive gray, horizontal bedding
20-9	1091.53—1092.00 m	0.47 m	Mudstone, gray, horizontal bedding
20-10	1092.00—1092.10 m	0.10 m	Tuff, light gray
20-11	1092.10—1092.30 m	0.20 m	Mudstone, olive gray, horizontal bedding
20-12	1092.30—1092.54 m	0.24 m	Mudstone, dark brown, horizontal bedding
21-1	1092.54—1093.10 m	0.56 m	Mudstone, grayish black, horizontal bedding
21-2	1093.10—1094.44 m	1.34 m	Mudstone, medium gray, horizontal bedding, 1cm thick calcspar and bioclast laminae at 1093.54 m
21-3	1094.44—1094.67 m	0.23 m	Oil shale, olive gray
21-4	1094.67—1094.72 m	0.05 m	Limestone, light gray, massive
21-5	1094.72—1095.44 m	0.72 m	Mudstone, olive-gray, horizontal bedding, 1cm thick recrystallized limestone at 1095.09 m, with much ostracoda
21-6	1095.44—1095.54 m	0.10 m	Oil shale, olive gray
21-7	1095.54—1095.74 m	0.20 m	Dolomite, light olive gray, wavy ripple bedding, with mudstone strips
21-8	1095.74—1096.38 m	0.64 m	Mudstone, gray, horizontal bedding
21-9	1096.38—1096.98 m	0.60 m	Mudstone, olive gray, horizontal bedding
21-10	1096.98—1097.12 m	0.14 m	Mudstone, dark brown, horizontal bedding
22-1	1097.12—1097.52 m	0.40 m	Mudstone, gray, horizontal bedding, ostracoda fossils locally
22-2	1097.52—1098.09 m	0.57 m	Oil shale, olive gray
22-3	1098.09—1100.57 m	2.48 m	Mudstone, light greenish gray, massive, ostracoda occasionally, fracture filled with calcspar
23-1	1100.57—1102.37 m	1.80 m	Mudstone, dark greenish-gray, unclear horizontal bedding (estimated by cake-like fragments, and flat fragment surface), few ostracoda fossils, calcareous concretions locally

23-2	1102.37—1103.23 m	0.86 m	Mudstone, dark greenish gray, horizontal bedding, with few ostracoda fossils, calcareous concretions locally
23-3	1103.23—1103.31 m	0.08 m	Mudstone, dark greenish gray, interbedded with light greenish gray limestone, with ostracoda fossils
23-4	1103.31—1103.92 m	0.61 m	Mudstone, dark greenish gray, horizontal bedding, with ostracoda fossils
24-1	1103.92—1104.59 m	0.67 m	Mudstone, dark greenish gray, horizontal bedding, with ostracoda fossils fragments
25-1	1105.24—1106.04 m	0.80 m	Mudstone, dark greenish gray, horizontal bedding, ostracoda fossils occasionally
25-2	1106.04—1108.44 m	2.40 m	Mudstone, slightly greenish-black, calcareous, unclear horizontal bedding (estimated by cake-like fragments, and flat fragment surface), few integral ostracoda fossils, with size of 0.5—1 mm
25-3	1108.44—1108.74 m	0.30 m	Mudstone, dark greenish gray, horizontal bedding, with ostracoda fossils
25-4	1108.74—1108.99 m	0.25 m	Mudstone, dark greenish gray, horizontal bedding, with ostracoda fossils
25-5	1108.99—1110.19 m	1.20 m	Mudstone, dark greenish gray, horizontal bedding
25-6	1110.19—1111.84 m	1.65 m	Mudstone, slightly greenish-black, interbedded with grayish-black thinly-bedded, banded and lenticular ostracoda clastics, unclear horizontal bedding (estimated by cake-like fragments, and flat fragment surface), lenticular bedding, ostracoda laminae, erosion surface at the base, large number of incomplete ostracoda in banded deposition, clastic (due to turbidite transportation)
25-7	1111.84—1114.91 m	3.07 m	Mudstone, olive gray, interbedded with light olive gray thinly-bedded, banded and lenticular ostracoda clastics, horizontal bedding and lenticular bedding, ripple bedding in ostracoda clastics, with much generally-incomplete ostracoda fossils, clastic, ostracoda clastics distributed in banded, lenticular and thinly-bedded manner in mudstone due to turbidite, oil-rich partially, 2 cm dark gray altered volcanic ash laminae at 1112.24 m, below which 0.5—1 cm grayish-black ostracoda is abundant, gastropod fossils at 1112.79 m and 1113.84 m
25-8	1114.91—1115.24 m	0.33 m	Mudstone, slightly greenish-gray, interbedded with dark gray thinly-bedded, banded and lenticular ostracoda clastics, horizontal bedding and lenticular bedding, ripple bedding in ostracoda clastics, with much generally-incomplete ostracoda fossils, clastic, ostracoda clastics distributed in banded, lenticular and thinly-bedded manner in mudstone due to turbidite, oil-rich partially, few conchostracon at the base
25-9	1115.24—1115.41 m	0.17 m	Dolomite, light olive gray, wavy ripple bedding
25-10	1115.41—1116.22 m	0.81 m	Mudstone, greenish-gray, interbedded with medium dark gray thinly bedded, banded or lenticular ostracoda clastics, horizontal or lenticular bedding, ripple bedding in ostracoda clastics, with ostracoda fossils, incomplete and fragmental, banded, lenticular or laminated distribution (due to turbidite deposition), some oil-bearing
26-1	1116.22—1116.39 m	0.17 m	Mudstone, dark brown, horizontal bedding, ostracoda fossils

26-2	1116.39—1116.83 m	0.44 m	Mudstone, calcareous, grayish-black, with horizontal bedding, 0.5 cm inferior oil shale at 1116.50 m, and ichthyolite and ostracoda-enriched layers at 1116.49 m
26-3	1116.83—1117.72 m	0.89 m	Mudstone, dark gray, horizontal bedding, with ostracoda fossils
26-4	1117.72—1119.19 m	1.47 m	Mudstone, calcareous, grayish-black, horizontal bedding, with ostracoda fossils at certain layers, and bivalve fossils at 1118.52 m
26-5	1119.19—1119.47 m	0.28 m	Mudstone, dark gray, horizontal bedding, with lots of ostracoda fossils
26-6	1119.47—1119.84 m	0.37 m	Mudstone, grayish black, horizontal bedding, with ostracoda fossils
26-7	1119.84—1120.19 m	0.35 m	Mudstone, calcareous, black, horizontal bedding, with ostracoda fossils at certain layers, and golden brown conchostracans fossils
26-8	1120.19—1120.77 m	0.58 m	Mudstone, black, horizontal bedding, with ostracoda fossils
26-9	1120.77—1121.12 m	0.35 m	Mudstone, grayish black, horizontal bedding, with ostracoda fossils
26-10	1121.12—1121.51 m	0.39 m	Mudstone, black, horizontal bedding, with ostracoda fossils
26-11	1121.51—1122.57 m	1.06 m	Mudstone, dark gray, horizontal bedding, with ostracoda fossils
26-12	1122.57—1122.72 m	0.15 m	Mudstone, black, horizontal bedding, with ostracoda fossils
26-13	1122.72—1123.52 m	0.80 m	Mudstone, dark gray, horizontal bedding, with ostracoda fossils
26-14	1123.52—1123.72 m	0.20 m	Mudstone, black, horizontal bedding, with ostracoda fossils
26-15	1123.72—1124.56 m	0.84 m	Mudstone, grayish black, horizontal bedding, with ostracoda fossils
26-16	1124.56—1125.04 m	0.48 m	Mudstone, grayish black, horizontal bedding, with ostracoda fossils occasionally
26-17	1125.04—1125.3 m	0.26 m	Mudstone, grayish black, horizontal bedding, with ostracoda fossils and conchostracon fossils
26-18	1125.30—1125.34 m	0.04 m	Oil shale, dark gray, with few ostracoda fossils
26-19	1125.34—1125.75 m	0.41 m	Mudstone, grayish black, horizontal bedding, with conchostracon and ostracoda fossils, pyrite
26-20	1125.75—1126.22 m	0.47 m	Mudstone, dark gray, interbedded with ostracoda clastic strips and lens, horizontal ripple and lenticular beddings developed, in ostracoda rocks, horizontal ripple bedding, and ostracoda fossils
26-21	1126.22—1126.74 m	0.52 m	Mudstone, grayish black, horizontal bedding, with ostracoda and conchostracon fossils
26-22	1126.74—1127.22 m	0.48 m	Mudstone, medium dark gray, interbedded with dark gray ostracoda clastic mudstone, horizontal bedding and horizontal ripple bedding developed, in calcareous bioclastic mudstone, horizontal ripple bedding and foreset laminae developed, with ostracoda lens, bioclast
26-23	1127.22—1127.33 m	0.11 m	Mudstone, dark brown, horizontal bedding, with ostracoda fossils
27-1	1127.33—1128.17 m	0.84 m	Mudstone, grayish black, horizontal bedding, with ostracoda fossils and conchostracon

5.1.6　Nenjiang Formation Members 3, 4 and 5

The overlying formation of Nenjiang Formation Member 5 is the Sifangtai Formation, the contacting relationship between them is unconformable contact; the underlying formation is the Nenjiang Formation Member 4, the contacting relationship between them is conformable contact (Figure 5.6).

The detailed description of the Nenjiang Formation Member 5:

248-2	1021.60—1021.94 m	0.34 m	Silty mudstone, light red, massive, with calcareous concretions
248-3	1021.94—1022.63 m	0.69 m	Mudstone, light reddish brown, massive, calcareous concretions occasionally
248-4	1022.63—1023.69 m	1.06 m	Silty mudstone, light reddish brown, massive, with light greenish gray siltstone strips locally, calcareous concretions occasionally
248-5	1023.69—1023.91 m	0.22 m	Silty mudstone, light red, massive bedding, light greenish-gray impregnated spots locally
248-6	1023.91—1024.31 m	0.40 m	Muddy siltstone, light greenish gray, massive
248-7	1024.31—1024.52 m	0.21 m	Siltstone, light greenish gray, massive
248-8	1024.52—1025.31 m	0.79 m	Muddy siltstone, light reddish brown, massive, calcareous concretions occasionally
248-9	1025.31—1025.81 m	0.50 m	Silty mudstone, light reddish brown, massive
248-10	1025.81—1026.52 m	0.71 m	Muddy siltstone, light reddish brown, massive, calcareous concretions occasionally
248-11	1026.52—1026.87 m	0.35 m	Mudstone, grayish red, massive
248-12	1026.87—1027.32 m	0.40 m	Muddy siltstone, light greenish gray, massive
248-13	1027.32—1027.57 m	0.25 m	Siltstone, grayish red, massive
248-14	1027.57—1027.77 m	0.20 m	Muddy siltstone, light reddish brown, massive
248-15	1027.77—1028.33 m	0.56 m	Siltstone, light reddish brown, massive
248-16	1028.33—1028.44 m	0.11 m	Mudstone, grayish red, massive
250-1	1028.70—1029.80 m	1.10 m	Mudstone, grayish red, massive
250-2	1029.80—1031.75 m	1.95 m	Mudstone, grayish red, massive, with calcareous concretions
250-3	1031.75—1032.40 m	0.65 m	Silty mudstone, grayish red, massive
250-4	1032.40—1032.86 m	0.46 m	Siltstone, light greenish gray and grayish red, slump structures
250-5	1032.86—1033.43 m	0.57 m	Muddy siltstone, grayish red, wave ripple bedding, with siltstone strips and mudstone strips
250-6	1033.43—1034.01 m	0.58 m	Siltstone, light greenish gray, wave ripple bedding, with fine-grained sandstone and mudstone laminae
251-1	1034.01—1034.49 m	0.48 m	Sandstone, fine-grained, light gray, oblique bedding in the upper part, swash cross bedding in the lower part, parallel bedding at the base
251-2	1034.49—1034.86 m	0.37 m	Muddy siltstone, light gray, wavy bedding, with mudstone laminae and strips

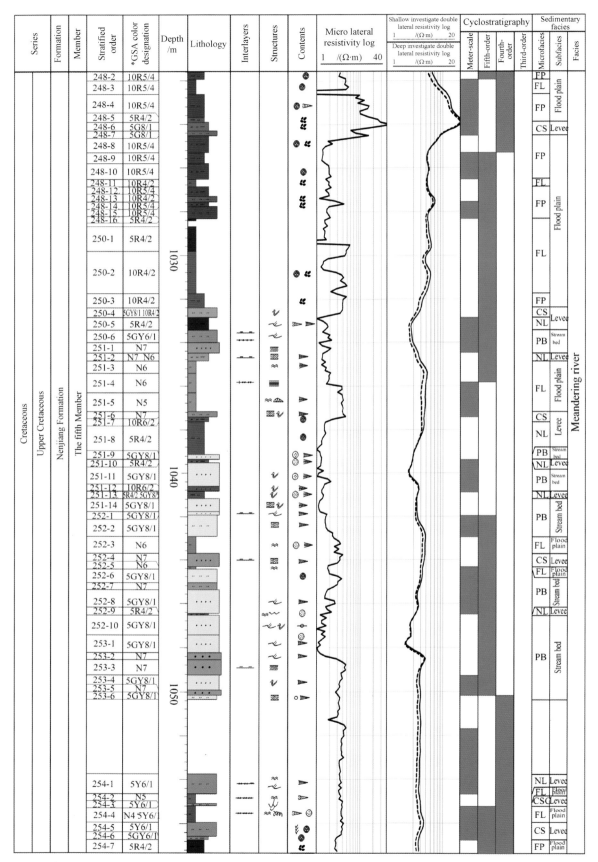

Figure 5.6 Stratigraphic chart of Nenjiang Formation Members 3, 4 and 5.

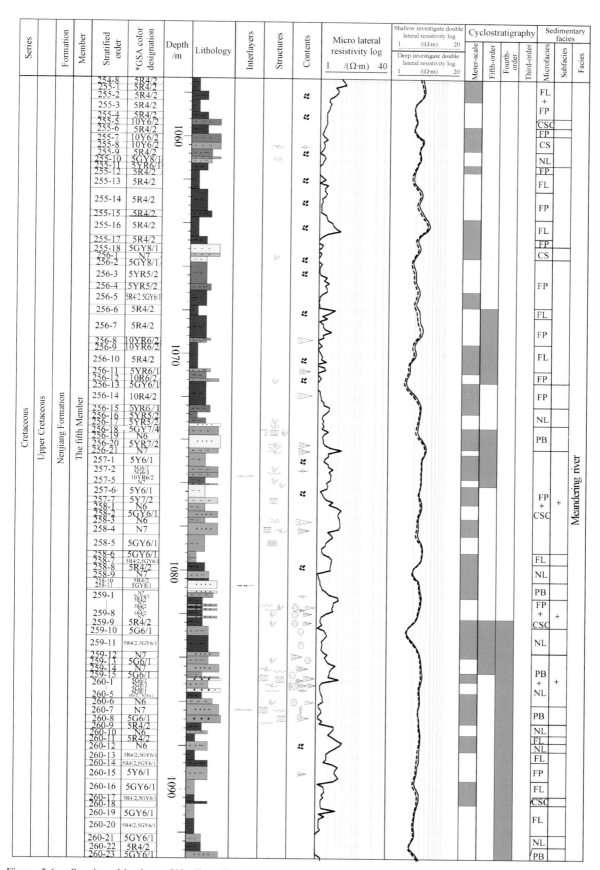

Figure 5.6　Stratigraphic chart of Nenjiang Formation Members 3, 4 and 5 (continued).

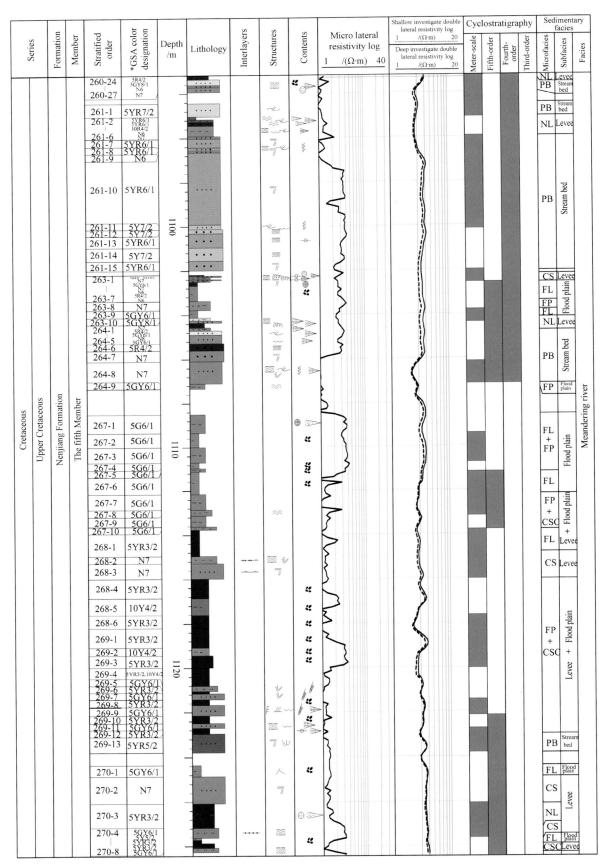

Figure 5.6　Stratigraphic chart of Nenjiang Formation Members 3, 4 and 5 (continued).

Section 5 SK-1 Core Description and Core Photographs

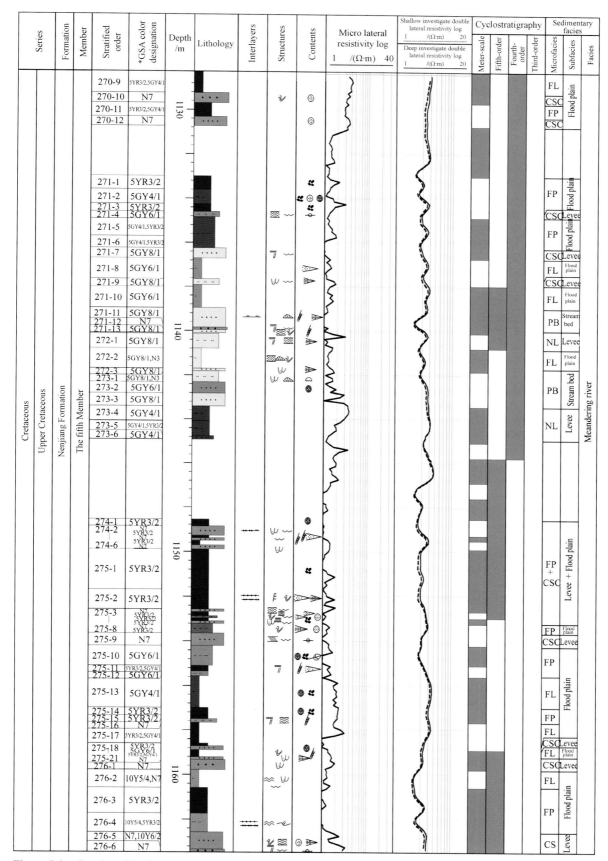

Figure 5.6 Stratigraphic chart of Nenjiang Formation Members 3, 4 and 5 (continued).

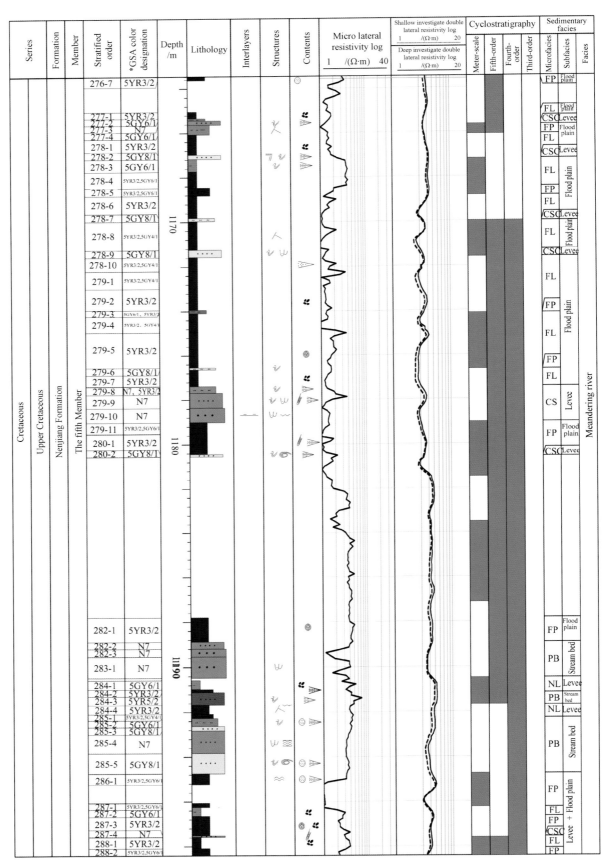

Figure 5.6 Stratigraphic chart of Nenjiang Formation Members 3, 4 and 5 (continued).

Section 5 SK-1 Core Description and Core Photographs

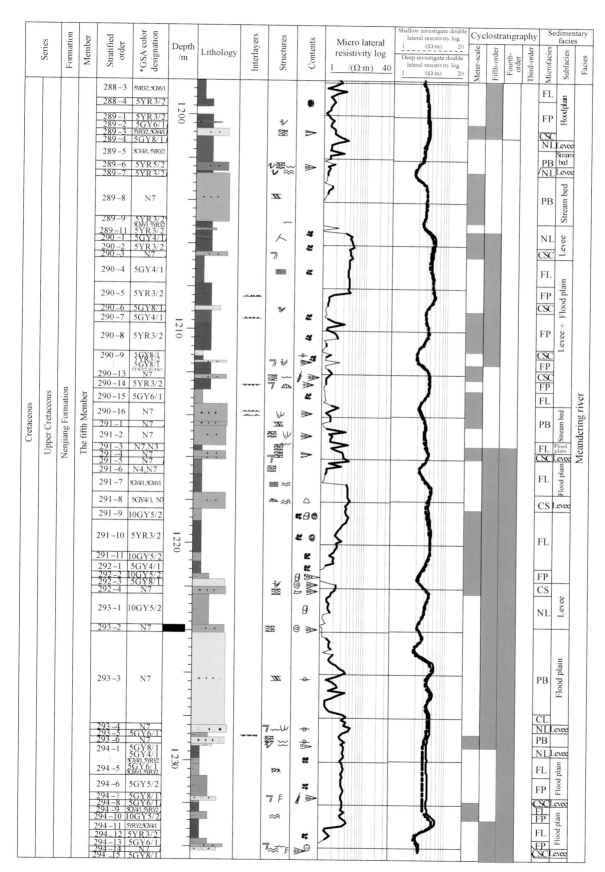

Figure 5.6 Stratigraphic chart of Nenjiang Formation Members 3, 4 and 5 (continued).

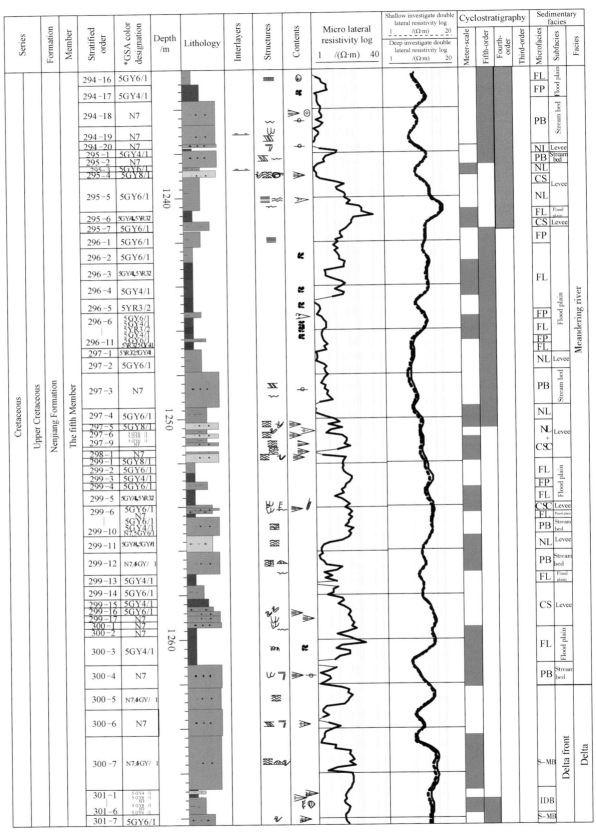

Figure 5.6 Stratigraphic chart of Nenjiang Formation Members 3, 4 and 5 (continued).

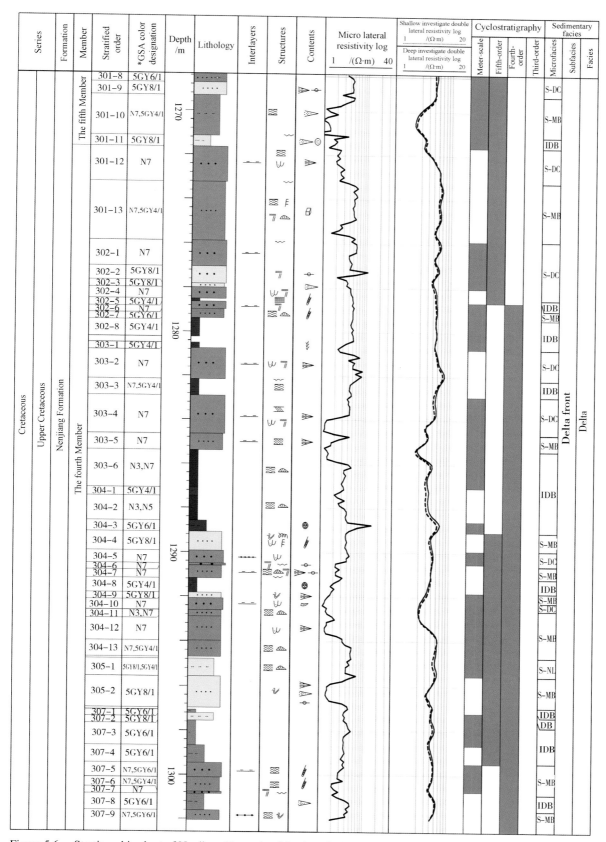

Figure 5.6　Stratigraphic chart of Nenjiang Formation Members 3, 4 and 5 (continued).

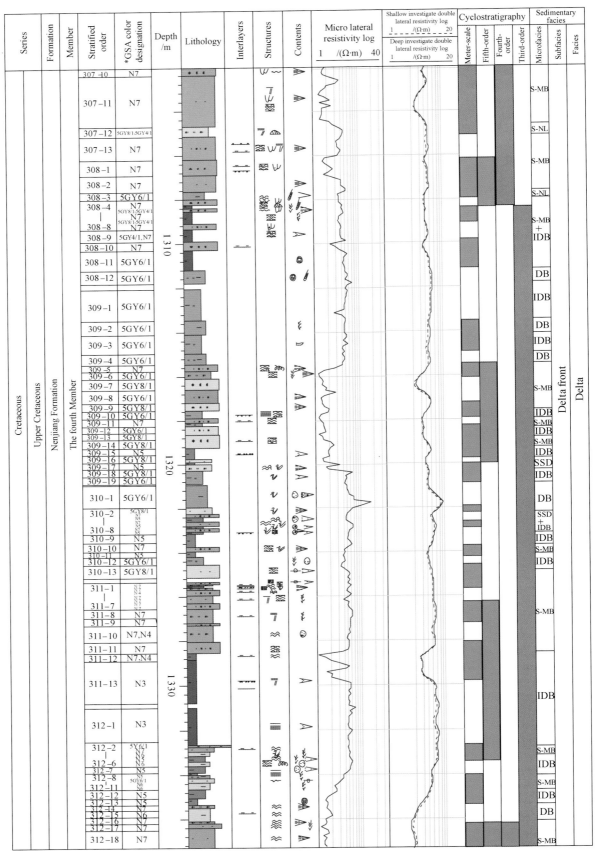

Figure 5.6 Stratigraphic chart of Nenjiang Formation Members 3, 4 and 5 (continued).

Section 5 SK-1 Core Description and Core Photographs | 237

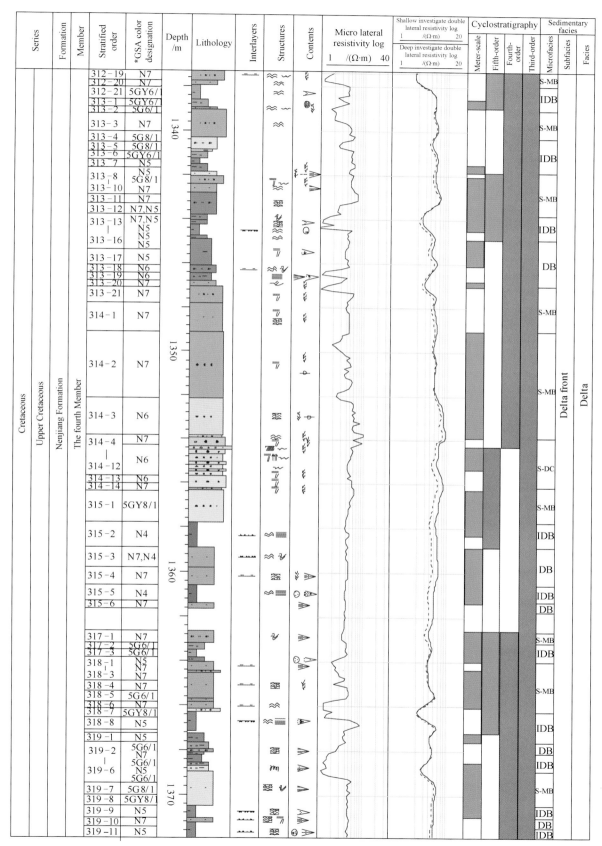

Figure 5.6 Stratigraphic chart of Nenjiang Formation Members 3, 4 and 5 (continued).

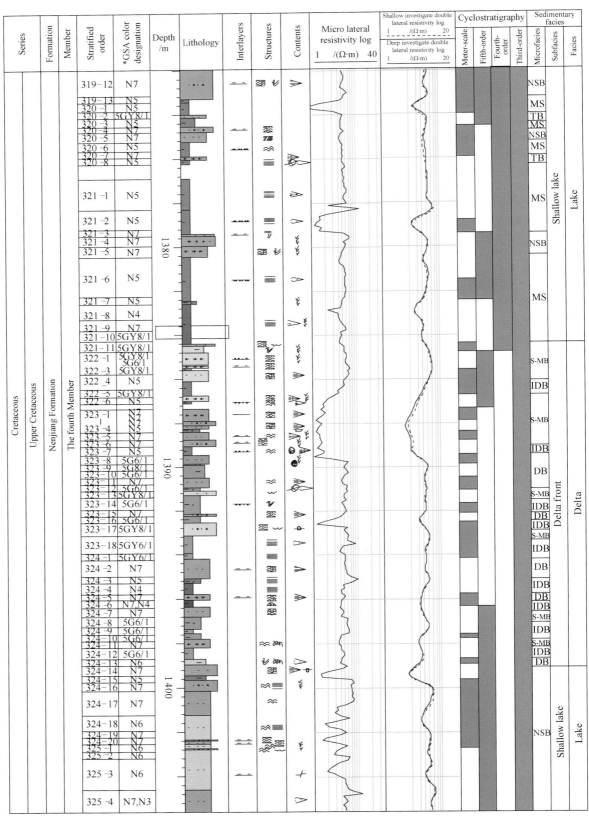

Figure 5.6   Stratigraphic chart of Nenjiang Formation Members 3, 4 and 5 (continued).

Section 5 SK-1 Core Description and Core Photographs

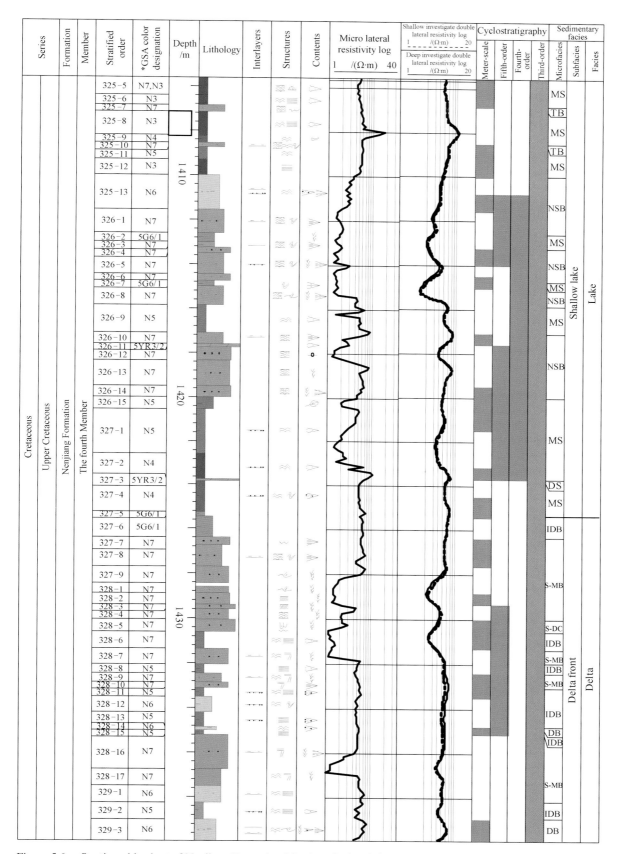

Figure 5.6 Stratigraphic chart of Nenjiang Formation Members 3, 4 and 5 (continued).

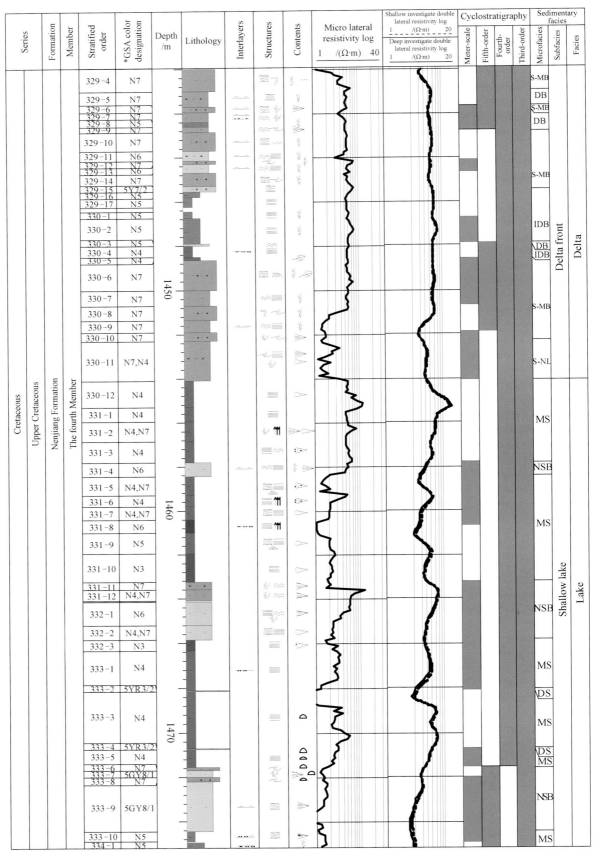

Figure 5.6　Stratigraphic chart of Nenjiang Formation Members 3, 4 and 5 (continued).

Section 5 SK-1 Core Description and Core Photographs

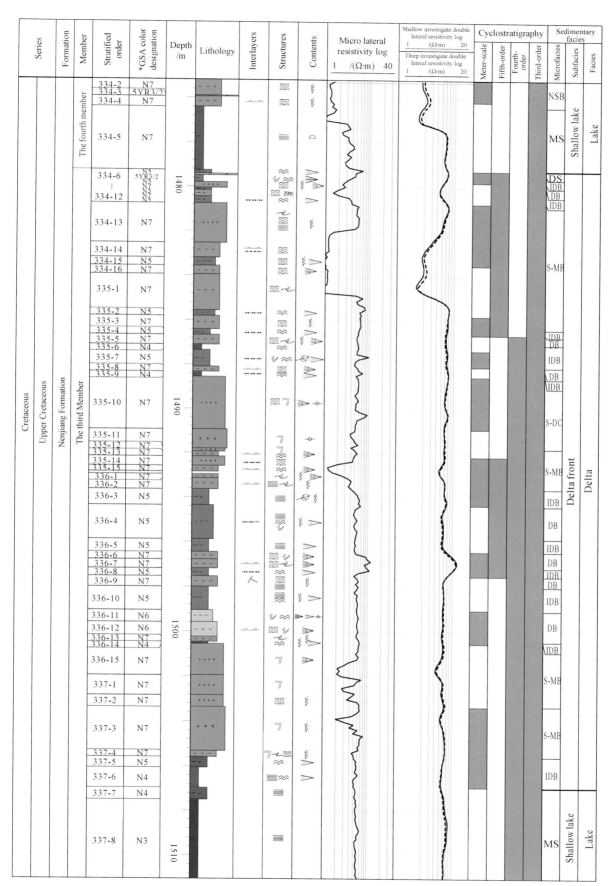

Figure 5.6 Stratigraphic chart of Nenjiang Formation Members 3, 4 and 5 (continued).

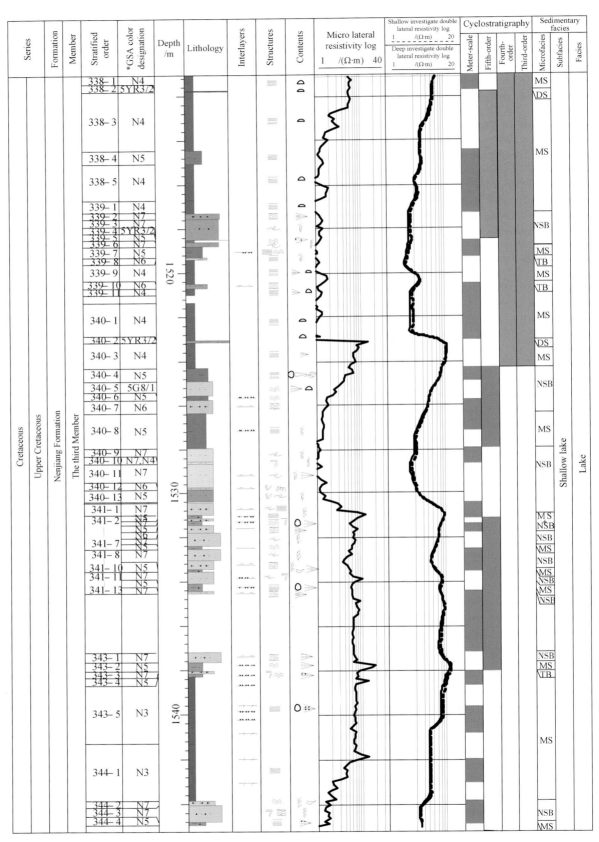

Figure 5.6 Stratigraphic chart of Nenjiang Formation Members 3, 4 and 5 (continued).

Section 5 SK-1 Core Description and Core Photographs

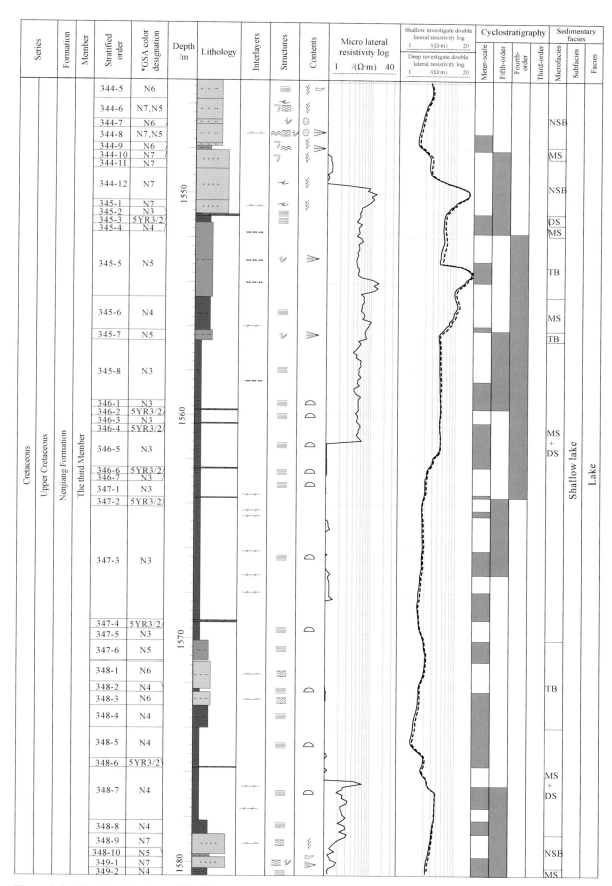

Figure 5.6　Stratigraphic chart of Nenjiang Formation Members 3, 4 and 5 (continued).

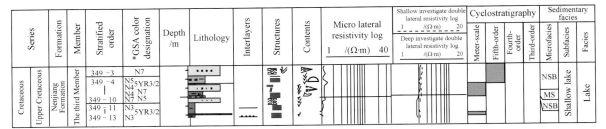

Figure 5.6 Stratigraphic chart of Nenjiang Formation Members 3, 4 and 5 (continued). Symbols are the same as those in Figure 5.1.

251-3	1034.86—1035.41 m	0.55 m	Mudstone, light gray, intercalated with light gray muddy siltstone, wavy bedding, siltstone strips
251-4	1035.41—1036.31 m	0.90 m	Mudstone, light gray, horizontal bedding, with muddy siltstone locally
251-5	1036.31—1037.16 m	0.85 m	Mudstone, gray, wavy bedding, with siltstone strips
251-6	1037.16—1037.41 m	0.25 m	Siltstone, light gray, wavy bedding, deformation bedding, with mudstone strips
251-7	1037.41—1037.63 m	0.22 m	Muddy siltstone, light reddish brown, massive, with few calcareous concretions
251-8	1037.63—1039.13 m	1.50 m	Silty mudstone, grayish red, massive, calcareous concretions occasionally
251-9	1039.13—1039.36 m	0.23 m	Fine-grained sandstone, light greenish gray, massive, grayish red silty mudstone strips
251-10	1039.36—1039.51 m	0.15 m	Muddy siltstone, grayish red, massive, with fine-grained sandstone, light greenish gray strips
251-11	1039.51—1040.61 m	1.10 m	Fine-grained sandstone, light greenish gray, deformation bedding, with grayish red silty mudstone strips
251-12	1040.61—1040.83 m	0.22 m	Sandstone, fine-grained, light red, massive bedding, interbedded with grayish-red silty mudstone laminae at 1040.66 m
251-13	1040.83—1041.16 m	0.33 m	Silty mudstone, grayish red, deformation bedding, with fine-grained sandstone, light greenish gray strips
251-14	1041.16—1041.77 m	0.61 m	Fine-grained sandstone, light greenish gray, wavy bedding, with grayish red silty mudstone strips
252-1	1041.77—1041.92 m	0.15 m	Fine-grained sandstone, light greenish gray, wave ripple bedding, with mudstone strips and laminaes
252-2	1041.92—1042.92 m	1.00 m	Siltstone, light greenish gray, wavy bedding, with mudstone strips
252-3	1042.92—1043.72 m	0.80 m	Mudstone, medium light gray, horizontal ripple bedding, fine lenticular and banded sandstones
252-4	1043.72—1044.32 m	0.60 m	Fine-grained sandstone, light gray, wavy bedding, with mudstone laminae and mudstone strips
252-5	1044.32—1044.42 m	0.10 m	Mudstone, light gray, wavy ripple bedding
252-6	1044.42—1045.07 m	0.65 m	Siltstone, light greenish gray, massive, with calcareous concretions

252-7	1045.07—1045.39 m	0.32 m	Siltstone, light gray, massive
252-8	1045.39—1046.45 m	1.06 m	Sandstone, fine-grained, light greenish-gray wavy cross bedding, erosion surface at the base, few mudstone laminae on the bedding surface
252-9	1046.45—1046.52 m	0.07 m	Muddy siltstone, grayish red, wavy ripple bedding, with light greenish gray sandstone balls
252-10	1046.52—1047.44 m	0.92 m	Fine-grained sandstone, light gray, wavy cross bedding, with mud gravels
253-1	1047.44—1048.24 m	0.80 m	Fine-grained sandstone, light greenish gray, wavy cross bedding, with mudstone strips
253-2	1048.24—1048.54 m	0.30 m	Medium-grained sandstone, light gray, wavy cross bedding, mudstone strips
253-3	1048.54—1049.24 m	0.70 m	Medium-grained sandstone, light gray, oblique bedding, with mudstone laminae
253-4	1049.24—1049.91 m	0.67 m	Fine-grained sandstone, light greenish gray, deformation bedding locally, with mudstone strips occasionally
253-5	1049.91—1050.14 m	0.23 m	Medium-grained sandstone, light gray, massive
253-6	1050.14—1050.36 m	0.22 m	Siltstone, light greenish gray, wavy bedding, with mudstone strips occasionally
254-1	1053.75—1054.64 m	0.89 m	Siltstone, light olive gray, with wavy ripple bedding and horizontal bedding, wavy cross bedding locally, with mudstone and silty mudstone laminae
254-2	1054.64—1055.09 m	0.45 m	Mudstone, gray, horizontal bedding, with siltstone strips, siltstone laminae in the lower part
254-3	1055.09—1055.19 m	0.10 m	Siltstone, light olive gray, with water-escape structure
254-4	1055.19—1055.96 m	0.77 m	Mudstone, medium gray, interbedded with light olive gray siltstone, wavy bedding, sandstone load at the base, sandstone laminae and bioturbation observed
254-5	1055.96—1056.57 m	0.61 m	Siltstone, light olive gray, massive, with calcareous concretions
254-6	1056.57—1056.73 m	0.16 m	Muddy siltstone, light greenish gray, massive, with few calcareous concretions
254-7	1056.73—1057.35 m	0.62 m	Silty mudstone, grayish-red, massive bedding, greenish-gray impregnated spots observed
254-8	1057.35—1057.55 m	0.20 m	Mudstone, grayish red, massive
255-1	1057.55—1057.95 m	0.40 m	Mudstone, grayish red, massive
255-2	1057.95—1058.35 m	0.40 m	Silty mudstone, grayish red, massive
255-3	1058.35—1058.91 m	0.56 m	Mudstone, grayish red, massive
255-4	1058.91—1059.23 m	0.32 m	Silty mudstone, grayish red, massive
255-5	1059.23—1059.51 m	0.28 m	Siltstone, light red, massive
255-6	1059.51—1059.91 m	0.40 m	Silty mudstone, grayish red, massive
255-7	1059.91—1060.31 m	0.40 m	Siltstone, light red, massive
255-8	1060.31—1060.60 m	0.29 m	Siltstone, light olive, wavy cross bedding, mud gravels occasionally

255-9	1060.60—1060.95 m	0.35 m	Muddy siltstone, grayish red, massive
255-10	1060.95—1061.05 m	0.10 m	Siltstone, greenish gray, wavy ripple bedding
255-11	1061.05—1061.25 m	0.20 m	Muddy siltstone, light brownish gray, massive
255-12	1061.25—1061.59 m	0.34 m	Silty mudstone, grayish red, massive
255-13	1061.59—1062.45 m	0.86 m	Mudstone, grayish red, massive
255-14	1062.45—1063.45 m	1.00 m	Silty mudstone, grayish red, massive
255-15	1063.45—1063.75 m	0.30 m	Muddy siltstone, grayish red, massive
255-16	1063.75—1064.63 m	0.88 m	Mudstone, grayish red, massive
255-17	1064.63—1064.98 m	0.35 m	Silty mudstone, grayish red, massive
255-18	1064.98—1065.40 m	0.42 m	Siltstone, light greenish gray, massive
256-1	1065.40—1065.55 m	0.15 m	Siltstone, light gray, massive, deformation bedding locally
256-2	1065.55—1065.80 m	0.25 m	Silty mudstone, light greenish gray, massive
256-3	1065.80—1066.80 m	1.00 m	Silty mudstone, light brown, massive
256-4	1066.80—1067.10 m	0.30 m	Muddy siltstone, light brown, massive
256-5	1067.10—1067.80 m	0.70 m	Silty mudstone, grayish red and greenish gray, massive
256-6	1067.80—1068.25 m	0.45 m	Mudstone, grayish red, massive
256-7	1068.25—1069.30 m	1.05 m	Silty mudstone, grayish red, massive
256-8	1069.30—1069.40 m	0.10 m	Muddy siltstone, light yellowish brown, massive, with few siltstone strips
256-9	1069.40—1069.50 m	0.10 m	Silty mudstone, light yellowish brown, massive
256-10	1069.50—1070.70 m	1.20 m	Mudstone, grayish red, massive
256-11	1070.70—1070.85 m	0.15 m	Muddy siltstone, light olive gray, massive, with grayish red mudstone strips
256-12	1070.85—1071.16 m	0.31 m	Silty mudstone, light red, massive
256-13	1071.16—1071.30 m	0.14 m	Siltstone, greenish gray, massive, with red mudstone spots
256-14	1071.30—1072.40 m	1.10 m	Silty mudstone, grayish red, massive, deformation bedding locally, with irregular sandstone strips
256-15	1072.40—1072.73 m	0.33 m	Muddy siltstone, light brown, massive, with bioturbation
256-16	1072.73—1073.00 m	0.27 m	Mudstone, light brown, massive
256-17	1073.00—1073.26 m	0.26 m	Silty mudstone, light brown, massive, deformation bedding locally
256-18	1073.26—1073.40 m	0.14 m	Fine-grained sandstone, yellowish green, with trough cross bedding
256-19	1073.40—1073.80 m	0.40 m	Siltstone, medium light gray, parallel bedding and wavy bedding locally, deformation bedding, 5 cm-thick sandstone-mudstone interbeds at the base
256-20	1073.80—1074.40 m	0.60 m	Fine-grained sandstone, gray orange pink, gentle wave-generated cross-bedding, wavy bedding and wavy cross bedding in the middle and lower parts, unclear erosion surface at the bottom, few mudstone laminae

256-21	1074.40—1074.63 m	0.23 m	Siltstone, light gray, wavy cross bedding, with few mudstone strips
257-1	1074.63—1075.25 m	0.62 m	Silty mudstone, light olive gray, massive
257-2	1075.25—1075.53 m	0.28 m	Siltstone, greenish gray, massive, deformation bedding locally, erosion surface at the base, mud gravels
257-3	1075.53—1075.75 m	0.22 m	Siltstone, greenish gray, massive, with mudstone laminae at the base
257-4	1075.75—1076.03 m	0.28 m	Silty mudstone, light reddish brown, massive, deformation bedding locally
257-5	1076.03—1076.09 m	0.06 m	Siltstone, light gray, slump structures
257-6	1076.09—1076.67 m	0.58 m	Silty mudstone, light olive gray, massive
257-7	1076.67—1076.94 m	0.27 m	Siltstone, yellowish gray, wavy bedding
258-1	1076.94—1077.29 m	0.35 m	Silty mudstone, light gray, massive
258-2	1077.29—1077.59 m	0.30 m	Siltstone, greenish gray, massive
258-3	1077.59—1077.84 m	0.25 m	Silty mudstone, light gray, wavy ripple bedding, with fine-grained sandstone strips
258-4	1077.84—1078.39 m	0.55 m	Siltstone, light gray, wavy bedding, with olive gray mudstone strips
258-5	1078.39—1079.14 m	0.75 m	Silty mudstone, greenish gray, horizontal bedding, wavy ripple bedding, with siltstone strips
258-6	1079.14—1079.44 m	0.30 m	Mudstone, greenish gray, massive
258-7	1079.44—1079.64 m	0.20 m	Mudstone, grayish red and greenish gray, massive
258-8	1079.64—1080.04 m	0.40 m	Silty mudstone, grayish red, massive
258-9	1080.04—1080.39 m	0.35 m	Muddy siltstone, light gray, massive
258-10	1080.39—1080.49 m	0.10 m	Muddy siltstone, grayish red, massive
258-11	1080.49—1080.85 m	0.36 m	Fine-grained sandstone, light greenish gray, massive, with silty mudstone laminae
259-1	1080.95—1081.05 m	0.10 m	Fine-grained sandstone, light gray, massive
259-2	1081.05—1081.25 m	0.20 m	Siltstone, greenish gray, massive
259-3	1081.25—1081.55 m	0.30 m	Silty mudstone, grayish red, massive, fine-grained sandstone, greenish gray
259-4	1081.55—1081.65 m	0.10 m	Fine-grained sandstone, light gray, massive, erosion surface at the base
259-5	1081.65—1081.75 m	0.10 m	Silty mudstone, grayish red, deformation bedding, with light gray sandstone balls and strips
259-6	1081.75—1081.85 m	0.10 m	Fine-grained sandstone, light gray, massive, erosion surface at the base, with grayish red mudstone strips
259-7	1081.85—1082.10 m	0.25 m	Silty mudstone, grayish red, massive, deformation bedding locally, with fine-grained sandstone balls and strips
259-8	1082.10—1082.20 m	0.10 m	Fine-grained sandstone, light gray, massive
259-9	1082.20—1082.55 m	0.35 m	Silty mudstone, grayish red, massive, deformation bedding locally, with fine-grained sandstone, light greenish gray balls and strips

259-10	1082.55—1083.00 m	0.45 m	Muddy siltstone, greenish gray, massive, with light gray fine-grained sandstone balls
259-11	1083.00—1083.75 m	0.75 m	Muddy siltstone, grayish red and greenish gray, massive, with fine-grained sandstone balls
259-12	1083.75—1083.95 m	0.20 m	Fine-grained sandstone, light gray, massive, with mudstone strips
259-13	1083.95—1084.30 m	0.35 m	Muddy siltstone, greenish gray, massive, fine-grained sandstone balls
259-14	1084.30—1084.65 m	0.35 m	Fine-grained sandstone, light gray, wavy cross bedding, erosion surface at the base, with grayish red mudstone strips
259-15	1084.65—1084.87 m	0.22 m	Silty mudstone, greenish gray, massive, with sandstone balls
260-1	1084.87—1084.94 m	0.07 m	Fine-grained sandstone, greenish gray, wavy bedding, mudstone strips
260-2	1084.94—1085.04 m	0.10 m	Medium-grained sandstone, light greenish gray, wavy bedding, mudstone strips at the base
260-3	1085.04—1085.39 m	0.35 m	Muddy siltstone, light olive gray, wavy cross bedding, fine-grained sandstone laminae and concretions, with carbon dust, grayish-red mudstone laminae at the base
260-4	1085.39—1085.57 m	0.18 m	Medium-grained sandstone, light greenish gray, massive, erosion surface and mud gravels at the base
260-5	1085.57—1085.87 m	0.30 m	Silty mudstone, grayish red and greenish gray, massive
260-6	1085.87—1086.12 m	0.25 m	Siltstone, light gray, massive, deformation locally, with fine-grained sandstone strips and mudstone laminae
260-7	1086.12—1086.62 m	0.50 m	Sandstone, fine-grained, light gray, wavy bedding, erosion surface at the base, thin argillaceous laminae with carbon dusts between bedding surfaces, 2cm-thick mudstone at the base, eroded at the top
260-8	1086.62—1086.97 m	0.35 m	Medium-grained sandstone, wavy bedding, erosion surface at the base, with mudstone strips
260-9	1086.97—1087.32 m	0.35 m	Silty mudstone, grayish red, massive
260-10	1087.32—1087.47 m	0.15 m	Muddy siltstone, light gray, massive
260-11	1087.47—1087.80 m	0.33 m	Mudstone, grayish red, massive
260-12	1087.80—1088.22 m	0.42 m	Muddy siltstone, light gray, massive
260-13	1088.22—1088.67 m	0.45 m	Mudstone, grayish red and greenish gray, massive
260-14	1088.67—1088.97 m	0.30 m	Muddy siltstone, grayish red and greenish gray, massive
260-15	1088.97—1089.57 m	0.60 m	Muddy siltstone, light olive gray, massive, with fine-grained sandstone strips
260-16	1089.57—1090.27 m	0.70 m	Mudstone, greenish gray, massive
260-17	1090.27—1090.57 m	0.30 m	Mudstone, grayish red and greenish gray, massive
260-18	1090.57—1090.67 m	0.10 m	Muddy siltstone, grayish red, massive
260-19	1090.67—1091.37 m	0.70 m	Mudstone, greenish gray, massive
260-20	1091.37—1092.07 m	0.70 m	Mudstone, grayish red and greenish gray, massive

Section 5　SK-1 Core Description and Core Photographs

260-21	1092.07—1092.47 m	0.40 m	Silty mudstone, greenish gray, massive
260-22	1092.47—1092.87 m	0.40 m	Silty mudstone, grayish red, massive
260-23	1092.87—1093.17 m	0.30 m	Fine-grained sandstone, greenish gray, massive, calcareous concretions occasionally
260-24	1093.17—1093.36 m	0.19 m	Muddy siltstone, grayish red, massive
260-25	1093.36—1093.62 m	0.26 m	Fine-grained sandstone, light greenish gray, massive, with red mudstone spots
260-26	1093.62—1093.72 m	0.10 m	Fine-grained sandstone, light gray, wavy cross bedding
260-27	1093.72—1093.94 m	0.22 m	Sandstone, fine-grained, light gray, massive bedding, a grayish-red mudstone laminae at 1093.77m, with sand balls and sand pillows
261-1	1094.44—1095.02 m	0.58 m	Fine-grained sandstone, grayish red, wavy cross bedding
261-2	1095.02—1095.15 m	0.13 m	Mudstone, light brown, wavy ripple bedding, with siltstone strips
261-3	1095.15—1095.26 m	0.11 m	Siltstone, light brown, wavy cross bedding, with mudstone strips
261-4	1095.26—1095.44 m	0.18 m	Siltstone, grayish red, wavy ripple bedding, with siltstone strips
261-5	1095.44—1095.91 m	0.47 m	Siltstone, medium light gray, wavy bedding and wavy cross bedding, weak erosion at the bottom, clear contact with the bottom lithology, silty mudstone laminae
261-6	1095.91—1096.02 m	0.11 m	Fine-grained sandstone, light gray, massive
261-7	1096.02—1096.42 m	0.40 m	Fine-grained sandstone, light brownish gray, parallel bedding
261-8	1096.42—1096.54 m	0.12 m	Fine-grained sandstone, light brownish gray, wavy bedding, cross bedding locally
261-9	1096.54—1096.64 m	0.10 m	Fine-grained sandstone, light gray, parallel bedding
261-10	1096.64—1099.90 m	3.26 m	Fine-grained sandstone, light brownish gray, parallel bedding
261-11	1099.90—1100.12 m	0.22 m	Fine-grained sandstone, light yellowish gray, wavy cross bedding, erosion surface at the base
261-12	1100.12—1100.31 m	0.19 m	Medium-grained sandstone, yellowish gray, oblique bedding
261-13	1100.31—1100.95 m	0.64 m	Medium-grained sandstone, light brownish gray, oblique bedding, mud gravels in the middle and lower parts
261-14	1100.95—1101.55 m	0.60 m	Medium-grained sandstone, yellowish gray, oblique bedding
261-15	1101.55—1101.99 m	0.44 m	Medium-grained sandstone, light brownish gray, parallel bedding, mud gravels in the lower part
263-1	1102.17—1102.22 m	0.05 m	Silty mudstone, grayish red and greenish gray, massive, with fine-grained sandstone balls
263-2	1102.22—1102.28 m	0.06 m	Fine-grained sandstone, light gray, massive, mudstone strips and mud gravels, with bioturbation
263-3	1102.28—1102.38 m	0.10 m	Muddy siltstone, greenish gray, wavy bedding, fine-grained sandstone strips and bioturbation
263-4	1102.38—1102.43 m	0.05 m	Fine-grained sandstone, light gray, massive, mud gravels

263-5	1102.43—1102.50 m	0.07 m	Muddy siltstone, light gray, parallel bedding
263-6	1102.50—1102.67 m	0.17 m	Mudstone, grayish red, massive, with calcareous concretions
263-7	1102.67—1103.37 m	0.70 m	Mudstone, light gray, massive
263-8	1103.37—1103.77 m	0.40 m	Muddy siltstone, light gray, massive
263-9	1103.77—1104.12 m	0.35 m	Mudstone, greenish gray, massive
263-10	1104.12—1104.25 m	0.13 m	Muddy siltstone, light greenish gray, massive
264-1	1104.25—1104.35 m	0.10 m	Fine-grained sandstone, light gray, wavy bedding, with grayish red mudstone strips
264-2	1104.35—1104.55 m	0.20 m	Silty mudstone, grayish red, massive, with sandstone strips and bioturbation
264-3	1104.55—1104.70 m	0.15 m	Muddy siltstone, greenish gray, massive, mudstone strips
264-4	1104.70—1104.88 m	0.18 m	Fine-grained sandstone, light gray, wavy cross bedding, mudstone strips
264-5	1104.88—1105.27 m	0.39 m	Fine-grained sandstone, light olive gray, massive, mudstone strips and fine-grained sandstone laminae at the base, erosion surface
264-6	1105.27—1105.60 m	0.33 m	Medium-grained sandstone, grayish red, oblique bedding
264-7	1105.60—1106.09 m	0.49 m	Medium-grained sandstone, light gray, parallel bedding
264-8	1106.09—1107.10 m	1.01 m	Sandstone, fine-grained, light gray, wavy bedding, wavy cross bedding locally, slight deformation locally, interbedded with greenish-gray mudstone laminae and carbon dust laminae
264-9	1107.10—1107.33 m	0.23 m	Silty mudstone, greenish gray, wavy ripple bedding
267-1	1108.49—1109.34 m	0.85 m	Silty mudstone, greenish gray, massive, siltstone strips at the top, calcareous concretions
267-2	1109.34—1109.99 m	0.65 m	Mudstone, greenish gray, massive
267-3	1109.99—1110.69 m	0.70 m	Silty mudstone, greenish gray, massive
267-4	1110.69—1110.95 m	0.26 m	Mudstone, greenish gray, massive
267-5	1110.95—1111.11 m	0.16 m	Silty mudstone, greenish gray, massive
267-6	1111.11—1112.14 m	1.03 m	Mudstone, greenish gray, massive
267-7	1112.14—1112.85 m	0.71 m	Silty mudstone, greenish gray, massive
267-8	1112.85—1113.11 m	0.26 m	Muddy siltstone, greenish gray, wavy ripple bedding
267-9	1113.11—1113.61 m	0.50 m	Silty mudstone, greenish gray, massive
267-10	1113.61—1113.75 m	0.14 m	Siltstone, greenish gray, massive
268-1	1113.75—1114.95 m	1.20 m	Mudstone, grayish brown, massive
268-2	1114.95—1115.27 m	0.32 m	Siltstone, light gray, wavy bedding, deformation structure locally, 5 layers of 1—3 cm thick olive gray silty mudstone
268-3	1115.27—1115.97 m	0.70 m	Sandstone, fine-grained, light gray, discontinuous parallel bedding, 1cm thick grayish-brown mudstone laminae at 1115.9 m

268-4	1115.97—1116.85 m	0.88 m	Silty mudstone, grayish brown, massive
268-5	1116.85—1117.60 m	0.75 m	Silty mudstone, grayish olive, massive
268-6	1117.60—1118.18 m	0.58 m	Silty mudstone, grayish-brown, massive bedding, grayish-olive (10Y4/2) spots locally
269-1	1118.18—1119.08 m	0.90 m	Silty mudstone, grayish-brown, massive bedding, grayish-olive (10Y4/2) spots locally
269-2	1119.08—1119.43 m	0.35 m	Silty mudstone, grayish olive, massive
269-3	1119.43—1119.98 m	0.55 m	Muddy siltstone, grayish brown, massive
269-4	1119.98—1120.78 m	0.80 m	Silty mudstone, grayish brown and grayish olive, massive
269-5	1120.78—1121.03 m	0.25 m	Siltstone, greenish gray, deformation bedding, trace fossils
269-6	1121.03—1121.13 m	0.10 m	Silty mudstone, grayish brown, massive
269-7	1121.13—1121.40 m	0.27 m	Fine-grained sandstone, greenish gray, deformation bedding, trace fossils
269-8	1121.40—1121.65 m	0.25 m	Silty mudstone, grayish brown, massive
269-9	1121.65—1122.13 m	0.48 m	Sandstone, fine grained, greenish-gray, climbing bedding, wedge-shaped cross bedding, erosion surface at the base, with mud gravels, and biogliph, grayish-brown mudstone laminae locally
269-10	1122.13—1122.48 m	0.35 m	Silty mudstone, grayish brown, massive
269-11	1122.48—1122.69 m	0.21 m	Fine-grained sandstone, greenish gray, wavy bedding, erosion surface at the base, mudstone strips locally
269-12	1122.69—1122.93 m	0.24 m	Silty mudstone, grayish-brown, massive bedding, 2 cm thick fine-grained sandstone laminae at 1122.78 m, with mud gravels
269-13	1122.93—1123.77 m	0.84 m	Fine-grained sandstone, light brown, parallel bedding, small trough cross bedding
270-1	1124.35—1124.85 m	0.50 m	Mudstone, greenish gray, massive, with irregular fracture
270-2	1124.85—1126.05 m	1.20 m	Fine-grained sandstone, light gray, parallel bedding
270-3	1126.05—1127.15 m	1.10 m	Muddy siltstone, grayish brown, massive, with greenish gray siltstone strips
270-4	1127.15—1127.61 m	0.46 m	Siltstone, greenish gray, wavy bedding, with fine-grained sandstone laminae
270-5	1127.61—1127.70 m	0.09 m	Mudstone, light olive gray, massive
270-6	1127.70—1127.85 m	0.15 m	Mudstone, grayish brown, massive
270-7	1127.85—1128.05 m	0.20 m	Muddy siltstone, grayish brown, massive
270-8	1128.05—1128.25 m	0.20 m	Siltstone, greenish gray, discontinuous wavy bedding
270-9	1128.25—1129.22 m	0.97 m	Mudstone, grayish brown and dark greenish gray, massive
270-10	1129.22—1129.67 m	0.45 m	Medium-grained sandstone, light gray, deformation bedding
270-11	1129.67—1130.29 m	0.62 m	Silty mudstone, grayish brown and dark greenish gray, massive
270-12	1130.29—1130.72 m	0.43 m	Fine-grained sandstone, light gray, massive

271-1	1132.98—1133.58 m	0.60 m	Silty mudstone, grayish brown, massive
271-2	1133.58—1134.23 m	0.65 m	Silty mudstone, dark greenish gray, massive, calcareous concretions at the base
271-3	1134.23—1134.58 m	0.35 m	Silty mudstone, grayish brown, massive
271-4	1134.58—1134.82 m	0.24 m	Siltstone, greenish gray, discontinuous wavy bedding, erosion surface at the base, mud gravels
271-5	1134.82—1135.68 m	0.86 m	Muddy siltstone, dark greenish gray and grayish brown, massive
271-6	1135.68—1136.23 m	0.55 m	Silty mudstone, dark greenish and grayish brown, massive
271-7	1136.23—1136.68 m	0.45 m	Fine-grained sandstone, light greenish gray, parallel bedding, erosion surface at the base
271-8	1136.68—1137.63 m	0.95 m	Mudstone, greenish gray, massive
271-9	1137.63—1137.90 m	0.27 m	Siltstone, light greenish gray, small trough cross bedding, erosion surface at the base, with mudstone strips
271-10	1137.90—1138.93 m	1.03 m	Mudstone, greenish gray, massive
271-11	1138.93—1139.83 m	0.90 m	Fine-grained sandstone, light greenish gray, wavy bedding, with mudstone strips and laminae, trace fossils
271-12	1139.83—1139.93 m	0.10 m	Medium-grained sandstone, light gray, parallel bedding
271-13	1139.93—1140.08 m	0.15 m	Siltstone, light greenish gray, wavy bedding, erosion surface at the base, deformation bedding at the top
272-1	1140.12—1140.77 m	0.65 m	Siltstone, light greenish gray, parallel bedding in the upper part, wavy bedding in the lower part, mudstone strips locally
272-2	1140.77—1141.67 m	0.90 m	Siltstone, light greenish gray, interbedded with dark gray mudstone, wavy bedding, deformation bedding locally
272-3	1141.67—1141.79 m	0.12 m	Fine-grained sandstone, light greenish gray, trough cross bedding, dark gray mudstone strips at the base
273-1	1141.79—1142.29 m	0.50 m	Siltstone, light greenish-gray, interbedded with dark gray mudstone, lenticular bedding, small trough cross bedding in siltstone, ostracoda fossils in mudstone
273-2	1142.29—1142.79 m	0.50 m	Fine-grained sandstone, greenish gray, massive, with lots of calcareous concretions
273-3	1142.79—1143.39 m	0.60 m	Fine-grained sandstone, light greenish gray
273-4	1143.39—1144.04 m	0.65 m	Silty mudstone, dark greenish gray, massive
273-5	1144.04—1144.74 m	0.70 m	Silty mudstone, dark greenish gray and grayish brown, massive
273-6	1144.74—1144.89 m	0.15 m	Muddy siltstone, dark greenish gray, massive
274-1	1148.53—1148.85 m	0.32 m	Silty mudstone, grayish brown, massive, with calcareous concretions
274-2	1148.85—1149.25 m	0.40 m	Fine-grained sandstone, light gray, trough cross bedding
274-3	1149.25—1149.37 m	0.12 m	Silty mudstone, grayish brown, massive, trace fossils
274-4	1149.37—1149.48 m	0.11 m	Fine-grained sandstone, light gray, trace fossils

274-5	1149.48—1149.68 m	0.20 m	Mudstone, grayish brown, massive
274-6	1149.68—1149.87 m	0.19 m	Fine-grained sandstone, light gray, trough cross bedding
275-1	1149.87—1151.66 m	1.79 m	Silty mudstone, grayish brown, massive
275-2	1151.66—1152.53 m	0.87 m	Silty mudstone, grayish brown, deformation bedding, climbing bedding
275-3	1152.53—1152.64 m	0.11 m	Fine-grained sandstone, light gray, discontinuous wavy bedding, erosion surface at the base
275-4	1152.64—1152.84 m	0.20 m	Silty mudstone, grayish-brown, massive bedding, 2 cm thick siltstone with wavy bedding
275-5	1152.84—1152.94 m	0.10 m	Siltstone, light brown, deformation bedding, with few fine-grained sandstone strips
275-6	1152.94—1153.07 m	0.13 m	Muddy siltstone, grayish brown, wavy bedding, erosion surface at the base
275-7	1153.07—1153.17 m	0.10 m	Fine-grained sandstone, light gray, trough cross bedding
275-8	1153.17—1153.62 m	0.45 m	Muddy siltstone, grayish brown, massive, deformation bedding
275-9	1153.62—1154.17 m	0.55 m	Fine-grained sandstone, light gray, tabular cross bedding, erosion surface and mud gravels at the base
275-10	1154.17—1155.07 m	0.90 m	Muddy siltstone, greenish gray, massive, calcareous concretions locally
275-11	1155.07—1155.32 m	0.25 m	Silty mudstone, grayish brown and dark greenish gray, horizontal bedding, with trace fossils, siltstone strips locally
275-12	1155.32—1155.53 m	0.21 m	Silty mudstone, greenish gray, massive
275-13	1155.53—1156.97 m	1.44 m	Mudstone, dark greenish gray, massive, with calcareous concretions locally
275-14	1156.97—1157.32 m	0.35 m	Silty mudstone, grayish brown, massive, with calcareous concretions locally
275-15	1157.32—1157.47 m	0.15 m	Silty mudstone, grayish brown, massive
275-16	1157.47—1157.63 m	0.16 m	Siltstone, light gray, wavy bedding, horizontal bedding, with trace fossils
275-17	1157.63—1158.57 m	0.94 m	Mudstone, grayish brown and greenish gray, massive
275-18	1158.57—1158.69 m	0.12 m	Silty mudstone, grayish brown, massive, calcareous concretions locally
275-19	1158.69—1158.87 m	0.18 m	Fine-grained sandstone, greenish gray, deformation bedding, with trace fossils
275-20	1158.87—1159.19 m	0.32 m	Mudstone, grayish brown and dark greenish gray, massive
275-21	1159.19—1159.30 m	0.11 m	Fine-grained sandstone, light gray, small trough cross bedding, with few grayish brown mudstone strips
276-1	1159.30—1159.75 m	0.45 m	Medium-grained sandstone, light gray, small trough cross bedding
276-2	1159.75—1160.55 m	0.80 m	Mudstone, light olive gray, interbedded with light gray siltstone, wavy ripple bedding, trough cross bedding, erosion surface at the base
276-3	1160.55—1161.70 m	1.15 m	Muddy siltstone, grayish brown, massive
276-4	1161.70—1162.54 m	0.84 m	Silty mudstone, light olive gray and grayish-brown, horizontal ripple bedding, wavy bedding, with some thinly bedded siltstone and fine-grained sandstone laminae

276-5	1162.54—1163.30 m	0.76 m	Fine-grained sandstone, light gray and light olive gray, deformation bedding, wavy bedding, with grayish brown mudstone strips
276-6	1163.30—1163.43 m	0.13 m	Medium-grained sandstone, light gray, parallel bedding, erosion surface at the base, with mud gravels
276-7	1163.43—1163.60 m	0.17 m	Muddy siltstone, grayish brown, massive
277-1	1165.00—1165.30 m	0.30 m	Mudstone, grayish brown, massive
277-2	1165.30—1165.40 m	0.10 m	Silty mudstone, greenish gray, massive
277-3	1165.40—1165.60 m	0.20 m	Fine-grained sandstone, light gray, deformation bedding, mudstone strips
277-4	1165.60—1166.03 m	0.43 m	Muddy siltstone, greenish gray, massive, irregular fracture at the top
278-1	1166.03—1166.95 m	0.92 m	Mudstone, grayish brown, massive
278-2	1166.95—1167.13 m	0.18 m	Fine-grained sandstone, light greenish gray, parallel bedding, water-escape structure, with greenish gray mudstone strips
278-3	1167.13—1167.71 m	0.58 m	Mudstone, greenish gray, massive, deformation bedding, with irregular fine-grained sandstone strips
278-4	1167.71—1168.43 m	0.72 m	Mudstone, grayish brown and greenish gray, massive
278-5	1168.43—1168.78 m	0.35 m	Muddy siltstone, grayish brown and greenish gray, massive
278-6	1168.78—1169.78 m	1.00 m	Mudstone, grayish brown, massive
278-7	1169.78—1169.94 m	0.16 m	Siltstone, light greenish gray, massive
278-8	1169.94—1171.21 m	1.27 m	Mudstone, grayish brown and greenish gray, massive, with irregular fracture in the upper part
278-9	1171.21—1171.53 m	0.32 m	Fine-grained sandstone, light greenish gray, deformation bedding, trough cross bedding
278-10	1171.53—1172.15 m	0.62 m	Mudstone, grayish brown and greenish gray, massive, siltstone strips at the base
279-1	1172.15—1172.95 m	0.80 m	Mudstone, grayish brown and greenish gray, massive
279-2	1172.95—1173.90 m	0.95 m	Mudstone, grayish brown, massive
279-3	1173.90—1174.05 m	0.15 m	Silty mudstone, greenish gray and grayish brown, massive
279-4	1174.05—1174.90 m	0.85 m	Mudstone, grayish brown and greenish gray, massive
279-5	1174.90—1176.50 m	1.60 m	Mudstone, grayish brown, massive, with calcareous concretions
279-6	1176.50—1176.60 m	0.10 m	Siltstone, light greenish gray, slump structures
279-7	1176.60—1177.35 m	0.75 m	Mudstone, grayish brown, massive
279-8	1177.35—1177.65 m	0.30 m	Siltstone, light gray, slump structures, with grayish brown mudstone strips
279-9	1177.65—1178.33 m	0.68 m	Fine-grained sandstone, light gray, trough cross bedding, deformation bedding locally, with lots of mudstone strips and trace fossils
279-10	1178.33—1178.98 m	0.65 m	Sandstone, medium-grained, light gray, trough cross bedding, 6cm thick grayish-brown mudstone laminae at 1178.70 m

279-11	1178.98—1179.49 m	0.51 m	Silty mudstone, greenish gray, massive
280-1	1179.49—1180.39 m	0.90 m	Silty mudstone, grayish brown, massive, trace fossils, fine-grained sandstone strips
280-2	1180.39—1180.50 m	0.11 m	Fine-grained sandstone, light greenish gray, deformation bedding, convolution bedding, with grayish brown mudstone strips
282-1	1187.75—1188.85 m	1.10 m	Silty mudstone, grayish-brown, massive, a calcareous concretion (diameter >9.5 cm) at 1188.50 m, and small calcareous concretions locally
282-2	1188.85—1189.15 m	0.30 m	Fine-grained sandstone, light gray
282-3	1189.15—1189.50 m	0.35 m	Medium-grained sandstone, light gray
283-1	1189.50—1190.42 m	0.92 m	Medium-grained sandstone, light gray, trough cross bedding
284-1	1190.53—1190.93 m	0.40 m	Mudstone, greenish gray, massive
284-2	1190.93—1191.08 m	0.15 m	Muddy siltstone, grayish brown, massive, fine-grained sandstone strips
284-3	1191.08—1191.63 m	0.55 m	Fine-grained sandstone, light brown, deformation bedding, with mudstone strips
284-4	1191.63—1192.03 m	0.40 m	Silty mudstone, grayish brown, massive, fracture
285-1	1192.03—1192.23 m	0.20 m	Muddy siltstone, grayish brown and greenish gray, massive
285-2	1192.23—1192.58 m	0.35 m	Siltstone, greenish gray, deformation bedding, deformation bedding, fine-grained sandstone strips
285-3	1192.58—1192.78 m	0.20 m	Fine-grained sandstone, light greenish gray
285-4	1192.78—1193.78 m	1.00 m	Fine-grained sandstone, light gray, trough cross bedding, wavy bedding
285-5	1193.78—1194.71 m	0.93 m	Sandstone, fine-grained, light greenish-gray, slump bedding, roll-up bedding, with irregular mudstone laminae and light gray medium-grained sandstone concretions
286-1	1194.71—1195.2 m	0.49 m	Silty mudstone, grayish-brown and greenish-gray, horizontal ripple bedding, with fine grained sandstone laminae and concretions, greenish-gray in surrounding
287-1	1196.01—1196.21 m	0.20 m	Silty mudstone, greenish gray, massive
287-2	1196.21—1196.61 m	0.40 m	Mudstone, greenish gray, massive
287-3	1196.61—1197.48 m	0.87 m	Silty mudstone, grayish brown, massive, calcareous concretions
287-4	1197.48—1197.53 m	0.05 m	Fine-grained sandstone, light gray, trace fossils
288-1	1197.53—1198.03 m	0.50 m	Mudstone, grayish brown, massive
288-2	1198.03—1198.36 m	0.33 m	Silty mudstone, greenish gray, massive
288-3	1198.36—1199.23 m	0.87 m	Mudstone, greenish gray, massive
288-4	1199.23—1199.61 m	0.38 m	Silty mudstone, grayish brown, massive, calcareous concretions
289-1	1199.61—1199.96 m	0.35 m	Silty mudstone, grayish brown, massive
289-2	1199.96—1200.01 m	0.05 m	Siltstone, greenish gray, deformation bedding

289-3	1200.01—1200.29 m	0.28 m	Silty mudstone, greenish gray, massive
289-4	1200.29—1200.61 m	0.32 m	Fine-grained sandstone, light greenish gray, wavy bedding, mudstone strips
289-5	1200.61—1201.81 m	1.20 m	Silty mudstone, grayish brown, massive
289-6	1201.81—1202.21 m	0.40 m	Fine-grained sandstone, light brown, wavy bedding, deformation bedding locally, erosion surface at the base, with mudstone strips
289-7	1202.21—1202.31 m	0.10 m	Silty mudstone, grayish brown, wavy ripple bedding, deformation bedding
289-8	1202.31—1204.61 m	2.30 m	Medium-grained sandstone, light gray, tabular cross bedding, erosion surface at the base
289-9	1204.61—1204.91 m	0.30 m	Silty mudstone, grayish brown, massive
289-10	1204.91—1205.01 m	0.10 m	Muddy siltstone, greenish gray and grayish brown, massive
289-11	1205.01—1205.23 m	0.22 m	Silty mudstone, grayish brown, massive
290-1	1205.23—1205.43 m	0.20 m	Silty mudstone, dark greenish-gray, massive bedding, with fractures filled with grayish-brown silty mudstone
290-2	1205.43—1206.03 m	0.60 m	Silty mudstone, grayish brown, massive
290-3	1206.03—1206.21 m	0.18 m	Fine-grained sandstone, light gray, parallel bedding, erosion surface at the base
290-4	1206.21—1207.48 m	1.27 m	Mudstone, dark greenish-gray, massive bedding, horizontal bedding locally, grayish-brown spots distributed evenly in dark greenish-gray mudstone
290-5	1207.48—1208.53 m	1.05 m	Silty mudstone, grayish-brown, massive bedding, two 5 cm-thick greenish-gray argillaceous siltstone laminaes at 1208.13 m and 1208.33 m
290-6	1208.53—1208.73 m	0.20 m	Siltstone, light greenish gray, deformation bedding locally
290-7	1208.73—1209.33 m	0.60 m	Silty mudstone, dark greenish-gray, massive bedding, with grayish-brown spots, fine-grained sandstone with slump structure at 1209.13 m
290-8	1209.33—1210.63 m	1.30 m	Silty mudstone, grayish-brown, massive bedding, greenish-gray spots (around sandy strips) locally
290-9	1210.63—1210.86 m	0.23 m	Mudstone, light greenish gray, with mud gravels
290-10	1210.86—1211.06 m	0.20 m	Mudstone, grayish brown, massive
290-11	1211.06—1211.18 m	0.12 m	Fine-grained sandstone, light greenish gray, parallel bedding, water-escape structure locally, with mudstone strips
290-12	1211.18—1211.73 m	0.55 m	Silty mudstone, grayish brown and greenish gray, massive
290-13	1211.73—1211.91 m	0.18 m	Fine-grained sandstone, light gray, wavy bedding, erosion surface at the base, mudstone strip and trace fossils
290-14	1211.91—1212.43 m	0.52 m	Silty mudstone, grayish brown, horizontal bedding, with fine-grained sandstone laminae and strips
290-15	1212.43—1213.08 m	0.65 m	Mudstone, greenish gray, massive
290-16	1213.08—1213.93 m	0.85 m	Sandstone, medium-grained, light gray, unclear trough cross bedding, two 5 cm-thick fine-grained sandstone laminae with tabular cross bedding at 1213.38 and 1213.48 m, mudstone laminae and laminae locally

291-1	1213.93—1214.13 m	0.20 m	Medium-grained sandstone, light gray, tabular cross bedding
291-2	1214.13—1215.03 m	0.90 m	Fine-grained sandstone, light gray, trough cross bedding, wavy bedding, mudstone strips
291-3	1215.03—1215.28 m	0.25 m	Mudstone, dark gray, interbedded with light gray fine-grained sandstone, wavy bedding
291-4	1215.28—1215.58 m	0.30 m	Fine-grained sandstone, light gray, wavy bedding, mudstone strips
291-5	1215.58—1215.68 m	0.10 m	Fine-grained sandstone, light gray, parallel bedding
291-6	1215.68—1216.43 m	0.75 m	Mudstone, dark gray, interbedded with light gray fine-grained sandstone, wavy bedding
291-7	1216.43—1217.28 m	0.85 m	Mudstone, dark greenish gray, interbedded with greenish gray siltstone, horizontal bedding, wavy ripple bedding
291-8	1217.28—1218.03 m	0.75 m	Mudstone, dark greenish-gray, interbedded with light gray fine-grained sandstone, lenticular bedding, ripple bedding, erosion at the base, with integral ostracoda fossils in sandstone
291-9	1218.03—1218.58 m	0.55 m	Mudstone, grayish green, massive, calcareous concretions and pyrite
291-10	1218.58—1220.03 m	1.45 m	Mudstone, grayish brown, massive, calcareous concretions occasionally
291-11	1220.03—1220.42 m	0.39 m	Mudstone, grayish green, massive
292-1	1220.42—1221.02 m	0.60 m	Mudstone, greenish gray, massive
292-2	1221.02—1221.27 m	0.25 m	Silty mudstone, grayish green, massive, with lots of pyrite, with siltstone strips
292-3	1221.27—1221.62 m	0.35 m	Fine-grained sandstone, light greenish gray, deformation bedding, with mudstone strips
292-4	1221.62—1221.93 m	0.31 m	Medium-grained sandstone, light gray, wavy bedding, with mudstone strips and fossil fragments
293-1	1222.25—1223.73 m	1.48 m	Silty mudstone, grayish green, massive, deformation bedding locally
293-2	1223.73—1224.05 m	0.32 m	Fine-grained sandstone, light gray, wavy bedding, tabular cross bedding, with mudstone strips occasionally
293-3	1224.05—1228.35 m	4.30 m	Medium-grained sandstone, light gray, tabular cross bedding, with mud gravels locally
293-4	1228.35—1228.72 m	0.37 m	Conglomeratic sandstones, light gray, with mud gravels, light gray, parallel bedding, trough cross bedding, erosion surface at the base, a black elliptical substance at 1228.55 m to be possibly ostracoda limestone formed after intensive ostracoda death, due to highly hydraulic movement
293-5	1228.72—1228.90 m	0.18 m	Silty mudstone, grayish-green, massive, 5 cm thick light gray siltstone laminae at 1228.75 m
293-6	1228.90—1229.20 m	0.30 m	Medium-grained sandstone, light gray, wavy bedding, erosion surface and mud gravels at the base
294-1	1229.20—1229.30 m	0.10 m	Muddy siltstone, light greenish gray, deformation bedding, erosion surface at the base

294-2	1229.30—1229.60 m	0.30 m	Mudstone, dark greenish gray, massive
294-3	1229.60—1229.85 m	0.25 m	Mudstone, dark greenish gray and grayish brown, massive
294-4	1229.85—1230.10 m	0.25 m	Mudstone, greenish gray, massive
294-5	1230.10—1230.65 m	0.55 m	Mudstone, greenish gray and grayish brown, massive
294-6	1230.65—1231.60 m	0.95 m	Silty mudstone, dark yellowish green, massive
294-7	1231.60—1231.80 m	0.20 m	Siltstone, light greenish gray, parallel bedding, climbing bedding locally, with trace fossils and mudstone strips
294-8	1231.80—1231.90 m	0.10 m	Mudstone, greenish gray, massive
294-9	1231.90—1232.30 m	0.40 m	Mudstone, dark greenish gray and grayish brown, massive
294-10	1232.30—1232.70 m	0.40 m	Silty mudstone, grayish green, massive, with wavy ripple bedding
294-11	1232.70—1233.30 m	0.60 m	Mudstone, grayish brown and dark greenish gray, massive
294-12	1233.30—1233.65 m	0.35 m	Mudstone, grayish brown, massive
294-13	1233.65—1233.90 m	0.25 m	Silty mudstone, greenish gray, massive
294-14	1233.90—1234.05 m	0.15 m	Fine-grained sandstone, light gray, parallel bedding, erosion surface at the base
294-15	1234.05—1234.20 m	0.15 m	Siltstone, light greenish gray, with trace fossils, wavy ripple bedding, climbing bedding, mudstone strips
294-16	1234.20—1234.85 m	0.65 m	Mudstone, greenish gray, horizontal bedding
294-17	1234.85—1235.60 m	0.75 m	Silty mudstone, dark greenish gray, massive
294-18	1235.60—1236.70 m	1.10 m	Fine-grained sandstone, light gray, trough cross bedding, erosion surface and mud gravels at the base, mudstone strips locally
294-19	1236.70—1237.55 m	0.85 m	Sandstone, fine-grained, light gray, trough cross bedding in the upper part, tabular cross bedding in the lower part, 2 cm thick greenish-gray mudstone at the top
294-20	1237.55—1237.70 m	0.15 m	Medium-grained sandstone, light gray, parallel bedding, mud gravels
295-1	1237.73—1237.83 m	0.10 m	Mudstone, dark greenish gray, massive
295-2	1237.83—1238.58 m	0.75 m	Medium-grained sandstone, light gray, tabular cross bedding, erosion surface at the base
295-3	1238.58—1238.75 m	0.17 m	Silty mudstone, greenish gray, massive, with mudstone laminae
295-4	1238.75—1239.03 m	0.28 m	Fine-grained sandstone, light greenish gray, deformation bedding, convolution bedding, wavy bedding, with mudstone strips
295-5	1239.03—1240.53 m	1.50 m	Silty mudstone, greenish-gray, horizontal bedding and horizontal ripple bedding, 3 medium gritstone layers (3 cm, 10 cm and 4 cm thick) at 1239.75 m, 1240.13 m and 1240.28 m, erosion surfaces at the base
295-6	1240.53—1241.03 m	0.50 m	Mudstone, dark greenish gray and grayish brown, massive
295-7	1241.03—1241.43 m	0.40 m	Siltstone, greenish gray, massive
296-1	1241.50—1242.20 m	0.70 m	Silty mudstone, greenish gray, massive, horizontal bedding locally

296-2	1242.20—1242.90 m	0.70 m	Mudstone, greenish gray, massive
296-3	1242.90—1243.65 m	0.75 m	Mudstone, grayish brown and greenish gray, massive
296-4	1243.65—1244.50 m	0.85 m	Mudstone, dark greenish gray, massive
296-5	1244.50—1245.15 m	0.65 m	Mudstone, grayish brown, massive
296-6	1245.15—1245.40 m	0.25 m	Silty mudstone, greenish gray, massive, deformation bedding locally
296-7	1245.40—1245.85 m	0.45 m	Mudstone, dark greenish gray, massive
296-8	1245.85—1246.10 m	0.25 m	Mudstone, grayish brown, massive
296-9	1246.10—1246.35 m	0.25 m	Mudstone, dark greenish gray, massive
296-10	1246.35—1246.50 m	0.15 m	Muddy siltstone, greenish gray, massive
296-11	1246.50—1246.68 m	0.18 m	Mudstone, grayish brown and dark greenish gray, massive
297-1	1246.73—1246.83 m	0.10 m	Mudstone, grayish brown and greenish gray, massive
297-2	1246.83—1247.83 m	1.00 m	Silty mudstone, greenish gray, massive
297-3	1247.83—1249.43 m	1.60 m	Medium-grained sandstone, light gray, tabular cross bedding, mud gravels at the base
297-4	1249.43—1250.13 m	0.70 m	Muddy siltstone, greenish gray, massive
297-5	1250.13—1250.43 m	0.30 m	Fine-grained sandstone, light greenish gray, wavy bedding, deformation bedding locally, with mudstone strips
297-6	1250.43—1250.63 m	0.20 m	Muddy siltstone, greenish gray, massive, deformation bedding
297-7	1250.63—1250.83 m	0.20 m	Fine-grained sandstone, light greenish gray, wavy bedding, mudstone strips
297-8	1250.83—1251.07 m	0.24 m	Muddy siltstone, greenish gray, massive
297-9	1251.07—1251.17 m	0.10 m	Fine-grained sandstone, light gray, wavy bedding, mudstone strips
298-1	1251.42—1251.57 m	0.15 m	Fine-grained sandstone, light gray, wavy bedding, mudstone strips
299-1	1251.57—1251.92 m	0.35 m	Fine-grained sandstone, light greenish gray, wavy bedding, deformation bedding, mudstone strips
299-2	1251.92—1252.47 m	0.55 m	Mudstone, greenish gray, massive
299-3	1252.47—1252.87 m	0.40 m	Mudstone, greenish gray, massive
299-4	1252.87—1253.19 m	0.32 m	Muddy siltstone, greenish gray, massive
299-5	1253.19—1253.92 m	0.73 m	Mudstone, grayish brown, massive
299-6	1253.92—1254.05 m	0.13 m	Siltstone, greenish gray, trough cross bedding, climbing bedding, trace fossils
299-7	1254.05—1254.17 m	0.12 m	Fine-grained sandstone, light gray, trough cross bedding, mudstone strips
299-8	1254.17—1254.29 m	0.12 m	Siltstone, greenish gray
299-9	1254.29—1254.47 m	0.18 m	Mudstone, greenish gray, massive
299-10	1254.47—1255.32 m	0.85 m	Fine-grained sandstone, light gray, interbedded with greenish gray mudstone, wavy bedding

299-11	1255.32—1256.07 m	0.75 m	Siltstone, light greenish gray, interbedded with greenish gray mudstone, wavy bedding
299-12	1256.07—1257.07 m	1.00 m	Sandstone, fine-grained, light gray, interbedded with dark greenish-gray mudstone, wavy bedding and lenticular bedding, erosion surface at each sandstone base
299-13	1257.07—1257.57 m	0.50 m	Mudstone, greenish gray, massive
299-14	1257.57—1258.17 m	0.60 m	Silty mudstone, greenish gray, massive
299-15	1258.17—1258.52 m	0.35 m	Muddy siltstone, greenish gray, massive
299-16	1258.52—1258.72 m	0.20 m	Siltstone, greenish gray, deformation bedding
299-17	1258.72—1258.91 m	0.19 m	Fine-grained sandstone, light gray, deformation bedding, mudstone strips
300-1	1258.91—1259.18 m	0.27 m	Fine-grained sandstone, light gray, trough cross bedding, mudstone strips
300-2	1259.18—1259.47 m	0.29 m	Sandstone, medium-grained, light gray, trough cross bedding, erosion surface at the base, a lot of mudstone laminae, pyrites as cements infilling in grains
300-3	1259.47—1261.11 m	1.64 m	Mudstone, dark greenish gray, massive, bioturbation locally
300-4	1261.11—1262.13 m	1.02 m	Medium-grained sandstone, light gray, trough cross bedding, parallel bedding, mudstone strips, mud gravels
300-5	1262.13—1263.13 m	1.00 m	Fine-grained sandstone, light gray, interbedded with dark greenish gray mudstone, wavy bedding
300-6	1263.13—1264.31 m	1.18 m	Sandstone, medium-grained, light gray, tabular cross bedding, parallel bedding, fine-grained sandstone laminae with mudstone laminae locally, organic laminae at the base
300-7	1264.31—1266.74 m	2.43 m	Sandstone, fine-grained, light gray, interbedded with dark gray mudstone, wavy bedding, lenticular bedding, slump bedding, sand liquefaction at 1266.21—1266.56 m
301-1	1266.74—1266.77 m	0.03 m	Mudstone, greenish gray, massive
301-2	1266.77—1266.89 m	0.12 m	Mudstone, light yellowish gray, massive
301-3	1266.89—1267.04 m	0.15 m	Mudstone, dark gray, horizontal bedding, fine-grained sandstone strips
301-4	1267.04—1267.10 m	0.06 m	Mudstone, light yellowish gray, massive
301-5	1267.10—1267.13 m	0.03 m	Mudstone, dark gray, horizontal bedding, fine-grained sandstone strips
301-6	1267.13—1267.84 m	0.71 m	Silty mudstone, greenish gray, massive, conchostracon fossils occasionally
301-7	1267.84—1268.34 m	0.50 m	Siltstone, greenish gray, deformation bedding locally, mudstone strips
301-8	1268.34—1268.74 m	0.40 m	Fine-grained sandstone, greenish gray, with lots of mud gravels
301-9	1268.74—1269.32 m	0.58 m	Fine-grained sandstone, light greenish gray, mudstone strips and mud gravels directionally aligned
301-10	1269.32—1271.14 m	1.82 m	Fine-grained sandstone, light gray, interbedded with dark greenish gray mudstone, wavy bedding
301-11	1271.14—1271.64 m	0.50 m	Muddy siltstone, light greenish gray, massive, with sandstone strips.

The overlying formation is the Nenjiang Formation Member 5, the contacting relationship between them is conformable contact; the underlying formation is the Nenjiang Formation Member 3, the contacting relationship between them is conformable contact (Figure 5.6).

The detailed description of the Nenjiang Formation Member 4:

301-12	1271.64—1273.24 m	1.60 m	Medium-grained sandstone, light gray, trough cross bedding, with mud laminae and strips
301-13	1273.24—1275.88 m	2.64 m	Sandstone, fine-grained, light gray, interbedded with dark greenish-gray mudstone, wavy bedding, climbing, parallel and lenticular beddings, pyrite crystals locally
302-1	1275.88—1277.03 m	1.15 m	Medium-grained sandstone, light gray, with lots of mudstone, dark gray
302-2	1277.03—1277.83 m	0.80 m	Medium-grained sandstone, light greenish gray, parallel bedding, with few mud gravels in the lower part
302-3	1277.83—1278.00 m	0.17 m	Fine-grained sandstone, light greenish gray
302-4	1278.00—1278.50 m	0.50 m	Medium-grained sandstone, light gray, trough cross bedding, parallel bedding, mudstone strips in the lower part, with trace fossils
302-5	1278.50—1278.63 m	0.13 m	Mudstone, dark greenish gray, horizontal bedding, trace fossils
302-6	1278.63—1278.98 m	0.35 m	Medium-grained sandstone, light gray, parallel bedding, with mudstone laminae
302-7	1278.98—1279.38 m	0.40 m	Fine-grained sandstone, greenish gray, interbedded with mudstone, wavy bedding, trace fossils
302-8	1279.38—1280.24 m	0.86 m	Mudstone, greenish gray, massive
303-1	1280.50—1280.75 m	0.25 m	Mudstone, greenish gray, massive
303-2	1280.75—1282.15 m	1.40 m	Medium-grained sandstone, light gray, trough cross bedding, parallel bedding
303-3	1282.15—1282.90 m	0.75 m	Mudstone, dark greenish gray, interbedded with light gray fine-grained sandstone, wavy bedding
303-4	1282.90—1284.65 m	1.75 m	Sandstone, medium-grained, light gray, with tabular cross bedding in the upper section and trough cross bedding and parallel bedding in the lower section, argillaceous thin layers and strips locally
303-5	1284.65—1285.38 m	0.73 m	Fine-grained sandstone, light gray, wavy bedding, with lots of mudstone strips
303-6	1285.38—1287.09 m	1.71 m	Mudstone, dark gray, interbedded with fine-grained sandstone, wavy bedding
304-1	1287.09—1287.47 m	0.38 m	Mudstone, dark greenish gray, massive
304-2	1287.47—1288.59 m	1.12 m	Mudstone, dark gray, interbedded with gray fine-grained sandstone, wavy bedding
304-3	1288.59—1289.09 m	0.50 m	Silty mudstone, greenish gray, massive, calcareous concretions occasionally
304-4	1289.09—1289.94 m	0.85 m	Sandstone, fine-grained, light greenish-gray, deformation bedding in the upper part, trough cross bedding in the lower part, climbing bedding, biogliph, bioturbation
304-5	1289.94—1290.51 m	0.57 m	Sandstone, medium-grained, light gray, unclear trough cross bedding, two greenish-gray fine sandstone laminae

304-6	1290.51—1290.61 m	0.10 m	Coarse-grained sandstone, light gray, parallel bedding, mud gravels at the base
304-7	1290.61—1291.19 m	0.58 m	Sandstone, fine-grained, light gray, wavy bedding, lenticular bedding, parallel bedding locally, a lot of mudstone laminae and strips, mud gravels at the base
304-8	1291.19—1291.84 m	0.65 m	Mudstone, greenish gray, massive, calcareous concretions occasionally
304-9	1291.84—1292.09 m	0.25 m	Fine-grained sandstone, light greenish gray, deformation bedding, mudstone strips
304-10	1292.09—1292.62 m	0.53 m	Sandstone, medium-grained, light gray, trough cross bedding, 3 cm thick dark gray mudstone laminae at 1292.47 m, large fossils
304-11	1292.62—1292.89 m	0.27 m	Fine-grained sandstone, light gray, interbedded with dark gray mudstone, wavy bedding
304-12	1292.89—1293.99 m	1.10 m	Fine-grained sandstone, light gray, trough cross bedding, mudstone strips occasionally
304-13	1293.99—1294.71 m	0.72 m	Fine-grained sandstone, light gray, interbedded with dark greenish gray mudstone, wavy bedding
305-1	1294.78—1295.56 m	0.78 m	Siltstone, light greenish gray, interbedded with dark greenish gray fine-grained sandstone, wavy bedding
305-2	1295.56—1296.98 m	1.42 m	Fine-grained sandstone, light greenish gray, deformation bedding, mudstone strips, mud gravels
307-1	1297.07—1297.22 m	0.15 m	Mudstone, greenish gray, massive
307-2	1297.22—1297.55 m	0.33 m	Siltstone, light greenish gray, massive
307-3	1297.55—1298.67 m	1.12 m	Mudstone, greenish gray, massive
307-4	1298.67—1299.47 m	0.80 m	Silty mudstone, greenish gray, massive
307-5	1299.47—1300.12 m	0.65 m	Medium-grained sandstone, light gray, wavy bedding, trace fossils, with lots of greenish gray mudstone laminae
307-6	1300.12—1300.77 m	0.65 m	Fine-grained sandstone, light gray, interbedded with dark greenish gray mudstone, wavy bedding, trace fossils
307-7	1300.77—1300.87 m	0.10 m	Medium-grained sandstone, light gray, parallel bedding
307-8	1300.87—1301.62 m	0.75 m	Silty mudstone, greenish gray, massive
307-9	1301.62—1302.05 m	0.43 m	Fine-grained sandstone, light gray, interbedded with greenish gray mudstone, wavy bedding, deformation bedding
307-10	1302.05—1302.40 m	0.35 m	Medium-grained sandstone, light gray, trough cross bedding, mudstone strips
307-11	1302.40—1304.67 m	2.27 m	Fine-grained sandstone, light gray, parallel bedding, trough cross bedding, wavy bedding, mudstone strips
307-12	1304.67—1305.12 m	0.45 m	Siltstone, light greenish gray, interbedded with dark greenish gray mudstone, horizontal bedding
307-13	1305.12—1306.21 m	1.09 m	Fine-grained sandstone, light gray, trough cross bedding, parallel bedding, wavy bedding

308-1	1306.21—1306.91 m	0.70 m	Sandstone, medium-grained, light gray, wavy bedding, trough cross bedding, dark gray mudstone laminae, light gray fine-grained sandstone laminae, a 0.5 cm-thick light grayish olive mudstone laminae at 1306.70 m
308-2	1306.91—1307.66 m	0.75 m	Fine-grained sandstone, light gray, mudstone strips
308-3	1307.66—1308.01 m	0.35 m	Muddy siltstone, greenish gray, massive, wavy ripple bedding, trace fossils, siltstone strips
308-4	1308.01—1308.23 m	0.22 m	Medium-grained sandstone, light gray, trough cross bedding, erosion surface at the base
308-5	1308.23—1308.38 m	0.15 m	Mudstone, dark greenish gray, interbedded with fine-grained sandstone, wavy bedding, trace fossils
308-6	1308.38—1308.56 m	0.18 m	Medium-grained sandstone, light gray, trough cross bedding, erosion surface at the base, with few mudstone strips
308-7	1308.56—1309.11 m	0.55 m	Mudstone, dark greenish gray, interbedded with light greenish gray fine-grained sandstone, wavy bedding
308-8	1309.11—1309.37 m	0.26 m	Fine-grained sandstone, light gray, trough cross bedding
308-9	1309.37—1309.91 m	0.54 m	Mudstone, dark greenish gray, interbedded with light gray fine-grained sandstone, wavy bedding, pyrite locally
308-10	1309.91—1310.31 m	0.40 m	Medium-grained sandstone, light gray
308-11	1310.31—1311.21 m	0.90 m	Mudstone, greenish gray, massive
308-12	1311.21—1311.79 m	0.58 m	Muddy siltstone, greenish gray, massive, with trace fossils
309-1	1312.07—1313.47 m	1.40 m	Silty mudstone, greenish gray, massive
309-2	1313.47—1314.08 m	0.61 m	Muddy siltstone, greenish gray, massive
309-3	1314.08—1314.95 m	0.87 m	Silty mudstone, greenish gray, massive, fossil fragments
309-4	1314.95—1315.47 m	0.52 m	Muddy siltstone, greenish gray, massive
309-5	1315.47—1315.75 m	0.28 m	Fine-grained sandstone, light gray, wavy bedding, mudstone strips, bioturbation
309-6	1315.75—1316.09 m	0.34 m	Siltstone, greenish gray, wavy bedding, mudstone strips
309-7	1316.09—1316.62 m	0.53 m	Medium-grained sandstone, light greenish gray
309-8	1316.62—1317.22 m	0.60 m	Fine-grained sandstone, greenish gray, mudstone strips
309-9	1317.22—1317.57 m	0.35 m	Medium-grained sandstone, light greenish gray, mudstone strips
309-10	1317.57—1317.92 m	0.35 m	Silty mudstone, greenish gray, wavy ripple bedding, wavy bedding
309-11	1317.92—1318.12 m	0.20 m	Fine-grained sandstone, light gray, wavy bedding
309-12	1318.12—1318.24 m	0.12 m	Silty mudstone, greenish gray, massive
309-13	1318.24—1318.63 m	0.39 m	Fine-grained sandstone, light greenish gray, wavy bedding
309-14	1318.63—1319.25 m	0.62 m	Medium-grained sandstone, light greenish gray, wavy bedding
309-15	1319.25—1319.77 m	0.52 m	Mudstone, gray, massive, siltstone laminae and strips
309-16	1319.77—1319.90 m	0.13 m	Siltstone, light greenish gray, massive

309-17	1319.90—1320.02 m	0.12 m	Mudstone, gray, massive
309-18	1320.02—1320.31 m	0.29 m	Siltstone, light greenish gray, wavy ripple bedding, deformation bedding, mudstone strips occasionally
309-19	1320.31—1320.90 m	0.59 m	Silty mudstone, greenish gray, massive, deformation bedding
310-1	1320.94—1321.94 m	1.00 m	Muddy siltstone, greenish gray, deformation bedding locally
310-2	1321.94—1322.24 m	0.30 m	Siltstone, light greenish gray, deformation bedding, mudstone strips
310-3	1322.24—1322.39 m	0.15 m	Fine-grained sandstone, light gray
310-4	1322.39—1322.59 m	0.20 m	Silty mudstone, gray, wavy ripple bedding
310-5	1322.59—1322.82 m	0.23 m	Fine-grained sandstone, light gray, wavy ripple bedding, deformation bedding locally, mudstone strips
310-6	1322.82—1322.92 m	0.10 m	Silty mudstone, gray, deformation bedding
310-7	1322.92—1323.03 m	0.11 m	Fine-grained sandstone, light gray, slump structures, mudstone strips
310-8	1323.03—1323.14 m	0.11 m	Silty mudstone, gray, wavy bedding
310-9	1323.14—1323.74 m	0.60 m	Mudstone, gray, massive
310-10	1323.74—1323.94 m	0.20 m	Fine-grained sandstone, light gray, wavy bedding, mudstone strips
310-11	1323.94—1324.22 m	0.28 m	Mudstone, gray, massive
310-12	1324.22—1324.54 m	0.32 m	Silty mudstone, greenish gray, massive
310-13	1324.54—1325.14 m	0.60 m	Fine-grained sandstone, light greenish gray, wavy bedding, mud gravels and sandstone strips locally
311-1	1325.37—1325.43 m	0.06 m	Fine-grained sandstone, light gray, discontinuous wavy bedding, with mudstone strips
311-2	1325.43—1325.52 m	0.09 m	Siltstone, light gray, wavy bedding, mudstone laminae
311-3	1325.52—1325.59 m	0.07 m	Silty mudstone, dark gray, wavy ripple bedding, convolution bedding locally
311-4	1325.59—1325.69 m	0.10 m	Siltstone, light gray, wavy bedding, deformation bedding locally, mudstone laminae and strips
311-5	1325.69—1325.95 m	0.26 m	Sandstone, fine-grained, light gray, horizontal ripple bedding, mudstone laminae locally, discontinuous mudstone laminae, some mudstone laminae not produced
311-6	1325.95—1326.30 m	0.35 m	Siltstone, light gray, parallel bedding, wavy bedding in the lower section, generally mudstone laminae in the upper part, carbon dust concentrated on the bedding surface, few carbon dusts in the lower part
311-7	1326.30—1326.57 m	0.27 m	Fine-grained sandstone, light gray, parallel bedding
311-8	1326.57—1327.23 m	0.66 m	Siltstone, light gray, parallel bedding, abundant carbon dusts on both upper and lower bedding surfaces, less on middle surface, mudstone laminae in the lower part (interbed locally)
311-9	1327.23—1327.31 m	0.08 m	Fine-grained sandstone, light gray, parallel bedding
311-10	1327.31—1328.09 m	0.78 m	Siltstone, light gray, interbedded with medium dark gray mudstone, sand-mudstone with horizontal wavy bedding, mixed sandstone structure in the lower part, sand balls occasionally

Section 5 SK-1 Core Description and Core Photographs

311-11	1328.09—1328.59 m	0.50 m	Fine-grained sandstone, light gray, discontinuous wavy bedding
311-12	1328.59—1328.86 m	0.27 m	Mudstone, dark gray, interbedded with siltstone, light gray, wavy ripple bedding
311-13	1328.86—1330.83 m	1.97 m	Mudstone, dark gray, parallel bedding, siltstone laminae and few thin interbeds, light olive gray marl interbeds at the base
312-1	1331.02—1332.65 m	1.63 m	Mudstone, dark gray, horizontal bedding, siltstone strips
312-2	1332.65—1332.76 m	0.11 m	Marl, light olive gray, horizontal bedding
312-3	1332.76—1332.96 m	0.20 m	Siltstone, light gray, ripple bedding, 3 cm-thick dark gray mudstone at the top, mudstone laminae, vermicular sand-veins in mudstone
312-4	1332.96—1333.17 m	0.21 m	Muddy siltstone, medium light gray with deformation structures, massive bedding, plentiful carbon dusts, 1.5 cm mudstone laminae at the top
312-5	1333.17—1333.41 m	0.24 m	Silty mudstone, gray, wavy ripple bedding, with few siltstone strips
312-6	1333.41—1333.61 m	0.20 m	Muddy siltstone, light gray, wavy bedding, bioturbation, irregular siltstone
312-7	1333.61—1333.92 m	0.31 m	Silty mudstone, gray, horizontal bedding, with siltstone strips
312-8	1333.92—1334.02 m	0.10 m	Mudstone, gray, horizontal bedding
312-9	1334.02—1334.16 m	0.14 m	Muddy siltstone, greenish-gray, unclear horizontal bedding, 4 cm-thick dark greenish-gray mud powder at the top, plentiful carbon dusts
312-10	1334.16—1334.35 m	0.19 m	Siltstone, light gray, mud gravels
312-11	1334.35—1334.68 m	0.33 m	Muddy siltstone, light gray, massive
312-12	1334.68—1335.05 m	0.37 m	Silty mudstone, gray, massive
312-13	1335.05—1335.35 m	0.30 m	Mudstone, gray, calcareous concretions
312-14	1335.35—1335.51 m	0.16 m	Siltstone, light gray, wavy ripple bedding, mudstone strips
312-15	1335.51—1336.10 m	0.59 m	Muddy siltstone, light gray, wavy ripple bedding, with lots of mudstone laminae
312-16	1336.10—1336.20 m	0.10 m	Siltstone, light gray, wavy ripple bedding, mudstone strips at the top
312-17	1336.20—1336.44 m	0.24 m	Fine-grained sandstone, light gray, wavy ripple bedding, mudstone strips
312-18	1336.44—1337.34 m	0.90 m	Siltstone, light gray, wavy ripple bedding, mudstone strips
312-19	1337.34—1337.76 m	0.42 m	Sandstone, fine-grained, light gray, ripple bedding, erosion surface at the base, mudstone laminae, fallen sand-veins, large carbon dust on the erosion surface
312-20	1337.76—1337.99 m	0.23 m	Siltstone, light gray, wavy ripple bedding, mudstone strips
312-21	1337.99—1338.60 m	0.61 m	Mudstone, greenish gray, wavy ripple bedding, siltstone strips
313-1	1338.60—1338.97 m	0.37 m	Silty mudstone, greenish gray, massive, calcareous concretions
313-2	1338.97—1339.07 m	0.10 m	Siltstone, greenish gray, wavy ripple bedding, weak erosion surface at the base
313-3	1339.07—1340.34 m	1.27 m	Fine-grained sandstone, light gray, parallel bedding
313-4	1340.34—1340.86 m	0.52 m	Siltstone, light greenish gray, wavy ripple bedding
313-5	1340.86—1340.96 m	0.10 m	Muddy siltstone, light greenish gray, massive

313-6	1340.96—1341.28 m	0.32 m	Silty mudstone, greenish gray, massive
313-7	1341.28—1341.50 m	0.22 m	Mudstone, gray, horizontal ripple bedding
313-8	1341.50—1341.85 m	0.35 m	Silty mudstone, gray, horizontal ripple bedding
313-9	1341.85—1342.07 m	0.22 m	Siltstone, light greenish gray, massive, weak erosion surface at the base, with mudstone strips
313-10	1342.07—1342.43 m	0.36 m	Fine-grained sandstone, light gray, parallel bedding, weak erosion surface at the base
313-11	1342.43—1342.79 m	0.36 m	Siltstone, light gray, wavy ripple bedding, mudstone strips occasionally
313-12	1342.79—1343.80 m	1.01 m	Siltstone, light gray, interbedded with gray mudstone, wavy bedding
313-13	1343.80—1344.05 m	0.25 m	Silty mudstone, gray, deformation bedding
313-14	1344.05—1344.26 m	0.21 m	Silty mudstone, medium gray, interbedded with light gray siltstone, wavy bedding, thickness-varying sandstone, wedge-shaped sandstone laminae locally
313-15	1344.26—1344.78 m	0.52 m	Mudstone, gray, wavy ripple bedding
313-16	1344.78—1344.90 m	0.12 m	Silty mudstone, gray, wavy ripple bedding
313-17	1344.90—1346.13 m	1.23 m	Muddy siltstone, gray, parallel bedding, with siltstone strips
313-18	1346.13—1346.48 m	0.35 m	Siltstone, medium light gray, ripple bedding, foreset bedding locally, liquefied deformation bedding, mudstone laminae
313-19	1346.48—1346.83 m	0.35 m	Muddy siltstone, light gray, horizontal bedding, siltstone strips and mudstone strips locally, trace fossils occasionally
313-20	1346.83—1347.05 m	0.22 m	Siltstone, light gray, wavy cross bedding
313-21	1347.05—1347.84 m	0.79 m	Fine-grained sandstone, light gray, parallel bedding
314-1	1347.84—1349.10 m	1.26 m	Fine-grained sandstone, light gray, parallel bedding, wavy bedding at the base
314-2	1349.10—1352.14 m	3.04 m	Medium-grained sandstone, light gray, parallel bedding, mud gravels occasionally
314-3	1352.14—1353.83 m	1.69 m	Sandstone, medium-grained, medium light gray, wavy bedding, with carbon dusts, subangular mud gravels and sand gravels at 1353.56 m and 1353.68 m
314-4	1353.83—1353.92 m	0.09 m	Fine-grained sandstone, light gray, discontinuous, wavy ripple bedding
314-5	1353.92—1354.34 m	0.42 m	Coarse-grained sandstone, gray, parallel bedding
314-6	1354.34—1354.47 m	0.13 m	Fine-grained conglomerate, oblique bedding, with erosion surface
314-7	1354.47—1354.70 m	0.23 m	Coarse-grained sandstone, with fine gravels, medium light gray, parallel bedding, some fine-grained quartz
314-8	1354.70—1355.01 m	0.31 m	Medium-grained sandstone, light gray, parallel bedding, normal graded bedding
314-9	1355.01—1355.16 m	0.15 m	Coarse-grained sandstone, light gray, massive
314-10	1355.16—1355.37 m	0.21 m	Medium-grained sandstone, light gray, massive
314-11	1355.37—1355.49 m	0.12 m	Coarse-grained sandstone, light gray, massive
314-12	1355.49—1355.69 m	0.20 m	Medium-grained sandstone, light gray, parallel bedding

314-13	1355.69—1356.21 m	0.52 m	Coarse-grained sandstone, parallel bedding
314-14	1356.21—1356.31 m	0.10 m	Fine-grained sandstone, light gray, parallel bedding
315-1	1356.37—1357.75 m	1.38 m	Sandstone, medium-grained, light greenish-gray, unclear bedding, quartz gravels occasionally, white concretions at 0.98 m not bubbled with acid, surrounded by black carbonaceous substance
315-2	1357.75—1358.87 m	1.12 m	Mudstone, dark brown, horizontal bedding, wavy ripple bedding, with sandstone strips and laminae locally
315-3	1358.87—1359.77 m	0.90 m	Siltstone, light gray, ripple bedding, deformation structure locally, siltstone laminae interbedded with mudstone
315-4	1359.77—1360.59 m	0.82 m	Siltstone, light gray, wavy bedding, with mudstone laminae and strips
315-5	1360.59—1361.37 m	0.78 m	Mudstone, dark brown, horizontal bedding, wavy ripple bedding, siltstone strips locally
315-6	1361.37—1361.62 m	0.25 m	Siltstone, light gray, massive, mudstone strips
317-1	1362.64—1363.19 m	0.55 m	Siltstone, light gray, massive, deformation bedding, mudstone strips at 1363.14 m
317-2	1363.19—1363.49 m	0.30 m	Silty mudstone, greenish gray, massive
317-3	1363.49—1363.81 m	0.32 m	Mudstone, greenish gray, massive
318-1	1363.81—1364.03 m	0.22 m	Silty mudstone, gray, massive
318-2	1364.03—1364.41 m	0.38 m	Siltstone, light gray, massive, with few mudstone strips and laminae
318-3	1364.41—1364.51 m	0.10 m	Fine-grained sandstone, light gray, parallel bedding, mudstone strips occasionally
318-4	1364.51—1365.65 m	1.14 m	Siltstone, light gray, wavy bedding, with mud laminae
318-5	1365.65—1365.85 m	0.20 m	Siltstone, greenish gray, massive
318-6	1365.85—1366.13 m	0.28 m	Siltstone, light gray, wavy ripple bedding
318-7	1366.13—1366.25 m	0.12 m	Fine-grained sandstone, light greenish gray
318-8	1366.25—1367.11 m	0.86 m	Mudstone, medium gray, horizontal bedding, horizontal ripple bedding, slight deformation locally, a lot of siltstone laminae and thin interbeds
319-1	1367.24—1367.64 m	0.40 m	Silty mudstone, gray, massive
319-2	1367.64—1367.92 m	0.28 m	Muddy siltstone, greenish gray, massive
319-3	1367.92—1368.09 m	0.17 m	Siltstone, light gray, wavy bedding, mudstone strips occasionally
319-4	1368.09—1368.19 m	0.10 m	Silty mudstone, greenish gray, massive
319-5	1368.19—1368.52 m	0.33 m	Silty mudstone, gray, massive
319-6	1368.52—1368.66 m	0.14 m	Muddy siltstone, greenish gray, massive
319-7	1368.66—1368.92 m	0.26 m	Muddy siltstone, light greenish gray, massive, bioturbation
319-8	1368.92—1370.48 m	1.56 m	Siltstone, light greenish gray, wavy bedding, deformation bedding, mudstone strips
319-9	1370.48—1371.02 m	0.54 m	Mudstone, gray, interbedded with light gray siltstone, wavy bedding, with siltstone strips and laminae

319-10	1371.02—1371.24 m	0.22 m	Siltstone, light gray, parallel bedding, wavy bedding, fine-grained sandstone laminae and mudstone strips
319-11	1371.24—1371.96 m	0.72 m	Mudstone, gray, wavy bedding, fine-grained sandstone laminae and strips
319-12	1371.96—1373.26 m	1.30 m	Fine-grained sandstone, light gray, wavy bedding, deformation bedding
319-13	1373.26—1373.44 m	0.18 m	Mudstone, gray, massive
320-1	1373.44—1373.99 m	0.55 m	Mudstone, gray, massive
320-2	1373.99—1374.14 m	0.15 m	Siltstone, light gray, massive
320-3	1374.14—1374.51 m	0.37 m	Silty mudstone, gray, massive
320-4	1374.51—1374.79 m	0.28 m	Fine-grained sandstone, light gray, wavy bedding, with mudstone strips
320-5	1374.79—1375.22 m	0.43 m	Siltstone, light gray, oblique bedding
320-6	1375.22—1375.89 m	0.67 m	Mudstone, gray, wavy ripple bedding, with siltstone laminaes
320-7	1375.89—1376.02 m	0.13 m	Siltstone, light gray, mudstone strips
320-8	1376.02—1376.24 m	0.22 m	Mudstone, medium gray, horizontal bedding, interbedded with siltstone laminae locally, plant fragments occasionally on the section
321-1	1376.88—1378.28 m	1.40 m	Mudstone, gray, horizontal bedding, with siltstone strips
321-2	1378.28—1379.38 m	1.10 m	Mudstone, gray, horizontal bedding, with siltstone strips and laminae
321-3	1379.38—1379.48 m	0.10 m	Fine-grained sandstone, light gray, parallel bedding
321-4	1379.48—1379.91 m	0.43 m	Medium-grained sandstone, light gray
321-5	1379.91—1380.41 m	0.50 m	Siltstone, light gray, deformation bedding, wavy bedding
321-6	1380.41—1382.43 m	2.02 m	Mudstone, gray, horizontal bedding, siltstone laminae
321-7	1382.43—1382.58 m	0.15 m	Silty mudstone, gray, massive
321-8	1382.58—1384.38 m	1.80 m	Mudstone, dark brown, parallel bedding
321-9	1384.38—1384.48 m	0.10 m	Fine-grained sandstone, light gray, wavy bedding
321-10	1384.48—1384.78 m	0.30 m	Muddy siltstone, light greenish gray, deformation bedding locally
321-11	1384.78—1385.31 m	0.53 m	Siltstone, light greenish gray, wavy bedding, with fine-grained sandstone laminae
322-1	1385.37—1385.47 m	0.10 m	Siltstone, light greenish gray, wavy bedding, with mudstone laminae
322-2	1385.47—1385.57 m	0.10 m	Mudstone, greenish gray, massive
322-3	1385.57—1386.07 m	0.50 m	Siltstone, light greenish gray, wavy bedding, mudstone strips
322-4	1386.07—1386.72 m	0.65 m	Mudstone, gray, massive, wavy ripple bedding
322-5	1386.72—1387.00 m	0.28 m	Siltstone, light greenish gray, wavy bedding, with mudstone strips
322-6	1387.00—1387.10 m	0.10 m	Mudstone, gray, wavy ripple bedding, with sandstone strips and laminae
323-1	1387.32—1387.87 m	0.55 m	Siltstone, light gray, wavy bedding, with mudstone laminae and strips
323-2	1387.87—1388.06 m	0.19 m	Fine-grained sandstone, light gray

323-3	1388.06—1388.22 m	0.16 m	Siltstone, light gray, wavy bedding, with mudstone strips
323-4	1388.22—1388.38 m	0.16 m	Silty mudstone, gray, massive, fossil fragments occasionally
323-5	1388.38—1388.75 m	0.37 m	Siltstone, light gray, wavy ripple bedding
323-6	1388.75—1389.02 m	0.27 m	Fine-grained sandstone, light gray, wavy bedding, with mudstone laminae
323-7	1389.02—1389.46 m	0.44 m	Mudstone, gray, wavy ripple bedding, with fine-grained sandstone laminae and strips
323-8	1389.46—1389.83 m	0.37 m	Siltstone, greenish gray, massive
323-9	1389.83—1389.88 m	0.05 m	Mudstone, yellowish gray, massive, calcareous concretions occasionally
323-10	1389.88—1390.47 m	0.59 m	Muddy siltstone, greenish gray, massive
323-11	1390.47—1390.94 m	0.47 m	Siltstone, light gray, wavy ripple bedding, mudstone strips
323-12	1390.94—1391.06 m	0.12 m	Silty mudstone, greenish gray, massive, with siltstone strips
323-13	1391.06—1391.30 m	0.24 m	Fine-grained sandstone, light greenish gray
323-14	1391.30—1392.13 m	0.83 m	Muddy siltstone, greenish-gray, deformation bedding locally, with fine-grained sandstone laminae
323-15	1392.13—1392.23 m	0.10 m	Fine-grained sandstone, light gray, wavy bedding, mudstone strips
323-16	1392.23—1392.51 m	0.28 m	Silty mudstone, greenish gray, massive
323-17	1392.51—1393.12 m	0.61 m	Fine-grained sandstone, light greenish gray, wavy bedding, mud gravels at 1392.43 m
323-18	1393.12—1393.92 m	0.80 m	Mudstone, greenish gray, horizontal bedding, siltstone strips
324-1	1393.92—1394.18 m	0.26 m	Mudstone, greenish gray, horizontal bedding
324-2	1394.18—1395.09 m	0.91 m	Siltstone, light gray, wavy bedding, mudstone laminae and strips
324-3	1395.09—1395.29 m	0.20 m	Silty mudstone, gray, horizontal bedding
324-4	1395.29—1395.77 m	0.48 m	Mudstone, dark brown, horizontal bedding, mudstone strips occasionally
324-5	1395.77—1396.05 m	0.28 m	Siltstone, light gray, wavy bedding in the middle and upper parts, with mudstone laminae and strips
324-6	1396.05—1396.35 m	0.30 m	Mudstone, dark gray, interbedded with light gray siltstone, lenticular bedding in the middle and upper part, flaser bedding in the lower part, with sand balls
324-7	1396.35—1396.76 m	0.41 m	Siltstone, light gray, flaser-wavy composite bedding
324-8	1396.76—1397.35 m	0.59 m	Silty mudstone, greenish gray, massive
324-9	1397.35—1397.55 m	0.20 m	Mudstone, greenish gray, massive
324-10	1397.55—1397.80 m	0.25 m	Silty mudstone, greenish gray, massive
324-11	1397.80—1398.19 m	0.39 m	Siltstone, light gray, wavy ripple bedding, with weak bioturbation
324-12	1398.19—1398.71 m	0.52 m	Mudstone, greenish gray, massive
324-13	1398.71—1398.92 m	0.21 m	Muddy siltstone, light gray, deformation bedding, siltstone strips, bioturbation
324-14	1398.92—1399.43 m	0.51 m	Fine-grained sandstone, light gray, wavy bedding, mudstone strips occasionally

324-15	1399.43—1399.63 m	0.20 m	Muddy siltstone, gray, massive, with siltstone strips occasionally
324-16	1399.63—1400.12 m	0.49 m	Fine-grained sandstone, light gray, parallel bedding, wavy ripple bedding
324-17	1400.12—1401.25 m	1.13 m	Siltstone, light gray, ripple bedding, silty mudstone laminae at 1400.15 m and 1400.25 m, generally carbon dust in the middle and lower part, fining downward
324-18	1401.25—1402.31 m	1.06 m	Siltstone, medium light gray, horizontal ripple and horizontal bedding at the top, sandstone-mudstone interbeds with horizontal wavy bedding below 1401.42 m
324-19	1402.31—1402.40 m	0.09 m	Fine-grained sandstone, light gray, wavy bedding, with mudstone laminae
324-20	1402.40—1402.55 m	0.15 m	Siltstone, light gray, wavy bedding, with mudstone laminae
325-1	1402.55—1402.72 m	0.17 m	Siltstone, light gray, wavy ripple bedding
325-2	1402.72—1402.79 m	0.07 m	Fine-grained sandstone, light gray, wavy ripple bedding, weak erosion surface at the face
325-3	1402.79—1404.57 m	1.78 m	Siltstone, medium light gray, gentle wavy bedding, dark gray mudstone laminae and sand balls in mudstone, step normal fault at 1403.65 m
325-4	1404.57—1405.61 m	1.04 m	Siltstone, light gray, interbedded with dark gray mudstone, horizontal wavy ripple bedding, siltstone strips
325-5	1405.61—1406.41 m	0.80 m	Mudstone, dark gray, interbedded with light gray siltstone, wavy bedding, lenticular bedding, deformation structure locally, siltstone laminae, lens and sand balls
325-6	1406.41—1406.85 m	0.44 m	Mudstone, dark gray, horizontal bedding and wavy ripple bedding, with siltstone strips
325-7	1406.85—1407.15 m	0.30 m	Siltstone, light gray, wavy bedding in the upper part, massive in the lower part
325-8	1407.15—1408.25 m	1.10 m	Mudstone, dark gray, horizontal bedding and horizontal ripple bedding, siltstone laminae, lenticular limestone at 1407.95 m
325-9	1408.25—1408.55 m	0.30 m	Mudstone, dark brown, wavy ripple bedding, with fossil fragments
325-10	1408.55—1408.76	0.21 m	Siltstone, light gray, wavy bedding, wavy ripple bedding, deformation bedding, with mudstone laminae
325-11	1408.76—1409.30 m	0.54 m	Mudstone, gray, wavy ripple bedding, with siltstone strips
325-12	1409.30—1410.00 m	0.70 m	Mudstone, dark gray, horizontal bedding
325-13	1410.00—1411.57 m	1.57 m	Muddy siltstone, light gray, wavy ripple bedding, with mudstone strips and laminae
326-1	1411.57—1412.59 m	1.02 m	Siltstone, light gray, wavy bedding, deformation bedding
326-2	1412.59—1413.02 m	0.43 m	Silty mudstone, greenish gray, massive
326-3	1413.02—1413.25 m	0.23 m	Silty mudstone, light gray, wavy bedding, deformation bedding locally, with dark gray mudstone laminae and strips
326-4	1413.25—1413.52 m	0.27 m	Fine-grained sandstone, light gray

326-5	1413.52—1414.47 m	0.95 m	Siltstone, light gray, wavy bedding, deformation bedding locally, with mudstone strips and fine-grained sandstone strips
326-6	1414.47—1414.75 m	0.28 m	Fine-grained sandstone, light gray
326-7	1414.75—1415.02 m	0.27 m	Silty mudstone, greenish gray, massive, with fine-grained sandstone strips
326-8	1415.02—1415.87 m	0.85 m	Siltstone, light gray, wavy bedding, with few mudstone strips
326-9	1415.87—1417.12 m	1.25 m	Mudstone, medium gray, ripple bedding, evenly interbedded with light gray silty mudstone or siltstone laminae, 2 cm-thick dolomite laminae at 1416.73 m
326-10	1417.12—1417.60 m	0.48 m	Siltstone, light gray, wavy bedding, with mudstone strips and laminae
326-11	1417.60—1417.75 m	0.15 m	Dolomite, olive gray, massive, with siltstone strips
326-12	1417.75—1418.37 m	0.62 m	Medium-grained sandstone, light gray, wavy bedding, with mud gravels
326-13	1418.37—1419.49 m	1.12 m	Fine-grained sandstone, light gray, wavy bedding
326-14	1419.49—1420.01 m	0.52 m	Medium-grained sandstone, light gray, wavy bedding, with few mudstone strips
326-15	1420.01—1420.55 m	0.54 m	Silty mudstone, medium gray, massive bedding, a lot of plant fragments, some siltstone laminae intersected with horizontal level at 1420.27 m
327-1	1420.57—1422.57 m	2.00 m	Mudstone, medium gray, horizontal ripple bedding, plentiful light gray siltstone laminae, in much layers locally
327-2	1422.57—1423.72 m	1.15 m	Mudstone, dark brown, wavy ripple bedding, with siltstone strips and laminae
327-3	1423.72—1423.80 m	0.08 m	Dolomite, olive black, massive
327-4	1423.80—1425.20 m	1.40 m	Mudstone, dark brown, wavy ripple bedding, with siltstone strips and laminae
327-5	1425.20—1425.42 m	0.22 m	Mudstone, greenish gray, massive
327-6	1425.42—1426.37 m	0.95 m	Silty mudstone, greenish gray, massive
327-7	1426.37—1426.72 m	0.35 m	Medium-grained sandstone, light gray, water-escape structure, mudstone strips occasionally
327-8	1426.72—1427.67 m	0.95 m	Siltstone, light gray, wavy bedding, deformation bedding, with mudstone strips and laminae in the lower part
327-9	1427.67—1428.44 m	0.77 m	Fine-grained sandstone, light gray
328-1	1428.57—1428.87 m	0.30 m	Fine-grained sandstone, light gray
328-2	1428.87—1429.35 m	0.48 m	Medium-grained sandstone, light gray, oblique bedding
328-3	1429.35—1429.56 m	0.21 m	Coarse-grained sandstone, light gray, oblique bedding
328-4	1429.56—1430.02 m	0.46 m	Fine-grained sandstone, light gray, discontinuous wavy bedding
328-5	1430.02—1430.59 m	0.57 m	Coarse-grained sandstone, light gray, small trough cross bedding
328-6	1430.59—1431.34 m	0.75 m	Mudstone, medium gray, ripple bedding, horizontal bedding, deformation structure locally, siltstone laminae, sand balls, increasing and thickening upward
328-7	1431.34—1432.00 m	0.66 m	Mudstone laminae. Sandstone, fine-grained, light gray, ripple bedding, discontinuous ripple bedding, parallel bedding locally, a lot of carbon dusts and few mudstone laminae on the bedding surface

328-8	1432.00—1432.43 m	0.43 m	Mudstone, gray, horizontal bedding, with siltstone strips
328-9	1432.43—1432.80 m	0.37 m	Siltstone, light gray, wavy ripple bedding, parallel bedding locally, with mudstone laminae
328-10	1432.80—1433.07 m	0.27 m	Fine-grained sandstone, light gray, parallel bedding, wavy ripple bedding, mudstone strips
328-11	1433.07—1433.47 m	0.40 m	Mudstone, gray, horizontal bedding, wavy ripple bedding, with light gray siltstone strips and laminae
328-12	1433.47—1434.18 m	0.71 m	Silty mudstone, light gray, wavy ripple bedding, deformation bedding locally, siltstone laminae
328-13	1434.18—1434.82 m	0.64 m	Mudstone, gray, horizontal bedding and wavy ripple bedding, with siltstone strips and laminaes
328-14	1434.82—1434.97 m	0.15 m	Muddy siltstone, light gray, wavy ripple bedding, deformation bedding locally, with siltstone strips
328-15	1434.97—1435.17 m	0.20 m	Mudstone, gray, horizontal bedding
328-16	1435.17—1436.73 m	1.56 m	Fine-grained sandstone, light gray, parallel bedding, water-escape structure locally, with mudstone strips and laminae
328-17	1436.73—1437.46 m	0.73 m	Siltstone, light gray, parallel bedding, wavy ripple bedding, with mudstone laminae
329-1	1437.46—1438.21 m	0.75 m	Siltstone, medium light gray, horizontal ripple bedding, horizontal bedding locally, mudstone laminae, with sand balls
329-2	1438.21—1438.97 m	0.76 m	Mudstone, gray, horizontal bedding, wavy ripple bedding locally, siltstone strips and laminae
329-3	1438.97—1439.96 m	0.99 m	Muddy siltstone, medium light gray, horizontal ripple bedding in the middle and upper parts, deformation at and below 1439.76 m, cross bedding, carbon dusts between layers in the upper part, mudstone or siltstone laminae and thin interbeds, sand increasing downward
329-4	1439.96—1440.98 m	1.02 m	Fine-grained sandstone, light gray, parallel bedding, wavy bedding
329-5	1440.98—1441.66 m	0.68 m	Siltstone, light gray, wavy bedding, with mudstone laminae
329-6	1441.66—1441.91 m	0.25 m	Fine-grained sandstone, light gray, wavy bedding, deformation bedding at the base, mudstone strips and laminae at the base
329-7	1441.91—1442.38 m	0.47 m	Siltstone, light gray, wavy ripple bedding in the middle and upper parts, deformation bedding in the lower part
329-8	1442.38—1442.55 m	0.17 m	Siltstone, gray, wavy bedding
329-9	1442.55—1442.82 m	0.27 m	Siltstone, light gray, with wavy ripple bedding, horizontal bedding and wavy cross bedding locally, mudstone strips
329-10	1442.82—1443.69 m	0.87 m	Fine-grained sandstone, light gray, wavy bedding, discontinuous, wavy bedding
329-11	1443.69—1444.13 m	0.44 m	Siltstone, light gray, wavy ripple bedding, horizontal bedding and weak erosion surface locally
329-12	1444.13—1444.28 m	0.15 m	Fine-grained sandstone, light gray, wavy cross bedding

329-13	1444.28—1444.67 m	0.39 m	Siltstone, light gray, wavy ripple bedding, horizontal bedding and deformation bedding locally, with mudstone laminae
329-14	1444.67—1445.26 m	0.59 m	Fine-grained sandstone, light gray, with wavy bedding
329-15	1445.26—1445.52 m	0.26 m	Fine-grained sandstone, yellowish gray, wavy bedding
329-16	1445.52—1445.79 m	0.27 m	Silty mudstone, gray, wavy ripple bedding
329-17	1445.79—1446.22 m	0.43 m	Mudstone, gray, horizontal bedding
330-1	1446.44—1446.74 m	0.30 m	Mudstone, gray, horizontal bedding
330-2	1446.74—1447.88 m	1.14 m	Silty mudstone, gray, horizontal bedding
330-3	1447.88—1447.99 m	0.11 m	Siltstone, light gray, wavy ripple bedding
330-4	1447.99—1448.48 m	0.49 m	Mudstone, medium dark gray, horizontal bedding, elliptic dolomite at 1448.30—1448.34 m, siltstone laminae at the top
330-5	1448.48—1448.61 m	0.13 m	Silty mudstone, dark gray, horizontal bedding
330-6	1448.61—1449.99 m	1.38 m	Fine-grained sandstone, light gray, wavy bedding, discontinuous wavy bedding
330-7	1449.99—1450.65 m	0.66 m	Siltstone, light gray, wavy ripple bedding, horizontal bedding, deformation bedding locally
330-8	1450.65—1451.35 m	0.70 m	Fine-grained sandstone, light gray, parallel bedding, wavy cross bedding
330-9	1451.35—1451.89 m	0.54 m	Siltstone, light gray, horizontal bedding, horizontal ripple bedding (low-angle cross bedding), mudstone laminae, few carbon dust on bedding surface
330-10	1451.89—1452.29 m	0.40 m	Fine-grained sandstone, light gray, deformation bedding, wavy bedding at the base, irregular mudstone strips
330-11	1452.29—1454.08 m	1.79 m	Siltstone, light gray, interbedded with medium gray mudstone, wavy bedding, horizontal bedding, wavy cross bedding locally, deformation bedding, siltstone laminae, mudstone laminae
330-12	1454.08—1455.30 m	1.22 m	Mudstone, dark brown, horizontal bedding, siltstone strips in the upper part
331-1	1455.30—1455.91 m	0.61 m	Mudstone, dark brown, horizontal bedding
331-2	1455.91—1456.82 m	0.91 m	Mudstone, dark gray, interbedded with light gray siltstone, deformation bedding, normal graded bedding, with siltstone strips and mudstone strips
331-3	1456.82—1457.78 m	0.96 m	Mudstone, dark brown, horizontal bedding, wavy ripple bedding, with siltstone strips
331-4	1457.78—1458.45 m	0.67 m	Siltstone, light gray, wavy ripple bedding, horizontal bedding, water-escape structure locally, with mudstone strips and laminae
331-5	1458.45—1459.30 m	0.85 m	Mudstone, dark gray, interbedded with light gray siltstone
331-6	1459.30—1459.80 m	0.50 m	Mudstone, medium dark gray, horizontal bedding, normal graded bedding in sandstone, siltstone laminae [horizontal ripple bedding or normal graded bedding, slight erosion in the lower part (turbidite)]
331-7	1459.80—1460.40 m	0.60 m	Mudstone, dark gray, interbedded with light gray siltstone
331-8	1460.40—1461.00 m	0.60 m	Mudstone, dark brown, horizontal bedding, normal graded bedding in the sandstone

331-9	1461.00—1461.95 m	0.95 m	Mudstone, medium dark gray, interbedded with light gray siltstone, horizontal wavy bedding, lenticular bedding locally, siltstone laminae (horizontal ripple bedding or normal graded bedding, slight erosion in the lower part)
331-10	1461.95—1463.25 m	1.30 m	Mudstone, dark gray, horizontal bedding
331-11	1463.25—1463.60 m	0.35 m	Siltstone, light gray, deformation bedding, discontinuous wavy ripple bedding, with mudstone strips
331-12	1463.60—1464.07 m	0.47 m	Siltstone, light gray, intercalated with dark gray mudstone, wavy ripple bedding, deformation bedding locally, mudstone strips
332-1	1464.12—1465.16 m	1.04 m	Siltstone, light gray, horizontal bedding, wavy ripple bedding, deformation bedding locally, mudstone strips and laminae
332-2	1465.16—1465.82 m	0.66 m	Siltstone, light gray, interbedded with dark gray mudstone, siltstone strips
332-3	1465.82—1466.35 m	0.53 m	Mudstone, dark gray, horizontal bedding, siltstone strips at the top
333-1	1466.35—1468.07 m	1.72 m	Mudstone, dark brown, horizontal bedding, silty mudstone laminae
333-2	1468.07—1468.15 m	0.08 m	Dolomite, olive gray, massive
333-3	1468.15—1470.70 m	2.55 m	Mudstone, dark brown, horizontal bedding, ostracoda fossils occasionally
333-4	1470.70—1470.75 m	0.05 m	Dolomite, olive gray, massive, ostracoda fossils
333-5	1470.75—1471.55 m	0.80 m	Mudstone, dark brown, horizontal bedding, ostracoda fossils occasionally
333-6	1471.55—1471.70 m	0.15 m	Fine-grained sandstone, light gray, wavy cross bedding, with ostracoda
333-7	1471.70—1472.00 m	0.30 m	Siltstone, light greenish gray, deformation bedding, with few ostracoda fossils in the siltstone, mudstone strips
333-8	1472.00—1472.20 m	0.20 m	Fine-grained sandstone, light gray, wavy cross bedding, ostracoda fossils
333-9	1472.20—1474.45 m	2.25 m	Siltstone, light greenish gray, wavy bedding, with lots of mudstone laminae and strips
333-10	1474.45—1474.88 m	0.43 m	Mudstone, gray, wavy ripple bedding, with silty mudstone and siltstone strips and laminae
334-1	1474.96—1475.21 m	0.25 m	Silty mudstone, gray, massive, with few siltstone laminae
334-2	1475.21—1475.91 m	0.70 m	Siltstone, light gray, wavy bedding
334-3	1475.91—1475.96 m	0.05 m	Dolomite, olive gray, massive
334-4	1475.96—1476.41 m	0.45 m	Siltstone, light gray, wavy bedding, with mudstone laminaes
334-5	1476.41—1479.26 m	2.85 m	Mudstone, dark brown, horizontal bedding, ostracoda fossils occasionally

The overlying formation of the Nenjiang Formation Member 3 is the Nenjiang Formation Member 4, the contacting relationship between them is conformable contact; the underlying formation is the Nenjiang Formation Member 2, the contacting relationship between them is conformable contact (Figure 5.6).

The detailed description of the Nenjiang Formation Member 3:

334-6	1479.26—1479.42 m	0.16 m	Silty mudstone, gray, wavy ripple bedding, siltstone strips
334-7	1479.42—1479.50 m	0.08 m	Dolomite, olive gray, massive
334-8	1479.50—1479.80 m	0.30 m	Mudstone, gray, wavy ripple bedding, with fine-grained sandstone strips, deformation bedding
334-9	1479.80—1480.06 m	0.26 m	Fine-grained sandstone, light gray, wavy bedding, mudstone strips
334-10	1480.06—1480.16 m	0.10 m	Silty mudstone, gray, massive, with dolomite strips at the base
334-11	1480.16—1480.46 m	0.30 m	Silty mudstone, light gray, wavy bedding, bioturbation
334-12	1480.46—1480.76 m	0.30 m	Silty mudstone, gray, wavy ripple bedding, with lots of siltstone laminae and strips
334-13	1480.76—1482.54 m	1.78 m	Sandstone, fine-grained, light gray, wave-generated cross-bedding in the upper part, wavy bedding in the middle part, parallel bedding in the lower part, carbon dust enriched in local layers
334-14	1482.54—1483.21 m	0.67 m	Siltstone, light gray, wavy bedding (hummocky), with argillaceous and fine grained sandstone laminae, 1 cm-thick dolomite interlayer at 1483.11 m
334-15	1483.21—1483.56 m	0.35 m	Muddy siltstone, gray, wavy bedding, siltstone strips
334-16	1483.56—1483.96 m	0.40 m	Siltstone, light gray, wavy bedding, mudstone strips
335-1	1483.96—1485.56 m	1.60 m	Dolomite, olive gray, siltstone, light gray, wavy bedding, wave-generated cross-bedding, with mudstone laminae locally, three olive gray dolomite laminae at 1484.64 m, 1484.86 m and 1484.88 m
335-2	1485.56—1485.81 m	0.25 m	Mufddy siltstone, gray, wavy ripple bedding, with siltstone laminae and strips
335-3	1485.81—1486.41 m	0.60 m	Siltstone, light gray, wavy bedding
335-4	1486.41—1486.66 m	0.25 m	Silty mudstone, gray, wavy ripple bedding, with siltstone strips and laminae
335-5	1486.66—1487.11 m	0.45 m	Siltstone, light gray, wavy bedding, deformation bedding locally, with few mudstone strips
335-6	1487.11—1487.36 m	0.25 m	Mudstone, dark brown, wavy ripple bedding
335-7	1487.36—1488.11 m	0.75 m	Silty mudstone, medium gray, ripple bedding, deformation structure locally, siltstone laminae, thin, 10 cm plant fragments reserved in silty mudstone near-vertically at 1487.56 m
335-8	1488.11—1488.34 m	0.23 m	Siltstone, light gray, wavy bedding, mudstone strips and laminae in the lower part
335-9	1488.34—1488.56 m	0.22 m	Mudstone, dark brown, wavy ripple bedding, siltstone strips and laminae
335-10	1488.56—1490.96 m	2.40 m	Fine-grained sandstone, light gray, parallel bedding and wavy bedding, mudstone strips and mud gravels occasionally
335-11	1490.96—1491.83 m	0.87 m	Medium-grained sandstone, light gray, parallel bedding, mud gravels occasionally

335-12	1491.83—1491.96 m	0.13 m	Fine-grained sandstone, light gray, parallel bedding
335-13	1491.96—1492.17 m	0.21 m	Siltstone, light gray, wavy ripple bedding, deformation bedding locally, with mudstone strips laminae
335-14	1492.17—1492.60 m	0.43 m	Fine-grained sandstone, light gray, wavy bedding
335-15	1492.60—1492.86 m	0.26 m	Siltstone, light gray, wavy ripple bedding, with mudstone laminae and strips
336-1	1492.98—1493.16 m	0.18 m	Siltstone, light gray, wavy cross bedding, with mudstone strips
336-2	1493.16—1493.65 m	0.49 m	Siltstone, light gray, horizontal bedding, wave-generated cross-bedding locally, two 3 cm-thick mudstone laminae at 1493.43—1493.51 m, with plentiful carbon dusts
336-3	1493.65—1494.37 m	0.72 m	Silty mudstone, gray, horizontal bedding, with siltstone laminae and strips
336-4	1494.37—1495.88 m	1.51 m	Muddy siltstone, medium gray, horizontal bedding and horizontal ripple bedding in the middle and upper parts, deformation bedding at the base, carbon dust and silty mudstone laminae as well as siltstone laminae occasionally
336-5	1495.88—1496.46 m	0.58 m	Silty mudstone, gray, horizontal bedding, siltstone strips
336-6	1496.46—1496.78 m	0.32 m	Siltstone, light gray, massive, with mudstone strips
336-7	1496.78—1497.18 m	0.40 m	Siltstone, light gray, wavy bedding, with mudstone strips and laminaes
336-8	1497.18—1497.53 m	0.35 m	Silty mudstone, gray, wavy ripple bedding, with mudstone and siltstone strips
336-9	1497.53—1498.01 m	0.48 m	Siltstone, light gray, discontinuous wavy bedding and sand balls in the upper part, horizontal bedding and carbon dusts in the middle and lower part, high-angle fracture in the upper part, filled with muds
336-10	1498.01—1499.07 m	1.06 m	Silty mudstone, medium gray, horizontal bedding and horizontal ripple bedding and siltstone laminae in the middle and upper parts, weak deformation bedding, vermiform and irregular siltstones, and few carbon dusts in the lower part
336-11	1499.07—1499.60 m	0.53 m	Muddy siltstone, light gray, discontinuous wavy ripple bedding, weak deformation bedding locally, with siltstone strips, irregular mud gravels at the top
336-12	1499.60—1500.25 m	0.65 m	Siltstone, light gray, discontinuous wavy bedding in the lower part, deformation bedding in the middle and upper parts, wavy bedding locally, with mudstone strips and laminaes
336-13	1500.25—1500.48 m	0.23 m	Siltstone, light gray, wavy cross bedding
336-14	1500.48—1500.58 m	0.10 m	Silty mudstone, dark gray, wavy ripple bedding, with few siltstone strips
336-15	1500.58—1501.97 m	1.39 m	Fine-grained sandstone, light gray, parallel bedding, with a mudstone laminae at 1500.81 m, gray, horizontal bedding
337-1	1501.97—1502.85 m	0.88 m	Fine-grained sandstone, light gray, parallel bedding
337-2	1502.85—1503.37 m	0.52 m	Fine-grained sandstone, light gray, wavy bedding
337-3	1503.37—1505.32 m	1.95 m	Medium-grained sandstone, light gray, parallel bedding

337-4	1505.32—1505.59 m	0.27 m	Siltstone, light gray, parallel bedding in the upper part, wavy cross bedding in the middle part, wavy bedding in the lower part
337-5	1505.59—1506.07 m	0.48 m	Silty mudstone, gray, wavy ripple bedding, with siltstone strips
337-6	1506.07—1506.97 m	0.90 m	Mudstone, dark brown, horizontal bedding, wavy ripple bedding, with siltstone strips
337-7	1506.97—1507.50 m	0.53 m	Silty mudstone, dark gray, horizontal bedding
337-8	1507.50—1511.06 m	3.56 m	Mudstone, dark gray, horizontal bedding, two dolomite interlayers at 1508.81 m and 1509.29 m, 0.8 cm-thick dolomite laminae at 1510.13 m
338-1	1511.06—1511.76 m	0.70 m	Mudstone, dark brown, horizontal bedding, ostracoda fossils occasionally
338-2	1511.76—1511.80 m	0.04 m	Dolomite, olive gray, massive, with ostracoda fossils
338-3	1511.80—1514.46 m	2.66 m	Mudstone, dark brown, horizontal bedding, ostracoda fossils occasionally
338-4	1514.46—1515.06 m	0.60 m	Silty mudstone, gray, horizontal bedding
338-5	1515.06—1516.68 m	1.62 m	Mudstone, dark brown, horizontal bedding, ostracoda fossils occasionally
339-1	1516.68—1517.28 m	0.60 m	Mudstone, dark brown, horizontal bedding, ostracoda fossils occasionally
339-2	1517.28—1517.53 m	0.25 m	Medium-grained sandstone, light gray, wavy bedding, mudstone strips at the top
339-3	1517.53—1518.48 m	0.95 m	Fine-grained sandstone, light gray, wavy cross bedding
339-4	1518.48—1518.50 m	0.02 m	Dolomite, olive gray, massive
339-5	1518.50—1518.58 m	0.08 m	Mudstone, gray, wavy ripple bedding, siltstone strips
339-6	1518.58—1518.81 m	0.23 m	Fine-grained sandstone, light gray, wavy cross bedding
339-7	1518.81—1519.28 m	0.47 m	Silty mudstone, gray, wavy ripple bedding, horizontal bedding, with muddy siltstone laminae
339-8	1519.28—1519.38 m	0.10 m	Muddy siltstone, light gray, horizontal bedding
339-9	1519.38—1520.48 m	1.10 m	Mudstone, dark brown, horizontal bedding, with few muddy siltstone strips, ostracoda fossils occasionally
339-10	1520.48—1520.68 m	0.20 m	Muddy siltstone, light gray, wavy bedding, with mudstone laminae
339-11	1520.68—1521.00 m	0.32 m	Mudstone, dark brown, horizontal bedding, siltstone strips, with ostracoda fossils occasionally
340-1	1521.35—1523.05 m	1.70 m	Mudstone, dark brown, horizontal bedding, ostracoda fossils occasionally
340-2	1523.05—1523.15 m	0.10 m	Dolomite, olive gray, massive, with few ostracoda
340-3	1523.15—1524.28 m	1.13 m	Mudstone, dark brown, horizontal bedding, with muddy siltstone strips occasionally
340-4	1524.28—1524.90 m	0.62 m	Muddy siltstone, gray, wavy bedding
340-5	1524.90—1525.55 m	0.65 m	Siltstone, light greenish gray, wavy ripple bedding, with few mudstone strips, ostracoda fossils

340-6	1525.55—1525.73 m	0.18 m	Muddy siltstone, gray, wavy ripple bedding, with siltstone laminae
340-7	1525.73—1526.35 m	0.62 m	Siltstone, light gray, wavy bedding, with mudstone laminae
340-8	1526.35—1527.97 m	1.62 m	Silty mudstone, gray, horizontal bedding, siltstone laminae
340-9	1527.97—1528.55 m	0.58 m	Siltstone, light gray, wavy cross bedding
340-10	1528.55—1528.65 m	0.10 m	Siltstone, light gray, interbedded with dark gray siltstone, parallel bedding
340-11	1528.65—1529.75 m	1.10 m	Siltstone, light gray, wave-generated cross-bedding, 4 cm-thick fine-grained sandstone at 1529.05 m, carbon dust layer and mudstone laminae occasionally
340-12	1529.75—1529.85 m	0.10 m	Siltstone, light gray, deformation bedding, bioturbation, Mudstone laminae
340-13	1529.85—1530.39 m	0.54 m	Siltstone, light gray, deformation bedding at the top, wave-generated cross-bedding in the lower part, carbon dust layers distributed along the bedding surface
341-1	1530.46—1530.98 m	0.52 m	Siltstone, light gray, wavy cross bedding, parallel bedding, mudstone laminae
341-2	1530.98—1531.16 m	0.18 m	Silty mudstone, gray, horizontal bedding, siltstone strips
341-3	1531.16—1531.26 m	0.10 m	Siltstone, light gray, parallel bedding
341-4	1531.26—1531.43 m	0.17 m	Silty mudstone, gray, wavy ripple bedding, siltstone strips and laminae
341-5	1531.43—1531.81 m	0.38 m	Siltstone, light gray, wavy bedding, with mudstone strips and laminae
341-6	1531.81—1532.41 m	0.60 m	Fine-grained sandstone, light gray, wavy cross bedding
341-7	1532.41—1532.56 m	0.15 m	Silty mudstone, gray, wavy ripple bedding, with siltstone laminae
341-8	1532.56—1533.06 m	0.50 m	Fine-grained sandstone, light gray, wavy cross bedding
341-9	1533.06—1533.46 m	0.40 m	Siltstone, light gray, wavy bedding, parallel bedding locally, with medium dark gray mudstone laminae, and rich carbon dust
341-10	1533.46—1533.56 m	0.10 m	Silty mudstone, gray, with few siltstone strips
341-11	1533.56—1534.11 m	0.55 m	Siltstone, light gray, wavy cross bedding,parallel bedding, with siltstone strips
341-12	1534.11—1534.43 m	0.32 m	Silty mudstone, gray, horizontal bedding, with siltstone strips and laminae
341-13	1534.43—1534.54 m	0.11 m	Siltstone, light gray, wavy bedding
343-1	1537.23—1537.63 m	0.40 m	Fine-grained sandstone, light gray, wavy bedding, with mudstone strips and laminae
343-2	1537.63—1538.03 m	0.40 m	Silty mudstone, gray, wavy ripple bedding, with siltstone laminae and strips
343-3	1538.03—1538.13 m	0.10 m	Siltstone, light gray, parallel bedding
343-4	1538.13—1538.33 m	0.20 m	Silty mudstone, gray, wavy ripple bedding, with siltstone laminae and strips
343-5	1538.33—1541.96 m	3.63 m	Mudstone, dark gray, horizontal bedding
344-1	1541.31—1543.84 m	2.53 m	Mudstone, dark gray, horizontal bedding, dolomite interlayers at 1542.02 m and 1543.02 m

344-2	1543.84—1544.05 m	0.21 m	Siltstone, light gray, wavy ripple bedding
344-3	1544.05—1544.78 m	0.73 m	Fine-grained sandstone, light gray, parallel bedding, wavy bedding
344-4	1544.78—1544.94 m	0.16 m	Muddy siltstone, gray, horizontal bedding, siltstone strips
344-5	1544.94—1545.79 m	0.85 m	Siltstone, light gray, horizontal bedding, with lots of fossil fragments, mudstone laminae locally
344-6	1545.79—1546.72 m	0.93 m	Siltstone, light gray, intercalated with medium gray mudstone laminae, parallel bedding, wavy bedding, horizontal ripple bedding, wave-generated cross-bedding at the top, few carbon dusts on the siltstone surface
344-7	1546.72—1546.91 m	0.19 m	Siltstone, light gray, wavy bedding, deformation bedding, with mudstone laminae
344-8	1546.91—1547.77 m	0.86 m	Siltstone, light gray, intercalated with gray mudstone, wavy bedding, wavy ripple bedding, deformation bedding locally, with siltstone strips
344-9	1547.77—1547.90 m	0.13 m	Siltstone, light gray, parallel bedding
344-10	1547.90—1548.09 m	0.19 m	Silty mudstone, medium gray, horizontal ripple bedding, possibly dolomite at 1547.93 m, siltstone laminae at the base
344-11	1548.09—1548.91 m	0.82 m	Fine-grained sandstone, light gray, parallel bedding
344-12	1548.91—1550.34 m	1.43 m	Sandstone, fine-grained, light gray, wave-generated cross-bedding, a lot of carbon dusts at 1549.21 m and 1549.61 m
345-1	1550.34—1550.92 m	0.58 m	Fine-grained sandstone, light gray, wavy cross bedding, mudstone laminae
345-2	1550.92—1550.96 m	0.04 m	Mudstone, dark gray, horizontal bedding
345-3	1550.96—1551.04 m	0.08 m	Dolomite, olive black, massive
345-4	1551.04—1551.34 m	0.30 m	Silty mudstone, dark gray, horizontal bedding
345-5	1551.34—1554.69 m	3.35 m	Muddy siltstone, gray, massive, with deformation bedding fine-grained sandstone strips
345-6	1554.69—1556.19 m	1.50 m	Silty mudstone, dark gray, with horizontal bedding
345-7	1556.19—1556.64 m	0.45 m	Muddy siltstone, gray, massive, with deformation bedding fine-grained sandstone
345-8	1556.64—1559.36 m	2.72 m	Mudstone, dark gray, horizontal bedding
346-1	1559.36—1559.73 m	0.37 m	Mudstone, dark gray, horizontal bedding, few ostracoda shells
346-2	1559.73—1559.78 m	0.05 m	Dolomite, olive black, massive
346-3	1559.78—1560.35 m	0.57 m	Mudstone, dark gray, horizontal bedding, few ostracoda shells
346-4	1560.35—1560.39 m	0.04 m	Dolomite, olive black, massive
346-5	1560.39—1562.37 m	1.98 m	Mudstone, dark gray, horizontal bedding, few ostracoda shells
346-6	1562.37—1562.41 m	0.04 m	Dolomite, olive black, massive

346-7	1562.41—1562.62 m	0.21 m	Mudstone, dark gray, horizontal bedding, few ostracoda shells
347-1	1562.62—1563.72 m	1.10 m	Mudstone, dark gray, horizontal bedding, few ostracoda shells
347-2	1563.72—1563.75 m	0.03 m	Dolomite, olive black, massive
347-3	1563.75—1569.20 m	5.45 m	Mudstone, dark gray, horizontal bedding, few ostracoda shells
347-4	1569.20—1569.30 m	0.10 m	Dolomite, olive black, massive
347-5	1569.30—1570.12 m	0.82 m	Mudstone, dark gray, horizontal bedding, few ostracoda shells
347-6	1570.12—1571.00 m	0.88 m	Silty mudstone, gray, horizontal bedding
348-1	1571.06—1572.26 m	1.20 m	Muddy siltstone, light gray, wavy bedding, mudstone strips locally
348-2	1572.26—1572.41 m	0.15 m	Mudstone, dark gray, horizontal bedding, few ostracoda shells
348-3	1572.41—1573.02 m	0.61 m	Muddy siltstone, light gray, wavy bedding, mudstone strips locally
348-4	1573.02—1574.01 m	0.99 m	Silty mudstone, dark gray, horizontal bedding
348-5	1574.01—1575.76 m	1.75 m	Mudstone, dark gray, horizontal bedding, few ostracoda shells
348-6	1575.76—1575.80 m	0.04 m	Dolomite, olive black, massive
348-7	1575.80—1578.16 m	2.36 m	Mudstone, dark gray, horizontal bedding, few ostracoda shells
348-8	1578.16—1578.76 m	0.60 m	Silty mudstone, dark gray, horizontal bedding
348-9	1578.76—1579.64 m	0.88 m	Fine-grained sandstone, light gray, wavy bedding
348-10	1579.64—1579.77 m	0.13 m	Muddy siltstone, gray, massive
349-1	1579.77—1580.27 m	0.50 m	Fine-grained sandstone, light gray, wavy bedding, with mudstone strips
349-2	1580.27—1580.62 m	0.35 m	Silty mudstone, dark gray, horizontal bedding
349-3	1580.62—1581.07 m	0.45 m	Fine-grained sandstone, light gray, wavy bedding, mudstone strips
349-4	1581.07—1581.39 m	0.32 m	Muddy siltstone, gray, massive in the upper part, horizontal bedding in the lower part
349-5	1581.39—1581.41 m	0.02 m	Dolomite, olive black, massive
349-6	1581.41—1581.47 m	0.06 m	Mudstone, dark gray, horizontal bedding, few ostracoda shells
349-7	1581.47—1581.77 m	0.30 m	Fine-grained sandstone, light gray, parallel bedding, with mudstone strips
349-8	1581.77—1581.97 m	0.20 m	Silty mudstone, dark gray, horizontal bedding
349-9	1581.97—1582.07 m	0.10 m	Muddy siltstone, gray, massive, with plant fossils fragments
349-10	1582.07—1582.27 m	0.20 m	Fine-grained sandstone, light gray, wavy bedding
349-11	1582.27—1582.77 m	0.50 m	Mudstone, dark gray, horizontal bedding, with fine-grained sandstone laminae
349-12	1582.77—1582.83 m	0.06 m	Dolomite, olive black, massive
349-13	1582.83—1582.93 m	0.10 m	Mudstone, dark gray, horizontal bedding, laminae

5.1.7 Sifangtai Formation

The overlying formation of Sifangtai Formation is the Mingshui Formation Member 1, the contacting relationship between them is parallel conformable contact; the underlying formation is the Nenjiang Formation Member 5, the contacting relationship between them is unconformable contact (Figure 5.7).

The detailed description of the Sifangtai Formation:

207-6	807.12—807.34 m	0.22 m	Muddy siltstone, light greenish gray, massive, erosion surface at the base, fossil fragments
207-7	807.34—807.47 m	0.13 m	Mudstone, grayish brown, wavy ripple bedding
208-1	807.47—807.67 m	0.20 m	Mudstone, brownish gray, wavy ripple bedding, with trace fossils
208-2	807.67—807.81 m	0.14 m	Mudstone, grayish green, horizontal bedding, with 1 cm-thick siltstone laminae and erosion surface at the base
208-3	807.81—807.97 m	0.16 m	Silty mudstone, greenish gray, massive, fossil fragments, with ostracoda fossils occasionally
208-4	807.97—808.57 m	0.60 m	Silty mudstone, grayish red, massive, with ostracoda fossils
208-5	808.57—809.52 m	0.95 m	Mudstone, gray, massive, with irregular mudstone strips, calcareous concretions occasionally
208-6	809.52—809.92 m	0.40 m	Mudstone, greenish gray, massive, calcareous concretions
208-7	809.92—811.13 m	1.21 m	Mudstone, gray, wavy ripple bedding, with fractures filled with calcspar at the top, fossil fragments
208-8	811.13—811.61 m	0.48 m	Silty mudstone, greenish gray, massive, fossil fragments
208-9	811.61—812.97 m	1.36 m	Mudstone, brown, massive, with few calcareous concretions
208-10	812.97—813.51 m	0.54 m	Mudstone, greenish gray, massive, fossil fragments, irregular fracture filled with mudstone
208-11	813.51—813.89 m	0.38 m	Silty mudstone, grayish brown and greenish gray, massive, with irregular fracture
208-12	813.89—814.67 m	0.78 m	Mudstone, brown, massive, with few calcareous concretions and fracture
208-13	814.67—815.08 m	0.41 m	Mudstone, light brown, massive
209-1	815.08—816.38 m	1.30 m	Silty mudstone, greenish gray, massive, with few calcareous concretions and greenish gray spots
209-2	816.39—816.63 m	0.24 m	Silty mudstone, greenish gray, massive, with calcareous concretions
209-3	816.63—817.26 m	0.63 m	Mudstone, light brown, massive, with few calcareous concretions and greenish gray spots
209-4	817.26—817.78 m	0.52 m	Silty mudstone, greenish gray, massive, with light brown spots
209-5	817.78—818.03 m	0.25 m	Muddy siltstone, light olive gray, massive, with few ostracoda fossils

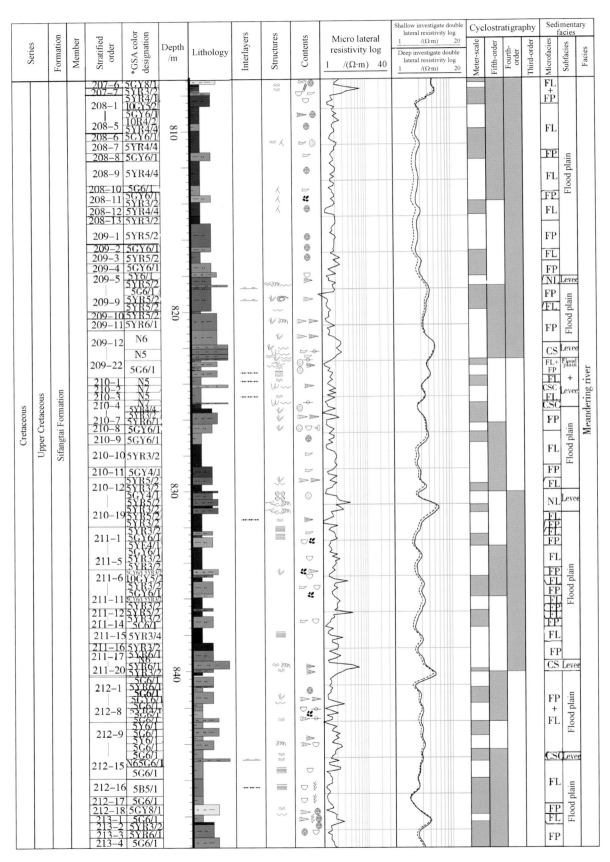

Figure 5.7　Stratigraphic chart of Sifangtai Formation.

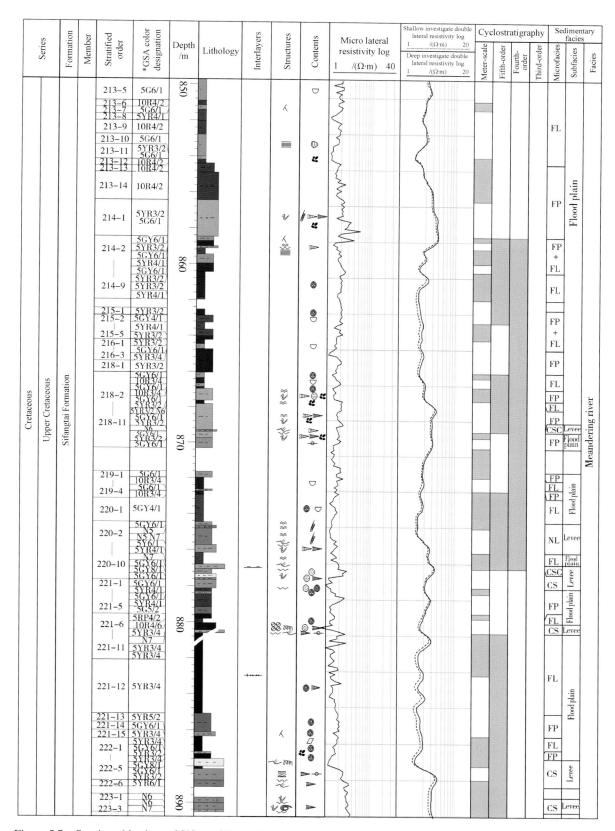

Figure 5.7 Stratigraphic chart of Sifangtai Formation (continued).

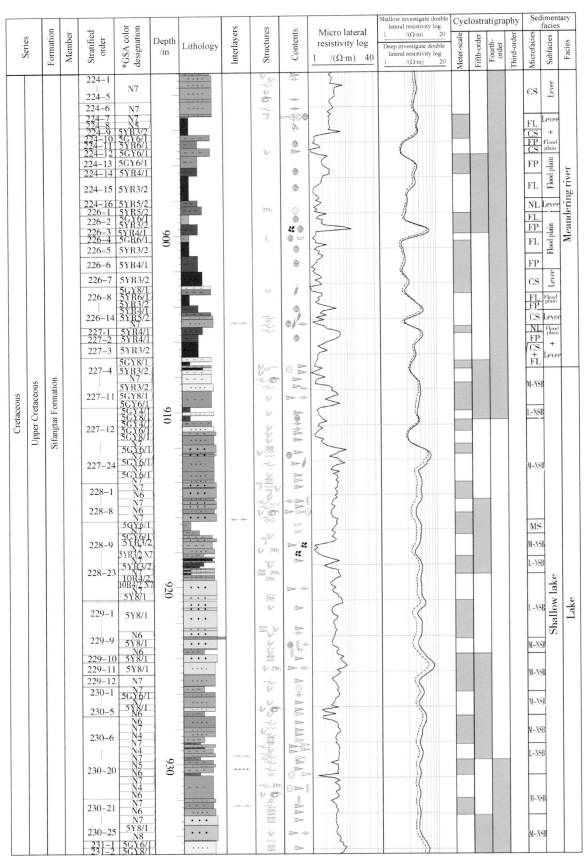

Figure 5.7　Stratigraphic chart of Sifangtai Formation (continued).

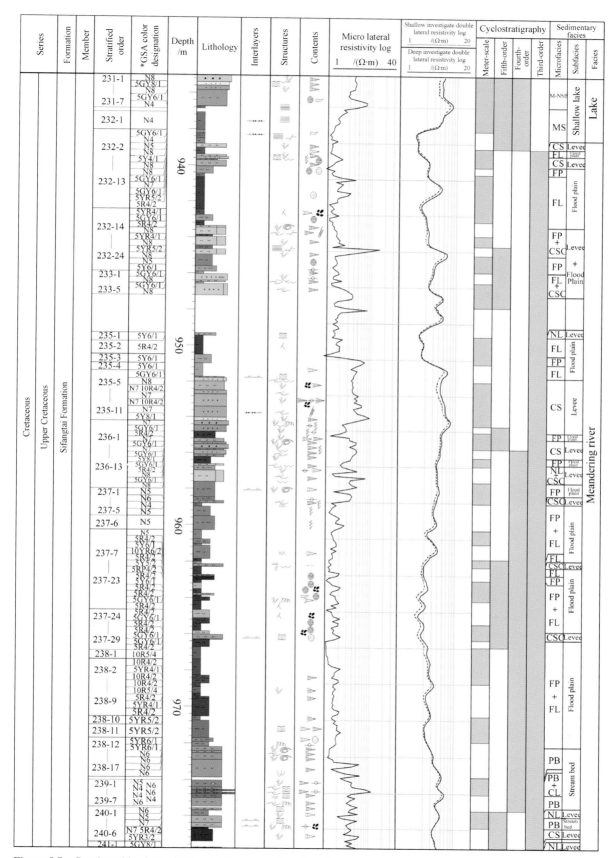

Figure 5.7　Stratigraphic chart of Sifangtai Formation (continued).

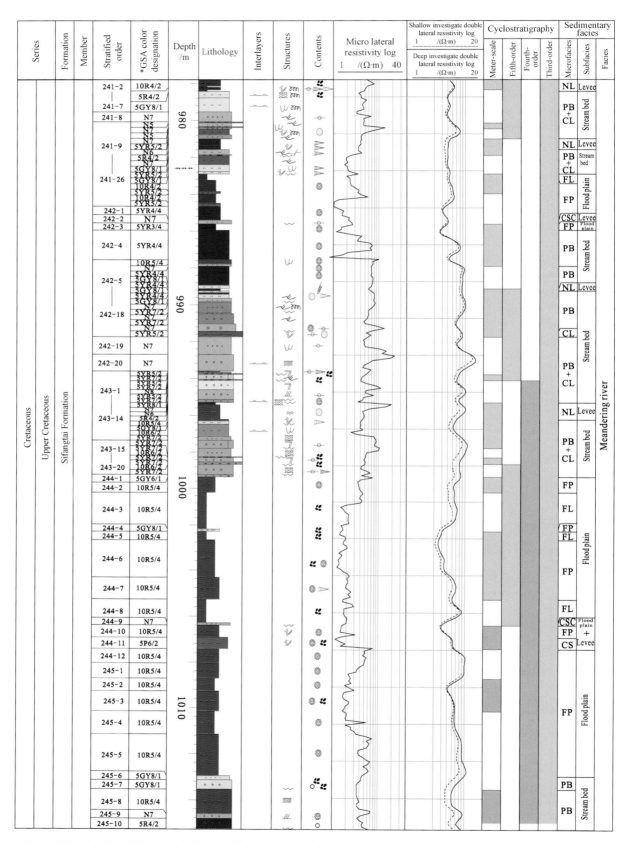

Figure 5.7 Stratigraphic chart of Sifangtai Formation (continued).

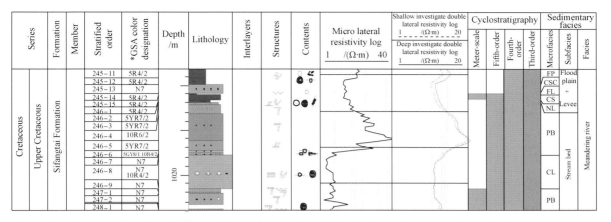

Figure 5.7　Stratigraphic chart of Sifangtai Formation (continued). Symbols are the same as those in Figure 5.1.

209-6	818.03—818.48 m	0.45 m	Silty mudstone, light brown, massive, with greenish gray muddy siltstone strips at the base
209-7	818.48—818.58 m	0.10 m	Siltstone, greenish gray, discontinuous wavy ripple bedding, deformation bedding locally, with bioturbation, weak erosion surface at the base
209-8	818.58—818.69 m	0.11 m	Silty mudstone, light brown, horizontal bedding, 1cm-thick mudstone laminae at the base
209-9	818.69—819.92 m	1.23 m	Silty mudstone, light brown, slump structures, with siltstone strips and mudstone laminae
209-10	819.92—820.03 m	0.11 m	Mudstone, light brown, massive
209-11	820.03—820.99 m	0.96 m	Muddy siltstone, light gray, slump structures, with irregular mudstone and siltstone strips, bioturbation locally
209-12	820.99—821.80 m	0.81 m	Muddy siltstone, light gray, slump structures, with lots of irregular siltstone and mudstone strips
209-13	821.80—822.03 m	0.23 m	Siltstone, light gray, massive, deformation bedding locally, weak erosion surface at the base
209-14	822.03—822.34 m	0.31 m	Siltstone, gray, sand ripple bedding, weak erosion surface at the base, fossil fragments
209-15	822.34—822.57 m	0.23 m	Siltstone, gray, slump structures in the middle and upper parts, discontinuous wavy ripple bedding in the lower part, normal graded bedding at the base, erosion surface at the base, with fossil fragments
209-16	822.57—822.65 m	0.08 m	Siltstone, gray, discontinuous wavy ripple bedding in the upper part, normal graded bedding in the lower part, erosion surface at the base
209-17	822.65—822.74 m	0.09 m	Mudstone, greenish gray, massive, siltstones laminae locally
209-18	822.74—823.01 m	0.27 m	Muddy siltstone, medium gray, deformation bedding in the middle and upper parts, wave-generated cross-bedding in the lower part, erosion surface at the base, a lot of irregular calcareous siltstone strips
209-19	823.01—823.08 m	0.07 m	Mudstone, greenish-gray, massive bedding, with siltstone locally

209-20	823.08—823.19 m	0.11 m	Mudstone, gritty with siltstone, medium gray, 4 cm wave-generated sand ripple bedding in the upper part, 5 cm wavy bedding in the middle part, 2 cm normal graded bedding at the base, weak erosion at the base.
209-21	823.19—823.24 m	0.05 m	Mudstone, greenish gray, wavy ripple bedding, siltstone laminae locally
209-22	823.24—823.61 m	0.37 m	Muddy siltstone, greenish gray, intercalated with siltstone, wavy bedding and wavy cross bedding, with large number of fossil fragments at the base
210-1	823.61—824.05 m	0.44 m	Mudstone, medium gray, interbedded with medium light gray siltstone, ripple bedding in mudstone, wave-generated sand ripple bedding in siltstone, siltstone laminae developed at 823.73—823.76 m, 823.86—823.93 m and 824.00—824.02 m, weak erosion at the base of siltstone
210-2	824.05—824.19 m	0.14 m	Siltstone, light gray, small trough cross bedding, weak erosion surface at the base, mudstone strips
210-3	824.19—825.05 m	0.86 m	Mudstone, medium gray, interbedded with light gray siltstone, horizontal bedding in mudstone, ripple bedding in siltstone, a mudstone layer with gravels at 824.24—0.64 m, siltstone laminae interbedded at 824.32 m, 824.73 m and 824.96 m
210-4	825.05—825.11 m	0.06 m	Siltstone, light gray, sand ripple bedding, weak erosion surface at the base, mud gravels occasionally
210-5	825.11—825.36 m	0.25 m	Mudstone, brown, massive, siltstone occasionally
210-6	825.36—825.61 m	0.25 m	Silty mudstone, grayish brown, massive, deformation bedding at the base
210-7	825.61—826.23 m	0.62 m	Muddy siltstone, light brownish gray, slump structures, with irregular siltstone and mudstone strips locally
210-8	826.23—826.73 m	0.50 m	Silty mudstone, greenish gray, slump structures, irregular siltstone strips, with mussel and ostracoda fossils
210-9	826.73—827.33 m	0.60 m	Mudstone, greenish gray, massive, with few calcareous concretions
210-10	827.33—828.58 m	1.25 m	Mudstone, grayish brown, massive, with greenish gray spots and fossil fragments locally
210-11	828.58—829.15 m	0.57 m	Silty mudstone, dark greenish gray, massive, few fossil fragments
210-12	829.15—829.61 m	0.46 m	Silty mudstone, light brown, slump structures, weak erosion surface at the base, with siltstone and mudstone strips
210-13	829.61—829.95 m	0.34 m	Mudstone, grayish brown, massive
210-14	829.95—830.06 m	0.11 m	Muddy siltstone, dark greenish gray, massive
210-15	830.06—830.43 m	0.37 m	Silty mudstone, light brown, deformation bedding, with siltstone balls
210-16	830.43—830.79 m	0.36 m	Muddy siltstone, light brown, sand ripple bedding, intensive bioturbation, weak erosion surface at the base, with lots of siltstone strips and balls
210-17	830.79—830.90 m	0.11 m	Silty mudstone, grayish brown, massive
210-18	830.90—831.23 m	0.33 m	Siltstone, light brown, sand ripple bedding, intensive bioturbation, weak erosion surface at the base

210-19	831.23—831.92 m	0.69 m	Mudstone, grayish brown, horizontal bedding, with muddy siltstone strips
211-1	831.92—832.20 m	0.28 m	Mudstone, grayish brown, horizontal bedding
211-2	832.20—832.42 m	0.22 m	Silty mudstone, greenish gray, massive, with few fossil fragments
211-3	832.42—832.55 m	0.13 m	Mudstone, brown, horizontal bedding
211-4	832.55—833.09 m	0.54 m	Silty mudstone, grayish brown and greenish gray, massive, ostracoda fossils occasionally
211-5	833.09—834.34 m	1.25 m	Mudstone, grayish brown, massive, with ostracoda fossils occasionally
211-6	834.34—834.64 m	0.30 m	Muddy siltstone, grayish brown and greenish gray, slump structures, with irregular siltstone strips
211-7	834.64—834.78 m	0.14 m	Silty mudstone, grayish green, massive, with ostracoda fossils occasionally
211-8	834.78—835.01 m	0.23 m	Mudstone, grayish brown, massive, with few greenish gray spots
211-9	835.01—835.72 m	0.71 m	Silty mudstone, gray, massive, with few fossil fragments and ostracoda fossils
211-10	835.72—835.92 m	0.20 m	Silty mudstone, grayish brown and greenish gray, massive
211-11	835.92—836.51 m	0.59 m	Mudstone, grayish brown, massive
211-12	836.51—836.68 m	0.17 m	Silty mudstone, light brown, massive
211-13	836.68—836.96 m	0.28 m	Mudstone, grayish brown, massive
211-14	836.96—837.57 m	0.61 m	Silty mudstone, greenish gray, massive, with ostracoda fossils and fossil fragments occasionally
211-15	837.57—838.42 m	0.85 m	Mudstone, brown, horizontal bedding
211-16	838.42—838.84 m	0.42 m	Silty mudstone, grayish brown, massive
211-17	838.84—839.42 m	0.58 m	Silty mudstone, light brownish gray, massive
211-18	839.42—839.86 m	0.44 m	Siltstone, light gray, discontinuous wavy ripple bedding, intensive bioturbation, with silty mudstone strips
211-19	839.86—840.03 m	0.17 m	Silty mudstone, light brownish gray, wavy ripple bedding, with muddy siltstone strips
211-20	840.03—840.19 m	0.16 m	Mudstone, grayish brown, massive
212-1	840.29—840.77 m	0.48 m	Silty mudstone, greenish gray, massive
212-2	840.77—840.87 m	0.10 m	Silty mudstone, light brownish gray, massive
212-3	840.87—841.09 m	0.22 m	Mudstone, greenish gray, massive, calcareous concretions occasionally
212-4	841.09—841.74 m	0.65 m	Silty mudstone, greenish gray, slump structures, weak erosion surface at the base, with mudstone and siltstone strips
212-5	841.74—841.92 m	0.18 m	Mudstone, greenish gray, massive
212-6	841.92—842.14 m	0.22 m	Mudstone, greenish gray, massive, with mud gravels and ostracoda fossils
212-7	842.14—842.56 m	0.42 m	Mudstone, greenish gray and brownish gray, massive
212-8	842.56—842.75 m	0.19 m	Muddy siltstone, greenish gray, discontinuous wavy ripple bedding, siltstone strips locally, few mud gravels at the base

212-9	842.75—843.09 m	0.34 m	Mudstone, lightolive gray, massive
212-10	843.09—843.54 m	0.45 m	Muddy siltstone, greenish gray, massive
212-11	843.54—843.79 m	0.25 m	Mudstone, light olive gray massive, with muddy siltstone laminae
212-12	843.79—844.29 m	0.50 m	Silty mudstone, greenish-gray, massive bedding, bioturbation structures, siltstone laminae and ostracoda fossils developed in 843.89—843.99 m
212-13	844.29—844.77 m	0.48 m	Mudstone, greenish gray, wavy ripple bedding
212-14	844.77—844.94 m	0.17 m	Mudstone, greenish gray, interbedded with light gray siltstone, horizontal bedding and wavy ripple bedding, deformation bedding locally
212-15	844.94—845.89 m	0.95 m	Mudstone, greenish gray, horizontal bedding, with lots of ostracoda fossils
212-16	845.89—846.87 m	0.98 m	Mudstone, gray, horizontal bedding, with ostracoda fossils
212-17	846.87—847.32 m	0.45 m	Mudstone, greenish gray, massive, with ostracoda fossils and charcoals
212-18	847.32—847.91 m	0.59 m	Muddy siltstone, light greenish gray, discontinuous wavy ripple bedding, deformation bedding locally, weak erosion surface at the base, with lots of siltstone strips, calcareous concretions
213-1	847.91—848.31 m	0.40 m	Mudstone, greenish gray, massive, with ostracoda fossils and calcareous concretions
213-2	848.31—848.44 m	0.13 m	Silty mudstone, grayish brown, massive, with calcareous concretions
213-3	848.44—849.17 m	0.73 m	Silty mudstone, light brown, massive, with lots of ostracoda fossils and few calcareous concretions
213-4	849.17—849.71 m	0.54 m	Muddy siltstone, greenish gray, massive, with few grayish brown spots
213-5	849.71—850.91 m	1.20 m	Mudstone, greenish gray, massive, with ostracoda fossils
213-6	850.91—851.16 m	0.25 m	Mudstone, grayish red, massive
213-7	851.16—851.39 m	0.23 m	Mudstone, greenish gray, massive, with calcspar-filled irregular fracture
213-8	851.39—852.01 m	0.62 m	Mudstone, brownish gray, massive
213-9	852.01—852.78 m	0.77 m	Mudstone, grayish red, massive
213-10	852.78—854.00 m	1.22 m	Mudstone, greenish gray, horizontal bedding, with conchostracon fossils occasionally
213-11	854.00—854.16 m	0.16 m	Mudstone, grayish brown and greenish gray, massive
213-12	854.16—854.38 m	0.22 m	Mudstone, grayish red, massive
213-13	854.38—854.88 m	0.50 m	Silty mudstone, grayish red, massive
213-14	854.88—856.44 m	1.56 m	Muddy siltstone, grayish red, massive
214-1	856.44—858.47 m	2.03 m	Muddy siltstone, grayish brown and greenish gray, deformation structures, with bioturbation, siltstone and mudstone strips
214-2	858.47—858.69 m	0.22 m	Silty mudstone, greenish gray, massive, with mudstone-filled irregular fracture.
214-3	858.69—859.00 m	0.31 m	Silty mudstone, grayish brown, massive

214-4	859.00—859.12 m	0.12 m	Muddy siltstone, greenish gray, discontinuous wavy ripple bedding, weak erosion surface at the base, siltstone strips
214-5	859.12—859.45 m	0.33 m	Mudstone, brownish gray, horizontal bedding
214-6	859.45—859.97 m	0.52 m	Silty mudstone, greenish gray, massive
214-7	859.97—860.44 m	0.47 m	Silty mudstone, grayish brown, massive
214-8	860.44—861.86 m	1.42 m	Mudstone, grayish brown, massive, calcareous concretions
214-9	861.86—861.99 m	0.13 m	Mudstone, brownish gray, massive
215-1	862.51—862.86 m	0.35 m	Mudstone, grayish brown, massive, calcareous concretions
215-2	862.86—863.26 m	0.40 m	Silty mudstone, dark greenish gray, massive, with ostracoda fossils
215-3	863.26—863.63 m	0.37 m	Mudstone, brownish gray, massive
215-4	863.63—864.01 m	0.38 m	Silty mudstone, greenish gray, massive
215-5	864.01—864.23 m	0.22 m	Silty mudstone, grayish brown, massive
216-1	864.23—864.46 m	0.23 m	Mudstone, grayish brown, massive
216-2	864.46—864.73 m	0.27 m	Mudstone, greenish gray, massive, with ostracoda fossils
216-3	864.73—864.74 m	0.01 m	Mudstone, brown, massive
218-1	864.74—866.04 m	1.30 m	Silty mudstone, grayish brown, massive
218-2	866.04—866.34 m	0.30 m	Mudstone, greenish gray, massive, with calcareous concretions
218-3	866.34—866.64 m	0.30 m	Mudstone, dark reddish brown, massive, with ostracoda fossils
218-4	866.64—866.84 m	0.20 m	Mudstone, greenish gray, massive
218-5	866.84—866.94 m	0.10 m	Mudstone, dark reddish brown, massive, with calcareous concretions occasionally
218-6	866.94—867.60 m	0.66 m	Silty mudstone, greenish-gray and grayish-brown, with slump structures, discontinuous ripple bedding at the base, siltstone laminae and sand balls
218-7	867.60—867.81 m	0.21 m	Mudstone, grayish brown, intercalated with light gray muddy siltstone, wavy ripple bedding
218-8	867.81—869.31 m	1.50 m	Silty mudstone, greenish gray and grayish brown, slump structures, discontinuous wavy ripple bedding at the base, with irregular siltstone and mudstone strips
218-9	869.31—869.41 m	0.10 m	Muddy siltstone, light gray, wavy cross bedding, weak erosion surface at the base, with siltstone strips
218-10	869.41—869.74 m	0.33 m	Silty mudstone, greenish gray and grayish brown, slump structures, discontinuous wavy ripple bedding at the base, with irregular siltstone and mudstone strips
218-11	869.74—870.32 m	0.58 m	Silty mudstone, greenish gray, massive, mud gravels locally
219-1	871.64—871.99 m	0.35 m	Silty mudstone, greenish gray, massive
219-2	871.99—872.64 m	0.65 m	Mudstone, dark reddish brown, massive, ostracoda fossils occasionally

219-3	872.64—872.74 m	0.10 m	Silty mudstone, greenish gray, massive
219-4	872.74—872.94 m	0.20 m	Mudstone, reddish brown, massive
220-1	872.94—874.46 m	1.52 m	Mudstone, dark greenish gray, horizontal bedding, with ostracoda fossils, siltstone strips locally
220-2	874.46—874.65 m	0.19 m	Silty mudstone, greenish gray, wavy ripple bedding
220-3	874.65—874.82 m	0.17 m	Muddy siltstone, gray, wavy ripple bedding, with trace fossils
220-4	874.82—875.71 m	0.89 m	Silty mudstone, medium gray, interbedded with light gray siltstone, ripple bedding, wave-generated sand ripple bedding and horizontal bedding, deformation bedding locally, erosion at the base, some trace fossils, intensive bioturbation
220-5	875.71—876.14 m	0.43 m	Muddy siltstone, light olive gray, slump structures, siltstone balls and mudstone strips locally
220-6	876.14—876.79 m	0.65 m	Mudstone, brownish gray, wavy ripple bedding
220-7	876.79—877.04 m	0.25 m	Siltstone, light gray, interbedded with mudstone, ripple bedding, wave-generated sand ripple bedding and erosion at the base, gravels occasionally, brownish-gray mudstone at 876.88—876.91 m
220-8	877.04—877.32 m	0.28 m	Silty mudstone, greenish gray, massive, with siltstone laminae
220-9	877.32—877.54 m	0.22 m	Muddy siltstone, light greenish gray, slump structures, siltstone balls in the middle and upper parts, mudstone strips at the top
220-10	877.54—877.64 m	0.10 m	Muddy siltstone, greenish gray, massive
221-1	877.64—878.12 m	0.48 m	Muddy siltstone, greenish gray, massive, weak erosion surface at the base, with lots of siltstone, calcareous concretions at the base
221-2	878.12—878.30 m	0.18 m	Silty mudstone, brownish gray, massive, with lots of calcareous concretions
221-3	878.30—878.44 m	0.14 m	Silty mudstone, greenish gray, massive
221-4	878.44—879.14 m	0.70 m	Silty mudstone, brownish gray, massive, with greenish gray spots
221-5	879.44—879.54 m	0.10 m	Silty mudstone, grayish green, massive, with grayish brown spots
221-6	879.54—879.81 m	0.27 m	Silty mudstone, grayish red, massive
221-7	879.81—880.05 m	0.24 m	Mudstone, reddish brown, massive, with calcareous concretions
221-8	880.05—880.44 m	0.39 m	Muddy siltstone, grayish brown, deformation bedding, trace fossils, with siltstone and irregular mudstone strips
221-9	880.44—880.61 m	0.17 m	Siltstone, light gray, parallel bedding in the middle and lower parts, sand ripple bedding in the upper part, erosion at the base, few mud gravels, intensive bioturbation, <1 cm greenish-gray mudstone on the bottom surface
221-10	880.61—881.74 m	1.13 m	Mudstone, brown, massive, with greenish gray mudstone strips at the top
221-11	881.74—882.09 m	0.35 m	Mudstone, brownish gray, massive
221-12	882.09—885.07 m	2.98 m	Mudstone, brown, massive, with siltstone strips and calcareous concretions
221-13	885.07—886.09 m	1.02 m	Silty mudstone, light brown, massive, with calcareous concretions occasionally

221-14	886.09—886.34 m	0.25 m	Muddy siltstone, greenish gray, massive, with irregular fracture, calcareous concretions occasionally
221-15	886.34—886.68 m	0.34 m	Mudstone, brown, massive, with calcspar
222-1	886.68—887.18 m	0.50 m	Mudstone, brown, massive, with calcareous concretions
222-2	887.18—887.31 m	0.13 m	Silty mudstone, greenish gray and grayish brown, massive
222-3	887.31—887.58 m	0.27 m	Silty mudstone, brown, massive, with few calcareous concretions
222-4	887.58—888.00 m	0.42 m	Siltstone, light greenish gray, slump structures, wave ripple bedding locally, bioturbation
222-5	888.00—888.76 m	0.76 m	Siltstone, greenish-gray, interbedded with grayish-brown mudstone, wavy bedding and wavy cross bedding, deformation bedding at the base, erosion on the bottom surface, 2 cm mud gravels deposited on the erosion surface, irregular mudstone laminae
222-6	888.76—889.12 m	0.36 m	Siltstone, light brownish gray, wave ripple bedding, deformation structures locally, with lots of mudstone strips
223-1	889.71—890.01 m	0.30 m	Siltstone, light gray, slump structures, weak erosion surface at the base
223-2	890.01—890.40 m	0.39 m	Siltstone, light gray, convolution bedding, deformation bedding, with irregular mudstone strips
223-3	890.40—890.53 m	0.13 m	Siltstone, light gray, deformation bedding, weak erosion surface at the base
223-4	890.53—890.79 m	0.26 m	Siltstone, light gray, deformation bedding, with irregular mudstone strips
224-1	890.79—890.89 m	0.10 m	Siltstone, light gray, deformation bedding
224-2	890.89—891.24 m	0.35 m	Fine-grained sandstone, light gray, deformation bedding, mud gravels locally
224-3	891.24—891.59 m	0.35 m	Fine-grained sandstone, light gray, wavy cross bedding
224-4	891.59—892.24 m	0.65 m	Fine-grained sandstone, light gray, deformation bedding
224-5	892.24—892.39 m	0.15 m	Fine-grained sandstone, light gray, convolution bedding, mud gravels locally
224-6	892.39—893.09 m	1.70 m	Fine-grained sandstone, light gray, wavy cross bedding in the upper part, deformation bedding in the lower part, mud gravels at the base
224-7	893.09—893.24 m	0.15 m	Fine-grained sandstone, light gray, parallel bedding
224-8	893.24—893.34 m	0.10 m	Mudstone, gray, horizontal bedding, with lots of fossil fragments and charcoal, with calcareous concretions
224-9	893.34—894.29 m	0.95 m	Mudstone, grayish brown, massive, with calcspar vein at 894.04 m
224-10	894.29—894.59 m	0.30 m	Siltstone, greenish gray, massive
224-11	894.59—895.05 m	0.46 m	Silty mudstone, light brownish gray, massive, with lots of calcareous concretions
224-12	895.05—895.49 m	0.44 m	Siltstone, greenish gray, massive, deformation bedding locally, with few mudstone strips at the base
224-13	895.49—896.14 m	0.65 m	Silty mudstone, greenish gray, massive
224-14	896.14—896.59 m	0.45 m	Silty mudstone, brownish gray, massive, with calcareous concretions

224-15	896.59—897.99 m	1.40 m	Mudstone, grayish brown, massive, with calcareous concretions
224-16	897.99—898.32 m	0.33 m	Silty mudstone, light brown, massive
226-1	898.32—898.77 m	0.45 m	Muddy siltstone, light brown, massive, with bioturbation and sandstone balls
226-2	898.77—899.32 m	0.55 m	Mudstone, greenish gray and grayish brown, massive, with calcareous concretions
226-3	899.32—899.92 m	0.60 m	Silty mudstone, brownish gray, massive, with calcareous concretions
226-4	899.92—900.32 m	0.40 m	Mudstone, greenish gray, massive, with calcareous concretions and fracture filled with mudstone
226-5	900.32—901.12 m	0.80 m	Mudstone, grayish brown, massive, with calcareous concretions
226-6	901.12—902.02 m	0.90 m	Silty mudstone, brownish gray, massive, with lots of calcareous concretions
226-7	902.02—902.82 m	0.80 m	Muddy siltstone, grayish brown, massive
226-8	902.82—903.02 m	0.20 m	Siltstone, light greenish gray, massive
226-9	903.02—903.32 m	0.30 m	Siltstone, light brown, wavy ripple bedding, with lots of trace fossils
226-10	903.32—903.92 m	0.60 m	Mudstone, grayish brown, massive
226-11	903.92—904.32 m	0.40 m	Silty mudstone, brownish gray, massive, with few calcareous concretions
226-12	904.32—904.52 m	0.20 m	Siltstone, light brown, massive
226-13	904.52—904.72 m	0.20 m	Fine-grained sandstone, light gray, wavy cross bedding, with lots of trace fossils
226-14	904.72—905.17 m	0.45 m	Mudstone, brownish gray, interbedded with light gray sandstone, wavy ripple bedding, wavy cross bedding in the lower part, with lots of calcareous concretions and trace fossils, mud gravels locally
227-1	905.17—905.57 m	0.40 m	Muddy siltstone, brownish gray, wave ripple bedding, deformation bedding locally
227-2	905.57—906.02 m	0.45 m	Silty mudstone, brownish gray, massive, with lots of calcareous concretions
227-3	906.02—906.87 m	0.85 m	Silty mudstone, grayish brown, massive
227-4	906.87—907.12 m	0.25 m	Siltstone, light greenish gray, massive
227-5	907.12—907.32 m	0.20 m	Mudstone, grayish brown, massive
227-6	907.32—907.47 m	0.15 m	Siltstone, light gray, deformation bedding, erosion surface at the base, with mudstone strips
227-7	907.47—907.57 m	0.10 m	Muddy siltstone, grayish brown, wavy ripple bedding, sandstone balls at the base
227-8	907.57—907.82 m	0.25 m	Siltstone, light greenish gray, deformation bedding
227-9	907.82—908.37 m	0.55 m	Siltstone, light greenish gray, wavy cross bedding, deformation bedding locally, with mudstone strips
227-10	908.37—908.77 m	0.40 m	Fine-grained sandstone, light greenish gray, wavy cross bedding, with lots of mudstone strips

227-11	908.77—909.70 m	0.93 m	Siltstone, greenish gray, deformation bedding locally, with mud gravels and mudstone strips
227-12	909.70—909.97 m	0.27 m	Mudstone, dark greenish gray, massive
227-13	909.97—910.17 m	0.20 m	Fine-grained sandstone, light greenish gray, deformation bedding, with lots of mudstone strips
227-14	910.17—910.47 m	0.30 m	Mudstone, dark greenish gray, massive, with few calcareous concretions
227-15	910.47—910.77 m	0.30 m	Siltstone, greenish gray, directionally aligned mud gravels
227-16	910.77—911.07 m	0.30 m	Fine-grained sandstone, light greenish gray, deformation bedding, mud gravels locally
227-17	911.07—911.27 m	0.20 m	Medium-grained sandstone, light gray, deformation bedding, with mudstone strips
227-18	911.27—911.82 m	0.55 m	Fine-grained sandstone, greenish gray, deformation bedding, mudstone strips occasionally
227-19	911.82—912.27 m	0.45 m	Medium-grained sandstone, light gray, oblique bedding, erosion surface at the base
227-20	912.27—912.37 m	0.10 m	Fine-grained sandstone, greenish gray, deformation bedding locally, mudstone strips
227-21	912.37—912.47 m	0.10 m	Medium-grained sandstone, light gray, massive, erosion surface at the base, with lots of calcareous concretions
227-22	912.47—913.37 m	0.90 m	Fine-grained sandstone, greenish gray, wavy cross bedding, deformation bedding locally, with mudstone strips and trace fossils
227-23	913.37—913.87 m	0.50 m	Fine-grained sandstone, greenish gray, deformation bedding, with lots of mudstone strips
227-24	913.87—914.01 m	0.14 m	Fine-grained sandstone, light gray, parallel bedding
228-1	914.01—914.30 m	0.29 m	Medium-grained sandstone, light gray, parallel bedding, erosion surface at the base
228-2	914.30—914.35 m	0.05 m	Fine-grained sandstone, light gray, deformation bedding, water-escape structure, irregular mudstone strips
228-3	914.35—914.91 m	0.56 m	Medium-grained sandstone, light gray, parallel bedding
228-4	914.91—915.03 m	0.12 m	Fine-grained sandstone, light gray, deformation bedding
228-5	915.03—915.25 m	0.22 m	Medium-grained sandstone, light gray, wavy cross bedding, erosion surface at the base, above which there are conglomerate and mudstone strips
228-6	915.25—915.51 m	0.26 m	Medium-grained sandstone, light gray, wavy cross bedding, with lots of mudstone strips in the lower part
228-7	915.51—915.78 m	0.27 m	Fine-grained sandstone, light gray, deformation bedding, convolution bedding locally, with sandstone balls and irregular mudstone strips
228-8	915.78—916.21 m	0.43 m	Medium-grained sandstone, light gray, wavy cross bedding, erosion surface at the base, above which there are 3cm-thick conglomerate laminae

228-9	916.21—916.76 m	0.55 m	Mudstone, greenish gray, massive, fossil fragments locally
228-10	916.76—917.01 m	0.25 m	Silty mudstone, greenish gray, massive, fossil fragments occasionally
228-11	917.01—917.21 m	0.20 m	Siltstone, light gray, deformation bedding
228-12	917.21—917.39 m	0.18 m	Medium-grained sandstone, light gray, parallel bedding
228-13	917.39—917.81 m	0.42 m	Siltstone, grayish brown and greenish gray, massive, with mudstone strips occasionally
228-14	917.81—918.10 m	0.29 m	Fine-grained sandstone, light gray, deformation bedding, with trace fossils
228-15	918.10—918.26 m	0.16 m	Muddy siltstone, grayish brown and greenish gray, massive
228-16	918.26—918.36 m	0.10 m	Fine-grained sandstone, light gray, massive
228-17	918.36—918.50 m	0.14 m	Siltstone, grayish brown, massive
228-18	918.50—918.58 m	0.08 m	Fine-grained sandstone, light gray, massive, erosion surface at the base, with few mudstone strips
228-19	918.58—918.81 m	0.23 m	Muddy siltstone, grayish red, massive, deformation bedding locally
228-20	918.81—919.43 m	0.62 m	Silty mudstone, grayish red, interbedded with light gray fine-grained sandstone, wavy bedding, deformation bedding locally, with bioturbation
228-21	919.43—919.61 m	0.18 m	Fine-grained sandstone, light gray, massive, bioturbation occasionally
228-22	919.61—920.46 m	0.85 m	Medium-grained sandstone, yellowish gray, massive, erosion surface at the base, above which there are mud gravels strips
228-23	920.46—920.69 m	0.23 m	Medium-grained sandstone, yellowish gray, massive, bioturbation
229-1	920.79—921.14 m	0.35 m	Medium-grained sandstone, yellowish gray, massive
229-2	921.14—921.29 m	0.15 m	Medium-grained sandstone, yellowish gray, slump structures, with gray mudstone strips
229-3	921.29—922.26 m	0.97 m	Medium-grained sandstone, yellowish gray, massive
229-4	922.26—922.42 m	0.16 m	Medium-grained sandstone, yellowish gray, wavy bedding, mudstone strips
229-5	922.42—922.75 m	0.33 m	Medium-grained sandstone, yellowish gray, massive
229-6	922.75—922.89 m	0.14 m	Fine-grained conglomerate, gray, massive
229-7	922.89—923.42 m	0.53 m	Medium-grained sandstone, yellowish gray, slump structure, erosion surface at the base, with lots of calcareous concretions and mud gravels
229-8	923.42—923.49 m	0.07 m	Medium-grained sandstone, yellowish gray, massive, mud gravels
229-9	923.49—923.75 m	0.26 m	Muddy siltstone, gray, massive at the top, deformation bedding in the middle and lower parts, with charcoal occasionally
229-10	923.75—924.19 m	0.44 m	Medium-grained sandstone, yellowish gray, wavy bedding, with mudstone strips occasionally
229-11	924.19—924.89 m	0.70 m	Fine-grained sandstone, yellowish gray, slump structures, bioturbation, with lots of mudstone strips and mud gravels

229-12	924.89—925.54 m	0.65 m	Fine-grained sandstone, light gray, wave ripple bedding, mudstone strips occasionally
230-1	925.56—925.89 m	0.33 m	Muddy siltstone, light gray, defomation structure, with irregular gray mudstone strips
230-2	925.89—926.16 m	0.27 m	Siltstone, light gray, massive, with lots of mud gravels
230-3	926.16—926.76 m	0.60 m	Fine-grained sandstone, light gray, deformation bedding, with irregular mudstone strips
230-4	926.76—926.99 m	0.23 m	Fine-grained sandstone, light gray, deformation bedding, convolution bedding in the upper part, with few mudstone strips
230-5	926.99—927.26 m	0.27 m	Muddy siltstone, light gray, deformation bedding, with siltstone strips
230-6	927.26—927.90 m	0.64 m	Sandstone, fine-grained, light gray, small wave-generated cross-bedding, erosion at the base, mudstone laminae occasionally
230-7	927.90—928.13 m	0.23 m	Fine-grained sandstone, light gray, deformation bedding, with few mudstone strips
230-8	928.13—928.43 m	0.30 m	Fine-grained sandstone, light gray, wavy cross bedding, with few mudstone strips
230-9	928.43—928.76 m	0.33 m	Muddy siltstone, light gray, deformation bedding, wavy bedding locally, with siltstone strips
230-10	928.76—928.86 m	0.10 m	Silty mudstone, dark gray, wavy ripple bedding
230-11	928.86—929.01 m	0.15 m	Fine-grained sandstone, light gray, wavy cross bedding, with mudstone strips
230-12	929.01—929.10 m	0.09 m	Muddy siltstone, dark gray, deformation bedding, with siltstone strips
230-13	929.10—929.36 m	0.26 m	Siltstone, light gray, deformation structure, erosion surface at the base, with few mudstone strips
230-14	929.36—929.56 m	0.20 m	Sandstone, fine-grained, light gray, wave-generated cross-bedding, 0.8 cm-thick mudstone laminae at 928.46 m and 929.36 m
230-15	929.56—929.74 m	0.18 m	Silty mudstone, gray, with wavy ripple bedding, siltstone strips
230-16	929.74—929.98 m	0.24 m	Siltstone, light gray, wavy cross bedding, mudstone strips
230-17	929.98—930.13 m	0.15 m	Fine-grained sandstone, light gray, wavy cross bedding
230-18	930.13—930.46 m	0.33 m	Siltstone, light gray, wavy cross bedding, with mudstone strips
230-19	930.46—930.62 m	0.16 m	Silty mudstone, dark gray, wavy ripple bedding, deformation bedding locally, with few siltstone balls and strips
230-20	930.62—931.91 m	1.29 m	Siltstone, light gray, wavy cross bedding, deformation bedding locally, bioturbation, convolution bedding occasionally
230-21	931.91—932.11 m	0.20 m	Fine-grained sandstone, light gray, oblique bedding, charcoal occasionally
230-22	932.11—932.46 m	0.35 m	Siltstone, light gray, wavy cross bedding, deformation bedding locally, with mudstone strips and laminae
230-23	932.46—932.86 m	0.40 m	Fine-grained sandstone, light gray, wavy bedding, mudstone strips

230-24	932.86—933.41 m	0.55 m	Medium-grained sandstone, yellowish gray, wavy bedding, with mudstone strips locally
230-25	933.41—934.24 m	0.83 m	Medium-grained sandstone, light gray, massive and deformation bedding, with mudstone laminae and mud gravels
231-1	934.24—934.29 m	0.05 m	Fine-grained sandstone, greenish gray, wavy bedding, erosion surface at the base
231-2	934.29—935.02 m	0.73 m	Fine-grained sandstone, light greenish gray, wavy bedding, with mudstone strips
231-3	935.02—935.39 m	0.37 m	Medium-grained sandstone, light gray, deformation bedding, erosion surface at the base, with mudstone strips
231-4	935.39—935.59 m	0.20 m	Fine-grained sandstone, light greenish gray, wavy bedding, with mudstone strips
231-5	935.59—935.84 m	0.25 m	Fine-grained sandstone, light gray, deformation bedding, mudstone strips
231-6	935.84—936.74 m	0.90 m	Siltstone, greenish gray, massive, calcareous concretions and fossil fragments
231-7	936.74—936.86 m	0.12 m	Mudstone, dark brown, horizontal bedding
232-1	937.06—938.06 m	1.00 m	Mudstone, dark brown, horizontal bedding, intercalated with siltstone
232-2	938.06—938.21 m	0.15 m	Mudstone, greenish gray, massive, with fossil fragments
232-3	938.21—938.51 m	0.30 m	Mudstone, dark brown, horizontal bedding
232-4	938.51—938.81 m	0.30 m	Mudstone, gray, massive
232-5	938.81—939.01 m	0.20 m	Siltstone, light gray, deformation bedding, with mudstone strips
232-6	939.01—939.41 m	0.40 m	Mudstone, olive gray, massive, siltstone strips
232-7	939.41—939.51 m	0.10 m	Siltstone, light gray, deformation bedding, with mudstone strips
232-8	939.51—939.63 m	0.12 m	Fine-grained sandstone, light gray, wavy bedding, erosion surface at the base, with lots of charcoals
232-9	939.63—939.73 m	0.10 m	Muddy siltstone, greenish gray, massive, with calcareous concretions occasionally
232-10	939.73—940.06 m	0.33 m	Siltstone, light gray, wavy ripple bedding, with mudstone strips
232-11	940.06—940.46 m	0.40 m	Silty mudstone, greenish gray, massive, calcareous concretions locally
232-12	940.46—940.66 m	0.20 m	Mudstone, light brown, massive
232-13	940.66—942.46 m	1.80 m	Mudstone, grayish red, massive
232-14	942.46—942.76 m	0.30 m	Mudstone, greenish gray and brownish gray, massive, siltstone strips locally
232-15	942.76—943.11 m	0.35 m	Silty mudstone, greenish gray, massive, calcareous concretions occasionally
232-16	943.11—943.36 m	0.25 m	Mudstone, grayish red, massive
232-17	943.36—943.46 m	0.10 m	Siltstone, light gray, deformation bedding and convolution bedding, with mudstone strips

232-18	943.46—943.86 m	0.40 m	Siltstone, light gray, wavy cross bedding, mudstone strips locally
232-19	943.86—943.96 m	0.10 m	Silty mudstone, brownish gray, massive
232-20	943.96—944.46 m	0.50 m	Siltstone, light gray, deformation bedding, with light brown mudstone strips
232-21	944.46—944.66 m	0.20 m	Silty mudstone, light brown, horizontal bedding
232-22	944.66—944.96 m	0.30 m	Siltstone, light gray, deformation bedding, with mudstone strips
232-23	944.96—945.61 m	0.65 m	Silty mudstone, gray, massive, with calcareous concretions
232-24	945.61—945.89 m	0.28 m	Muddy siltstone, light olive gray, massive, with greenish gray mud gravels
233-1	945.89—946.04 m	0.15 m	Mudstone, greenish gray, massive, with siltstone strips
233-2	946.04—946.39 m	0.35 m	Fine-grained sandstone, light gray, slump structures, erosion surface at the base, with mud gravels
233-3	946.39—946.49 m	0.10 m	Siltstone, light gray, wavy cross bedding, erosion surface at the base, mudstone strips
233-4	946.49—946.57 m	0.08 m	Mudstone, greenish gray, massive
233-5	946.57—947.22 m	0.65 m	Sandstone, calcareous and fine grained, very light gray, wave-generated cross-bedding, 2 cm-thick mudstone laminae at 947.09 m
235-1	949.30—949.45 m	0.15 m	Muddy siltstone, light olive gray, wavy bedding
235-2	949.45—950.50 m	1.05 m	Mudstone, grayish red, massive, with irregular fracture
235-3	950.50—950.98 m	0.48 m	Silty mudstone, light olive gray, massive
235-4	950.98—951.71 m	0.73 m	Mudstone, light olive gray, massive, with siltstone strips
235-5	951.71—951.78 m	0.07 m	Siltstone, greenish gray, wavy bedding, mudstone laminae
235-6	951.78—951.98 m	0.20 m	Sandstone, light gray, wavy cross bedding, erosion surface at the base
235-7	951.98—952.48 m	0.50 m	Siltstone, light gray and grayish red, deformation bedding, irregular mudstone strips
235-8	952.48—952.70 m	0.22 m	Fine-grained sandstone, light gray, wavy cross bedding
235-9	952.70—953.40 m	0.70 m	Fine-grained sandstone, light gray and grayish red, slump structures, mud gravels and mudstone strips in the middle and lower part, with fine-grained sandstone strips
235-10	953.40—954.02 m	0.62 m	Sandstone, fine-grained, light gray, wave-generated cross-bedding, synsedimentary deformation structures locally, biogliph, grayish-red silty mudstone laminae at 953.60—953.65 m
235-11	954.02—954.22 m	0.20 m	Fine-grained sandstone, yellowish gray, deformation bedding, with irregular mudstone strips
236-1	954.22—954.45 m	0.23 m	Medium-grained sandstone, light gray, deformation bedding, with bioturbation, erosion surface at the base, with charcoal and mud gravels
236-2	954.45—954.82 m	0.37 m	Fine-grained sandstone, greenish gray, deformation bedding, with charcoal, mud gravels occasionally

236-3	954.82—955.19 m	0.37 m	Muddy siltstone, grayish red, deformation bedding, with siltstone strips and mudstone strips
236-4	955.19—955.40 m	0.21 m	Siltstone, light gray, convolution bedding, with mudstone strips
236-5	955.40—955.52 m	0.12 m	Fine-grained sandstone, greenish gray, deformation bedding, water-escape structure, with mud gravels and mudstone strips occasionally
236-6	955.52—955.76 m	0.24 m	Medium-grained sandstone, light gray, massive, mudstone strips occasionally
236-7	955.76—955.84 m	0.08 m	Siltstone, greenish gray, massive, charcoal occasionally
236-8	955.84—956.07 m	0.23 m	Fine-grained sandstone, yellowish gray, wavy cross bedding, with mudstone strips
236-9	956.07—956.22 m	0.15 m	Siltstone, greenish gray, slump structures, with mudstone strips
236-10	956.22—956.62 m	0.40 m	Silty mudstone, grayish red, massive
236-11	956.62—956.73 m	0.11 m	Siltstone, light gray, wavy cross bedding, erosion surface at the base, mudstone strips
236-12	956.73—956.99 m	0.26 m	Muddy siltstone, greenish gray, deformation bedding, with mud gravels and siltstone strips
236-13	956.99—957.62 m	0.63 m	Siltstone, light gray, wavy cross bedding, erosion surface at the base, with few mudstone strips
237-1	957.62—958.49 m	0.87 m	Muddy siltstone, gray, deformation bedding in the upper part, convolution bedding in the middle and lower part, with lots of siltstone strips
237-2	958.49—958.67 m	0.18 m	Muddy siltstone, gray, massive, with few mud gravels at the top
237-3	958.67—958.80 m	0.13 m	Siltstone, light gray, wavy cross bedding, deformation structure locally, with mudstone strips
237-4	958.80—958.99 m	0.19 m	Mudstone, dark brown, wavy ripple bedding, charcoal
237-5	958.99—959.52 m	0.53 m	Silty mudstone, gray, massive, with charcoal
237-6	959.52—960.29 m	0.77 m	Muddy siltstone, gray, massive, with charcoal
237-7	960.29—960.77 m	0.48 m	Mudstone, gray, massive
237-8	960.77—961.02 m	0.25 m	Silty mudstone, gray, massive
237-9	961.02—961.12 m	0.10 m	Mudstone, gray, massive
237-10	961.12—961.84 m	0.72 m	Silty mudstone, gray, massive, wavy ripple bedding locally, with siltstone strips occasionally
237-11	961.84—962.00 m	0.16 m	Mudstone, grayish red, massive
237-12	962.00—962.12 m	0.12 m	Mudstone, light olive gray, massive
237-13	962.12—962.29 m	0.17 m	Siltstone, light yellowish brown, parallel bedding, erosion surface at the base
237-14	962.29—962.72 m	0.43 m	Mudstone, grayish red, massive
237-15	962.72—962.82 m	0.10 m	Muddy siltstone, light olive gray, massive, with calcareous concretions occasionally

237-16	962.82—963.02 m	0.20 m	Muddy siltstone, grayish red, massive
237-17	963.02—963.18 m	0.16 m	Mudstone, grayish red, massive, with calcareous concretions occasionally
237-18	963.18—963.52 m	0.34 m	Mudstone, light olive gray and grayish red, massive, calcareous concretions
237-19	963.52—963.76 m	0.24 m	Mudstone, grayish red, massive
237-20	963.76—963.87 m	0.11 m	Muddy siltstone, greenish gray, massive, with fossil fragments and calcareous concretions occasionally
237-21	963.87—963.97 m	0.10 m	Mudstone, grayish red, massive
237-22	963.97—964.15 m	0.18 m	Silty mudstone, grayish red, massive, deformation structure, with bioturbation structure, siltstone strips
237-23	964.15—964.67 m	0.52 m	Mudstone, grayish red, massive, with irregular fracture
237-24	964.67—965.12 m	0.45 m	Silty mudstone, grayish red and greenish gray, massive, with irregular fracture
237-25	965.12—965.42 m	0.30 m	Silty mudstone, grayish red, massive, with calcareous concretions occasionally
237-26	965.42—965.82 m	0.40 m	Mudstone, grayish red, massive, with calcareous concretions occasionally
237-27	965.82—966.02 m	0.20 m	Silty mudstone, grayish red and greenish gray, massive
237-28	966.02—966.30 m	0.28 m	Siltstone, greenish gray, interbedded with grayish red mudstone, wavy bedding
237-29	966.30—966.73 m	0.43 m	Silty mudstone, grayish red, massive
238-1	966.73—967.41 m	0.68 m	Mudstone, light reddish brown, massive
238-2	967.41—968.28 m	0.87 m	Mudstone, grayish red, massive
238-3	968.28—968.47 m	0.19 m	Silty mudstone, brownish gray, massive, with grayish green mudstone strips
238-4	968.47—968.73 m	0.26 m	Silty mudstone, grayish red, massive, with light gray strips
238-5	968.73—968.93 m	0.20 m	Mudstone, grayish red, massive
238-6	968.93—969.28 m	0.35 m	Mudstone, light reddish brown, deformation bedding in the upper part, massive in the middle part, and wavy ripple bedding in the lower part, with light gray muddy siltstone strips
238-7	969.28—969.46 m	0.18 m	Mudstone, grayish red, massive
238-8	969.46—969.58 m	0.12 m	Silty mudstone, brownish gray, massive, with light gray muddy siltstone strips locally
238-9	969.58—970.53 m	0.95 m	Silty mudstone, grayish red, massive
238-10	970.53—970.76 m	0.23 m	Mudstone, light brown, massive, with light gray muddy siltstone strips
238-11	970.76—971.72 m	0.96 m	Silty mudstone, light brown, wavy bedding, deformation bedding locally, with light gray, siltstone strips and grayish red mudstone strips
238-12	971.72—971.93 m	0.21 m	Muddy siltstone, light brown, massive, siltstone strips and sandstone balls
238-13	971.93—972.18 m	0.25 m	Silty mudstone, light brown, massive, bioturbation, with mud gravels
238-14	972.18—972.42 m	0.24 m	Siltstone, light gray, slump structures, with convolution bedding locally, with mudstone strips

238-15	972.42—972.53 m	0.11 m	Siltstone, light gray, wavy cross bedding, with mudstone strips
238-16	972.53—972.91 m	0.38 m	Siltstone, light gray, slump structures, convolution bedding, with irregular mudstone strips
238-17	972.91—973.86 m	0.95 m	Siltstone, light gray, trough cross bedding, with mudstone strips
239-1	974.02—974.42 m	0.40 m	Siltstone, gray, slump structures, with irregular mudstone strips and mud gravels
239-2	974.42—974.53 m	0.11 m	Siltstone, light gray, wavy cross bedding, erosion surface at the base, with mudstone strips and mud gravels locally
239-3	974.53—974.62 m	0.09 m	Fine-grained conglomerate, dark gray, parallel bedding in the lower part, wavy bedding in the upper part
239-4	974.62—974.69 m	0.07 m	Siltstone, light gray, with wavy cross bedding, mudstone strips, with mud gravels occasionally
239-5	974.69—974.81 m	0.12 m	Fine-grained conglomerate, dark gray, massive, deformation bedding locally, with mudstone strips
239-6	974.81—975.00 m	0.19 m	Siltstone, dark gray, wavy ripple bedding
239-7	975.00—975.49 m	0.49 m	Siltstone, light gray, wavy cross bedding, with bioturbation locally, mudstone strips
240-1	975.56—975.69 m	0.13 m	Siltstone, light gray, wavy bedding, with mudstone strips
240-2	975.69—976.02 m	0.33 m	Silty mudstone, gray, deformation bedding, siltstone strips
240-3	976.02—976.39 m	0.37 m	Siltstone, light gray, flaser bedding, erosion surface at the base, 5 mm and 1mm dark red mudstone laminae respectively at 976.18 m and 976.32 m, 5 cm siltstone with bioclasts at 976.32—976.37 m
240-4	976.39—976.67 m	0.28 m	Siltstone, light gray, interbedded with grayish red mudstone, slump structures, bioturbation, with mud gravels and mudstone strips
240-5	976.67—976.76 m	0.09 m	Silty mudstone, grayish brown, massive
240-6	976.76—977.42 m	0.66 m	Muddy siltstone, grayish brown, slump structures, with light gray siltstone strips
241-1	977.74—977.77 m	0.03 m	Siltstone, light greenish gray, massive
241-2	977.77—978.14 m	0.37 m	Muddy siltstone, grayish red, massive
241-3	978.14—978.20 m	0.06 m	Mudstone, grayish red, massive
241-4	978.20—978.35 m	0.15 m	Muddy siltstone, grayish red, slump structures, bioturbation, with irregular mud gravels and mudstone strips
241-5	978.35—978.42 m	0.07 m	Mudstone, grayish red, massive, with sandstone balls
241-6	978.42—978.77 m	0.35 m	Siltstone, light greenish-gray, intercalated with grayish-red silty mudstone, wavy bedding, deformation structure locally, bioturbation structures, grayish-red mudstone laminae

241-7	978.77—979.54 m	0.77 m	Siltstone, light greenish-gray, wavy bedding and wave-generated cross-bedding, bioturbation structures locally, mudstone laminae, sandstone in grayish-red locally
241-8	979.54—980.07 m	0.53 m	Fine-grained sandstone, light gray, wavy cross bedding, with mud gravels occasionally
241-9	980.07—980.14 m	0.07 m	Fine-grained conglomerate, gray, massive
241-10	980.14—980.34 m	0.20 m	Fine-grained sandstone, light gray, wavy cross bedding
241-11	980.34—980.38 m	0.04 m	Fine-grained conglomerate, gray, massive
241-12	980.38—980.46 m	0.08 m	Fine-grained sandstone, light gray, trough cross bedding, with mud gravels occasionally
241-13	980.46—980.78 m	0.32 m	Siltstone, light brown, deformation bedding, with sandstone balls and bioturbation
241-14	980.78—981.00 m	0.22 m	Fine-grained sandstone, light gray, wavy cross bedding
241-15	981.00—981.40 m	0.40 m	Muddy siltstone, grayish red, massive, with mudstone strips occasionally
241-16	981.40—981.57 m	0.17 m	Muddy siltstone, grayish red, slump structures, with irregular mudstone strips
241-17	981.57—981.86 m	0.29 m	Siltstone, light gray, wavy cross bedding, with wavy bedding locally, mudstone strips
241-18	981.86—981.89 m	0.03 m	Siltstone, light greenish gray, massive, erosion surface at the base
241-19	981.89—982.39 m	0.50 m	Siltstone, light brown, wavy cross bedding
241-20	982.39—982.76 m	0.37 m	Siltstone, light greenish gray, trough cross bedding, with fossil fragments and siltstone laminae
241-21	982.76—982.90 m	0.14 m	Siltstone, light greenish gray, deformation bedding, erosion surface at the base, mudstone strips
241-22	982.90—983.24 m	0.34 m	Mudstone, grayish brown, massive
241-23	983.24—983.84 m	0.60 m	Silty mudstone, grayish brown, massive, with lots of calcareous concretions
241-24	983.84—983.96 m	0.12 m	Muddy siltstone, light brown, massive
241-25	983.96—984.49 m	0.53 m	Silty mudstone, grayish brown, massive
241-26	984.49—984.64 m	0.15 m	Muddy siltstone, light brown, massive
242-1	984.64—985.39 m	0.75 m	Muddy siltstone, brown, massive, with deformation bedding locally, calcareous concretions
242-2	985.39—985.59 m	0.20 m	Sandstone, fine-grained, light gray, massive bedding, erosion surface at the base, mud gravels on the erosion surface
242-3	985.59—985.94 m	0.35 m	Silty mudstone, brown, massive, with calcareous concretions
242-4	985.94—987.54 m	1.60 m	Siltstone, brown, massive, with lots of calcareous concretions
242-5	987.54—987.79 m	0.25 m	Fine-grained sandstone, light brown, trough cross bedding, with calcareous concretions

242-6	987.79—987.89 m	0.10 m	Fine-grained sandstone, light gray, massive, calcareous concretions
242-7	987.89—988.94 m	1.05 m	Siltstone, brown, massive, with calcareous concretions locally
242-8	988.94—989.12 m	0.18 m	Siltstone, greenish gray, massive, with grayish brown spots
242-9	989.12—989.24 m	0.12 m	Muddy siltstone, brown, massive, with trace fossils
242-10	989.24—989.34 m	0.10 m	Siltstone, light greenish gray, massive
242-11	989.34—989.42 m	0.08 m	Siltstone, brown, massive
242-12	989.42—989.64 m	0.22 m	Siltstone, light greenish gray, massive, with brown siltstone strips
242-13	989.64—989.84 m	0.20 m	Fine-grained sandstone, light gray, wavy cross bedding, erosion surface at the base, with siltstone strips
242-14	989.84—990.39 m	0.55 m	Fine-grained sandstone, grayish red, wavy cross bedding, with bioturbation, erosion surface at the base
242-15	990.39—990.49 m	0.10 m	Fine-grained sandstone, light gray, wavy cross bedding
242-16	990.49—991.04 m	0.55 m	Medium-grained sandstone, grayish red, wavy cross bedding
242-17	991.04—991.44 m	0.40 m	Coarse-grained sandstone, light gray, massive, erosion surface at the base, mud gravels and calcareous concretions
242-18	991.44—991.74 m	0.30 m	Conglomerate, light brown, silty with a lot of angular mud gravels (size 2—20 mm), deformation structure locally, few gritty concretions
242-19	991.74—992.69 m	0.95 m	Fine-grained sandstone, light gray, wavy cross bedding, mud gravels at the base
242-20	992.69—993.59 m	0.90 m	Sandstone, medium-grained, light gray, oblique bedding, thin argillaceous layer at 993.14 m
243-1	993.59—993.67 m	0.08 m	Muddy siltstone, light brown, massive
243-2	993.67—993.73 m	0.06 m	Fine-grained sandstone, grayish red, parallel bedding, with mud gravels and mudstone strips
243-3	993.73—993.77 m	0.04 m	Fine-grained conglomerate, light brown and grayish red, massive, erosion surface at the base
243-4	993.77—994.04 m	0.27 m	Medium-grained sandstone, light gray, parallel bedding
243-5	994.04—994.09 m	0.05 m	Fine-grained conglomerate, with mud gravels, massive, erosion surface at the base
243-6	994.09—994.60 m	0.51 m	Medium-grained sandstone, grayish red, parallel bedding
243-7	994.60—995.14 m	0.54 m	Medium-grained sandstone, light gray, wedge-shaped cross bedding, erosion surface at the base, mud gravels
243-8	995.14—995.30 m	0.16 m	Medium-grained sandstone, grayish red, deformation bedding, wavy bedding locally, erosion surface at the base, irregular mudstone strips and calcareous concretions
243-9	995.30—995.43 m	0.13 m	Silty mudstone, grayish red, massive

243-10	995.43—996.21 m	0.78 m	Muddy siltstone, light reddish brown, wavy ripple bedding, with siltstone balls
243-11	996.21—996.31 m	0.10 m	Siltstone, light reddish brown, wavy ripple bedding, deformation bedding, erosion surface at the base, with light gray siltstone strips
243-12	996.31—996.49 m	0.18 m	Siltstone, light greenish gray, massive
243-13	996.49—997.14 m	0.65 m	Siltstone, light red, wavy bedding and wavy cross bedding
243-14	997.14—997.35 m	0.21 m	Fine-grained sandstone, wavy bedding
243-15	997.35—997.79 m	0.44 m	Medium-grained sandstone, parallel bedding, wavy bedding locally, erosion surface at the base
243-16	997.79—998.24 m	0.45 m	Fine-grained sandstone, grayish red, wavy bedding
243-17	998.24—998.29 m	0.05 m	Fine-grained conglomerate, light red, massive, erosion surface at the base
243-18	998.29—998.52 m	0.23 m	Medium-grained sandstone, parallel bedding, with mud gravels
243-19	998.52—998.67 m	0.15 m	Fine-grained conglomerate, light red, massive, erosion surface at the base
243-20	998.67—999.23 m	0.56 m	Medium-grained sandstone, light red, parallel bedding, wavy bedding, with mud gravels strips and laminae, erosion surface at the base, with lots of mud gravels
244-1	999.23—999.36 m	0.13 m	Medium-grained sandstone, greenish gray, massive
244-2	999.36—1000.21 m	0.85 m	Silty mudstone, light reddish brown, massive, calcareous concretions occasionally
244-3	1000.21—1002.21 m	2.00 m	Mudstone, light brown, massive
244-4	1002.21—1002.23 m	0.02 m	Muddy siltstone, light greenish gray, massive
244-5	1002.23—1002.63 m	0.40 m	Mudstone, light reddish-brown, massive bedding, greenish-gray spots with carbon-based concretions
244-6	1002.63—1004.33 m	1.70 m	Silty mudstone, light reddish-brown, massive bedding, calcareous concretions and greenish-gray spots occasionally
244-7	1004.33—1005.29 m	0.96 m	Muddy siltstone, light reddish-brown, massive bedding, few calcareous concretions, with light greenish-gray siltstone at 1006.63 m and 1005.23 m
244-8	1005.29—1006.29 m	1.00 m	Mudstone, light reddish brown, massive
244-9	1006.29—1006.39 m	0.10 m	Fine-grained sandstone, light gray, massive, erosion surface at the base
244-10	1006.39—1006.93 m	0.54 m	Muddy siltstone, light reddish brown, massive, calcareous concretions
244-11	1006.93—1007.43 m	0.50 m	Siltstone, light brown, massive, with lots of calcareous concretions
244-12	1007.43—1008.06 m	0.63 m	Muddy siltstone, light reddish brown, massive, with few calcareous concretions
245-1	1008.06—1008.71 m	0.65 m	Muddy siltstone, light reddish brown, massive, with lots of calcareous concretions
245-2	1008.71—1009.26 m	0.55 m	Silty mudstone, light reddish brown, massive, with lots of calcareous concretions

245-3	1009.26—1010.16 m	0.90 m	Muddy siltstone, light reddish brown, massive, calcareous concretions
245-4	1010.16—1011.16 m	1.00 m	Silty mudstone, light reddish brown, massive, calcareous concretions
245-5	1011.16—1012.96 m	1.80 m	Muddy siltstone, light reddish brown, massive, calcareous concretions
245-6	1012.96—1013.16 m	0.20 m	Fine-grained sandstone, light greenish gray, massive
245-7	1013.16—1013.56 m	0.40 m	Medium-grained sandstone, light greenish gray, massive, erosion surface at the base
245-8	1013.56—1014.61 m	1.05 m	Medium-grained sandstone, light reddish brown, oblique bedding
245-9	1014.61—1014.81 m	0.20 m	Medium-grained sandstone, light gray, cross bedding, with calcareous concretions occasionally
245-10	1014.81—1015.26 m	0.45 m	Medium-grained sandstone, grayish red, massive
245-11	1015.26—1015.66 m	0.40 m	Silty mudstone, grayish red, deformation bedding
245-12	1015.66—1015.96 m	0.30 m	Silty mudstone, grayish red, massive, with mud gravels and calcareous concretions
245-13	1015.96—1016.36 m	0.40 m	Medium-grained sandstone, light gray, massive, erosion surface at the base
245-14	1016.36—1016.41 m	0.05 m	Mudstone, grayish red, massive
245-15	1016.41—1016.66 m	0.25 m	Fine-grained sandstone, grayish red, massive
246-1	1016.66—1016.77 m	0.11 m	Muddy siltstone, grayish red, slump structures, with light greenish gray siltstone balls, with few calcareous concretions
246-2	1016.77—1016.84 m	0.07 m	Siltstone, grayish red, cross bedding
246-3	1016.84—1017.26 m	0.42 m	Fine-grained sandstone, grayish red, parallel bedding
246-4	1017.26—1018.46 m	1.20 m	Medium-grained sandstone, light red, parallel bedding
246-5	1018.46—1018.90 m	0.44 m	Medium-grained sandstone, grayish red, wedge-shaped cross bedding, massive locally, with mud gravels strips
246-6	1018.90—1019.03 m	0.13 m	Medium-grained sandstone, light grayish red and greenish gray, massive
246-7	1019.03—1019.16 m	0.13 m	Medium-grained sandstone, light gray, oblique bedding, with mud gravels
246-8	1019.16—1020.66 m	1.50 m	Conglomerate, fine grained and medium gritty with mud gravels, light gray and grayish-red, oblique bedding and massive bedding, deformation structure locally, erosion surface at the base, a 6 cm × 2 cm rhyolite gravel at 1019.71 m, wrapped by sandstone, calcareous concretions occasionally
246-9	1020.66—1020.76 m	0.10 m	Fine-grained sandstone, light gray, horizontal bedding
247-1	1020.76—1020.86 m	0.10 m	Sandstone, light gray, parallel bedding
247-2	1020.86—1021.32 m	0.46 m	Sandstone, medium grained with gravels, light gray, parallel bedding in the upper part, massive bedding in the middle and lower part, erosion surface at 1021.16 m
248-1	1021.57—1021.60 m	0.03 m	Sandstone, medium-grained, light gray, parallel bedding, erosion at the base, calcareous concretions (residual deposits) on the erosion surface

5.1.8 Mingshui Formation

The overlying formation of the Mingshui Formation Member 2 is Taikang Formation, the contacting relationship between them is unconformable contact; the underlying formation is the Mingshui Formation Member 1, the contacting relationship between them is conformable contact (Figure 5.8).

The detailed description of the Mingshui Formation Member 2:

58-1	210.66—211.44 m	0.78 m	Mudstone, olive gray, massive
60-1	212.16—213.03 m	0.87 m	Mudstone, olive gray, massive
62-1	213.03—214.62 m	1.59 m	Silty mudstone, olive gray, massive
64-1	215.08—216.48 m	1.40 m	Mudstone, yellowish brown, massive
65-1	216.48—217.18 m	0.70 m	Silty mudstone, yellowish brown, massive
66-1	217.18—218.58 m	1.40 m	Mudstone, yellowish brown, massive
67-1	218.58—218.95 m	0.37 m	Mudstone, olive gray, massive
67-2	218.95—219.56 m	0.61 m	Fine-grained sandstone, yellowish gray, parallel bedding, erosion surface at the base
67-3	219.56—219.73 m	0.17 m	Mudstone, olive gray, massive
68-1	219.89—220.49 m	0.60 m	Mudstone, dark greenish gray, massive
68-2	220.49—221.29 m	0.80 m	Mudstone, light yellowish brown, massive
69-1	221.29—222.85 m	1.56 m	Mudstone, yellowish gray, massive
71-1	223.22—224.78 m	1.56 m	Silty mudstone, olive gray, massive
72-1	224.78—226.36 m	1.58 m	Mudstone, olive gray, massive
73-1	226.36—227.59 m	1.23 m	Silty mudstone, olive gray, massive
74-1	227.59—229.19 m	1.60 m	Mudstone, olive gray, massive
75-1	229.19—230.19 m	1.00 m	Silty mudstone, olive gray, massive
76-1	230.82—232.09 m	1.27 m	Mudstone, olive gray, massive
79-1	233.71—234.24 m	0.53 m	Mudstone, gray, massive
80-1	234.24—234.66 m	0.42 m	Mudstone, gray, massive
81-1	234.66—235.78 m	1.12 m	Silty mudstone, gray, massive
82-1	235.78—236.88 m	1.10 m	Mudstone, gray, massive
83-1	236.88—237.35 m	0.47 m	Mudstone, gray, massive
84-1	237.35—237.60 m	0.25 m	Mudstone, gray, massive
84-2	237.60—238.08 m	0.48 m	Siltstone, light gray, discontinuous wavy bedding
85-1	238.15—238.65 m	0.50 m	Siltstone, light gray, discontinuous wavy bedding. few subangular quartzose conglomerates, 2—4 mm
86-1	239.18—239.38 m	0.20 m	Siltstone, light gray, discontinuous wavy bedding
86-2	239.38—239.88 m	0.50 m	Mudstone, gray, discontinuous wavy bedding

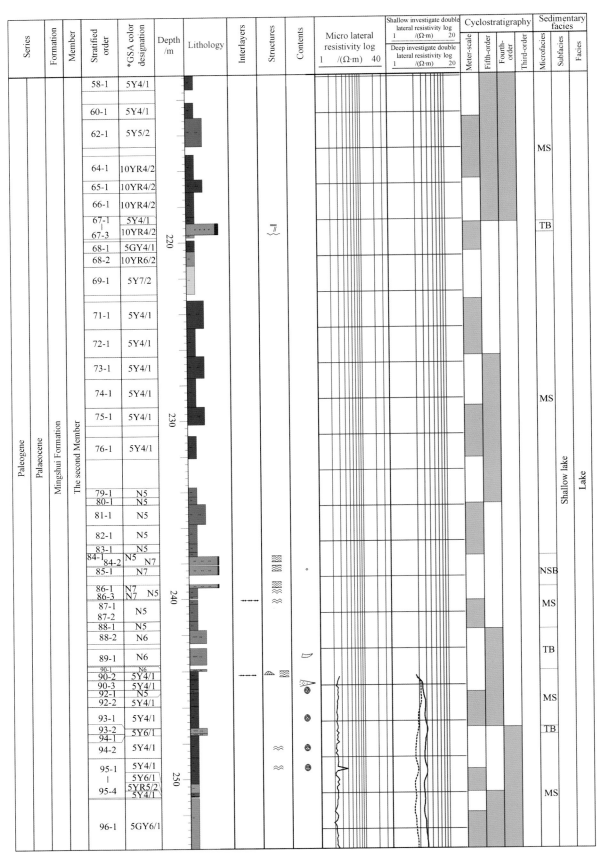

Figure 5.8　Stratigraphic chart of Mingshui Formation.

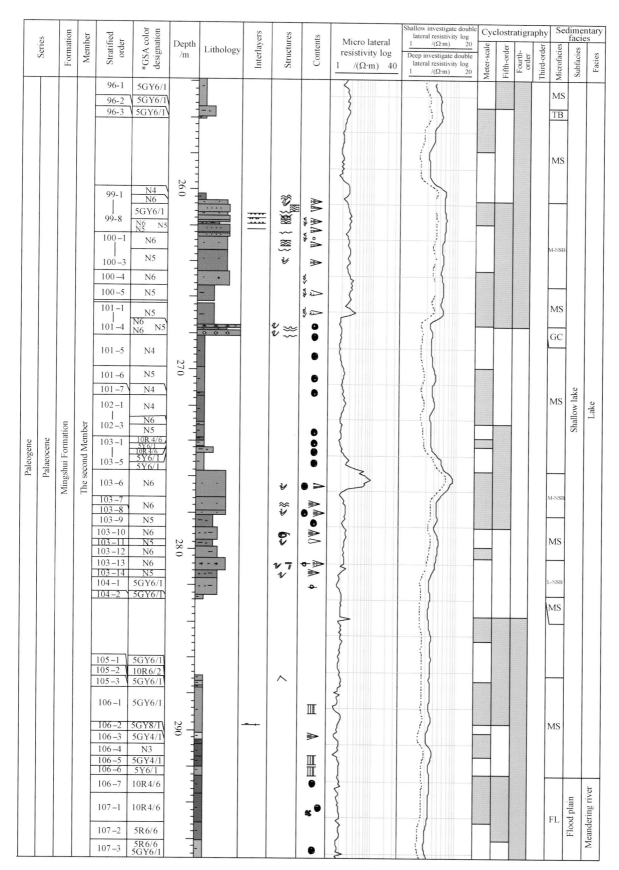

Figure 5.8 Stratigraphic chart of Mingshui Formation (continued).

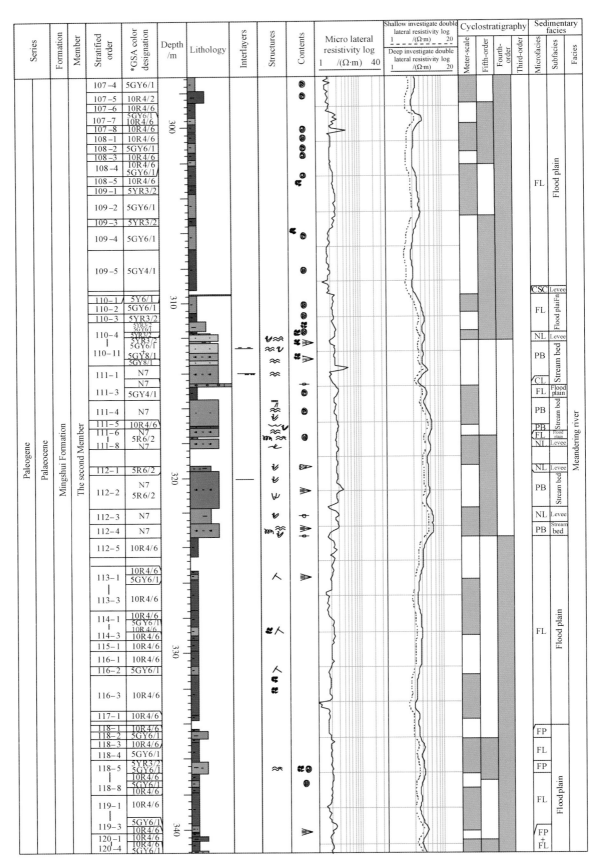

Figure 5.8 Stratigraphic chart of Mingshui Formation (continued).

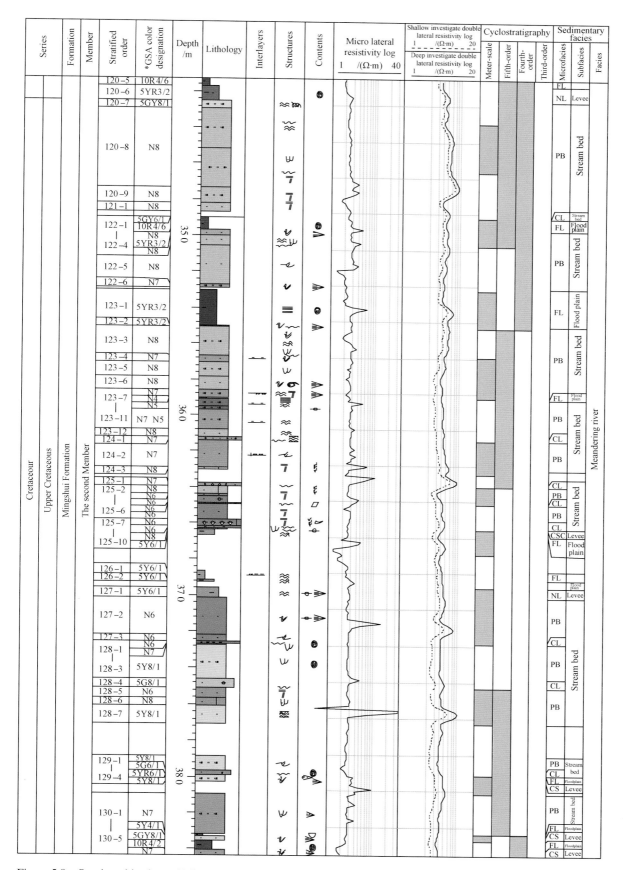

Figure 5.8 Stratigraphic chart of Mingshui Formation (continued).

312 | Initial Report of Continental Scientific Drilling Project of the Cretaceous Songliao Basin (SK-1)

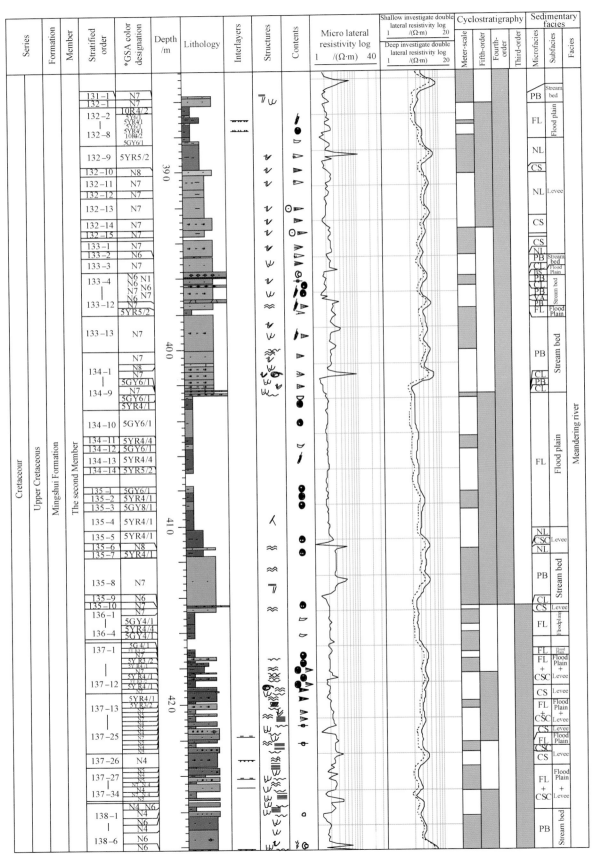

Figure 5.8 Stratigraphic chart of Mingshui Formation (continued).

Section 5 SK-1 Core Description and Core Photographs

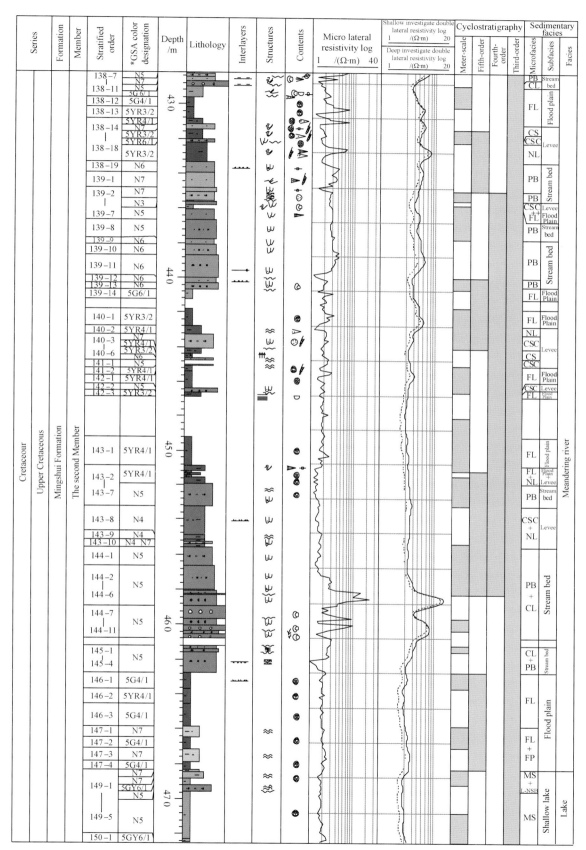

Figure 5.8 Stratigraphic chart of Mingshui Formation (continued).

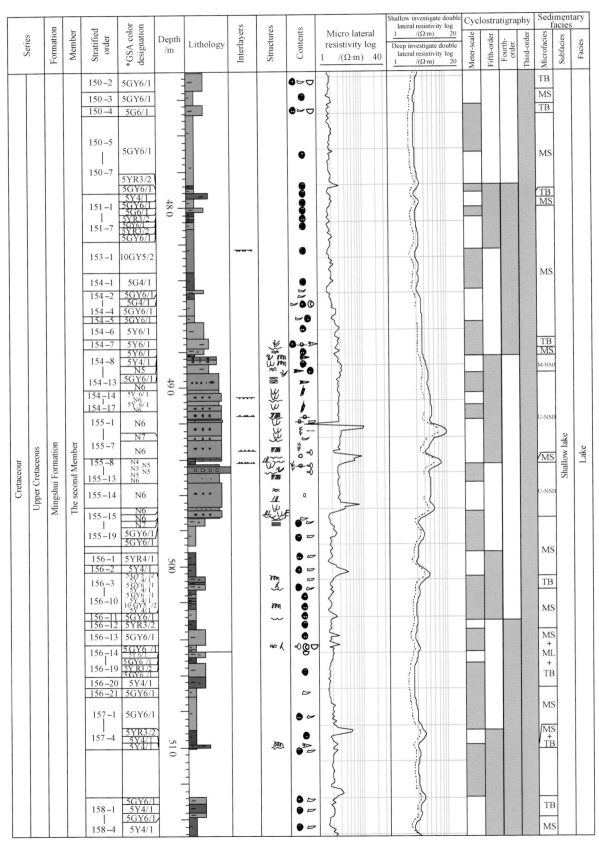

Figure 5.8 Stratigraphic chart of Mingshui Formation (continued).

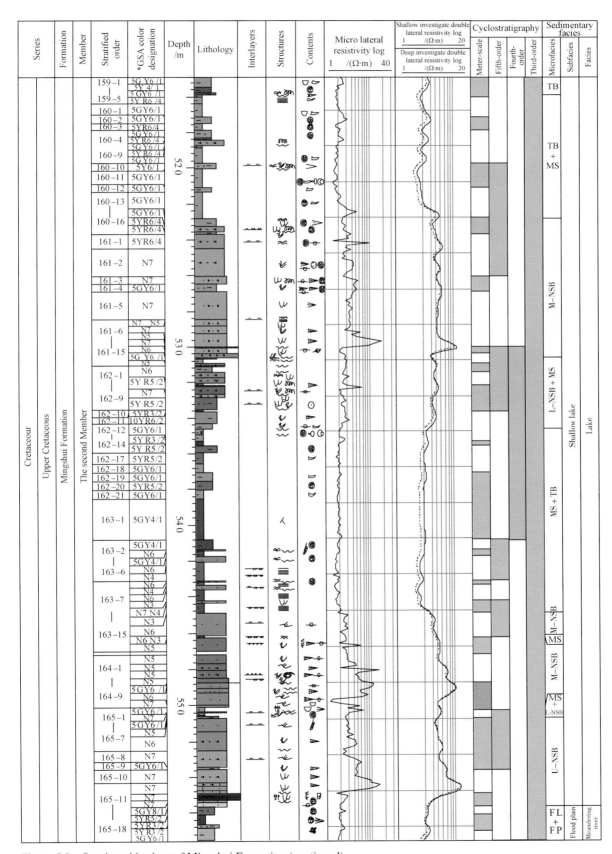

Figure 5.8 Stratigraphic chart of Mingshui Formation (continued).

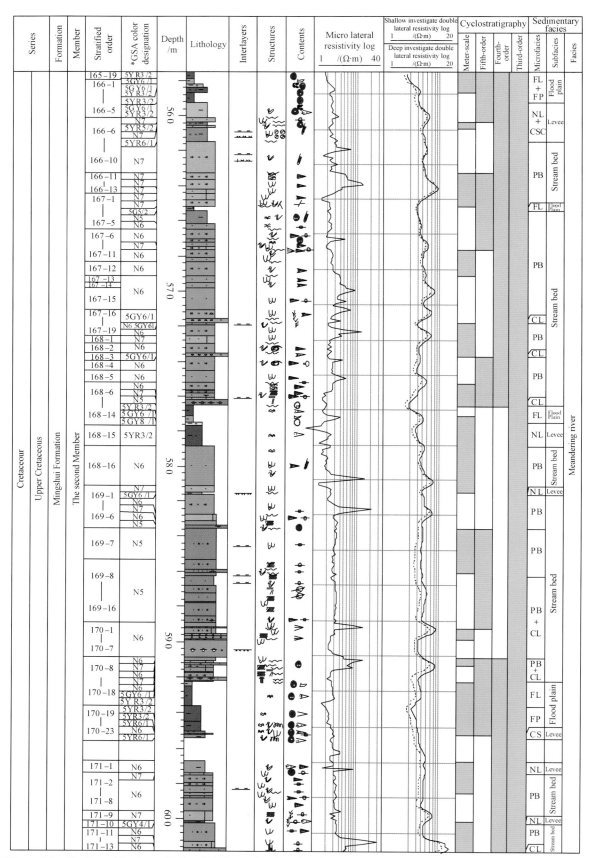

Figure 5.8 Stratigraphic chart of Mingshui Formation (continued).

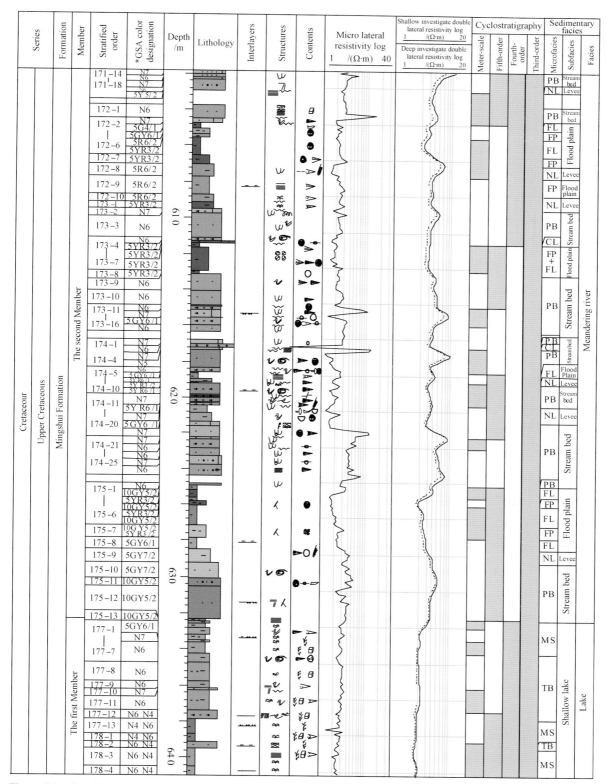

Figure 5.8 Stratigraphic chart of Mingshui Formation (continued).

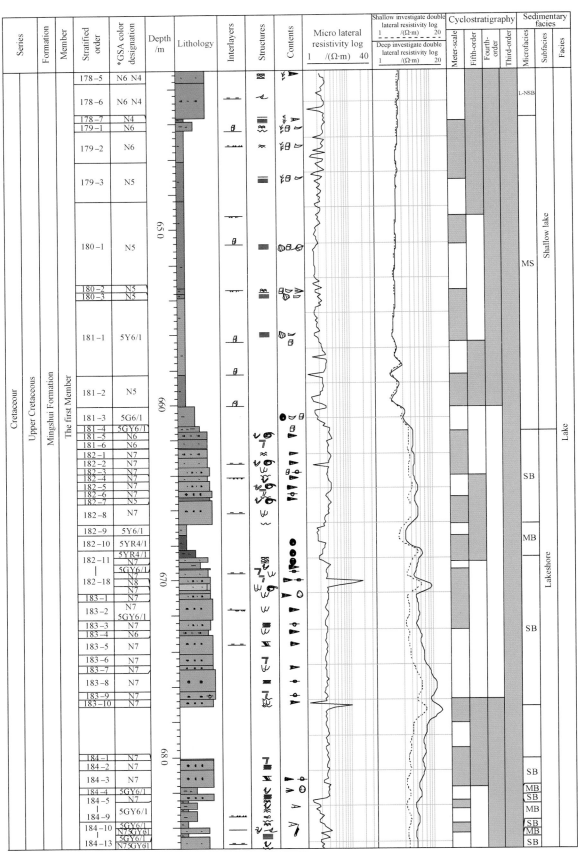

Figure 5.8 Stratigraphic chart of Mingshui Formation (continued).

Section 5 SK-1 Core Description and Core Photographs

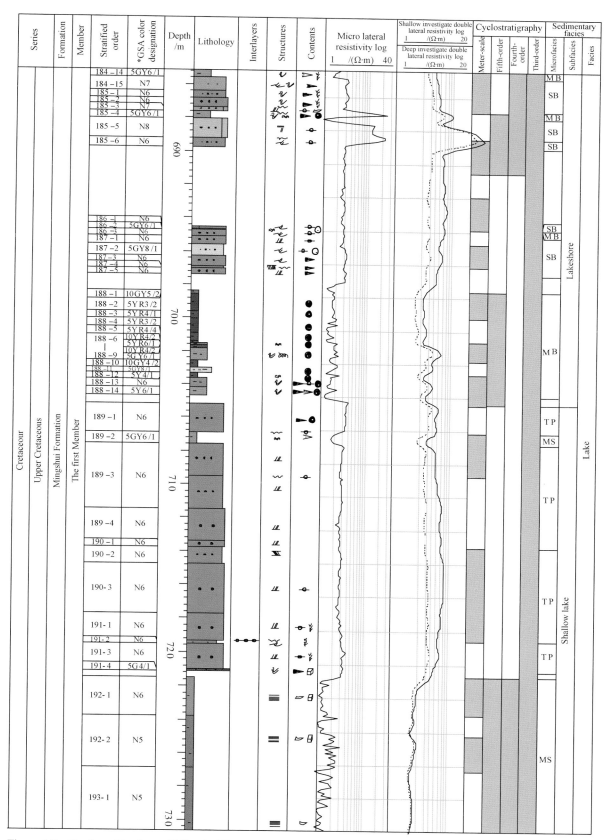

Figure 5.8 Stratigraphic chart of Mingshui Formation (continued).

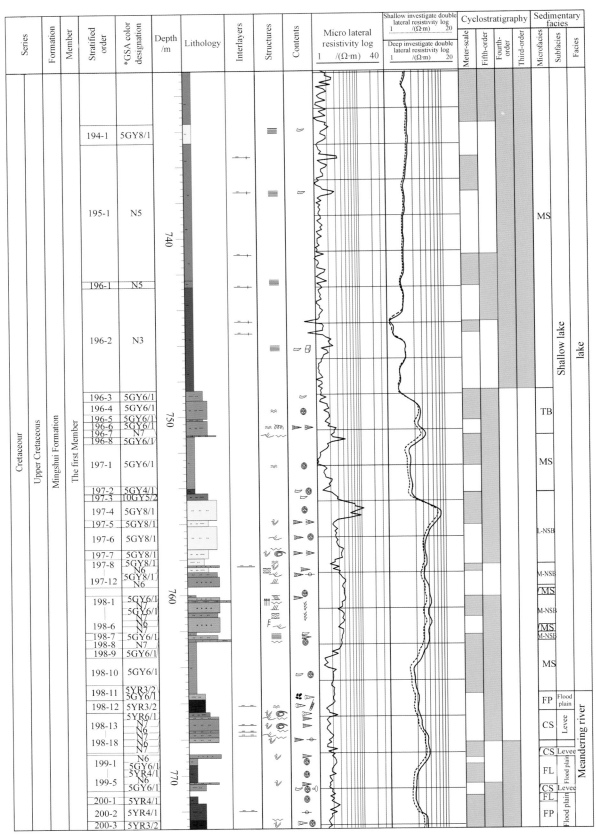

Figure 5.8　Stratigraphic chart of Mingshui Formation (continued).

Section 5 SK-1 Core Description and Core Photographs | 321

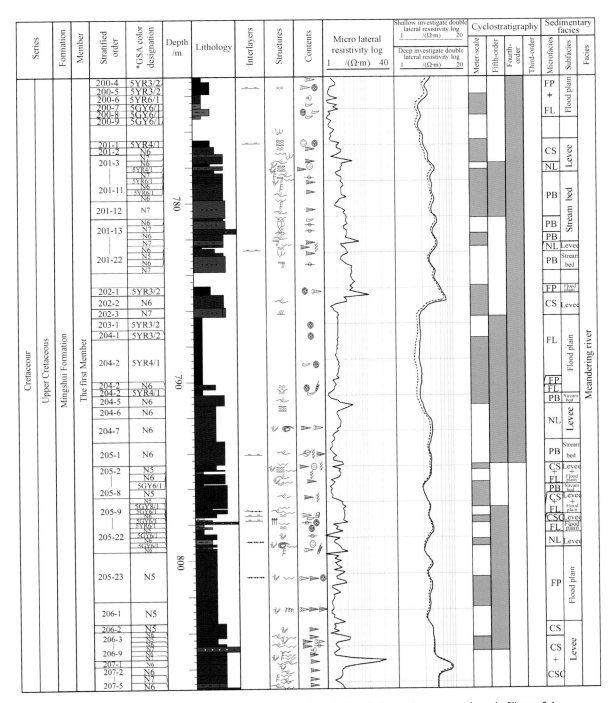

Figure 5.8 Stratigraphic chart of Mingshui Formation (continued). Symbols are the same as those in Figure 5.1.

86-3	239.88—239.98 m	0.10 m	Mudstone, medium dark gray, interbedded with light gray siltstone, wavy bedding
87-1	240.04—240.29 m	0.25 m	Mudstone, medium dark gray, interbedded with light gray siltstone, wavy bedding
87-2	240.29—241.23 m	0.94 m	Mudstone, gray, massive
88-1	241.23—241.73 m	0.50 m	Mudstone, gray, massive
88-2	241.73—242.46 m	0.73 m	Silty mudstone, light gray, massive
89-1	242.73—243.63 m	0.90 m	Silty mudstone, light gray, massive, bivalve fossils
90-1	243.85—243.98 m	0.13 m	Silty mudstone, light gray, massive
90-2	243.98—244.45 m	0.47 m	Mudstone, olive gray, interbedded with light gray siltstone, wavy bedding, lenticular beddings
90-3	244.45—244.98 m	0.53 m	Mudstone, olive gray, massive, with light gray siltite strips locally
92-1	244.98—245.17 m	0.19 m	Mudstone, gray, massive, much calcareous concretions
92-2	245.17—245.92 m	0.75 m	Mudstone, olive gray, massive
93-1	245.92—247.11 m	1.19 m	Mudstone, olive gray, massive, with calcareous crumbs at 246.42—246.57 m
93-2	247.11—247.41 m	0.30 m	Silty mudstone, olive gray, massive
94-1	247.42—247.52 m	0.10 m	Silty mudstone, olive gray, massive
94-2	247.52—248.78 m	1.26 m	Mudstone, olive gray, with ripple bedding, calcareous crumbs at the base, up to 2.5 cm diameter
95-1	248.79—250.25 m	1.46 m	Mudstone, olive gray, horizontal bedding, calcareous concretions at the base
95-2	250.25—250.75 m	0.50 m	Mudstone, olive gray, massive
95-3	250.75—250.92 m	0.17 m	Mudstone, light brown, massive
95-4	250.92—250.97 m	0.05 m	Mudstone, olive gray, massive
96-1	251.04—255.40 m	4.36 m	Mudstone, light greenish gray, massive
96-2	255.40—255.94 m	0.54 m	Silty mudstone, light greenish gray, massive
96-3	255.94—256.08 m	0.14 m	Mudstone, light greenish gray, massive
99-1	260.24—260.59 m	0.35 m	Mudstone, dark gray, horizontal bedding
99-2	260.59—260.85 m	0.26 m	Siltstone, medium light gray, with deformation structures, irregular mudstone laminae developed
99-3	260.85—261.24 m	0.39 m	Sandstone, fine-grained, light greenish-gray, wavy cross bedding, deformation bedding locally, erosion surface at the base
99-4	261.24—261.44 m	0.20 m	Siltstone, light greenish-gray, small wedge-shaped cross bedding and deformation bedding, scour structure at the base, irregular mudstone laminae and fine-grained sandstone strips, a few plant fragments
99-5	261.44—261.72 m	0.28 m	Sandstone, fine-grained, light greenish-gray, with wavy bedding, and cross bedding locally, mudstone and argillaceous siltstone laminae developed
99-6	261.72—261.83 m	0.11 m	Sandstone, fine-grained, medium light gray, massive structure, irregular mudstone laminae and charcoal developed

99-7	261.83—261.91 m	0.08 m	Muddy siltstone, medium gray, deformation bedding or discontinuous wavy bedding, charcoal, irregular mudstone laminae and fine-grained sandstone strips developed
99-8	261.91—262.40 m	0.49 m	Sandstone, fine-grained, medium gray, mainly deformation bedding, a few wavy bedding, and scour structure at the base
100-1	262.43—262.63 m	0.20 m	Sandstone, fine-grained, medium light gray, wavy bedding, with charcoal, gravels at the bas
100-2	262.63—263.35 m	0.72 m	Siltstone, medium light gray, thin mudstone in ripple bedding developed in the upper part of the section, massive structure in the middle and lower section, and scour structure at the base
100-3	263.35—264.51 m	1.16 m	Siltstone, medium gray, with fine-grained sandstone locally, argillaceous interbeds increasing upward, slump structures
100-4	264.51—265.34 m	0.83 m	Siltstone, fine-grained, medium light gray, parallel bedding, with deformation structures locally, large number of charcoals on the rock surface
100-5	265.34—266.23 m	0.89 m	Silty mudstone, medium gray, massive bedding, charcoals occasionally and a few siltstone strips
101-1	266.35—267.55 m	1.20 m	Silty mudstone, medium gray, massive structure, and deformation structure locally, charcoal largely, with siltstone locally
101-2	267.55—267.71 m	0.16 m	Conglomerate, medium-grained, medium light gray, silty, calcareous or concretionary, with poorly-sorted subangular gravels, erosion surface at the base, a large number of calcareous concretions
101-3	267.71—267.79 m	0.08 m	Siltstone, medium gray, deformation structures, erosion surface at the base, irregular mudstone laminae, a few charcoal
101-4	267.79—268.15 m	0.36 m	Medium-grained conglomerate, medium light gray, with poorly-sorted subangular gravels, deformation structure, erosion surface at the base, a large number of calcareous concretions
101-5	268.15—269.85 m	1.70 m	Mudstone, medium dark gray, massive bedding, with plentiful calcareous concretions
101-6	269.85—271.25 m	1.40 m	Mudstone, gray, massive, lots of calcareous concretions
101-7	271.25—271.45 m	0.20 m	Mudstone, medium dark gray, massive bedding, with plentiful calcareous concretions
102-1	271.49—272.99 m	1.50 m	Mudstone, dark gray, massive
102-2	272.99—273.14 m	0.15 m	Mudstone, light gray, massive
102-3	273.14—273.79 m	0.65 m	Mudstone, gray, massive, calcareous concretions occasionally
103-1	273.83—273.98 m	0.15 m	Mudstone, reddish brown, massive
103-2	273.98—274.31 m	0.33 m	Mudstone, light olive gray, massive, calcareous concretions occasionally
103-3	274.31—274.35 m	0.04 m	Mudstone, reddish brown, massive
103-4	274.35—274.70 m	0.35 m	Silty mudstone, light olive gray, massive, calcareous concretions occasionally
103-5	274.70—275.67 m	0.97 m	Mudstone, light olive gray, massive, calcareous concretions locally

103-6	275.67—277.16 m	1.49 m	Siltstone, medium light gray, slump structure, with preexisting beddings in local siltstone and few mudstone, irregular mudstone laminae developed, much calcareous concretions locally
103-7	277.16—277.88 m	0.72 m	Siltstone, medium light gray, discontinuous ripple bedding, few mudstone laminae
103-8	277.88—278.13 m	0.25 m	Siltstone, medium light gray, with deformation structures, irregular mudstone laminae developed calcareous concretions
103-9	278.13—278.83 m	0.70 m	Silty mudstone, medium gray, massive bedding, few calcareous concretions
103-10	278.83—279.53 m	0.70 m	Muddy siltstone, medium light gray, massive bedding, roll-up structure locally, with mudstone laminae
103-11	279.53—279.89 m	0.36 m	Silty mudstone, medium gray, massive bedding, roll-up structure locally, with siltstone concretions
103-12	279.89—280.47 m	0.58 m	Muddy siltstone, medium light gray, massive bedding
103-13	280.47—281.18 m	0.71 m	Siltstone, medium light gray, with deformation structures, parallel bedding locally, few mud gravels and mudstone laminae
103-14	281.18—281.61 m	0.43 m	Muddy siltstone, gray, deformation structures, with irregular mudstone laminae
104-1	281.61—282.51 m	0.90 m	Muddy siltstone, greenish gray, massive, with lots of mud gravels
104-2	282.51—282.81 m	0.30 m	Mudstone, greenish gray, massive
105-1	286.96—287.29 m	0.33 m	Mudstone, greenish gray, massive, with fractures
105-2	287.29—287.56 m	0.27 m	Mudstone, light red, massive, greenish-gray reduced spots
105-3	287.56—287.66 m	0.10 m	Mudstone, greenish gray, massive
106-1	287.66—290.16 m	2.50 m	Mudstone, greenish-gray, massive bedding, gypsum at the base, 1cm-thick red mudstone at 289.66 m
106-2	290.16—290.48 m	0.32 m	Mudstone, light greenish gray, massive, with greenish gray mudstone laminae and spots
106-3	290.48—290.76 m	0.28 m	Mudstone, dark greenish gray, massive
106-4	290.76—291.44 m	0.68 m	Mudstone, dark gray, massive
106-5	291.44—291.96 m	0.52 m	Mudstone, dark greenish gray, massive
106-6	291.96—292.46 m	0.50 m	Mudstone, olive gray, massive
106-7	292.46—293.49 m	1.03 m	Mudstone, reddish brown, massive, with calcareous concretions
107-1	293.49—295.09 m	1.60 m	Mudstone, reddish brown, massive, with calcareous concretions
107-2	295.09—296.09 m	1.00 m	Mudstone, light red, massive
107-3	296.09—297.09 m	1.00 m	Mudstone, light red and greenish gray, massive, with lots of calcareous concretions
107-4	297.09—297.88 m	0.79 m	Mudstone, greenish gray, massive, with lots of calcareous concretions
107-5	297.88—298.59 m	0.71 m	Silty mudstone, grayish red, massive, with lots of calcareous concretions
107-6	298.59—299.42 m	0.83 m	Mudstone, reddish brown, massive
107-7	299.42—299.79 m	0.37 m	Mudstone, reddish brown and greenish gray, massive

107-8	299.79—300.19 m	0.40 m	Mudstone, reddish brown, massive, with lots of calcareous concretions
108-1	300.28—300.83 m	0.55 m	Mudstone, reddish brown, massive, with lots of calcareous concretions
108-2	300.83—301.38 m	0.55 m	Mudstone, greenish gray, massive, with lots of calcareous concretions
108-3	301.38—301.81 m	0.43 m	Mudstone, reddish brown, massive, with lots of calcareous concretions
108-4	301.81—302.14 m	0.33 m	Mudstone, reddish brown and greenish gray, massive
108-5	302.14—303.25 m	1.11 m	Mudstone, reddish brown, massive, with few greenish gray mudstone spots and calcareous concretions
109-1	303.25—303.75 m	0.50 m	Mudstone, grayish brown, massive, with lots of greenish gray mudstone spots
109-2	303.75—305.07 m	1.32 m	Mudstone, greenish gray, massive, with grayish brown mudstone spots
109-3	305.07—305.51 m	0.44 m	Mudstone, grayish gray, massive, greenish gray mudstone spots
109-4	305.51—306.92 m	1.41 m	Mudstone, greenish gray, massive, grayish brown spots in the upper part, with few calcareous concretions locally
109-5	306.92—309.25 m	2.33 m	Mudstone, dark greenish gray, massive, with few calcareous concretions locally
110-1	309.51—309.59 m	0.08 m	Calcareous conglomerate, yellowish gray, massive
110-2	309.59—310.51 m	0.92 m	Mudstone, dark greenish gray, massive, with few calcareous concretions
110-3	310.51—311.06 m	0.55 m	Mudstone, grayish brown, massive, with few greenish gray spots and calcareous concretions
110-4	311.06—311.62 m	0.56 m	Silty mudstone, greenish gray and grayish brown, massive
110-5	311.62—311.70 m	0.08 m	Mudstone, grayish brown, massive, with few calcareous concretions
110-6	311.70—311.81 m	0.11 m	Muddy siltstone, light brownish-gray and light greenish-gray, massive bedding, irregular mudstone laminae developed
110-7	311.81—312.14 m	0.33 m	Siltstone, light greenish-gray, discontinuous ripple bedding, deformation bedding, roll-up structure locally, irregular mudstone laminae developed
110-8	312.14—312.28 m	0.14 m	Muddy siltstone, light brownish gray and light greenish gray, massive, with irregular mudstone laminae
110-9	312.28—312.85 m	0.57 m	Siltstone, light greenish-gray, discontinuous ripple bedding, deformation bedding, roll-up structure locally, with irregular mudstone laminae, and 3 cm mudstone at 312.41 m
110-10	312.85—313.23 m	0.38 m	Muddy siltstone, light brownish gray and light greenish gray, massive, with irregular mudstone laminars
110-11	313.23—313.53 m	0.30 m	Siltstone, light greenish gray, massive, with few irregular mudstone laminars
111-1	313.53—314.58 m	1.05 m	Siltstone, light gray, discontinuous wavy bedding and deformation bedding locally, with muddy siltstone strips
111-2	314.58—314.75 m	0.17 m	Medium-grained conglomerate, massive, few mud gravels
111-3	314.75—315.51 m	0.76 m	Mudstone, dark greenish gray, massive, few mud gravels
111-4	315.51—316.97 m	1.46 m	Siltstone, light gray, parallel bedding, discontinuous wavy bedding, with deformation structures locally, large calcareous concretions, <3 cm calcareous concretions observed respectively at 315.98 m, 316.16 m and 316.28 m, erosion at the base of each concretion layer

111-5	316.97—317.08 m	0.11 m	Siltstone, light gray, deformation bedding, with siltstone strips and balls
111-6	317.08—317.58 m	0.50 m	Siltstone, light gray, discontinuous wavy bedding, with grayish brown oxidized sports locally
111-7	317.58—317.71 m	0.13 m	Mudstone, light red, discontinuous ripple bedding, with calcareous concretions and bioturbation structures occasionally
111-8	317.71—318.30 m	0.59 m	Siltstone, light gray, wavy cross bedding, deformation bedding locally
112-1	319.26—319.56 m	0.30 m	Muddy siltstone, light red, deformation bedding, with irregular siltstone strips
112-2	319.56—321.76 m	2.20 m	Siltstone, light gray and light red, with deformation bedding, wavy bedding and occasional cross bedding, mudstone laminae, and mudstone laminae at 320.71 m
112-3	321.76—322.59 m	0.83 m	Muddy siltstone, light gray, massive bedding, deformation bedding locally, with mud gravels especially at 322.16 m
112-4	322.59—323.43 m	0.84 m	Siltstone, light gray, discontinuous ripple bedding, deformation bedding locally, bioturbation structures and irregular mudstone laminae and mud gravels
112-5	323.43—324.52 m	1.09 m	Mudstone, reddish brown, massive
113-1	325.31—325.51 m	0.20 m	Mudstone, reddish brown, massive
113-2	325.51—325.81 m	0.30 m	Mudstone, greenish gray, massive, with lots of irregular fractures and reddish brown mudstone strips
113-3	325.81—327.59 m	1.78 m	Mudstone, reddish brown, massive, with greenish-gray reduced spots at the top
114-1	327.59—328.49 m	0.90 m	Mudstone, reddish brown, massive
114-2	328.49—328.99 m	0.50 m	Mudstone, reddish brown and greenish gray, massive, with lots of irregular fractures
114-3	328.99—329.19 m	0.20 m	Mudstone, reddish brown, massive
115-1	329.19—329.87 m	0.68 m	Mudstone, reddish brown, massive
116-1	329.87—330.67 m	0.80 m	Mudstone, reddish brown, massive
116-2	330.67—331.20 m	0.53 m	Mudstone, greenish gray, massive, with lots of irregular fractures filled with reddish brown mudstone
116-3	331.20—333.49 m	2.29 m	Mudstone, reddish brown, massive, with greenish gray mudstone reduced spots
117-1	333.49—333.81 m	0.32 m	Mudstone, reddish brown, massive
118-1	334.28—334.41 m	0.13 m	Mudstone, reddish brown, massive
118-2	334.41—334.88 m	0.47 m	Silty mudstone, greenish gray, massive, with reddish brown oxidized spots
118-3	334.88—335.02 m	0.14 m	Mudstone, reddish brown, massive
118-4	335.02—336.14 m	1.12 m	Mudstone, greenish gray, massive
118-5	336.14—336.84 m	0.70 m	Siltstone, grayish brown and greenish gray, massive, discontinuous wavy bedding, few calcareous concretions
118-6	336.84—337.18 m	0.34 m	Mudstone, reddish brown, massive, with greenish gray reduced spots at base
118-7	337.18—337.58 m	0.40 m	Mudstone, greenish gray, massive, with few grayish brown oxidized spots and calcareous concretions

118-8	337.58—337.90 m	0.32 m	Mudstone, reddish brown, massive
119-1	337.90—339.76 m	1.86 m	Mudstone, reddish brown, massive
119-2	339.76—340.07 m	0.31 m	Silty mudstone, greenish gray, massive
119-3	340.07—340.17 m	0.10 m	Mudstone, reddish brown, massive, with irregular greenish gray mudstone laminae and spots
120-1	340.17—340.37 m	0.20 m	Mudstone, reddish brown, massive
120-2	340.37—340.58 m	0.21 m	Silty mudstone, reddish brown, massive
120-3	340.58—341.17 m	0.59 m	Mudstone, reddish brown, massive
120-4	341.17—341.31 m	0.14 m	Silty mudstone, reddish brown and greenish gray, massive
120-5	341.31—341.67 m	0.36 m	Mudstone, reddish brown, massive
120-6	341.67—342.52 m	0.85 m	Silty mudstone, grayish brown, massive, few calcareous concretions
120-7	341.67—342.93 m	1.26 m	Siltstone, light greenish-gray, massive bedding, discontinuous ripple bedding locally, with mud content increasing in the middle and upper parts, and also bioturbation structures
120-8	342.93—347.37 m	4.44 m	Siltstone, very light gray, with parallel bedding and trough cross bedding in the lower part, and discontinuous ripple bedding in the upper part, erosion surface at 343.6 m and 346.5 m
121-1	348.22—348.68 m	0.46 m	Siltstone, light gray, with parallel bedding
122-1	348.98—349.01 m	0.03 m	Conglomerate, medium grained, silty, greenish-gray, with massive structure, mainly mud gravel, sandstone gravel and calcareous concretions, subangular-subround
122-2	349.01—349.71 m	0.70 m	Mudstone, reddish brown, massive, with few calcareous concretions and sandstone locally
122-3	349.71—349.98 m	0.27 m	Siltstone, light gray and grayish brown, deformation bedding, with lots of irregular mudstone strips
122-4	349.98—350.52 m	0.54 m	Siltstone, light gray, with discontinuous wavy bedding and small trough cross bedding, erosion surface at the base
122-5	350.52—352.70 m	2.18 m	Siltstone, light gray, discontinuous wavy cross bedding
122-6	352.70—352.92 m	0.22 m	Siltstone, light gray, deformation bedding, with few mudstone strips
123-1	353.03—354.98 m	1.95 m	Silty mudstone, grayish brown, horizontal bedding, few calcareous concretions
123-2	354.98—355.08 m	0.10 m	Siltstone, grayish brown, deformation bedding, erosion surface and irregular mudstone strips at the base
123-3	355.08—356.67 m	1.59 m	Siltstone, very light gray, with discontinuous ripple bedding and deformation bedding at the top, small trough cross bedding in the middle and lower parts, erosion at the base
123-4	356.67—357.01 m	0.34 m	Siltstone, light gray, deformation bedding, 3 cm-thick gray mudstone laminae at the top
123-5	357.01—357.83 m	0.82 m	Siltstone, very light gray, with small wave-generated cross bedding
123-6	357.83—358.56 m	0.73 m	Siltstone, very light gray, with slump structures, roll-up structure locally, irregular mudstone laminae

123-7	358.56—359.03 m	0.47 m	Siltstone, light gray, discontinuous ripple bedding and parallel bedding, with argillaceous siltstone laminae and irregular mudstone laminae locally
123-8	359.03—359.08 m	0.05 m	Mudstone, medium dark gray, with horizontal bedding
123-9	359.08—359.47 m	0.39 m	Siltstone, medium gray, discontinuous horizontal ripple bedding, deformation structure locally, 8 mm-thick mudstone laminae at 359.23 m
123-10	359.47—359.70 m	0.23 m	Siltstone, medium gray, massive bedding, 3—4 cm-thick medium argillaceous conglomerate at the base
123-11	359.70—360.83 m	1.13 m	Siltstone, light gray, with medium gray mudstone laminae, ripple bedding, medium dark gray mudstone laminae at 359.98—360.03 m
123-12	360.83—361.27 m	0.44 m	Siltstone, light gray, with wavy bedding
124-1	361.27—361.40 m	0.13 m	Conglomerate, light gray, wavy bedding
124-2	361.40—362.90 m	1.50 m	Siltstone, light gray, wavy cross bedding extending upward, with 5 cm calcareous siltstone at 361.57 m
124-3	362.90—363.01 m	0.11 m	Siltstone, light gray, parallel bedding, with few charcoals
125-1	363.80—363.95 m	0.15 m	Conglomerate, light gray, massive, erosion surface at the base
125-2	363.95—364.40 m	0.45 m	Siltstone, light gray, parallel bedding, with charcoals
125-3	364.40—364.50 m	0.10 m	Siltstone, light gray, parallel
125-4	364.50—364.85 m	0.35 m	Siltstone, light gray, parallel bedding, with calcspar
125-5	364.85—364.95 m	0.10 m	Conglomerate, medium-grained, silty, medium light gray, with sand gravel and mud gravel, subround-subangular, massive bedding, erosion surface at the base
125-6	364.95—365.78 m	0.83 m	Siltstone, gray, parallel bedding
125-7	365.78—366.20 m	0.42 m	Conglomerate, fine-grained, calcareous, medium light gray, with mud gravels, parallel bedding, erosion surface at the base, calcareous bio-shell and charcoal occasionally
125-8	366.20—366.29 m	0.09 m	Conglomerate, light gray, with calcareous concretions, massive
125-9	366.29—366.34 m	0.05 m	Siltstone, light gray, few mudstone gravels, with small trough cross bedding
125-10	366.34—366.67 m	0.33 m	Silty mudstone, light olive gray, discontinuous wavy bedding, with few siltstone concretions
126-1	368.64—369.14 m	0.50 m	Mudstone, light olive gray, ripple bedding, with silty mudstone at 368.69 m
126-2	369.14—369.26 m	0.12 m	Silty mudstone, light olive gray, wavy bedding
127-1	369.56—370.16 m	0.60 m	Muddy siltstone, light olive gray, discontinuous wavy bedding, with lots of calcareous concretions
127-2	370.16—372.16 m	2.00 m	Siltstone, light gray, deformation bedding, with irregular mudstone strips and gravels
127-3	372.16—372.56 m	0.40 m	Siltstone, light gray, wavy cross bedding
128-1	372.56—372.66 m	0.10 m	Conglomerate, with silty and calcareous concretions, light gray, erosion surface at the base
128-2	372.66—372.76 m	0.10 m	Siltstone, light gray, with small trough cross bedding

128-3	372.76—374.61 m	1.85 m	Siltstone, yellowish-gray, trough cross bedding declining upward, calcareous concretions occasionally
128-4	374.61—375.06 m	0.45 m	Siltstone, light greenish-gray, with mud gravels, massive bedding, weak erosion surface at the base
128-5	375.06—375.66 m	0.60 m	Siltstone, light gray, parallel bedding
128-6	375.66—376.06 m	0.40 m	Siltstone, light gray, small trough cross bedding
128-7	376.06—377.09 m	1.03 m	Siltstone, yellowish gray, discontinuous wavy bedding
129-1	378.83—379.68 m	0.85 m	Siltstone, yellowish gray, with wave-ripple cross bedding
129-2	379.68—379.88 m	0.20 m	Siltstone, massive, erosion surface at the base, with few calcareous concretions
129-3	379.88—379.93 m	0.05 m	Mudstone, light brownish gray, deformation bedding, with few siltstone strips
129-4	379.93—380.29 m	0.36 m	Siltstone, yellowish gray, deformation bedding
130-1	380.96—383.21 m	2.25 m	Siltstone, light gray, trough cross bedding, with few mudstone strips
130-2	383.21—383.31 m	0.10 m	Mudstone, olive-gray, massive bedding, few ostracoda shell crumb and fine mineral grains
130-3	383.31—383.56 m	0.25 m	Siltstone, light greenish gray, deformation bedding, mudstone laminae at the top
130-4	383.56—384.06 m	0.50 m	Silty mudstone, grayish red, massive, with calcareous concretions
130-5	384.06—384.37 m	0.31 m	Siltstone, light gray, massive, deformation bedding in locally, with mudstone strips
131-1	385.68—385.83 m	0.15 m	Siltstone, light gray, massive
132-1	385.83—386.23 m	0.40 m	Siltstone, light gray, trough cross bedding
132-2	386.23—386.78 m	0.55 m	Mudstone, grayish red, massive
132-3	386.78—386.88 m	0.10 m	Mudstone, light olive gray, massive
132-4	386.88—387.18 m	0.30 m	Silty mudstone, brownish gray, with trace fossils and sandstone laminae
132-5	387.18—387.33 m	0.15 m	Mudstone, light olive gray, massive, with calcareous concretions
132-6	387.33—388.03 m	0.70 m	Mudstone, brownish gray, massive, with calcareous concretions, siltstone strips locally
132-7	388.03—388.23 m	0.20 m	Mudstone, grayish red, massive
132-8	388.23—388.43 m	0.20 m	Silty mudstone, greenish gray, massive, with bioclasts
132-9	388.43—389.83 m	1.40 m	Silty mudstone, light brown, deformation bedding, light gray siltstone strips
132-10	389.83—390.14 m	0.31 m	Siltstone, light gray, deformation bedding, with irregular mudstone strips
132-11	390.18—390.95 m	0.77 m	Muddy siltstone, light gray, deformation bedding, with light gray siltstone strips
132-12	390.95—391.43 m	0.48 m	Muddy siltstone, light gray, massive
132-13	391.43—392.58 m	1.15 m	Muddy siltstone, light gray, deformation bedding, with light gray siltstone strips
132-14	392.58—393.23 m	0.65 m	Muddy siltstone, light gray, deformation bedding, with light gray muddy siltstone strips
132-15	393.23—393.63 m	0.40 m	Muddy siltstone, light gray, deformation bedding, with light gray siltstone strips

133-1	393.82—394.72 m	0.90 m	Siltstone, light gray, deformation bedding, with light gray muddy siltstone strips
133-2	394.72—394.82 m	0.10 m	Silty mudstone, light gray, massive, siltstone strips locally
133-3	394.82—395.62 m	0.80 m	Siltstone, light gray, small trough cross bedding, with light gray silty mudstone strips
133-4	395.62—395.92 m	0.30 m	Conglomerate, silty and argillaceous, medium light gray, fine grained, massive bedding, with large number of bivalve fossils
133-5	395.92—395.93 m	0.01 m	Black seam
133-6	395.93—396.32 m	0.39 m	Siltstone, medium light gray, deformation bedding, with a 2cm-thick mud gravel layer at the top
133-7	396.32—396.34 m	0.02 m	Tuff, medium light gray
133-8	396.34—396.46 m	0.12 m	Conglomerate, silty and muddy, massive, with calcareous concretions
133-9	396.46—397.14 m	0.68 m	Siltstone, light gray, small trough cross bedding, with trace fossils and few calcareous concretions
133-10	397.14—397.32 m	0.18 m	Tuff, medium light gray
133-11	397.32—397.72 m	0.40 m	Siltstone, light gray, wavy bedding, with trace fossils and light gray brown mudstone strips
133-12	397.78—398.02 m	0.24 m	Mudstone, light brown, massive, with siltstone strips
133-13	398.02—400.05 m	2.03 m	Siltstone, light gray, small trough cross bedding, and deformation bedding locally, erosion surface at the base
134-1	400.05—400.69 m	0.64 m	Siltstone, light gray, small trough cross bedding, and deformation bedding locally
134-2	400.69—400.90 m	0.21 m	Calcareous siltstone, light gray, small trough cross bedding, with few irregular mudstone strips
134-3	400.90—401.25 m	0.35 m	Siltstone, light gray, small trough cross bedding
134-4	401.25—401.63 m	0.38 m	Siltstone, light gray, deformation bedding, convolution bedding locally, with few irregular mudstone strips
134-5	401.63—401.95 m	0.32 m	Siltstone, light gray, small trough cross bedding
134-6	401.95—402.15 m	0.20 m	Siltstone, light gray, deformation bedding
134-7	402.15—402.56 m	0.41 m	Fine-grained conglomerate, with calcareous concretions and mud gravels, greenish-gray, intercalated with light gray calcareous siltstone, 3 cycles containing conglomerate in the lower and calcareous siltstone in the upper, fining generally upward, massive bedding in conglomerate, small trough cross bedding in calcareous siltstone
134-8	402.56—402.81 m	0.25 m	Mudstone, greenish gray, massive, ostracoda occasionally
134-9	402.81—403.33 m	0.52 m	Mudstone, brownish gray, massive, calcareous concretions occasionally
134-10	403.23—404.95 m	1.72 m	Mudstone, greenish gray, massive, calcareous concretions occasionally
134-11	404.95—405.30 m	0.35 m	Mudstone, brownish gray, massive
134-12	405.30—405.65 m	0.35 m	Mudstone, greenish gray, massive, with fossil fragments

Section 5 SK-1 Core Description and Core Photographs

134-13	405.65—406.70 m	1.05 m	Mudstone, brownish gray, massive, with trace fossils
134-14	406.70—406.95 m	0.25 m	Mudstone, brownish gray, massive
135-1	407.71—408.11 m	0.40 m	Mudstone, greenish gray, massive, calcareous concretions occasionally
135-2	408.11—408.61 m	0.50 m	Mudstone, brownish gray, massive, calcareous concretions occasionally
135-3	408.61—409.11 m	0.50 m	Mudstone, greenish gray, massive, calcareous concretions occasionally
135-4	409.11—410.21 m	1.10 m	Mudstone, brownish-gray, massive bedding, with fractures filled with calcspar, greenish-gray reduced spots locally
135-5	410.21—411.19 m	0.98 m	Silty mudstone, brownish gray, massive, calcareous concretions locally
135-6	411.19—411.29 m	0.10 m	Siltstone, light gray, with small trough cross bedding
135-7	411.29—411.66 m	0.37 m	Silty mudstone, brownish gray, massive, calcareous concretions
135-8	411.66—414.51 m	2.85 m	Siltstone, light gray, parallel bedding and small trough cross bedding
135-9	414.51—414.53 m	0.02 m	Fine-grained conglomerate, with calcareous concretions and mud gravels, light gray, massive
135-10	414.53—414.61 m	0.08 m	Siltstone, light gray, massive
136-1	414.61—414.79 m	0.18 m	Siltstone, light gray, sand ripple bedding
136-2	414.79—415.51 m	0.72 m	Mudstone, dark greenish gray, massive
136-3	415.51—416.01 m	0.50 m	Mudstone, brown, massive
136-4	416.01—416.31 m	0.30 m	Mudstone, dark greenish gray, massive, with large number of fossil fragments
137-1	416.91—417.13 m	0.22 m	Mudstone, dark greenish gray, massive
137-2	417.13—417.36 m	0.23 m	Mudstone, grayish brown, massive, with few calcareous concretions
137-3	417.36—417.51 m	0.15 m	Siltstone, light gray, massive, erosion surface at the base
137-4	417.51—417.69 m	0.18 m	Mudstone, grayish brown, massive, with calcareous concretions
137-5	417.69—417.84 m	0.15 m	Muddy siltstone, massive, with calcareous concretions
137-6	417.84—418.21 m	0.37 m	Silty mudstone, brownish-gray, massive bedding, ripple bedding locally, deformation bedding at the base, argillaceous siltstone laminae locally, large number of calcareous concretions
137-7	418.21—418.31 m	0.10 m	Siltstone, light gray, wavy bedding, weak erosion surface at the base
137-8	418.31—418.48 m	0.17 m	Mudstone, brownish gray, wavy ripple bedding, with calcareous concretions
137-9	418.58—418.63 m	0.05 m	Siltstone, brownish gray, wavy ripple bedding, weak erosion surface and calcareous concretions at the base
137-10	418.63—419.01 m	0.38 m	Mudstone, grayish brown, massive, cross bedding at the top, muddy siltstone strips and calcareous concretions locally
137-11	419.01—419.28 m	0.27 m	Siltstone, brownish gray, wavy ripple bedding, convolution bedding at the base, mudstone strips and calcareous concretions locally
137-12	419.28—419.36 m	0.08 m	Fine-grained conglomerate, small trough cross bedding, erosion surface at the base
137-13	419.36—419.86 m	0.50 m	Siltstone, brownish gray, wavy ripple bedding, foreset laminae, with lots of mudstone strips, sandstone strips at the top

137-14	419.86—419.98 m	0.12 m	Mudstone, grayish brown, wavy ripple bedding
137-15	419.98—420.23 m	0.25 m	Siltstone, light gray, oblique bedding, erosion surface at the base
137-16	420.23—420.55 m	0.32 m	Mudstone, medium gray, massive bedding, deformation bedding at the base, mud gravels in the middle, irregularly and poorly sorted, argillaceous siltstone laminae locally
137-17	420.55—420.65 m	0.10 m	Siltstone, light gray, wavy ripple bedding
137-18	420.65—420.93 m	0.28 m	Mudstone, dark gray, horizontal bedding, muddy siltstone and siltstone strips locally
137-19	420.93—421.22 m	0.29 m	Siltstone, gray, wavy cross bedding, erosion surface at the base, with a mud gravels strip in the middle and lower part
137-20	421.22—421.29 m	0.07 m	Mudstone, dark gray, horizontal bedding
137-21	421.29—421.65 m	0.36 m	Fine-grained sandstone, light gray, massive, erosion surface at the base
137-22	421.65—421.95 m	0.30 m	Sandstone, fine-grained, light gray, trough cross bedding increasing upward, local mudstone laminae
137-23	421.95—422.12 m	0.17 m	Siltstone, medium gray, interbedded with medium dark gray mudstone, wavy bedding, mud gravels locally
137-24	422.12—422.34 m	0.22 m	Mudstone, gray, horizontal bedding
137-25	422.34—422.68 m	0.34 m	Fine-grained sandstone, massive, erosion surface at the base
137-26	422.68—423.45 m	0.77 m	Siltstone, medium dark gray, interbedded with dark gray mudstone, sand ripple bedding with scale increasing and mudstone decreasing upward, 2 cm thick fine gray sandstone laminae at 423.03—423.05 m
137-27	423.45—423.55 m	0.10 m	Mudstone, gray, horizontal bedding
137-28	423.55—423.93 m	0.38 m	Sandstone, fine-grained with gravels, medium gray, small trough cross bedding, and deformation bedding at the top, with larger scale and less mudstone upward
137-29	423.93—424.01 m	0.08 m	Silty mudstone, gray, massive, weak erosion surface and mud gravels at the base
137-30	424.01—424.31 m	0.30 m	Fine-grained sandstone, gray, interbedded with dark gray mudstone, trough cross bedding, erosion surface at the base
137-31	424.31—424.42 m	0.11 m	Mudstone, dark gray, wavy ripple bedding
137-32	424.42—424.87 m	0.45 m	Fine-grained sandstone, gray, interbedded with dark gray mudstone, trough cross bedding, erosion surface at the base
137-33	424.87—425.21 m	0.34 m	Mudstone, gray, horizontal bedding
137-34	425.21—425.35 m	0.14 m	Fine-grained sandstone, gray, trough cross bedding
138-1	425.44—425.67 m	0.23 m	Fine-grained sandstone, light gray, trough cross bedding, erosion surface at the base
138-2	425.44—425.67 m	0.23 m	Mudstone, dark gray, horizontal bedding
138-3	425.97—426.24 m	0.27 m	Fine-grained sandstone, light gray, trough cross bedding, erosion surface and 3 cm-thick conglomerate at the base

138-4	426.24—427.02 m	0.78 m	Siltstone, medium dark gray, trough cross bedding, with larger scale and less mudstone upward, dark gray mudstone developed at 426.30—426.35 m
138-5	427.02—428.11 m	1.09 m	Siltstone, medium light gray, with mud gravels, deformation bedding, cross bedding locally, mud gravels and mudstone laminae
138-6	428.11—428.19 m	0.08 m	Siltstone, medium gray, small wavy cross bedding, erosion surface at the base and 2 cm thick charcoal-bearing gritstone, shell fossil
138-7	428.19—428.25 m	0.06 m	Mudstone, medium gray, horizontal bedding, fill structure at the base, carbonized plant fossil
138-8	428.25—428.40 m	0.15 m	Siltstone, light gray, massive bedding, erosion surface at the base, laminae consisting of fine sandstone, gritstone and gravels in the middle, large bivalve fossil and 2 cm thick conglomerate with calcareous fossil fragments at the base
138-9	428.40—428.67 m	0.27 m	Siltstone, light gray, interbedded with medium dark gray mudstone, deformation bedding, roll-up bedding locally, 4 cm thick medium dark gray mudstone at the base
138-10	428.67—428.96 m	0.29 m	Polymictic conglomerate, medium gray, with sandstone, mud and calcareous concretionary gravels, size of 2 mm—2 cm or 1 cm averagely, poorly sorted, subangular-subround, massive bedding
138-11	428.96—429.62 m	0.66 m	Mudstone, greenish-gray, silty with fossils, unidirectional oblique bedding, small cross bedding locally, erosion surface at the base, large number of snail and ostracoda fossils, especially in certain layers, 3 cm thick mud gravels at the base, calcareous
138-12	429.62—430.14 m	0.52 m	Mudstone, dark greenish-gray, massive bedding, few calcareous concretions, grayish-brown secondary oxidized spots at the base, calcareous
138-13	430.14—430.82 m	0.68 m	Mudstone, grayish brown, massive, with calcareous concretions
138-14	430.82—431.14 m	0.32 m	Silty mudstone, brownish gray, massive, with few calcareous concretions and siltstone strips
138-15	431.14—431.44 m	0.30 m	Siltstone, light gray, deformation bedding, with large number of calcareous concretions, few mud gravels and trace fossils at the base
138-16	431.44—432.01 m	0.57 m	Silty mudstone, grayish-brown, massive bedding, with slump structure locally, few calcareous concretions, irregular siltstone laminae and biogliph
138-17	432.01—432.18 m	0.17 m	Siltstone, brownish-gray, small trough cross bedding increasing upward, erosion surface at the base, few calcareous concretions and biogliph
138-18	432.18—433.29 m	1.11 m	Muddy siltstone, grayish-brown, deformation bedding, large number of mudstone laminae and siltstone laminae, abundant siltstone and trace fossils at 432.34—432.44 m
138-19	433.29—433.93 m	0.64 m	Siltstone, medium light gray, small trough cross bedding, beaded mud gravels locally, fine sandstone laminae in the middle part
139-1	433.93—434.80 m	0.87 m	Siltstone, light gray, small trough cross bedding, mudstone laminae and few trace fossils locally
139-2	434.80—435.03 m	0.23 m	Sandstone, fine-grained, light gray, small trough cross bedding, mud gravels in the lower part along the horizon surface

139-3	435.03—435.53 m	0.50 m	Siltstone, light gray, deformation bedding in the upper part, wavy bedding in the middle part, small trough cross bedding in the lower part, erosion surface and fine sandstone ball at the base
139-4	435.53—435.61 m	0.08 m	Mudstone, dark gray, horizontal bedding
139-5	435.61—435.79 m	0.18 m	Fine-grained sandstone, gray, small trough cross bedding
139-6	435.79—436.05 m	0.26 m	Mudstone, conglomeratic, medium gray, massive bedding, creep deformation structure (due to intruding of fine grained sandstone into mudstone) at the top, large bivalve fossils, fine sandstone ball at the base
139-7	436.05—436.63 m	0.58 m	Siltstone, gray, small trough cross bedding, mudstone strips locally
139-8	436.63—437.77 m	1.14 m	Fine-grained sandstone, gray, small trough cross bedding
139-9	437.77—438.10 m	0.33 m	Siltstone, light gray, trough cross bedding
139-10	438.10—438.70 m	0.60 m	Fine-grained sandstone, light gray, trough cross bedding
139-11	438.70—440.03 m	1.33 m	Sandstone, light gray, trough cross bedding, erosion surface bedding, above which there are 5 cm-thick medium-grained sandstone
139-12	440.03—440.33 m	0.30 m	Sandstone, fine-grained, medium light gray, trough cross bedding, bivalve fossil and 2 cm thick gritstone with gravels at the base, the top, large bivalve fossils. fine sandstone ball at the base
139-13	440.33—440.66 m	0.33 m	Sandstone, medium-grained, medium light gray, trough cross bedding, erosion surface and 1 cm thick medium sandstone with gravels at the base, one with gravels at the base, the top, large bivalve fossils, fine sandstone ball at the base
139-14	440.66—441.23 m	0.57 m	Mudstone, greenish gray, massive
140-1	441.75—442.79 m	1.04 m	Mudstone, grayish brown, massive, with some calcareous concretions
140-2	442.79—443.25 m	0.46 m	Silty mudstone, brownish gray, wavy ripple bedding, deformation bedding locally, with siltstone strips and balls
140-3	443.25—444.09 m	0.84 m	Siltstone, light gray, trough cross bedding, erosion surface at the base, sandstone balls and trace fossils
140-4	444.09—444.38 m	0.29 m	Silty mudstone, brownish gray, deformation bedding and wavy ripple bedding, with siltstone strips
140-5	444.38—444.65 m	0.27 m	Mudstone, brownish gray, interbedded with siltstone, deformation bedding and graded bedding, erosion surface at the base
140-6	444.65—444.75 m	0.10 m	Siltstone, gray, wavy ripple bedding
141-1	445.03—445.08 m	0.05 m	Siltstone, gray, wavy ripple bedding
141-2	445.08—445.36 m	0.28 m	Mudstone, brownish gray, massive, with few calcareous concretions and trace fossils
142-1	445.38—446.38 m	1.00 m	Mudstone, brownish gray, massive, with few calcareous concretions
142-2	446.38—446.62 m	0.24 m	Sandstone, gray, small trough cross bedding, erosion surface at the base
142-3	446.62—446.88 m	0.26 m	Silty mudstone, bearing ostracoda, brownish gray, horizontal bedding

Section 5 SK-1 Core Description and Core Photographs

143-1	449.21—450.88 m	1.67 m	Mudstone, brownish gray, massive, with few calcareous concretions
143-2	450.88—451.21 m	0.33 m	Muddy siltstone, brownish-gray, massive bedding, with slump structure locally, mud gravels in the lower part and mudstone laminae in the upper part
143-3	451.21—451.33 m	0.12 m	Mudstone, brownish gray, massive
143-4	451.33—451.45 m	0.12 m	Silty mudstone, brownish gray, massive, with calcareous concretions
143-5	451.45—451.55 m	0.10 m	Mudstone, brownish gray, massive
143-6	451.55—451.93 m	0.38 m	Muddy siltstone, brownish gray, massive, with few calcareous concretions
143-7	451.93—453.21 m	1.28 m	Siltstone, medium gray, trough cross bedding, with deformation bedding and horizontal ripple bedding locally, few calcareous concretions. mudstone laminae in the upper part
143-8	453.21—454.93 m	1.72 m	Muddy siltstone, dark gray, interbedded with light gray siltstone, trough cross bedding
143-9	454.93—455.14 m	0.21 m	Muddy siltstone, dark gray, interbedded with light gray siltstone, wavy bedding
143-10	455.14—455.59 m	0.45 m	Muddy siltstone, dark gray, interbedded with light gray siltstone, trough cross bedding
144-1	455.59—456.62 m	1.03 m	Siltstone, medium gray, trough cross bedding, local bedding deformed and broken due to compression, medium dark gray and medium gray rhythmic beddings in fine grained layers, few calcareous concretions, mudstone laminae in the upper part
144-2	456.62—457.98 m	1.36 m	Fine-grained sandstone, medium gray, trough cross bedding, dark or light rhythmic beddings in fine grained layers. rhythmic beddings in fine grained layers, few calcareous concretions, mudstone laminae in the upper part
144-3	457.95—458.09 m	0.14 m	Fine-grained sandstone, gray, convolution bedding and trough cross bedding
144-4	458.09—458.27 m	0.18 m	Sandstone, gray, trough cross bedding
144-5	458.27—458.30 m	0.03 m	Polymictic conglomerate, medium gray, poorly sorted quartz, mud and calcareous concretionary gravels, with size of 0.2—1 cm, massive bedding, erosion surface at the base
144-6	458.30—458.92 m	0.62 m	Sandstone, gray, trough cross bedding
144-7	458.92—459.64 m	0.72 m	Polymictic conglomerate, medium gray, poorly sorted quartz, mud and calcareous concretionary gravels, with size of 0.2—3 cm, massive bedding, erosion surface at the base, bivalve fossil developed
144-8	459.64—460.03 m	0.39 m	Sandstone, gray, trough cross bedding
144-9	460.03—460.37 m	0.34 m	Polymictic conglomerate, medium gray, poorly sorted quartz, mud and calcareous concretionary gravels, with size of 0.2—2.5 cm, massive bedding, erosion surface at the base, bivathe base, bivalve fossil developed
144-10	460.37—460.52 m	0.15 m	Sandstone, gray, trough cross bedding
144-11	460.52—460.79 m	0.27 m	Polymictic conglomerate, medium gray, poorly sorted quartz, mud and calcareous concretionary gravels, with size of 0.2—2.5 cm, massive bedding, erosion surface at the base, bivathe base, bivalve fossil developed

145-1	461.17—462.22 m	1.05 m	Polymictic conglomerate, medium gray, with sandstone, mud and calcareous concretionary gravels, size of 2—2 cm or 1 cm averagely, poorly sorted, subangular-subround, massive bedding
145-2	461.22—461.36 m	0.14 m	Sandstone, medium grained, medium gray, interbedded with gritstone, wavy bedding, 5 cm thick gritstone in the upper 0 cm, erosion surface at the base, bivathe base, bivalve fossil developed, mas sive bedding, erosion surface at the base
145-3	461.36—461.59 m	0.23 m	Sandstone, medium grained, medium gray, interbedded with gritstone, mass bedding, 5cm thick gritstone in the upper 0 cm, erosion surface at the base, bivathe base, bivalve fossil developed, mas sive bedding, erosion surface at the base
145-4	461.59—462.73 m	1.14 m	Sandstone, medium-grained, medium gray, wavy bedding, a wedge-shaped conglomerate layer (3—10 cm) at 462.27 m
146-1	462.73—463.67 m	0.94 m	Mudstone, dark greenish-gray, massive bedding, few calcareous concretions, 3 cm thick siltstone laminae with parallel bedding at 463.03—463.06 m, conglome rate layer (3—10 cm) at 462.27 m
146-2	463.67—464.56 m	0.89 m	Mudstone, brownish gray, massive, with calcareous concretions
146-3	464.56—465.82 m	1.26 m	Mudstone, dark greenish gray, massive, with few calcareous concretions
147-1	465.82—466.54 m	0.72 m	Silty mudstone, light gray, wavy ripple bedding
147-2	466.54—467.10 m	0.56 m	Mudstone, greenish gray, wavy ripple bedding, calcareous concretions
147-3	467.10—467.82 m	0.72 m	Silty mudstone, light gray, wavy ripple bedding
147-4	467.82—468.32 m	0.50 m	Mudstone, greenish gray, wavy ripple bedding, calcareous concretions
149-1	468.32—468.45 m	0.13 m	Silty mudstone, light gray, massive, with calcareous concretions
149-2	468.45—468.79 m	0.34 m	Muddy siltstone, light gray, wavy ripple bedding, with calcareous concretions
149-3	468.79—469.19 m	0.40 m	Mudstone, greenish graym, massive
149-4	469.19—469.59 m	0.40 m	Siltstone, gray, trough cross bedding, erosion surface at the base
149-5	469.59—472.28 m	2.69 m	Mudstone, greenish-gray, massive bedding, large calcareous concretions at 470.75—470.92 m. with parallel bedding at 463.03—463.06 m, conglome rate layer (3—10 cm) at 462.27 m
150-1	472.28—472.49 m	0.21 m	Mudstone, greenish gray, massive, with few fossil fragments
150-2	472.49—473.47 m	0.98 m	Silty mudstone, greenish gray, massive, with fossil fragments and calcareous concretions
150-3	473.47—474.33 m	0.86 m	Mudstone, greenish gray, massive, with few calcareous concretions
150-4	474.33—474.88 m	0.55 m	Silty mudstone, greenish gray, massive, with large number fossil fragments, calcareous concretions and few ostracoda fossils
150-5	474.88—478.77 m	3.89 m	Mudstone, greenish gray, massive, with calcareous concretions
150-6	478.77—478.94 m	0.17 m	Mudstone, grayish brown, massive, with few calcareous concretions
150-7	478.94—479.21 m	0.27 m	Mudstone, greenish gray, massive, with few calcareous concretions
151-1	479.21—479.48 m	0.27 m	Muddy siltstone, olive gray, massive

151-2	479.48—479.98 m	0.50 m	Mudstone, greenish gray, massive, with few calcareous concretions
151-3	479.98—480.24 m	0.26 m	Silty mudstone, greenish gray, massive, with few calcareous concretions
151-4	480.24—480.45 m	0.21 m	Mudstone, grayish-brown, massive bedding, few large calcareous concretions (up to 3—4 cm diameter)
151-5	480.45—480.57 m	0.12 m	Mudstone, greenish gray, massive
151-6	480.57—480.83 m	0.26 m	Mudstone, grayish brown, massive, with few calcareous concretions
151-7	480.83—481.13 m	0.30 m	Mudstone, greenish gray, massive, with few calcareous concretions
153-1	481.13—483.59 m	2.46 m	Mudstone, grayish-green, massive bedding, calcareous concretions, light gray mudstone laminae at 481.23—481.26 m, 481.46—481.73 m and 482.33—482.41 m, siltstone laminae with wave-generated sand ripple bedding and horizontal ripple bedding
154-1	483.59—484.59 m	1.00 m	Mudstone, dark greenish gray, massive, with few calcareous concretions
154-2	484.59—484.75 m	0.16 m	Silty mudstone, greenish gray, massive, with few fossil fragments
154-3	484.75—485.09 m	0.34 m	Mudstone, dark greenish gray, massive, with few fossil fragments
154-4	485.09—485.97 m	0.88 m	Mudstone, clastizoic, greenish-gray, massive bedding, large number of fossil fragments, bivalve and gastropods fossils, calcareous concretions, plenty of mirror scratches
154-5	485.97—486.35 m	0.38 m	Mudstone, greenish gray, massive, with few calcareous concretions and fossil fragments
154-6	486.35—487.29 m	0.94 m	Silty mudstone, light olive gray, massive, with few calcareous concretions
154-7	487.29—487.84 m	0.55 m	Muddy siltstone, light olive gray, trough cross bedding, deformation bedding and weak erosion surface at the base, calcareous concretions and mud gravels, few siltstone laminae
154-8	487.84—488.12 m	0.28 m	Silty mudstone, light olive gray, ripple bedding, calcareous concretions
154-9	488.12—488.25 m	0.13 m	Mudstone, olive gray, massive bedding
154-10	488.25—488.42 m	0.17 m	Siltstone, medium gray, liquefied deformation bedding in the upper and middle parts, wavy bedding in the lower part, intensive bioturbation, mudstone laminae locally
154-11	488.42—488.68 m	0.26 m	Siltstone, medium gray and greenish-gray, trough cross bedding, intensive bioturbation, calcareous concretions and mudstone concretions occasionally
154-12	488.68—489.19 m	0.51 m	Muddy siltstone, greenish-gray, deformation bedding, discontinuous ripple bedding, mudstone laminae, and few calcareous concretions
154-13	489.19—490.15 m	0.96 m	Sandstone, fine-grained with mud gravels, medium light gray, deformation bedding, wavy bedding locally, cutting structure at 489.89 m
154-14	490.15—490.33 m	0.18 m	Sandstone, fine-grained, light olive gray, trough cross bedding, biogliph
154-15	490.33—490.77 m	0.44 m	Sandstone, medium-grained, medium light gray, trough cross bedding, erosion surface at the base, medium light gray fine grained sandstone laminae at 490.46—490.51 m
154-16	490.77—490.96 m	0.19 m	Sandstone, fine-grained, light olive gray, trough cross bedding, biogliph

154-17	490.96—491.33 m	0.37 m	Sandstone, medium-grained, medium light gray, trough cross bedding, erosion surface at the base
155-1	491.33—491.74 m	0.41 m	Sandstone, medium-grained, medium light gray, swash cross bedding, mud gravels on erosion surface at the base, 1cm thick medium dark gray argillaceous siltstone laminae at 491.53 m
155-2	491.74—491.80 m	0.06 m	Sandstone, fine-grained, medium light gray, trough cross bedding
155-3	491.80—491.88 m	0.08 m	Sandstone, medium grained, medium light gray, thinly bedded, swash cross bedding, erosion surface at the base, with conglomerates on the erosion surface, mud gravels and abundant fossil fragments
155-4	491.88—491.92 m	0.04 m	Sandstone, fine-grained, medium light gray, wave-generated sand ripple bedding
155-5	491.92—492.67 m	0.75 m	Sandstone, medium-grained, medium light gray, trough cross bedding at the top, swash cross beddings in the middle and lower parts, much micas and charcoals along the bedding plane
155-6	492.67—492.90 m	0.23 m	Sandstone, medium-grained, light gray, trough cross bedding, erosion surface at the base, much fossil fragments
155-7	492.90—493.89 m	0.99 m	Sandstone, medium-grained, medium light gray, swash cross bedding, 4 cm fine grained sandstone with trough cross bedding at the base, erosion surfaces at 493.03 m, 493.26 m, 493.58 m, 493.71 m and 493.92 m, gravels deposited on the erosion surfaces at 493.58 m, 493.71 m and 493.92 m, much charcoals on the bedding surface, mussel fossils at 493.25 m and the base
155-8	493.89—493.92 m	0.03 m	Sandstone (wedge-shaped bedding), medium grained, medium gray, interbedded with siltstone (deformation bedding), erosion surface at the base, much charcoals on the bedding surface
155-9	493.92—494.10 m	0.18 m	Mudstone, medium dark gray, horizontal bedding, 1cm-thick siltstone laminae in the middle part, ripple bedding, erosion at the base
155-10	494.10—494.26 m	0.16 m	Sandstone, fine-grained, medium gray, trough cross bedding, discontinuous horizontal ripple bedding, deformation bedding locally, filling at the base and in contact with the lower conglomerate, plentiful mussel fossils, mud gravels and charcoals
155-11	494.26—494.39 m	0.13 m	Conglomerate, polymictic, medium gray, deformation bedding in the matrix, mainly sand, mud and calcareous concretionary gravels, usually 0.5 cm diameter, erosion surface at the base, much mussel fossils
155-12	494.39—494.78 m	0.39 m	Sandstone, fine-grained, medium light gray, oblique bedding
155-13	494.78—495.30 m	0.52 m	Sandstone, medium-grained, medium light gray, moderate wedge-shaped cross bedding, trough cross bedding locally, few mud gravels
155-14	495.30—496.63 m	1.33 m	Sandstone, medium-grained, medium light gray, wedge-shaped cross bedding, trough cross bedding locally, erosion at the base, carbonized plant fragments at 497.11—497.15 m
155-15	496.63—496.88 m	0.25 m	Sandstone, fine-grained, light gray, wedge-shaped cross bedding, scour-fill structure at the base, greenish-gray argilla and bioclasts in bedding

155-16	496.88—497.19 m	0.31 m	Silty mudstone, greenish-gray, horizontal bedding, fossil fragments and calcareous concretions
155-17	497.19—497.26 m	0.07 m	Mudstone, greenish-gray, massive bedding, fossil fragments and few calcareous concretions
155-18	497.26—497.73 m	0.47 m	Mudstone, brownish-gray, massive bedding, calcareous concretions and fossil fragments
155-19	497.73—498.83 m	1.10 m	Mudstone, olive-gray, massive bedding, calcareous concretions, few ostracoda and snail fossils
156-1	499.20—499.86 m	0.66 m	Silty mudstone, grayish-green, massive bedding, bioturbation structures occasionally, fossil fragments
156-2	499.86—500.54 m	0.68 m	Silty mudstone, olive gray, massive bedding, with calcareous concretions, and vertical irregular fractures filled with calcspar
156-3	500.54—500.86 m	0.32 m	Silty mudstone, greenish-gray, massive bedding, with few fossil fragments and calcareous concretions
156-4	500.86—501.03 m	0.17 m	Mudstone, clastizoic, greenish-gray, ripple bedding, irregular calcspar-filled fractures, mussel, snail and ostracoda fossils
156-5	501.03—501.28 m	0.25 m	Lime mudstone, light olive gray, massive bedding, ostracoda fossils
156-6	501.28—501.51 m	0.23 m	Silty mudstone, greenish-gray, massive bedding, 2 mm gypsum filled in fractures at 504.72 m
156-7	501.51—501.77 m	0.26 m	Muddy siltstone, light olive gray, trough cross bedding, deformation bedding at the base, weak erosion at the base, calcareous concretions and mud gravels, few siltstone laminae
156-8	501.77—501.93 m	0.16 m	Silty mudstone, light olive gray, ripple bedding, calcareous concretions
156-9	501.93—502.02 m	0.09 m	Mudstone, olive-gray, massive bedding
156-10	502.02—502.35 m	0.33 m	Siltstone, medium gray, liquefied deformation bedding in the upper and middle parts, wavy bedding in the lower part, intensive bioturbation, mudstone laminae locally
156-11	502.35—502.94 m	0.59 m	Siltstone, medium gray and greenish-gray, trough cross bedding, intensive bioturbation, calcareous concretions and mudstone concretions occasionally
156-12	502.94—503.44 m	0.50 m	Muddy siltstone, greenish-gray, deformation bedding, discontinuous ripple bedding, mudstone laminae, and few calcareous concretions
156-13	503.44—504.31 m	0.87 m	Sandstone, fine-grained with mud gravels, medium light gray, deformation bedding, wavy bedding locally, cutting structure at 489.89 m
156-14	504.31—504.58 m	0.27 m	Sandstone, fine-grained, light olive gray, trough cross bedding, biogliph
156-15	504.58—504.68 m	0.10 m	Sandstone, medium-grained, medium light gray, trough cross bedding, erosion surface at the base, medium light gray fine grained sandstone laminae at 490.46—490.51 m
156-16	504.68—504.71 m	0.03 m	Sandstone, fine-grained, light olive gray, trough cross bedding, biogliph
156-17	504.71—504.98 m	0.27 m	Sandstone, medium-grained, medium light gray, trough cross bedding, erosion surface at the base

156-18	504.98—505.31 m	0.33 m	Mudstone, brownish gray, massive
156-19	505.31—506.07 m	0.76 m	Silty mudstone, greenish gray, massive, with few calcareous concretions
156-20	506.07—506.70 m	0.63 m	Silty mudstone, olive gray, massive
156-21	506.70—507.20 m	0.50 m	Mudstone, greenish gray, massive, with fossil fragments
157-1	507.20—509.02 m	1.82 m	Mudstone, greenish gray, massive, with calcareous concretions and fossil fragments
157-2	509.02—509.84 m	0.82 m	Mudstone, grayish brown, massive, with few calcareous concretions
157-3	509.84—510.01 m	0.17 m	Muddy siltstone, olive gray, deformation bedding, intensive bioturbation, erosion at the base, siltstone laminae locally
157-4	510.01—510.10 m	0.09 m	Mudstone, greenish gray, deformation bedding, with fossil fragments and few calcareous concretions
158-1	512.72—513.12 m	0.40 m	Silty mudstone, massive, with fossil fragments and calcareous concretions
158-2	513.12—513.62 m	0.50 m	Silty mudstone, olive gray, massive, with fossil fragments and calcareous concretions
158-3	513.62—513.82 m	0.20 m	Silty mudstone, greenish gray, massive, with fossil fragments and calcareous concretions
158-4	513.82—514.92 m	1.10 m	Mudstone, olive gray, massive, with few fossil fragments and calcareous concretions
159-1	514.92—515.49 m	0.57 m	Silty mudstone, greenish gray, massive, with ostracoda and few fossil fragments
159-2	515.49—515.75 m	0.26 m	Silty mudstone, olive gray, massive, with calcareous concretions occasionally
159-3	515.75—515.90 m	0.15 m	Muddy siltstone, olive gray, discontinuous wavy ripple bedding, intensive bioturbation, erosion surface at the base, with siltstone strips
159-4	515.90—516.04 m	0.14 m	Mudstone, greenish gray, horizontal bedding, fossil fragments
159-5	516.04—516.38 m	0.34 m	Mudstone, light brown, horizontal bedding
160-1	516.38—517.14 m	0.76 m	Mudstone, greenish gray, massive, with ostracoda fossils and fossil fragments
160-2	517.14—517.49 m	0.35 m	Silty mudstone, greenish gray, massive, calcareous concretions
160-3	517.49—517.88 m	0.39 m	Mudstone, light brown, massive, calcareous concretions
160-4	517.88—518.27 m	0.39 m	Silty mudstone, greenish gray, massive, calcareous concretions
160-5	518.27—518.36 m	0.09 m	Silty mudstone, light brown, massive, calcareous concretions
160-6	518.36—518.44 m	0.08 m	Silty mudstone, greenish gray, massive
160-7	518.44—518.71 m	0.27 m	Siltstone, greenish gray, massive, with bioturbation constructure, erosion surface at the base
160-8	518.71—518.88 m	0.17 m	Mudstone, light brown, massive
160-9	518.88—519.71 m	0.83 m	Mudstone, greenish gray, massive, calcareous concretion sand fossil fragments
160-10	519.71—520.05 m	0.34 m	Muddy siltstone, light olive gray, discontinuous horizontal ripple bedding, deformation bedding, dominant bioturbation structures, weak erosion at the base, siltstone laminae, sand balls, a mudstone laminae at 519.91—519.95 m
160-11	520.05—521.11 m	1.06 m	Mudstone, greenish gray, massive, with large number of fossil fragments, snail and mussel fossils, calcareous concretions

160-12	521.11—521.32 m	0.21 m	Silty mudstone, greenish gray, massive, fossil fragments occasionally
160-13	521.32—522.81 m	1.49 m	Mudstone, greenish gray, massive, calcareous concretion sand few fossil fragments
160-14	522.81—523.28 m	0.47 m	Muddy siltstone, greenish gray, discontinuous wavy ripple bedding and deformation bedding, with lots of calcareous concretions and siltstone strips
160-15	523.28—523.53 m	0.25 m	Siltstone, light brown, trough cross bedding, bioturbation structures, with muddy siltstone laminae and calcareous concretions
160-16	523.53—523.65 m	0.12 m	Muddy siltstone, light brown, deformation bedding
161-1	523.65—524.51 m	0.86 m	Siltstone, light brown, discontinuous ripple bedding, compressive deformation bedding, 4 irregular grayish-brown mudstone laminae, irregular mud gravels and calcareous concretions occasionally
161-2	524.51—526.02 m	1.51 m	Siltstone, greenish gray, deformation bedding, with sandstone balls and strips, mud gravels and few calcareous concretions
161-3	526.02—526.45 m	0.43 m	Sandstone, fine-grained, light gray, trough cross bedding, deformation bedding and bioturbation locally, mud gravels and mudstone laminae, few calcareous concretions
161-4	526.45—526.89 m	0.44 m	Siltstone, argillaceous with mud gravels, greenish-gray, deformation bedding, preexisting bedding reserved locally, scour-fill structure at the base, siltstone in the upper part, deformation ripple bedding, mud gravels, mudstone laminae and few calcareous concretions in the middle and lower parts
161-5	526.89—528.42 m	1.53 m	Fine-grained sandstone, light gray, trough cross bedding, with calcareous sandstone strips locally
161-6	528.42—528.49 m	0.07 m	Fine-grained sandstone, light gray, interbedded with gray mudstone, wavy bedding
161-7	528.49—528.85 m	0.36 m	Fine-grained sandstone, light gray, wavy ripple bedding
161-8	528.85—529.27 m	0.42 m	Fine-grained sandstone, light gray, deformation bedding, with irregular mudstone strips
161-9	529.27—529.99 m	0.72 m	Fine-grained sandstone, light gray, small trough cross bedding, weak deformation bedding locally, with some mudstone strips
161-10	529.99—530.04 m	0.05 m	Conglomerate, gray, massive, erosion surface at the base
161-11	530.04—530.37 m	0.33 m	Fine-grained sandstone, light gray, small trough cross bedding, with large number plate fossil fragments and mud gravels locally
161-12	530.37—530.59 m	0.22 m	Fine-grained conglomerate, light gray, wedge-shaped cross bedding, erosion surface at the base
161-13	530.59—530.68 m	0.09 m	Muddy siltstone, greenish gray, wavy ripple bedding, weak erosion surface at the base
161-14	530.68—530.90 m	0.22 m	Mudstone, greenish gray, massive, fill structure at the base
161-15	530.90—531.03 m	0.13 m	Siltstone, gray, massive
162-1	531.03—531.10 m	0.07 m	Siltstone, light gray, massive, erosion surface at the base
162-2	531.10—531.42 m	0.32 m	Silty mudstone, light gray, massive, fill structure at the base

162-3	531.42—531.72 m	0.30 m	Siltstone, light gray, deformation bedding and wavy ripple bedding
162-4	531.72—532.03 m	0.31 m	Siltstone, light gray, massive, with some mud gravels
162-5	532.03—532.13 m	0.10 m	Mudstone, light brown, massive
162-6	532.13—532.20 m	0.07 m	Siltstone, light brown, trough cross bedding, with mudstone strips and bioturbation
162-7	532.20—532.66 m	0.46 m	Siltstone, light brown, interbedded with mudstone, deformation bedding, erosion surface at the base, mud gravels locally, intensive bioturbation, 1 cm-thick mudstone at the base
162-8	532.66—532.83 m	0.17 m	Siltstone, light gray, trough cross bedding
162-9	532.83—533.54 m	0.71 m	Muddy siltstone, light brown, deformation bedding, 5 cm-thick wave-generated sand ripple bedding at the base, wavy bedding locally, irregular mudstone laminae, siltstone concretions
162-10	533.54—533.67 m	0.13 m	Mudstone, grayish brown, massive
162-11	533.67—534.33 m	0.66 m	Muddy siltstone, light yellowish brown, deformation bedding, with irregular mudstone strips
162-12	534.33—534.63 m	0.30 m	Silty mudstone, greenish gray, deformation bedding, weak erosion surface at the base, with fossil fragments and mud gravels
162-13	534.63—534.96 m	0.33 m	Mudstone, greenish gray, massive, with calcareous concretions and snail fossils
162-14	534.96—535.33 m	0.37 m	Mudstone, greenish gray, massive, calcareous concretions
162-15	535.33—535.57 m	0.24 m	Mudstone, grayish brown, massive
162-16	535.57—535.93 m	0.36 m	Muddy siltstone, light brown, massive
162-17	535.93—536.57 m	0.64 m	Mudstone, light brown, massive, fossil fragments
162-18	536.57—537.03 m	0.46 m	Mudstone, greenish gray, massive
162-19	537.03—537.53 m	0.50 m	Silty mudstone, greenish gray, massive, fossil fragments occasionally
162-20	537.53—538.03 m	0.50 m	Silty mudstone, light brown, massive, calcareous concretions
162-21	538.03—538.53 m	0.50 m	Mudstone, greenish gray, massive, with fossil fragments occasionally
163-1	538.53—540.84 m	2.31 m	Mudstone, dark greenish gray, massive, with fractures filled with calcspar, few ostracada fossils
163-2	540.84—541.38 m	0.54 m	Silty mudstone, dark greenish-gray, massive bedding, scour-fill structure at the base, few calcareous concretions
163-3	541.38—541.46 m	0.08 m	Siltstone, light gray, wavy ripple bedding, with trace fossils, erosion surface at the base
163-4	541.46—541.76 m	0.30 m	Mudstone, dark greenish gray, massive
163-5	541.76—542.05 m	0.29 m	Siltstone, medium light gray, horizontal bedding and ripple bedding, erosion surface at the base, few mud gravels above, few large calcareous concretions, up to 4 cm
163-6	542.05—543.11 m	1.06 m	Mudstone, medium dark gray, horizontal bedding, with 1cm and 3 cm thick light gray coarse siltstone laminae respectively at 542.58 m and 542.83 m, argillaceous siltstone laminae in 542.73—542.77 m

163-7	543.11—543.18 m	0.07 m	Siltstone, light gray, wavy ripple bedding, erosion surface at the base, calcareous concretions
163-8	543.18—543.33 m	0.15 m	Mudstone, dark brown, horizontal bedding
163-9	543.33—543.60 m	0.27 m	Siltstone, light gray, wavy ripple bedding, erosion surface at the base, with muddy siltstone strips
163-10	543.60—544.22 m	0.62 m	Mudstone, dark gray, horizontal bedding
163-11	544.22—544.36 m	0.14 m	Siltstone, light gray, interbedded with medium dark gray mudstone, wave-generated sand ripple bedding increasing upward
163-12	544.36—544.90 m	0.54 m	Mudstone, dark gray, horizontal bedding, siltstone laminae
163-13	544.90—546.16 m	1.26 m	Siltstone, medium light gray, wave-generated sand ripple bedding increasing upward, mudstone laminae, few mud gravels at 546.03 m
163-14	546.16—546.26 m	0.10 m	Mudstone, medium dark gray, interbedded with medium light gray siltstone, ripple bedding
163-15	546.26—546.98 m	0.72 m	Siltstone, medium gray, deformation bedding, large number of irregular mudstone laminae and charcoals, fine conglomerates at 546.33—546.36 m, erosion at the base, 2 cm-thick fine sandstone above
164-1	547.25—547.75 m	0.50 m	Siltstone, medium gray, deformation bedding, much mudstone laminae and few mud gravels
164-2	547.75—548.14 m	0.39 m	Siltstone, medium gray, interbedded with medium dark gray mudstone, wave-generated sand ripple bedding, a lot of mudstone laminae
164-3	548.14—548.53 m	0.39 m	Siltstone, medium gray, deformation bedding, roll-up structure locally, a lot of mudstone laminae in rip up manner, with few mud gravels, 2 cm-thick fine sandstone laminae at 548.35 m
164-4	548.53—548.78 m	0.25 m	Sandstone, fine-grained, medium gray, deformation bedding, argillaceous siltstone laminae
164-5	548.78—548.83 m	0.05 m	Conglomerate, sandy and fine-grained, greenish-gray, mainly calcareous concretionary gravels and sand gravels, massive bedding, strong erosion at the base
164-6	548.83—549.13 m	0.30 m	Fine-grained sandstone, light gray, small trough cross bedding, weak erosion surface at the base, with few mud gravels
164-7	549.13—549.37 m	0.24 m	Fine-grained sandstone, light gray, deformation bedding, erosion surface at the base, with few irregular mudstone strips
164-8	549.37—550.05 m	0.68 m	Fine-grained sandstone, light gray, trough cross bedding, mudstone gravels and strips locally, with ostracoda at the base
164-9	550.05—550.15 m	0.10 m	Fine-grained sandstone, light gray, wave ripple bedding, with ostracoda fossils
165-1	550.15—550.40 m	0.25 m	Mudstone, greenish gray, normal graded bedding, erosion surface at the base, with ostracoda fossils
165-2	550.40—550.49 m	0.09 m	Siltstone, light gray, interbedded with greenish gray mudstone, deformation bedding, erosion surface at the base

165-3	550.49—550.58 m	0.09 m	Mudstone, greenish gray, normal graded bedding, erosion surface at the base, with ostracoda fossils
165-4	550.58—550.65 m	0.07 m	Mudstone, gray, wavy ripple bedding
165-5	550.65—550.77 m	0.12 m	Muddy siltstone, gray, wavy ripple bedding deformation bedding, with few calcareous concretions
165-6	550.77—551.54 m	0.77 m	Siltstone, medium light gray, wave-generated sand ripple bedding, with wormholes, 3 mudstone laminae (<2 cm) at 551.20 m, 551.45 m and 551.51 m
165-7	551.51—552.56 m	1.05 m	Siltstone, medium gray, deformation bedding, erosion surface at the base, mudstone laminae at the top, irregular mudstone laminae and laminae in the middle and lower parts
165-8	552.56—553.51 m	0.95 m	Siltstone, light gray, wave-generated sand ripple bedding, mudstone laminae at 553.15 m, mud slump structure
165-9	553.51—553.65 m	0.14 m	Muddy siltstone, greenish-gray, deformation bedding, a lot of irregular mudstone laminae, 1 cm thick mudstone laminae at the top
165-10	553.65—554.37 m	0.72 m	Siltstone, light gray, wave-generated sand ripple bedding, a lot of irregular mudstone laminae
165-11	554.37—554.96 m	0.59 m	Siltstone, calcareous, light gray, small wave-generated cross-bedding, water-escape structure, weak erosion at the base, few mudstone laminae
165-12	554.96—555.36 m	0.40 m	4 sedimentary cycles: conglomerate at the base with unclear graded bedding, calcareous fine grained sandstone upward with wave-generated cross-bedding, sand-containing mudstone at the top with wavy bedding, 2 cm, 9 cm, 5 cm, 1 cm, 4 cm, 2 cm, 3 cm, 3 cm and 3 cm thick separately, a lot of carbonized plant fragments and pyrite grains in sand-containing mudstone
165-13	555.36—555.47 m	0.11 m	Siltstone, light gray, wavy cross bedding, erosion surface at the base, few calcareous concretions
165-14	555.47—555.71 m	0.24 m	Siltstone, light greenish gray, massive, mud gravels locally
165-15	555.71—556.15 m	0.44 m	Silty mudstone, light brown, massive, calcareous concretions
165-16	556.15—556.73 m	0.58 m	Mudstone, grayish brown, massive, calcareous concretions, muddy siltstone strips at 556.65 m
165-17	556.73—556.90 m	0.17 m	Muddy siltstone, greenish gray and grayish brown, massive
165-18	556.90—557.61 m	0.71 m	Silty mudstone, greenish gray and grayish brown, massive
165-19	557.61—557.93 m	0.32 m	Mudstone, grayish brown, massive, with few calcareous concretions
166-1	557.93—558.26 m	0.33 m	Mudstone, greenish gray, massive, with few fossil fragments
166-2	558.26—558.43 m	0.17 m	Silty mudstone, greenish gray and grayish brown, massive, with few calcareous concretions
166-3	558.43—558.77 m	0.34 m	Silty mudstone, grayish brown, massive, calcareous concretions
166-4	558.77—559.33 m	0.56 m	Mudstone, grayish brown, massive, calcareous concretions
166-5	559.33—560.17 m	0.84 m	Muddy siltstone, greenish gray and grayish brown, deformation bedding, fill structure at the base, with calcareous concretions and siltstone strips occasionally

166-6	560.17—560.44 m	0.27 m	Siltstone, light gray, trough cross bedding, with bioturbation and trace fossils, weak erosion surface at the base
166-7	560.44—560.60 m	0.16 m	Muddy siltstone, light brown, massive, deformation bedding locally, siltstone strips in the lower part, with calcareous concretions and siltstone strips occasionally
166-8	560.60—560.73 m	0.13 m	Siltstone, light gray, trough cross bedding, intensive bioturbation structures at the top, biogliph
166-9	560.73—561.53 m	0.80 m	Muddy siltstone, light brown, deformation bedding, a lot of irregular mudstone laminae, rip-up manner locally, few siltstone laminae
166-10	561.53—563.33 m	1.80 m	Siltstone, light gray, deformation bedding, slight erosion at the base, irregular mudstone laminae and calcareous siltstone strips, calcareous siltstone, with interbed at 562.27—562.41 m, containing biogliphs and light brown mudstone laminae
166-11	563.33—563.73 m	0.40 m	Siltstone, light gray, wavy bedding, deformation bedding, few greenish-gray mudstone laminae
166-12	563.73—564.24 m	0.51 m	Siltstone, light gray, small trough cross bedding, with few mudstone strips
166-13	564.24—564.54 m	0.30 m	Siltstone, light gray, small trough cross bedding
167-1	564.46—564.77 m	0.31 m	Siltstone, light gray, small trough cross bedding and deformation bedding, with few mudstone strips
167-2	564.77—565.16 m	0.39 m	Siltstone, light gray, deformation bedding in the upper, trough cross bedding in the middle, slump structures at the base, fractures, few mudstone laminae and mudstone laminae in the upper part
167-3	565.16—565.39 m	0.23 m	Mudstone, grayish green, massive, with siltstone strips and large number of fossil fragments
167-4	565.39—566.16 m	0.77 m	Muddy siltstone, gray, wavy ripple bedding, deformation bedding locally, with few calcareous concretions and trace fossils
167-5	566.16—566.44 m	0.28 m	Siltstone, light gray, wavy ripple bedding, erosion surface at the base, with mud gravels occasionally
167-6	566.44—567.01 m	0.57 m	Siltstone, light gray, deformation bedding, weak erosion surface at the base, irregular mudstone strips in the upper and middle parts, mud gravels in the lower part
167-7	567.01—567.23 m	0.22 m	Siltstone, light gray, wavy ripple bedding, weak deformation bedding locally, calcareous concretions occasionally
167-8	567.23—567.45 m	0.22 m	Siltstone, light gray, deformation bedding, with irregular mudstone strips
167-9	567.45—567.63 m	0.18 m	Siltstone, light gray, trough cross bedding, erosion surface at the base
167-10	567.63—568.09 m	0.46 m	Siltstone, medium light gray, deformation bedding, calcareous siltstone strips in the upper part, with roll-up characteristics, irregular mudstone laminae developed, few mud gravels at the base, 0.6 cm thick mudstone laminae at 567.82 m
167-11	568.09—568.31 m	0.22 m	Siltstone, medium light gray, small trough cross bedding, erosion surface at the base, above which there are directionally aligned mud gravels

167-12	568.31—569.13 m	0.82 m	Siltstone, light gray, deformation bedding, erosion surface at the base, with irregular mudstone strips
167-13	569.13—569.49 m	0.36 m	Siltstone, light gray, small wavy cross bedding, with few mudstone strips
167-14	569.49—569.82 m	0.33 m	Siltstone, light gray, deformation bedding, with irregular mudstone strips
167-15	569.82—571.01 m	1.19 m	Siltstone, light gray, small trough cross bedding, with mudstone strips and mud gravels locally
167-16	571.01—571.51 m	0.50 m	Siltstone, greenish-gray, with mud gravels, small trough cross bedding, deformation structure locally, erosion surface at the base, carbonized plant fragments and irregular mudstone laminae
167-17	571.51—571.77 m	0.26 m	Conglomerate, silty, greenish-gray, massive bedding, intensive erosion surface at the base, sandstone concretions and few carbonized plant fragments
167-18	571.77—571.93 m	0.16 m	Siltstone, gray, interbedded with greenish gray mudstone, deformation bedding.
167-19	571.93—572.53 m	0.60 m	Siltstone, gray, small trough cross bedding
168-1	572.54—572.94 m	0.40 m	Siltstone, calcareous, light gray, small trough cross bedding, erosion surface at the base, greenish-gray mudstone laminae (<1 cm) locally
168-2	572.94—573.53 m	0.59 m	Siltstone, light gray, deformation bedding, convolution bedding, weak erosion surface at the base, with few irregular mudstone strips
168-3	573.53—573.68 m	0.15 m	Mudstone, greenish gray, deformation bedding, with sandstone strips
168-4	573.68—574.54 m	0.86 m	Siltstone, light gray, deformation bedding, convolution bedding, weak erosion surface at the base, with few irregular mudstone strips and mussel fossils
168-5	574.54—575.22 m	0.68 m	Siltstone, light gray, trough cross bedding, with mudstone strips, erosion surface at the base
168-6	575.22—575.59 m	0.37 m	Siltstone, light gray, wavy cross bedding and wavy bedding, mudstone strips
168-7	575.59—575.68 m	0.09 m	Siltstone, light gray, oblique bedding
168-8	575.68—575.85 m	0.17 m	Siltstone, light gray, oblique bedding, with few mudstone strips
168-9	575.85—576.06 m	0.21 m	Siltstone, light gray, deformation bedding, with mud gravels and irregular mudstone strips
168-10	576.06—576.19 m	0.13 m	Siltstone, light gray, parallel bedding, with mudstone laminaes and mud gravels.
168-11	576.19—576.51 m	0.32 m	Fine-grained conglomerate, gray, massive, erosion surface at the base, with large number of sand balls
168-12	576.51—577.09 m	0.58 m	Mudstone, grayish brown, wavy ripple bedding, siltstone strips occasionally
168-13	577.09—577.49 m	0.40 m	Mudstone, greenish gray, massive, with lots of fossil fragments
168-14	577.49—577.64 m	0.15 m	Silty mudstone, light greenish gray, massive, with sand balls
168-15	577.64—578.85 m	1.21 m	Silty mudstone, grayish brown, wavy ripple bedding, with siltstone strips in the lower part
168-16	578.85—581.10 m	2.25 m	Siltstone, light gray, trough cross bedding, with irregular mudstone strips and trace fossils
169-1	581.10—581.50 m	0.40 m	Siltstone, light gray, trough cross bedding, erosion surface at the base

169-2	581.50—581.63 m	0.13 m	Silty mudstone, greenish gray, deformation bedding, with large number of siltstone strips
169-3	581.63—582.75 m	1.12 m	Siltstone, light gray, trough cross bedding, erosion surface at the base
169-4	582.75—583.07 m	0.32 m	Fine-grained sandstone, light gray, trough cross bedding, with mud gravels and mudstone strips occasionally
169-5	583.07—583.32 m	0.25 m	Fine-grained sandstone, light gray, oblique bedding
169-6	583.32—583.54 m	0.22 m	Fine-grained sandstone, gray, deformation bedding, erosion surface at the base, with calcareous concretions
169-7	583.54—585.31 m	1.77 m	Sandstone, fine-grained, medium gray, trough cross bedding, with mudstone laminae at 584.2—584.21 m, 584.8—584.84 m, 584.96—585 m, and 585.06—585.11 m, mud gravels locally
169-8	585.31—585.86 m	0.55 m	Sandstone, fine-grained, medium gray, unidirectional oblique bedding, mud gravels on the erosion surface at the base
169-9	585.86—586.07 m	0.21 m	Fine-grained sandstone, gray, deformation bedding, erosion surface at the base, with mud gravels
169-10	586.07—586.28 m	0.21 m	Fine-grained sandstone, gray, oblique bedding, with mud gravels locally
169-11	586.28—586.84 m	0.56 m	Siltstone, gray, deformation bedding, with lots of mud gravels and irregular mudstone strips
169-12	586.84—587.22 m	0.38 m	Fine-grained sandstone, gray, parallel bedding, with carbonized plant fragments
169-13	587.22—587.72 m	0.50 m	Fine-grained sandstone, gray, oblique bedding, with carbonized plant fragments
169-14	587.72—587.81 m	0.09 m	Fine-grained sandstone, gray, trough cross bedding
169-15	587.81—588.56	0.75 m	Fine-grained sandstone, gray, oblique bedding
169-16	588.56—588.90 m	0.34 m	Fine-grained sandstone, gray, deformation bedding, with mud gravels and mudstone strips
170-1	588.90—589.16 m	0.26 m	Fine-grained sandstone, light gray, trough cross bedding
170-2	589.16—589.23 m	0.07 m	Fine-grained conglomerate, light gray, massive, weak erosion surface at the base, with sandstone strips occasionally
170-3	589.23—589.50 m	0.27 m	Fine-grained sandstone, light gray, trough cross bedding
170-4	589.50—589.66 m	0.16 m	Fine-grained conglomerate, light gray, oblique bedding
170-5	589.66—589.84 m	0.18 m	Fine-grained conglomerate, light gray, massive, erosion surface at the base, with 3cm-thick sandstone laminae at the top
170-6	589.84—589.92 m	0.08 m	Siltstone, light gray, trough cross bedding.
170-7	589.92—590.90 m	0.98 m	Gompholite, light gray, massive, with fine-grained sandstone at 590.30—590.35 m
170-8	590.90—591.09 m	0.19 m	Fine-grained sandstone, light gray, trough cross bedding, erosion surface at the base
170-9	591.09—591.19 m	0.10 m	Siltstone, light gray, trough cross bedding, erosion surface at the base
170-10	591.19—591.34 m	0.15 m	Conglomerate, light gray, massive, erosion surface at the base, with calcareous concretions

170-11	591.34—591.46 m	0.12 m	Siltstone, light gray, oblique bedding, erosion surface at the base and conglomerate
170-12	591.46—591.70 m	0.24 m	Siltstone, light gray, trough cross bedding and parallel bedding
170-13	591.70—591.79 m	0.09 m	Siltstone, light gray, oblique bedding, erosion surface at the base
170-14	591.79—592.02 m	0.23 m	Siltstone, light gray, parallel bedding, with few mud gravels at the base
170-15	592.02—592.16 m	0.14 m	Gompholite, light gray, normal graded bedding, erosion surface at the base, above which there are directionally aligned mud gravels
170-16	592.16—592.29 m	0.13 m	Siltstone, light gray, massive, erosion surface at the base
170-17	592.29—592.50 m	0.21 m	Mudstone, greenish gray, massive
170-18	592.50—593.60 m	1.10 m	Mudstone, grayish brown, wavy ripple bedding, with calcareous concretions and few siltstone strips
170-19	593.60—594.47 m	0.87 m	Silty mudstone, grayish brown, massive, wavy ripple bedding in the upper part, with calcareous concretions and siltstone strips
170-20	594.47—595.00 m	0.53 m	Silty mudstone, grayish brown, wavy ripple bedding and deformation bedding, with bioturbation and large number siltstone strips and few calcareous concretions
170-21	595.00—595.14 m	0.14 m	Muddy siltstone, light brownish gray, deformation bedding, with irregular siltstone strips and calcareous concretions
170-22	595.14—595.24 m	0.10 m	Siltstone, medium gray, small trough cross bedding, 2 cm-thick fine gompholite at the base
170-23	595.24—595.40 m	0.16 m	Muddy siltstone, light brownish gray, deformation bedding, with irregular siltstone strips and calcareous concretions
171-1	596.72—597.41 m	0.69 m	Muddy siltstone, light gray, deformation bedding, with irregular mudstone and sandstone strips
171-2	597.41—597.49 m	0.08 m	Siltstone, light gray, trough cross bedding
171-3	597.49—597.92 m	0.43 m	Siltstone, light gray, deformation bedding, with irregular mudstone strips and mud gravels
171-4	597.92—598.12 m	0.20 m	Siltstone, light gray, trough cross bedding, mud gravels at the base
171-5	598.12—598.48 m	0.36 m	Siltstone, medium light gray, deformation bedding, intensive erosion surface at the base, irregular mudstone laminae and mud gravels, 1 cm-thick mudstone laminae at the top
171-6	598.48—598.67 m	0.19 m	Siltstone, light gray, trough cross bedding
171-7	598.67—598.79 m	0.12 m	Siltstone, light gray, deformation bedding, mud gravels and mudstone strips
171-8	598.79—599.52 m	0.73 m	Siltstone, medium light gray, trough cross bedding, mudstone laminae locally, irregular mudstone laminae at 599.36 m
171-9	599.52—600.09 m	0.57 m	Siltstone, light gray, oblique bedding, deformation bedding, with large number of charcoal and mud gravels
171-10	600.09—600.22 m	0.13 m	Muddy siltstone, dark greenish gray, deformation bedding, erosion surface at the base, with mud gravels, irregular siltstone strips and charcoal

Section 5 SK-1 Core Description and Core Photographs

171-11	600.22—601.29 m	1.07 m	Fine-grained sandstone, light gray, trough cross bedding, with mud gravels and irregular mudstone strips occasionally
171-12	601.29—601.66 m	0.37 m	Sandstone, light gray, deformation bedding, erosion surface at the base, mud gravels occasionally
171-13	601.66—601.84 m	0.18 m	Conglomerate, light gray, graded bedding, erosion surface at the base
171-14	601.84—601.99 m	0.15 m	Sandstone, light gray, trough cross bedding
171-15	601.99—602.19 m	0.20 m	Fine-grained sandstone, light gray, trough cross bedding
171-16	602.19—602.32 m	0.13 m	Sandstone, light gray, parallel bedding
171-17	602.32—602.92 m	0.60 m	Fine-grained sandstone, light gray, parallel bedding, erosion surface at the base, mud gravels
171-18	602.92—602.97 m	0.05 m	Silty mudstone, light olive gray, horizontal bedding
172-1	603.71—604.47 m	0.76 m	Sandstone, fine-grained, medium light gray, unidirectional oblique bedding, wavy bedding occasionally, weak scour structure at the base, pyrite crystals
172-2	604.47—604.71 m	0.24 m	Siltstone, light gray, deformation bedding, erosion surface at the base, with irregular siltstone strips
172-3	604.71—605.02 m	0.31 m	Mudstone, dark greenish gray, massive, fossil fragments occasionally
172-4	605.02—605.35 m	0.33 m	Silty mudstone, greenish gray, massive, with calcareous concretions at the base, calcareous concretions
172-5	605.35—605.54 m	0.19 m	Silty mudstone, light red, massive
172-6	605.54—606.51 m	0.97 m	Mudstone, grayish brown, massive, calcareous concretions
172-7	606.51—607.00 m	0.49 m	Silty mudstone, grayish brown, massive, calcareous concretions and light gray mudstone and siltstone strips
172-8	607.00—607.73 m	0.73 m	Muddy siltstone, light red, trough cross bedding, with a lot of sandstone laminae, few mica flake, biogliph
172-9	607.73—608.65 m	0.92 m	Silty mudstone, light red, horizontal bedding, with grayish-brown mudstone laminae at 607.79—607.83 m, and 608.34—608.39 m, argillaceous siltstone laminae at 607.91 m and 608.38 m
172-10	608.65—609.09 m	0.44 m	Muddy siltstone, light red, ripple bedding, with calcareous siltstone laminae at 608.68—608.86 m, erosion surface at the base
173-1	609.09—609.52 m	0.43 m	Muddy siltstone, grayish-brown, wavy bedding, siltstone laminae, mud gravels on erosion surface at the base
173-2	609.52—609.75 m	0.23 m	Siltstone, light gray, trough cross bedding, intensive bioturbation
173-3	609.75—611.09 m	1.34 m	Siltstone, light gray, trough cross bedding, deformation bedding locally
173-4	611.09—611.29 m	0.20 m	Siltstone, light gray, slump structures, convolution bedding locally, with irregular mudstone strips
173-5	611.29—611.39 m	0.10 m	Conglomerate, grayish brown, massive, erosion surface at the base, with calcareous concretions and few mud gravels
173-6	611.39—611.65 m	0.26 m	Mudstone, grayish brown, deformation bedding, with lots of muddy siltstone strips

173-7	611.65—612.98 m	1.33 m	Silty mudstone, grayish brown, deformation bedding, with lots of mudstone and siltstone strips, calcareous concretions occasionally
173-8	612.98—613.12 m	0.14 m	Mudstone, grayish brown, massive, siltstone strips occasionally
173-9	613.12—614.04 m	0.92 m	Siltstone, light gray, slump structures, with irregular siltstone and mudstone strips
173-10	614.04—614.84 m	0.80 m	Siltstone, light gray, trough cross bedding, irregular mudstone strips occasionally
173-11	614.84—614.92 m	0.08 m	Siltstone, bearsing conglomerate, light gray, massive, with calcareous concretions, erosion surface at the base
173-12	614.92—615.13 m	0.21 m	Siltstone, light gray, trough cross bedding, mudstone strips occasionally
173-13	615.13—615.51 m	0.38 m	Siltstone, calcareous, light gray, trough cross bedding, erosion surface at the base, 1 cm-thick greenish-gray argillaceous siltstone laminae with mud gravels at 615.39 m
173-14	615.51—615.56 m	0.05 m	Siltstone, greenish-gray, with mud gravels, massive bedding, erosion surface at the base, siltstone laminae with mud gravels
173-15	615.56—615.96 m	0.40 m	Siltstone, greenish-gray, with slump structures, erosion surface at the base, plentiful calcareous concretions and siltstone laminae, mud gravels locally
173-16	615.96—616.38 m	0.42 m	Siltstone, medium light gray, trough cross bedding, unidirectional oblique bedding at 616.06—616.12 m
174-1	616.78—617.03 m	0.25 m	Sandstone, light gray, trough cross bedding, erosion surface at the base, bearing conglomerate locally
174-2	617.03—617.24 m	0.21 m	Conglomerate, light gray, massive, erosion surface at the base
174-3	617.24—617.34 m	0.10 m	Sandstone, light gray, oblique bedding, erosion surface at the base
174-4	617.34—618.37 m	1.03 m	Siltstone, medium gray, with slump structures, roll-up bedding locally, erosion surface at the base, irregular mudstone laminaes and calcareous concretions developed
174-5	618.37—618.59 m	0.22 m	Siltstone, medium light gray, with slump structures, intensive erosion surface at the base, with mud gravels, irregular mudstone laminae, mussel fossils and calcareous concretions
174-6	618.59—618.63 m	0.04 m	Mudstone, greenish-gray, massive bedding
174-7	618.63—618.95 m	0.32 m	Mudstone, light brownish-gray, horizontal bedding, 5mm-thick calcareous siltstone at 618.88 m eroding the lower mudstone, siltstone laminae locally
174-8	618.95—619.17 m	0.22 m	Muddy siltstone, light brownish-gray, with mudstone laminae, trough cross bedding, ripple bedding, weak erosion at the base
174-9	619.17—619.22 m	0.05 m	Mudstone, grayish brown, wavy ripple bedding, muddy siltstone strips in the lower part
174-10	619.17—619.22 m	0.05 m	Siltstone, light brownish-gray, with slump structures, roll-up bedding locally, irregular mudstone laminae, about 2 cm-thick mudstone laminae at 619.34 m
174-11	619.83—619.88 m	0.05 m	Siltstone, light gray, trough cross bedding, convolution bedding occasionally, erosion surface at the base

174-12	619.88—619.93 m	0.05 m	Siltstone, light brownish gray, slump structures, convolution bedding locally, with irregular mudstone strips
174-13	619.93—620.02 m	0.09 m	Siltstone, light brownish gray, trough cross bedding, climbing bedding in the upper part, with mudstone strips
174-14	620.02—620.25 m	0.23 m	Siltstone, light brownish gray, slump structures, convolution bedding locally, with irregular mudstone strips
174-15	620.25—620.33 m	0.08 m	Muddy siltstone, light brownish gray, wavy ripple bedding, with lots of mudstone strips
174-16	620.33—620.47 m	0.14 m	Siltstone, light brownish gray, trough cross bedding, erosion surface at the base, with few mudstone strips and trace fossils
174-17	620.47—620.88 m	0.41 m	Silty mudstone, greenish gray, massive, with ostracoda fossils, snail fossils and few fossil fragments.
174-18	620.88—621.34 m	0.46 m	Muddy siltstone, greenish gray, slump structure, ostracoda fossils occasionally, calcareous concretions at the base
174-19	621.34—621.57 m	0.23 m	Siltstone, light gray, trough cross bedding, wavy bedding and climbing bedding in the upper part
174-20	621.57—622.21 m	0.64 m	Siltstone, light gray, trough cross bedding, erosion and large calcareous concretions at the base, mudstone laminae
174-21	622.21—622.38 m	0.17 m	Siltstone, light gray, with greenish-gray mudstone, trough cross bedding, mud gravels on the erosion surface at the base
174-22	622.38—622.86 m	0.48 m	Siltstone, medium light gray, trough cross bedding, mud gravels and mudstone laminae locally
174-23	622.86—623.31 m	0.45 m	Siltstone, light gray, trough cross bedding, with mudstone strips occasionally
174-24	623.31—623.81 m	0.50 m	Sanstone, light gray, trough cross bedding, mud gravels occasionally
174-25	623.81—624.38 m	0.57 m	Sandstone, bearing mud gravels, gray, oblique bedding, with mudstone strips
175-1	624.83—624.93 m	0.10 m	Fine-grained sandstone, light gray, trough cross bedding
175-2	624.93—625.23 m	0.30 m	Mudstone, grayish green, massive
175-3	625.23—625.63 m	0.40 m	Mudstone, grayish brown, massive
175-4	625.63—625.83 m	0.20 m	Mudstone, grayish green, massive
175-5	625.83—626.07 m	0.24 m	Silty mudstone, grayish brown, massive, calcareous concretions
175-6	626.07—627.17 m	1.10 m	Mudstone, greenish gray, massive
175-7	627.17—627.83 m	0.66 m	Silty mudstone, greenish gray and grayish brown, massive, with fractures
175-8	627.83—628.48 m	0.65 m	Mudstone, greenish gray, massive, with few mudstone strips
175-9	628.48—629.23 m	0.75 m	Muddy siltstone, grayish green, massive, with mudstone and siltstone strips, trace fossils
175-10	629.23—630.08 m	0.85 m	Siltstone, grayish green, with deformation bedding and convolution bedding
175-11	630.08—630.50 m	0.42 m	Sandstone, greenish gray, massive, with lots of calcareous concretions and fossil fragments, few mud gravels
175-12	630.50—632.33 m	1.83 m	Sandstone, fine-grained, grayish-green, parallel bedding, fractures, a layer of silty sandstone at 631.41—631.46 m

The overlying formation of Mingshui Formation Member 1 is the Mingshui Formation Member 2, the contacting relationship is conformable contact; the underlying formation is the Sifangtai Formation, the contacting relationship is parallel unconformable contact (Figure 5.8).

The detailed description of the Mingshui Formation Member 1.

175-13	632.33—632.45 m	0.12 m	Silty mudstone, grayish green, horizontal bedding
177-1	632.45—632.56 m	0.11 m	Muddy siltstone, greenish gray, wavy ripple bedding
177-2	632.56—632.81 m	0.25 m	Silty mudstone, greenish gray, wavy ripple bedding
177-3	632.81—633.16 m	0.35 m	Muddy siltstone, greenish gray, slump structures, with irregular siltstone and mudstone strips
177-4	633.16—633.48 m	0.32 m	Silty mudstone, greenish gray, wavy ripple bedding, with muddy siltstone laminae
177-5	633.48—633.61 m	0.13 m	Muddy siltstone, light gray, wavy ripple bedding and wavy cross bedding, with charcoal
177-6	633.61—634.21 m	0.60 m	Silty mudstone, light gray, wavy ripple bedding, with charcoal and pyrite
177-7	634.21—634.69 m	0.48 m	Muddy siltstone, light gray, slump structures, convolution bedding locally, with irregular mudstone strips and siltstone balls
177-8	634.69—635.80 m	1.11 m	Muddy siltstone, light gray, wavy ripple bedding, with lots of charcoal and few pyrite
177-9	635.80—636.14 m	0.34 m	Muddy siltstone, light gray, slump structures, with irregular siltstone strips
177-10	636.14—636.27 m	0.13 m	Siltstone, light gray, parallel bedding, weak erosion surface at the base
177-11	636.27—637.40 m	1.13 m	Muddy siltstone, light gray, wavy ripple bedding, deformation bedding locally, with lots of charcoal, pyrite and siltstone strips occasionally
177-12	637.40—637.90 m	0.50 m	Siltstone, medium light gray, interbedded with medium dark gray mudstone, with oblique bedding and wavy cross bedding, horizontal ripple bedding in mudstone, a lot of charcoals, few pyrite
177-13	637.90—638.65 m	0.75 m	Mudstone, medium dark gray, interbedded with medium light gray siltstone, wavy cross bedding, deformation structure locally, a lot of charcoals
178-1	638.65—639.11 m	0.46 m	Mudstone, medium dark gray, interbedded with medium light gray siltstone, horizontal bedding and horizontal ripple bedding in mudstone, wavy bedding in siltstone, charcoals and pyrites on the bedding surface
178-2	639.11—639.45 m	0.34 m	Siltstone intercalated with dark gray mudstone, wavy bedding, with large number of charcoal
178-3	639.45—640.47 m	1.02 m	Mudstone, dark gray, intercalated with light gray siltstone, horizontal bedding, wavy ripple bedding, with lots of pyrite and charcoals
178-4	640.47—640.98 m	0.51 m	Mudstone, dark gray, interbedded with light gray siltstone, horizontal bedding, wavy ripple bedding, with lots of pyrite and charcoals
178-5	640.98—641.63 m	0.65 m	Siltstone, light gray, intercalated with dark gray mudstone, wavy bedding, wavy cross bedding and wavy ripple bedding, with large number of charcoals

178-6	641.63—643.65 m	2.02 m	Siltstone interbedded with dark gray mudstone, wavy bedding and wavy cross bedding, with lots of charcoals
178-7	643.65—643.88 m	0.23 m	Mudstone, dark brown, horizontal bedding and wavy ripple bedding, with some siltstone strips and lots of charcoals
179-1	643.88—644.38 m	0.50 m	Silty mudstone, light gray, intercalated with muddy siltstone, wavy ripple bedding, weak erosion surface at the base, with pyrite and charcoals, few fossil fragments
179-2	644.38—646.12 m	1.74 m	Mudstone, light gray, intercalated with muddy siltstone, wavy ripple bedding, with pyrite and charcoals, few fossil fragments
179-3	646.12—648.38 m	2.26 m	Mudstone, gray, horizontal bedding, few fossil fragments and charcoal
180-1	648.38—653.30 m	4.92 m	Mudstone, medium gray, horizontal bedding, with plant fragments and pyrite interbedded, conchostracons occasionally, siltstone laminae at 650.80 m
180-2	653.30—653.60 m	0.30 m	Mudstone, gray, wavy ripple bedding, with Silty mudstone strips and pyrite, few fossil fragments
180-3	653.60—653.91 m	0.31 m	Mudstone, gray, horizontal bedding, with few conchostracon and fossil fragments
181-1	653.91—658.24 m	4.33 m	Mudstone, light olive gray, horizontal bedding, few conchostracon and fossil fragments, with 1.2 cm-thick pyrite laminae
181-2	658.24—660.07 m	1.83 m	Mudstone, gray, horizontal bedding, erosion surface at the base, with few pyrite and conchostracons fossils
181-3	660.07—661.14 m	1.07 m	Silty mudstone, greenish gray, massive, few pyrite and fossil fragments, calcareous concretions at the base
181-4	661.14—661.45 m	0.31 m	Muddy siltstone, greenish gray, massive, few pyrite
181-5	661.45—661.90 m	0.45 m	Siltstone, light gray, slump structures, mudstone strips, bioturbation locally
181-6	661.90—662.42 m	0.52 m	Siltstone, light gray, parallel bedding
182-1	662.48—662.98 m	0.50 m	Siltstone, light gray, wavy ripple bedding, with mudstone strips
182-2	662.98—663.51 m	0.53 m	Siltstone, light gray, interbedded with mudstone, with slump structures, irregular mudstone laminae and fine grained sandstone, mudstone laminae at 663.33—663.38 m
182-3	662.98—663.51 m	0.53 m	Fine-grained sandstone, light gray, trough cross bedding, erosion surface at the base, few pyrite and mud gravels
182-4	664.01—664.30 m	0.29 m	Siltstone, light gray, slump structures, with large number fo mudstone strips and sandstone strips
182-5	664.30—664.88 m	0.58 m	Fine-grained sandstone, light gray, slump structures, convolution bedding locally, few mudstone strips
182-6	664.88—665.21 m	0.33 m	Medium-grained sandstone, light gray, mud gravels, parallel bedding, erosion surface at the base
182-7	665.21—665.40 m	0.19 m	Fine-grained sandstone, gray, slump structures, convolution bedding locally, with irregular mudstone strips

182-8	665.40—666.78 m	1.38 m	Sandstone, medium-grained, light gray, trough cross bedding, erosion at the base, thickness-varying mudstone laminae or interbeds at 666.32 m, 666.42 m and 667.02 m
182-9	666.78—667.42 m	0.64 m	Mudstone, greenish gray, massive
182-10	667.42—668.28 m	0.86 m	Mudstone, brownish gray, massive, calcareous concretions occasionally
182-11	668.28—668.68 m	0.40 m	Silty mudstone, greenish gray, massive, with lots of calcareous concretions
182-12	668.68—668.94 m	0.26 m	Siltstone, light gray, wavy bedding, calcareous concretions
182-13	668.94—669.33 m	0.39 m	Muddy siltstone, greenish gray, wavy bedding, slump structure in the upper part, with irregular mudstone strips
182-14	669.33—669.40 m	0.07 m	Fine-grained sandstone, light gray, parallel bedding, with few mud gravels
182-15	669.40—669.80 m	0.40 m	Siltstone, light gray, trough cross bedding, mudstone laminae in the upper and lower parts
182-16	669.80—670.19 m	0.39 m	Fine-grained sandstone, light gray, parallel bedding, wavy bedding, with large number of mudstone strips and few mud gravels
182-17	670.19—670.53 m	0.34 m	Sandstone, light gray, trough cross bedding, with bioturbation
182-18	670.53—670.71 m	0.18 m	Fine-grained sandstone, light gray, trough cross bedding, bioturbation structure occasionally
183-1	670.71—671.03 m	0.32 m	Fine-grained sandstone, light gray, trough cross bedding, with siltstone balls and mudstone strips locally
183-2	671.03—672.24 m	1.21 m	Medium-grained sandstone, light gray, interbedded with greenish gray muddy siltstone, trough cross bedding, wavy ripple bedding in the lower part, with lots of mudstone strips
183-3	672.24—672.76 m	0.52 m	Fine-grained sandstone, light gray, oblique bedding, mud gravels
183-4	672.76—673.07 m	0.31 m	Siltstone, light gray, trough cross bedding, mudstone strips
183-5	673.07—674.14 m	1.07 m	Medium-grained sandstone, light gray, tabular cross bedding, mudstone laminae at 673.14—673.18 m
183-6	674.14—674.78 m	0.64 m	Medium-grained sandstone, light gray, parallel bedding
183-7	674.78—675.00 m	0.22 m	Medium-grained sandstone, light gray, trough cross bedding, mudstone strips
183-8	675.00—676.24 m	1.24 m	Coarse-grained sandstone, light gray, oblique bedding, mud gravels at the base
183-9	676.24—676.68 m	0.44 m	Coarse-grained sandstone, light gray, parallel bedding, erosion surface at the base and mud gravels
183-10	676.68—677.10 m	0.42 m	Medium-grained sandstone, light gray, trough cross bedding, mudstone strips
184-1	680.08—680.17 m	0.09 m	Medium-grained sandstone, light gray, parallel bedding
184-2	680.17—680.71 m	0.54 m	Medium-grained sandstone, light gray, oblique bedding
184-3	680.71—681.69 m	0.98 m	Medium-grained sandstone, light gray, tabular cross bedding, mud gravels and irregular mudstone strips

184-4	681.69—682.08 m	0.39 m	Silty mudstone, greenish gray, slump structures, with fine-grained sandstone strips and balls
184-5	682.08—682.42 m	0.34 m	Medium-grained sandstone, light gray, oblique bedding in the upper part, trough cross bedding in the lower part
184-6	682.42—682.52 m	0.10 m	Mudstone, greenish gray, wavy ripple bedding
184-7	682.52—682.91 m	0.39 m	Silty mudstone, greenish gray, slump structures, siltstone strips
184-8	682.91—683.03 m	0.12 m	Mudstone, greenish gray, wavy ripple bedding
184-9	683.03—683.59 m	0.56 m	Silty mudstone, greenish gray, horizontal bedding、wavy ripple bedding, 4 cm-thick muddy siltstone at the base
184-10	683.59—683.98 m	0.39 m	Mudstone, greenish gray, horizontal bedding and wavy ripple bedding. Siltstone in 683.74—683.78 m
184-11	683.98—684.31 m	0.33 m	Siltstone, light gray, interbedded with greenish gray mudstone, convolution bedding at the top, deformation bedding and wave ripple bedding in the lower part, with trace fossils
184-12	684.31—684.48 m	0.17 m	Mudstone, greenish gray, horizontal bedding
184-13	684.48—685.16 m	0.68 m	Siltstone, light gray, intercalated with mudstone, greenish gray, slump structures, erosion surface at the base
184-14	685.16—685.57 m	0.41 m	Silty mudstone, greenish gray, slump structures, with irregular siltstone strips and few charcoals
184-15	685.57—686.39 m	0.82 m	Fine-grained sandstone, light gray, wave ripple bedding, deformation bedding, convolution bedding locally, with irregular mudstone strips
185-1	686.39—686.79 m	0.40 m	Fine-grained sandstone, light gray, deformation bedding, with irregular mudstone strips and charcoal
185-2	686.79—687.33 m	0.54 m	Medium-grained sandstone, light gray, deformation bedding, weak erosion surface at the base, with irregular mudstone strips and charcoal
185-3	687.33—687.61 m	0.28 m	Medium-grained conglomerate, light gray, wedge-shaped cross bedding, erosion surface at the base, with mud gravels and fossil fragments
185-4	687.61—688.01 m	0.40 m	Silty mudstone, greenish gray, deformation bedding in the upperpart, horizontal bedding and wavy ripple bedding in the lower part, mudstone strips and calcareous concretions at the top
185-5	688.01—689.22 m	1.21 m	Medium-grained sandstone, light gray, parallel bedding, erosion surface at the base, mud gravels occasionally
185-6	689.22—689.76 m	0.54 m	Fine-grained sandstone, light gray, wavy cross bedding, mud gravels occasionally
186-1	694.53—694.63 m	0.10 m	Medium-grained sandstone, light gray, wavy cross bedding, mud gravels occasionally
186-2	694.63—694.73 m	0.10 m	Mudstone, greenish gray, wavy ripple bedding
186-3	694.73—695.08 m	0.35 m	Medium-grained sandstone, light gray, wavy cross bedding, mud gravels occasionally

187-1	695.08—695.56 m	0.48 m	Medium-grained sandstone, light gray, wedge-shaped cross bedding, mud gravels at the base
187-2	695.56—696.24 m	0.68 m	Fine-grained sandstone, light greenish gray, wavy cross bedding, deformation bedding in the middle part, erosion surface at 695.91 m, with directionally aligned mud gravels
187-3	696.24—696.88 m	0.64 m	Medium-grained sandstone, light gray, wavy cross bedding, with few mudstone strips
187-4	696.88—696.99 m	0.11 m	Fine-grained sandstone, light gray, oblique bedding, erosion surface at the base, with few mudstone strips
187-5	696.99—697.40 m	0.41 m	Medium-grained sandstone, light gray, wedge-shaped cross bedding, with few mudstone strips
188-1	698.39—698.55 m	0.16 m	Mudstone, grayish green, massive
188-2	698.55—699.59 m	1.04 m	Mudstone, grayish brown, massive, calcareous concretions occasionally
188-3	699.59—700.02 m	0.43 m	Mudstone, brownish gray, massive, with lots of calcareous concretions
188-4	700.02—700.89 m	0.87 m	Mudstone, grayish brown, massive, calcareous concretions locally
188-5	700.89—701.39 m	0.50 m	Mudstone, brown, massive, calcareous concretions locally
188-6	701.39—701.52 m	0.13 m	Mudstone, dark yellowish brown, massive
188-7	701.52—701.64 m	0.12 m	Silty mudstone, greenish gray, slump structures, with discontinuous wavy bedding locally, calcareous concretions
188-8	701.64—701.86 m	0.22 m	Silty mudstone, dark yellowish brown, massive
188-9	701.86—702.61 m	0.75 m	Silty mudstone, greenish gray, massive, bioturbation, deformation bedding at the base, calcareous concretions
188-10	702.61—702.99 m	0.38 m	Mudstone, dark yellowish brown, massive
188-11	702.99—703.34 m	0.35 m	Muddy siltstone, light greenish gray, massive, calcareous concretions
188-12	703.34—703.64 m	0.30 m	Mudstone, olive gray, wavy ripple bedding, calcareous concretions
188-13	703.64—704.15 m	0.51 m	Silty mudstone, light gray, slump structures, irregular mudstone strips, mud gravels and calcareous concretions
188-14	704.15—704.69 m	0.54 m	Silty mudstone, light olive gray, slump structures, irregular mudstone strips and muddy siltstone strips, few calcareous concretions
189-1	705.20—706.89 m	1.69 m	Medium-grained sandstone, light gray, massive, erosion surface at the base, calcareous concretions at the top, with few irregular mudstone strips
189-2	706.89—707.58 m	0.69 m	Mudstone, greenish gray, wavy ripple bedding, horizontal bedding, siltstone strips
189-3	707.58—711.45 m	3.87 m	Medium-grained sandstone, light gray, wedge-shaped cross bedding, few mud gravels, large number of mud gravels at 709.07 m, erosion surface at the base
189-4	711.45—713.43 m	1.98 m	Coarse-grained sandstone, light gray, wedge-shaped cross bedding
190-1	713.43—713.79 m	0.36 m	Coarse-grained sandstone, light gray, wedge-shaped cross bedding

190-2	713.79—714.79 m	1.00 m	Medium-grained sandstone, light gray, tabular cross bedding
190-3	714.79—717.72 m	2.93 m	Coarse-grained sandstone, light gray, wedge-shaped cross bedding, with conglomerate locally
191-1	717.72—719.31 m	1.59 m	Coarse-grained sandstone, light gray, wedge-shaped cross bedding, erosion surface at the base, with few charcoal, mud gravels locally
191-2	719.31—719.53 m	0.22 m	Siltstone, light gray, wavy bedding, small wavy cross bedding, with lots of charcoal
191-3	719.53—720.98 m	1.45 m	Sandstone, coarse-grained, medium light gray, wedge-shaped cross bedding, few charcoal, mud gravels occasionally, quartz gravels at 720.67 m
191-4	720.98—721.11 m	0.13 m	Medium-grained sandstone, dark greenish gray, slump structures, with mudstone strips and pyrite
192-1	721.46—723.76 m	2.30 m	Mudstone, light gray, horizontal bedding, with few fossil fragments and pyrite
192-2	723.76—726.84 m	3.08 m	Mudstone, gray, horizontal bedding, with few fossil fragments and pyrite
193-1	723.84—733.56 m	9.72 m	Mudstone, gray, horizontal bedding, fossil fragments occasionally
194-1	733.56—734.66 m	1.10 m	Mudstone, light greenish gray, horizontal bedding, fossil fragments occasionally
195-1	734.66—742.33 m	7.67 m	Mudstone, gray, horizontal bedding, fossil fragments occasionally
196-1	742.33—742.73 m	0.40 m	Mudstone, gray, horizontal bedding
196-2	742.73—748.51 m	5.78 m	Mudstone, dark gray, horizontal bedding, fossil fragments occasionally, 2 gray mudstone interbeds (about 2 cm thick) at 744.55 m and 744.63 m, 2 pyrite interbeds (about 2 cm thick) at 744.58 m and 745.21 m
196-3	748.51—749.08 m	0.57 m	Silty mudstone, greenish gray, massive, fossil fragments occasionally
196-4	749.08—750.12 m	1.04 m	Muddy siltstone, greenish gray, wavy ripple bedding, deformation bedding at the base, calcareous concretions
196-5	750.12—750.23 m	0.11 m	Silty mudstone, greenish gray, massive
196-6	750.23—751.01 m	0.78 m	Muddy siltstone, greenish gray, discontinuous wavy ripple bedding and deformation bedding, weak bioturbation, with irregular mudstone and siltstone strips
196-7	751.01—751.07 m	0.06 m	Siltstone, light gray, wavy cross bedding, erosion surface at the base
196-8	751.07—751.09 m	0.02 m	Mudstone, greenish gray, wavy ripple bedding
197-1	751.09—754.04 m	2.95 m	Mudstone, greenish gray, massive, calcareous concretions
197-2	754.04—754.30 m	0.26 m	Mudstone, greenish gray, massive, with fossil fragments and calcareous concretions
197-3	754.30—754.65 m	0.35 m	Muddy siltstone, grayish green, massive, fossil fragments
197-4	754.65—755.77 m	1.12 m	Siltstone, light greenish gray, massive, with lots of calcareous concretions
197-5	755.77—756.12 m	0.35 m	Muddy siltstone, light greenish gray, slump structures, with irregular siltstone and mudstone strips

197-6	756.12—757.43 m	1.31 m	Siltstone, light greenish, gray, wave-generated cross-bedding, weak erosion at the base, calcareous concretions and few mudstone laminae, 1 cm-thick grayish-brown mudstone at the base
197-7	757.43—757.94 m	0.51 m	Muddy siltstone, light greenish gray, slump structures, convolution bedding locally, with irregular mudstone and siltstone strips
197-8	757.94—758.32 m	0.38 m	Silty mudstone, light greenish gray, horizontal bedding, wavy ripple bedding at the base
197-9	758.32—758.42 m	0.10 m	Fine-grained sandstone, light gray, slump structures, with mudstone laminae and strips
197-10	758.42—758.70 m	0.28 m	Muddy siltstone, light greenish gray, wavy bedding, with irregular mudstone strips
197-11	758.70—758.97 m	0.27 m	Siltstone, light gray, wavy cross bedding, mudstone strips, mud gravels at the top
197-12	758.97—759.52 m	0.55 m	Fine-grained sandstone, light gray, wedge-shaped cross bedding
198-1	759.99—760.09 m	0.10 m	Mudstone, greenish gray, massive, with few calcareous concretions
198-2	760.09—760.29 m	0.20 m	Fine-grained sandstone, light gray, oblique bedding, with mudstone strips
198-3	760.29—760.39 m	0.10 m	Fine-grained conglomerate, greenish gray, normal graded bedding, erosion surface at the base
198-4	760.39—760.96 m	0.57 m	Fine-grained sandstone, light gray, parallel bedding, wedge-shaped cross bedding at the top, with few charcoal
198-5	760.96—761.23 m	0.27 m	Siltstone, light gray, parallel bedding, wavy bedding at the base, with large number of charcoal
198-6	761.23—762.03 m	0.80 m	Sandstone, fine-grained, light gray, unidirectional oblique bedding at the top, wavy bedding and climbing bedding in the middle section, wave-generated cross-bedding in the lower section
198-7	762.03—762.22 m	0.19 m	Mudstone, greenish gray, massive, with lots of fossil fragments
198-8	762.22—762.46 m	0.24 m	Siltstone, light gray, parallel bedding and wavy ripple bedding, mudstone strips
198-9	762.46—762.57 m	0.11 m	Fine-grained conglomerate, greenish gray, massive, erosion surface at the base, with calcareous concretions
198-10	762.57—765.46 m	2.89 m	Mudstone, greenish gray, massive, fossil fragments and few calcareous concretions
198-11	765.46—765.78 m	0.32 m	Silty mudstone, grayish brown and greenish gray, massive, with irregular siltstone strips
198-12	765.78—766.48 m	0.70 m	Silty mudstone, grayish brown, interbedded with mudstone, wavy ripple bedding, trace fossils, siltstone strips occasionally
198-13	766.48—766.72 m	0.24 m	Muddy siltstone, light brownish gray, slump structures, convolution bedding locally, weak erosion surface at the base, with irregular mudstone and silty mudstone strips

198-14	766.72—766.83 m	0.11 m	Siltstone, light gray, wave ripple bedding, weak erosion surface at the base, with lots of grayish brown mudstone strips
198-15	766.83—767.50 m	0.67 m	Siltstone, light gray, slump structures, convolution bedding locally, weak erosion surface at the base, with irregular mudstone strips, mudstone laminae occasionally
198-16	767.50—767.61 m	0.11 m	Siltstone, light gray, wavy cross bedding, deformation bedding locally, weak erosion surface at the base, mudstone laminae
198-17	767.61—767.88 m	0.27 m	Siltstone, light gray, wavy cross bedding, weak erosion surface at the base, mudstone laminae
198-18	767.88—768.14 m	0.26 m	Siltstone, light gray, slump structures, mudstone strips and few mud gravels
199-1	768.80—769.02 m	0.22 m	Fine-grained sandstone, light gray, slump structures, with irregular mudstone
199-2	769.02—769.40 m	0.38 m	Mudstone, greenish gray, massive, few calcareous concretions
199-3	769.40—770.36 m	0.96 m	Mudstone, brownish gray, massive, few calcareous concretions
199-4	770.36—770.46 m	0.10 m	Siltstone, light gray, slump structures, irregular mudstone
199-5	770.46—770.82 m	0.36 m	Silty mudstone, greenish gray, massive, with calcareous concretions and fossil fragments, snail fossils
200-1	771.13—771.53 m	0.40 m	Mudstone, brownish gray, massive, calcareous concretions occasionally
200-2	771.53—772.48 m	0.95 m	Silty mudstone, brownish-gray, massive bedding, weak erosion at the base, 4 cm-thick greenish-gray mudstone laminae at the base, greenish-gray mud gravels occasionally
200-3	772.48—772.96 m	0.48 m	Silty mudstone, grayish brown, massive, slump structure in the middle part, siltstone strips and calcareous concretions
200-4	772.96—773.22 m	0.26 m	Mudstone, grayish brown, massive, with siltstone strips occasionally
200-5	773.22—773.93 m	0.71 m	Silty mudstone, grayish-brown, ripple bedding, siltstone laminae and calcareous concretions occasionally, with mudstone laminae at 773.47—773.53 m
200-6	773.93—774.43 m	0.50 m	Mudstone, light brownish gray, massive, with massive fossils
200-7	774.43—774.73 m	0.30 m	Mudstone, greenish gray, massive, with fossil fragments
200-8	774.73—775.09 m	0.36 m	Silty mudstone, greenish gray, massive, with calcareous concretions and fossil fragments
200-9	775.09—775.28 m	0.19 m	Mudstone, greenish gray, massive, fossil fragments
201-1	776.49—776.84 m	0.35 m	Silty mudstone, brownish gray, interbedded with mudstone, wavy bedding and wavy ripple bedding, deformation bedding locally, weak erosion surface at the base, with siltstone strips and calcareous concretions locally
201-2	776.84—777.24 m	0.40 m	Muddy siltstone, gray, wavy ripple bedding, weak erosion surface at the base, mudstone strips
201-3	777.24—777.43 m	0.19 m	Calcareous siltstone, light gray, deformation structure and water-escape structure at the top, climbing bedding in the middle, wave-generated cross-bedding in the lower part

201-4	777.43—777.77 m	0.34 m	Siltstone, light gray, wave ripple bedding, mudstone strips
201-5	777.77—777.93 m	0.16 m	Silty mudstone, light gray, slump structures
201-6	777.93—778.13 m	0.20 m	Siltstone, light gray, deformation bedding in the upper part, climbing bedding in the middle part, wavy bedding in the lower part, weak erosion surface at the base, with few mudstone strips
201-7	778.13—778.32 m	0.19 m	Muddy siltstone, light brownish gray, slump structures, with irregular mudstone strips
201-8	778.32—778.51 m	0.19 m	Siltstone, light gray, deformation bedding in the upper pat, wave ripple bedding and climbing bedding in the middle and lower part, with mud gravels at the top
201-9	778.51—778.67 m	0.16 m	Siltstone, light gray, parallel bedding, mud gravels on the erosion surface at the base
201-10	778.67—778.84 m	0.17 m	Siltstone, light brownish gray, slump structures, with lots of mudstone strips
201-11	778.84—779.82 m	0.98 m	Siltstone, light gray, wavy cross bedding, climbing bedding and deformation bedding locally, weak erosion surface at the base, with mudstone strips
201-12	779.82—780.82 m	1.00 m	Fine-grained sandstone, light gray, wavy cross bedding, climbing bedding locally, parallel bedding at the base, few mudstone strips
201-13	780.82—780.96 m	0.14 m	Siltstone, light gray, wavy ripple bedding, wavy cross bedding at the base, with fossil fragments
201-14	780.96—781.42 m	0.46 m	Fine-grained sandstone, light gray, wavy cross bedding, climbing bedding locally, mud gravels occasionally
201-15	781.42—781.69 m	0.27 m	Fine-grained conglomerate, light gray, wedge-shaped cross bedding, directionally aligned mud gravels, erosion surface at the base
201-16	781.69—782.03 m	0.34 m	Siltstone, light gray, wave ripple bedding
201-17	781.03—782.33 m	1.30 m	Siltstone, light gray, parallel bedding, weak erosion surface at the base, mudstone strips occasionally
201-18	782.33—782.44 m	0.11 m	Siltstone, light gray, parallel bedding, with lots of charcoal and mudstone strips.
201-19	782.44—782.52 m	0.08 m	Mudstone, gray, wavy ripple bedding, with few siltstone
201-20	782.52—782.75 m	0.23 m	Muddy siltstone, light gray, intercalated with gray mudstone, wavy bedding
201-21	782.75—782.94 m	0.19 m	Siltstone, light gray, climbing bedding, mudstone strips
201-22	782.94—783.84 m	0.90 m	Sandstone, fine-grained, light gray, wave-generated sand ripple bedding at the top, parallel beddings in the middle and lower parts, mud gravels at the base, weak erosion
202-1	784.64—785.12 m	0.48 m	Silty mudstone, grayish brown, massive, with few calcareous concretions and siltstone
202-2	785.12—785.83 m	0.71 m	Siltstone, light gray, wave ripple bedding and wavy bedding
202-3	785.83—786.37 m	0.54 m	Sandstone, light gray, wedge-shaped cross bedding
203-1	786.37—787.33 m	0.96 m	Mudstone, grayish brown, massive, with calcareous concretions
204-1	787.33—787.56 m	0.23 m	Mudstone, grayish brown, massive, with calcareous concretions

Section 5 SK-1 Core Description and Core Photographs

204-2	787.56—790.12 m	2.56 m	Mudstone, brownish gray, massive, with fossil fragments and calcareous concretions
204-3	790.12—790.40 m	0.28	Muddy siltstone, light gray, wavy ripple bedding, erosion surface at the base, trace fossils, calcareous concretions
204-4	790.40—790.69 m	0.29 m	Mudstone, brownish gray, wavy ripple bedding, with few fossil fragments
204-5	790.69—791.30 m	0.61 m	Siltstone, light gray, wavy cross bedding
204-6	791.30—791.97 m	0.67 m	Muddy siltstone, light gray, horizontal bedding, wavy bedding locally
204-7	791.97—793.32 m	1.35 m	Muddy siltstone, medium light gray with slump structures, roll-up bedding locally, irregular mudstone laminae and sandstone strips
205-1	793.32—794.64 m	1.32 m	Siltstone, medium light gray, with charcoals, wavy cross bedding, ripple bedding locally, weak erosion at the base, carbonized fossil fragments on the bedding surface, calcareous siltstone interlayer locally, 3—4 cm mudstone laminae at the top, small synsedimentary normal fault
205-2	794.64—794.92 m	0.28 m	Muddy siltstone, medium gray, with slump structures, irregular mudstone laminae and few siltstone laminae, a lot of charcoals locally
205-3	794.92—794.99 m	0.07 m	Siltstone, light gray, wavy bedding, with some charcoal
205-4	794.99—795.17 m	0.18 m	Mudstone, greenish gray, horizontal bedding
205-5	795.17—795.52 m	0.35 m	Siltstone, gray, wavy cross bedding, with lots of charcoal
205-6	795.52—795.70 m	0.18 m	Siltstone, gray, slump structures, weak erosion surface at the base, with few irregular mudstone strips
205-7	795.70—795.78 m	0.08 m	Mudstone, gray, wavy ripple bedding
205-8	795.78—797.07 m	1.29 m	Sandstone, fine-grained, medium gray, wave-generated sand ripple bedding, deformation structure locally, erosion at the base, mudstone laminae, thin interlayer at the base, 10 cm calcareous siltstone, a lot of charcoals on the layer surface
205-9	797.07—797.17 m	0.10 m	Silty mudstone, gray, wavy bedding, erosion surface at the base, with mud gravels
205-10	797.17—797.38 m	0.21 m	Fine-grained sandstone, gray, wave ripple bedding, erosion surface at the base, with 4 cm-thick corse-grained sandstone
205-11	797.38—797.60 m	0.22 m	Muddy siltstone, medium gray, interbedded with mudstone, ripple bedding, and wavy bedding locally, weak erosion and also 3 cm thin siltstone at the base, pyrites occasionally
205-12	797.60—797.74 m	0.14 m	Mudstone, clastizoic, light greenish-gray, massive bedding, 1 cm medium dark gray mudstone at the base, with fossil fragments
205-13	797.74—797.83 m	0.09 m	Mudstone, silty with bioclasts, greenish-gray, massive bedding, 1 cm dark gray mudstone at the base, horizontal bedding
205-14	797.83—797.99 m	0.16 m	Conglomerate, fine grained, calcareous concretionary, medium light gray, unclear normal graded bedding, weak erosion, few mud gravels
205-15	797.99—798.14 m	0.15 m	Mudstone, greenish gray, massive, with few calcareous concretions

205-16	798.14—798.26 m	0.12 m	Silty mudstone, greenish gray, wavy ripple bedding, few calcareous concretions
205-17	798.26—798.55 m	0.29 m	Mudstone, light brownish gray, massive, with greenish gray irregular mudstone strips
205-18	798.55—798.82 m	0.27 m	Muddy siltstone, gray, slump structures, with few mud gravels
205-19	798.82—799.09 m	0.27 m	Silty mudstone, light olive gray, massive, with few siltstone
205-20	799.09—799.24 m	0.15 m	Muddy siltstone, light gray, deformation bedding, convolution bedding locally, erosion surface
205-21	799.24—799.35 m	0.11 m	Silty mudstone, light olive gray, massive
205-22	799.35—799.50 m	0.15 m	Muddy siltstone, light gray, climbing bedding in the upper and middle parts, wavy cross bedding in the lower part, with siltstone strips, trace fossils locally
205-23	799.50—802.21 m	2.71 m	Muddy siltstone, gray, slump structures, with irregular mudstone and siltstone strips, calcareous concretions and mudstone laminae
206-1	802.21—803.46 m	1.25 m	Muddy siltstone, gray, slump structures, with irregular mudstone and silty mudstone strips, bioturbation
206-2	803.46—804.17 m	0.71 m	Siltstone, gray, slump structures, mudstone strips
206-3	804.17—804.27 m	0.10 m	Siltstone, light gray, wave ripple bedding, weak erosion surface at the base, mudstone strips
206-4	804.27—804.61 m	0.34 m	Muddy siltstone, gray, slump structures, irregular mudstone strips, siltstone strips
206-5	804.61—804.74 m	0.13 m	Siltstone, light gray, wave ripple bedding, weak erosion surface at the base, mudstone strips, mud gravels at the base
206-6	804.74—805.04 m	0.30 m	Fine-grained conglomerate, light gray, massive, erosion surface at the base, with mussel fossils and fossil fragments
206-7	805.04—805.27 m	0.23 m	Siltstone, light gray, wavy ripple bedding and deformation bedding at the top, climbing bedding in the middle and lower parts, erosion surface at the base, siltstone strips
206-8	805.27—805.41 m	0.14 m	Siltstone, light gray, slump structures, erosion surface at the base
206-9	805.41—805.46 m	0.05 m	Siltstone, light gray, climbing bedding
207-1	805.46—805.73 m	0.27 m	Siltstone, medium light gray, with slump structures, 7 cm wave-generated sand ripple bedding and weak erosion at the base, irregular mudstone laminae
207-2	805.73—806.35 m	0.62 m	Siltstone, medium light gray, with slump structures, wave-generated sand ripple bedding, 3 cm-thick calcareous siltstone and weak erosion at the base, irregular mudstone laminae
207-3	806.35—806.76 m	0.41 m	Siltstone, medium light gray, ripple bedding and climbing bedding at the top, wave-generated sand ripple bedding in the middle and lower parts, weak erosion surface and irregular mudstone laminae at the base
207-4	806.76—807.02 m	0.26 m	Calcareous siltstone, light gray, wave-generated sand ripple bedding, deformation structure locally, irregular mudstone laminae, weak erosion surface at the surface
207-5	807.02—807.12 m	0.10 m	Fine-grained conglomerate, light gray, massive, weak erosion surface at the base

5.1.9 Taikang Formation

The detailed description of Taikang Formation.

1-1	164.77—165.32 m	0.55 m	Sandy conglomerate, light gray
2-1	165.33—165.73 m	0.4 m	Sandy conglomerate, light gray
3-1	165.87—166.47 m	0.6 m	Coarse conglomerate, light gray
4-1	166.97—167.89 m	0.92 m	Conglomeratic sandstones, light gray
5-1	167.89—168.39 m	0.5 m	Conglomeratic sandstones, light gray
6-1	168.70—169.32 m	0.62 m	Conglomeratic sandstones, light gray
7-1	169.36—170.19 m	0.83 m	Conglomeratic sandstones, light gray
8-1	170.19—170.83 m	0.64 m	Sandy conglomerate, light gray
9-1	171.09—171.69 m	0.6 m	Conglomeratic sandstones, light gray
10-1	172.27—172.97 m	0.7 m	Sandy conglomerate, light gray
11-1	173.01—173.53 m	0.52 m	Conglomerate, light gray
12-1	173.53—174.70 m	1.17 m	Sandy conglomerate, light gray
13-1	174.70—175.20 m	0.5 m	Conglomeratic sandstones, light gray
14-1	175.25—175.50 m	0.25 m	Conglomeratic sandstones, light gray
14-2	175.50—176.18 m	0.68 m	Sandy conglomerate, light gray
16-1	176.18—176.68 m	0.5 m	Sandy conglomerate, light gray
16-2	176.68—177.96 m	1.28 m	Conglomeratic sandstones, light gray
17-1	177.96—178.56 m	0.6 m	Sandy conglomerate, light gray
18-1	179.16—179.72 m	0.56 m	Sandy conglomerate, light gray
19-1	179.72—179.92 m	0.2 m	Sandy conglomerate, light gray
19-2	179.92—180.17 m	0.25 m	Fine-grained sandstone, light gray
20-1	180.30—180.57 m	0.27 m	Medium-grained sandstone, light gray
21-1	180.57—181.26 m	0.66 m	Conglomeratic sandstones, light gray
22-1	181.26—182.32 m	1.06 m	Conglomeratic sandstones, light gray
23-1	182.32—183.62 m	1.3 m	Conglomeratic sandstones, light gray
24-1	183.62—184.07 m	0.45 m	Conglomeratic sandstones, light gray
24-2	184.07—184.52 m	0.45 m	Fine-grained conglomerate, light gray
24-3	184.52—184.71 m	0.19 m	Conglomeratic sandstones, light gray
25-1	184.71—185.09 m	0.38 m	Fine-grained conglomerate, light gray
26-1	186.08—187.31 m	1.23 m	Fine-grained conglomerate, light gray
27-1	187.35—187.81 m	0.46 m	Conglomeratic sandstones, light gray
29-1	188.80—189.45 m	0.65 m	Fine-grained conglomerate, light gray

31-1	189.76—190.22 m	0.45 m	Fine-grained conglomerate, light gray
32-1	190.22—191.11 m	0.89 m	Fine-grained conglomerate, light gray
33-1	191.22—191.42 m	0.2 m	Fine-grained conglomerate, light gray
33-2	191.42—191.72 m	0.3 m	Conglomeratic sandstones, light gray
33-3	191.72—191.82 m	0.1 m	Fine-grained conglomerate, light gray
34-1	192.2—192.3 m	0.1 m	Fine-grained conglomerate, light gray
34-2	192.3—192.7 m	0.4 m	Sandy conglomerate, medium light gray, interbedded with medium-grained sandstone
34-3	192.7—193 m	0.3 m	Fine-grained sandstone, light gray
35-1	193—193.2 m	0.2 m	Medium-grained sandstone, light gray
35-2	193.2—193.5 m	0.3 m	Fine-grained conglomerate, light gray
36-1	193.54—193.98 m	0.44 m	Medium-grained sandstone, light gray
38-1	194.2—194.70 m	0.5 m	Conglomeratic sandstones, light gray
38-2	194.7—195.52 m	1.32 m	Sandy conglomerate, light gray
39-1	195.52—196.15 m	0.63 m	Sandy conglomerate, light gray
40-1	196.55—196.87 m	0.32 m	Sandy conglomerate, light gray
41-1	197.52—198.05 m	0.53 m	Conglomeratic sandstones, light gray
42-1	198.05—198.5 m	0.45 m	Conglomeratic sandstones, light gray
42-1	198.5—198.69 m	0.19 m	Sandy conglomerate, light gray
43-1	198.69—199.44 m	0.49 m	Fine-grained sandstone, light gray
44-1	199.44—199.49 m	0.05 m	Mudstone, olive gray
44-2	199.49—200.22 m	0.73 m	Conglomeratic sandstones, light gray
44-3	200.22—200.32 m	0.1 m	Coarse sandstone, light gray
45-1	200.36—200.79 m	0.43 m	Conglomeratic sandstones, light gray
45-2	200.79—201.19 m	0.4 m	Sandy conglomerate, light gray
45-3	201.19—201.29 m	0.1 m	Conglomeratic sandstones, light gray
46-1	201.34—201.63 m	0.29 m	Conglomeratic sandstones, light gray
46-2	201.63—201.73 m	0.1 m	Sandy conglomerate, light gray
47-1	201.88—201.98 m	0.1 m	Sandy conglomerate, light gray
47-2	201.98—202.77 m	0.79 m	Conglomeratic sandstones, light gray
48-1	202.77—202.90 m	0.13 m	Sandy conglomerate, light gray
48-2	202.90—203.14 m	0.24 m	Coarse sandstone, light gray
48-3	203.14—203.32 m	0.18 m	Sandy conglomerate, light gray
48-4	203.32—203.74 m	0.42 m	Sandy conglomerate, light gray
49-1	203.74—204.54 m	0.8 m	Sandy conglomerate, light gray
50-1	204.81—204.99 m	0.18 m	Sandy conglomerate, light gray

50-2	204.99—205.41 m	0.42 m	Siltstone, light gray
50-3	205.41—205.64 m	0.23 m	Sandy conglomerate, light gray
50-4	205.64—205.79 m	0.15 m	Conglomeratic sandstones, light gray
50-5	205.79—205.88 m	0.09 m	Sandy conglomerate, light gray
51-1	205.88—206.24 m	0.36 m	Sandy conglomerate, light gray
51-2	206.24—206.72 m	0.48 m	Sandy conglomerate, light gray
51-3	206.72—207.07 m	0.35 m	Sandy conglomerate, light gray
52-1	207.07—207.63 m	0.56 m	Sandy conglomerate, light gray
52-2	207.63—208.24 m	0.61 m	Conglomeratic sandstones, light gray
53-1	208.24—208.40 m	0.16 m	Conglomeratic sandstones, light gray
53-2	208.40—208.61 m	0.21 m	Sandy conglomerate, light gray
53-3	208.61—209.01 m	0.4 m	Coarse sandstone, light gray
54-1	209.43—209.68 m	0.25 m	Coarse sandstone, light gray.
56-1	209.88—210.57 m	0.69 m	Sandy conglomerate, light gray

5.2 SK-I Core Photographs

5.2.1 SK-I s Core Photographs

Section 5 SK-1 Core Description and Core Photographs | 367

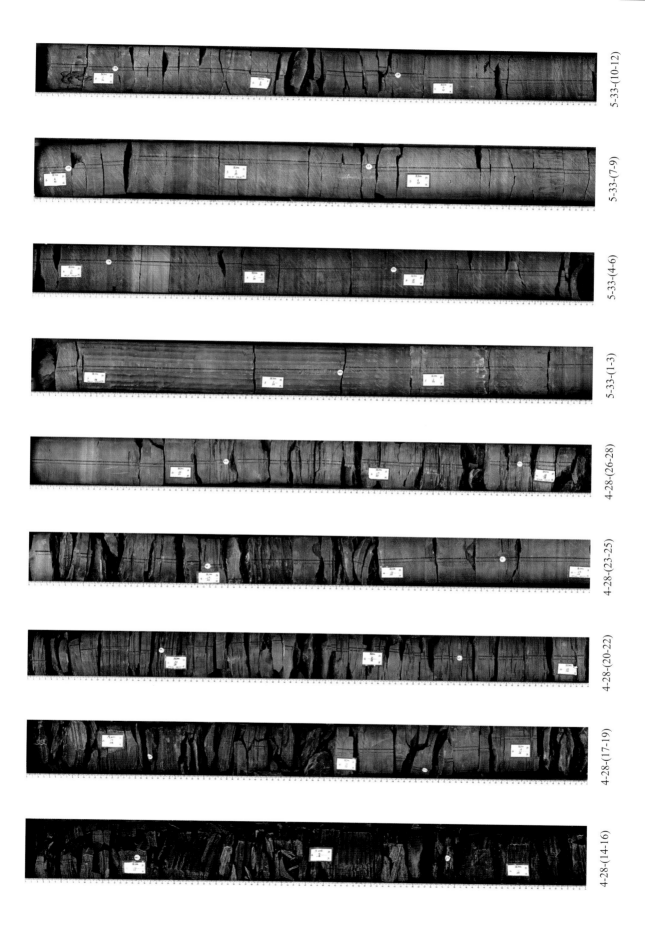

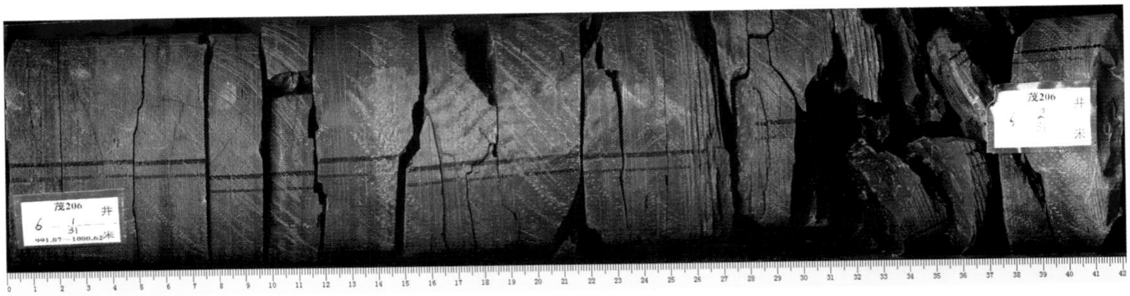

368 | Initial Report of Continental Scientific Drilling Project of the Cretaceous Songliao Basin (SK-1)

6-31-(1-2)

5-33-(31-33)

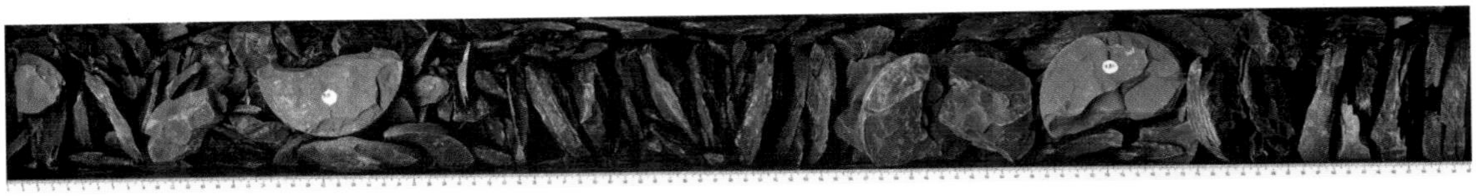

5-33-(28-30)

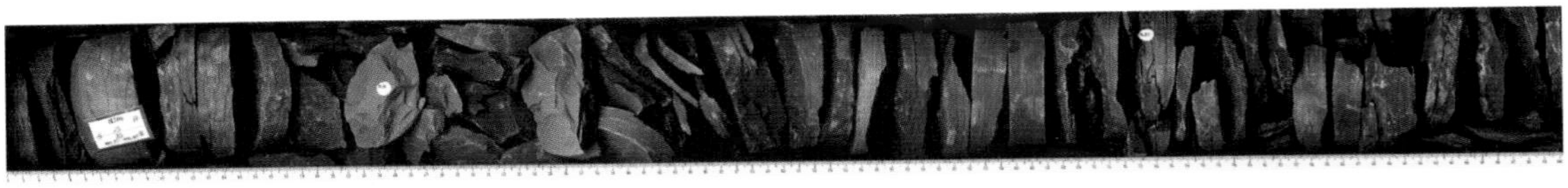

5-33-(25-27)

5-33-(22-24)

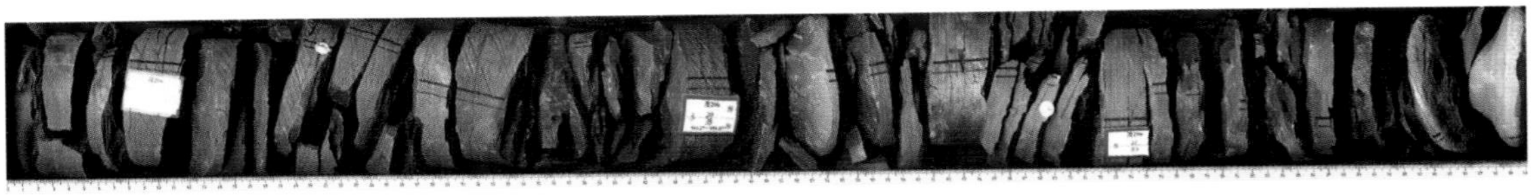

5-33-(19-21)

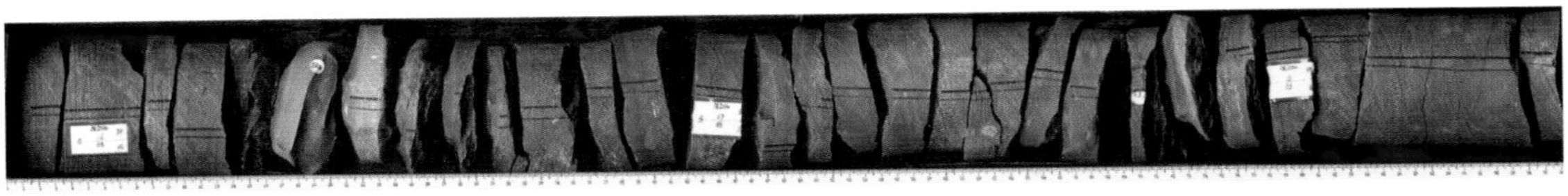

5-33-(16-18)

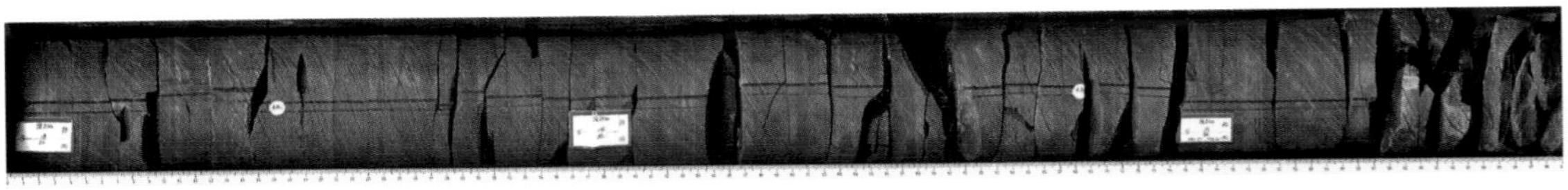

5-33-(13-15)

Section 5 SK-1 Core Description and Core Photographs

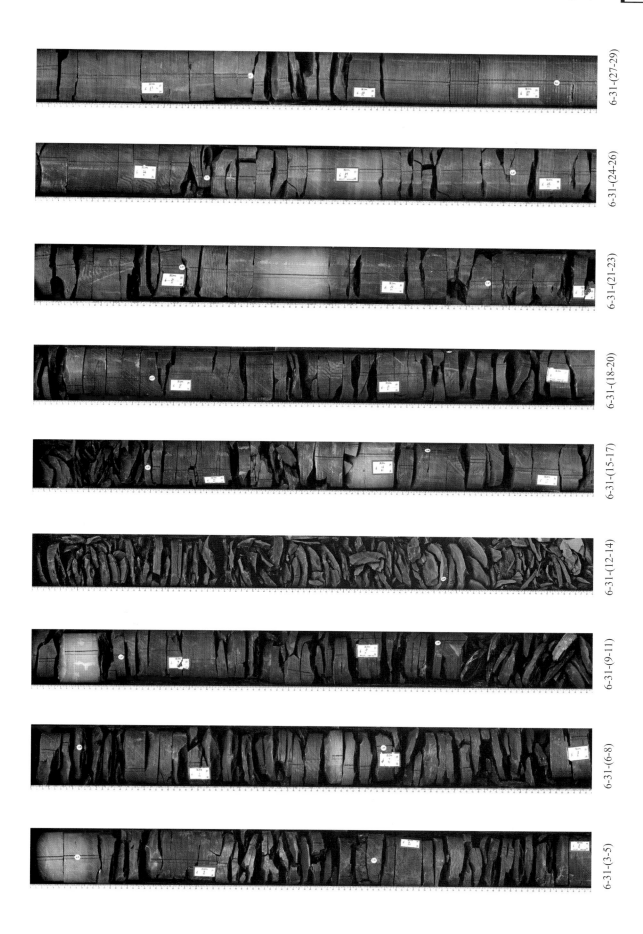

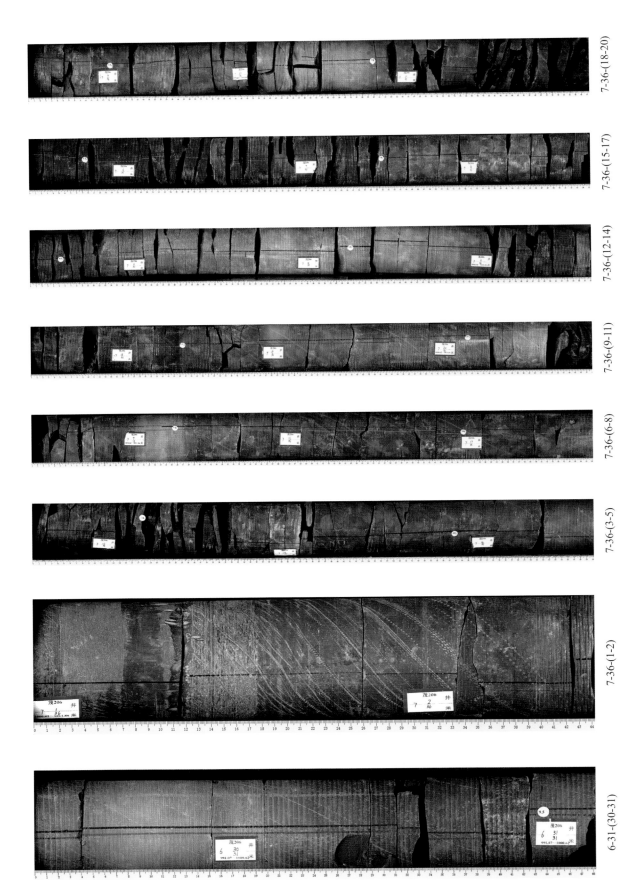

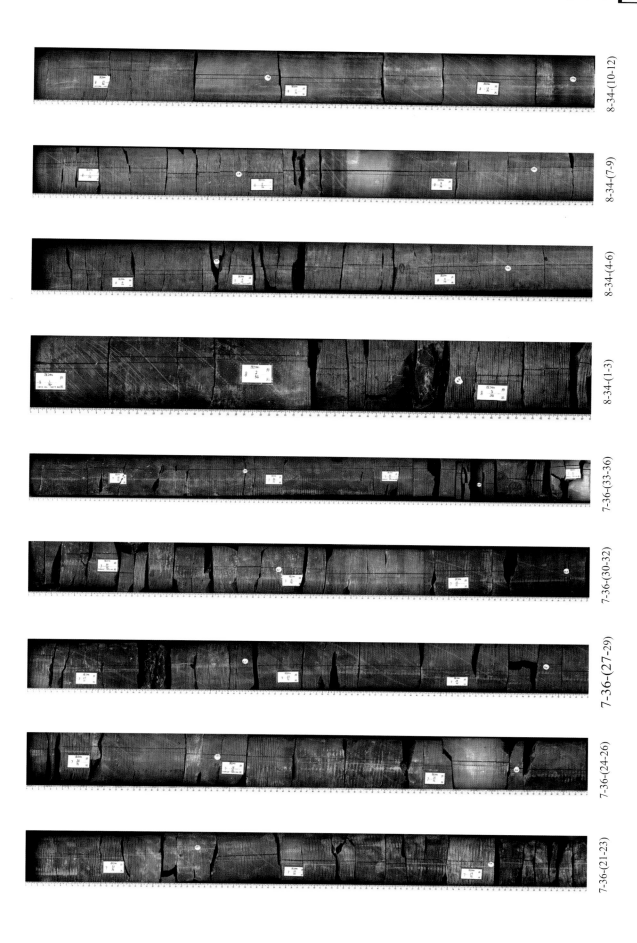

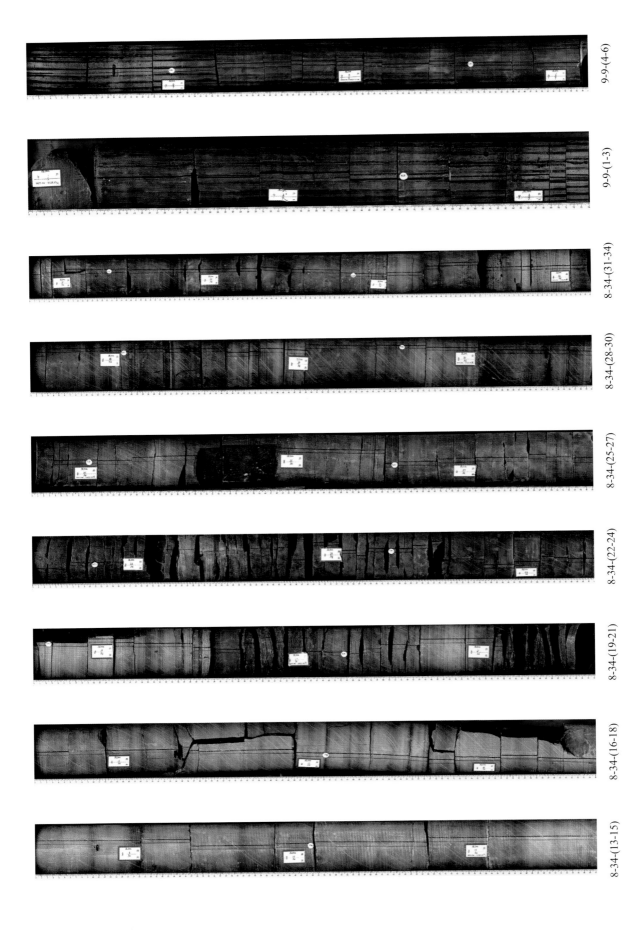

Section 5 SK-1 Core Description and Core Photographs

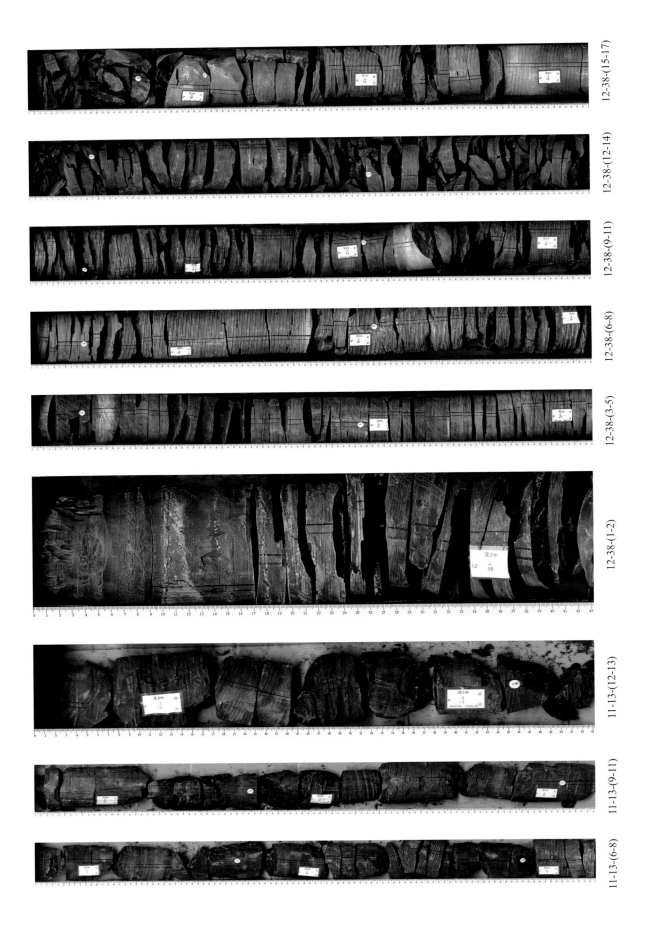

Section 5 SK-1 Core Description and Core Photographs | 375

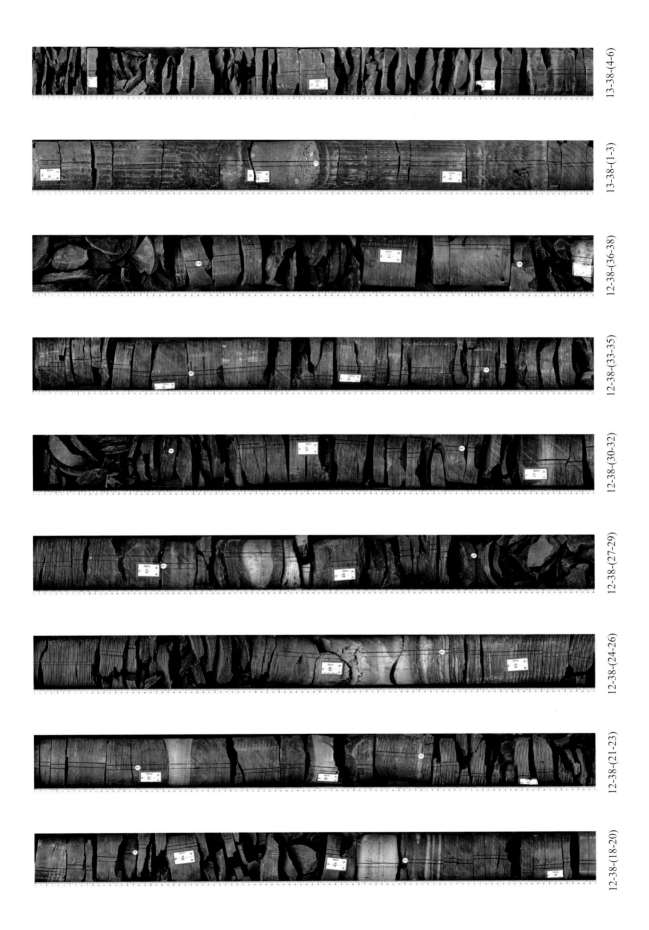

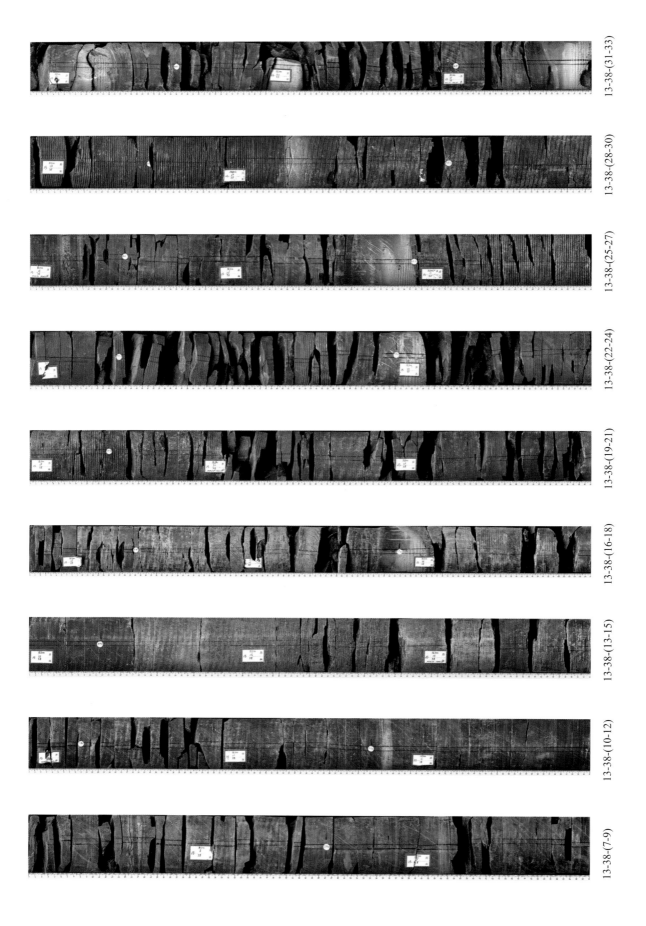

Section 5 SK-1 Core Description and Core Photographs | *377*

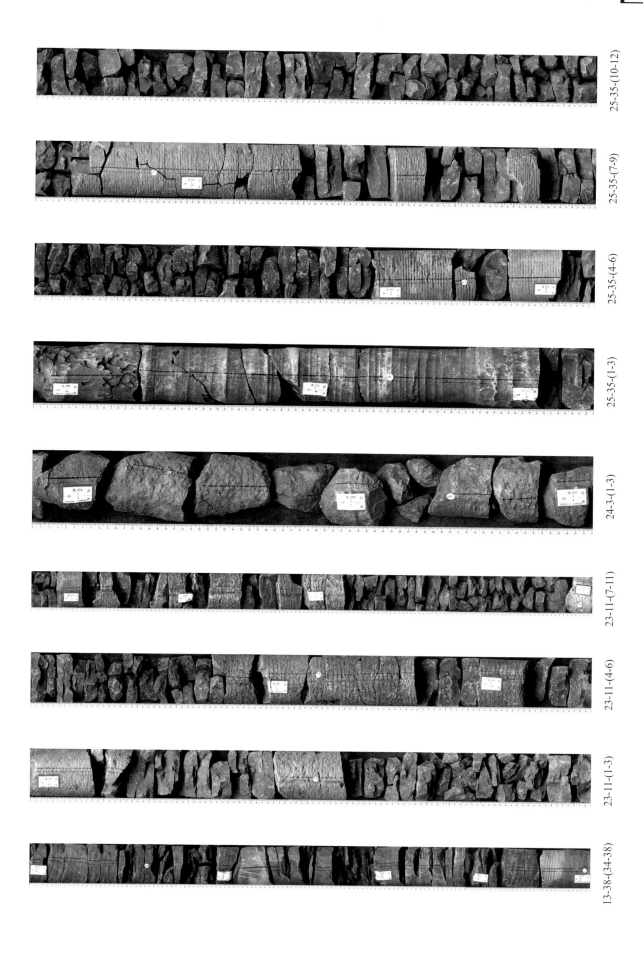

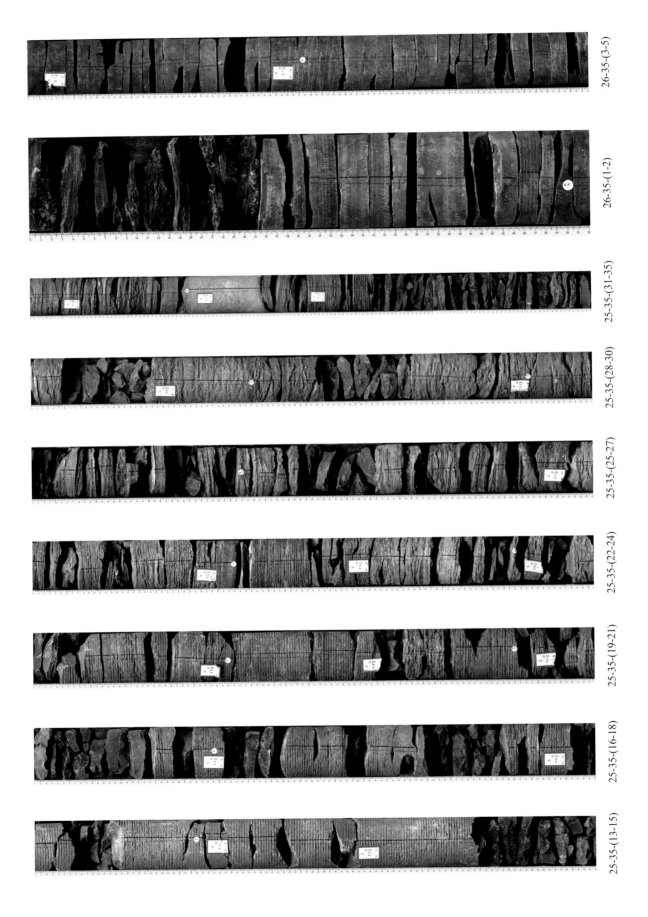

Section 5 SK-1 Core Description and Core Photographs | 379

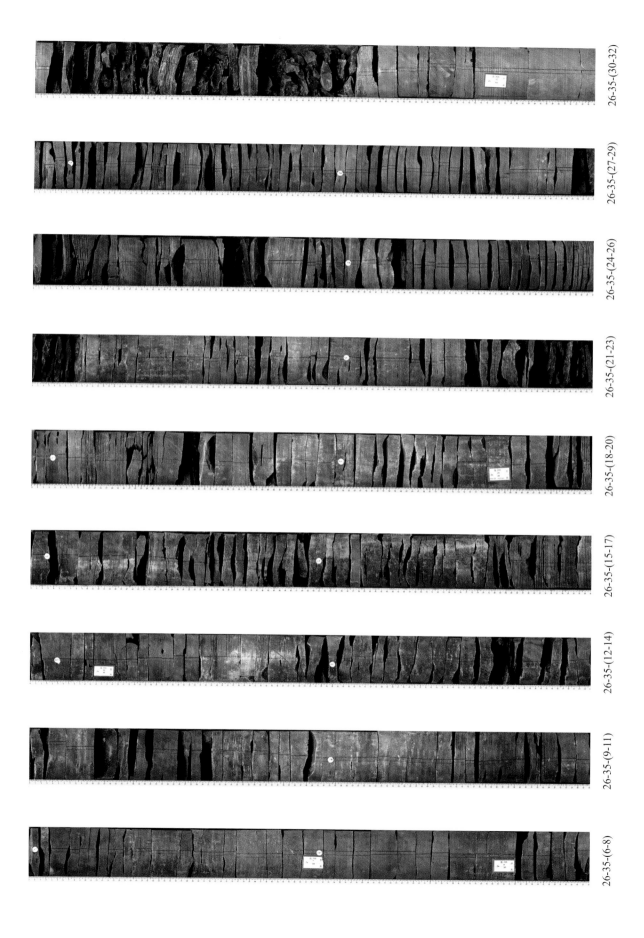

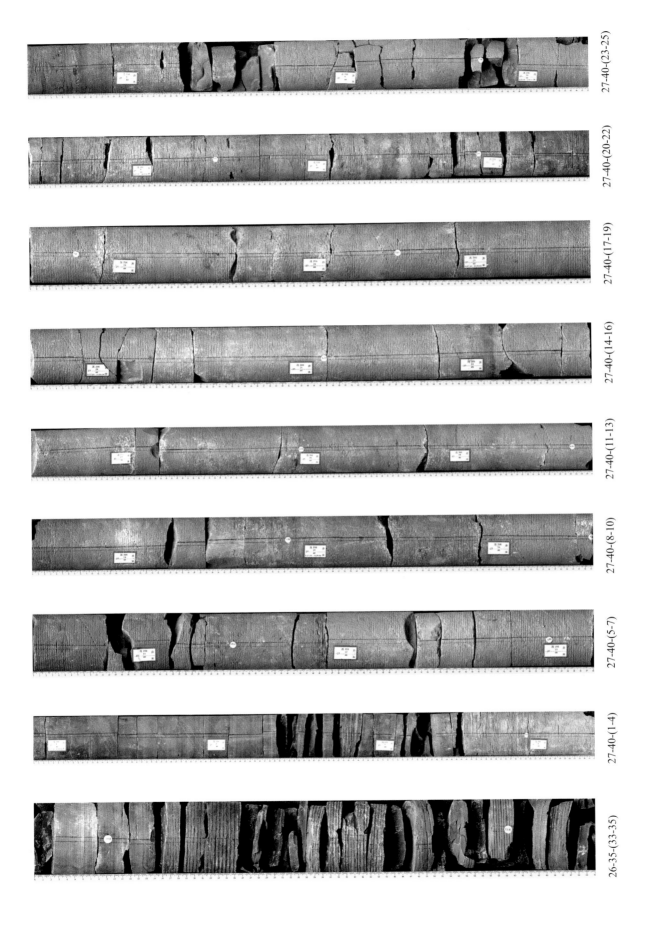

Section 5 SK-1 Core Description and Core Photographs | 381

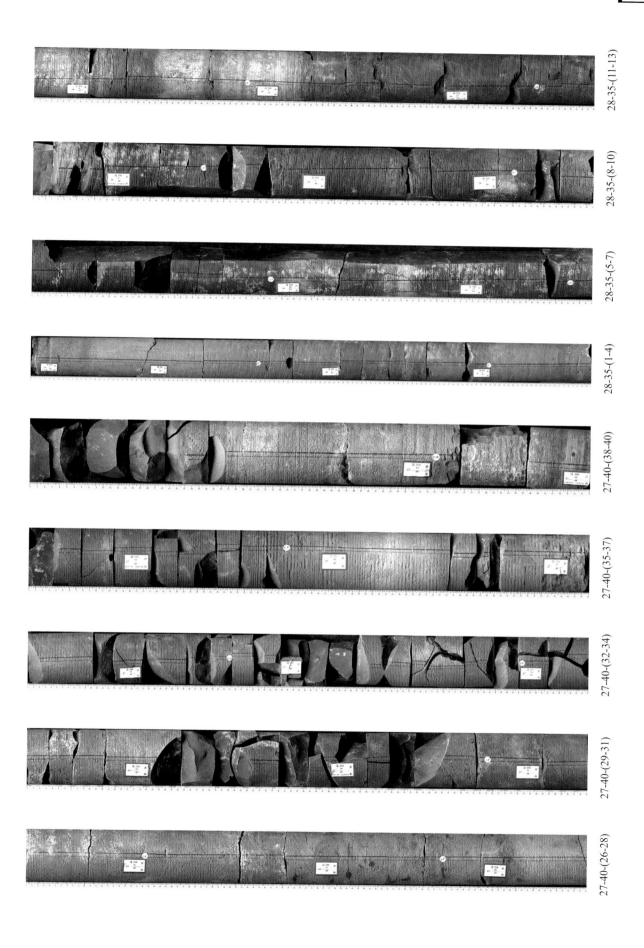

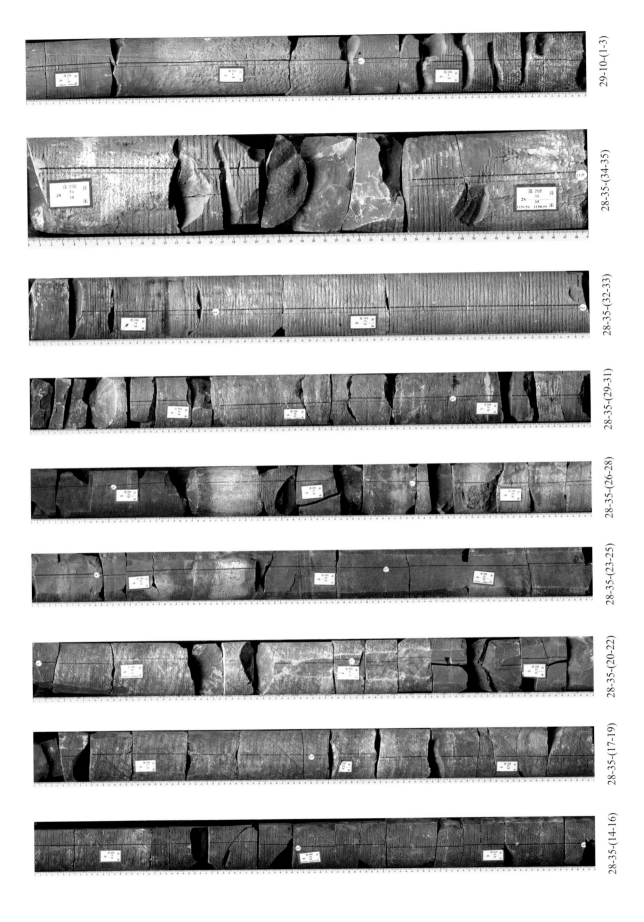

Section 5 SK-1 Core Description and Core Photographs

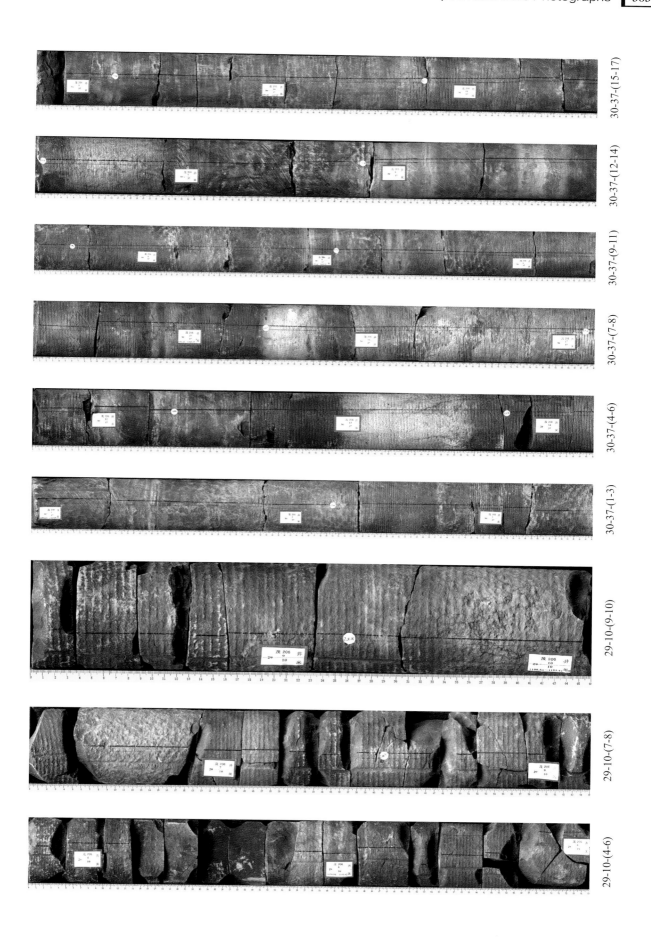

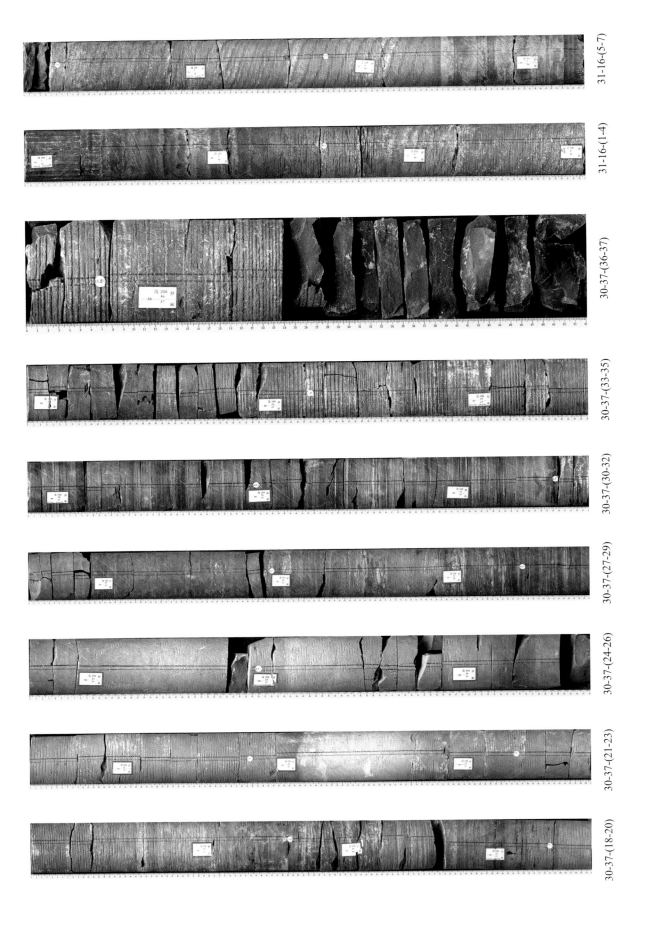

Section 5 SK-1 Core Description and Core Photographs | 385

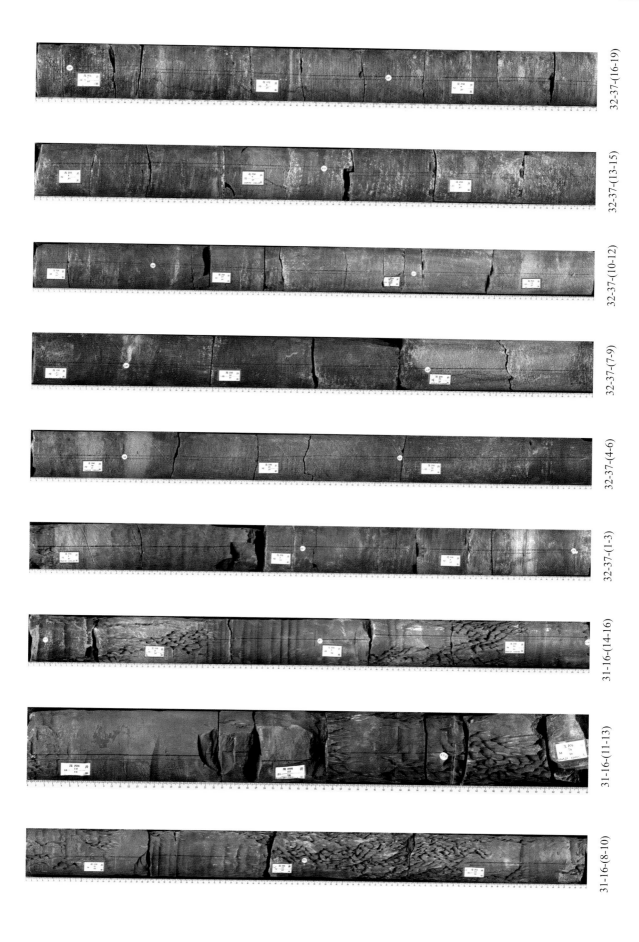

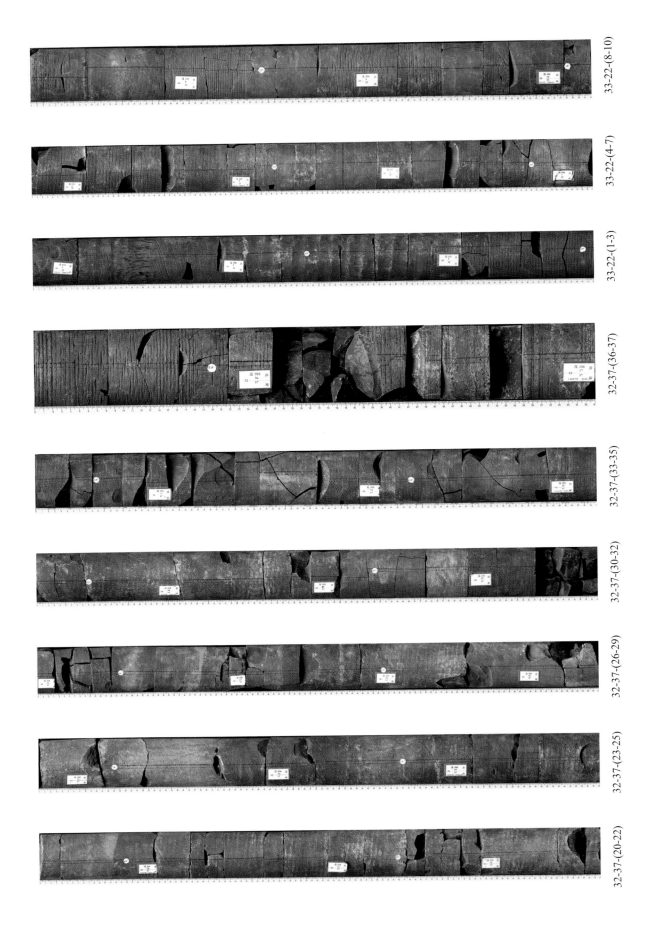

Section 5 SK-1 Core Description and Core Photographs

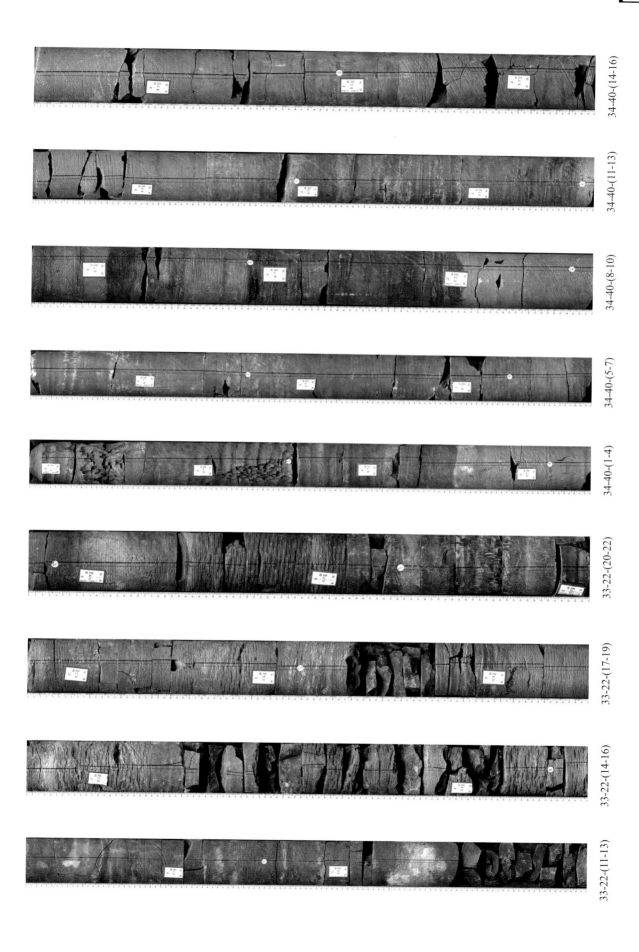

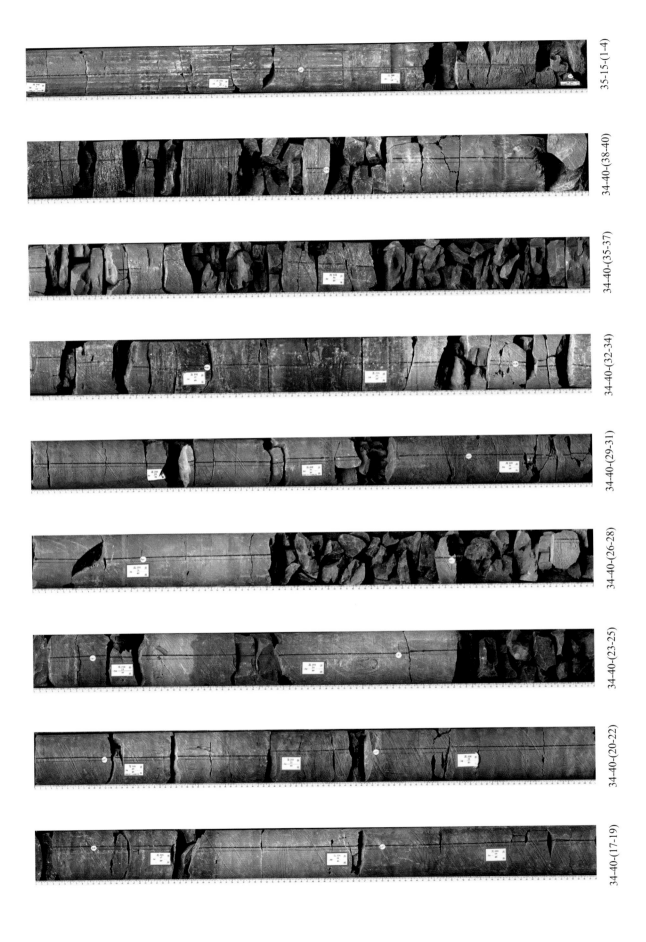

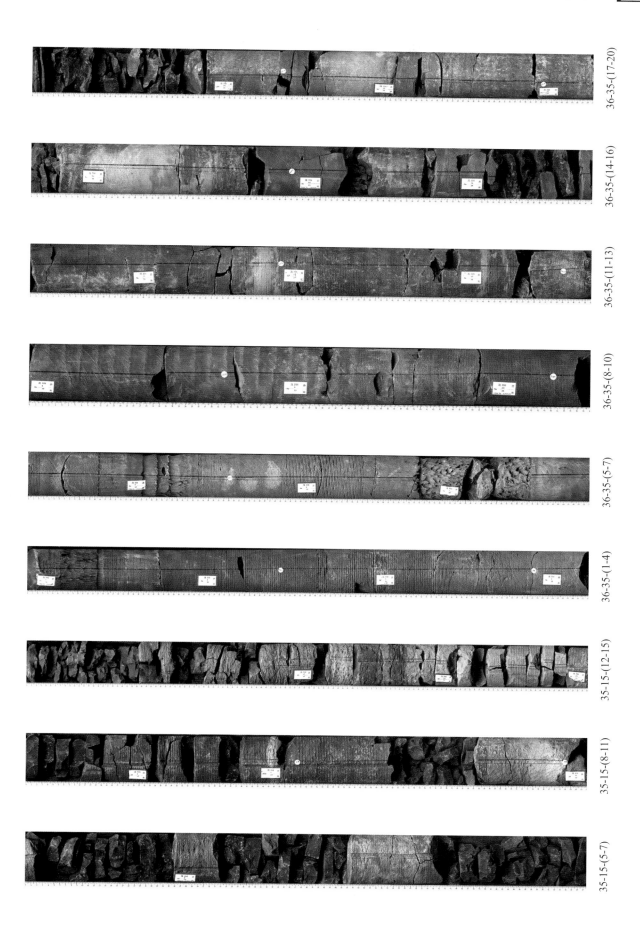

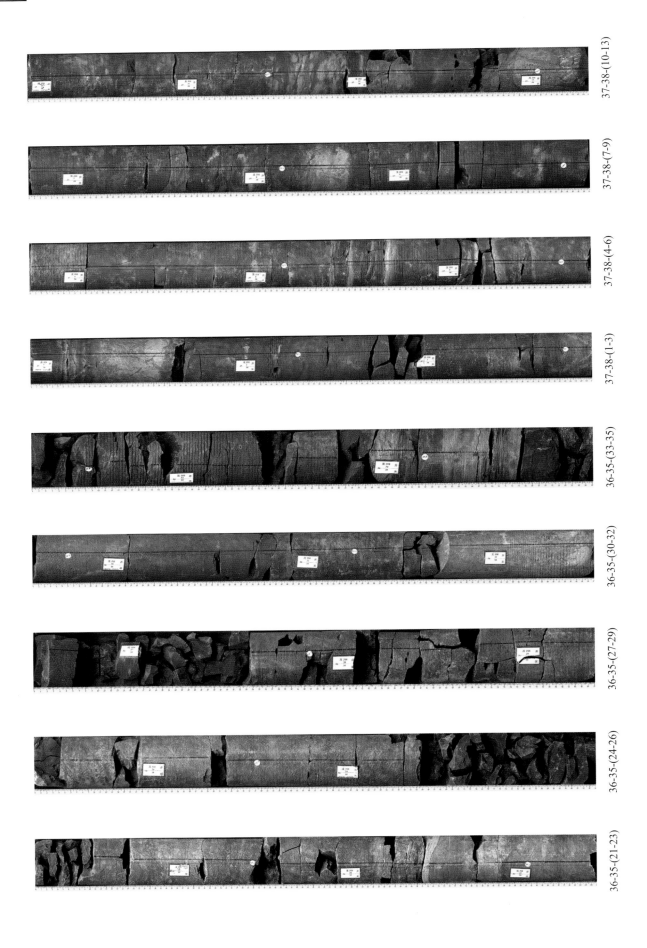

Section 5 SK-1 Core Description and Core Photographs | *391*

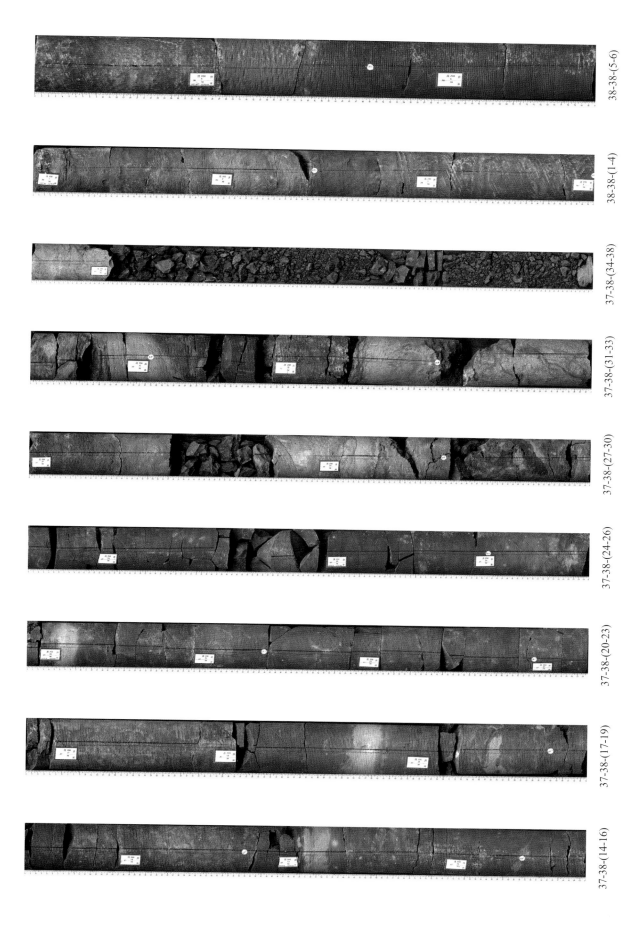

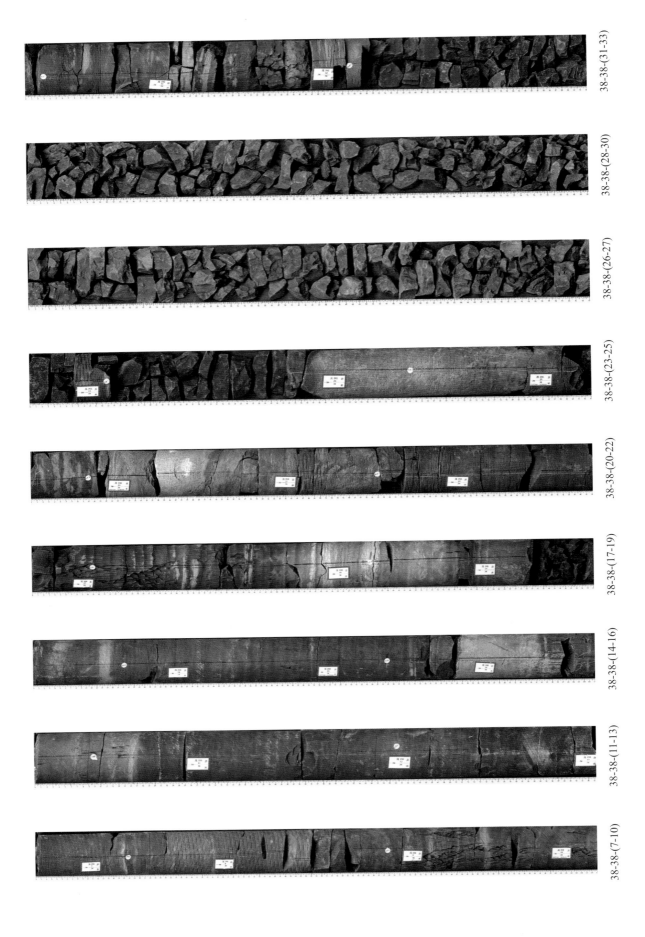

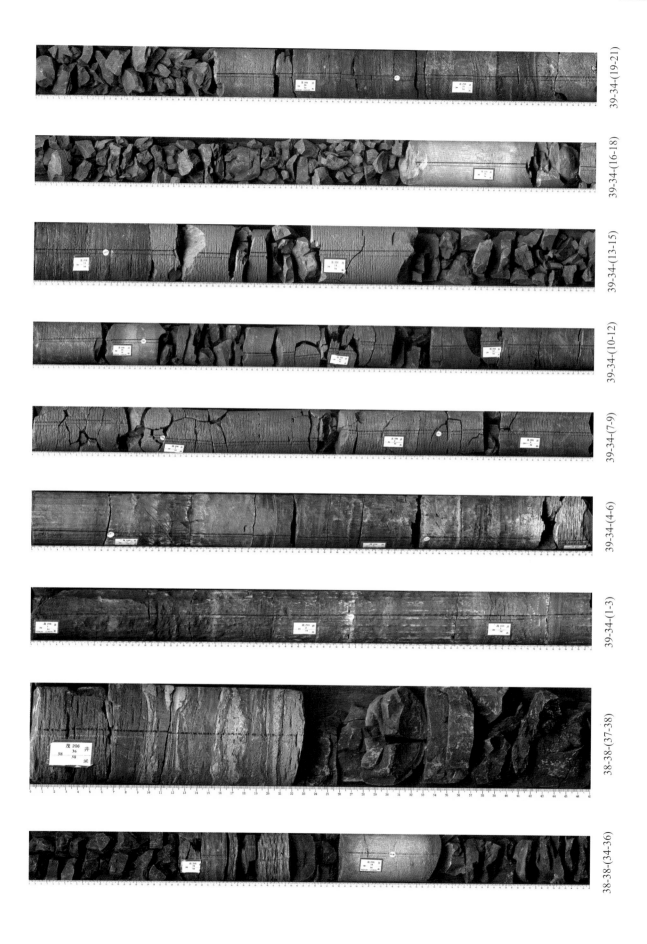

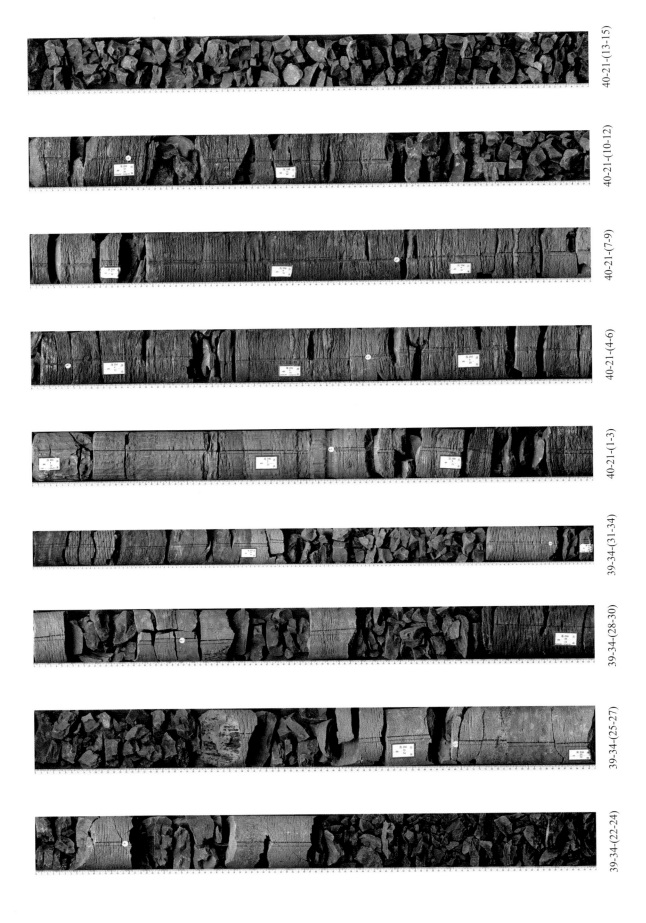

Section 5 SK-1 Core Description and Core Photographs

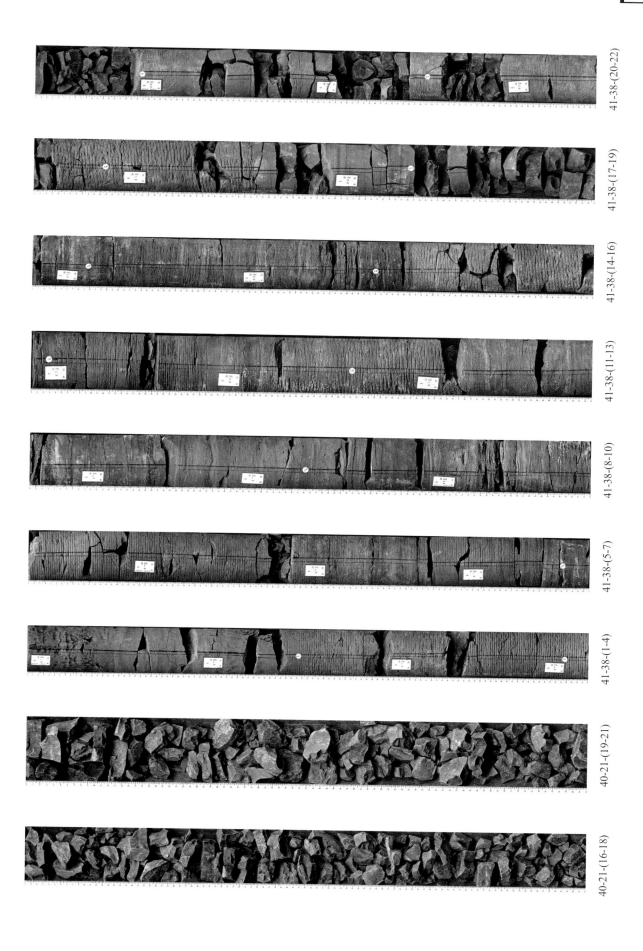

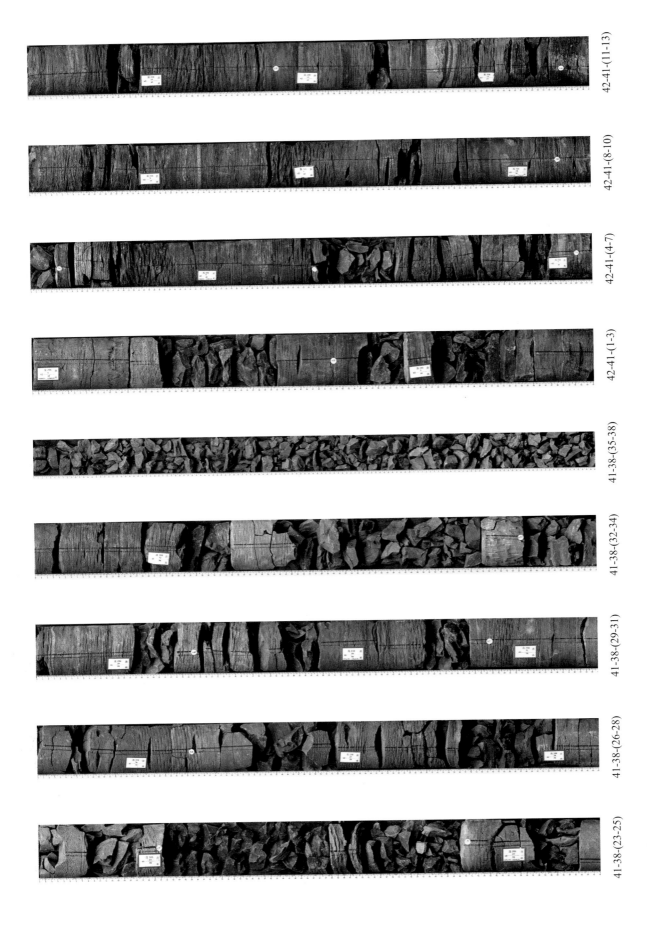

Section 5 SK-1 Core Description and Core Photographs

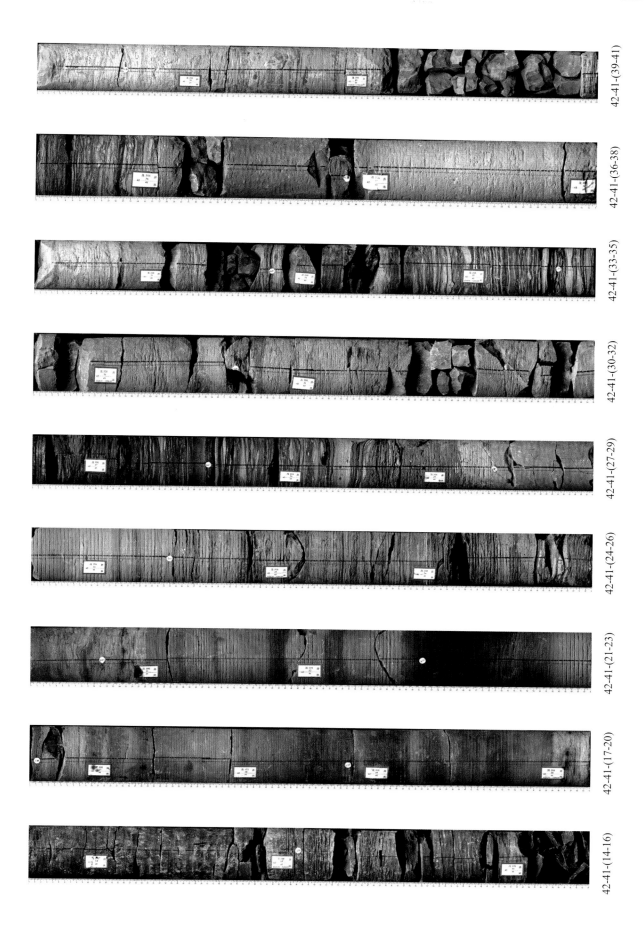

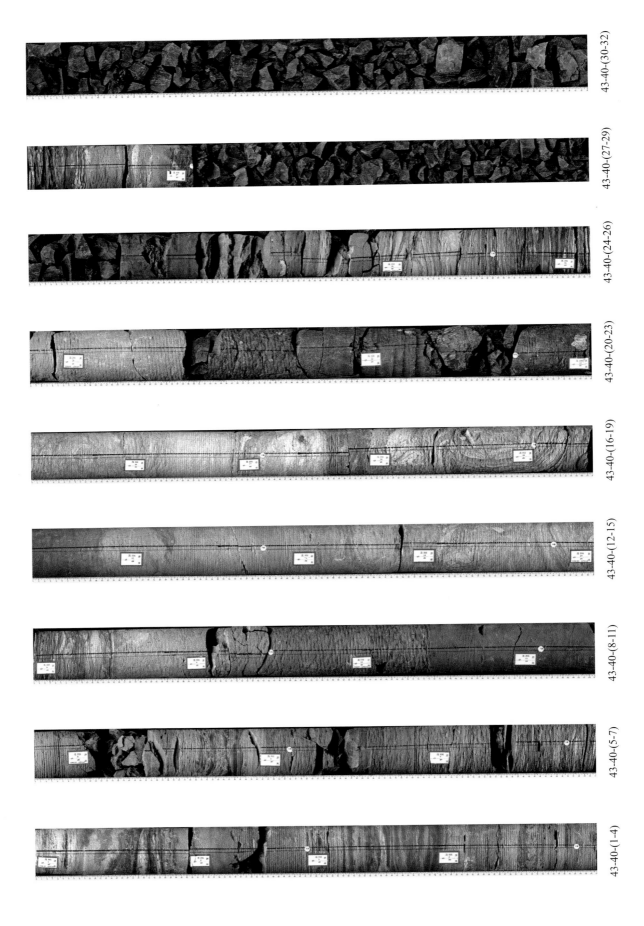

Section 5 SK-1 Core Description and Core Photographs | *399*

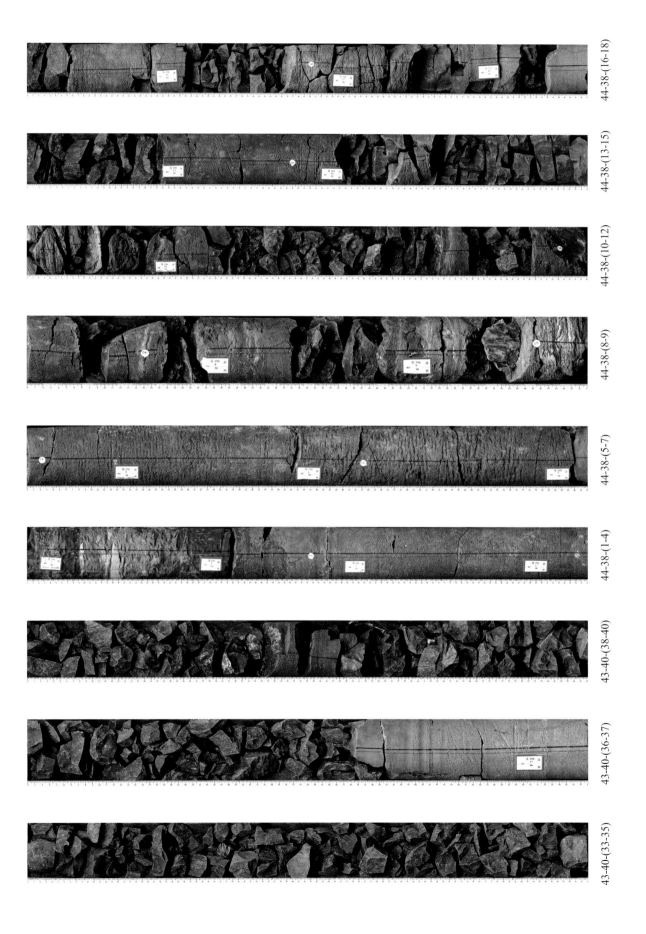

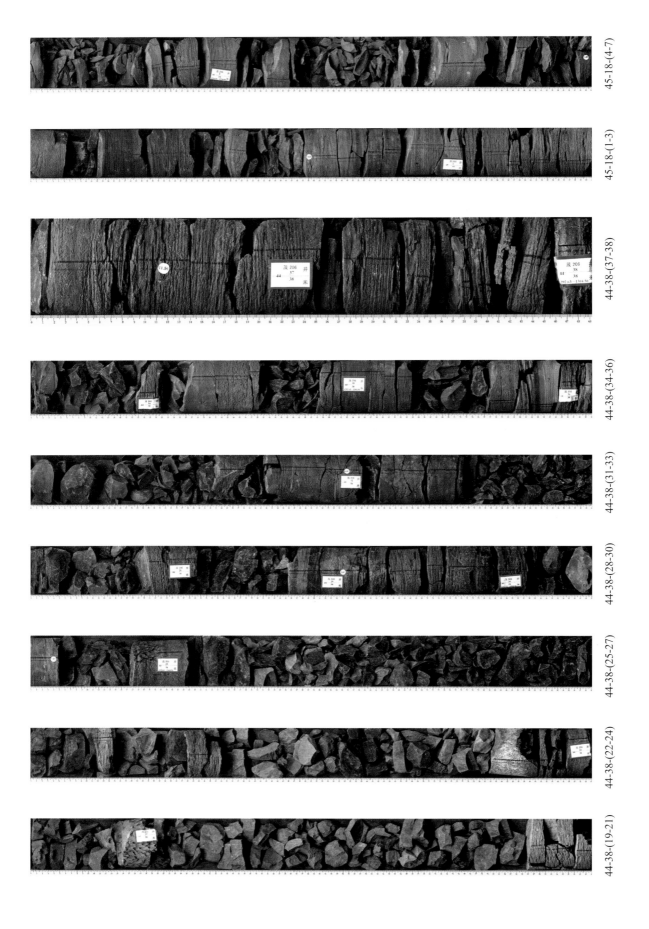

Section 5 SK-1 Core Description and Core Photographs

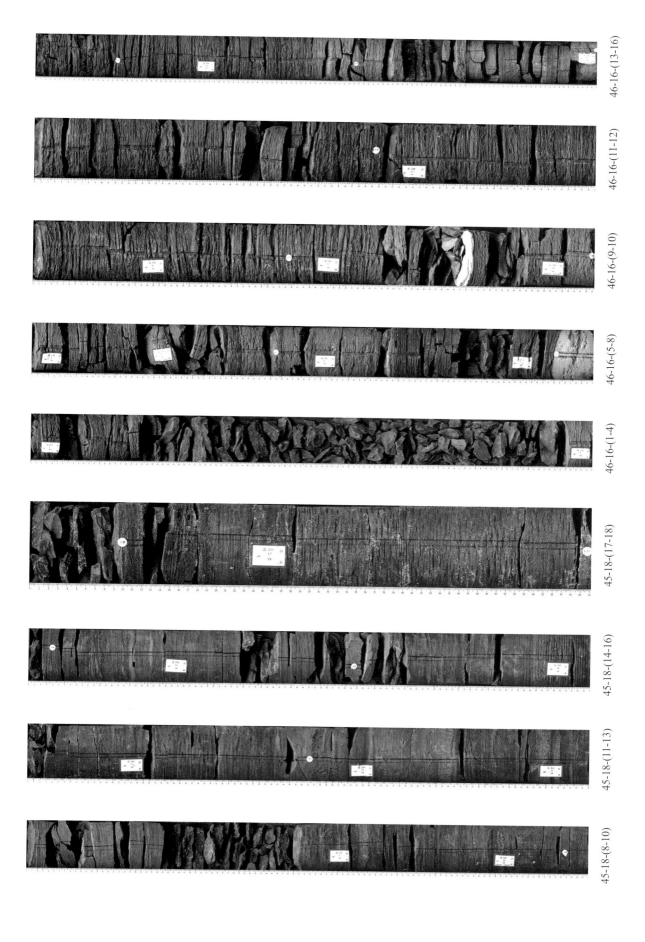

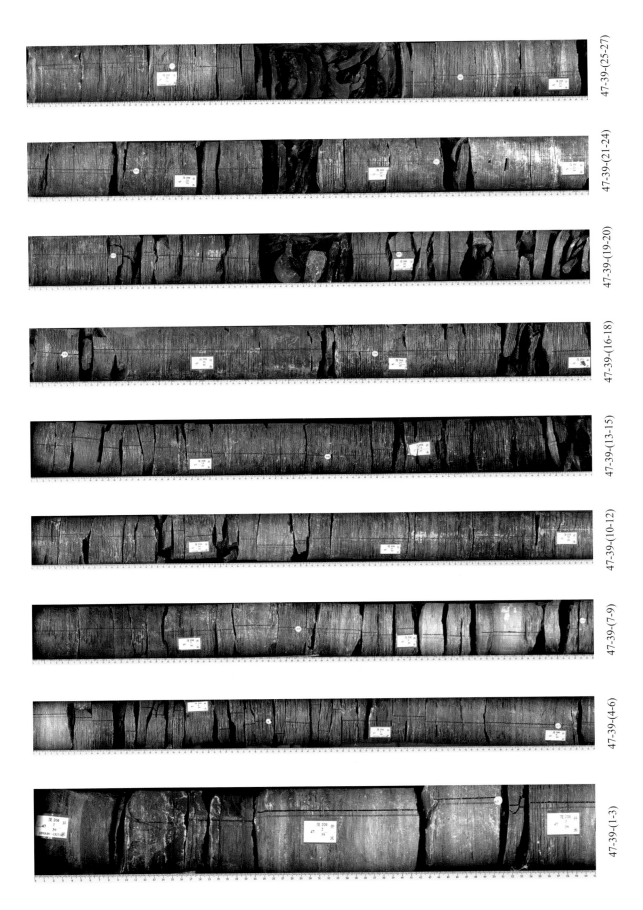

Section 5 SK-1 Core Description and Core Photographs | 403

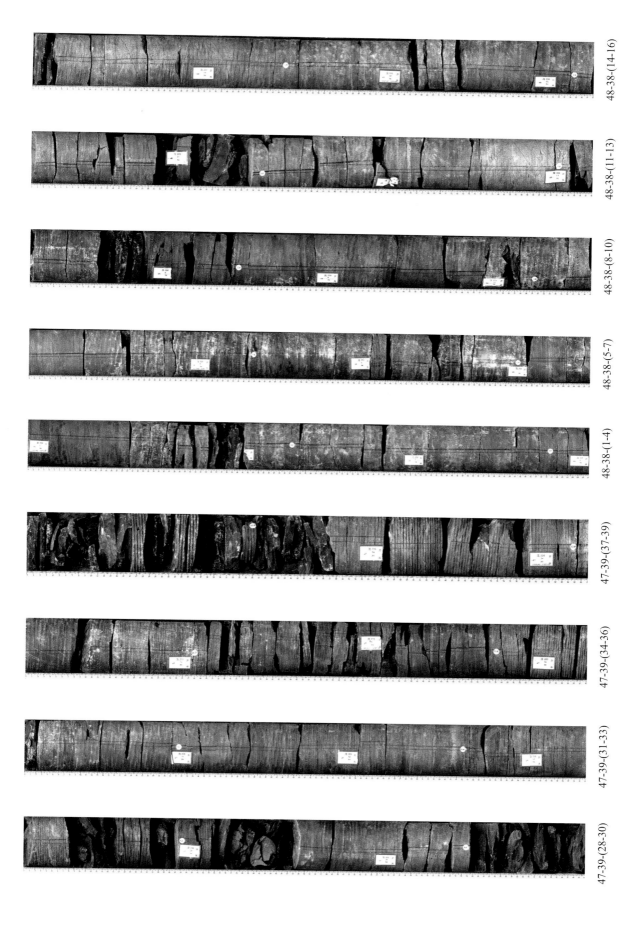

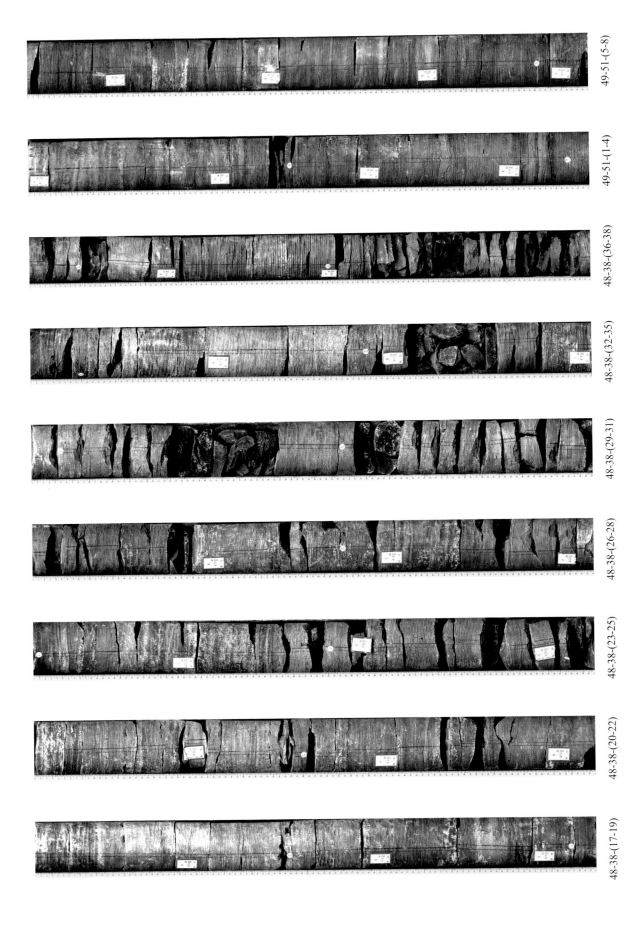

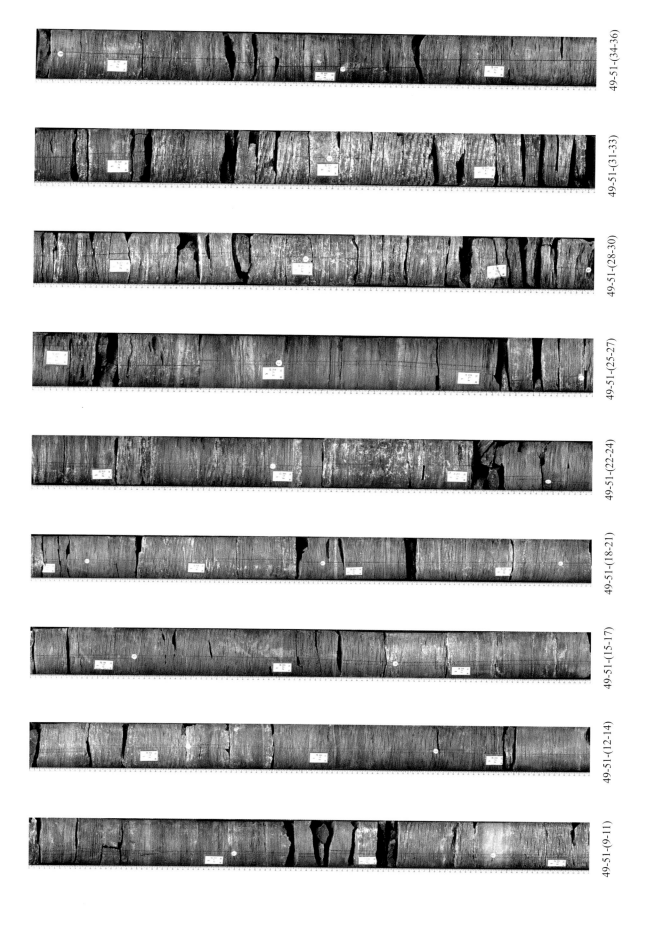

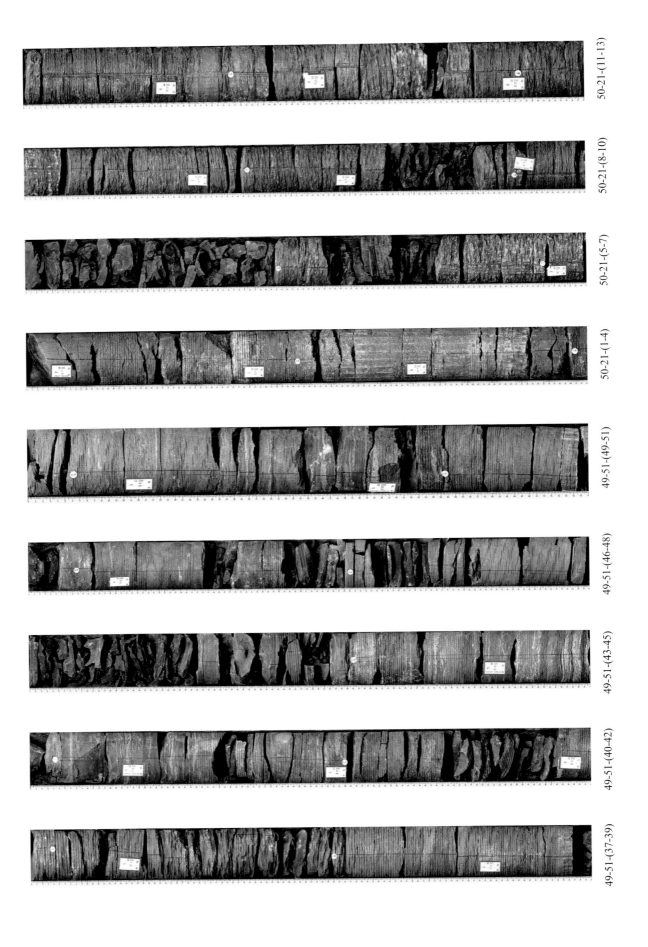

Section 5 SK-1 Core Description and Core Photographs 407

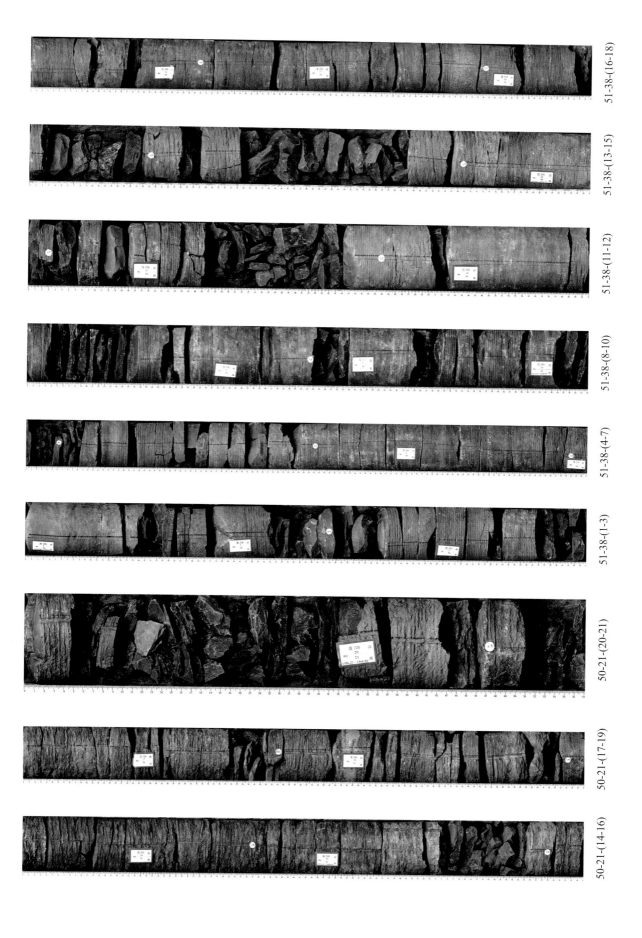

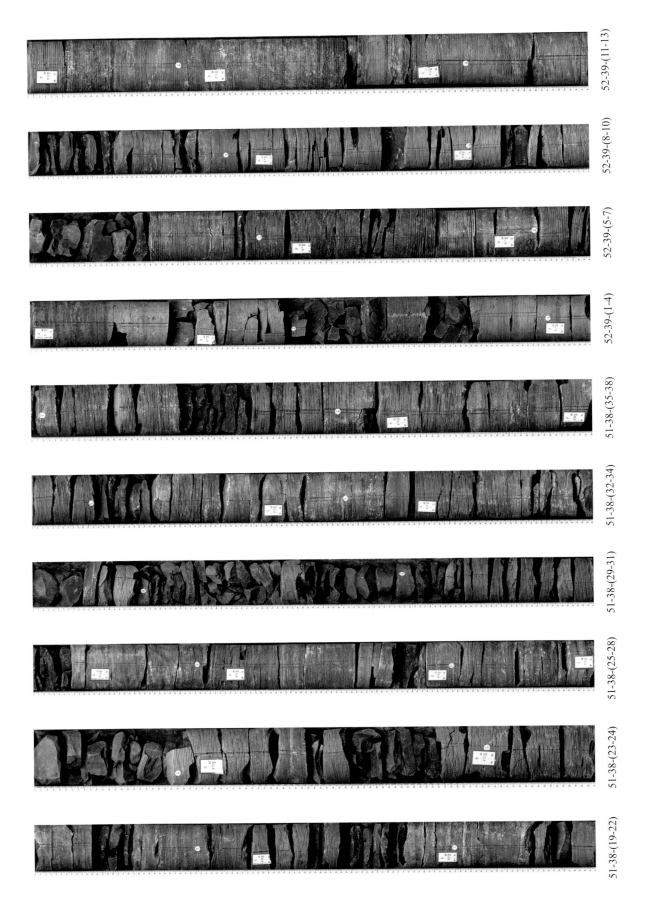

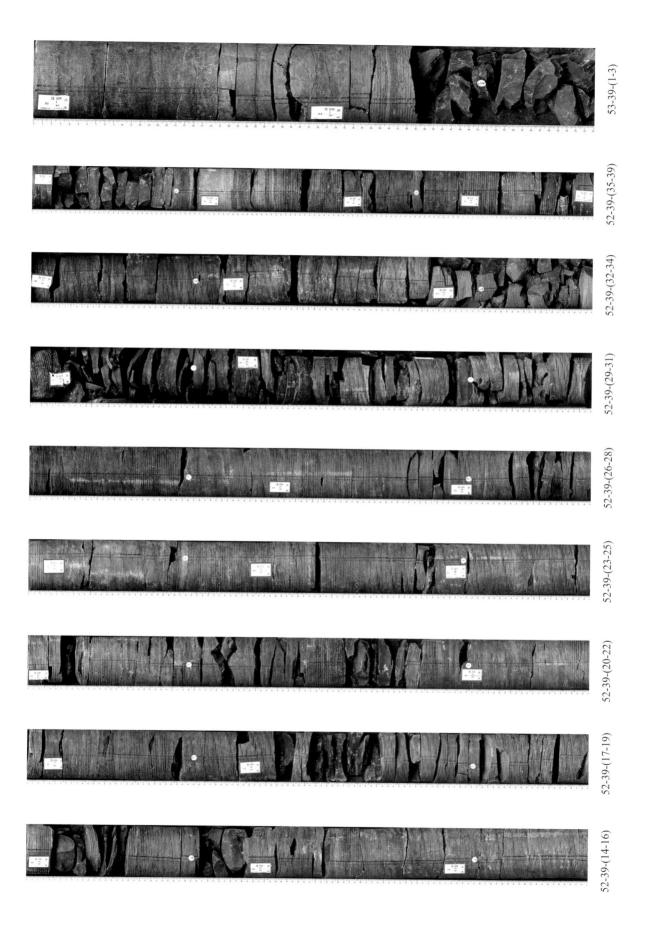

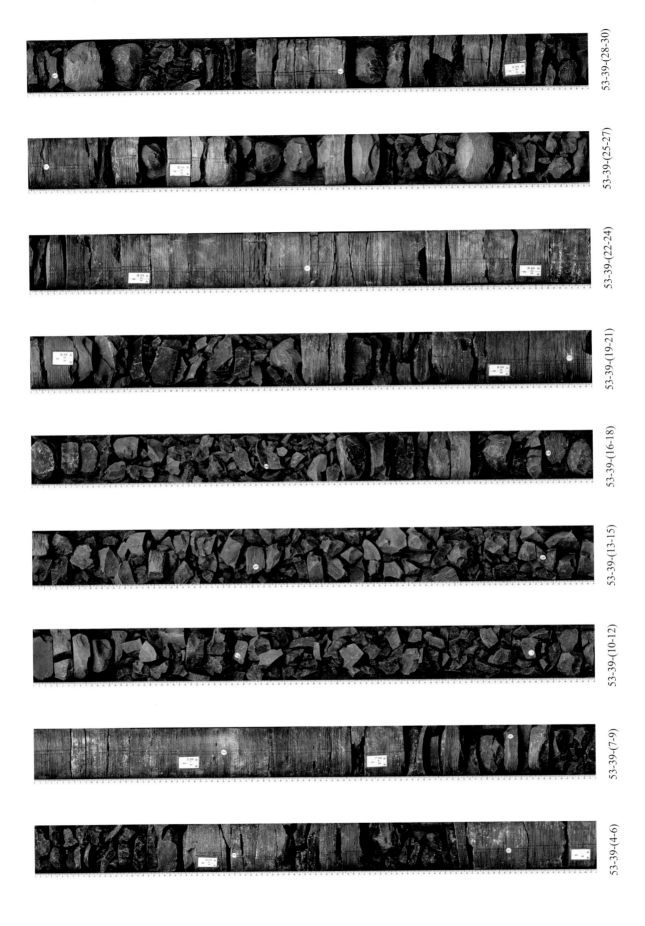

Section 5 SK-1 Core Description and Core Photographs | *411*

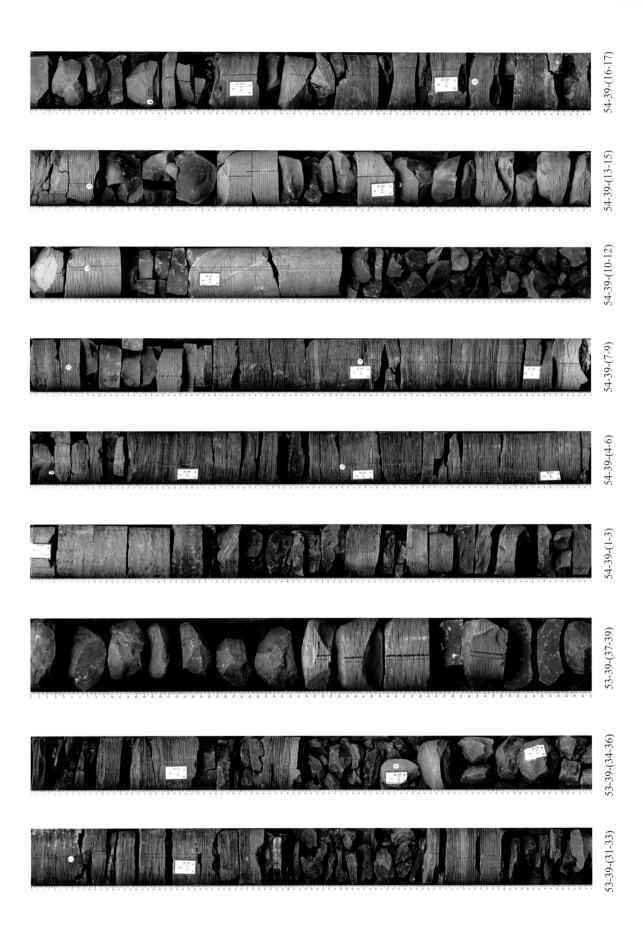

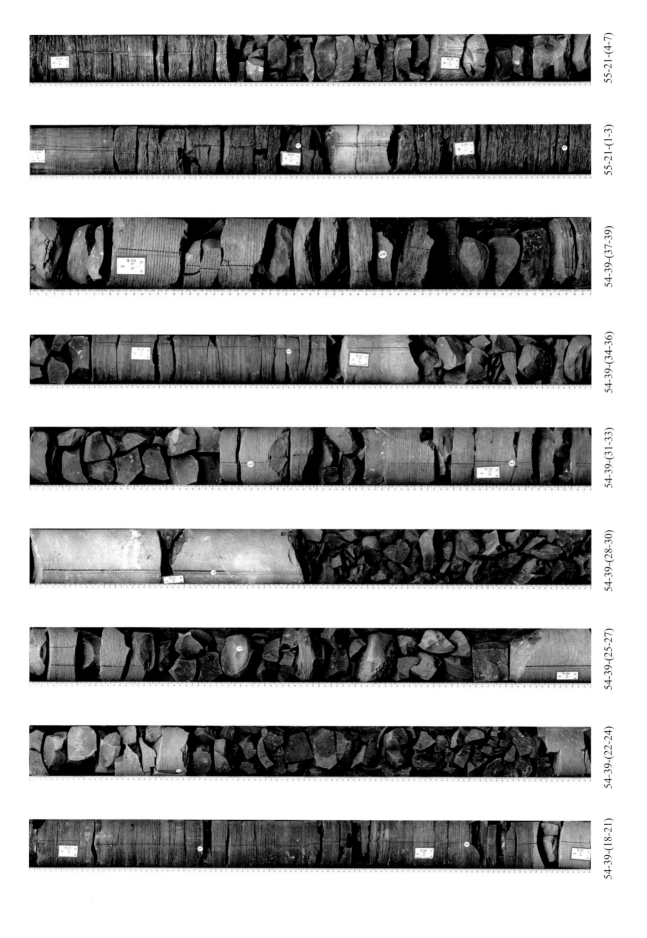

Section 5 SK-1 Core Description and Core Photographs

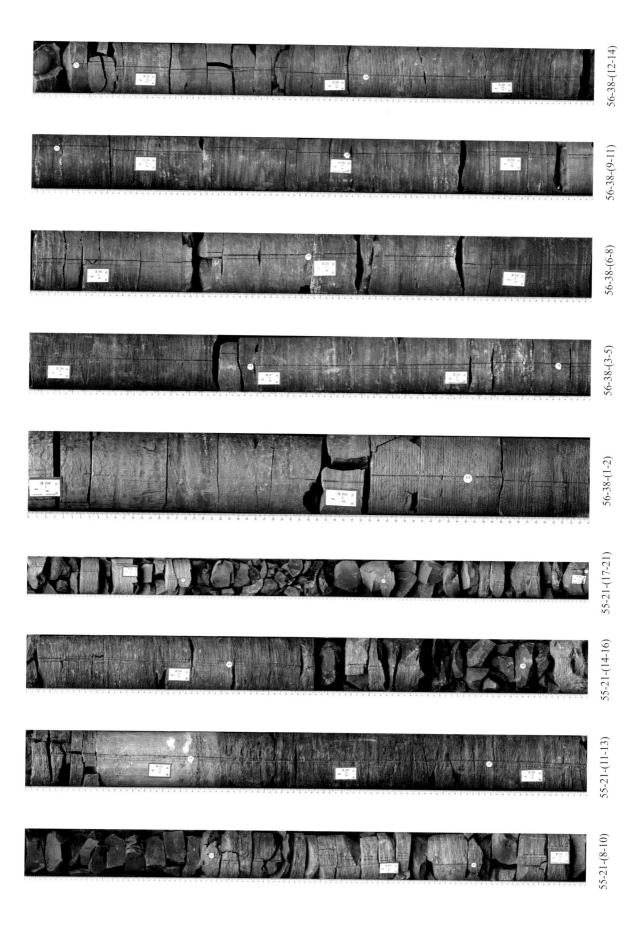

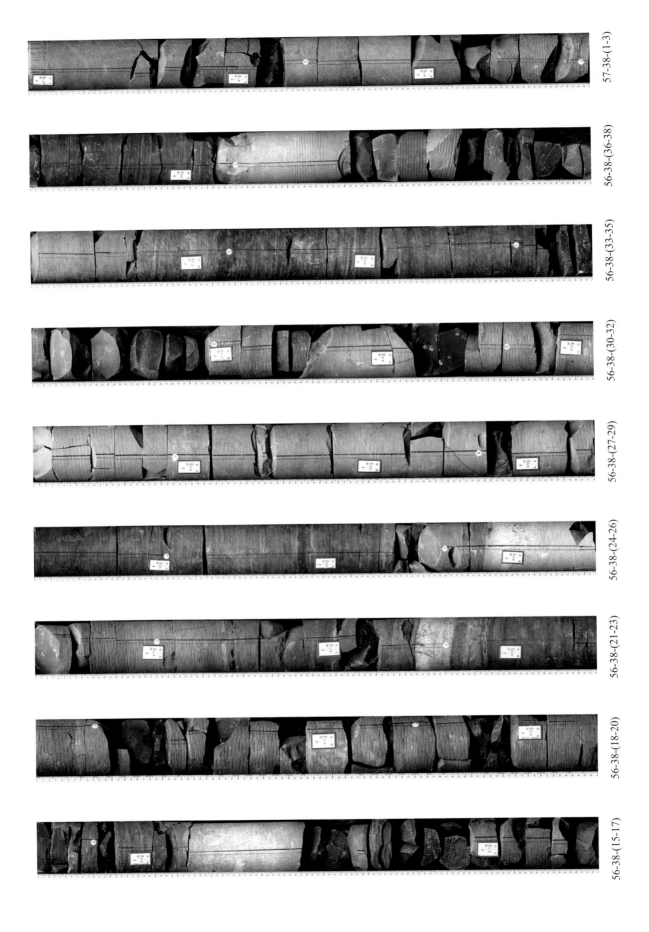

Section 5 SK-1 Core Description and Core Photographs

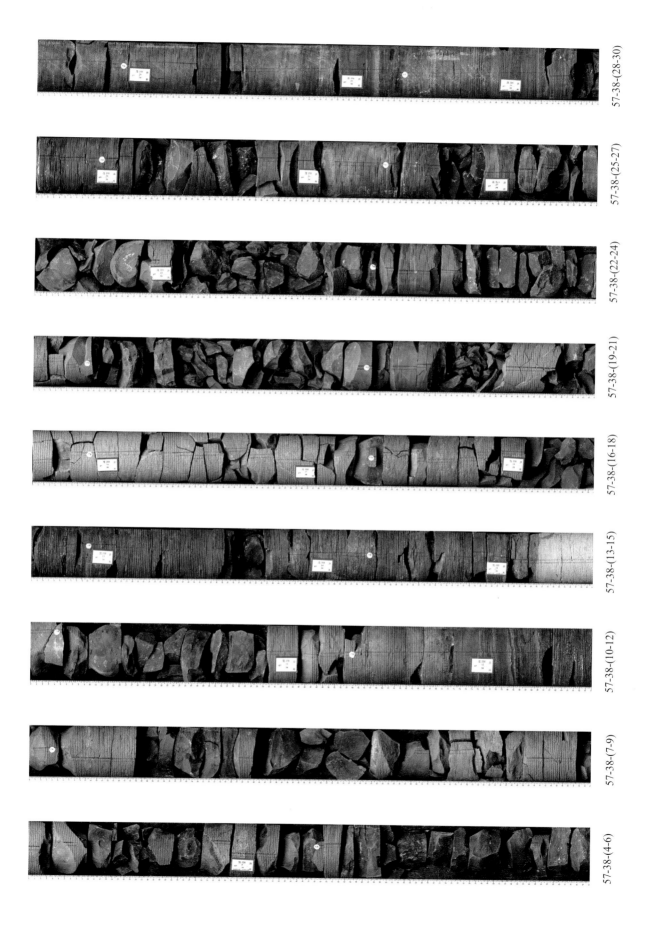

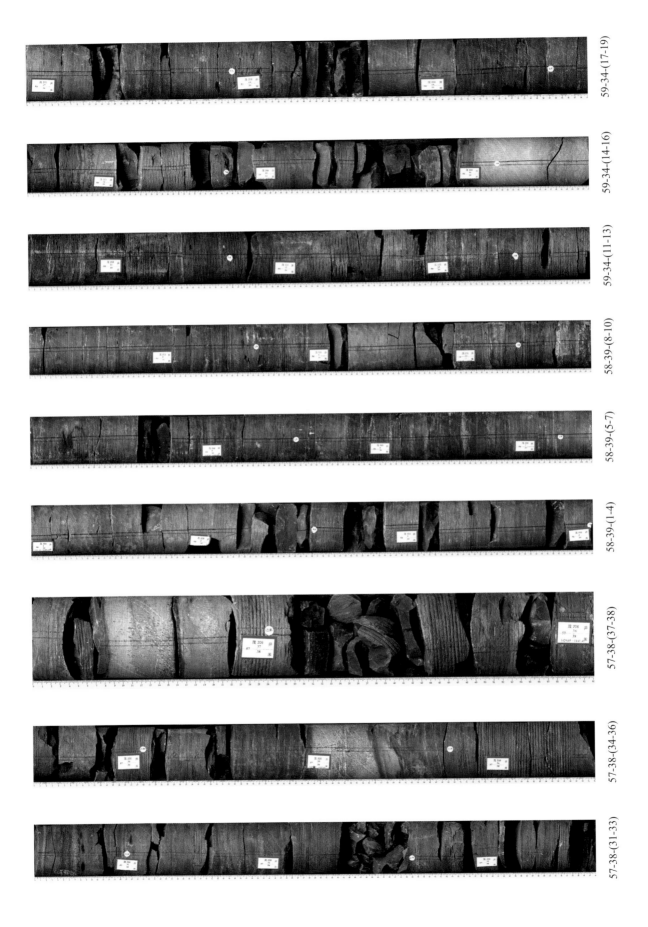

Section 5 SK-1 Core Description and Core Photographs | 417

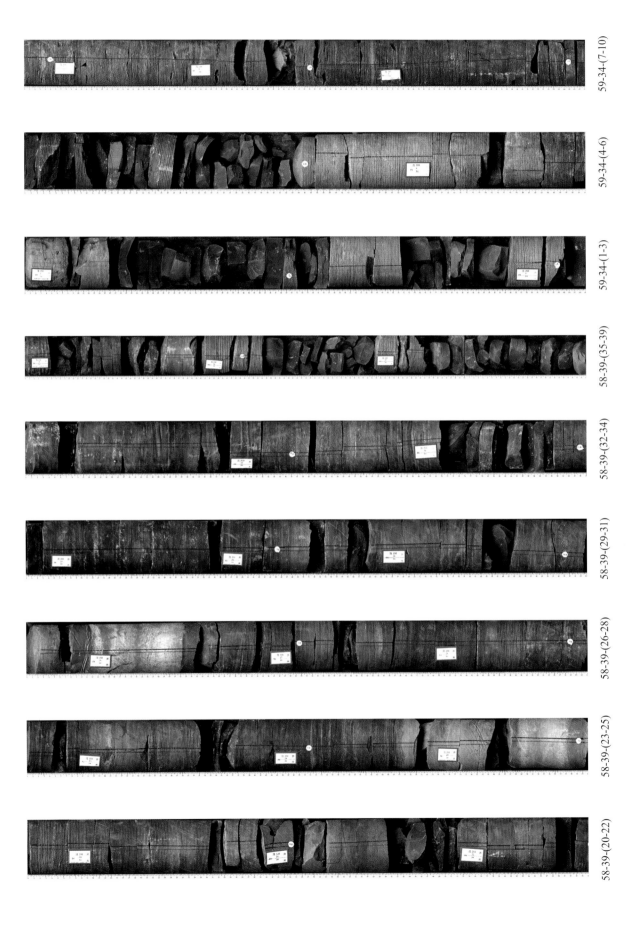

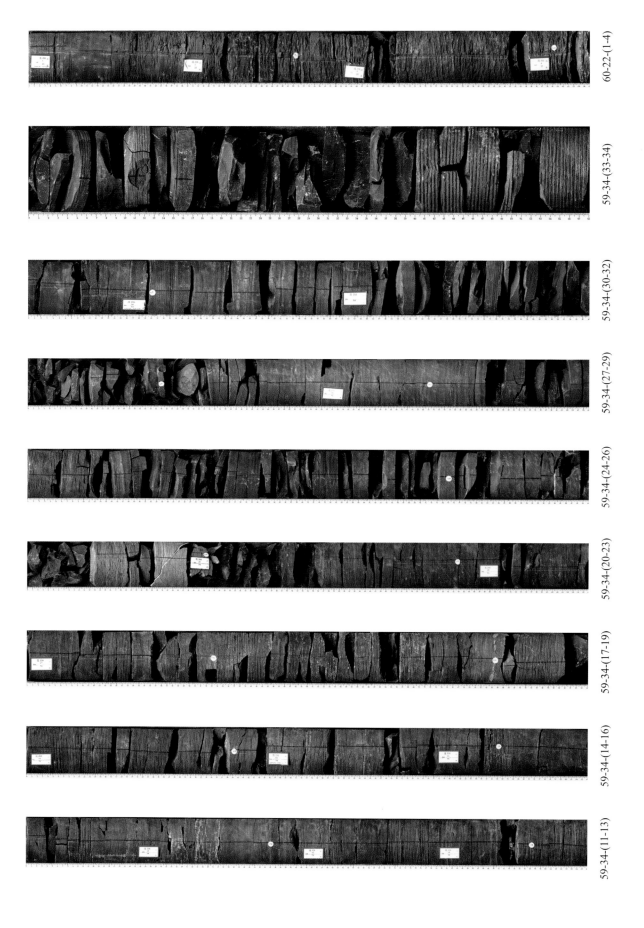

Section 5 SK-1 Core Description and Core Photographs | *419*

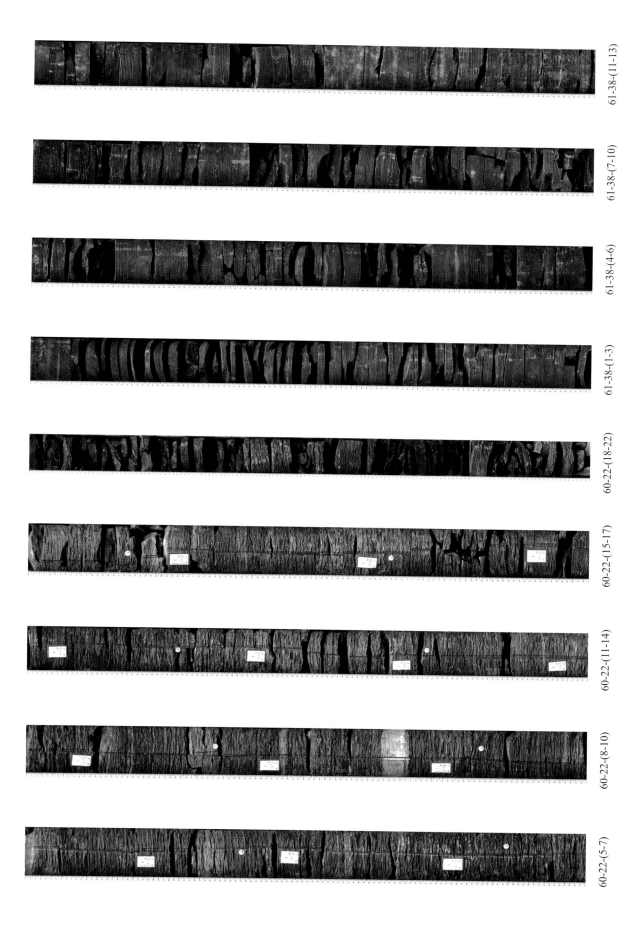

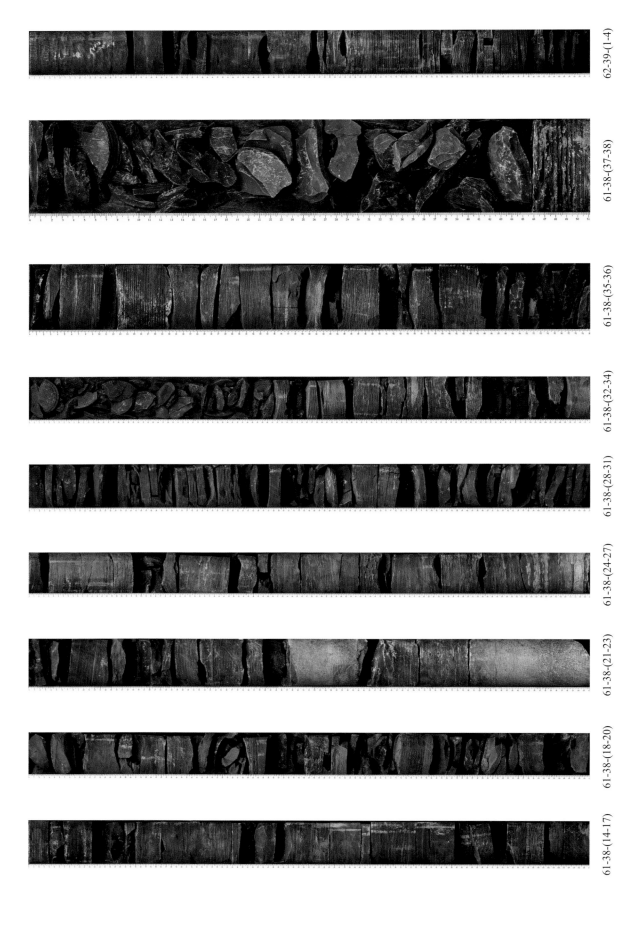

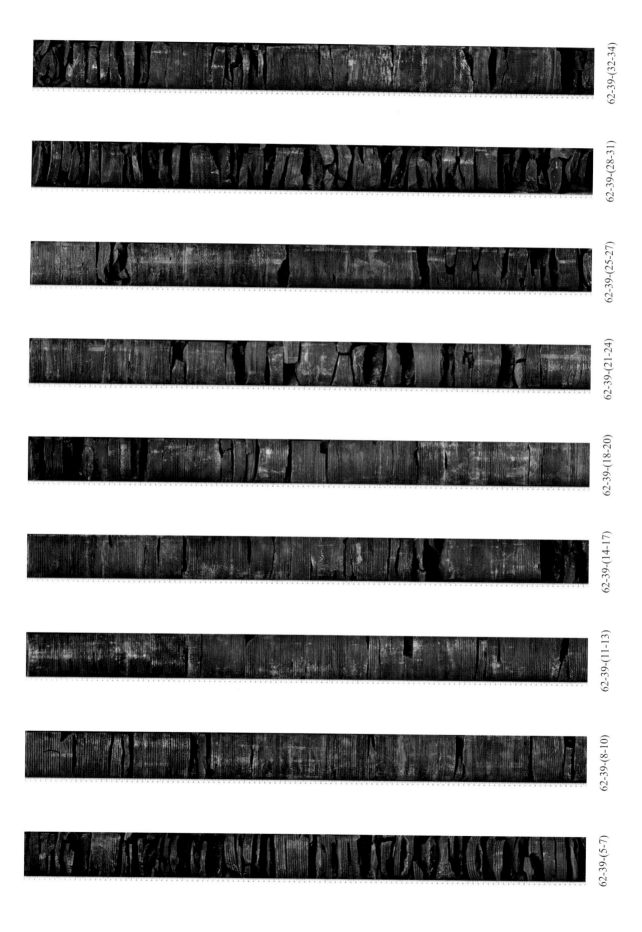

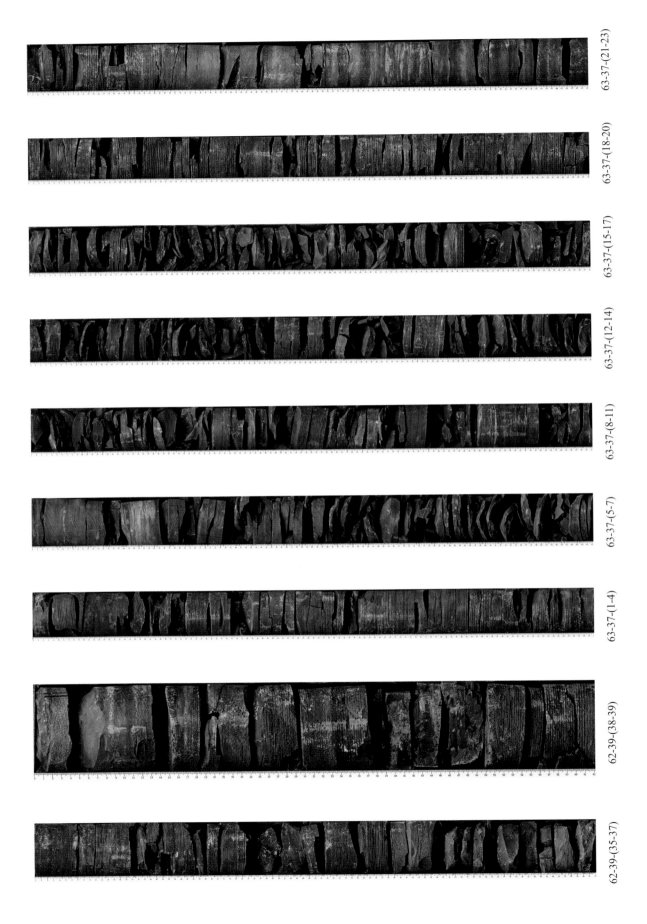

Section 5 SK-1 Core Description and Core Photographs | 423

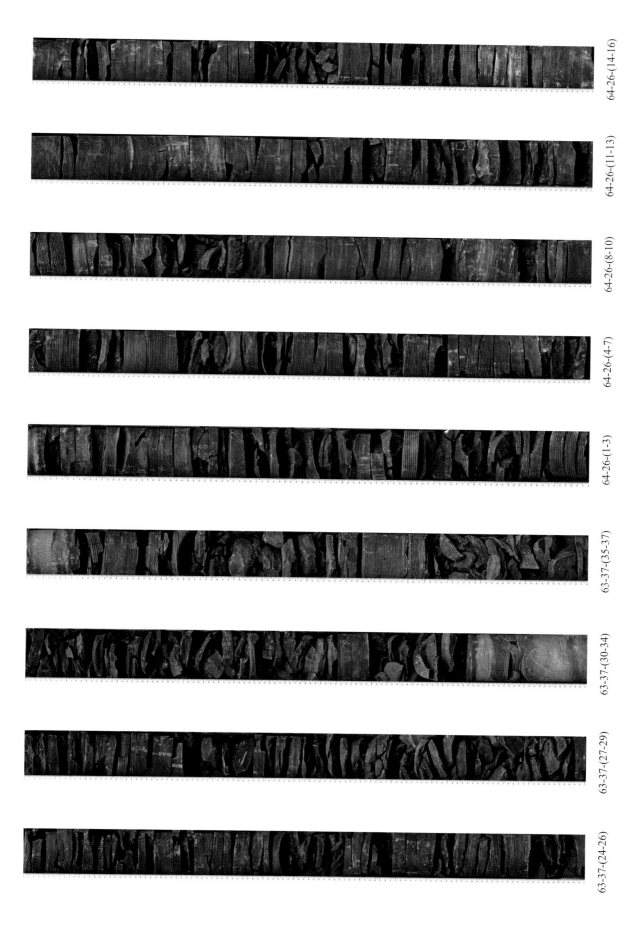

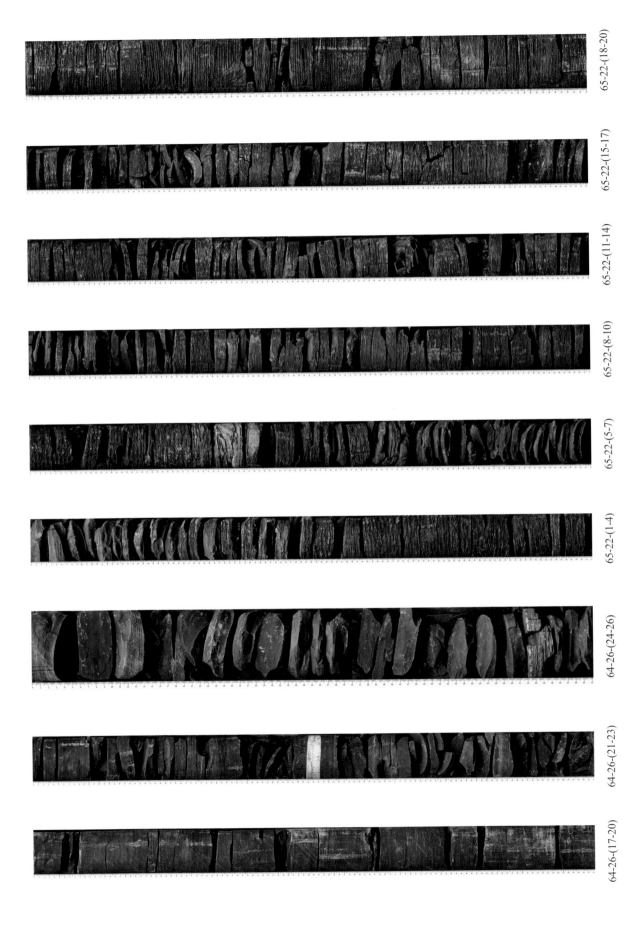

Section 5 SK-1 Core Description and Core Photographs

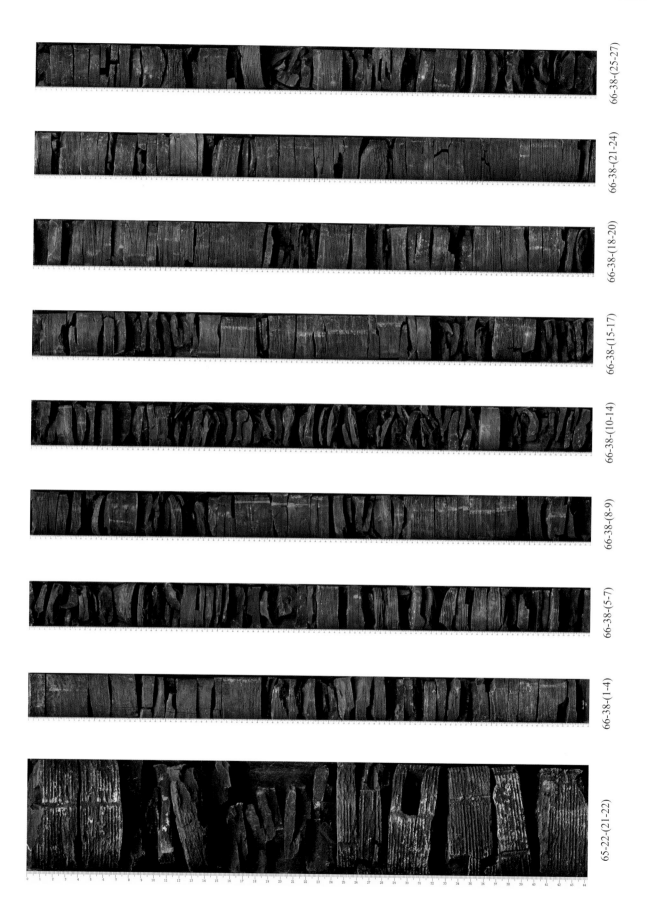

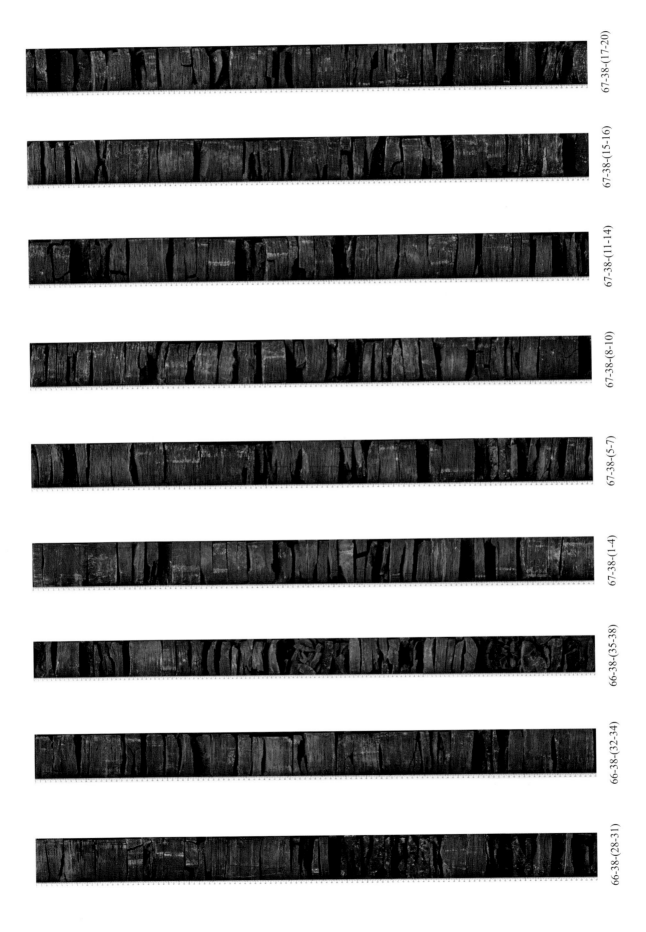

Section 5 SK-1 Core Description and Core Photographs | *427*

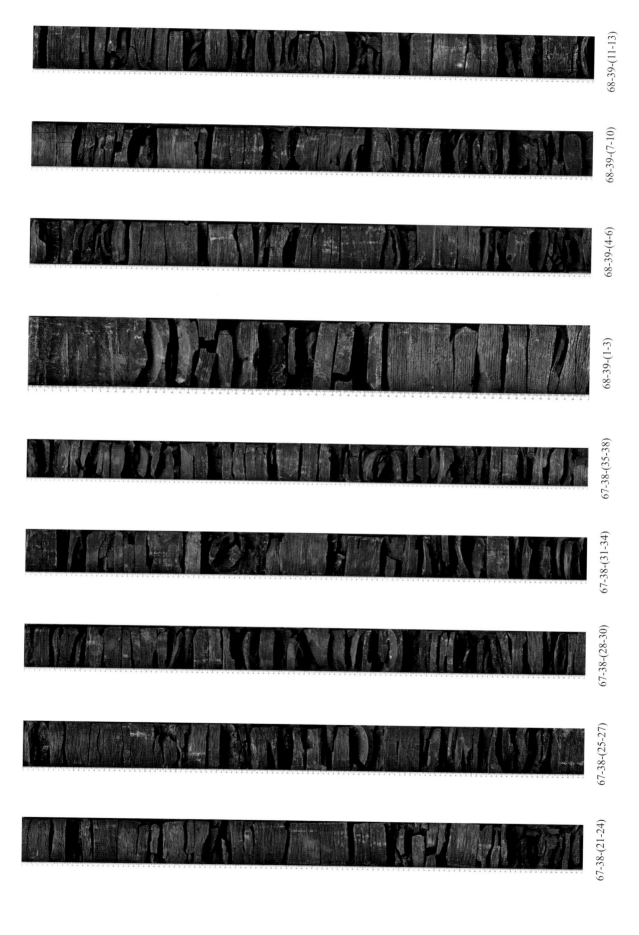

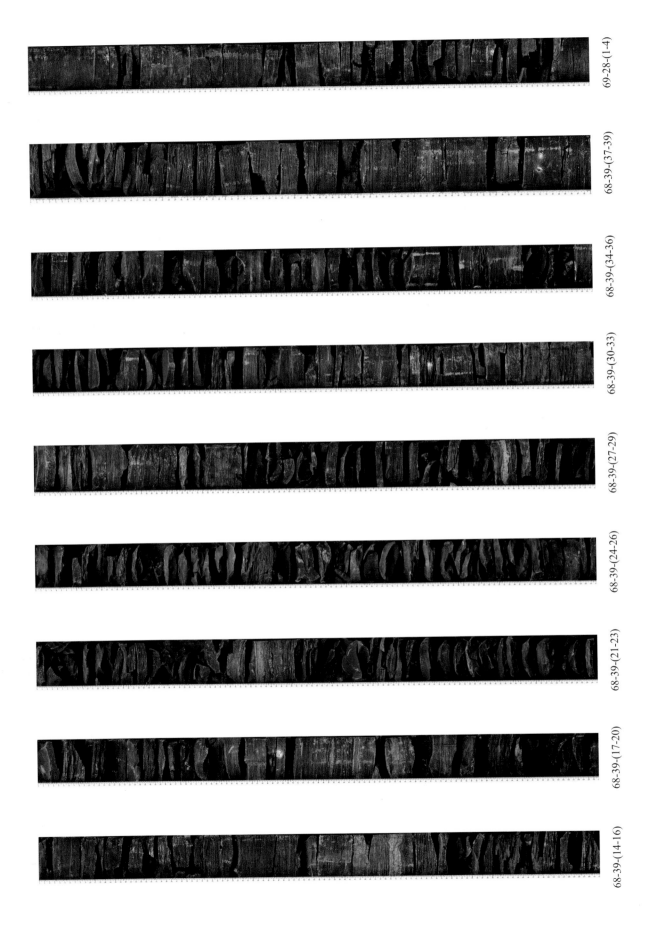

Section 5 SK-1 Core Description and Core Photographs

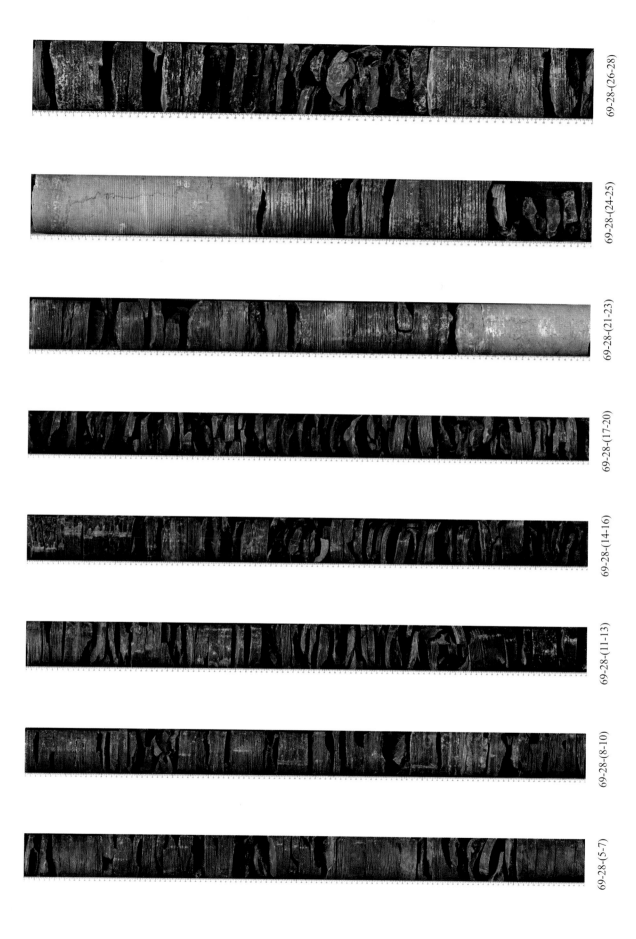

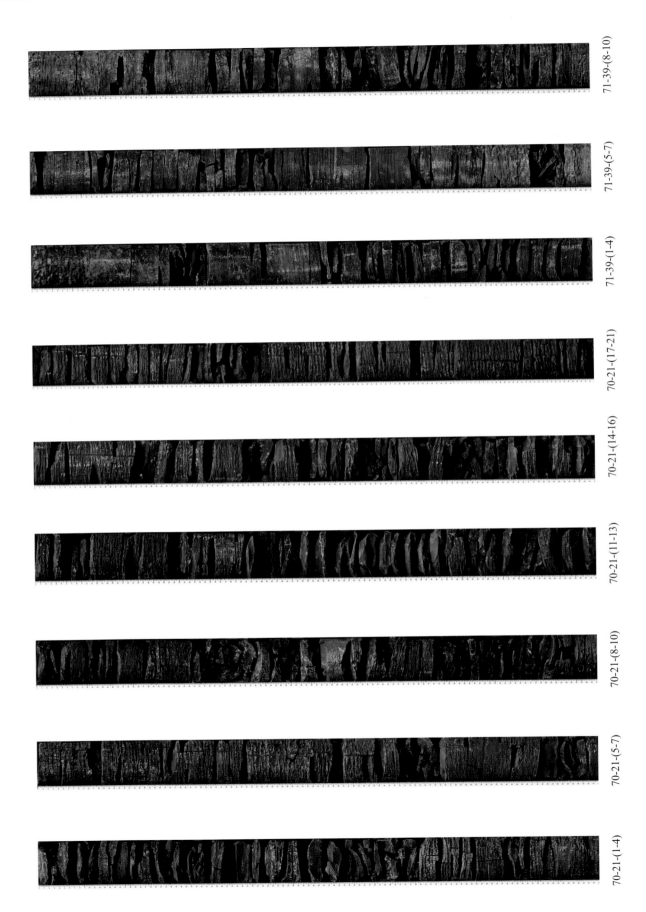

Section 5 SK-1 Core Description and Core Photographs | *431*

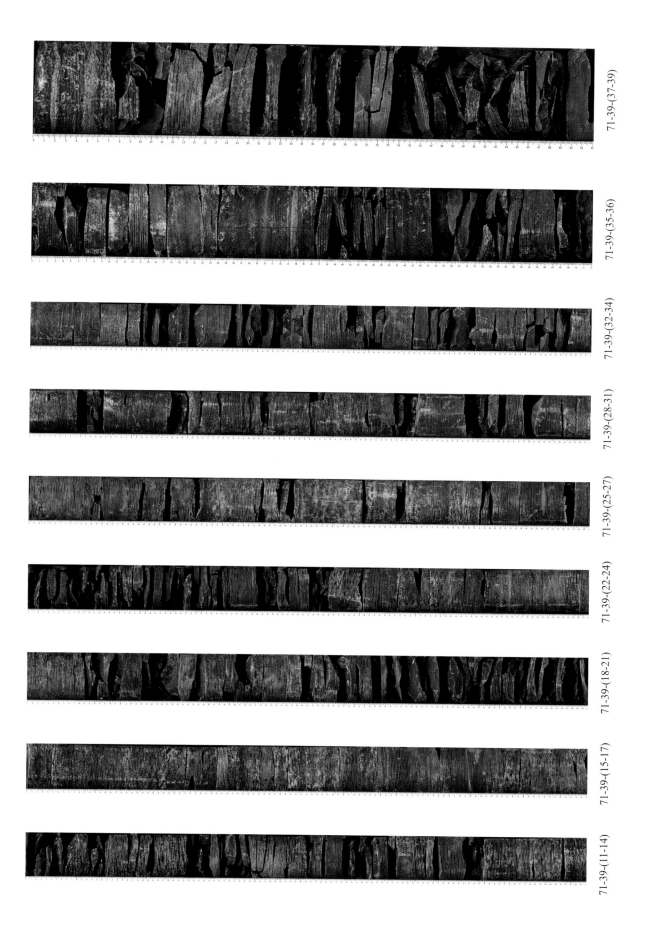

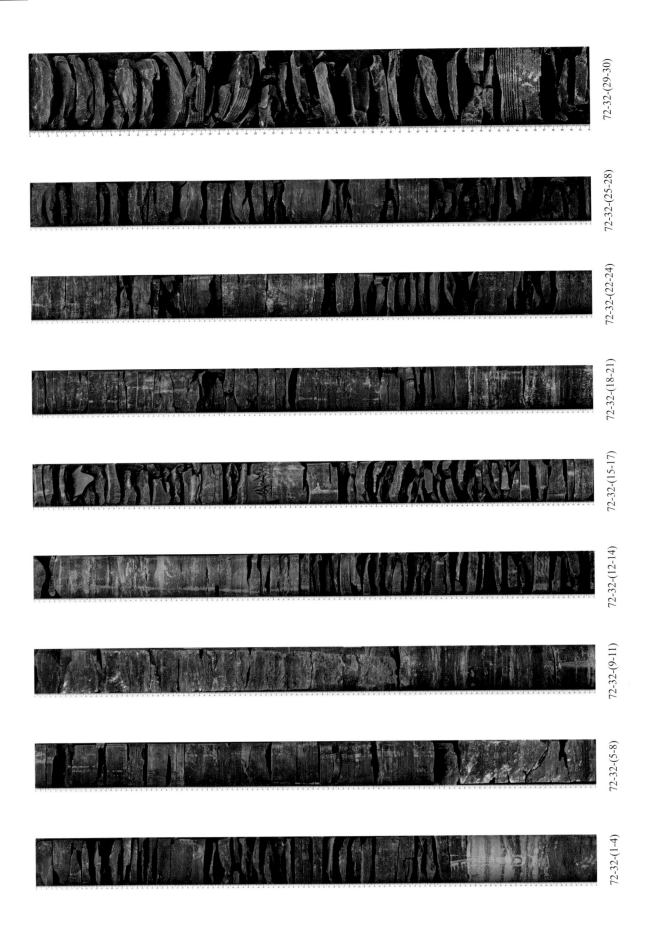

Section 5 SK-1 Core Description and Core Photographs | *433*

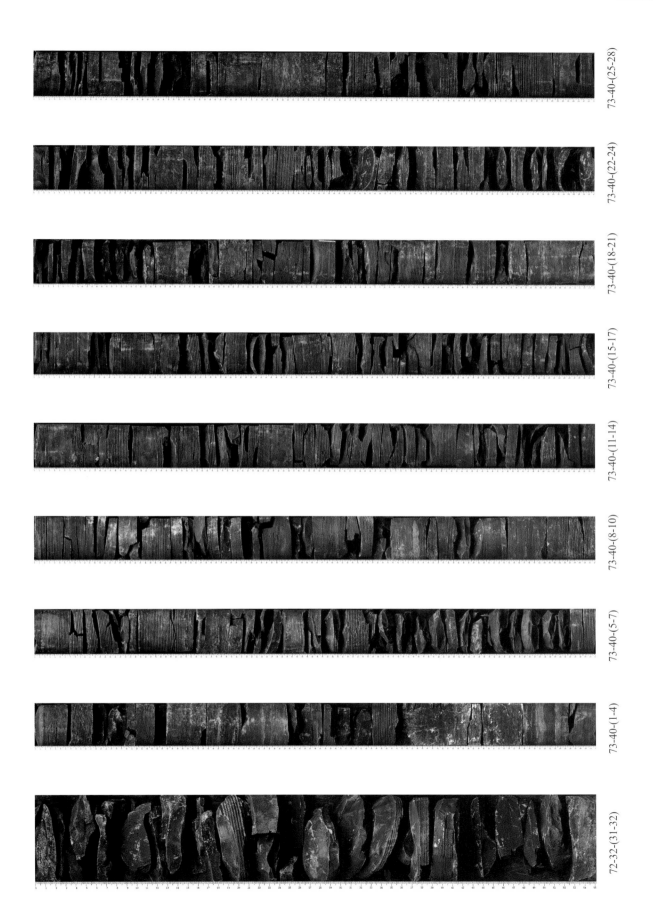

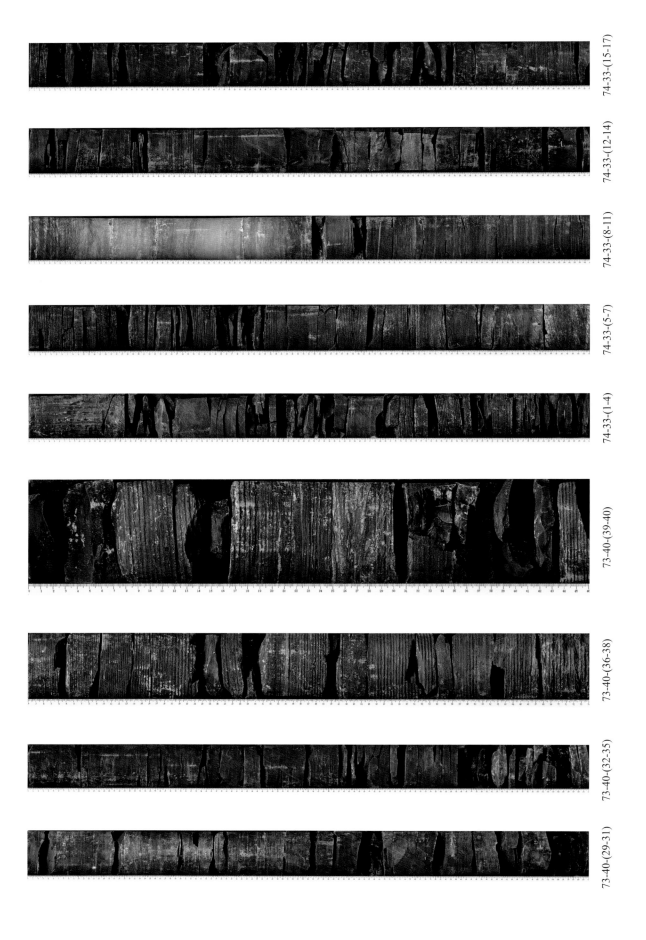

Section 5 SK-1 Core Description and Core Photographs | 435

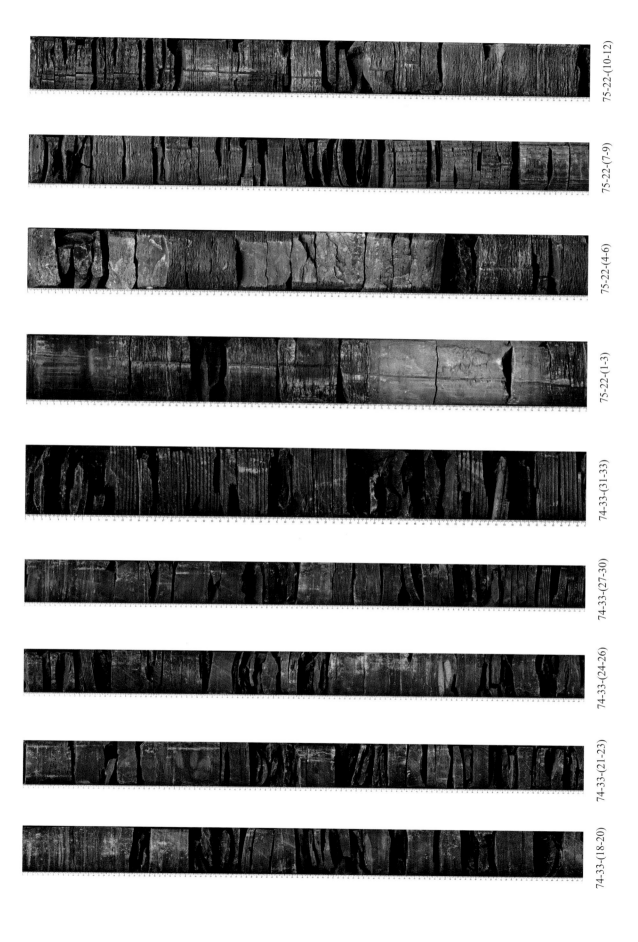

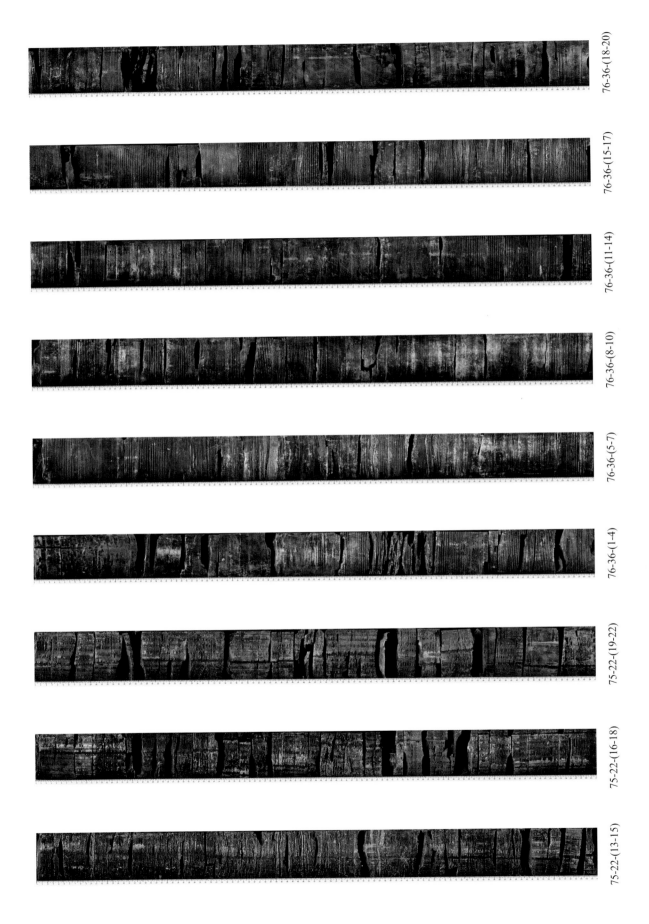

Section 5 SK-1 Core Description and Core Photographs | *437*

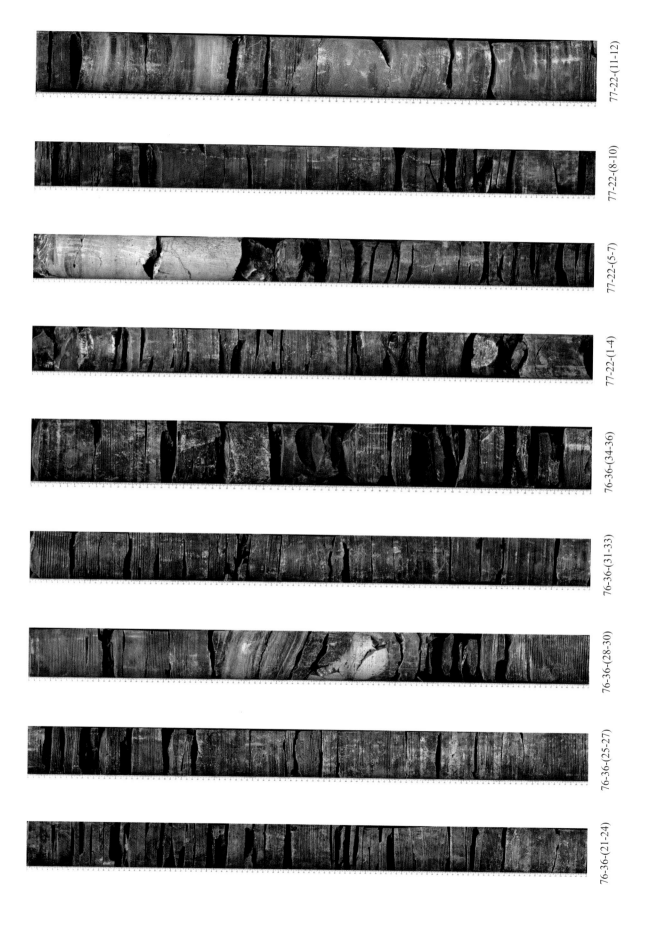

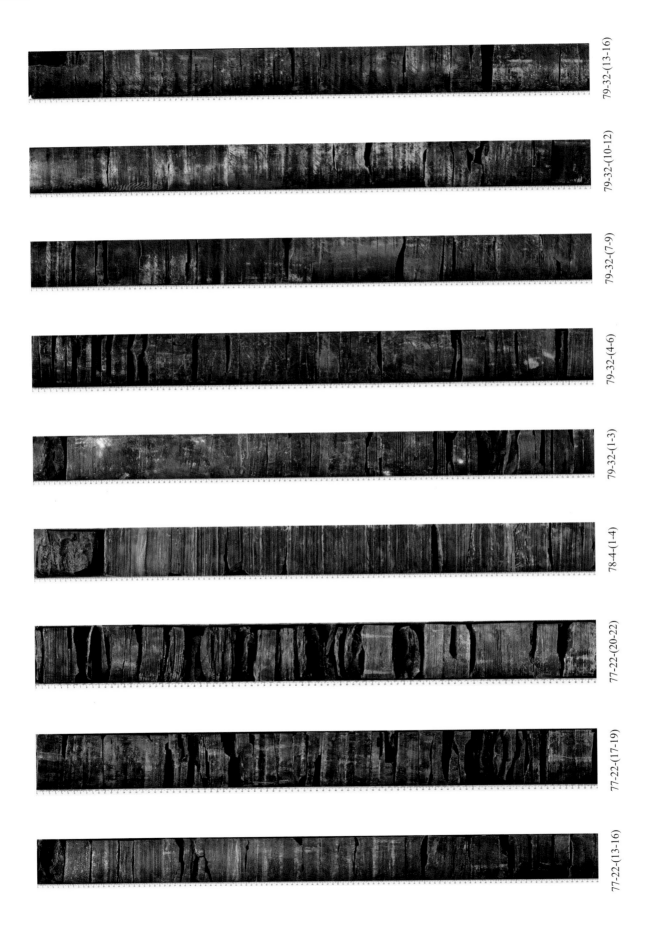

Section 5 SK-1 Core Description and Core Photographs

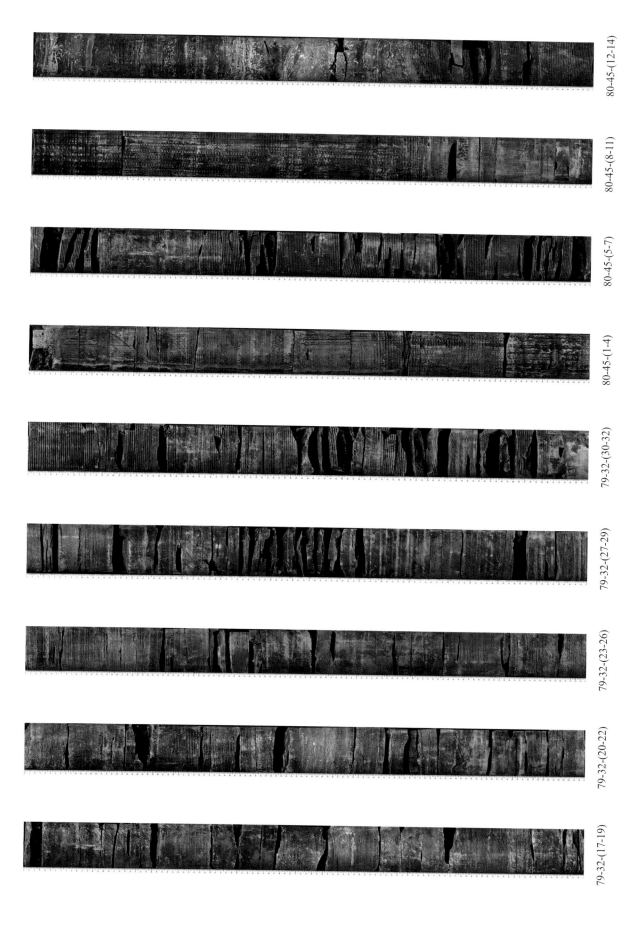

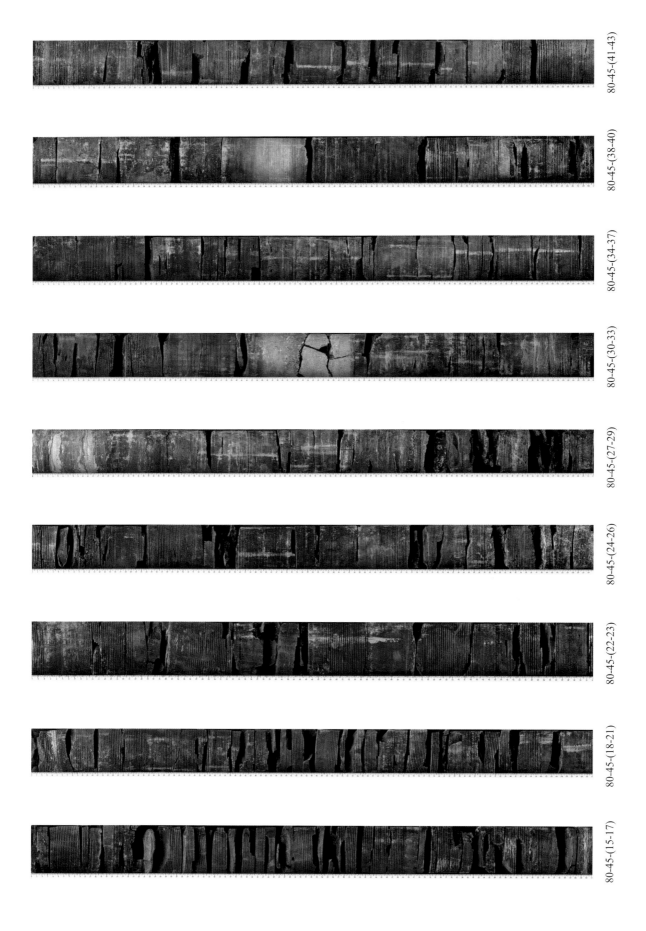

Section 5 SK-1 Core Description and Core Photographs | *441*

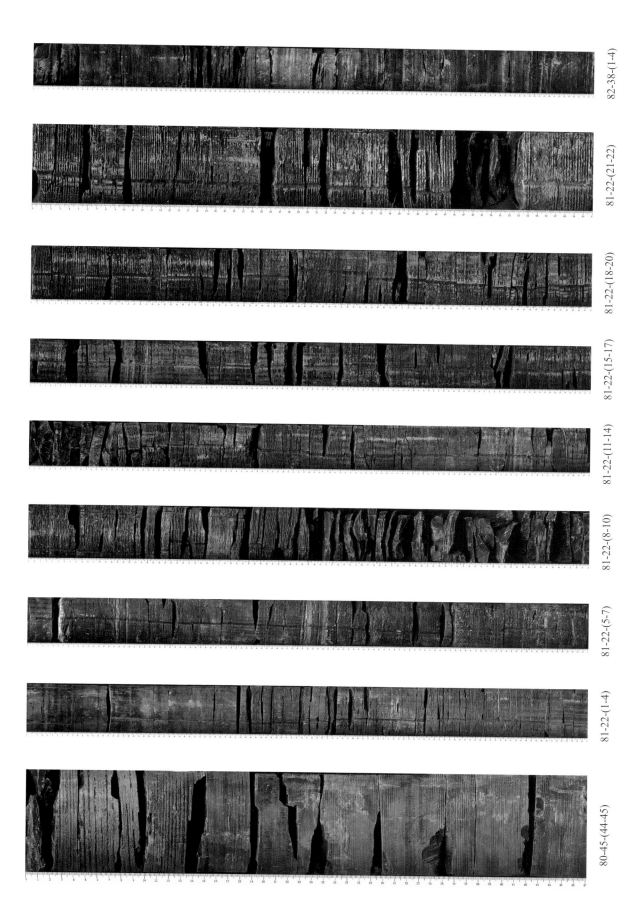

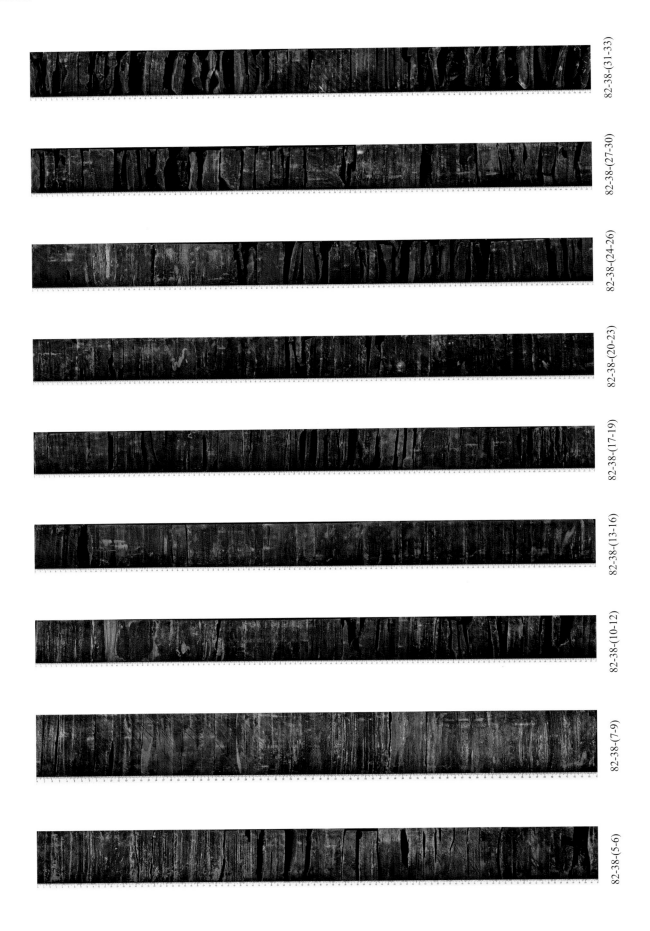

Section 5 SK-1 Core Description and Core Photographs | *443*

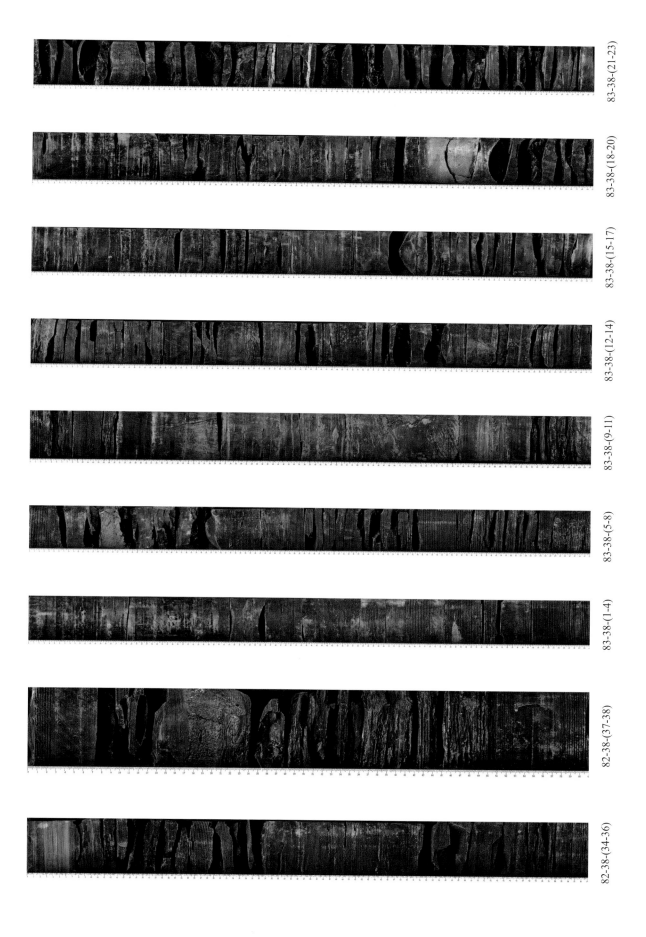

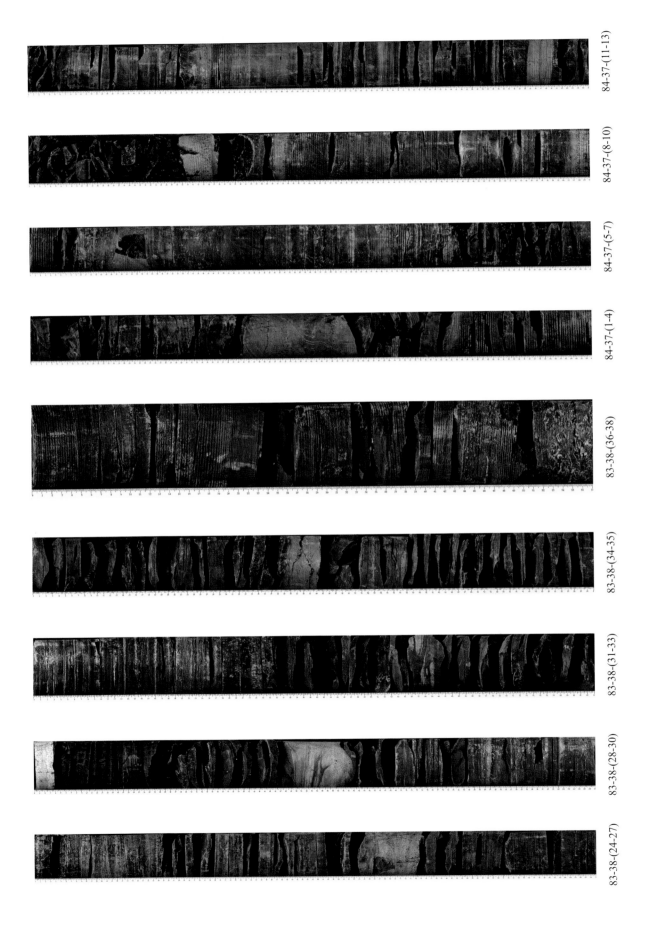

Section 5 SK-1 Core Description and Core Photographs

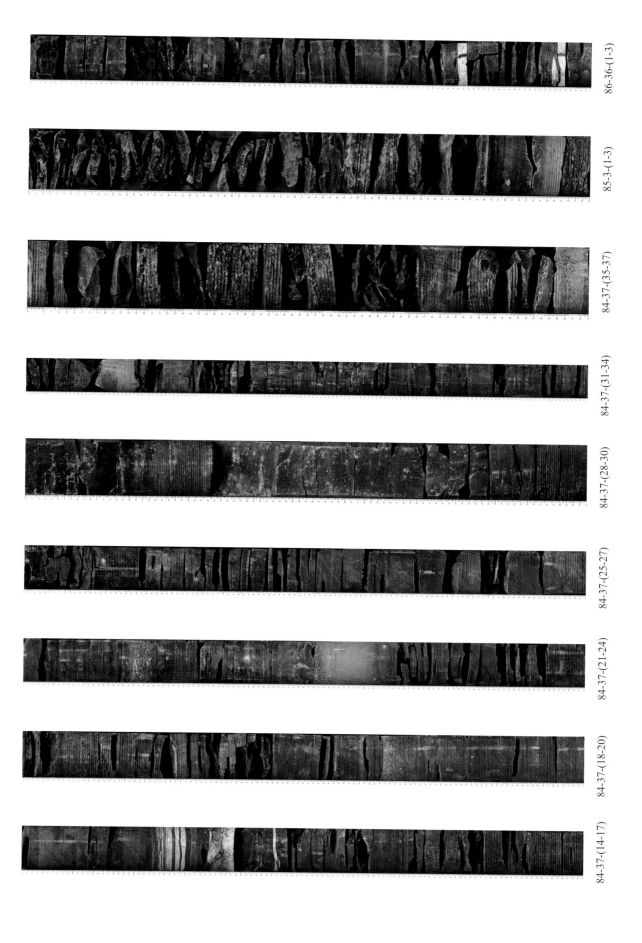

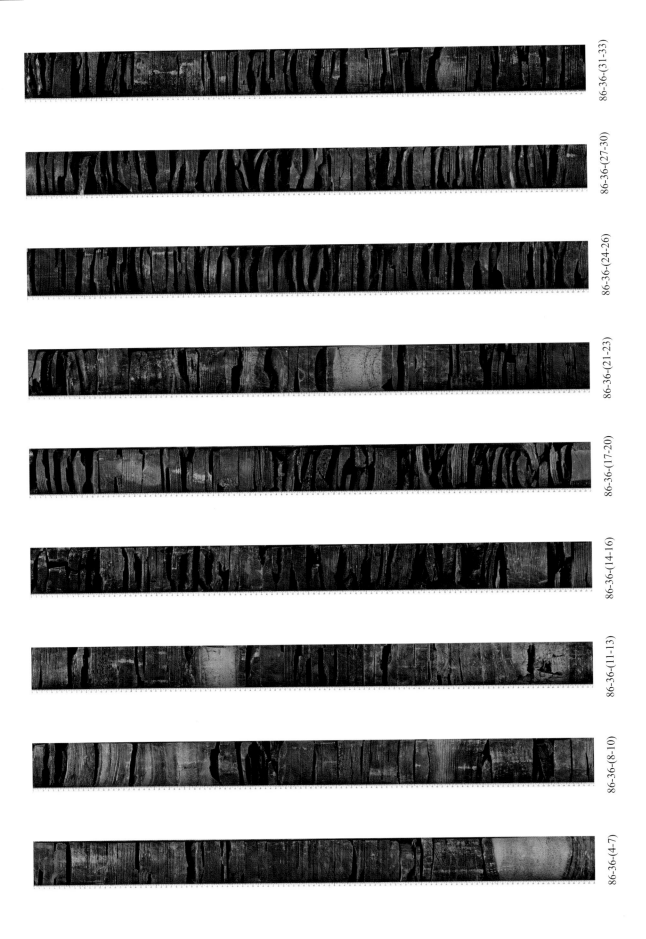

Section 5 SK-1 Core Description and Core Photographs | *447*

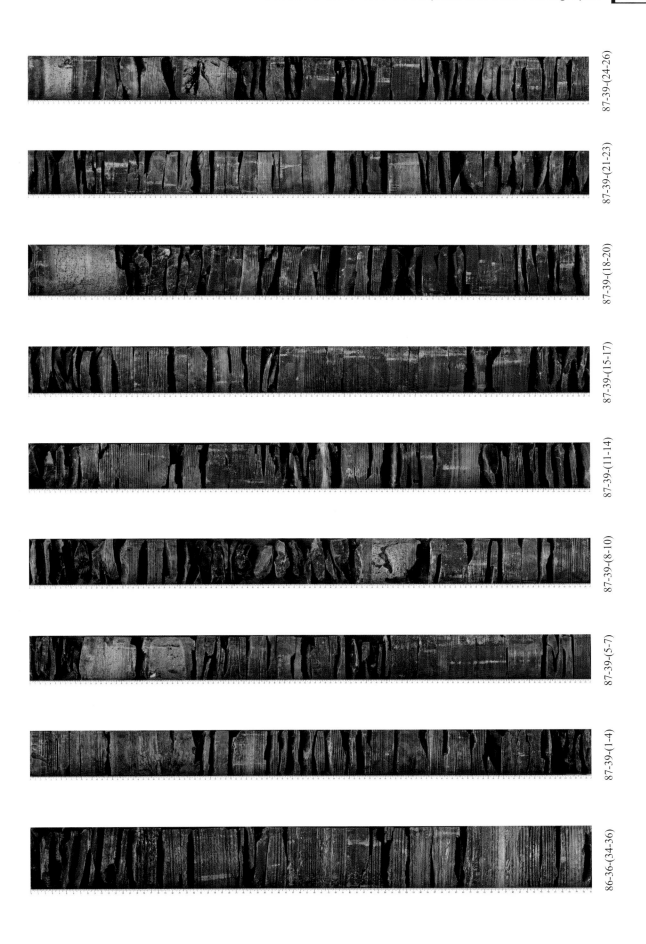

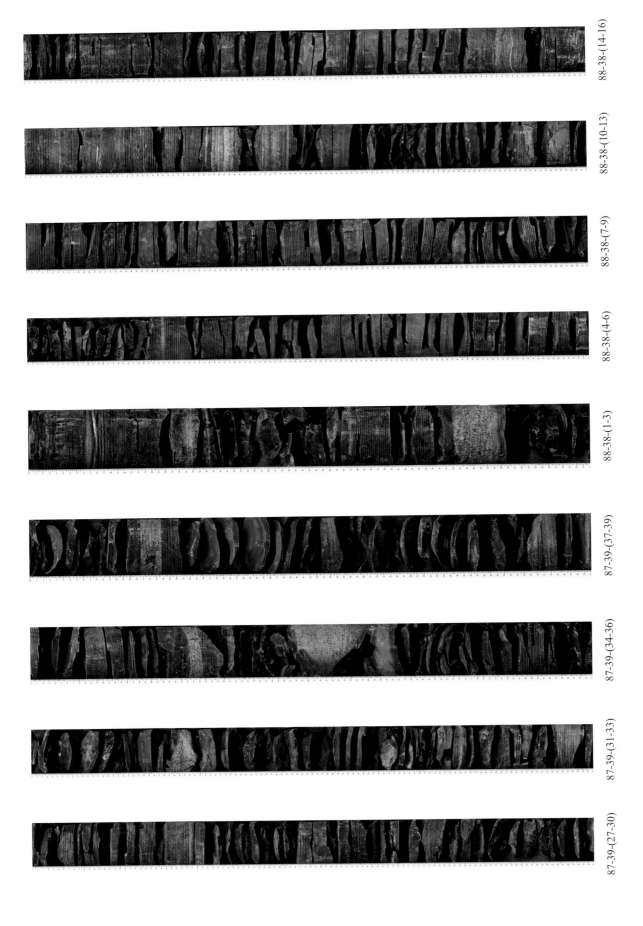

Section 5 SK-1 Core Description and Core Photographs | 449

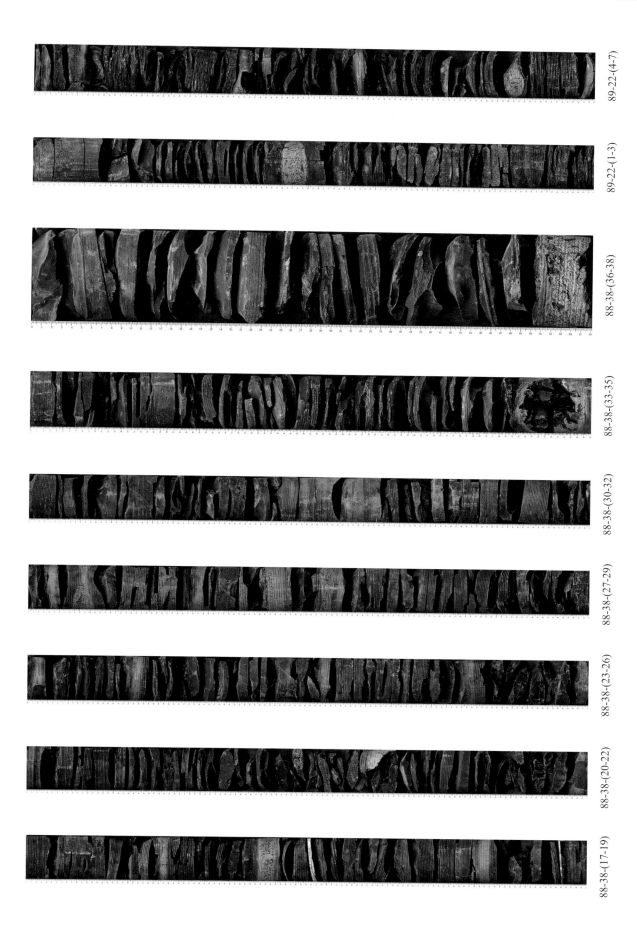

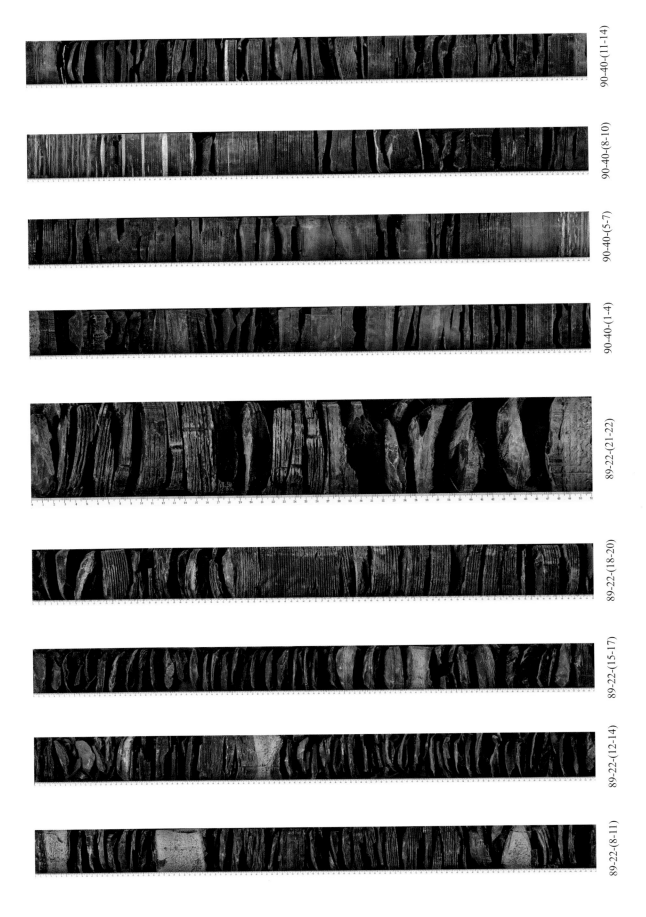

Section 5 SK-1 Core Description and Core Photographs

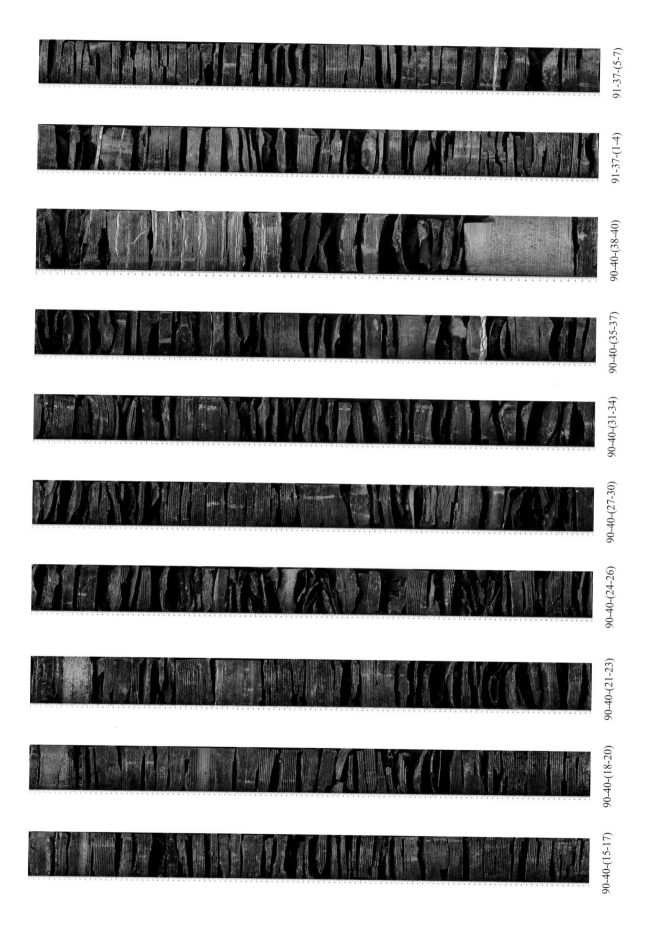

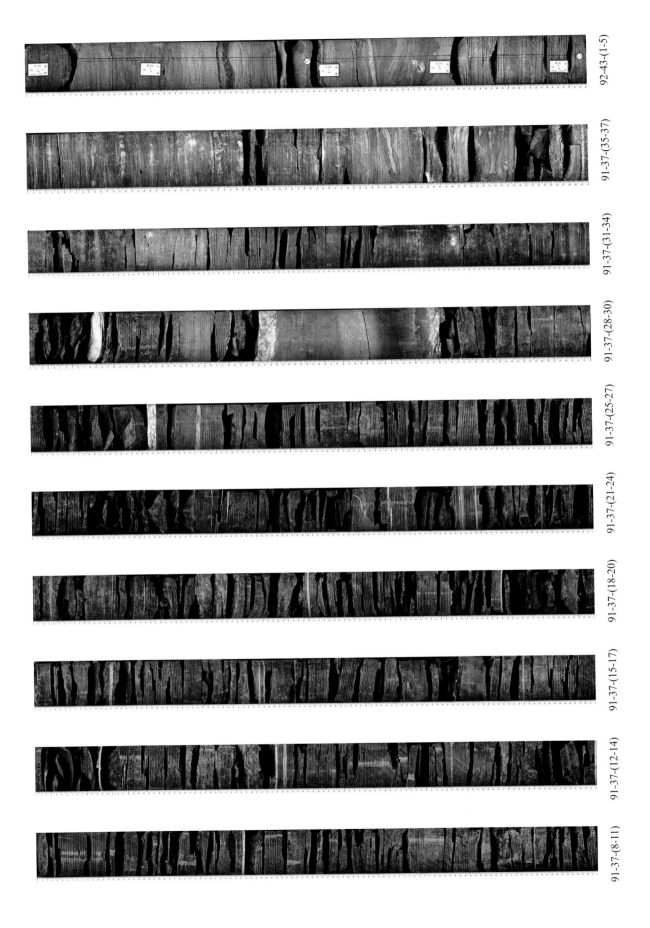

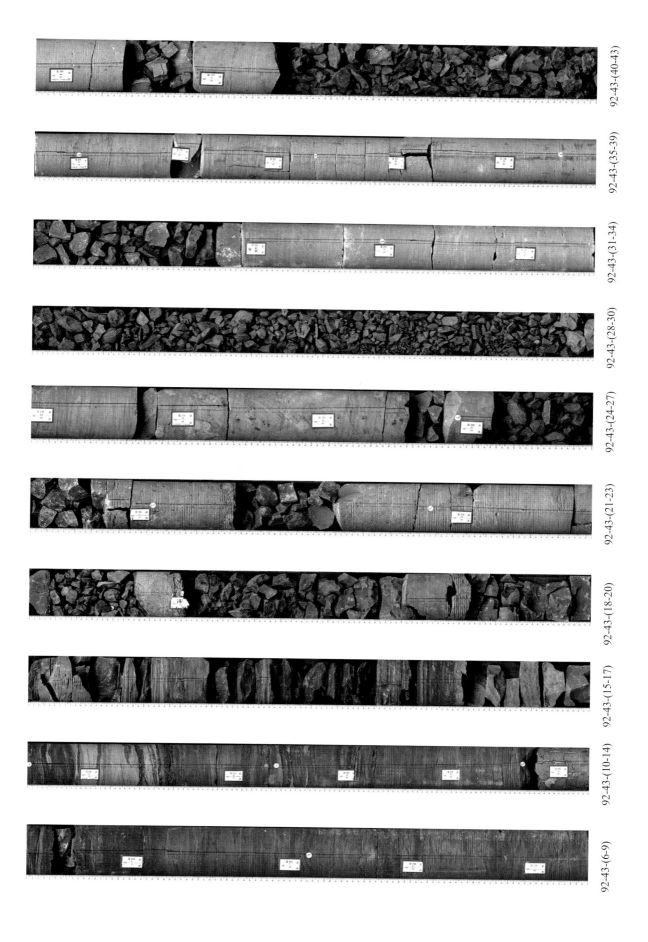

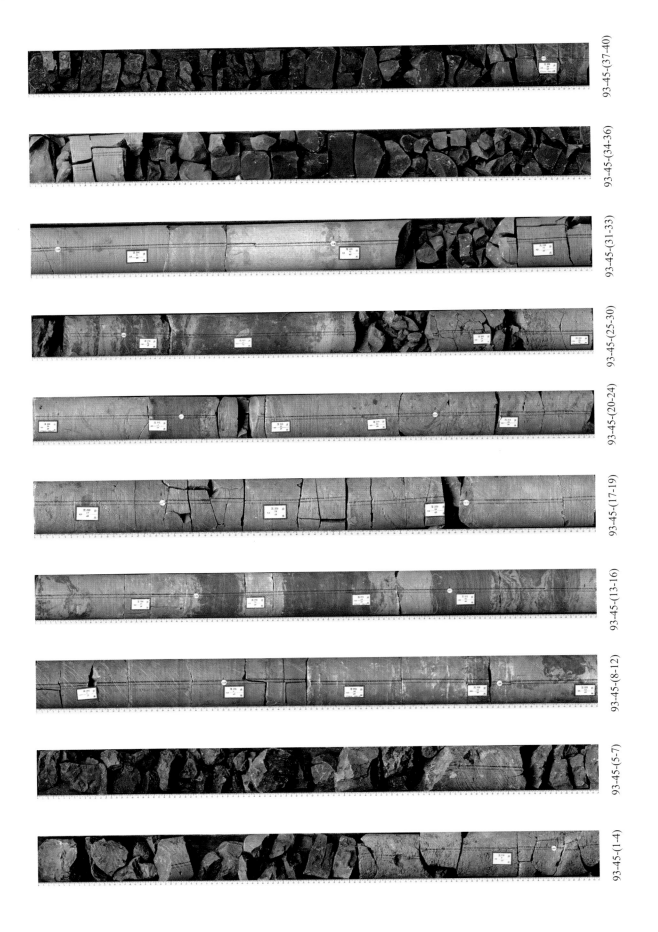

Section 5 SK-1 Core Description and Core Photographs | *455*

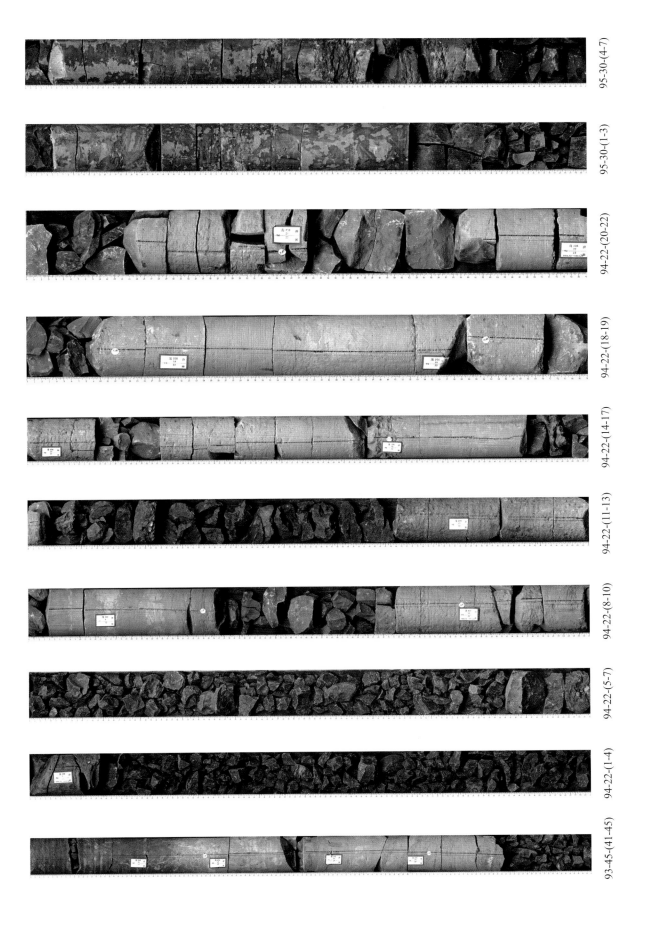

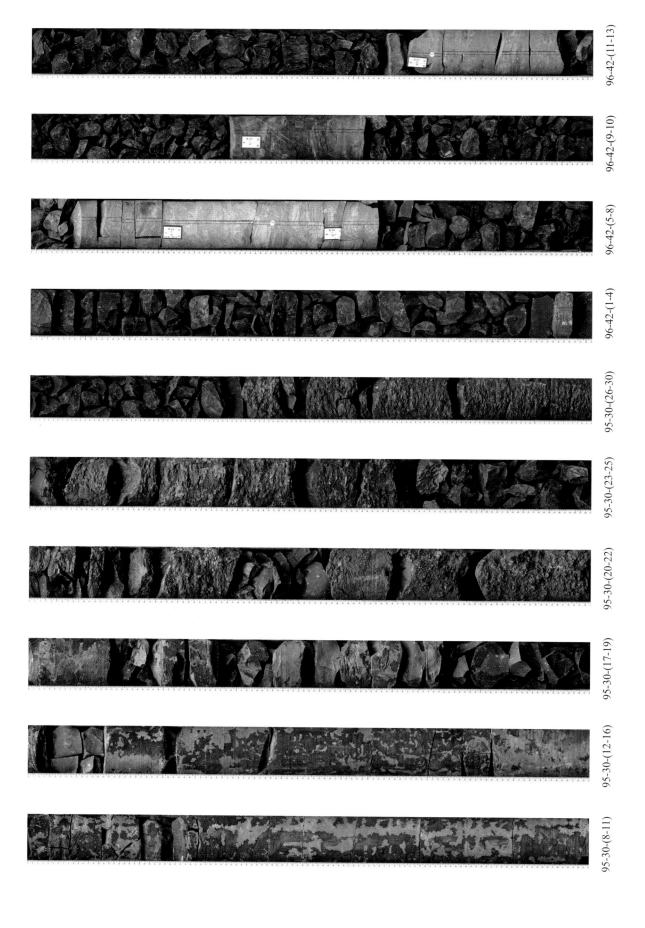

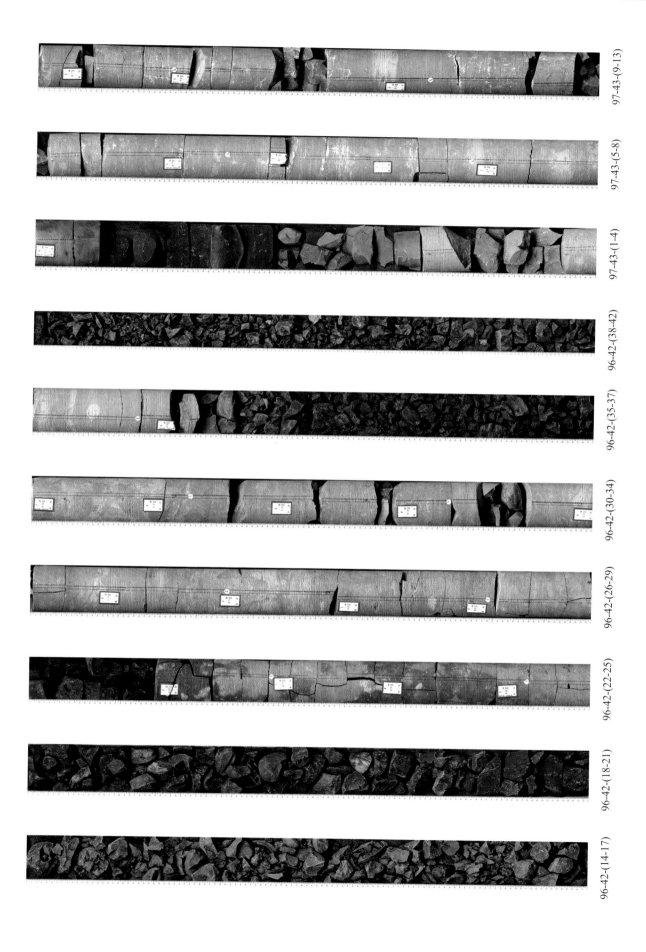

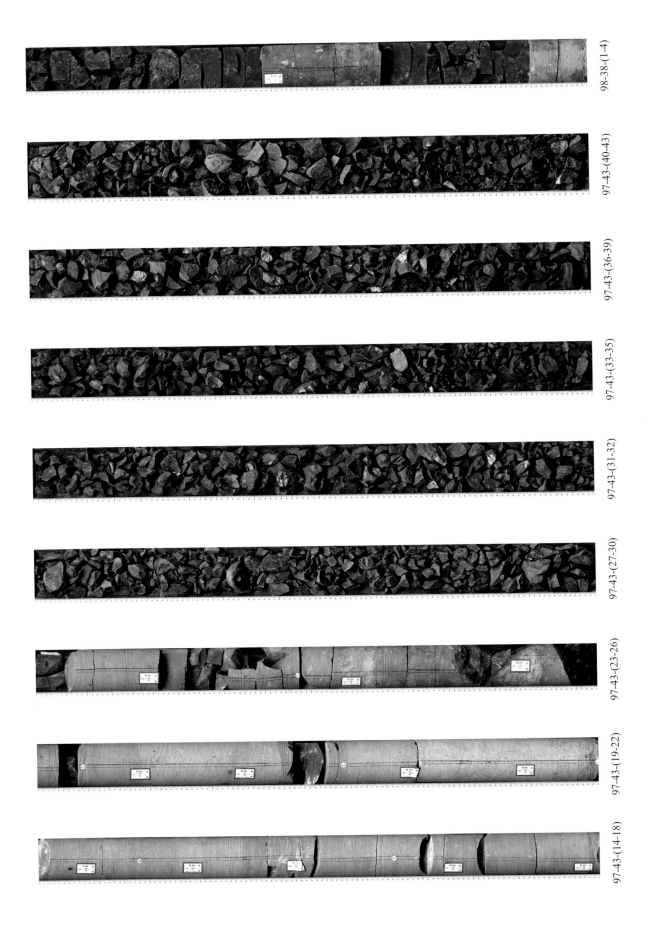

Section 5 SK-1 Core Description and Core Photographs

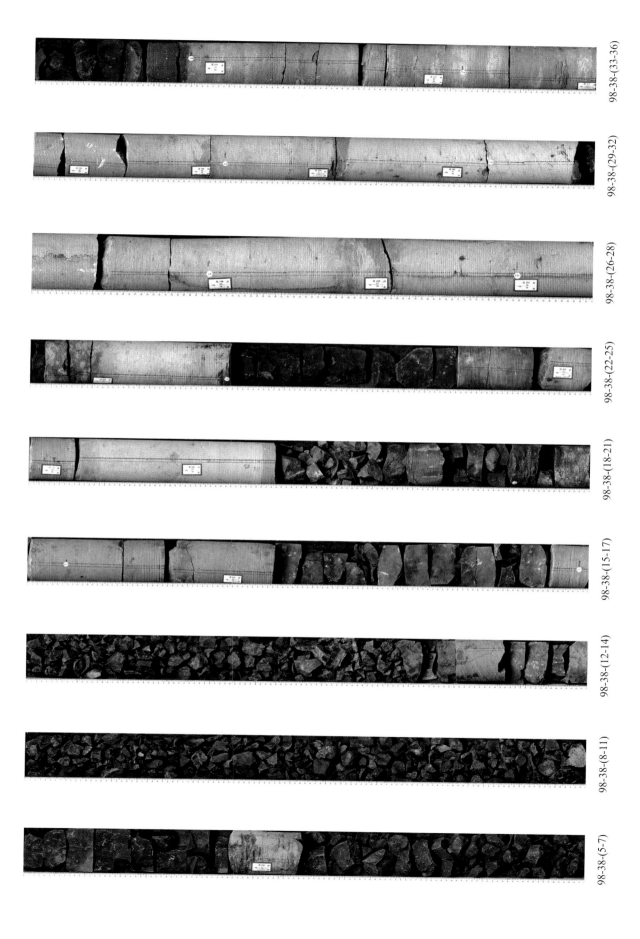

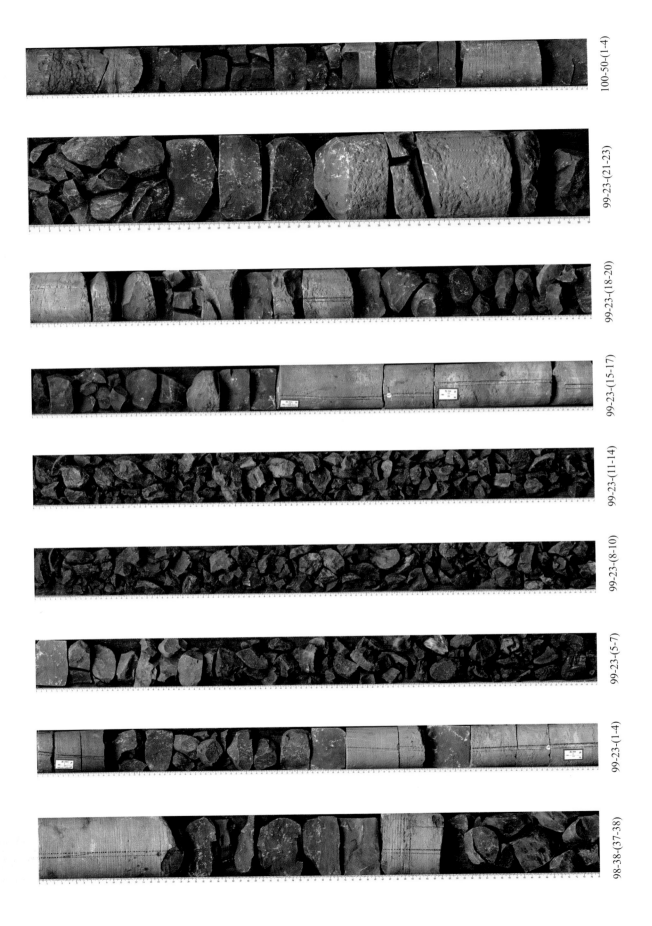

Section 5 SK-1 Core Description and Core Photographs

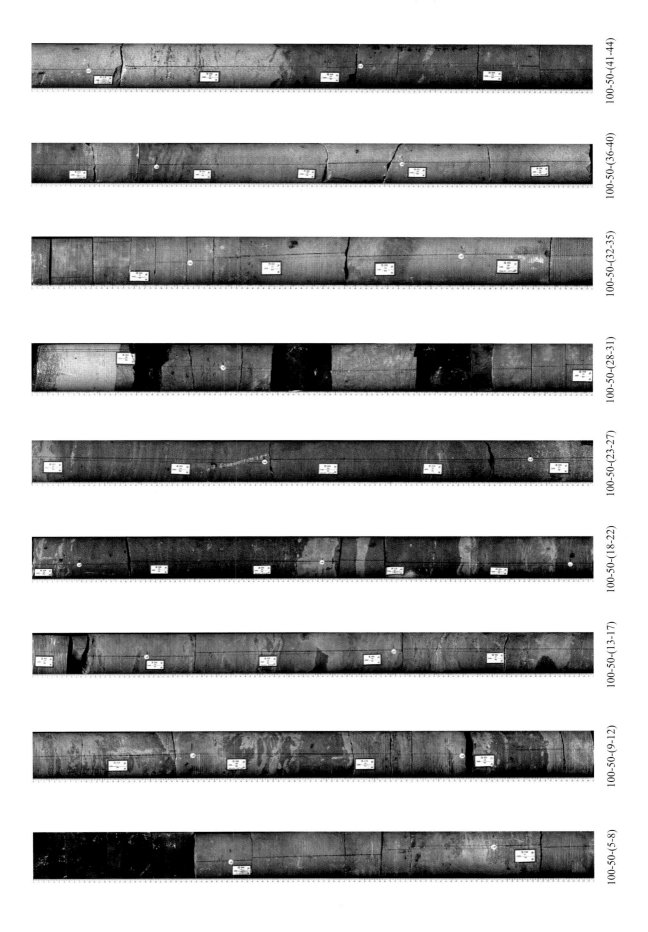

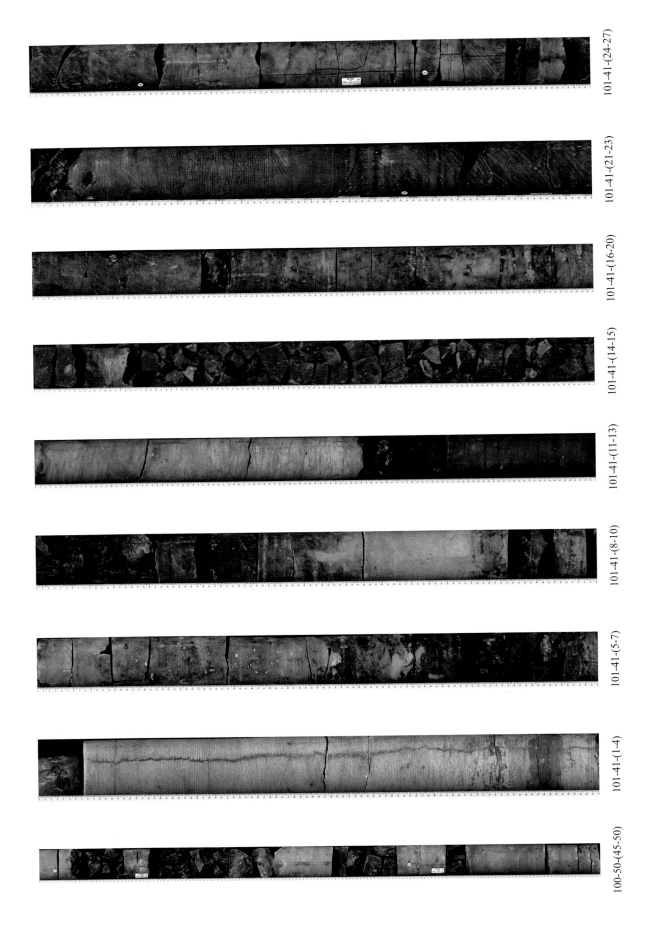

Section 5 SK-1 Core Description and Core Photographs 463

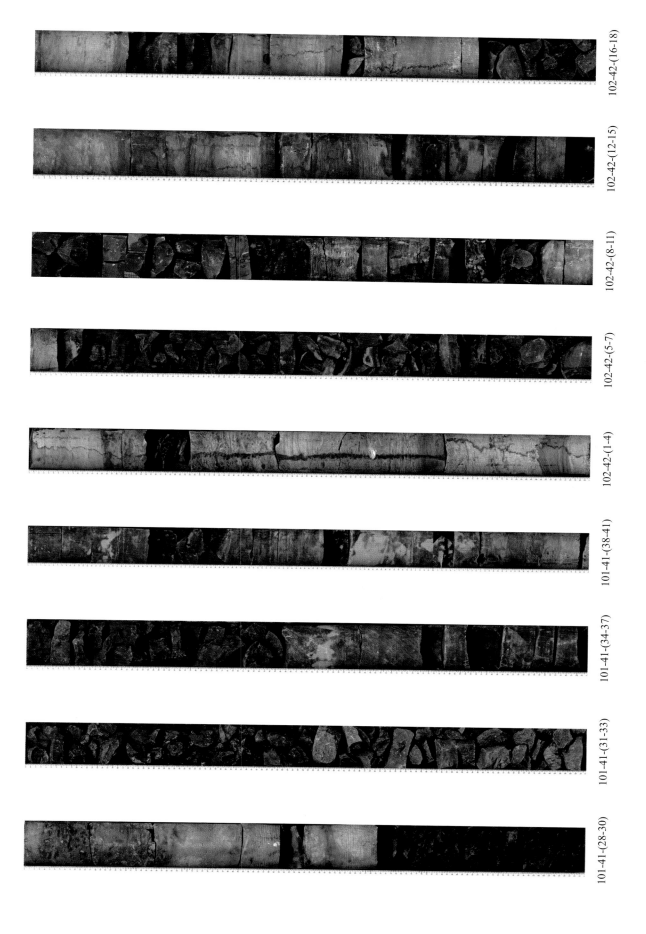

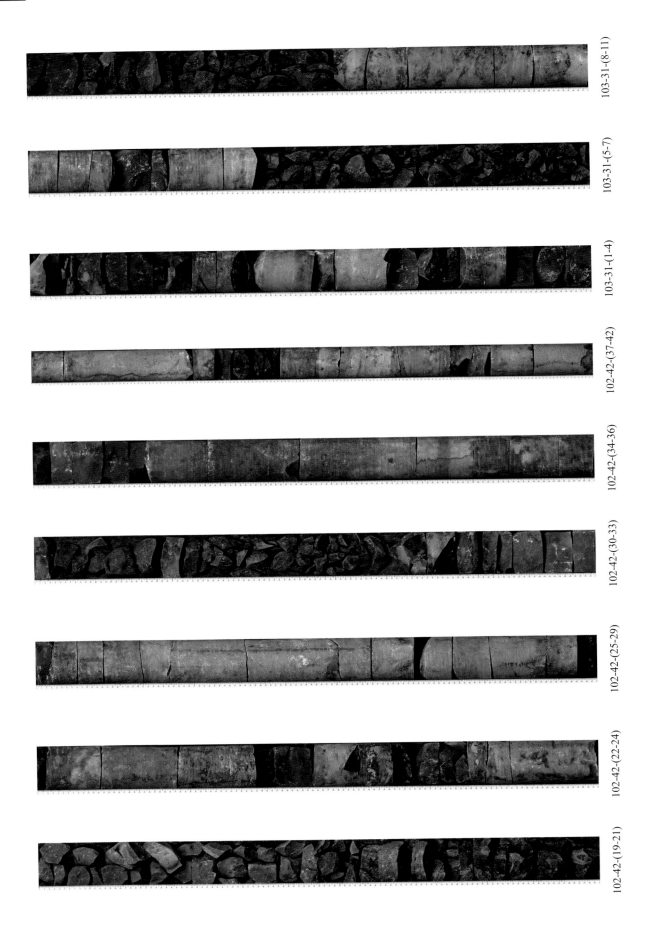

Section 5 SK-1 Core Description and Core Photographs | *465*

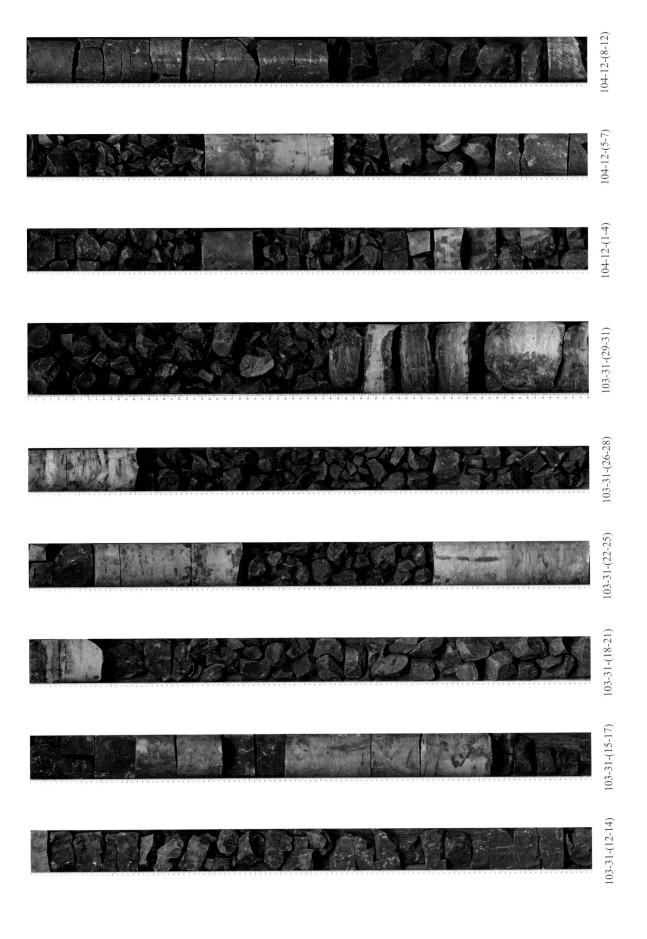

5.2.2 SK-1n Core Photographs

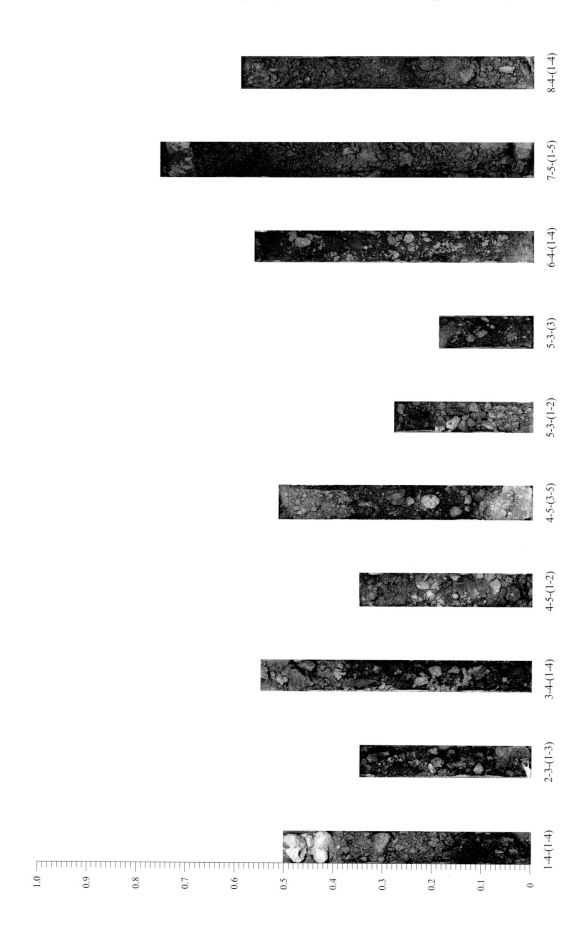

Section 5 SK-1 Core Description and Core Photographs

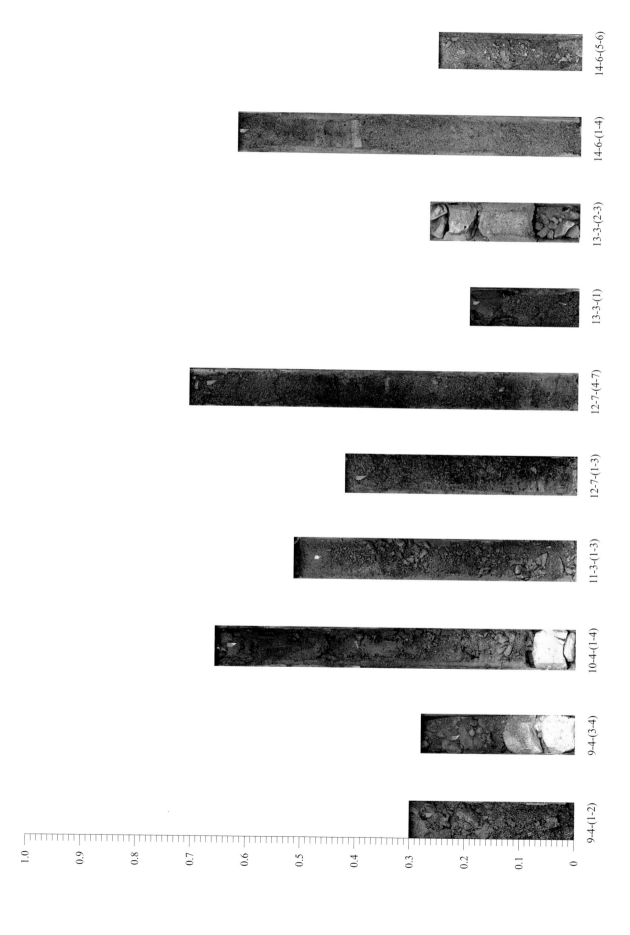

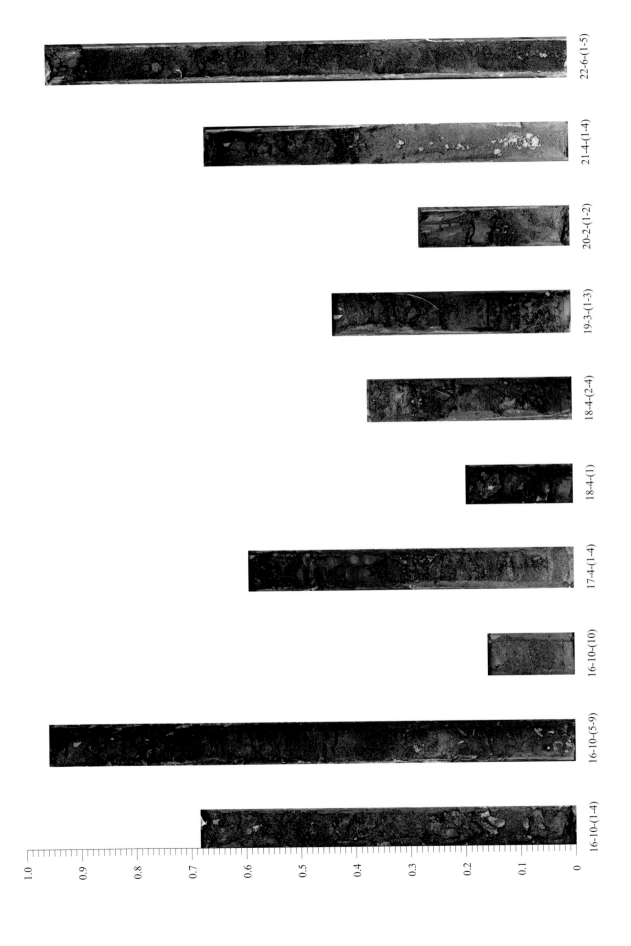

Section 5 SK-1 Core Description and Core Photographs | *469*

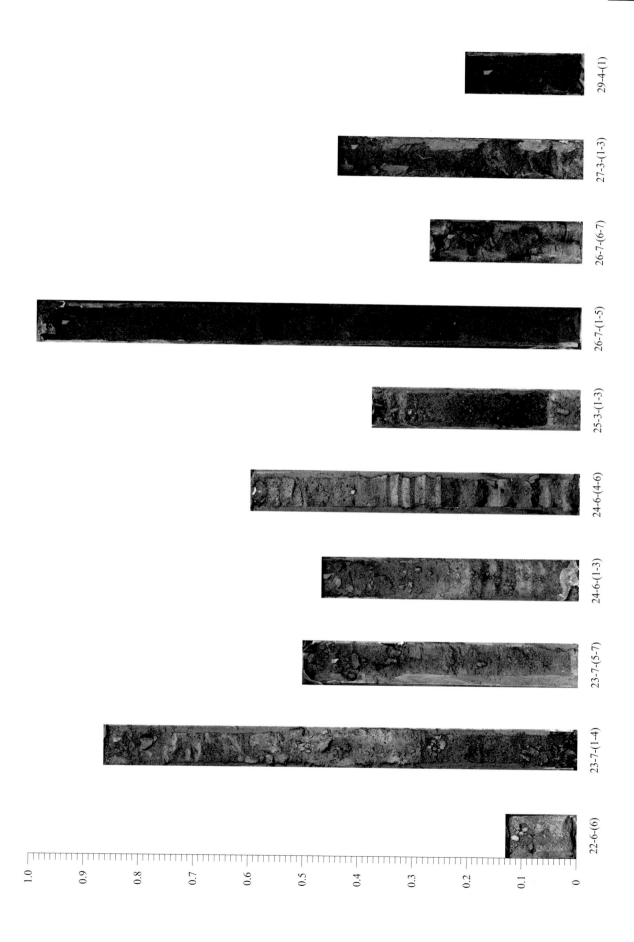

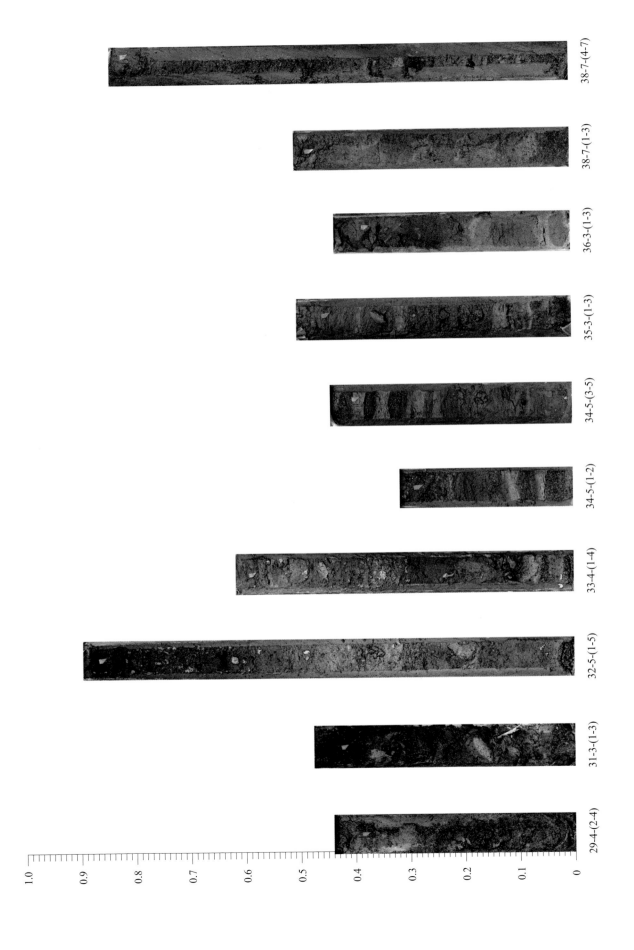

Section 5 SK-1 Core Description and Core Photographs

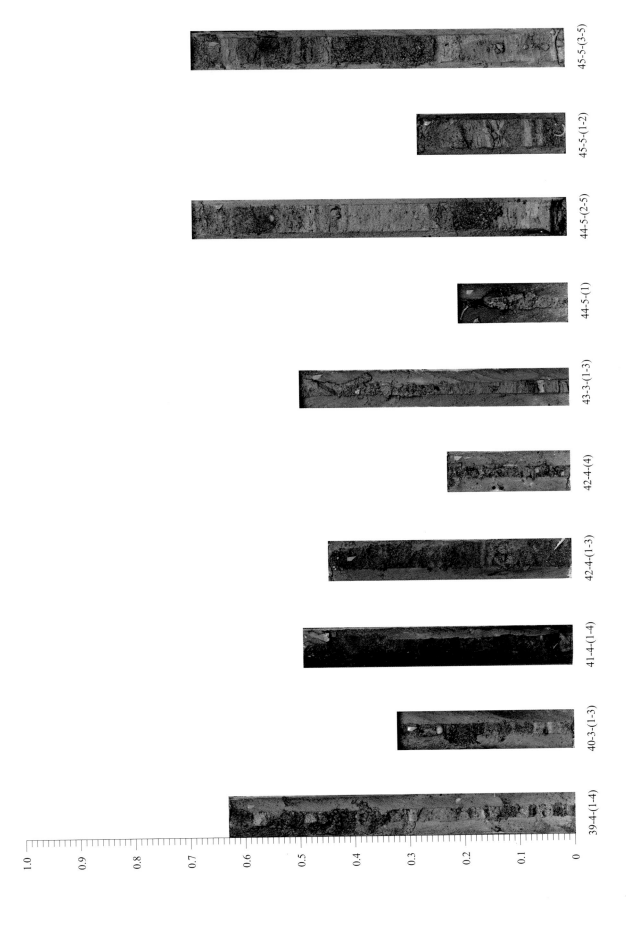

Section 5 SK-1 Core Description and Core Photographs | *473*

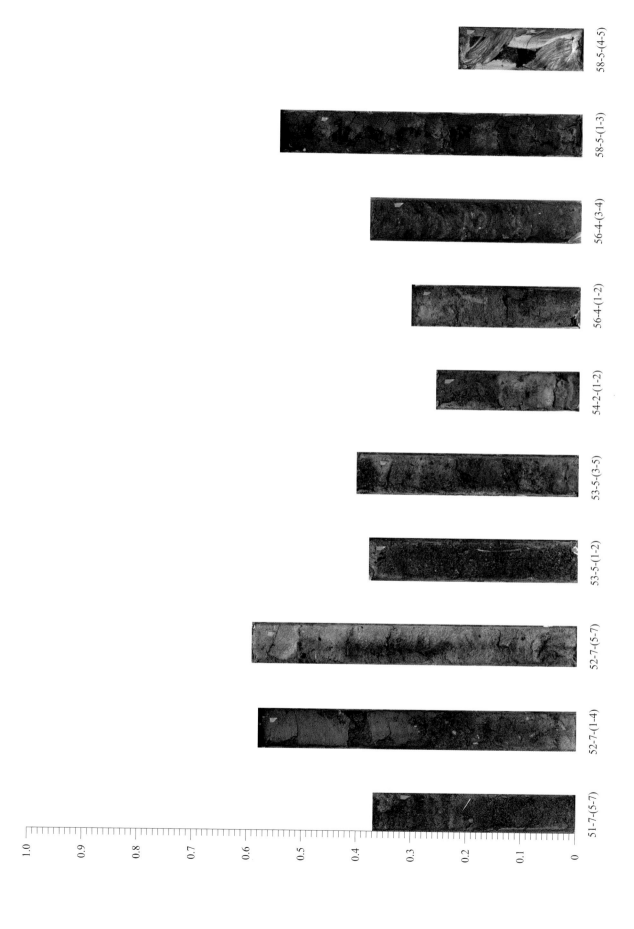

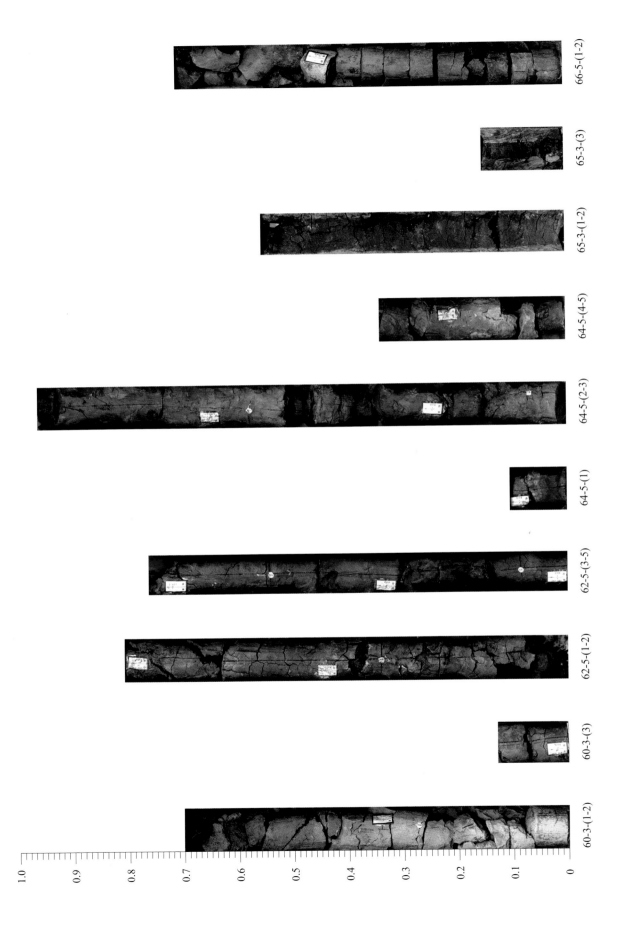

Section 5 SK-1 Core Description and Core Photographs | *475*

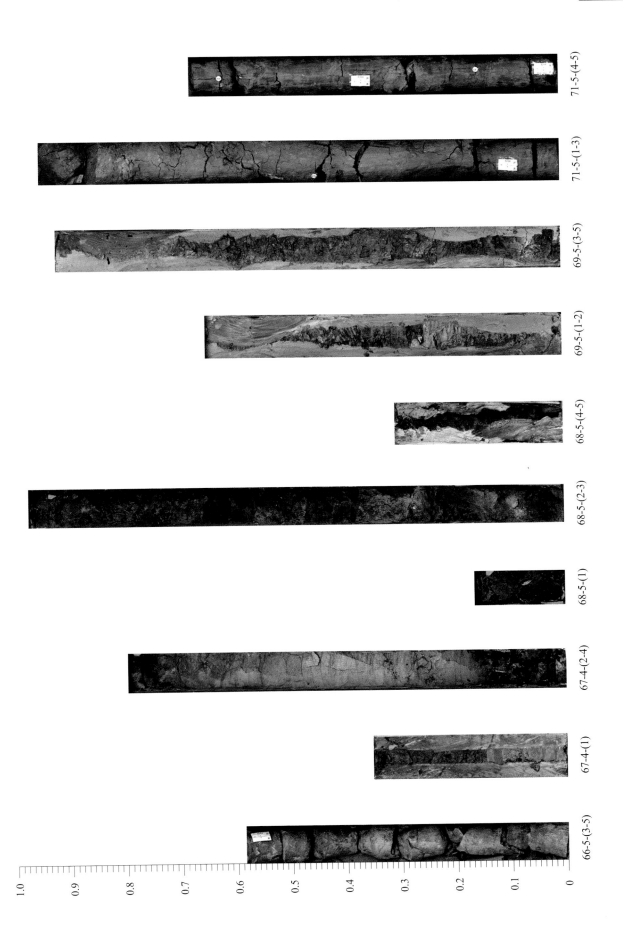

Section 5 SK-1 Core Description and Core Photographs

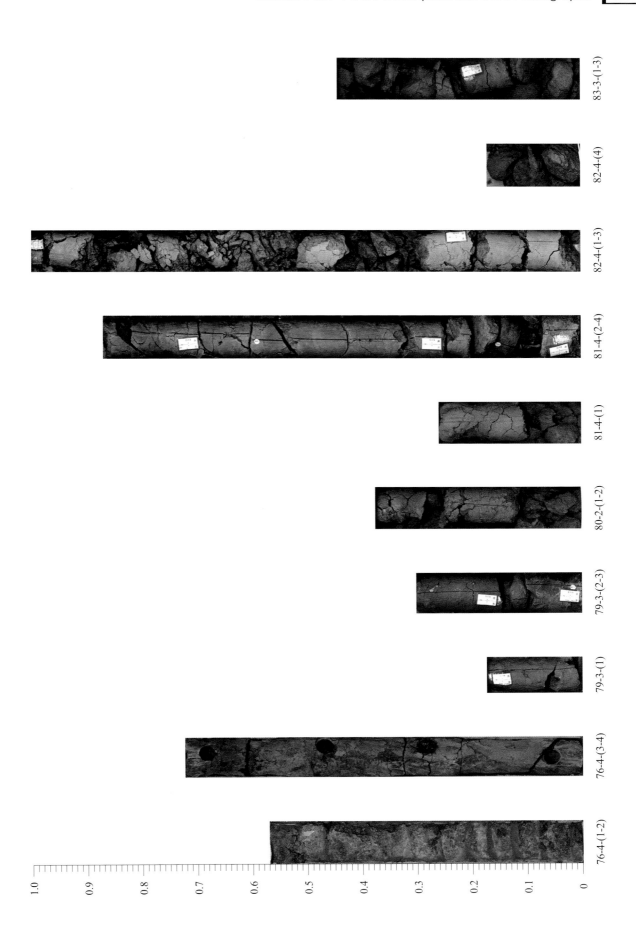

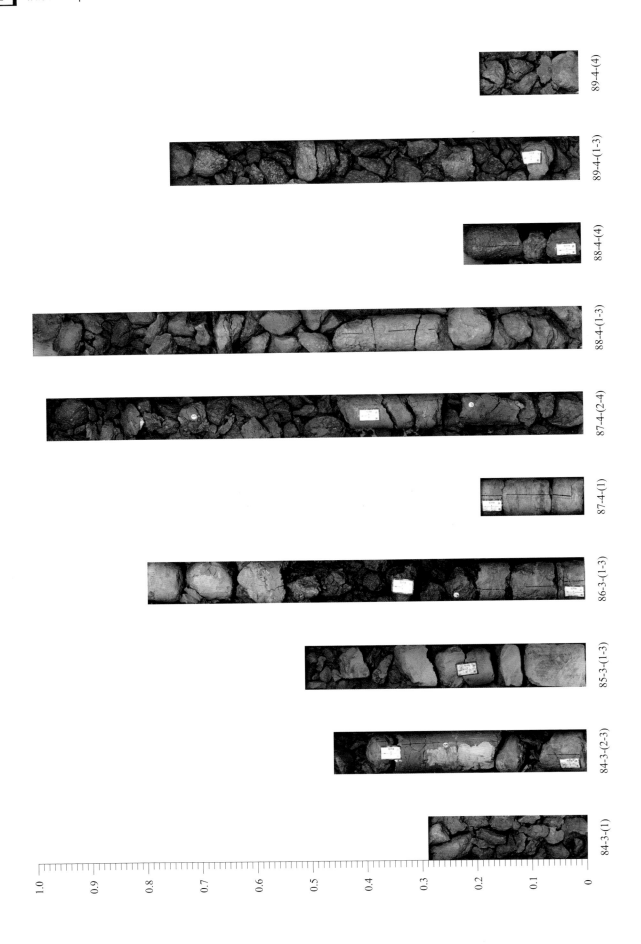

Section 5 SK-1 Core Description and Core Photographs

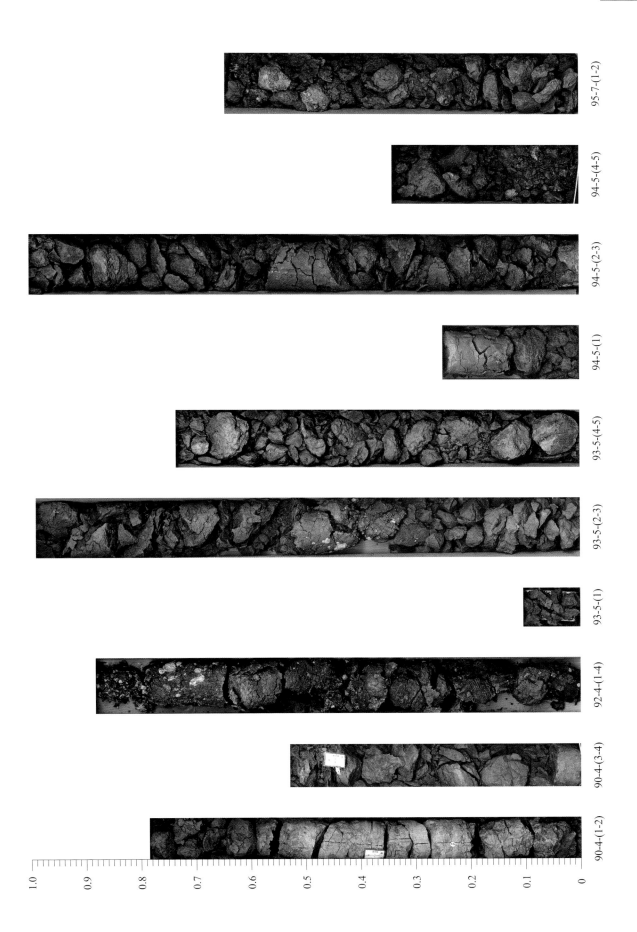

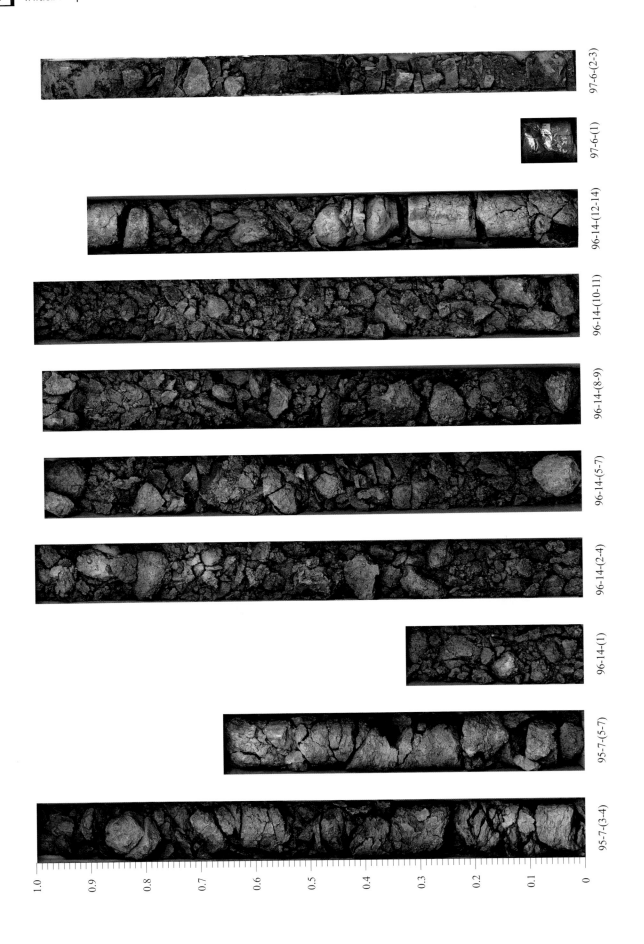

Section 5 SK-1 Core Description and Core Photographs | *481*

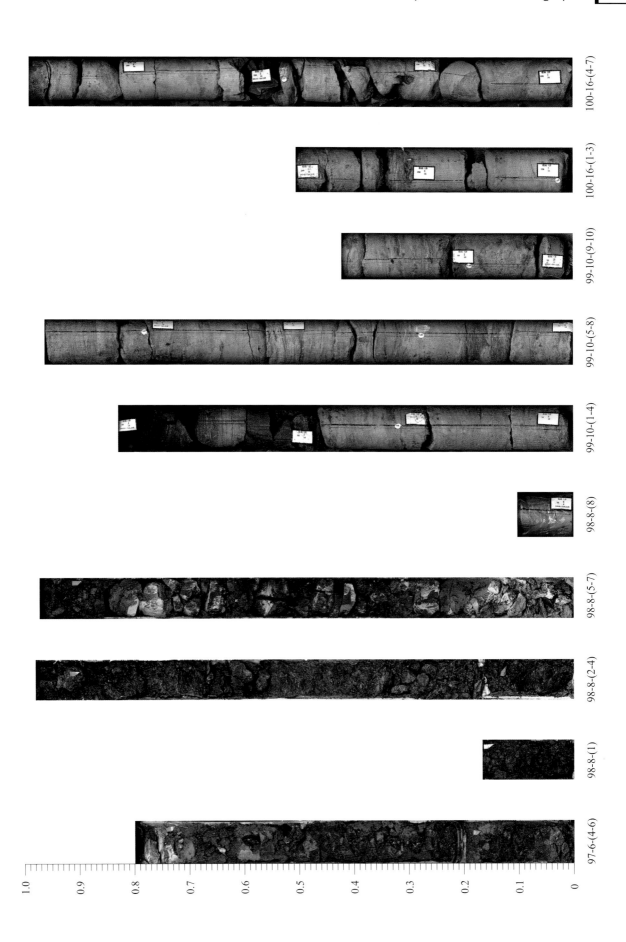

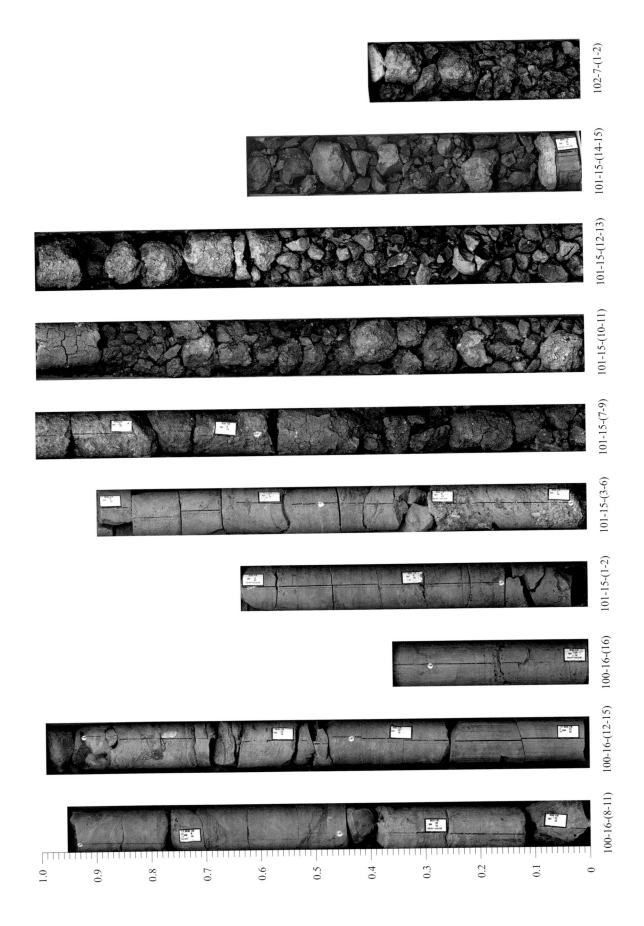

Section 5 SK-1 Core Description and Core Photographs | 483

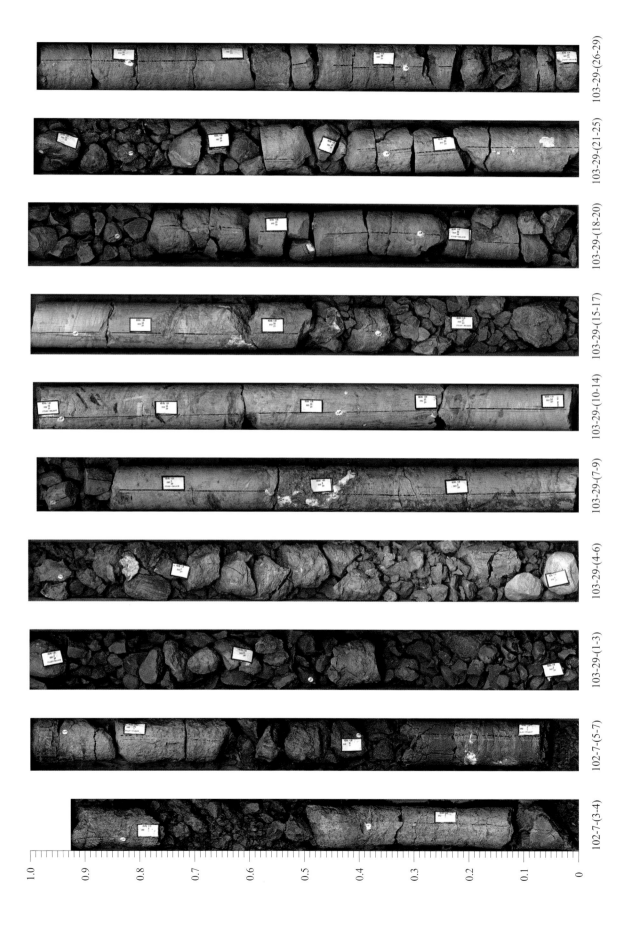

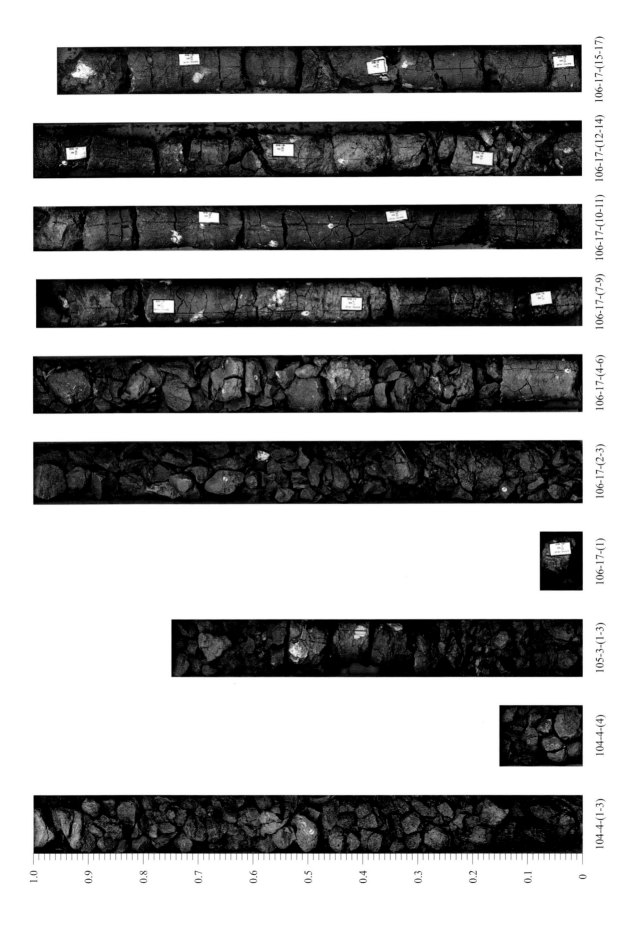

Section 5 SK-1 Core Description and Core Photographs | *485*

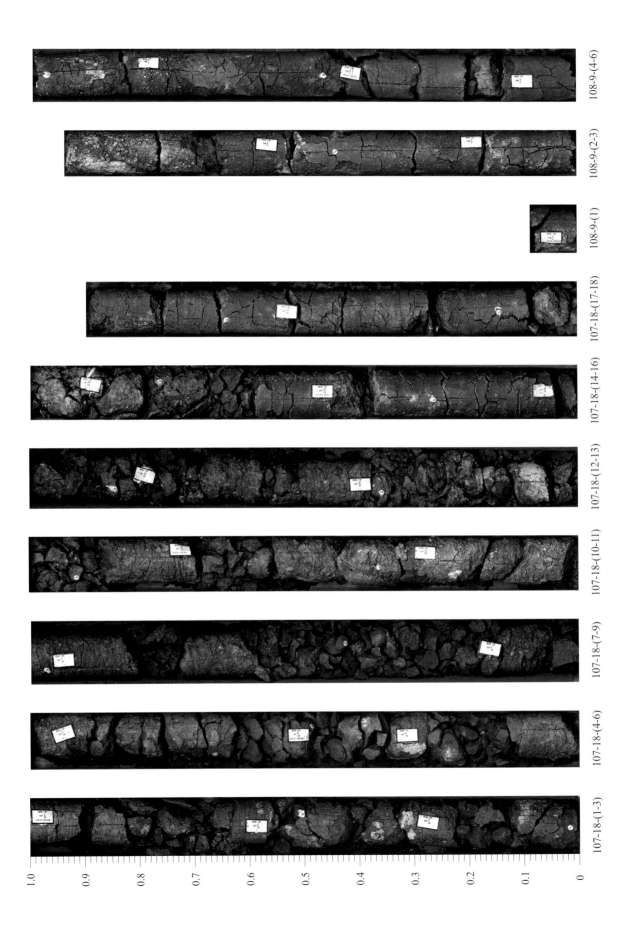

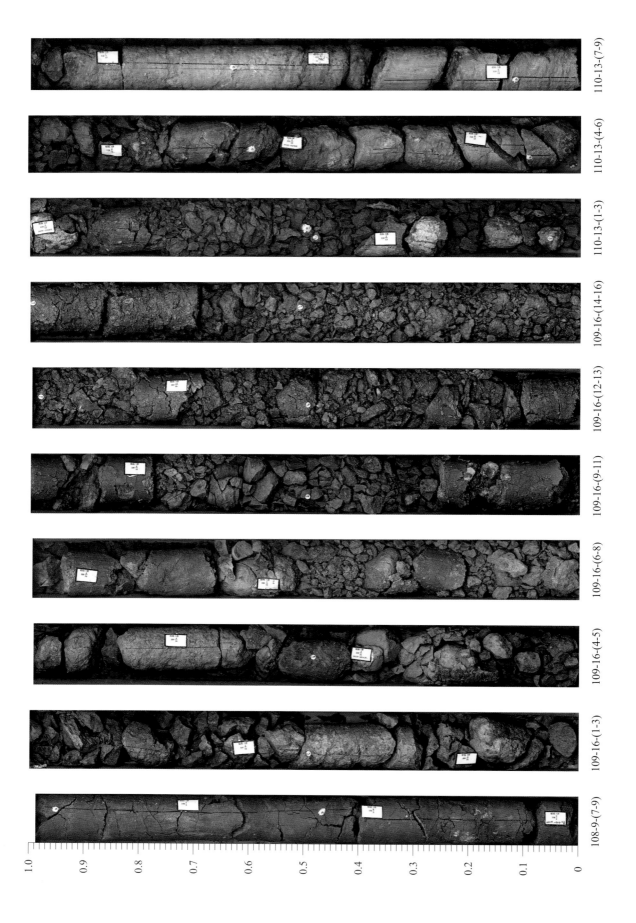

Section 5 SK-1 Core Description and Core Photographs | *487*

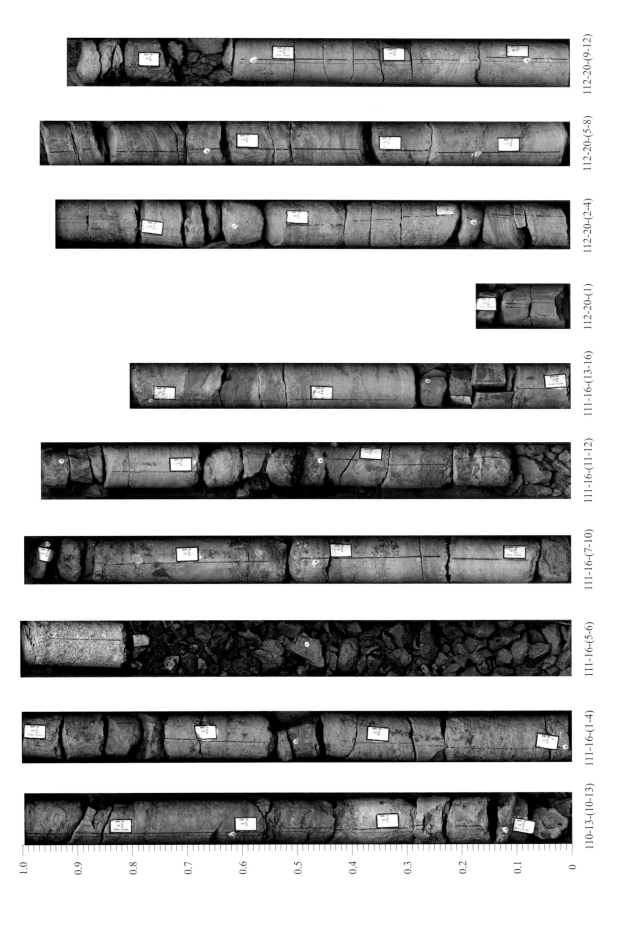

Initial Report of Continental Scientific Drilling Project of the Cretaceous Songliao Basin (SK-1)

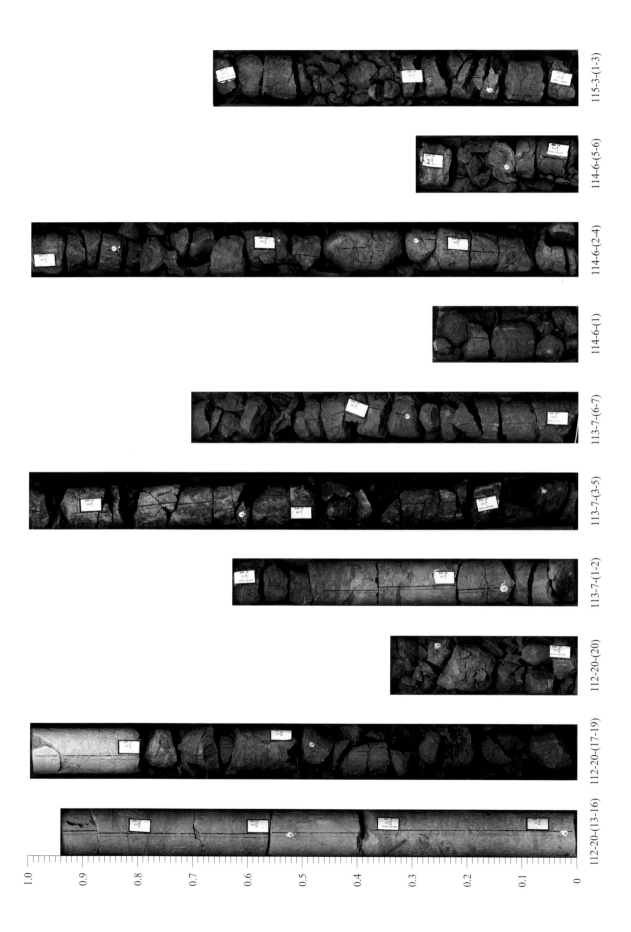

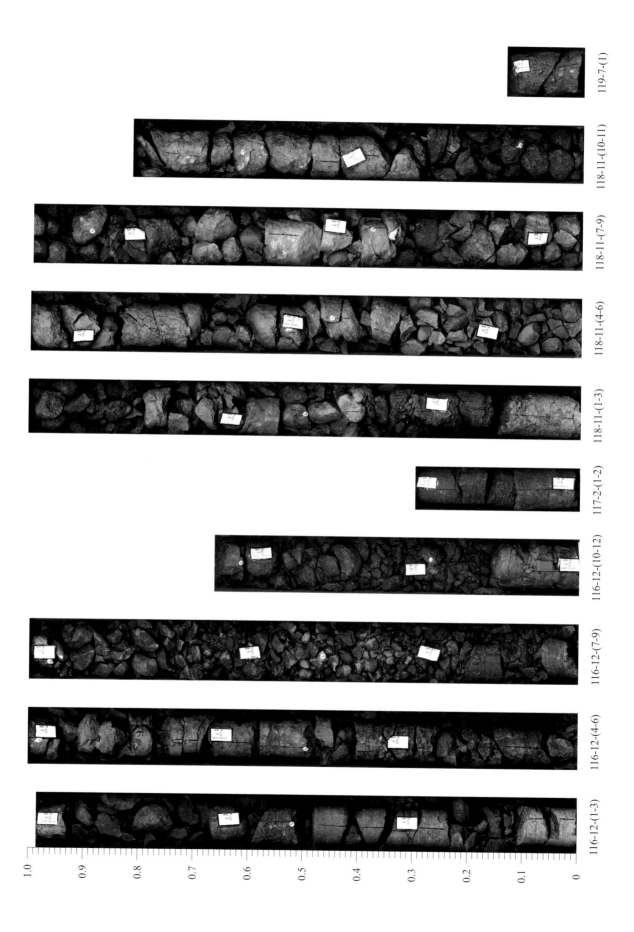

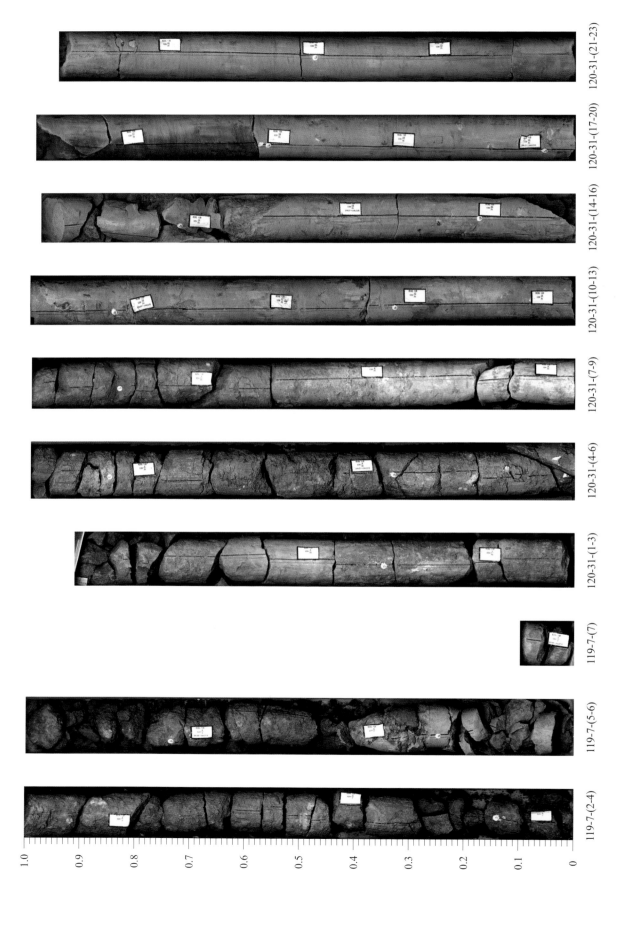

Section 5 SK-1 Core Description and Core Photographs

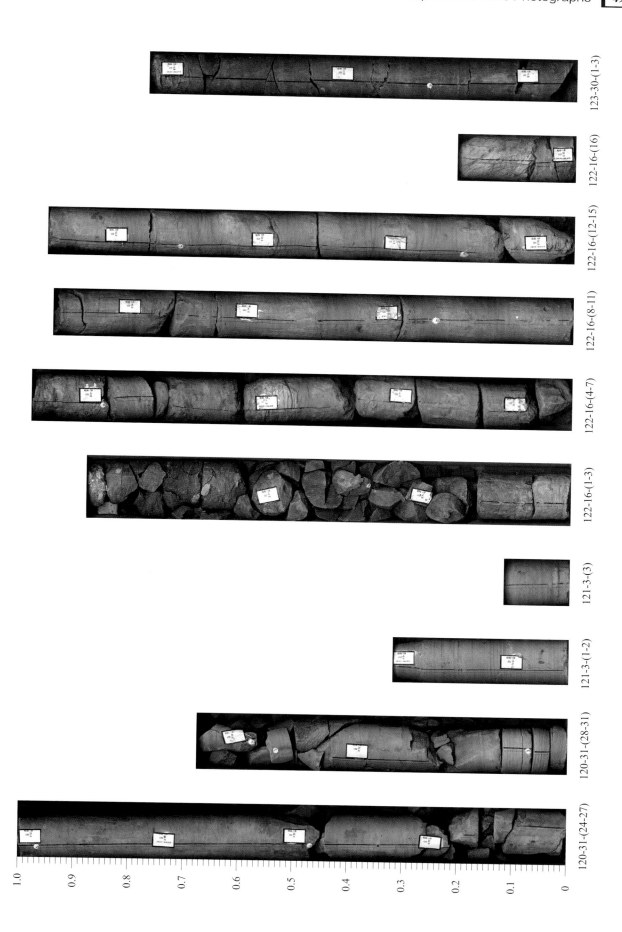

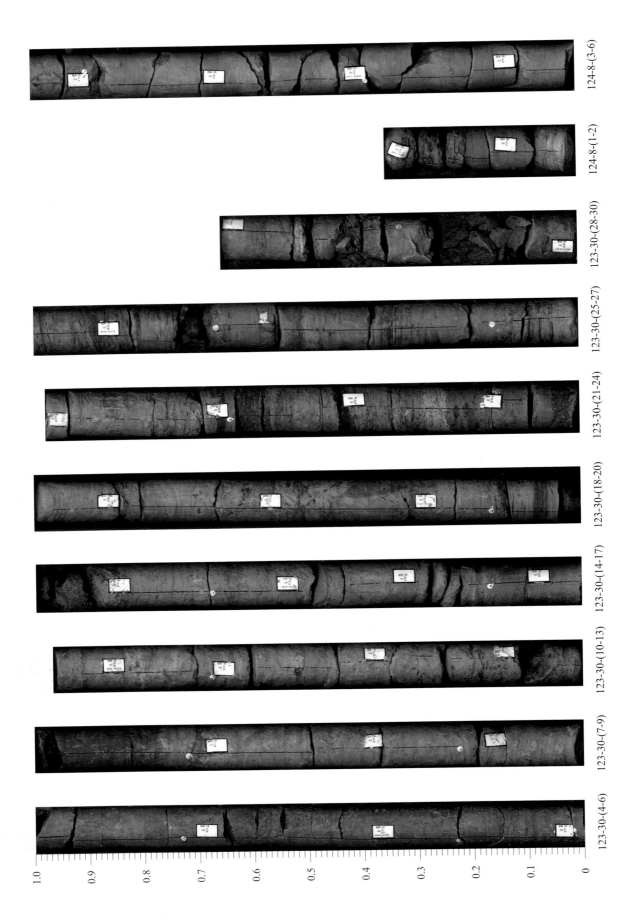

Section 5 SK-1 Core Description and Core Photographs

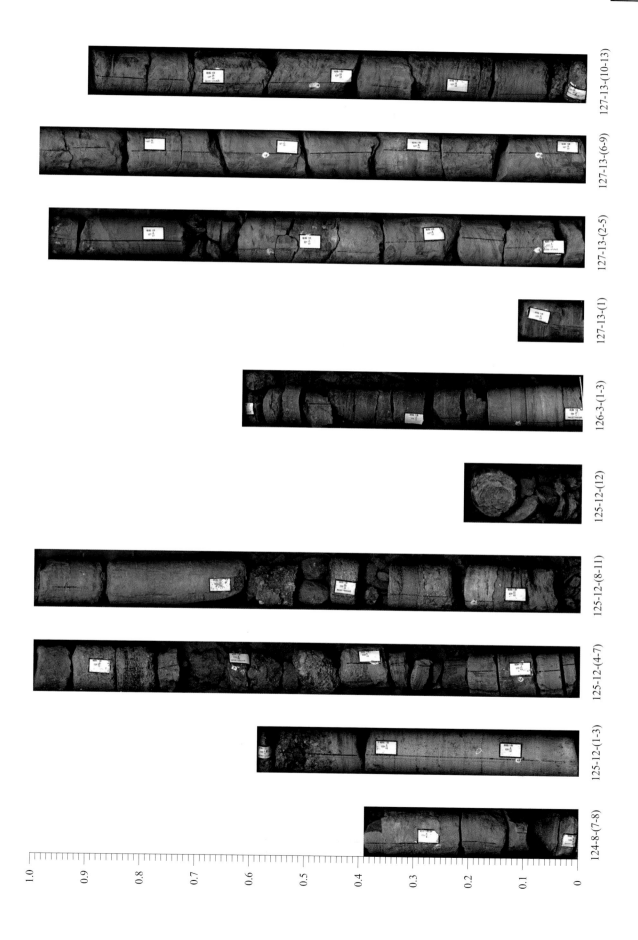

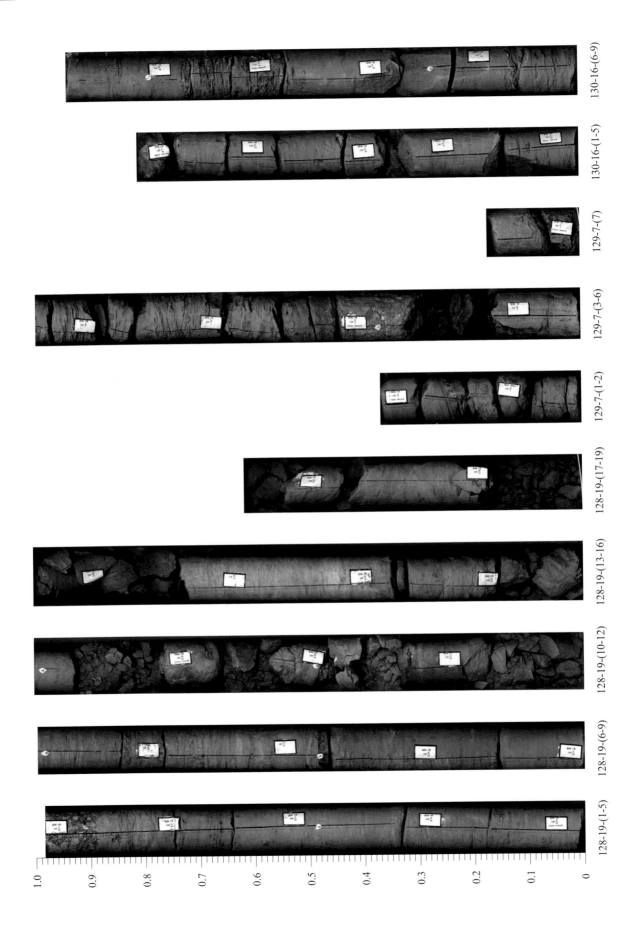

Section 5 SK-1 Core Description and Core Photographs

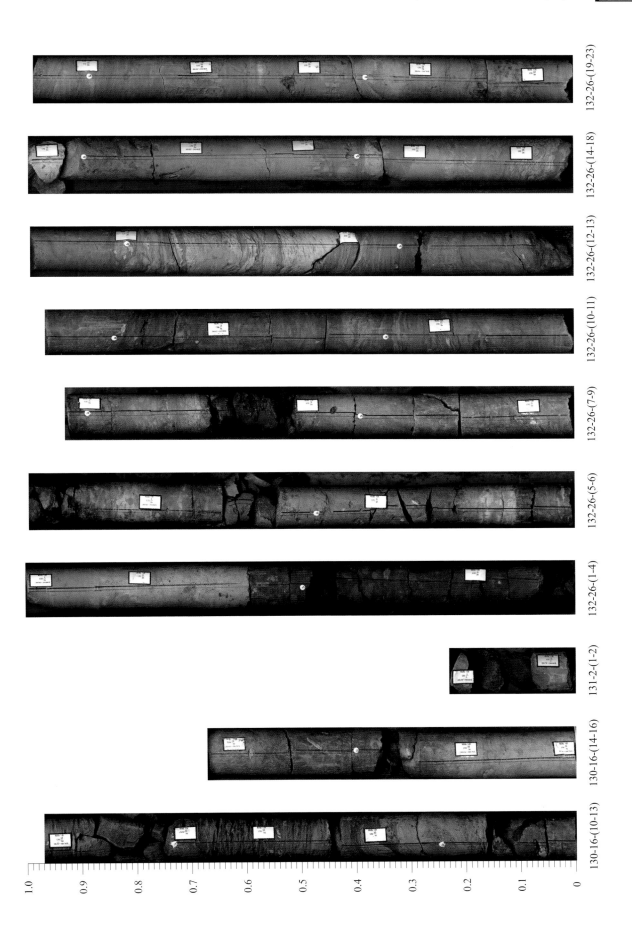

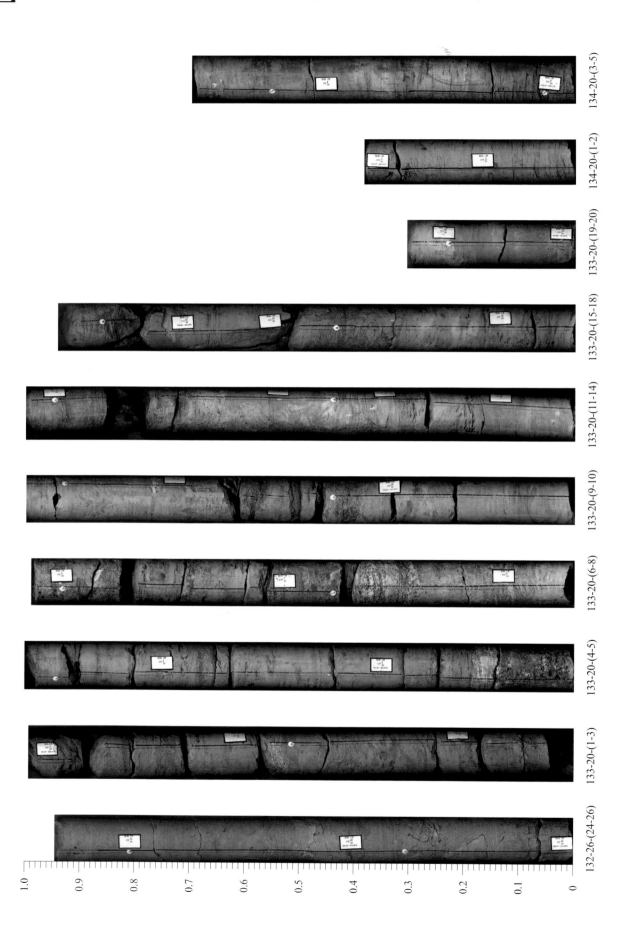

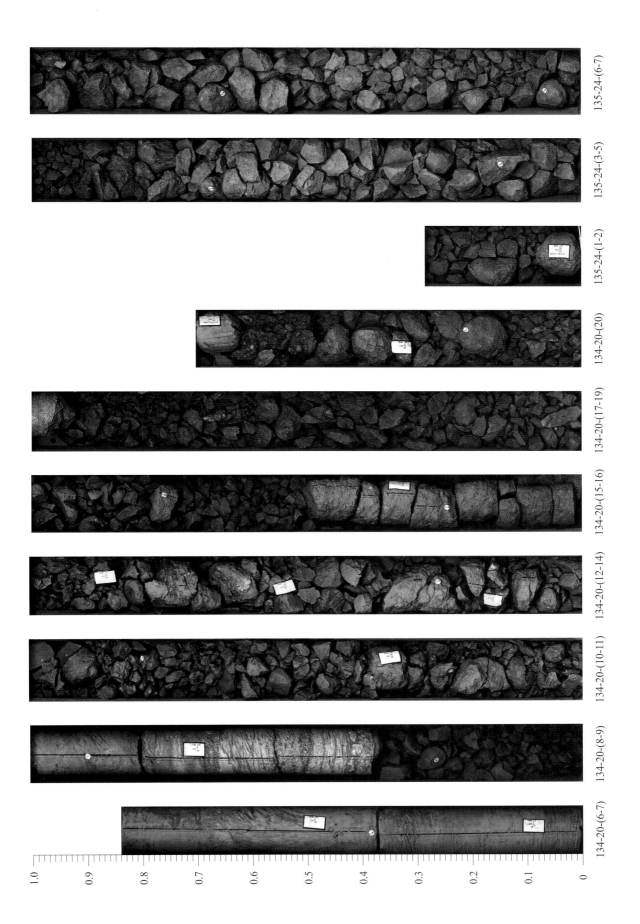

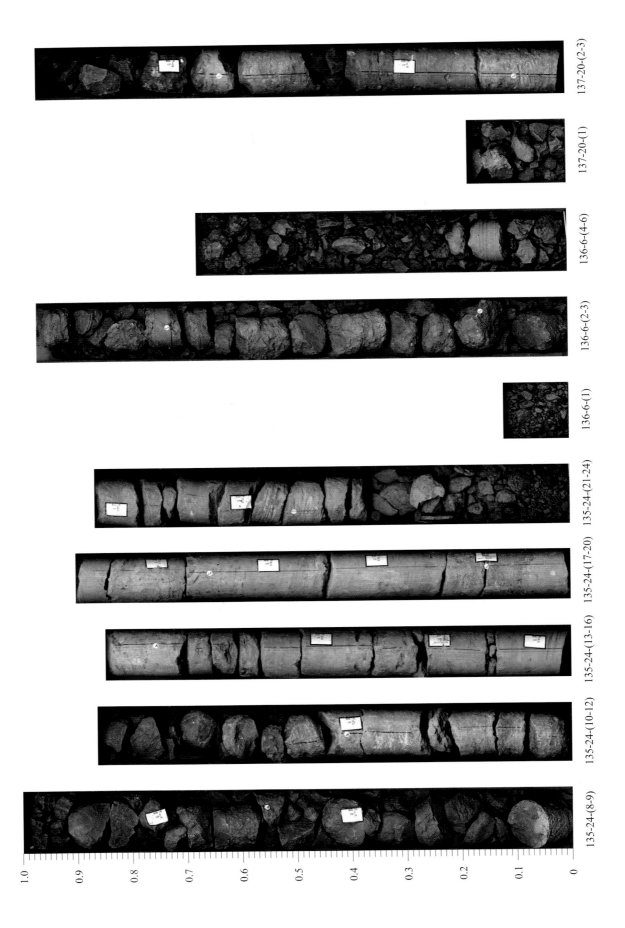

Section 5 SK-1 Core Description and Core Photographs | *499*

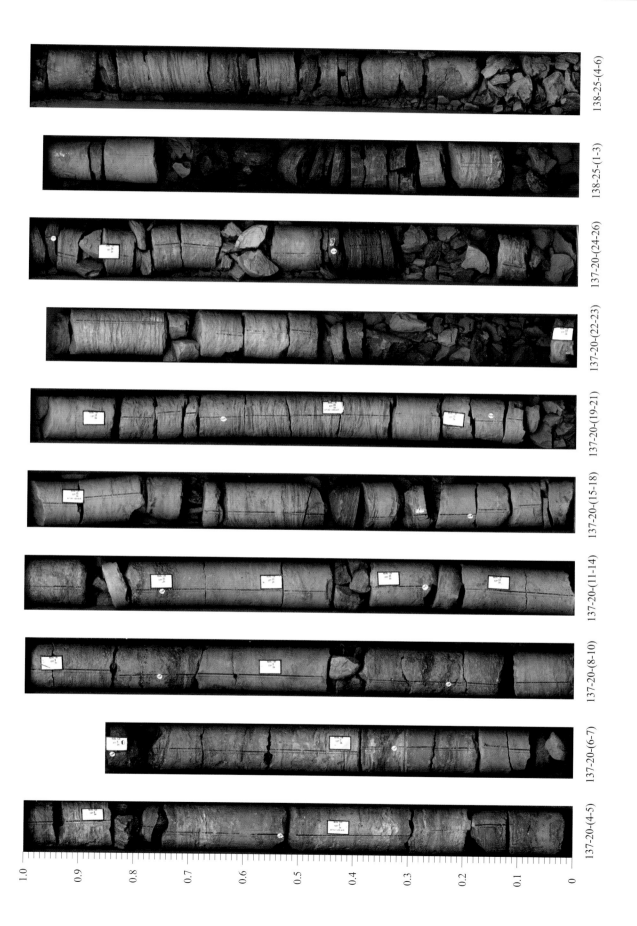

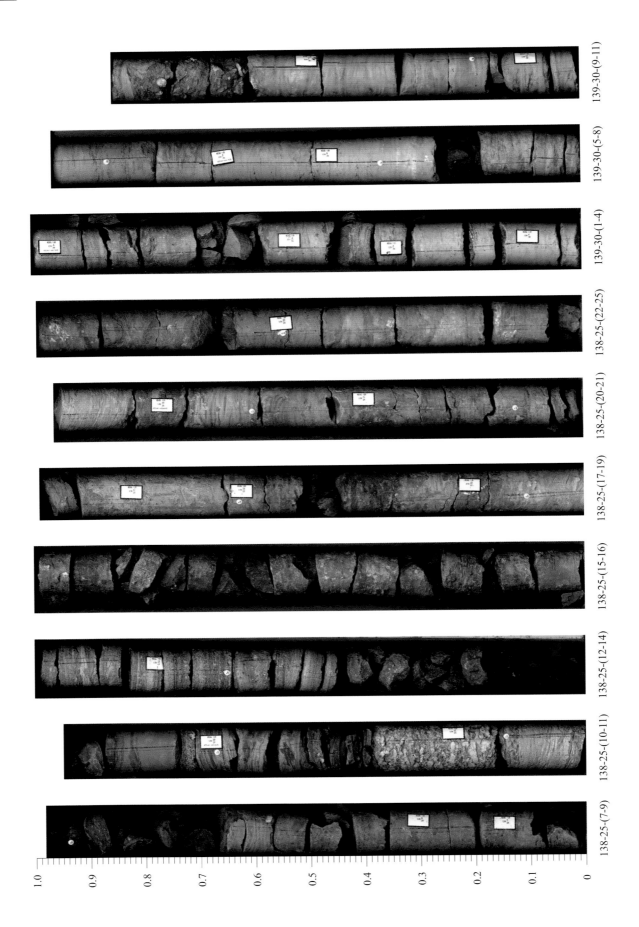

Section 5 SK-1 Core Description and Core Photographs

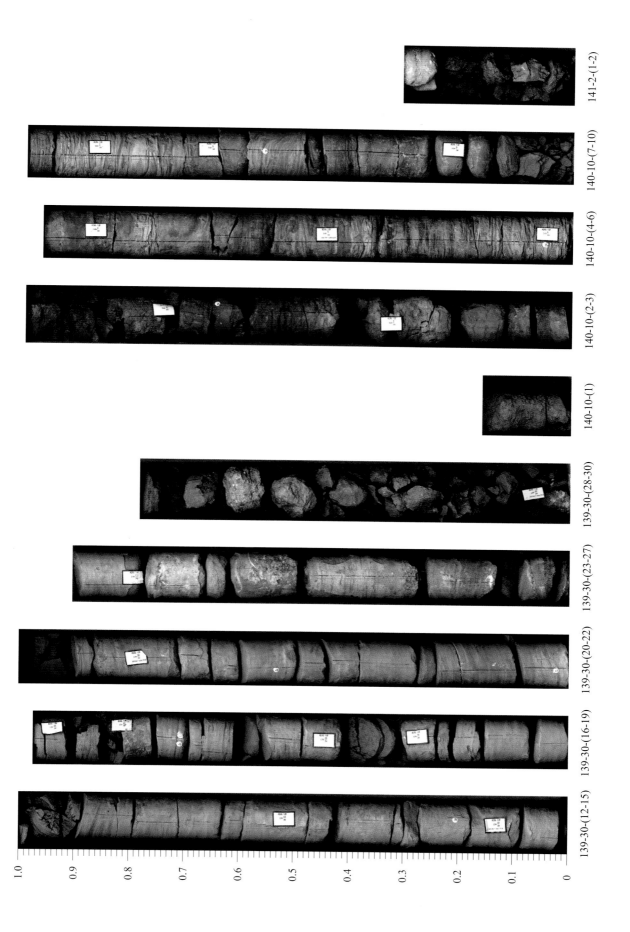

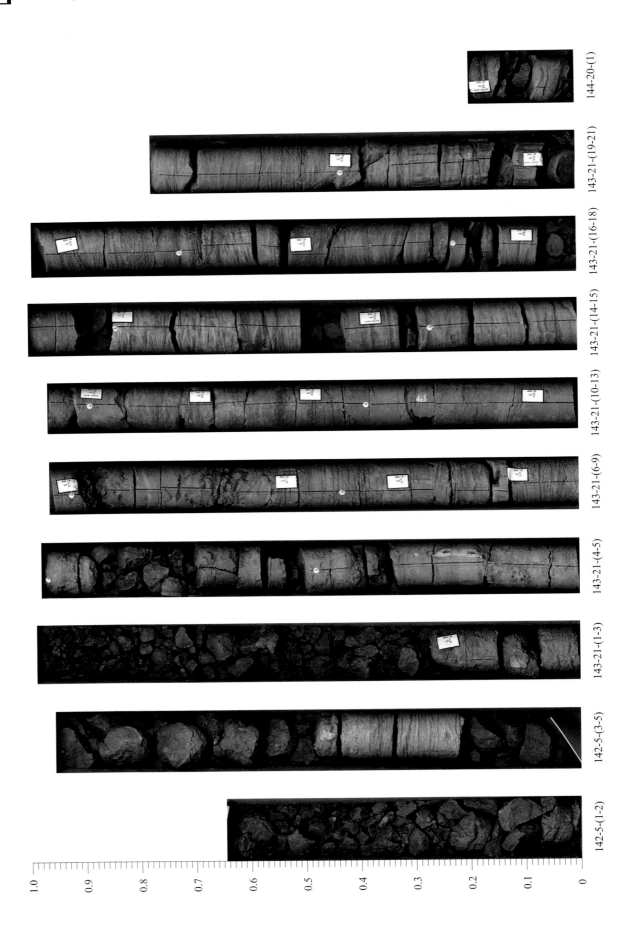

Section 5 SK-1 Core Description and Core Photographs | 503

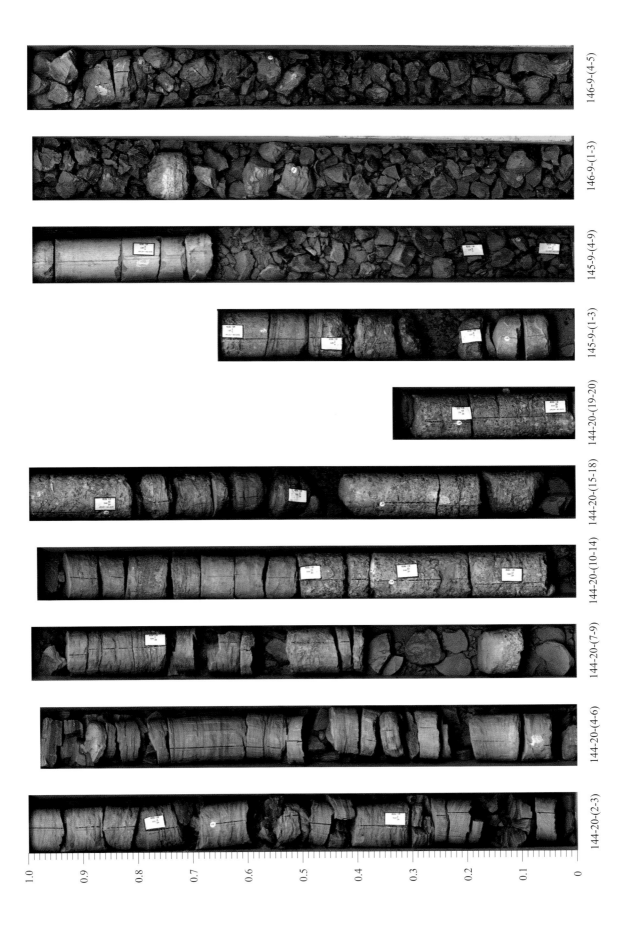

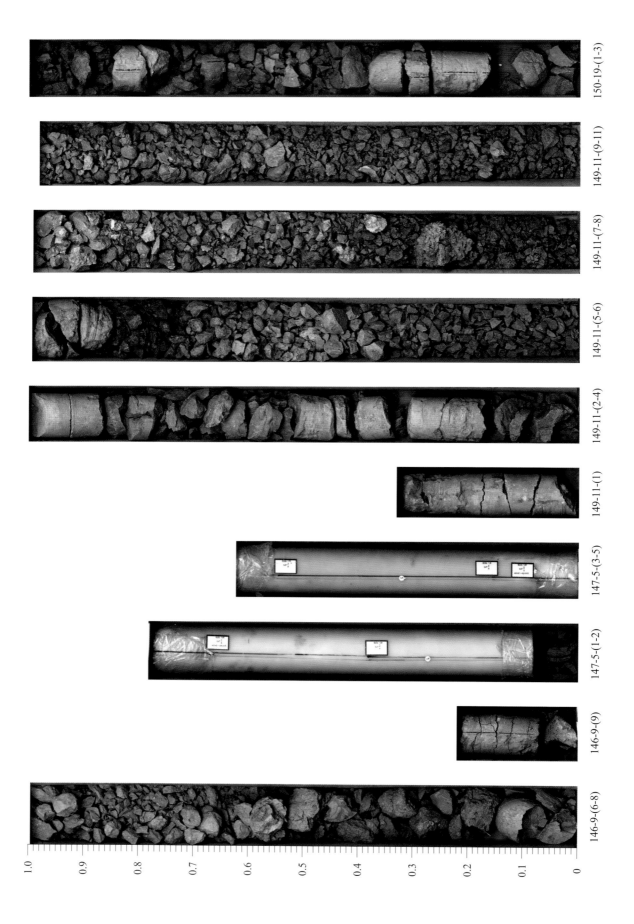

Section 5 SK-1 Core Description and Core Photographs

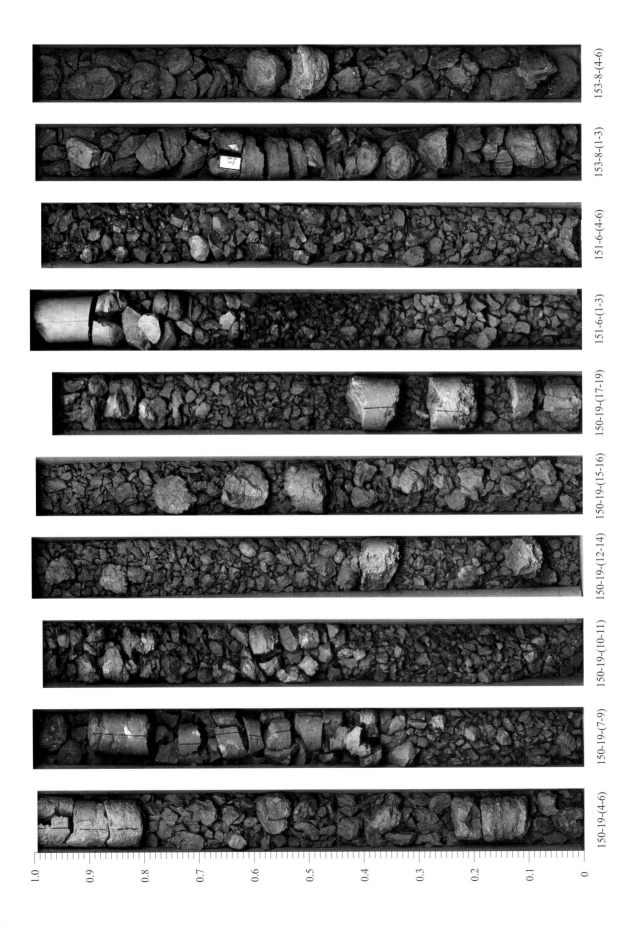

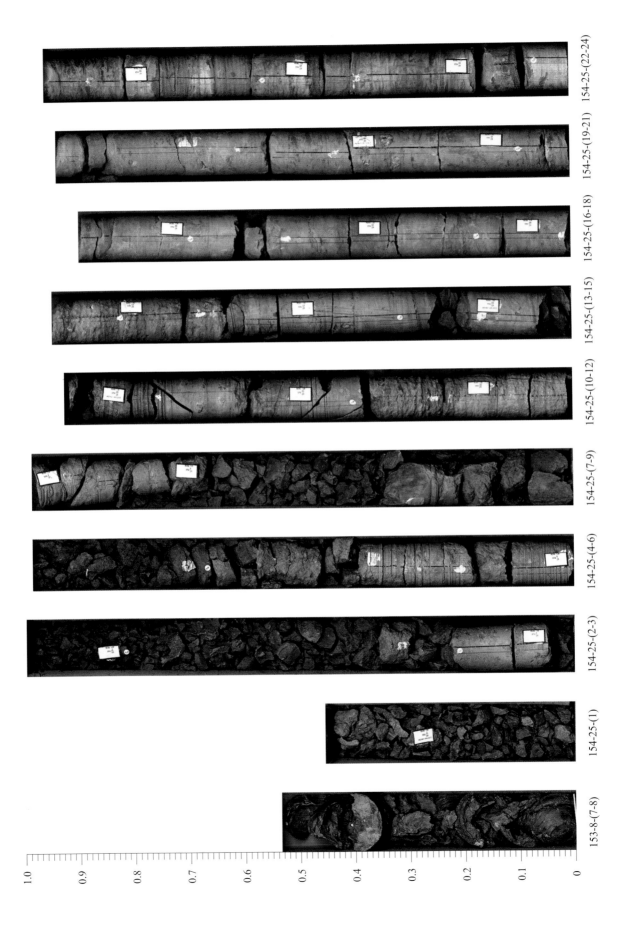

Section 5 SK-1 Core Description and Core Photographs

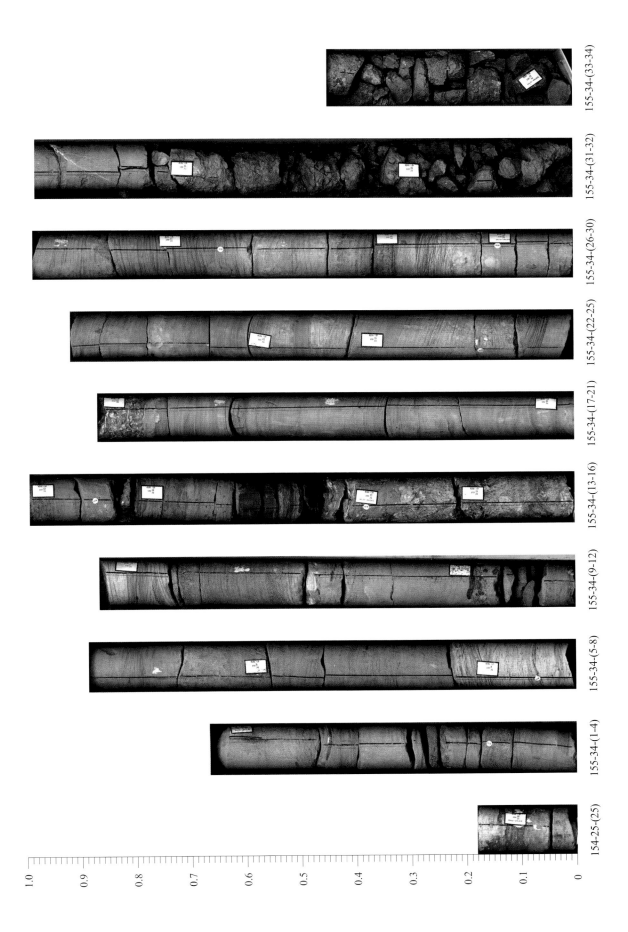

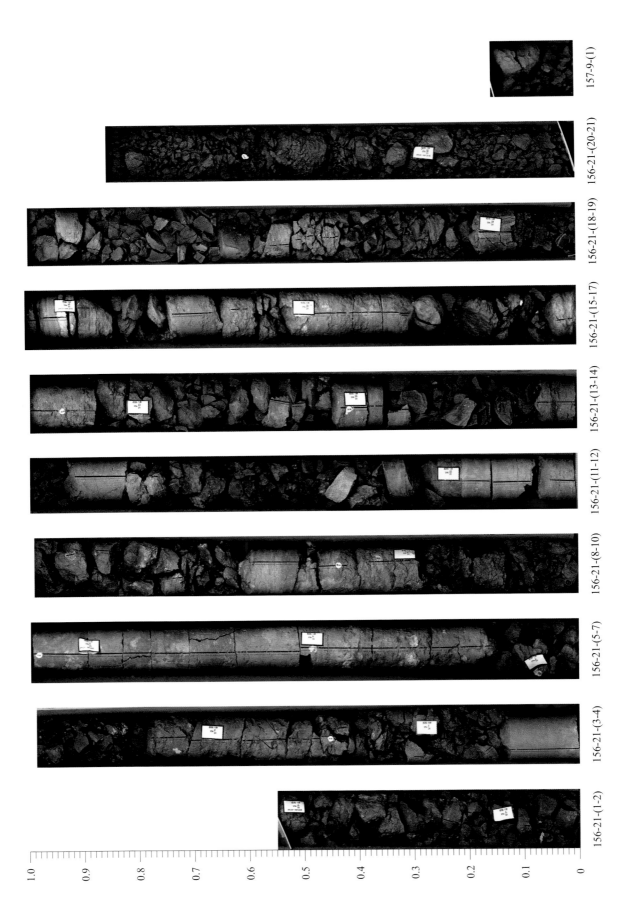

Section 5 SK-1 Core Description and Core Photographs | *509*

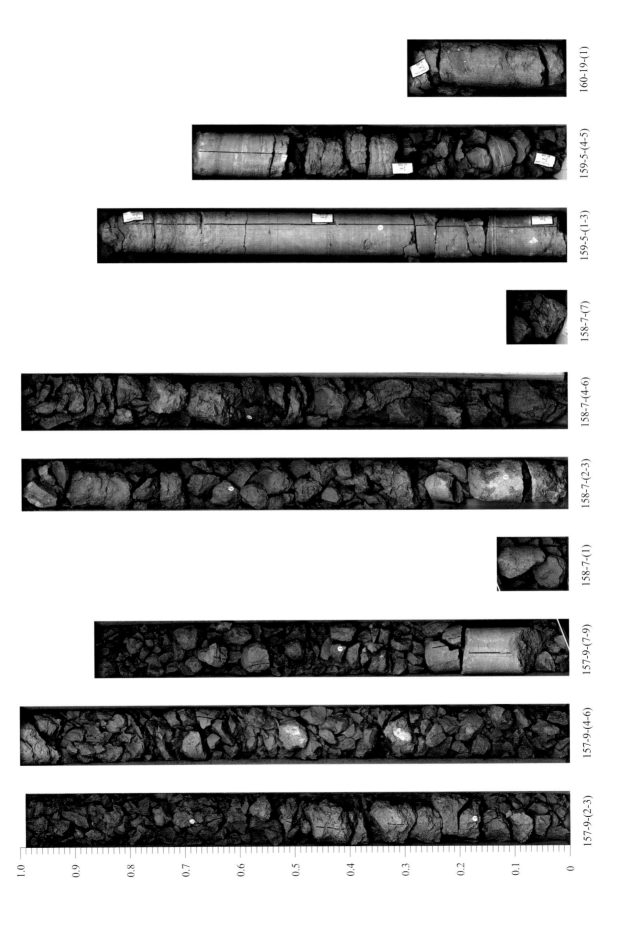

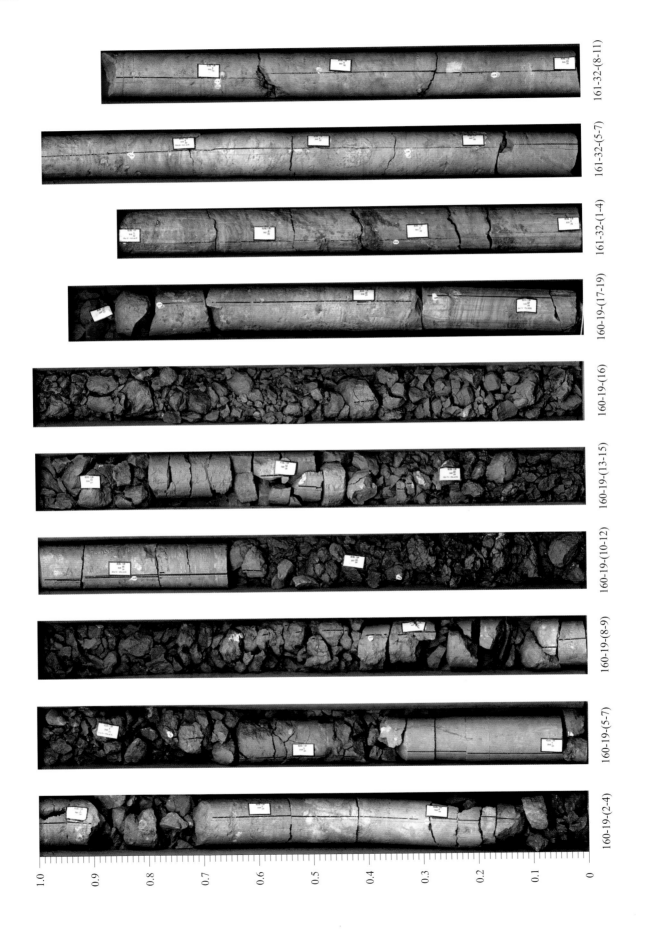

Section 5 SK-1 Core Description and Core Photographs

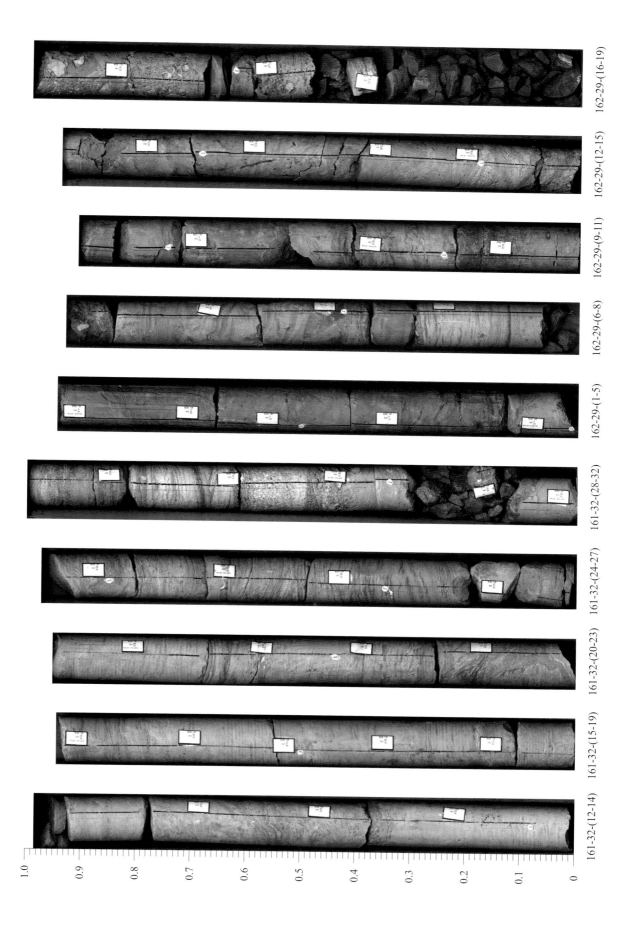

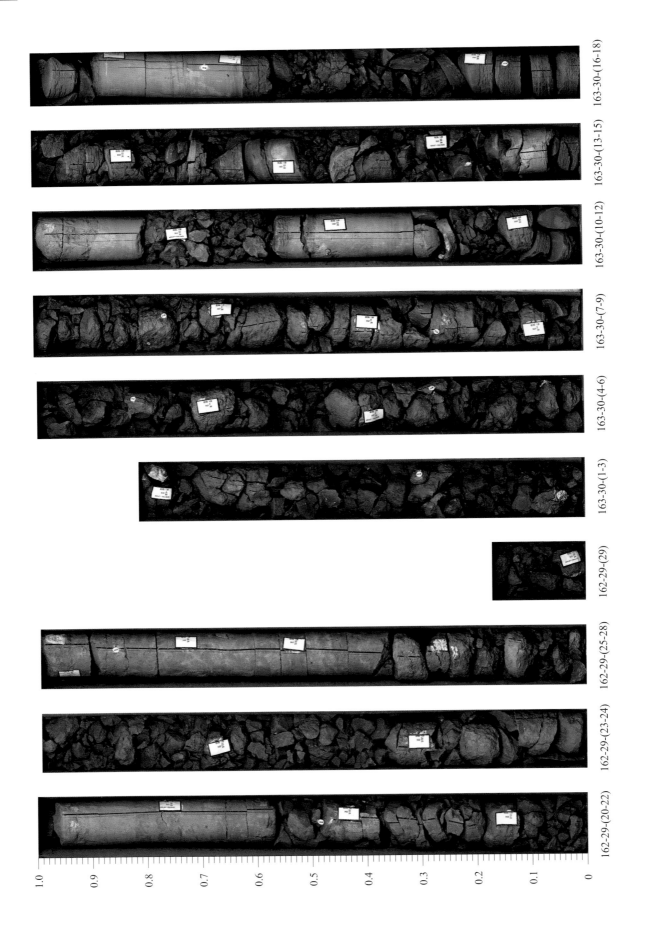

Section 5 SK-1 Core Description and Core Photographs

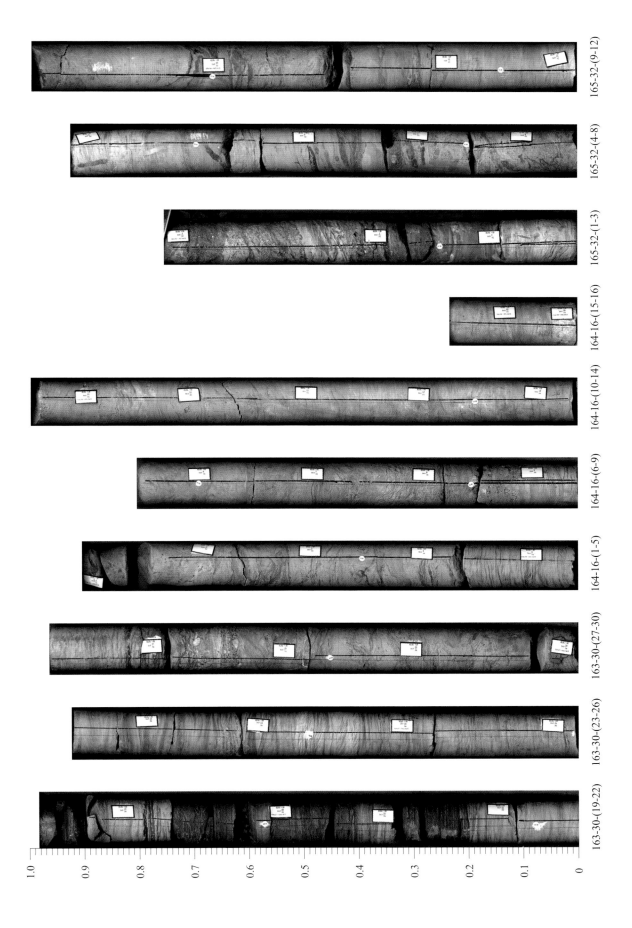

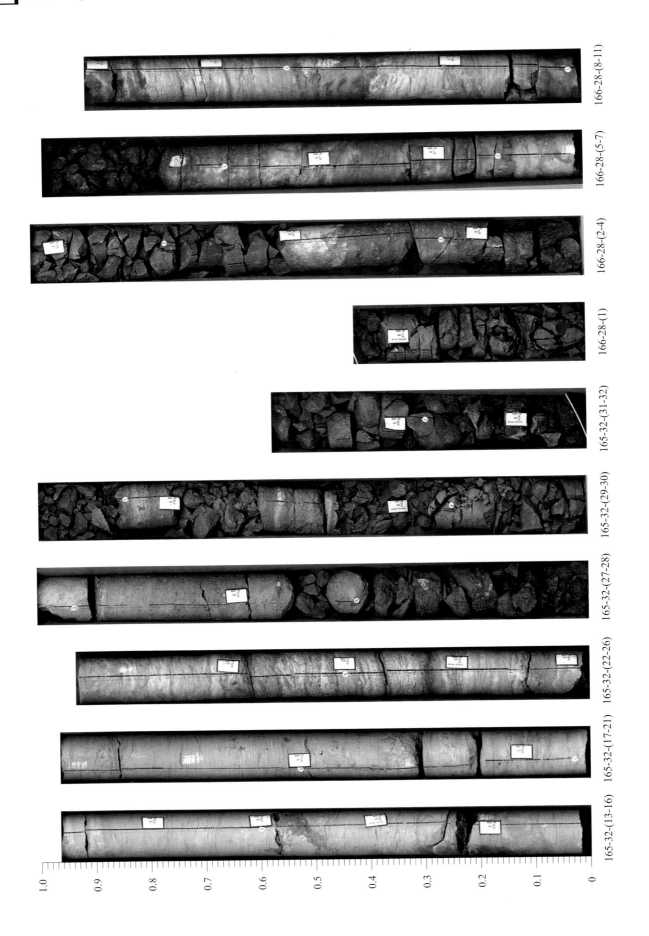

Section 5 SK-1 Core Description and Core Photographs

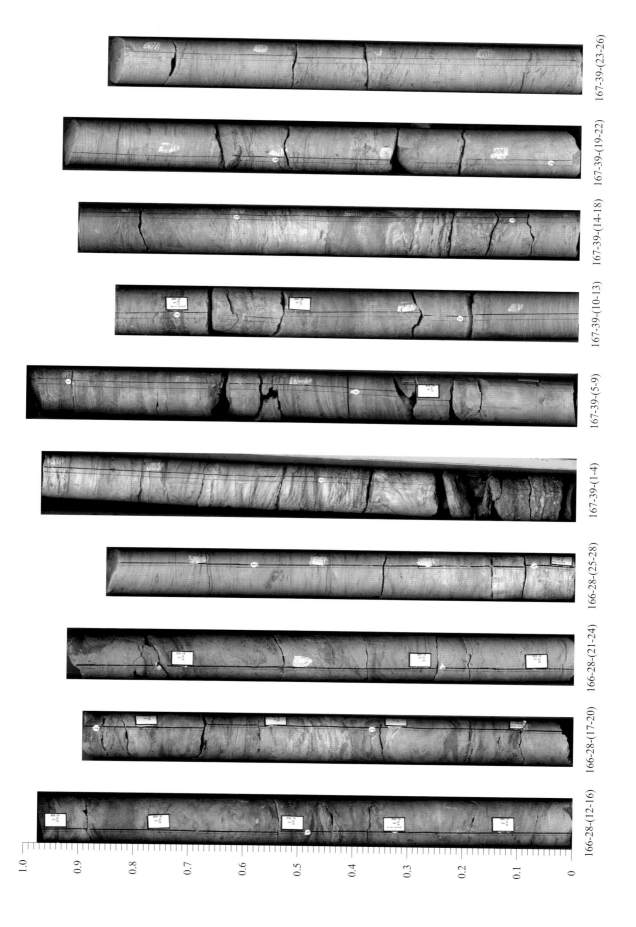

Initial Report of Continental Scientific Drilling Project of the Cretaceous Songliao Basin (SK-1)

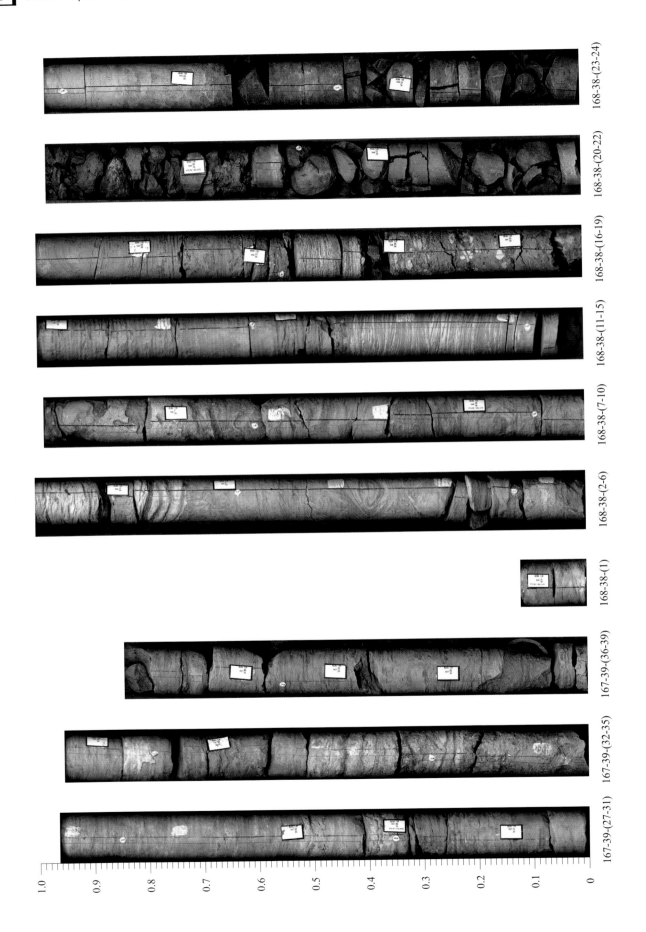

Section 5 SK–1 Core Description and Core Photographs

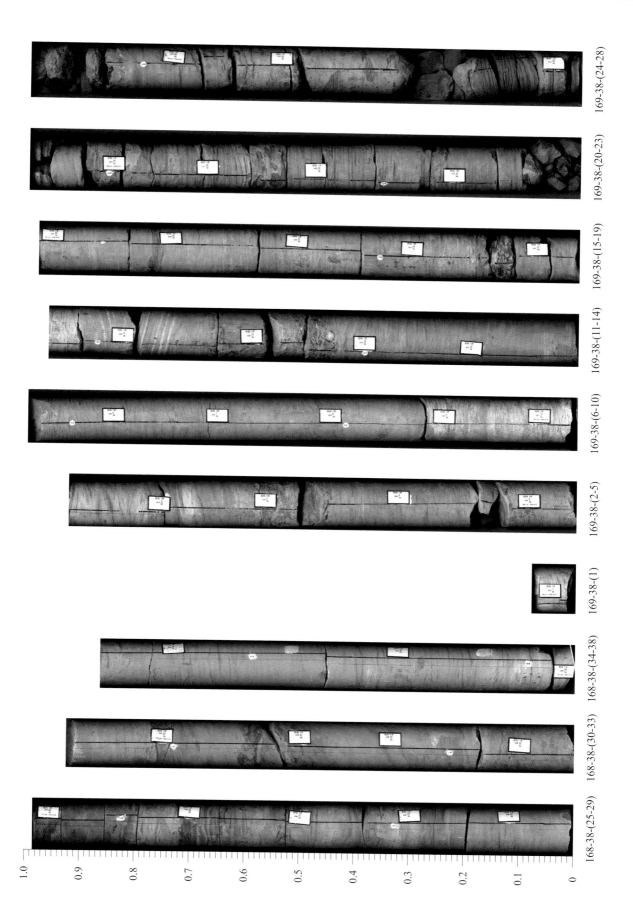

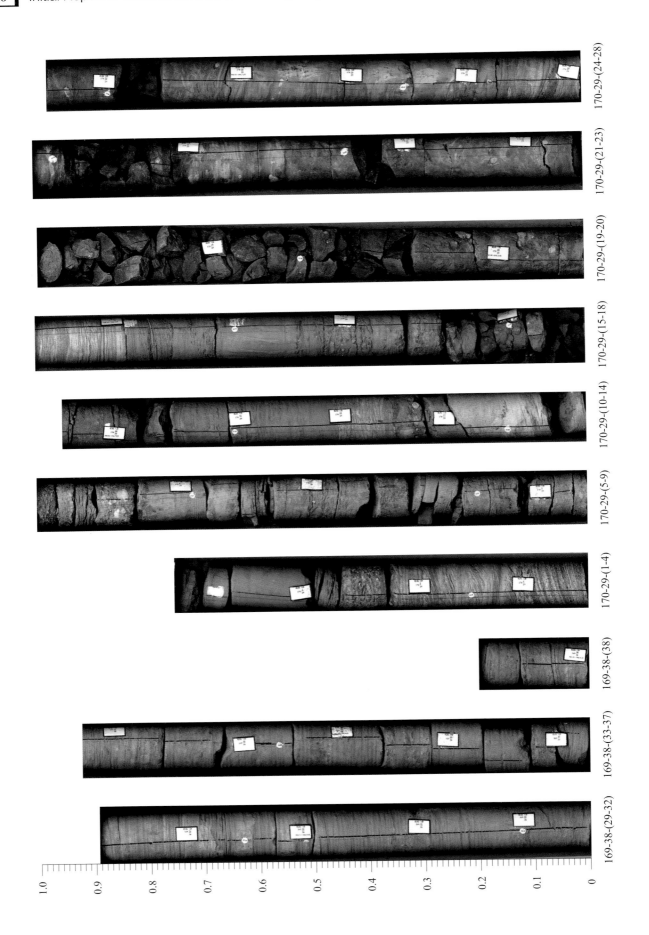

Section 5 SK-1 Core Description and Core Photographs

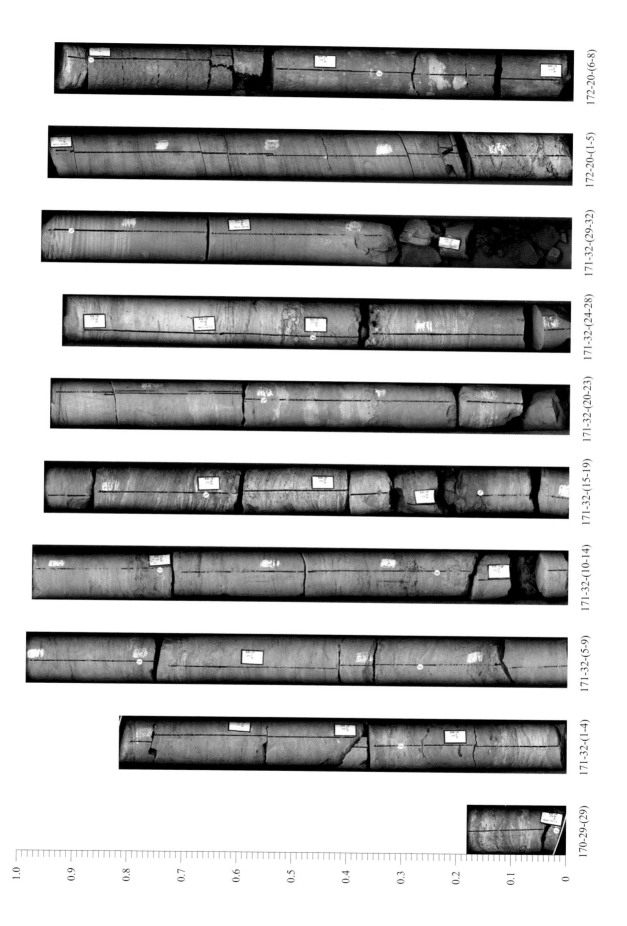

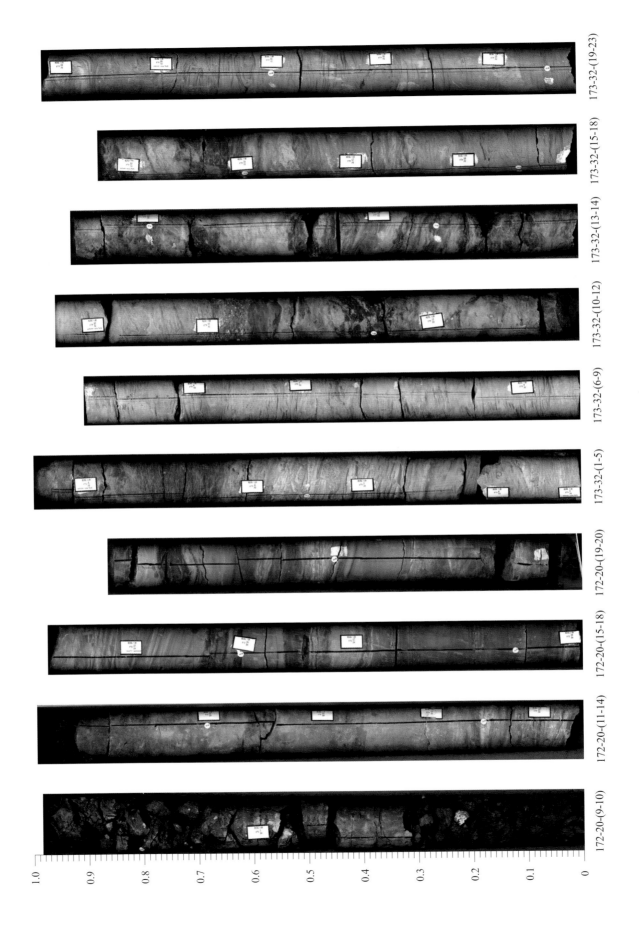

Section 5 SK-1 Core Description and Core Photographs | *521*

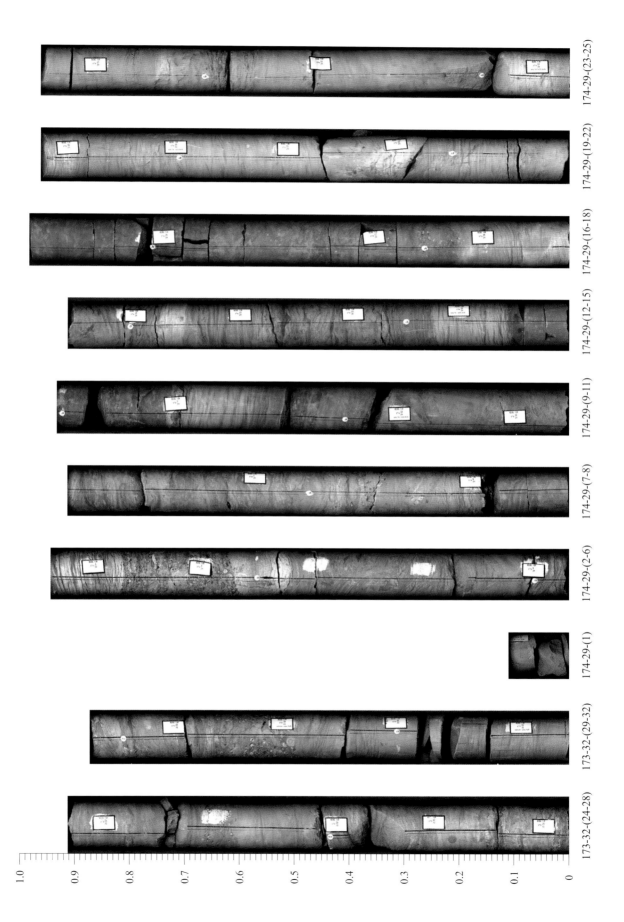

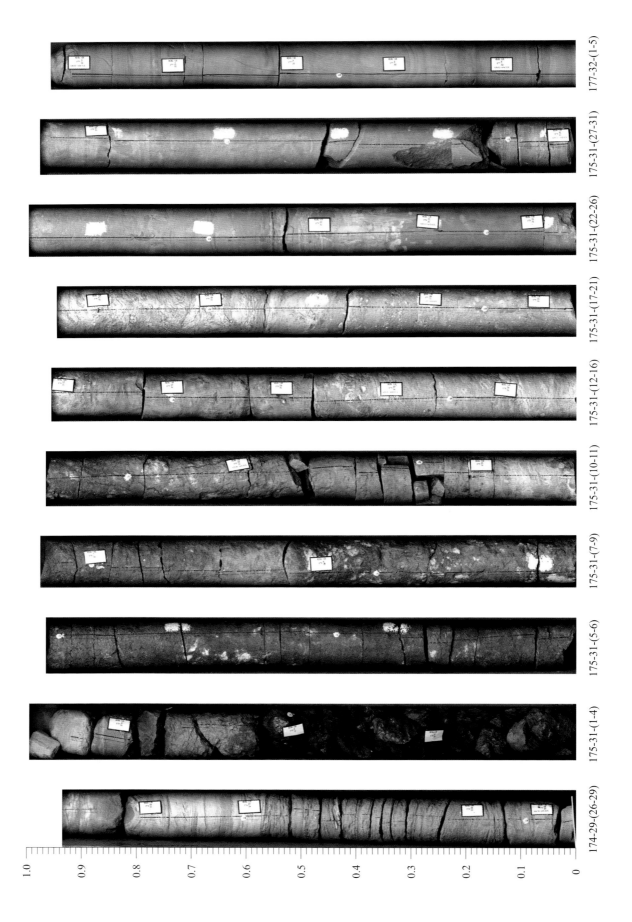

Section 5 SK–1 Core Description and Core Photographs | 523

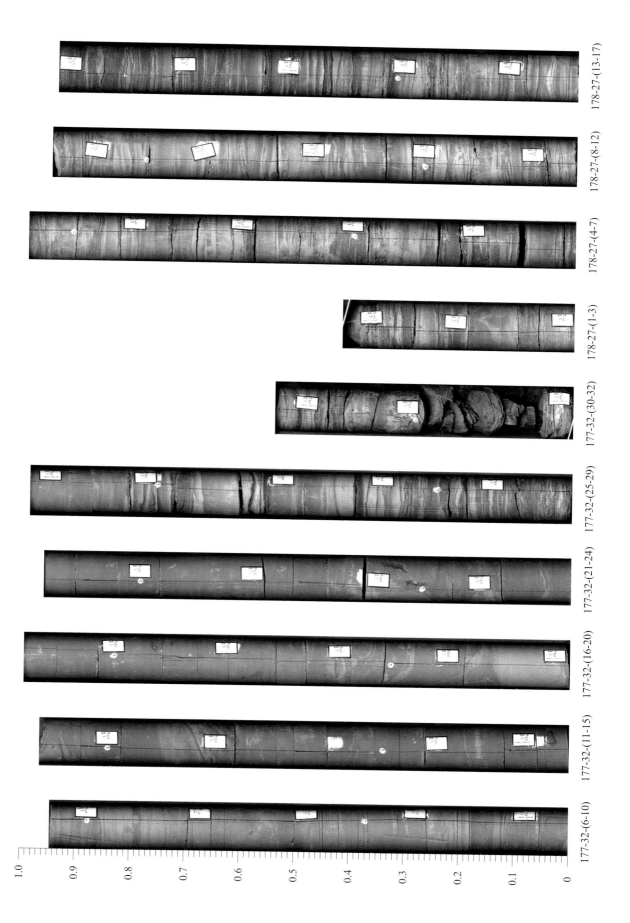

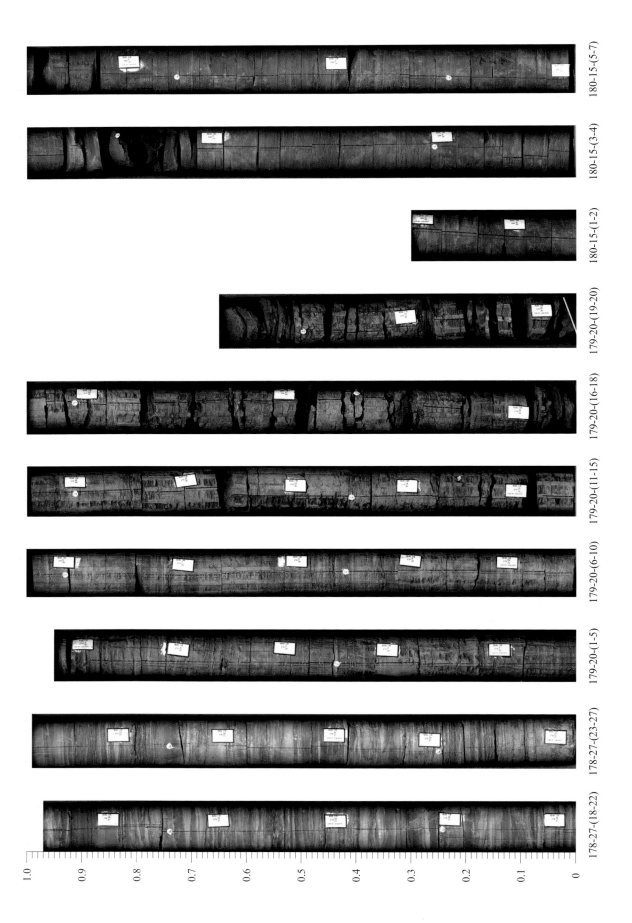

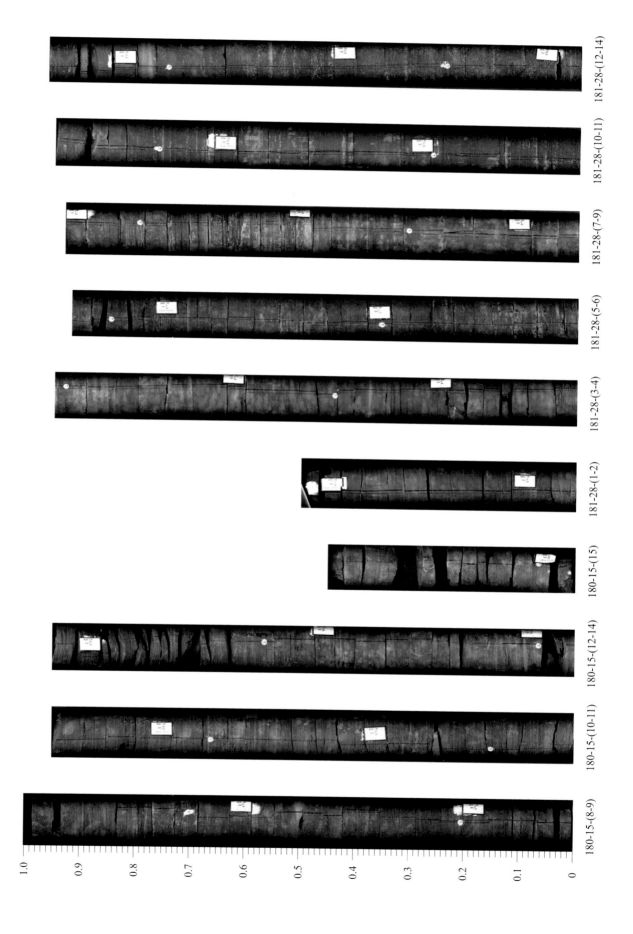

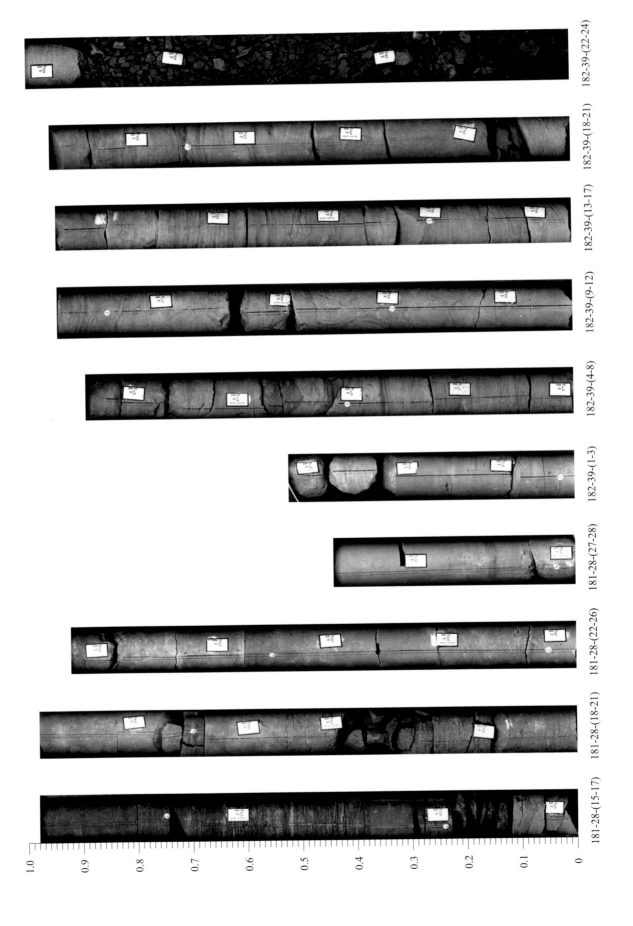

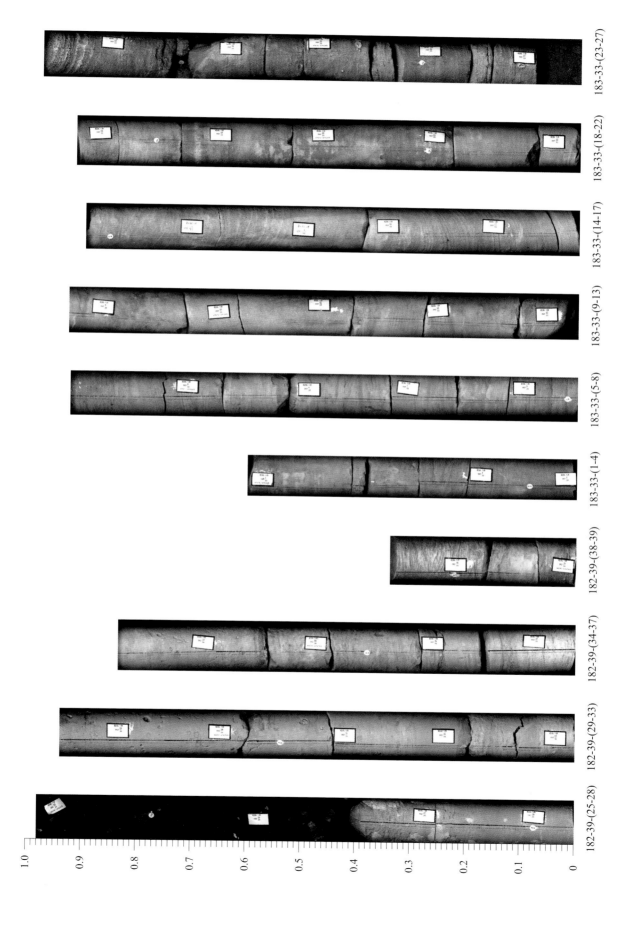

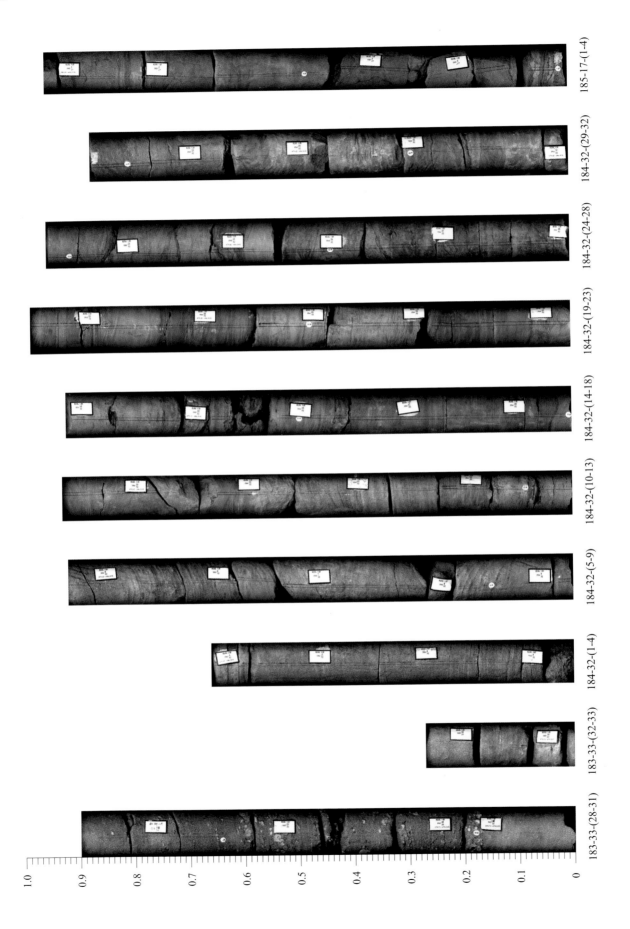

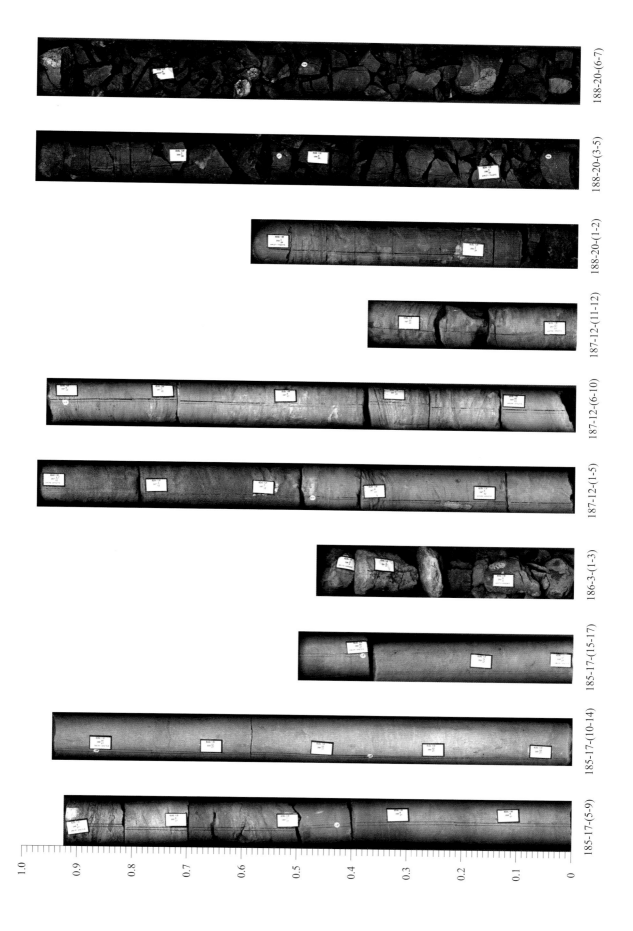

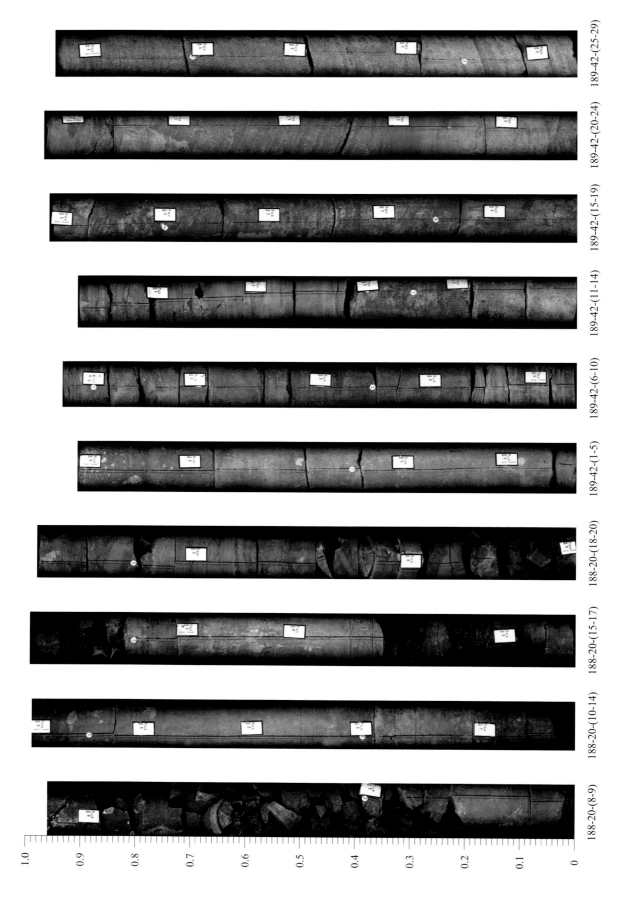

Section 5 SK-1 Core Description and Core Photographs

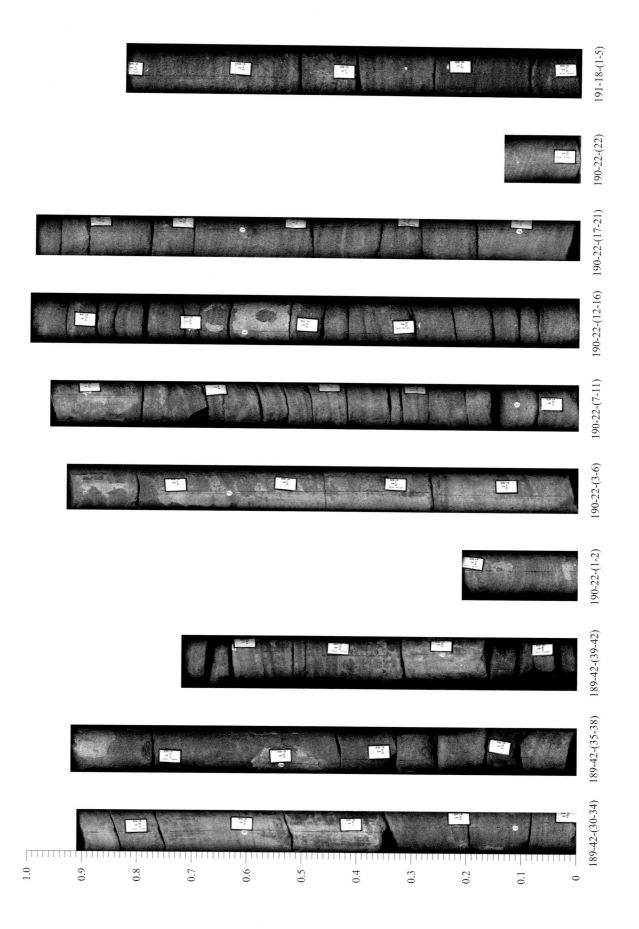

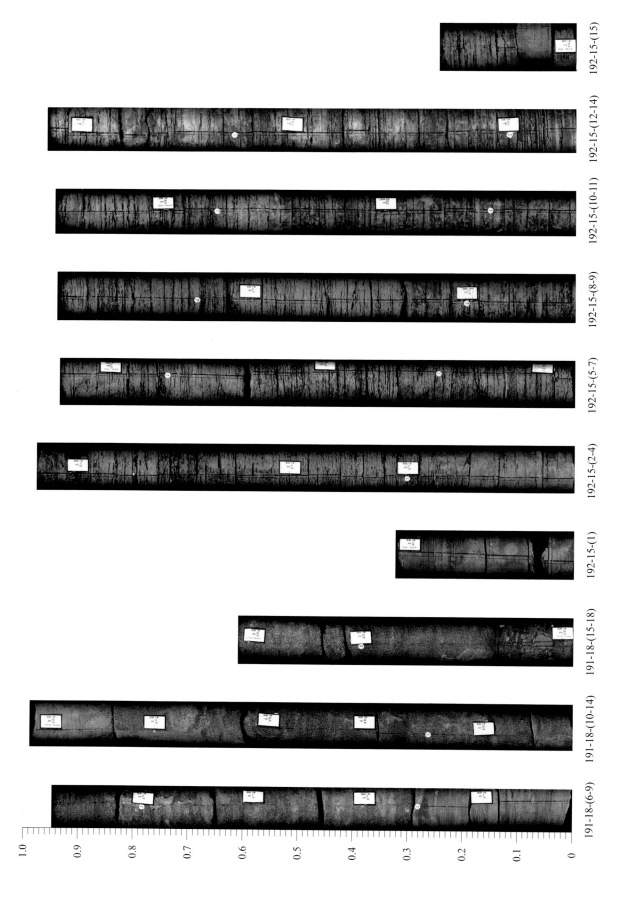

Section 5 SK-1 Core Description and Core Photographs

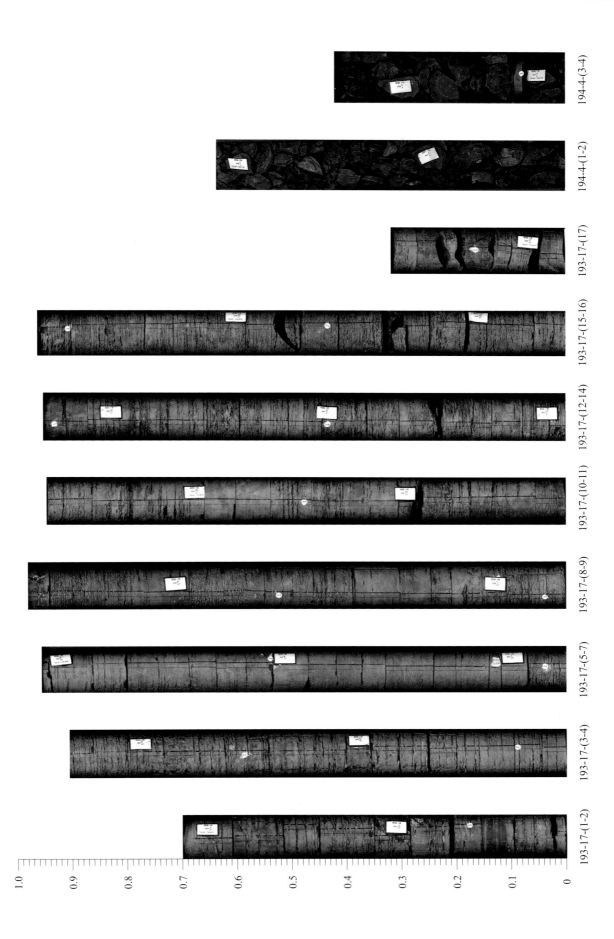

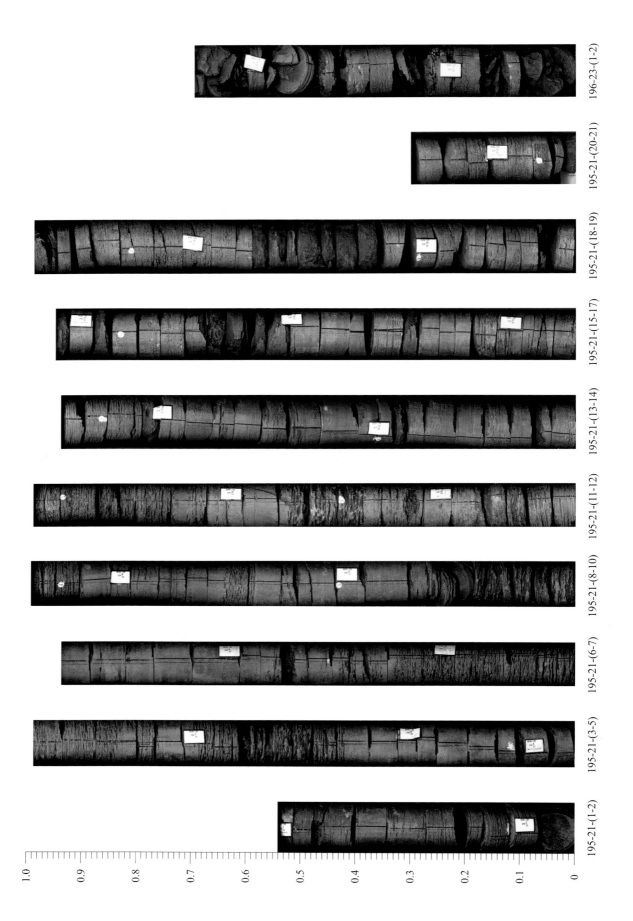

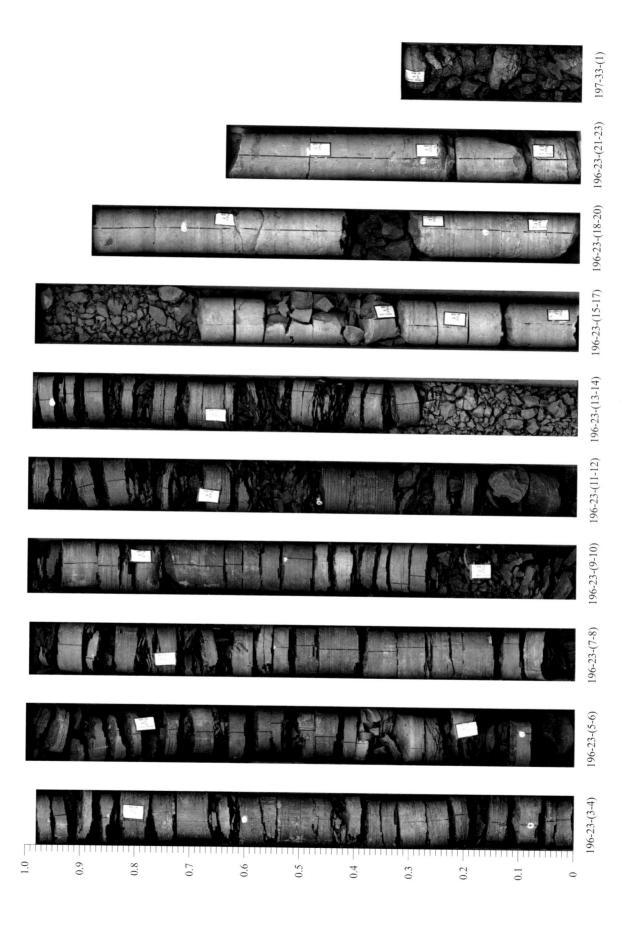

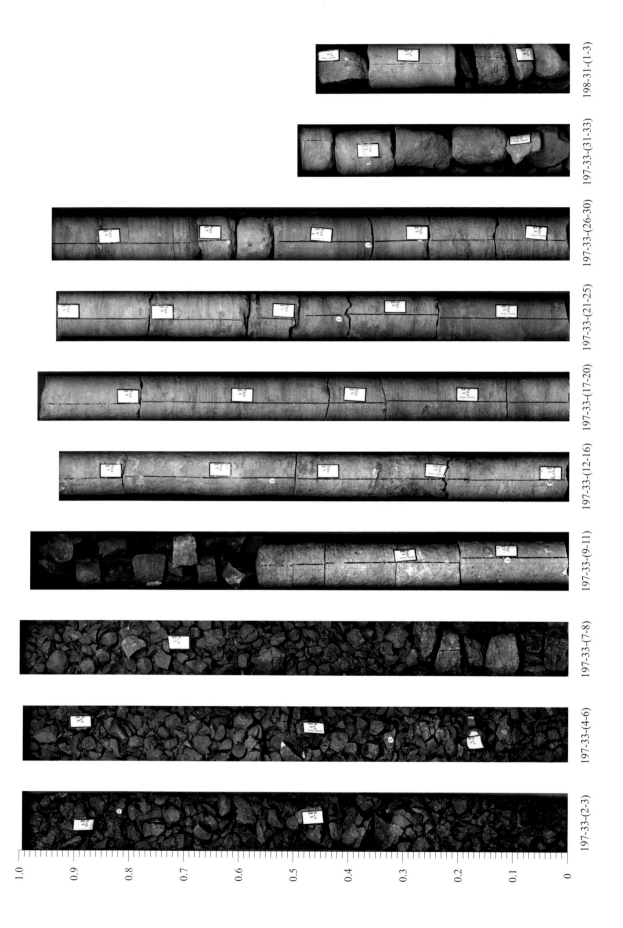

Section 5 SK-1 Core Description and Core Photographs | 537

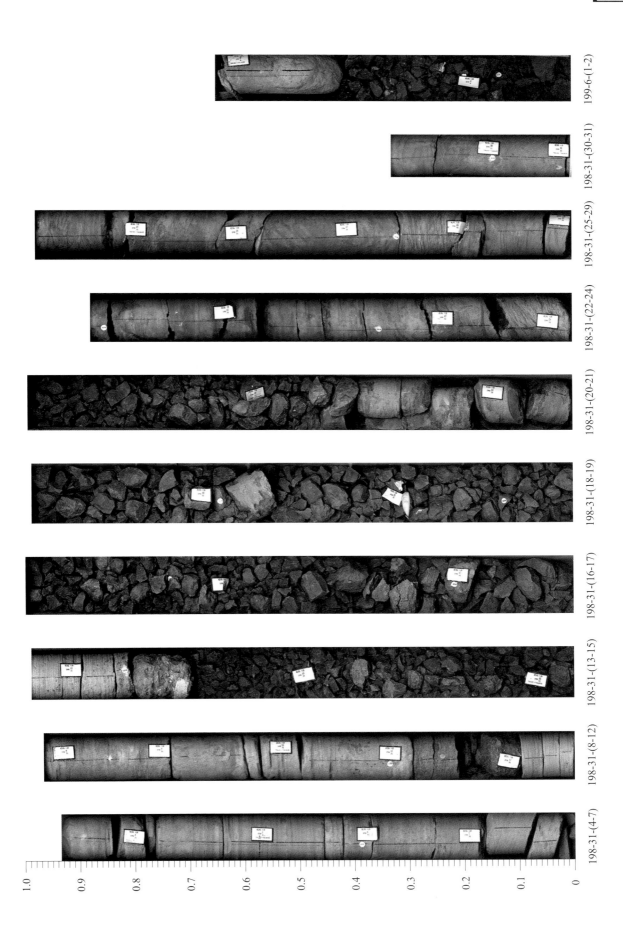

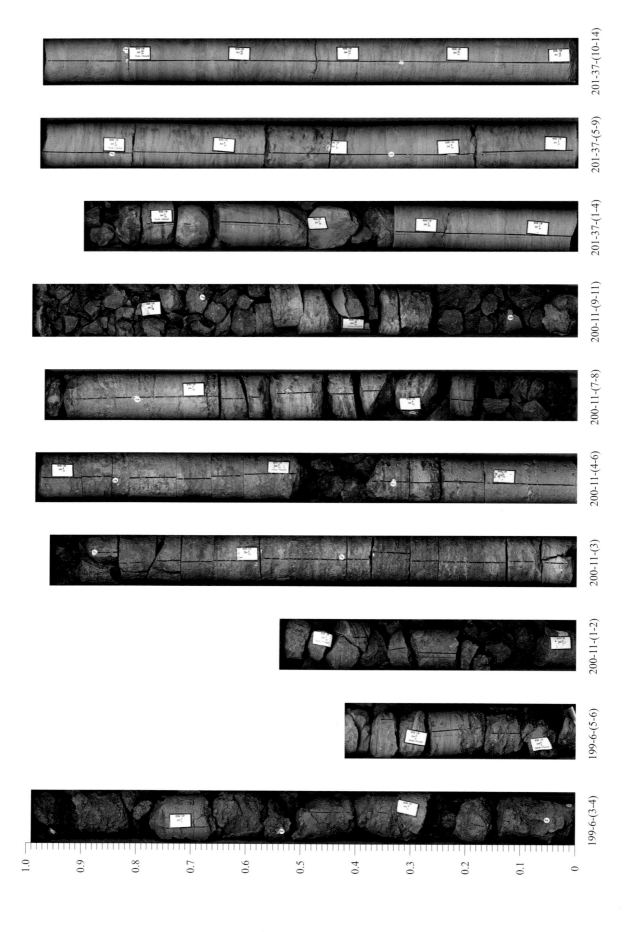

Section 5 SK-1 Core Description and Core Photographs

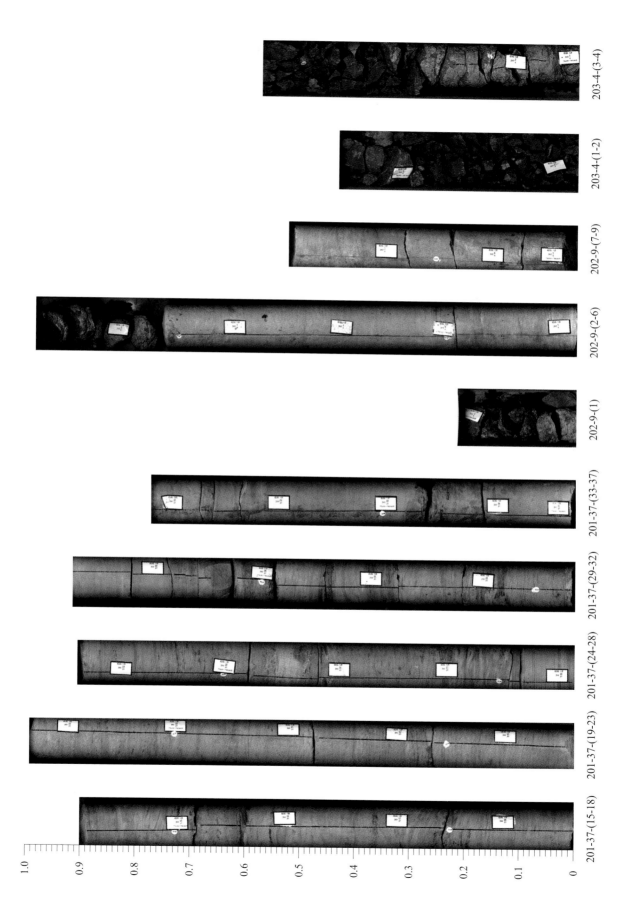

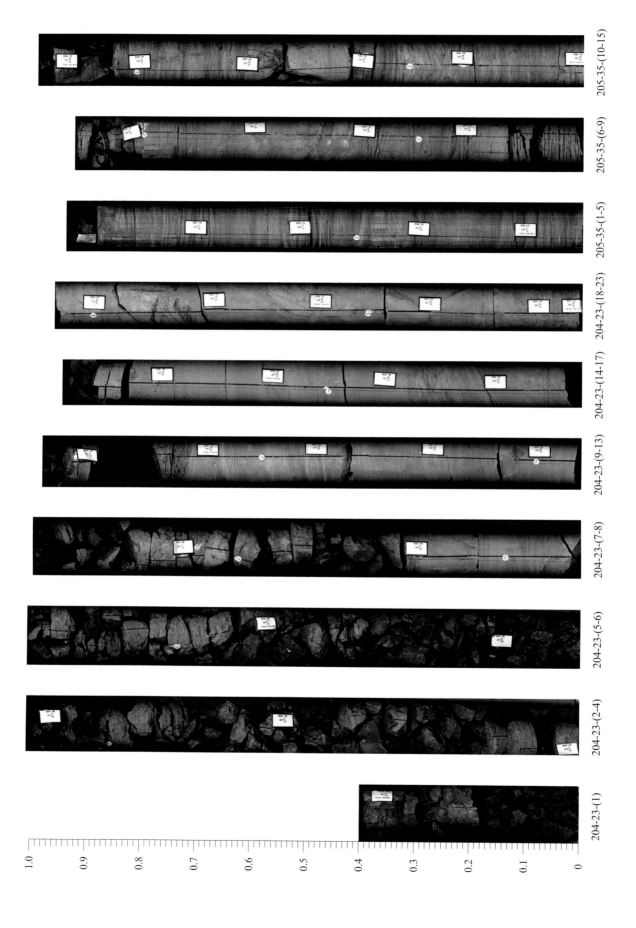

Section 5 SK-1 Core Description and Core Photographs | *541*

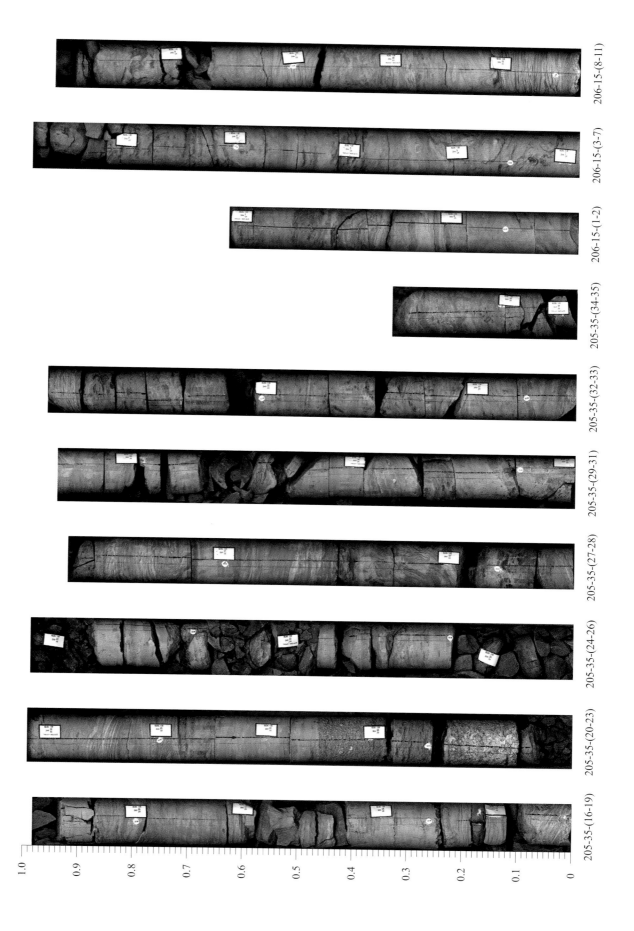

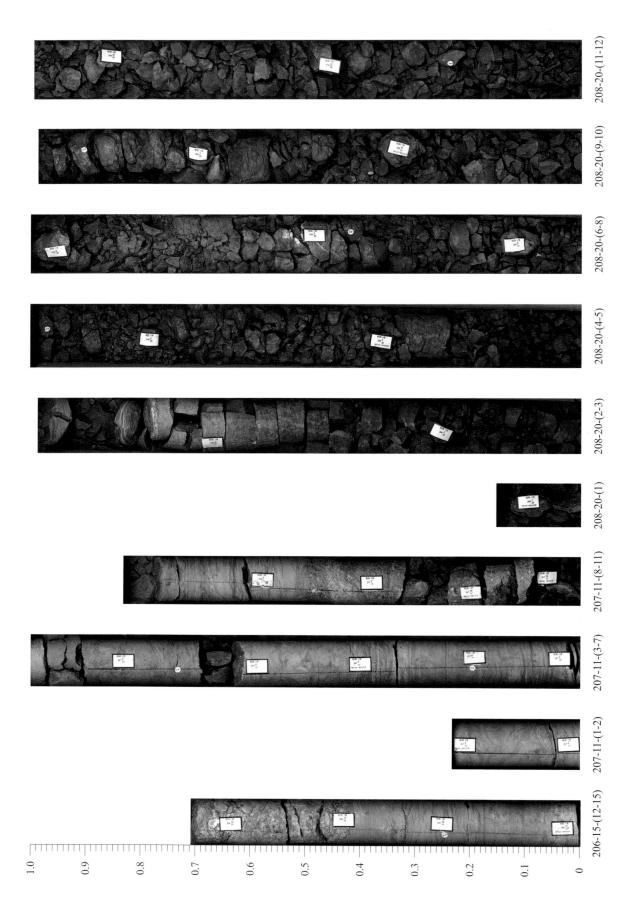

Section 5 SK-1 Core Description and Core Photographs

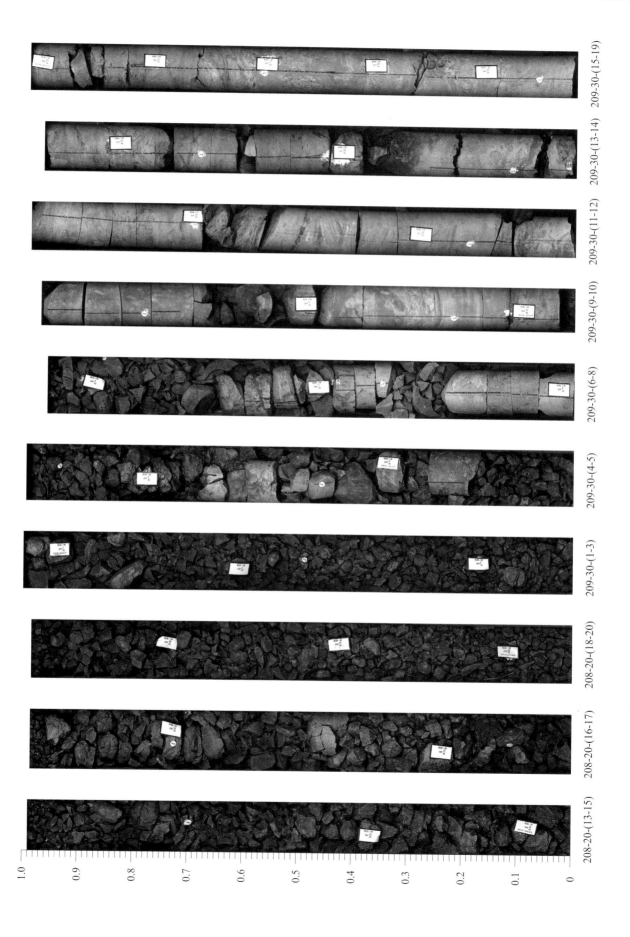

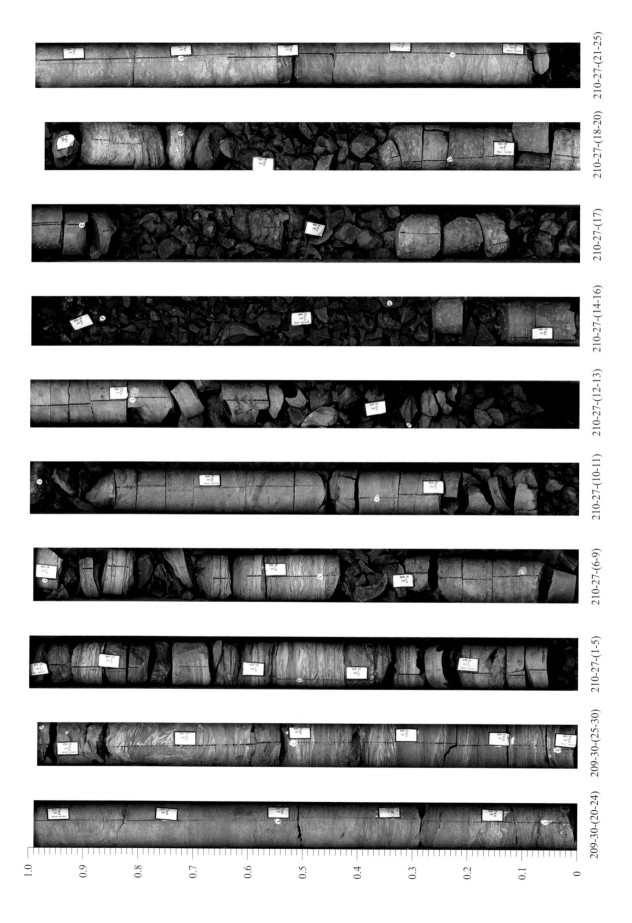

Section 5 SK-1 Core Description and Core Photographs

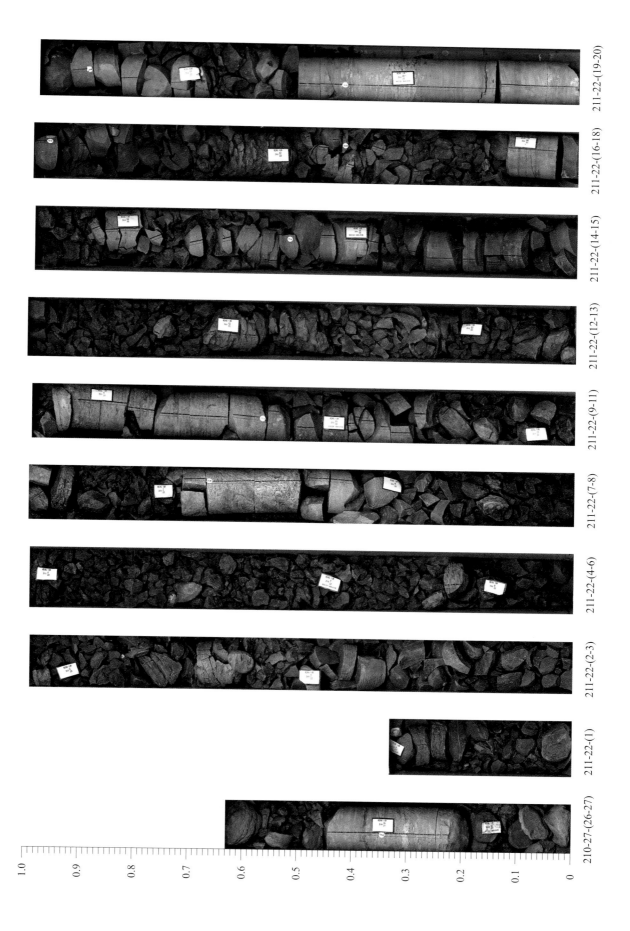

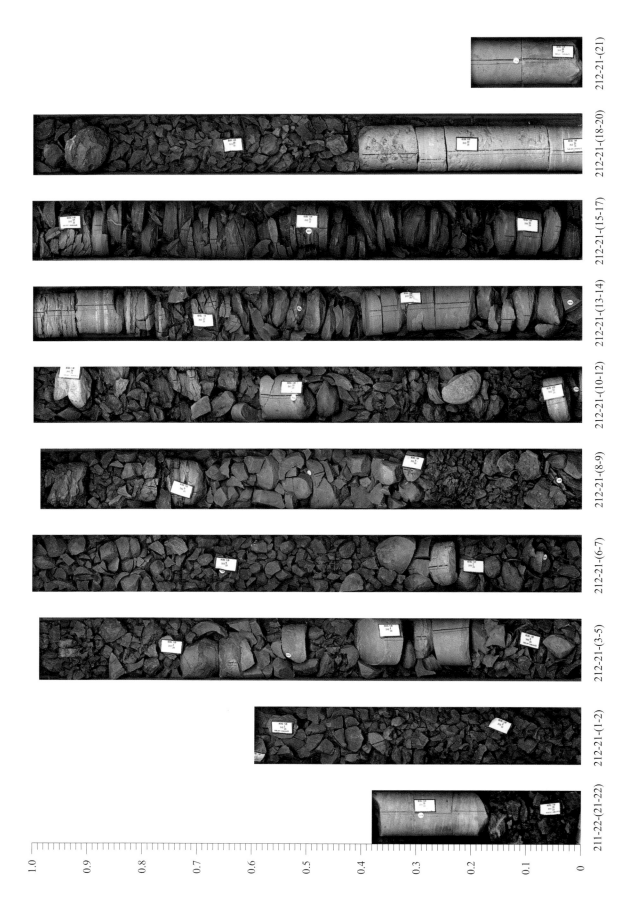

Section 5 SK-1 Core Description and Core Photographs | *547*

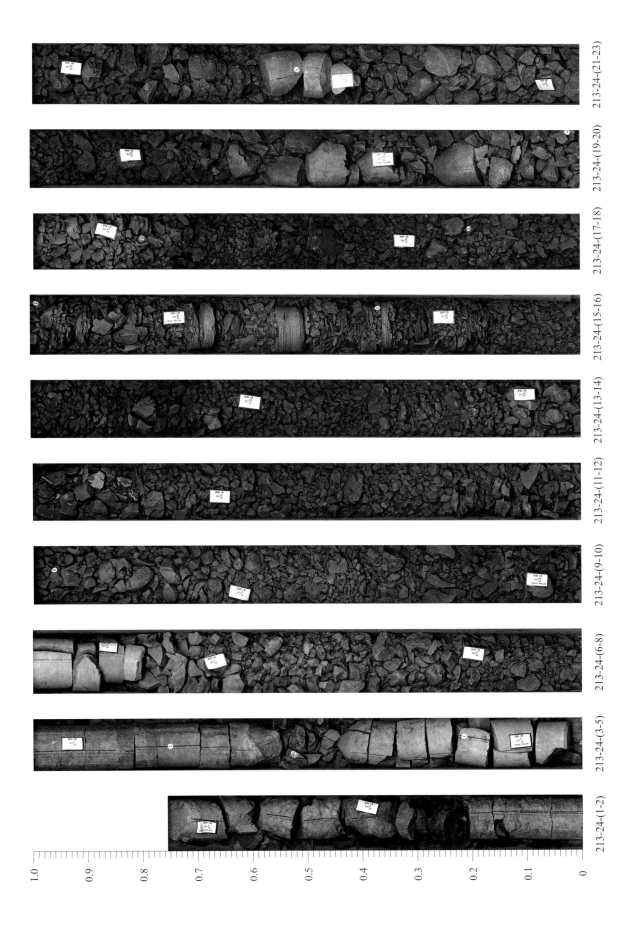

548 | Initial Report of Continental Scientific Drilling Project of the Cretaceous Songliao Basin (SK-1)

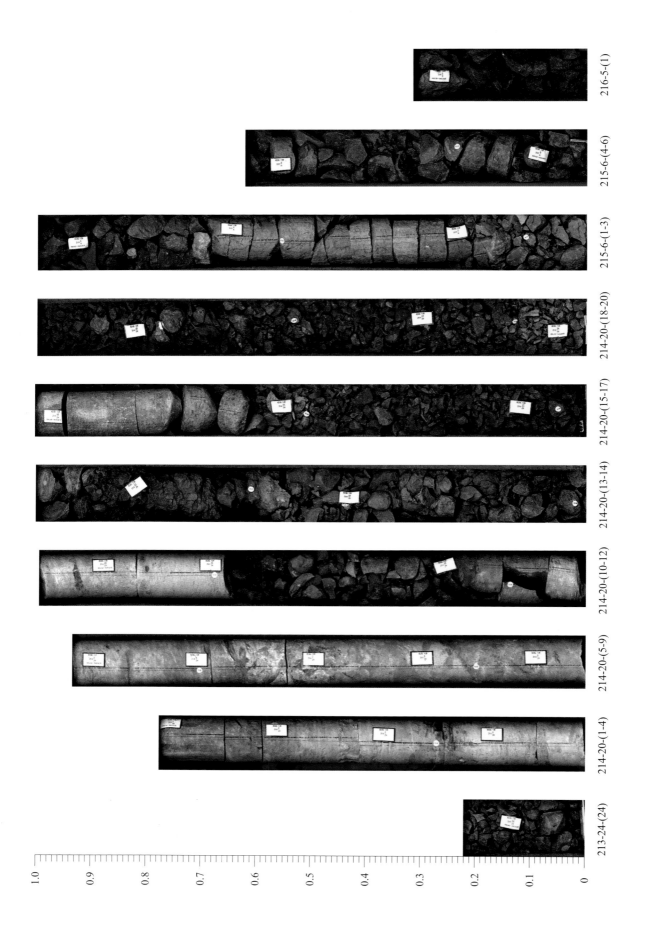

Section 5 SK-1 Core Description and Core Photographs

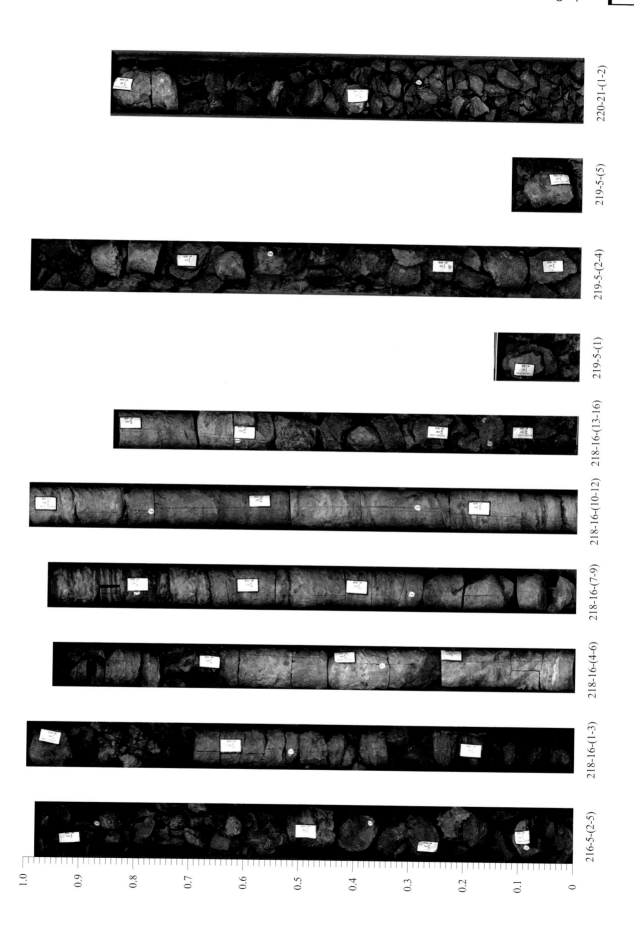

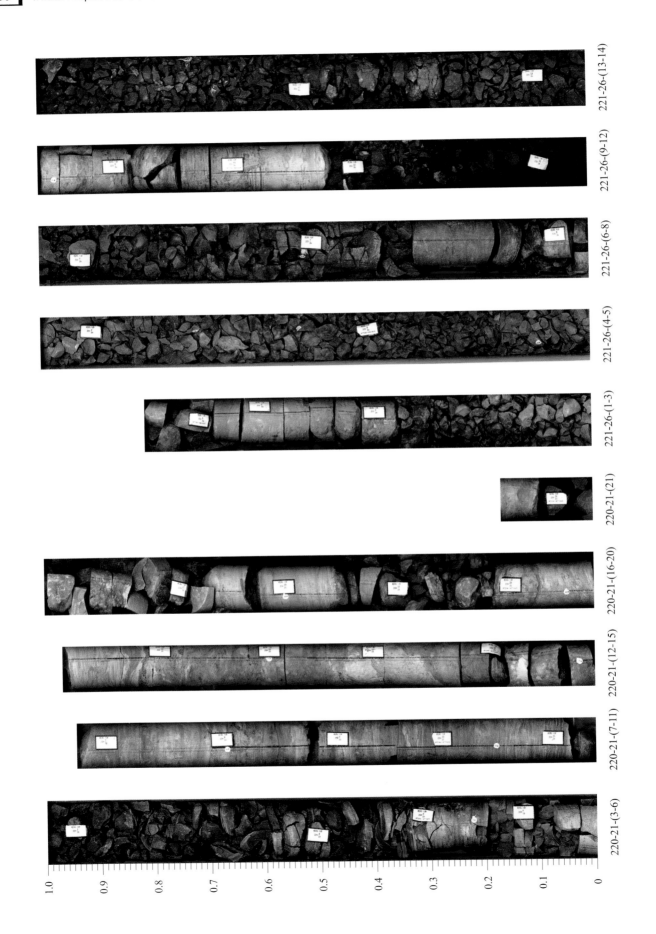

Section 5 SK-1 Core Description and Core Photographs 551

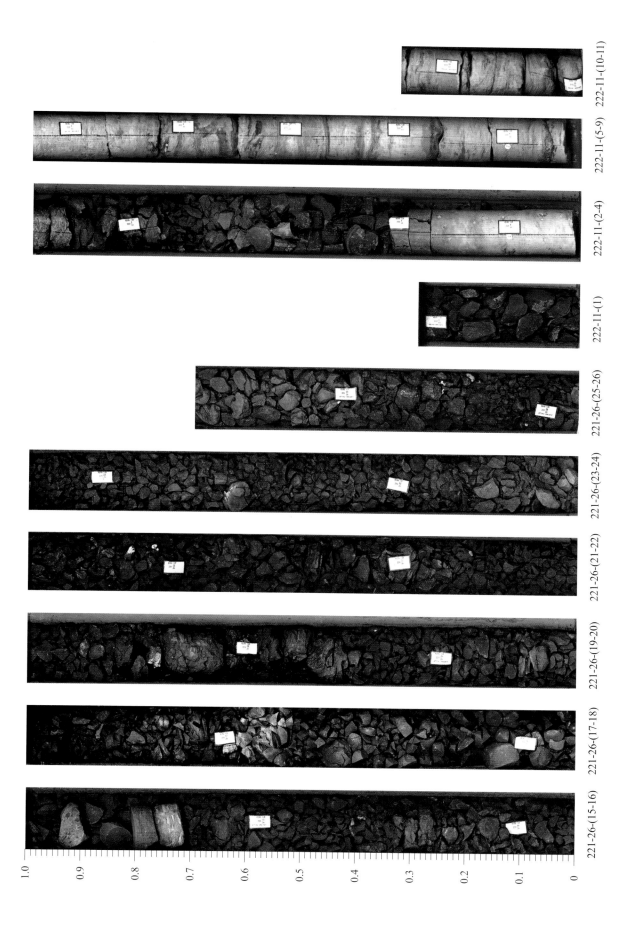

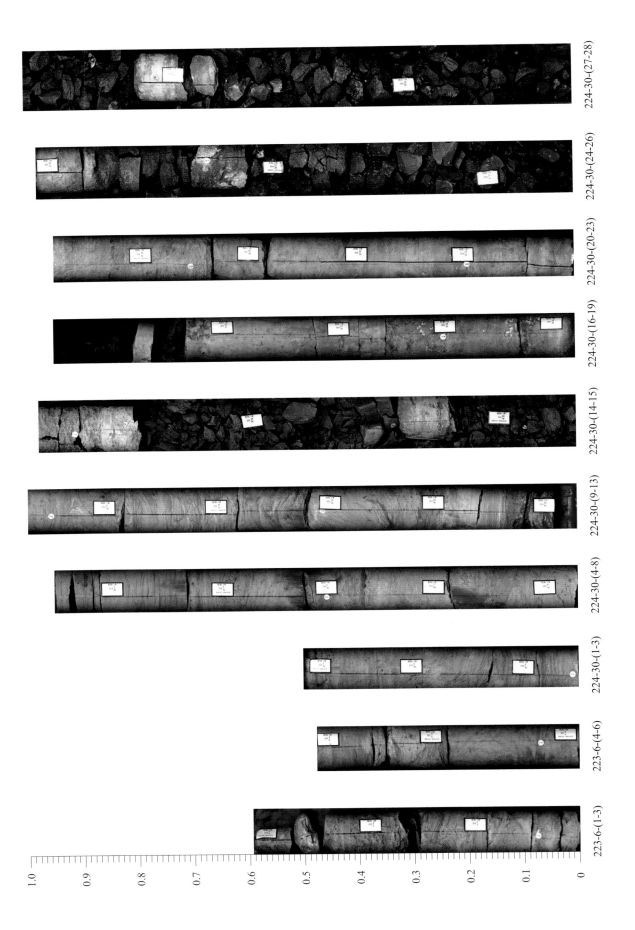

Section 5 SK-1 Core Description and Core Photographs

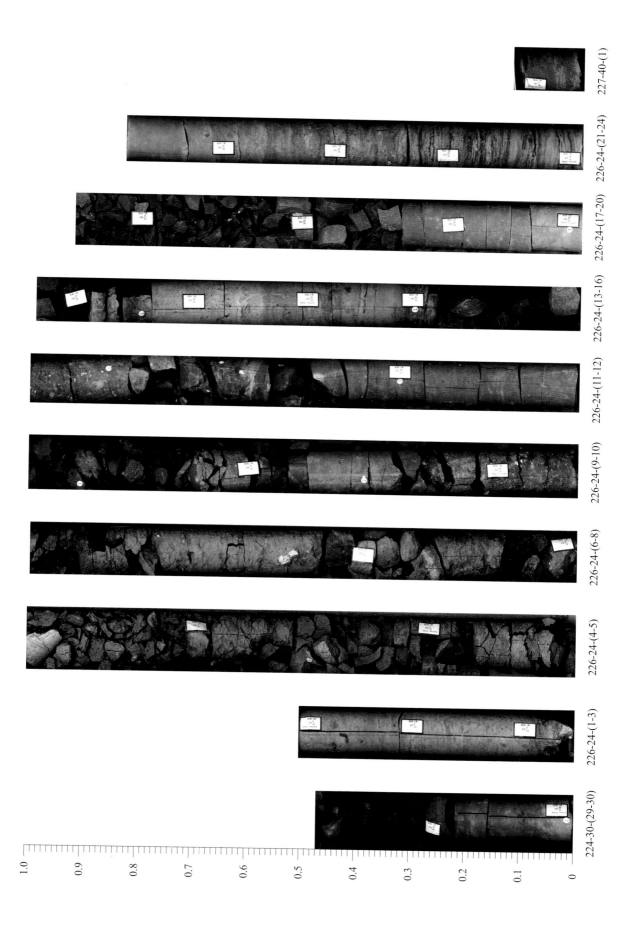

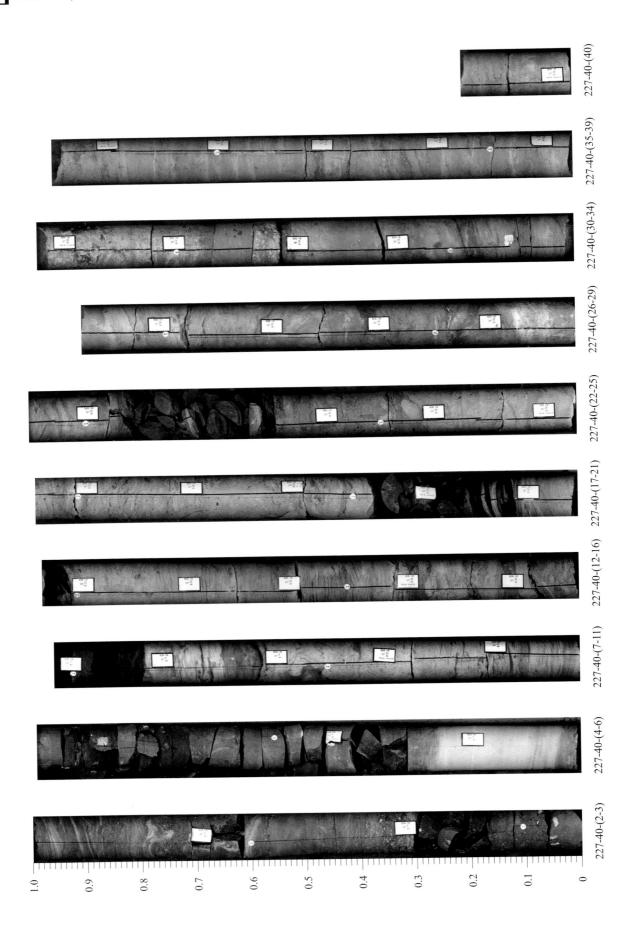

Section 5 SK-1 Core Description and Core Photographs

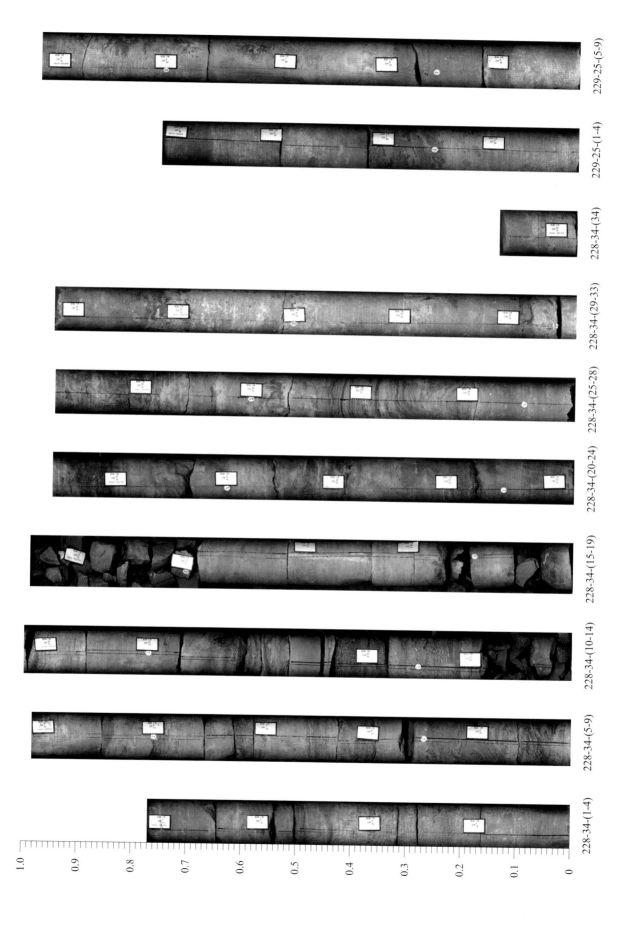

556 Initial Report of Continental Scientific Drilling Project of the Cretaceous Songliao Basin (SK-1)

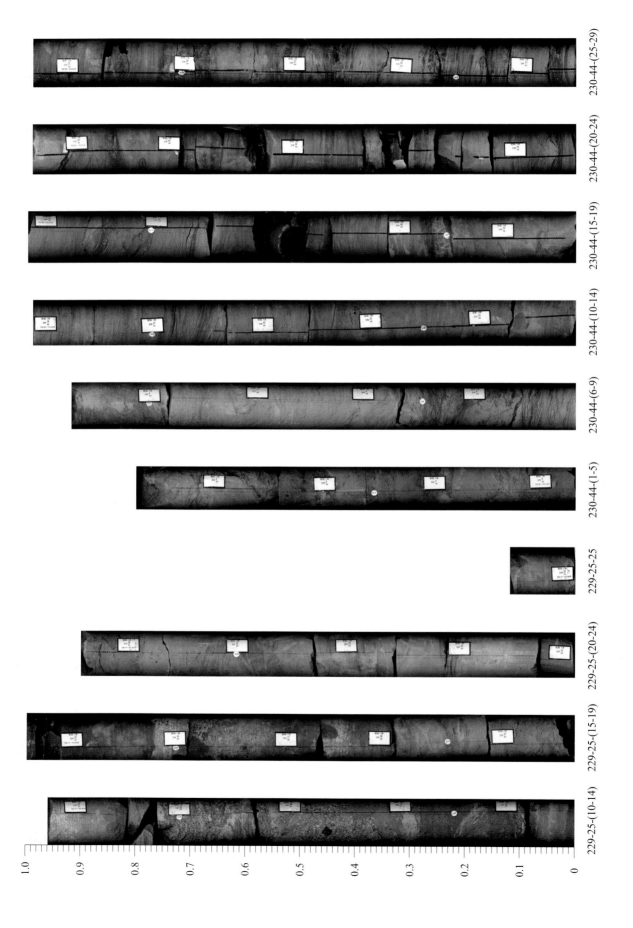

Section 5 SK-1 Core Description and Core Photographs

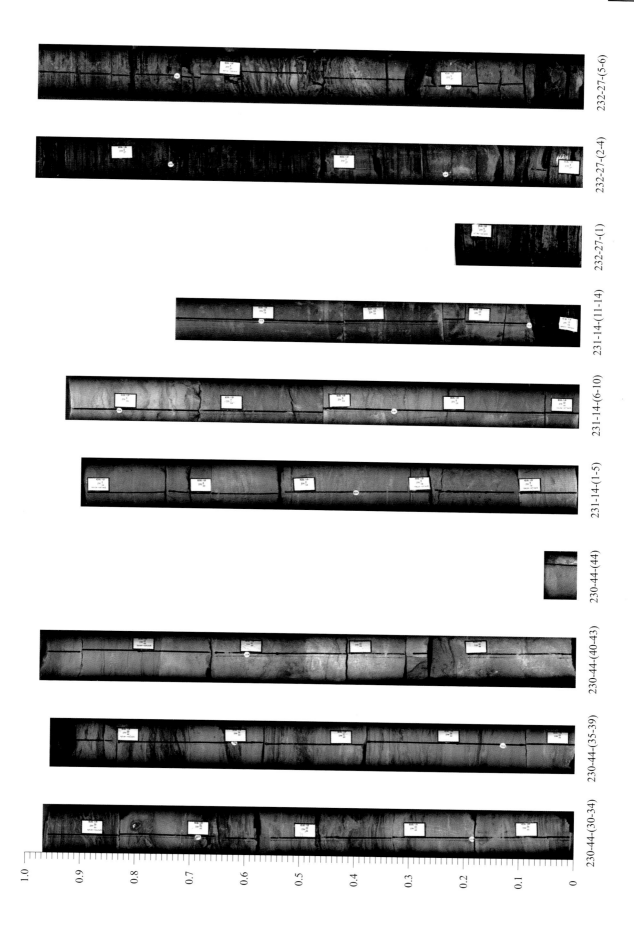

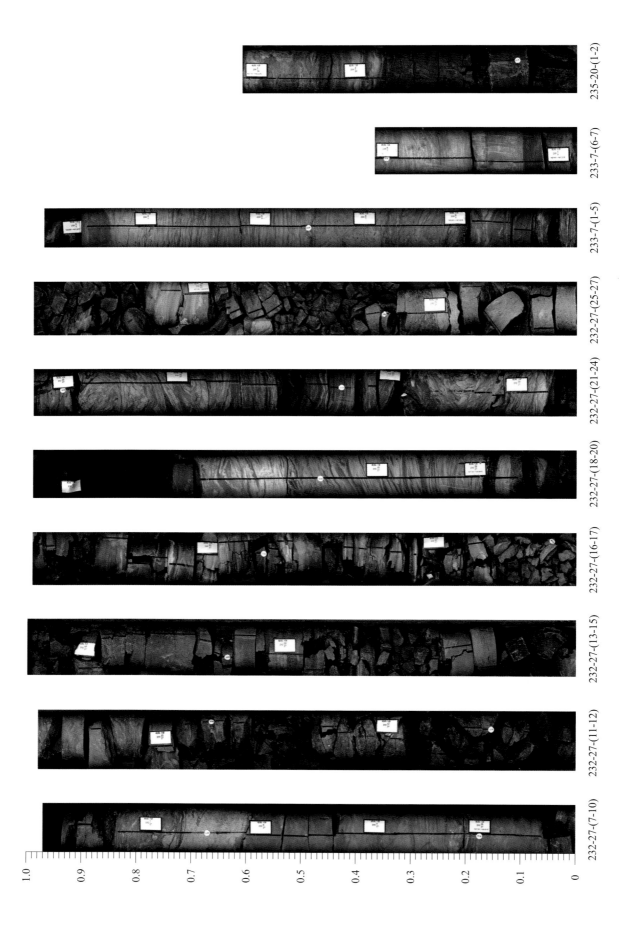

Section 5 SK–1 Core Description and Core Photographs

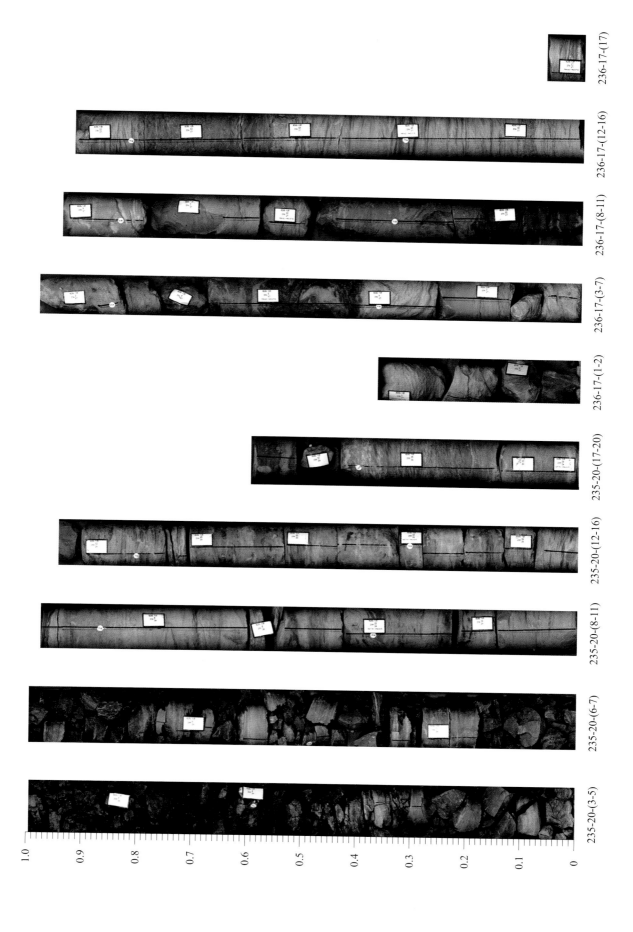

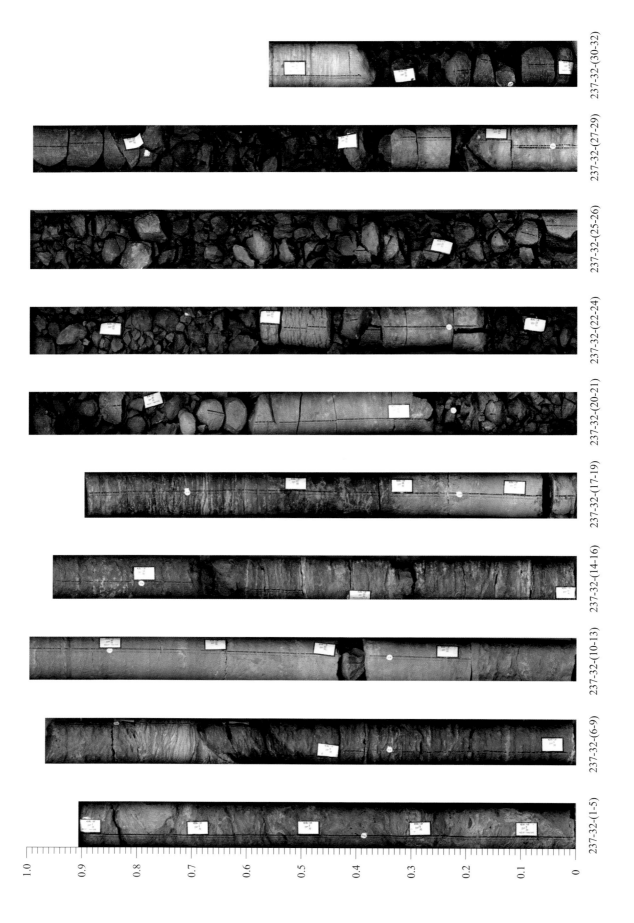

Section 5 SK-1 Core Description and Core Photographs | *561*

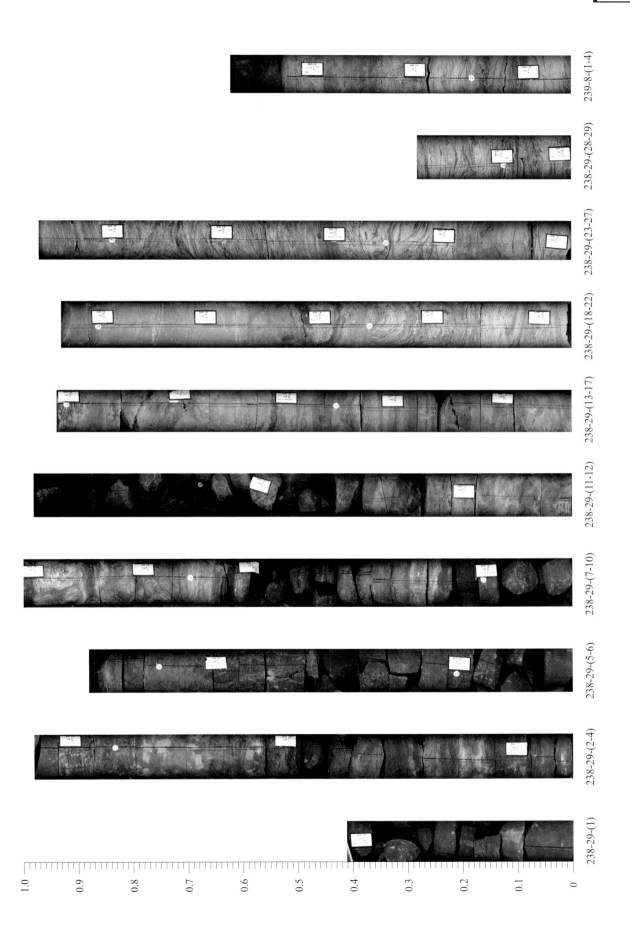

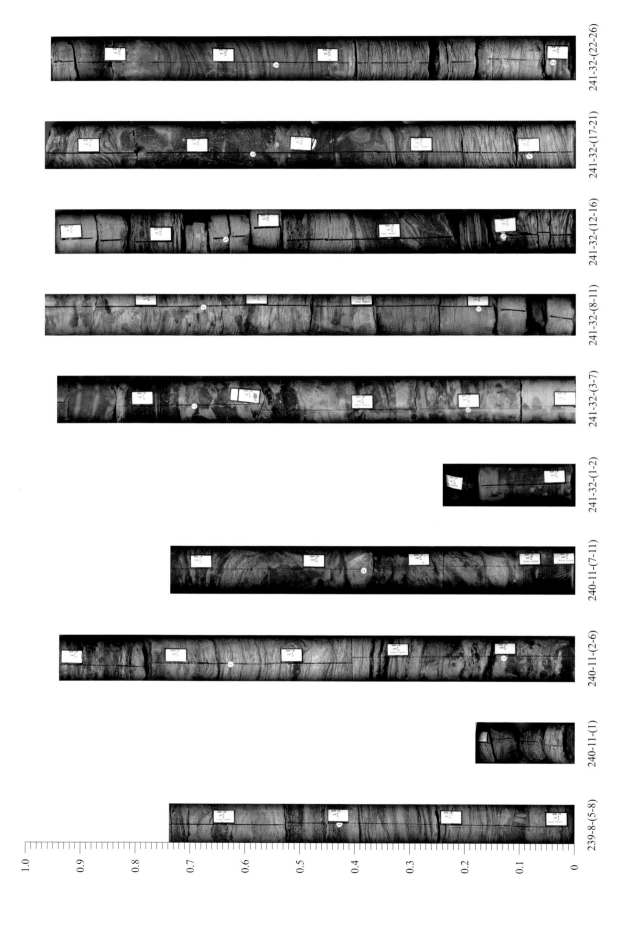

Section 5 SK-1 Core Description and Core Photographs | *563*

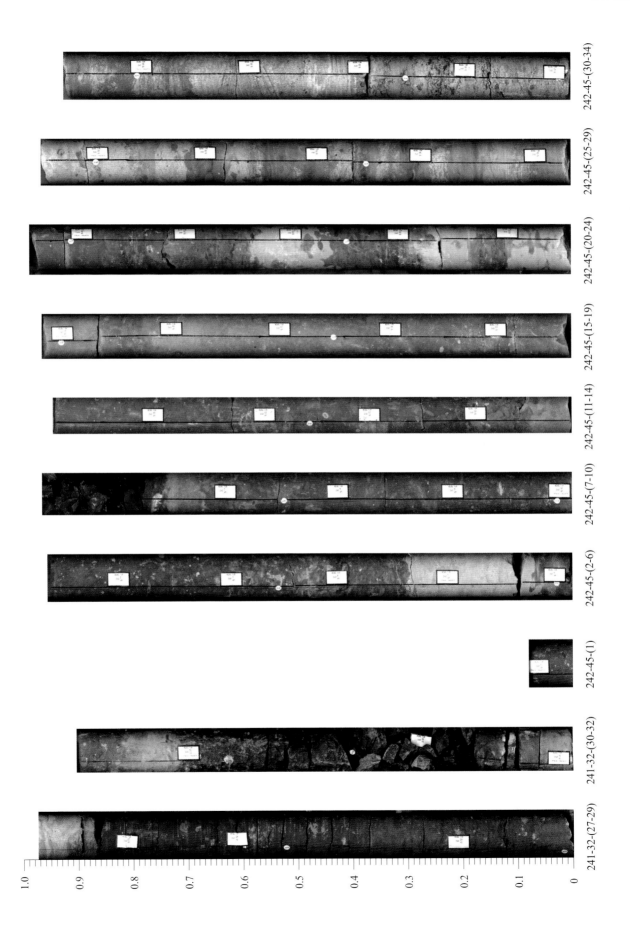

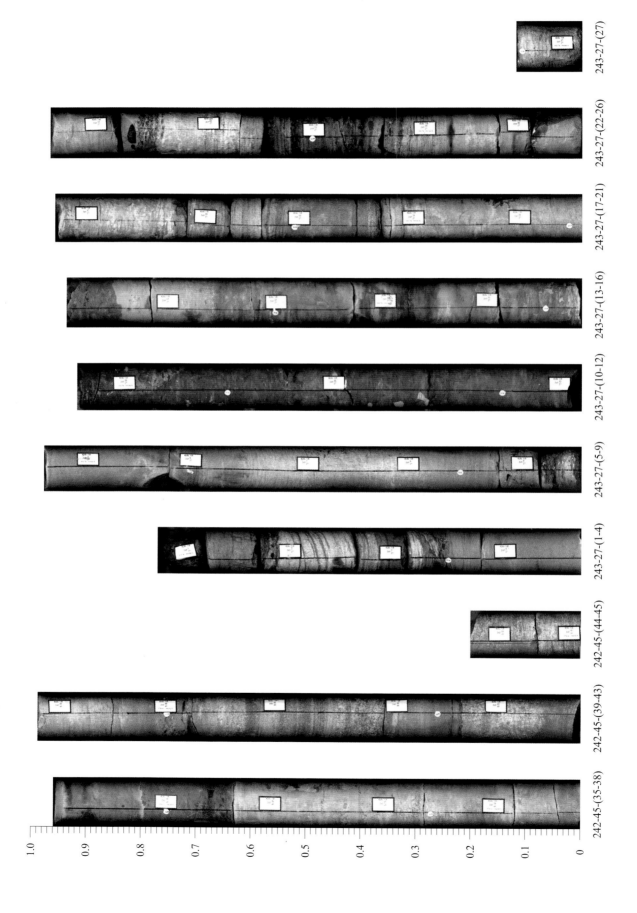

Section 5 SK-1 Core Description and Core Photographs

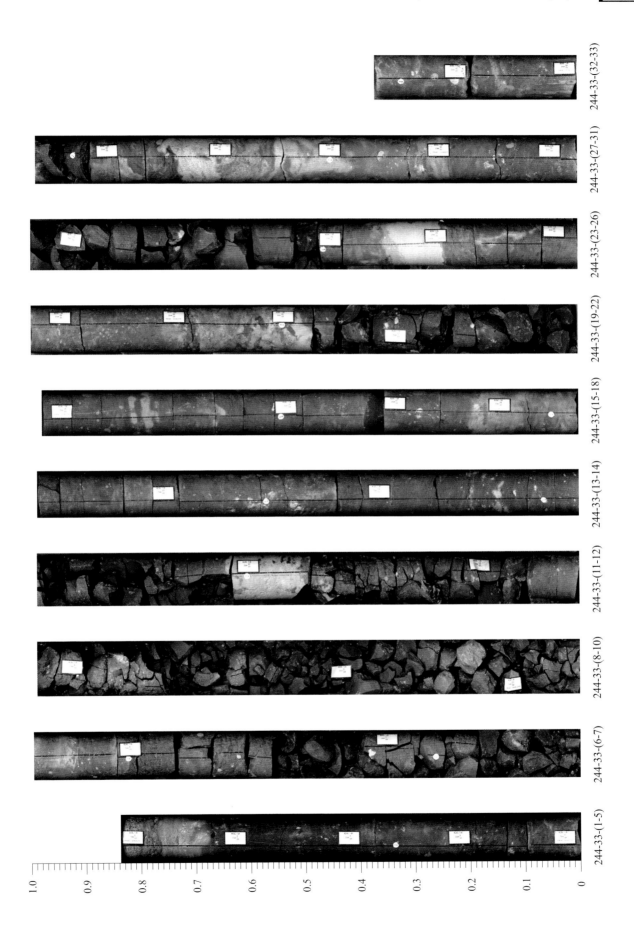

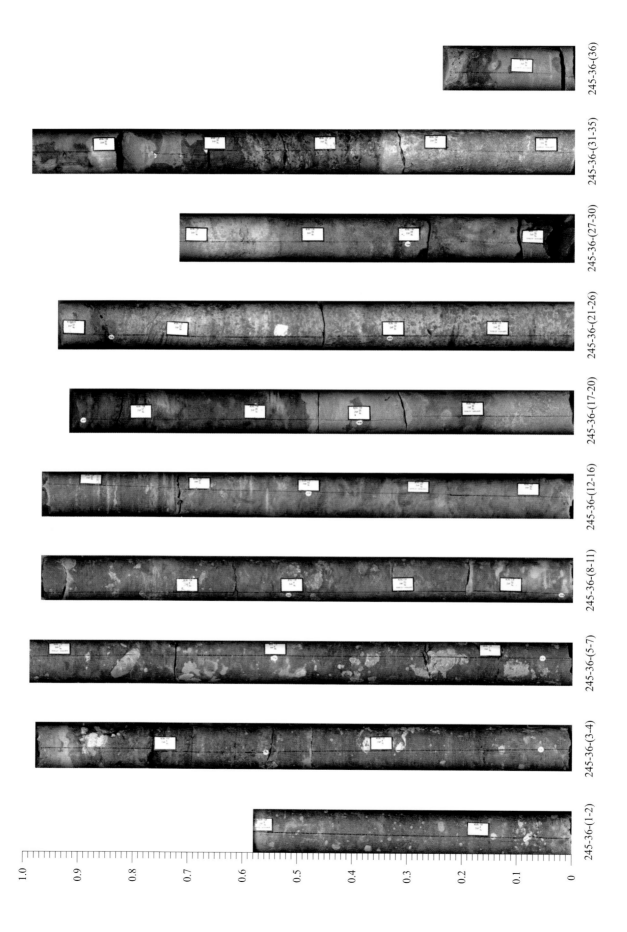

Section 5 SK-1 Core Description and Core Photographs

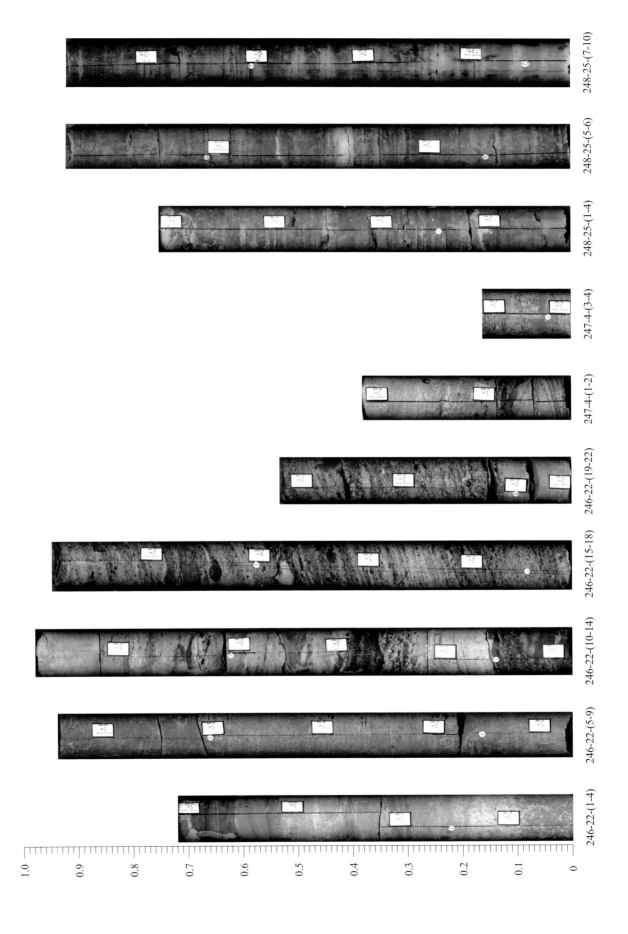

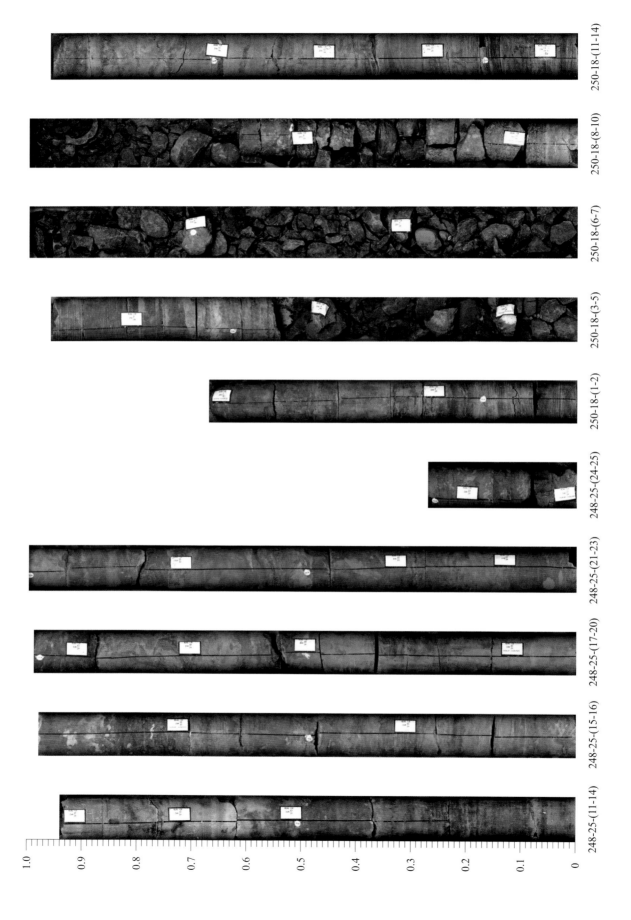

Section 5 SK-1 Core Description and Core Photographs | 569

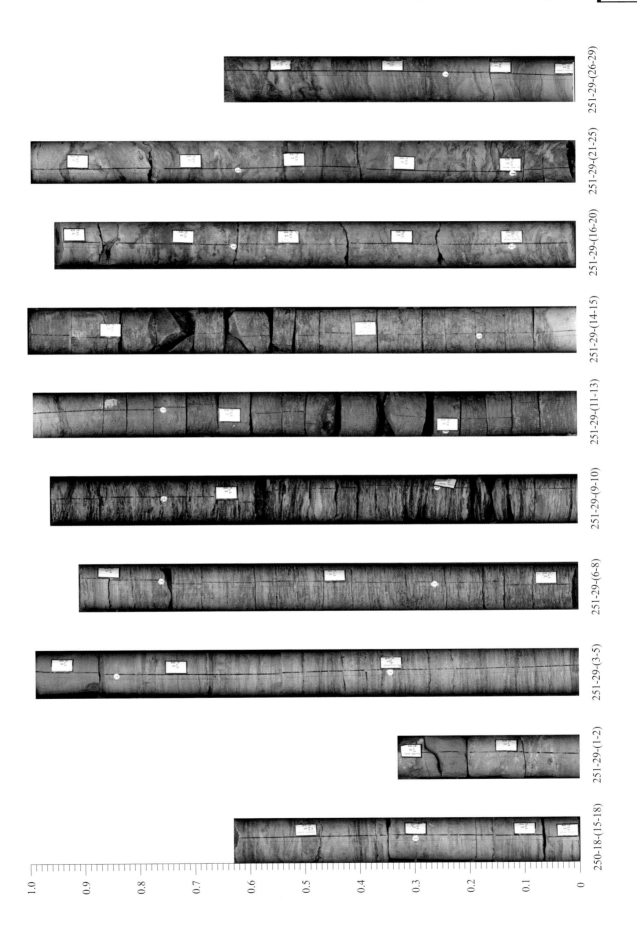

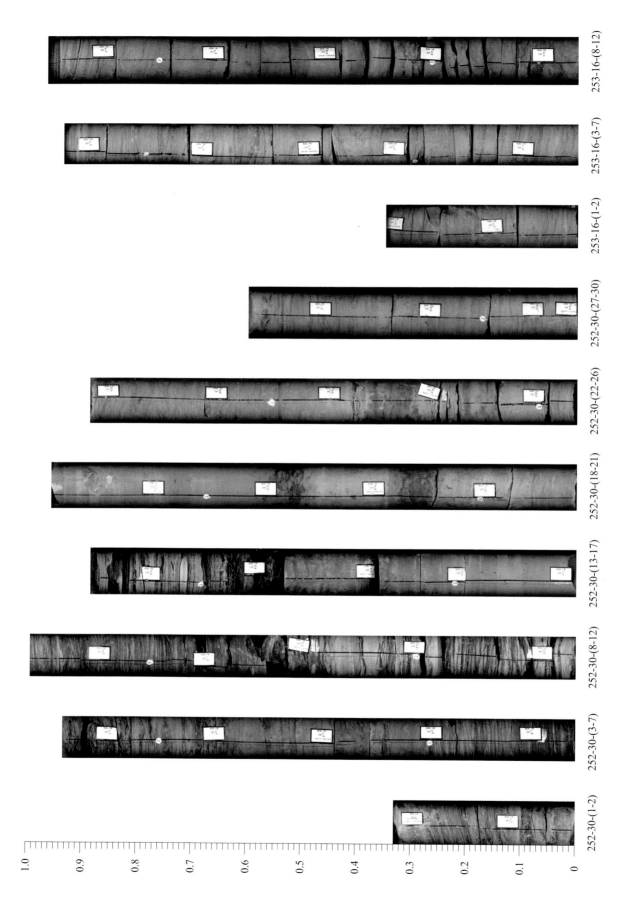

Section 5 SK-1 Core Description and Core Photographs | *571*

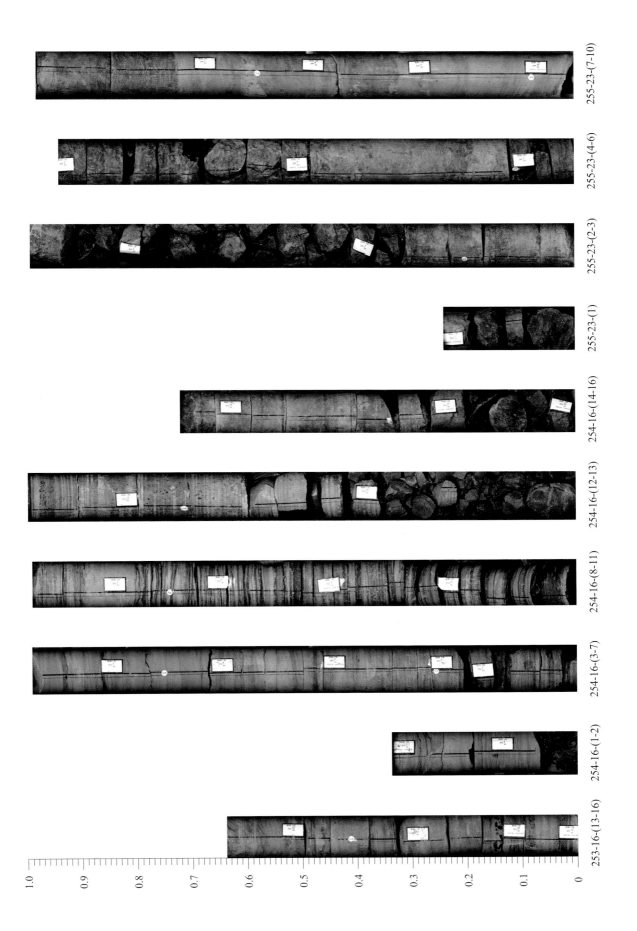

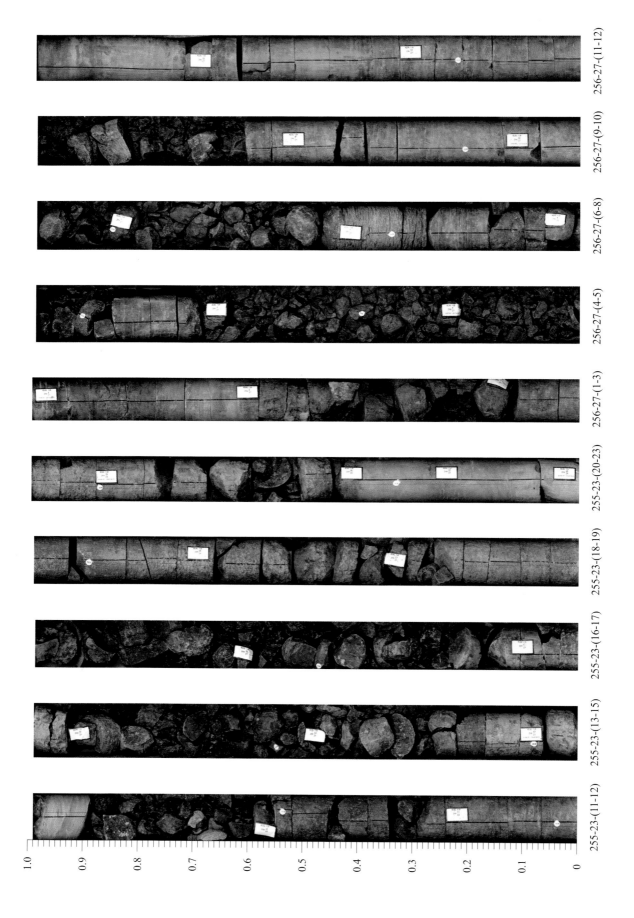

Section 5 SK-1 Core Description and Core Photographs

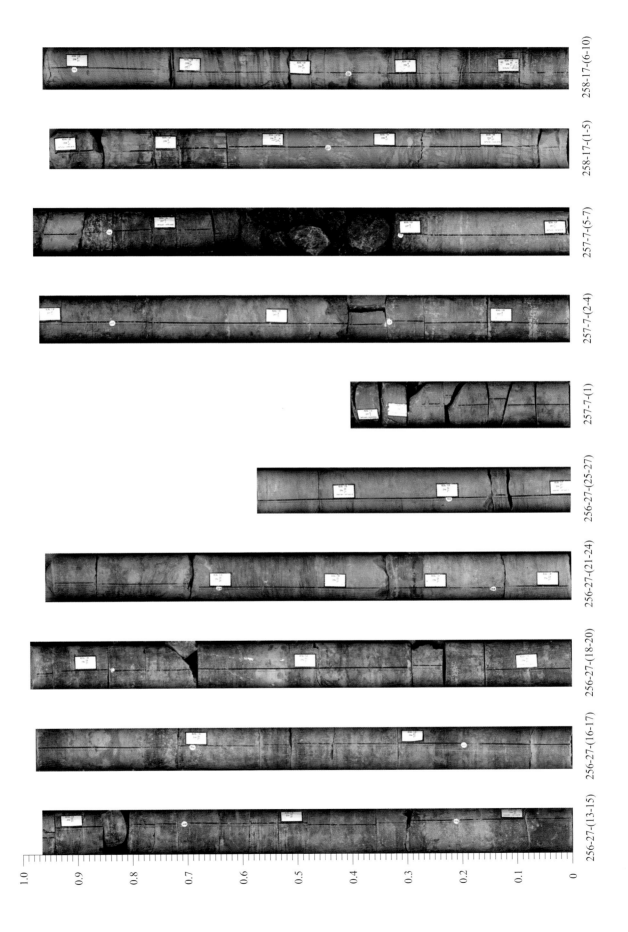

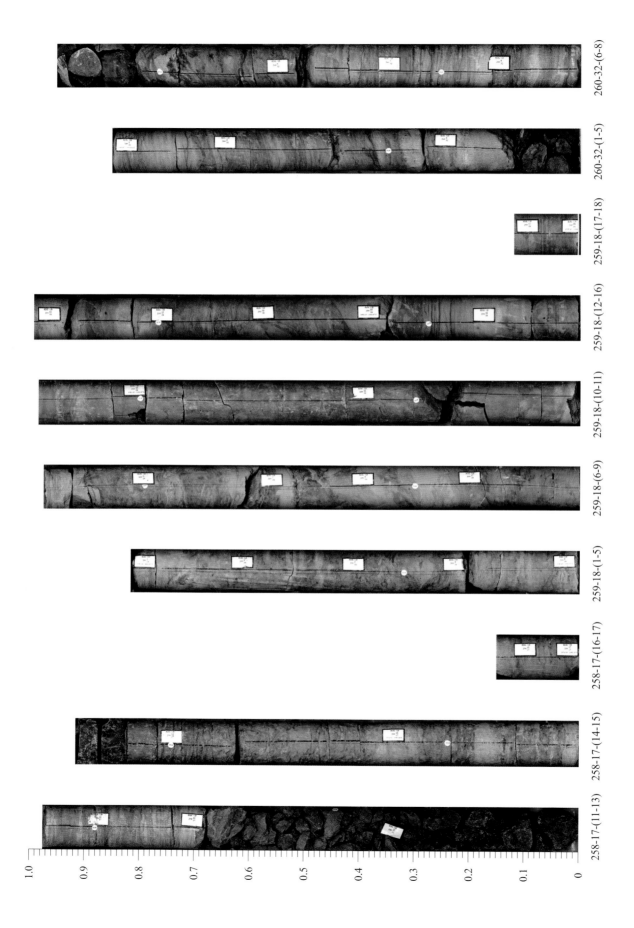

Section 5 SK-1 Core Description and Core Photographs

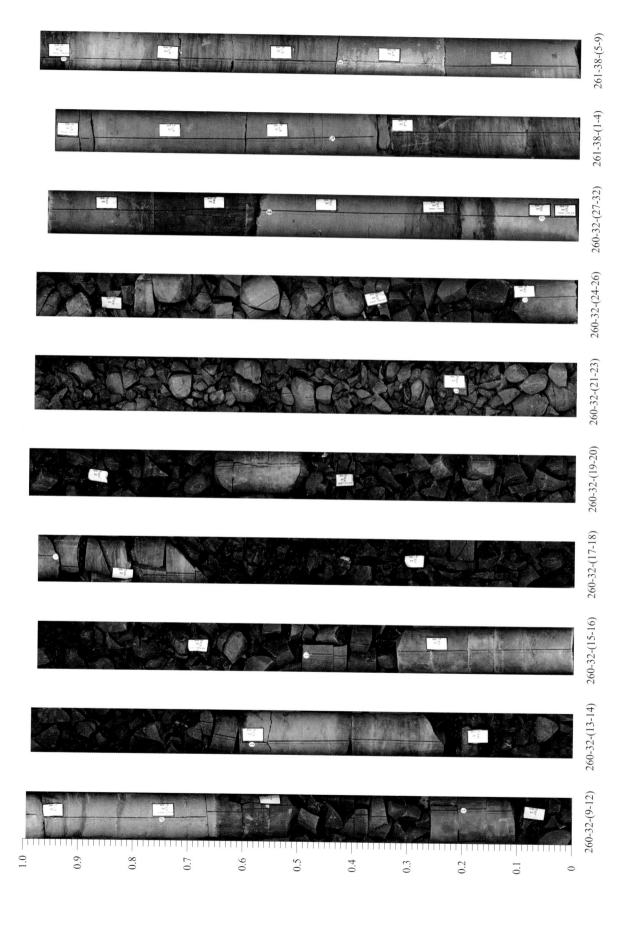

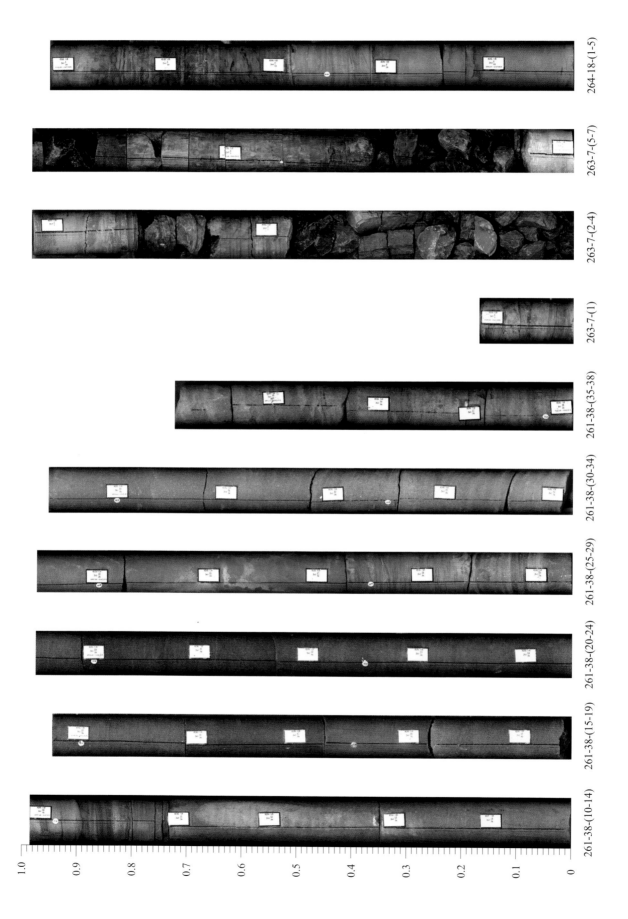

Section 5 SK-1 Core Description and Core Photographs | 577

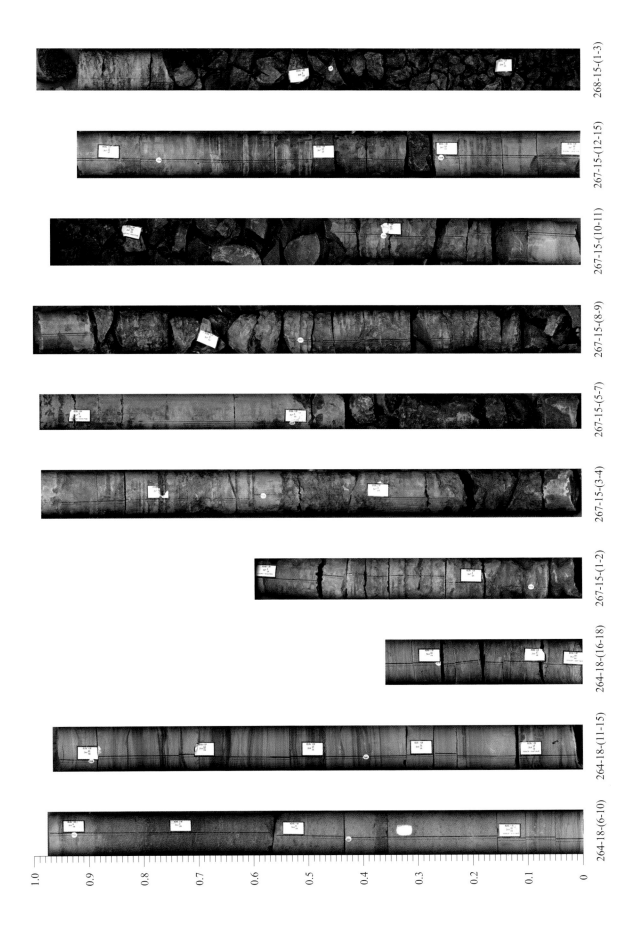

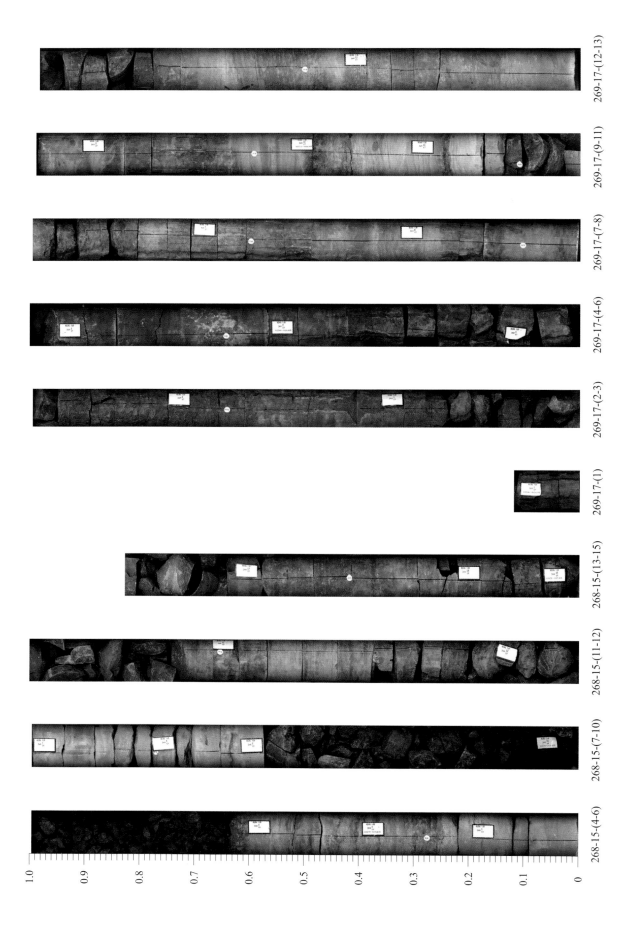

Section 5 SK-1 Core Description and Core Photographs

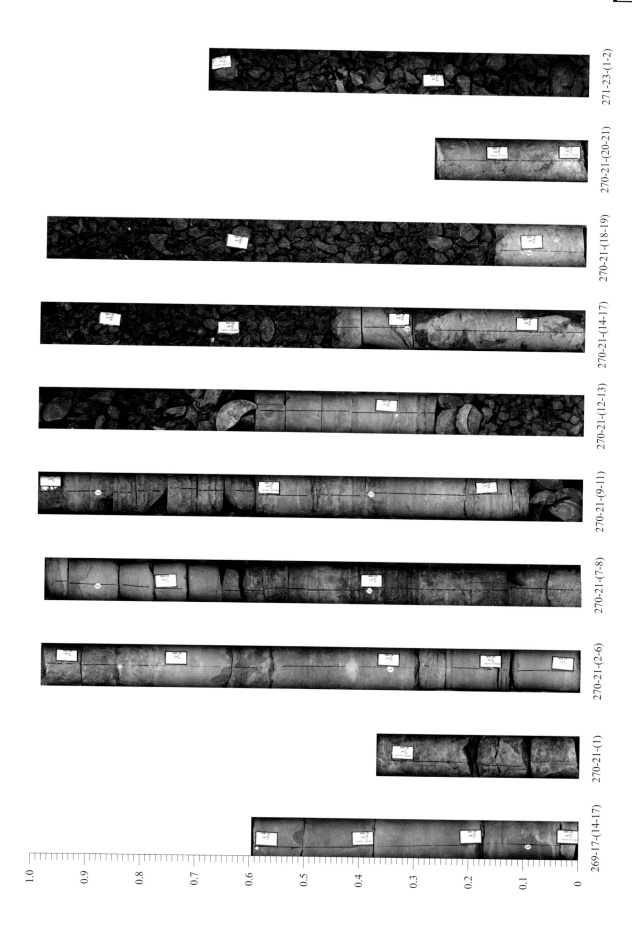

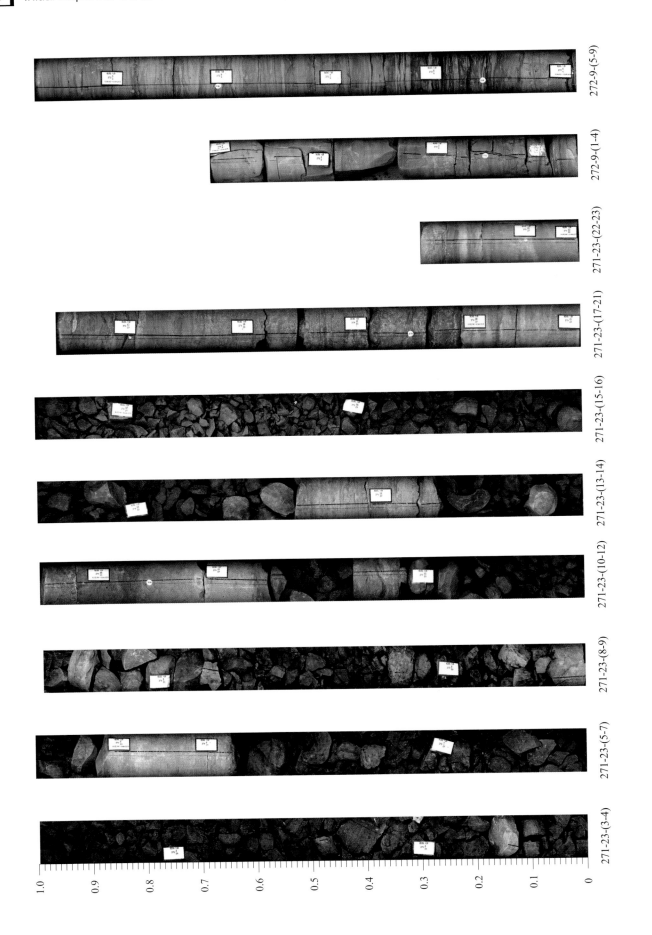

Section 5 SK-1 Core Description and Core Photographs

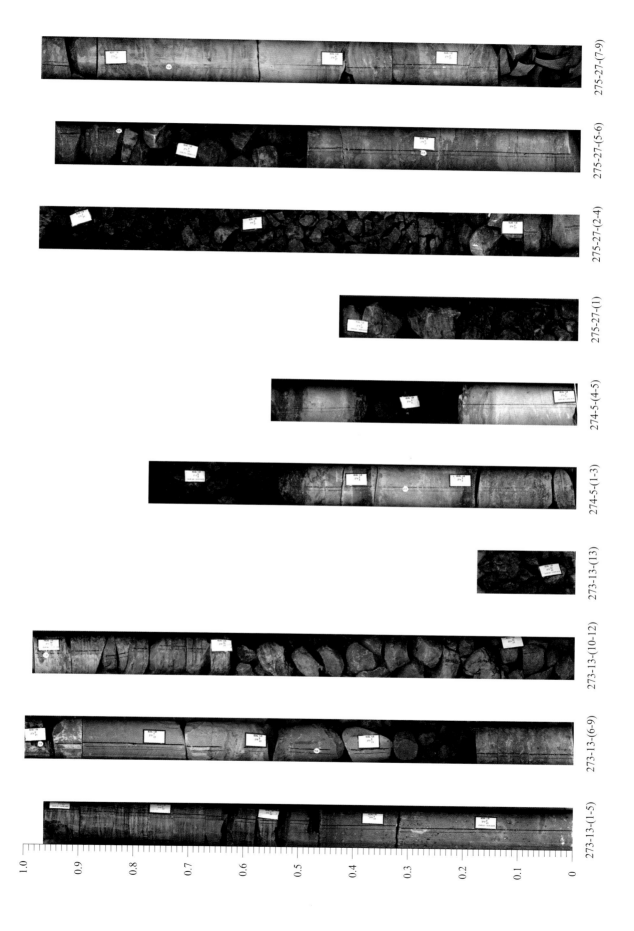

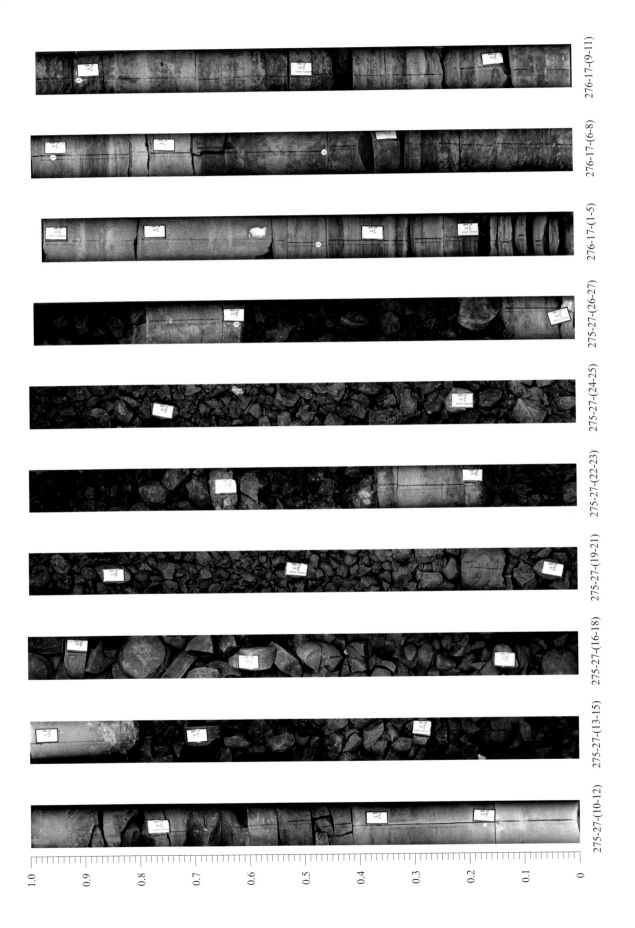

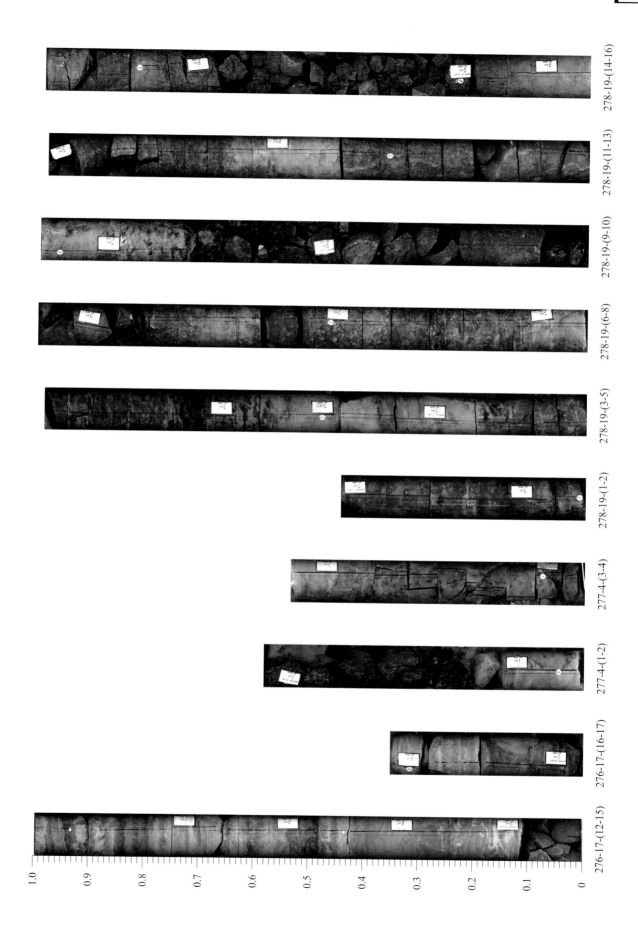

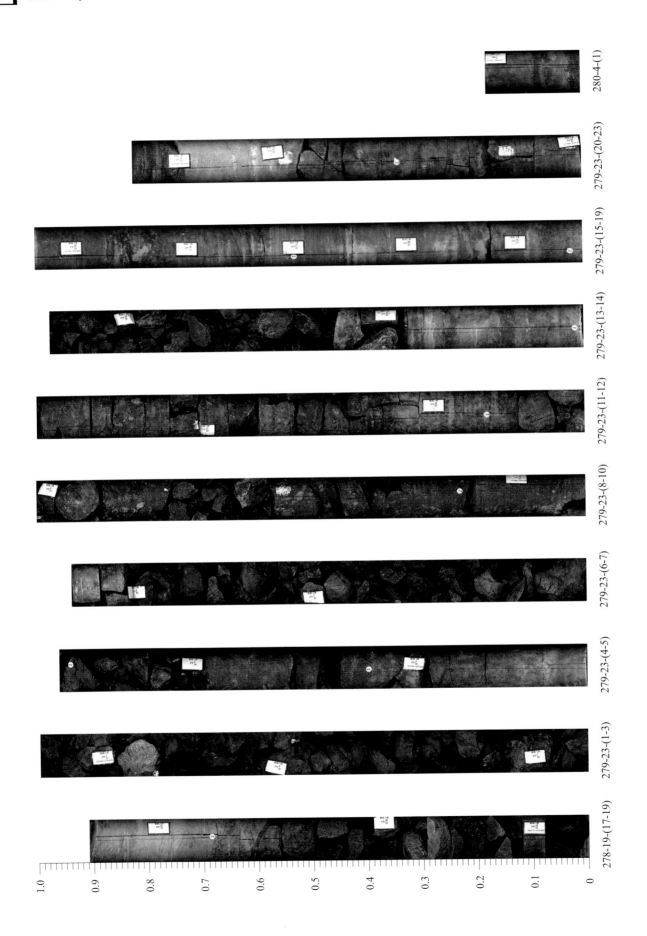

Section 5 SK-1 Core Description and Core Photographs

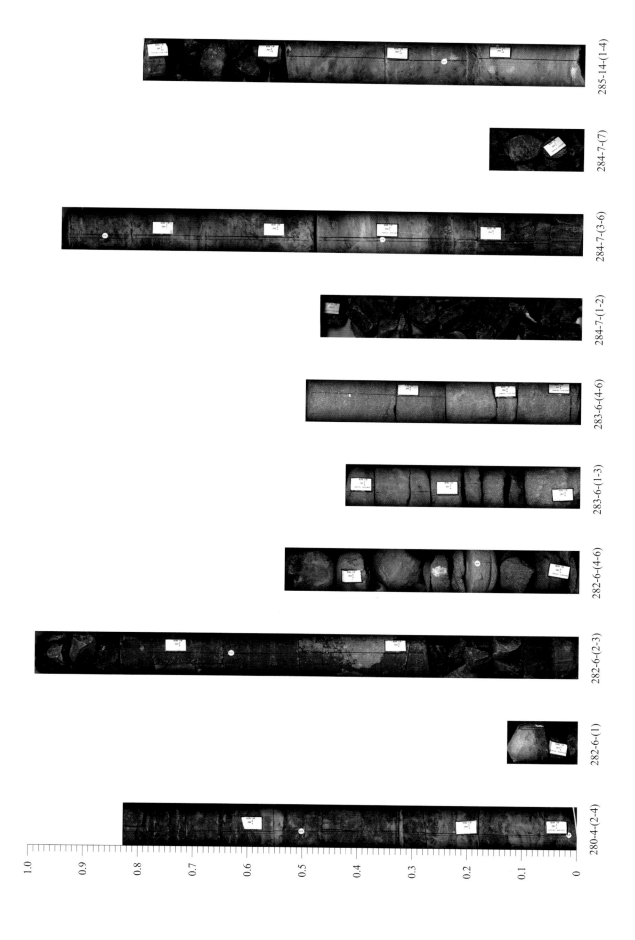

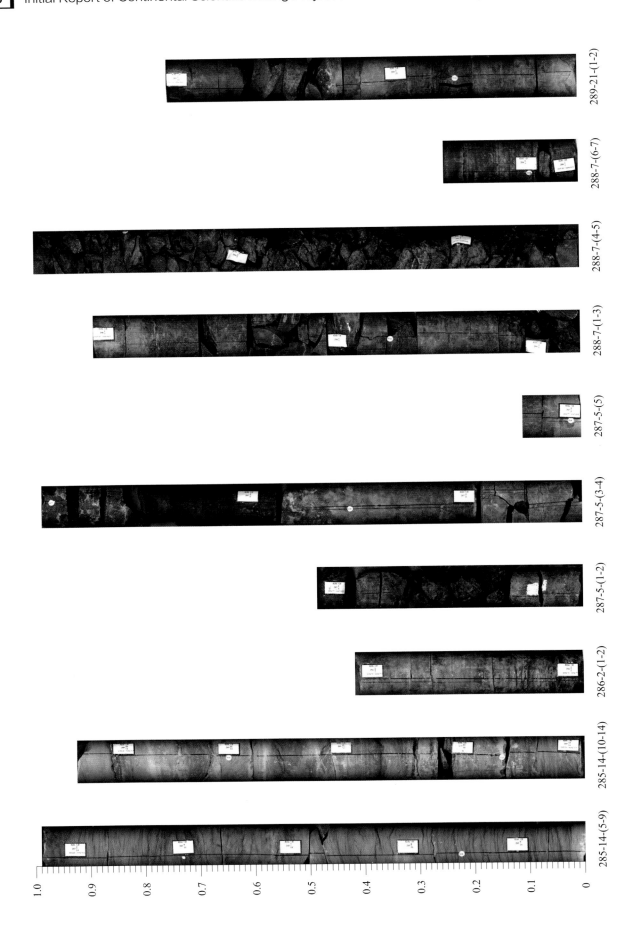

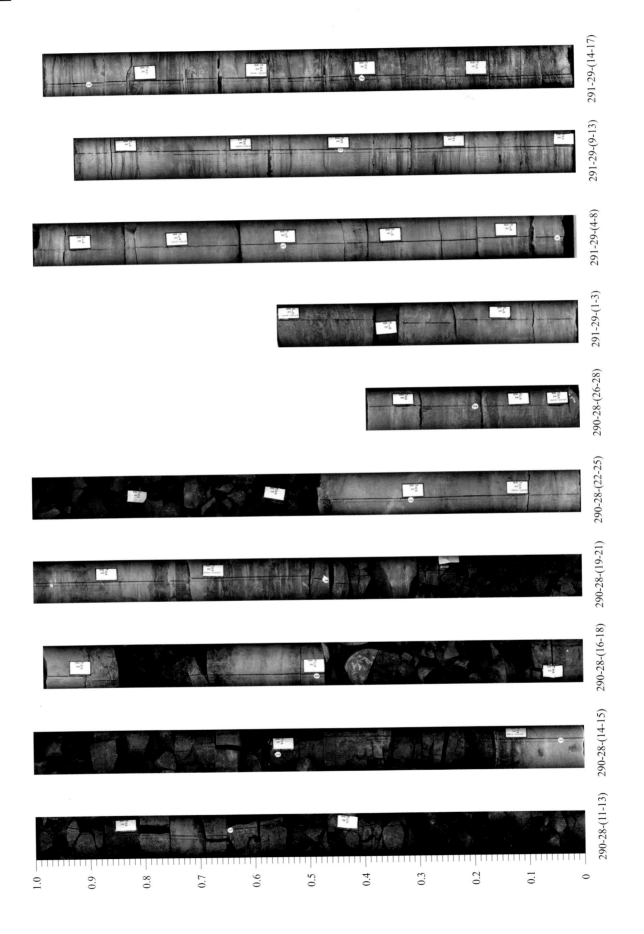

Section 5 SK-1 Core Description and Core Photographs

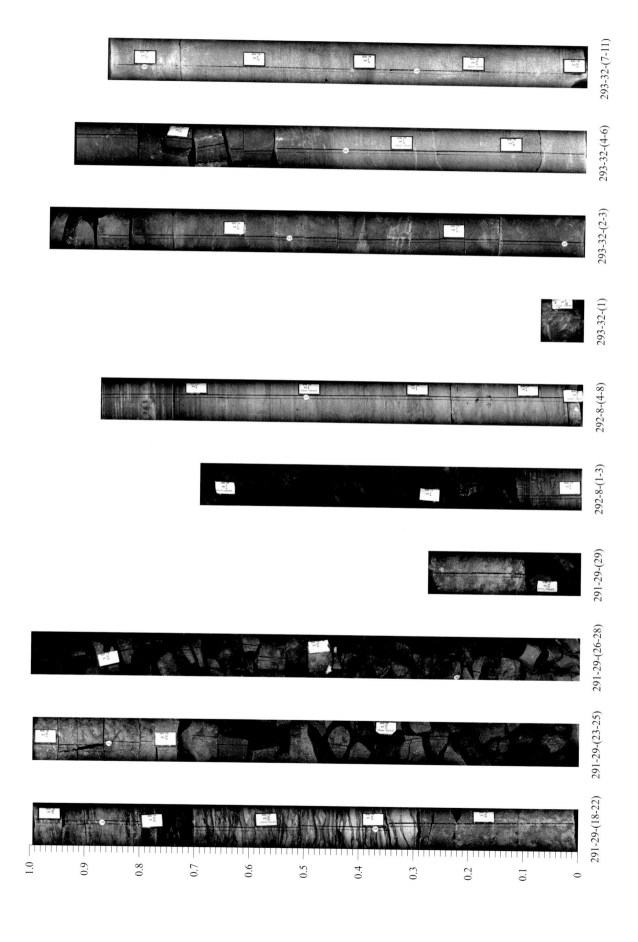

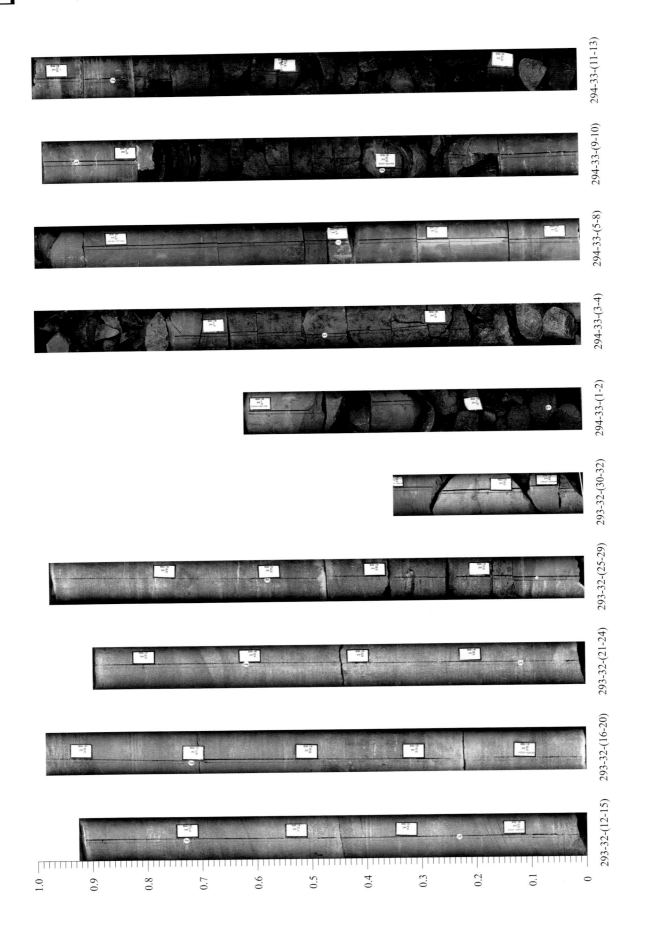

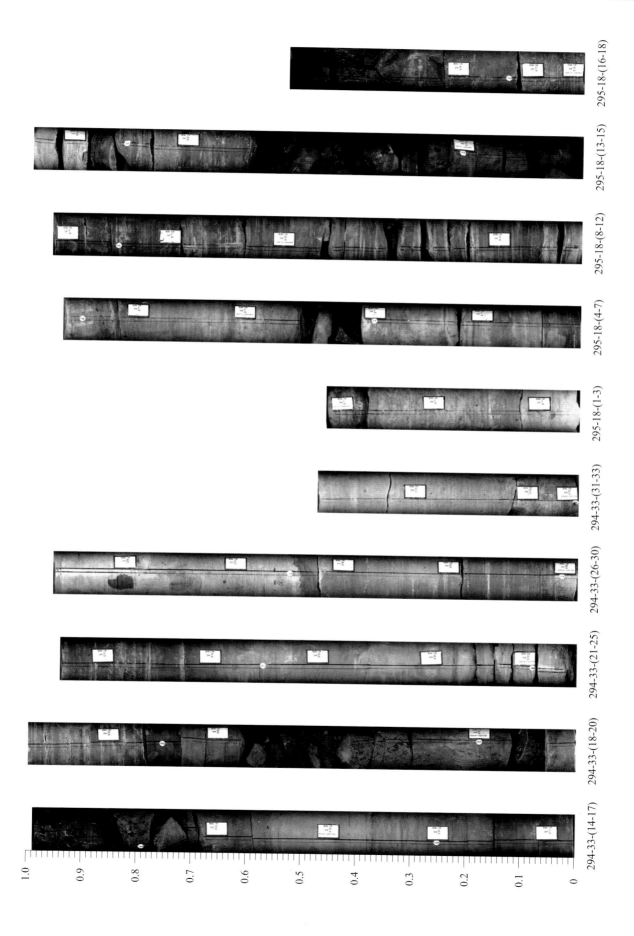

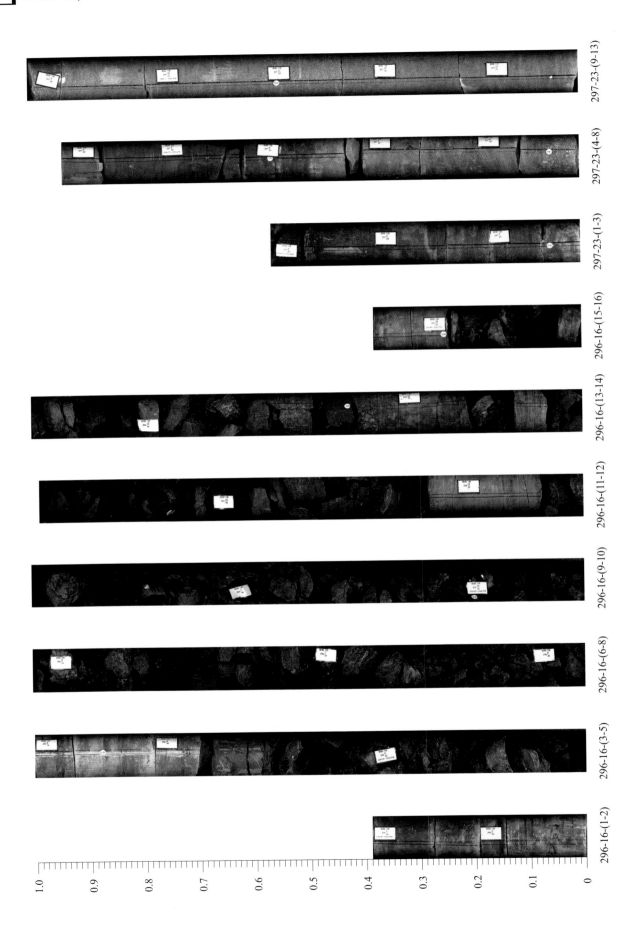

Section 5 SK-1 Core Description and Core Photographs | 593

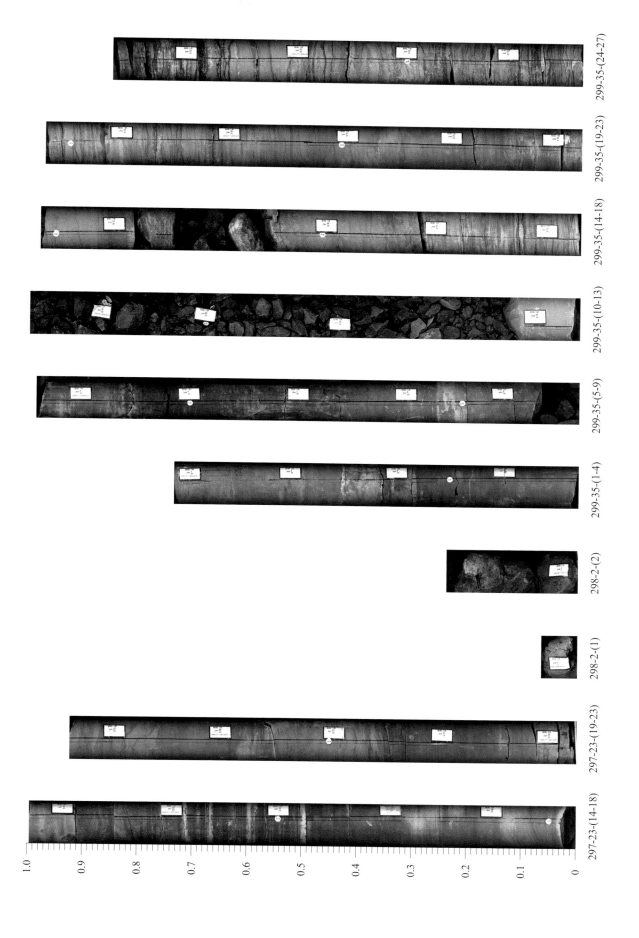

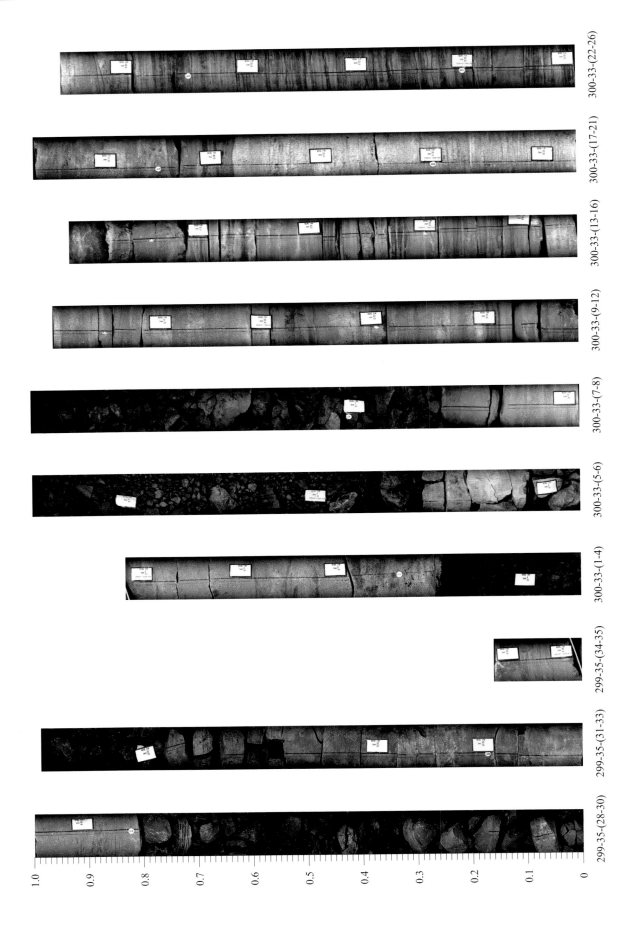

Section 5 SK-1 Core Description and Core Photographs

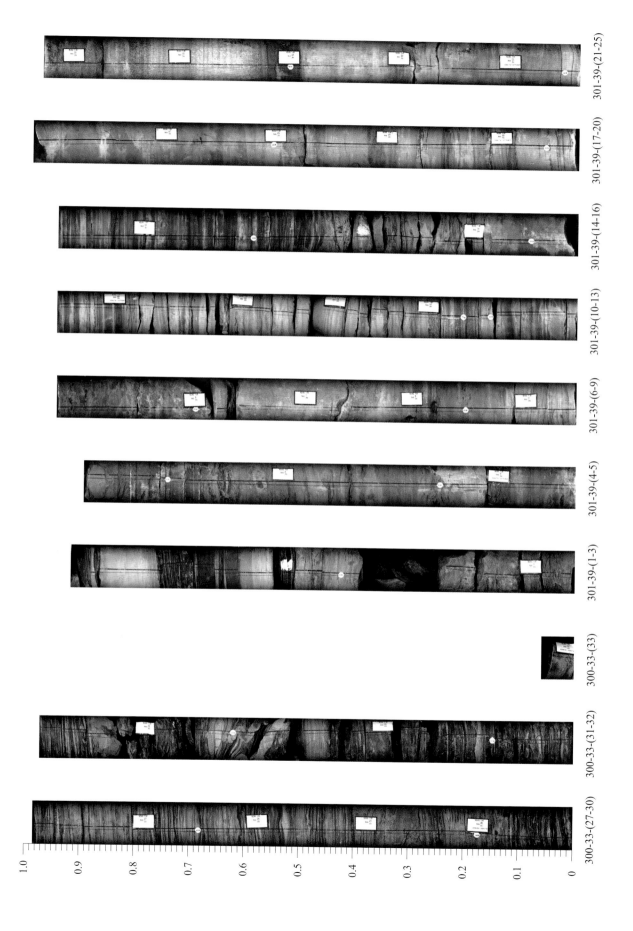

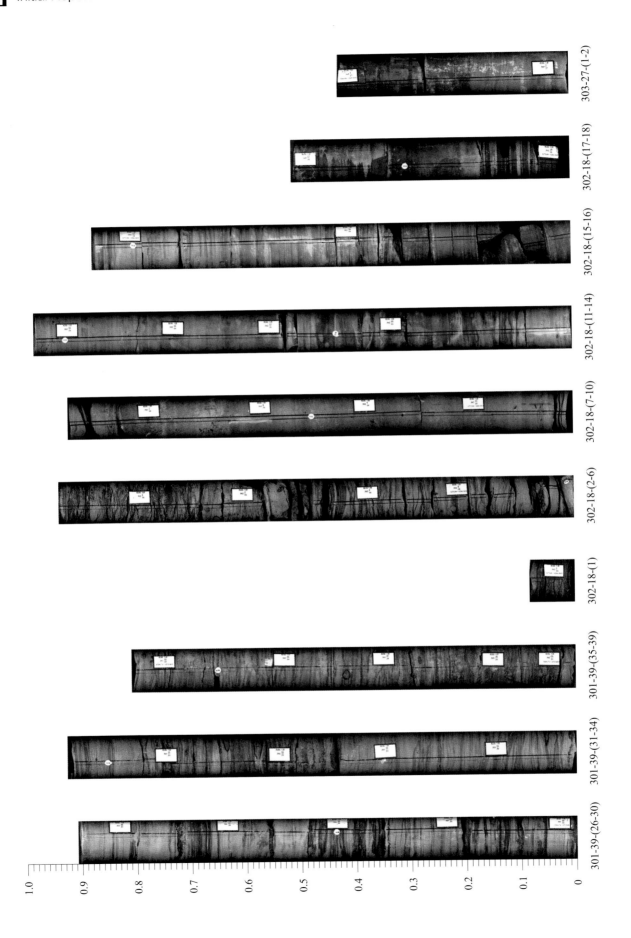

Section 5 SK-1 Core Description and Core Photographs

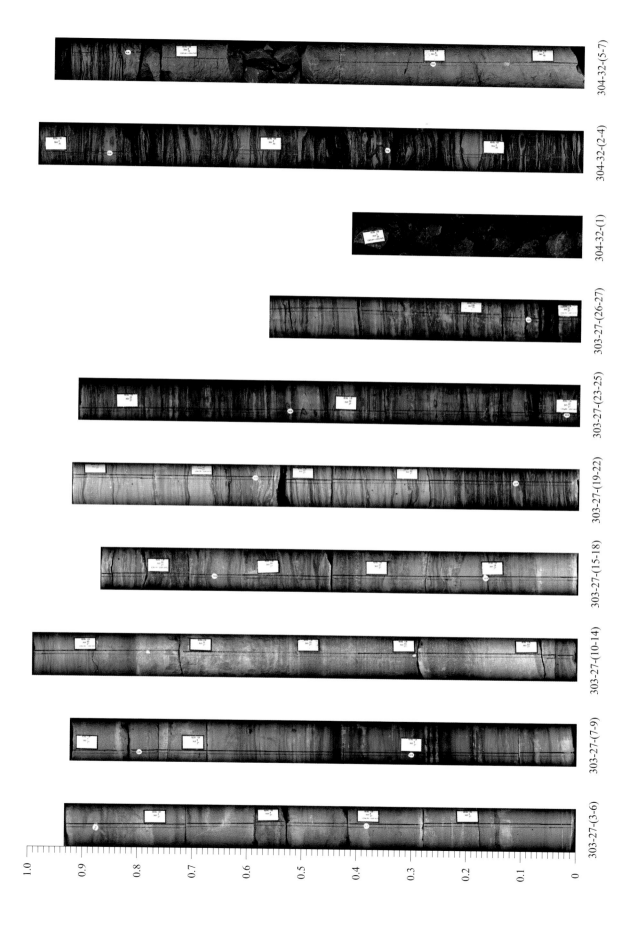

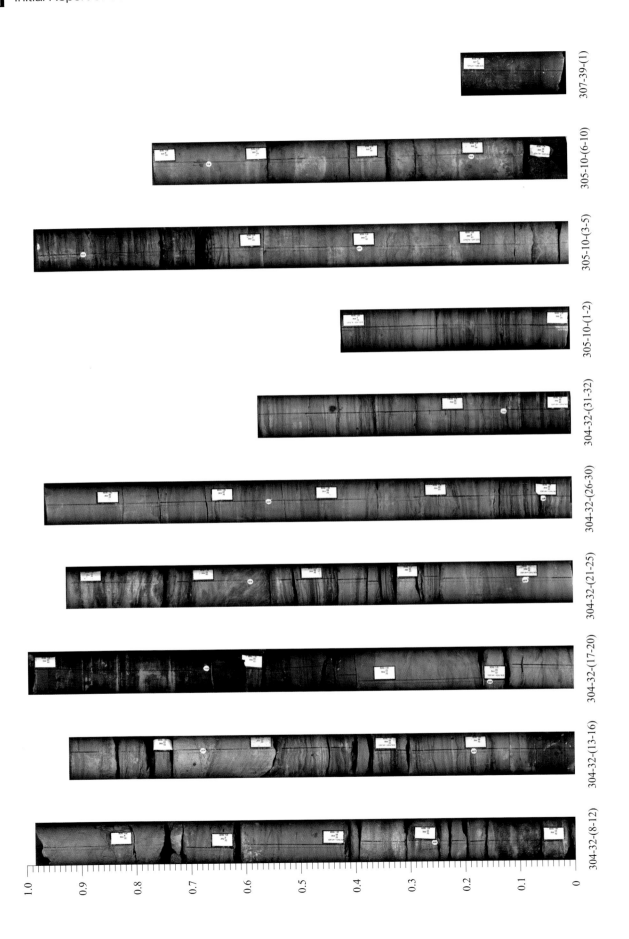

Section 5 SK-1 Core Description and Core Photographs

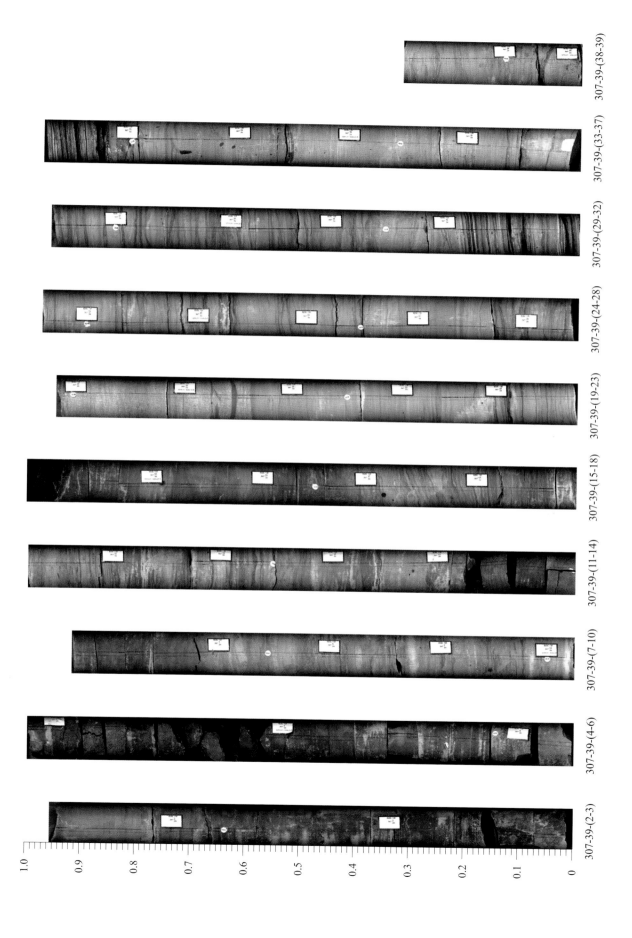

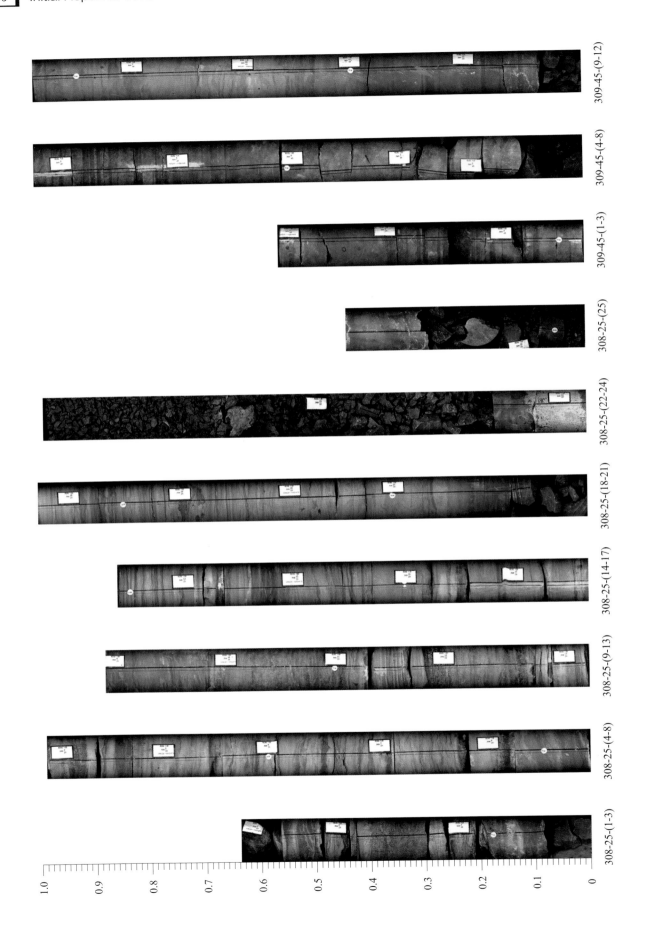

Section 5 SK-1 Core Description and Core Photographs | *601*

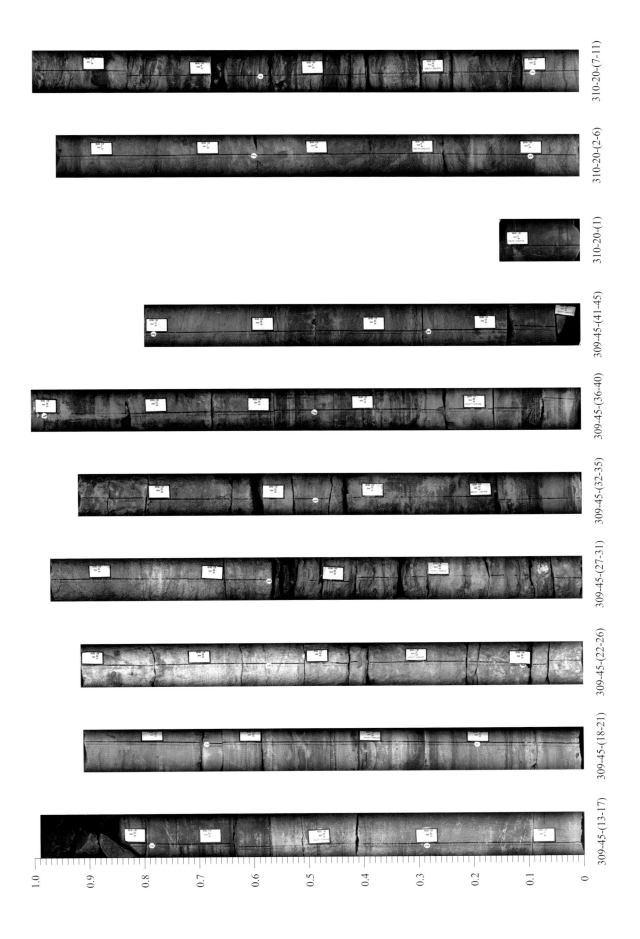

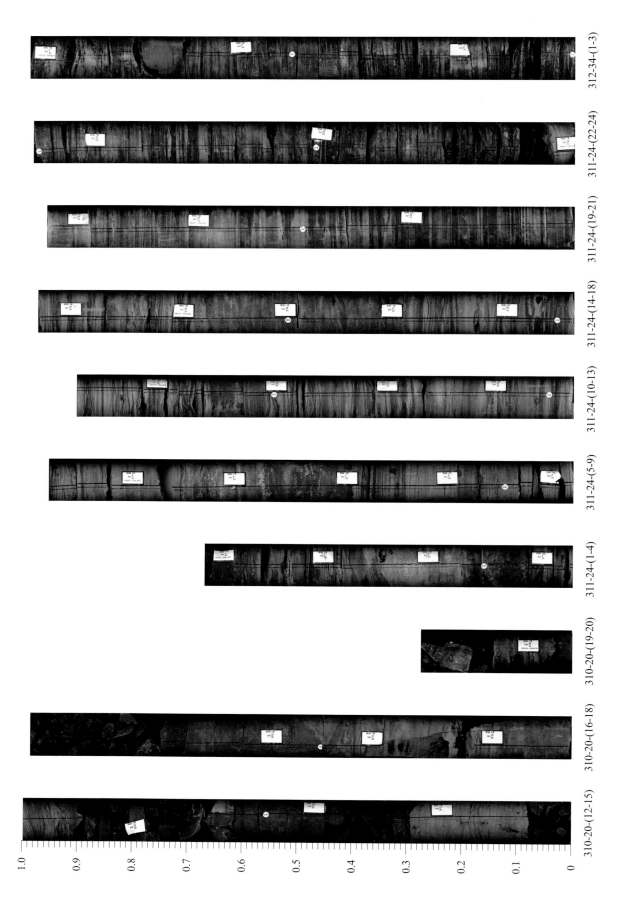

Section 5 SK-1 Core Description and Core Photographs | 603

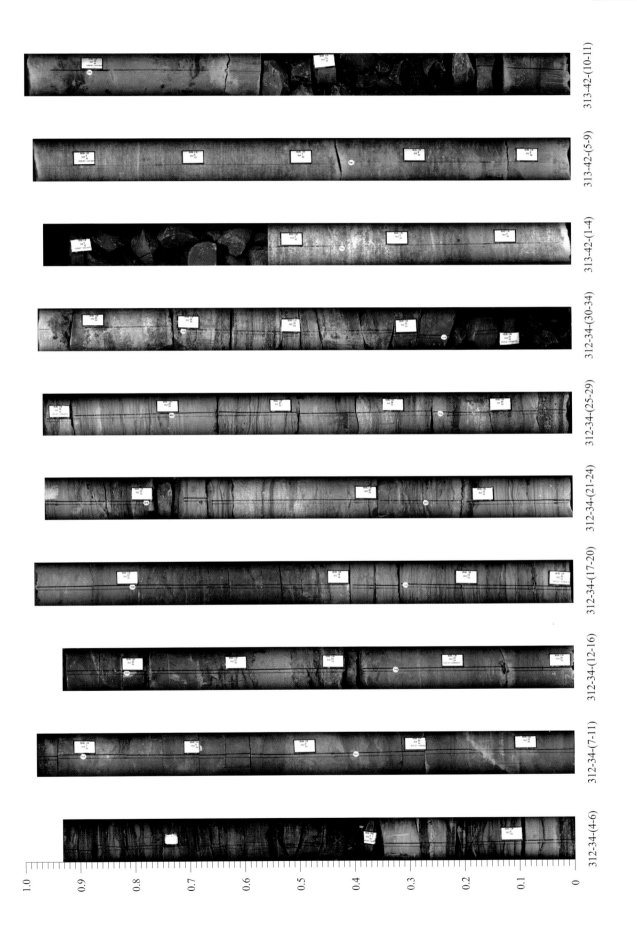

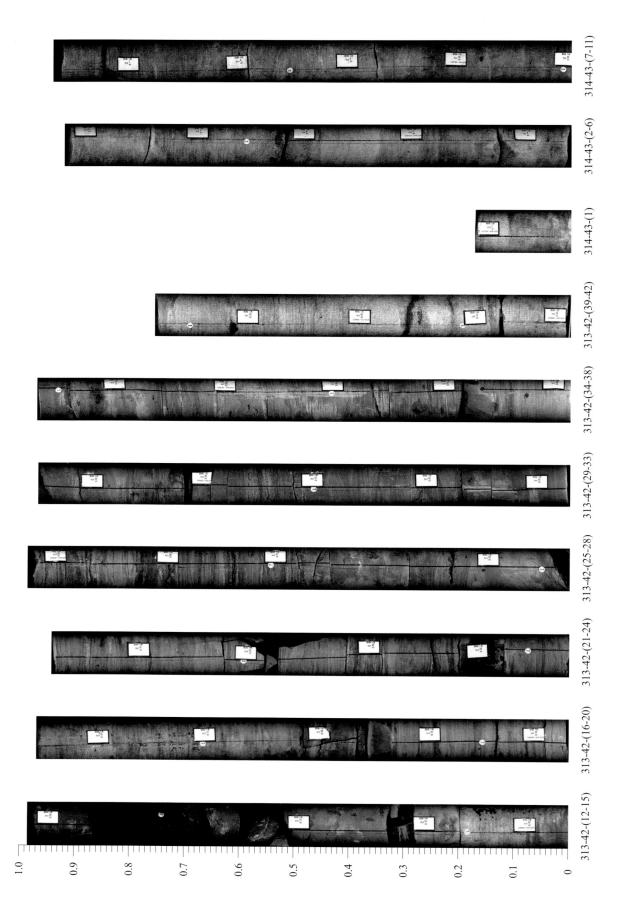

Section 5 SK-1 Core Description and Core Photographs | 605

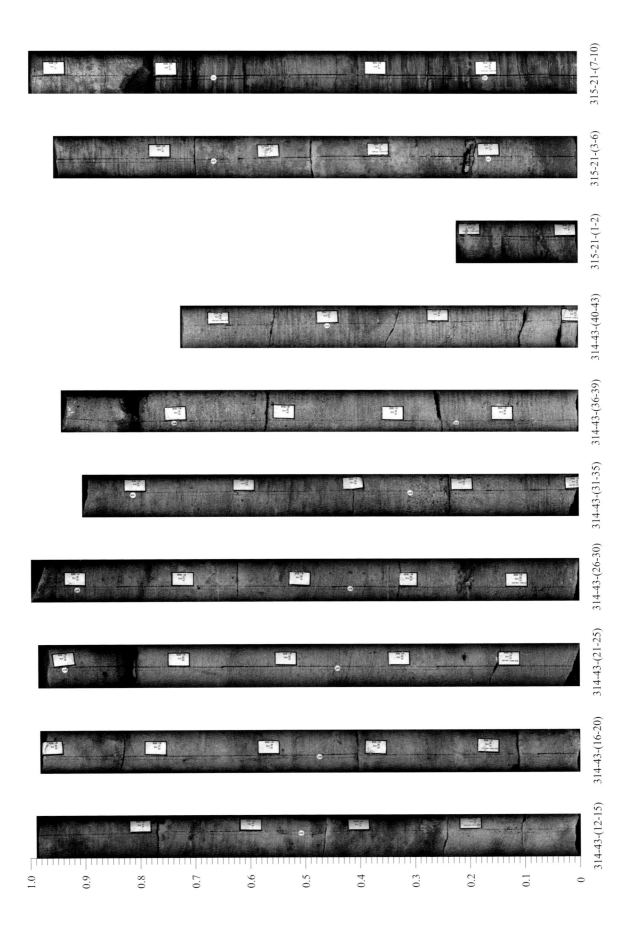

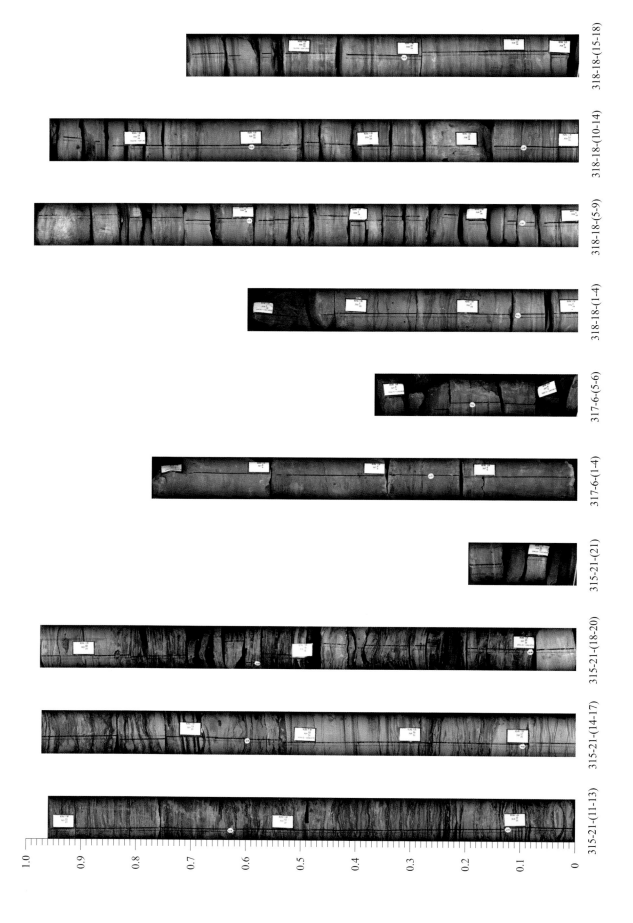

Section 5 SK-1 Core Description and Core Photographs

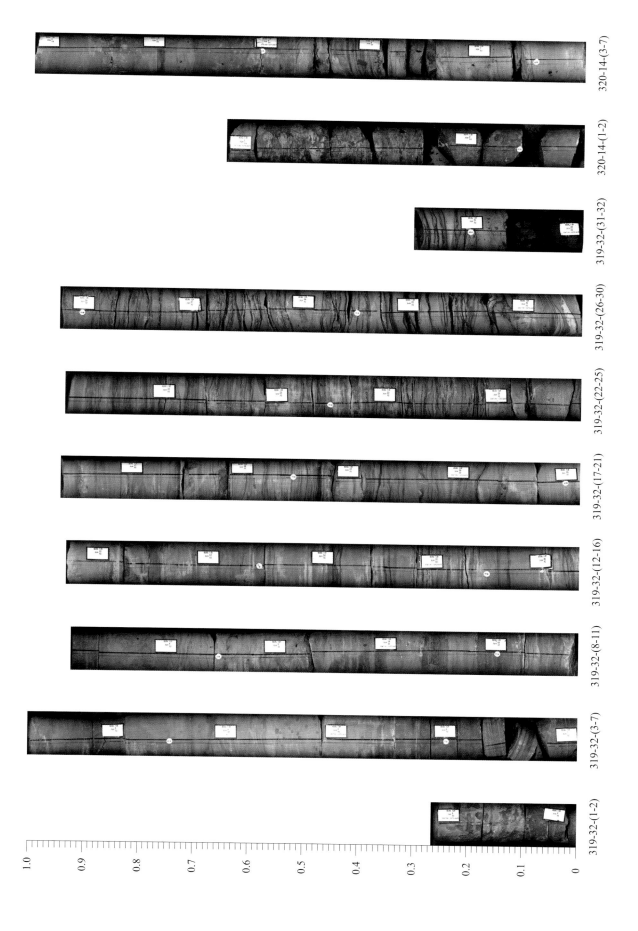

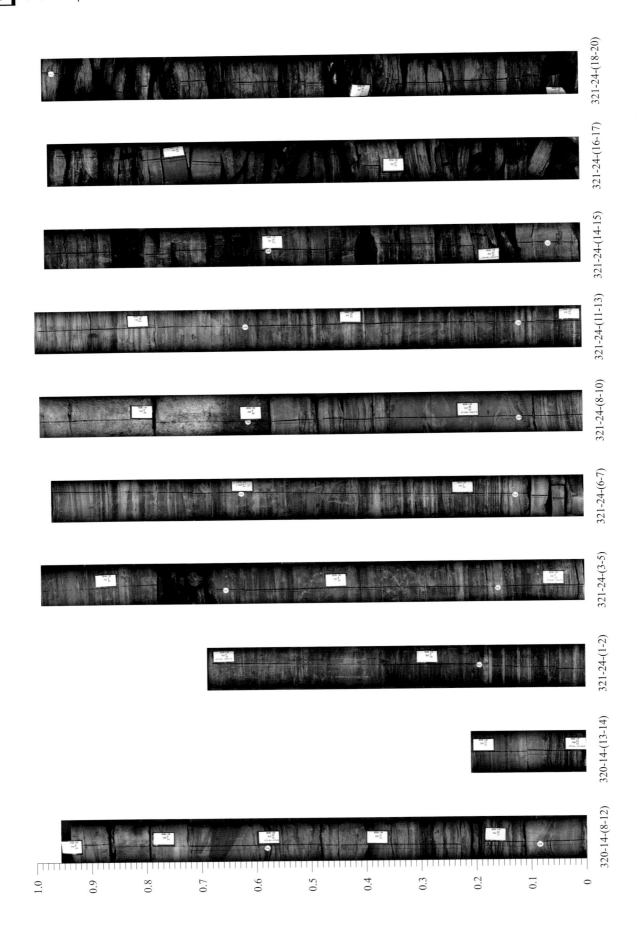

Section 5 SK-1 Core Description and Core Photographs | 609

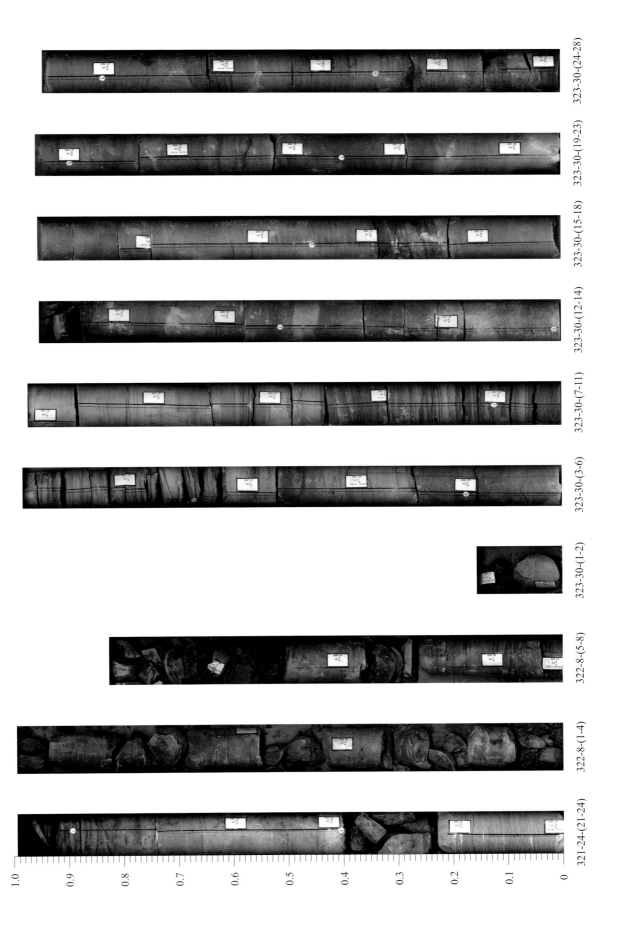

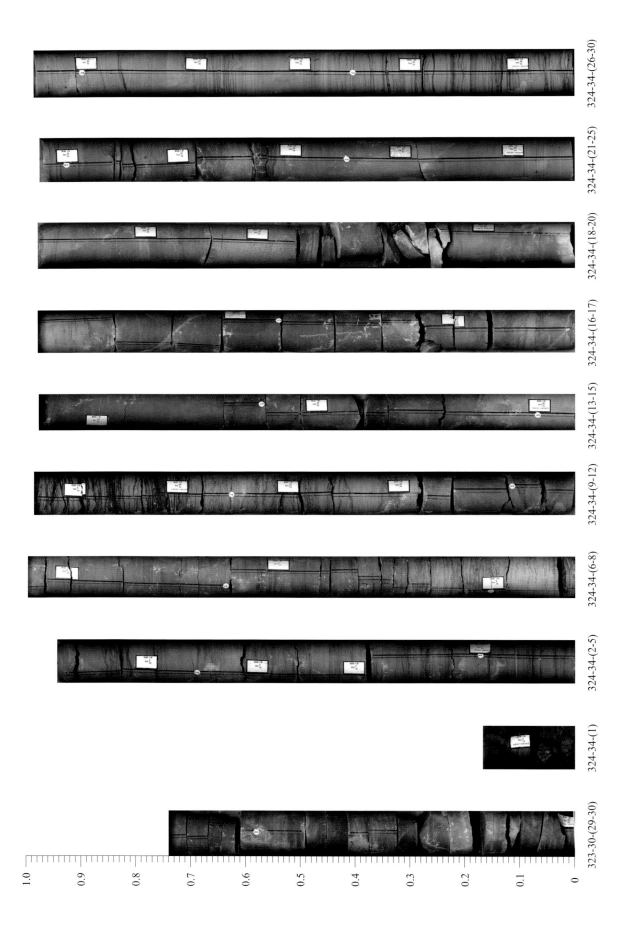

Section 5 SK-1 Core Description and Core Photographs | 611

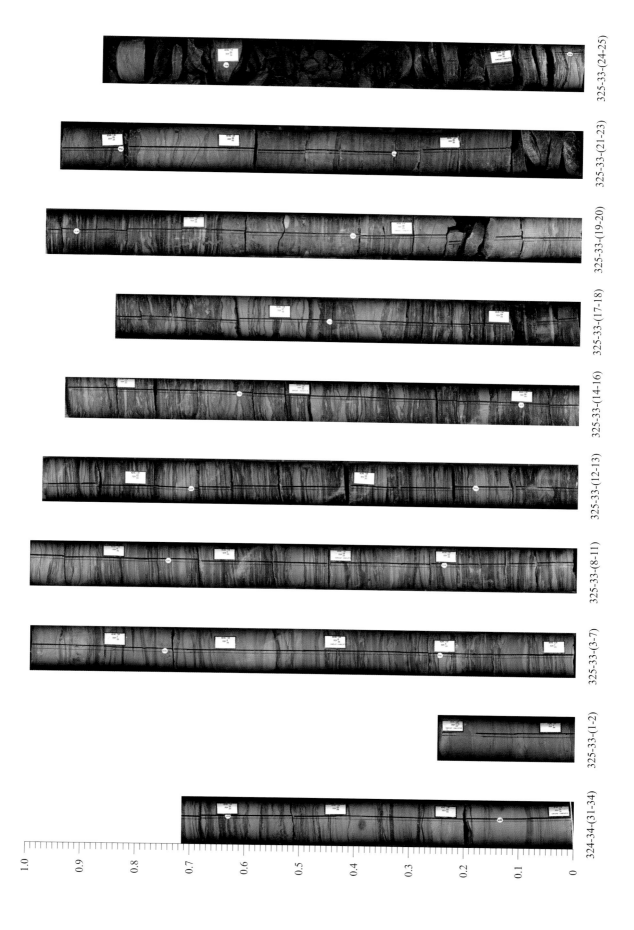

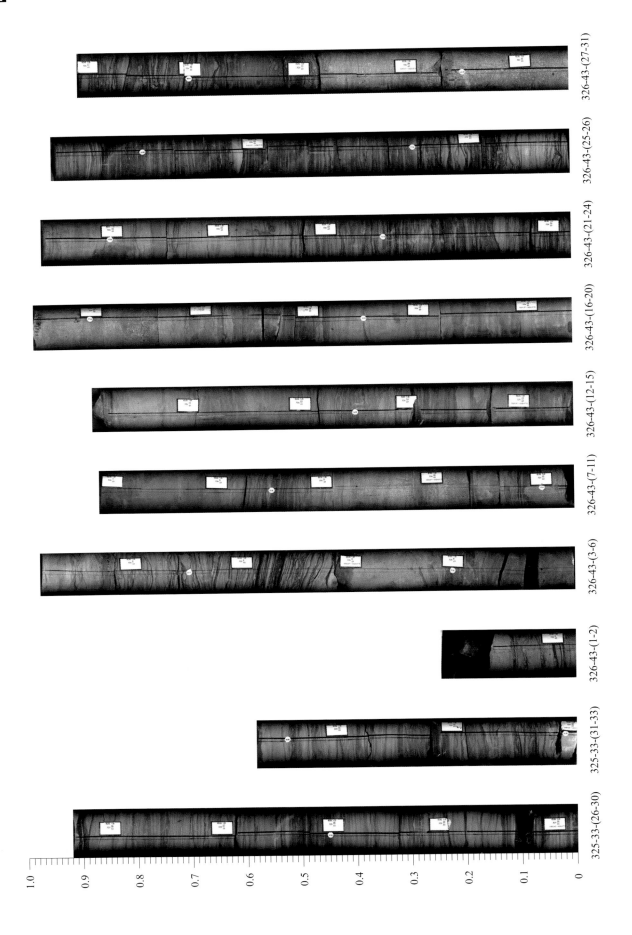

Section 5 SK–1 Core Description and Core Photographs | 613

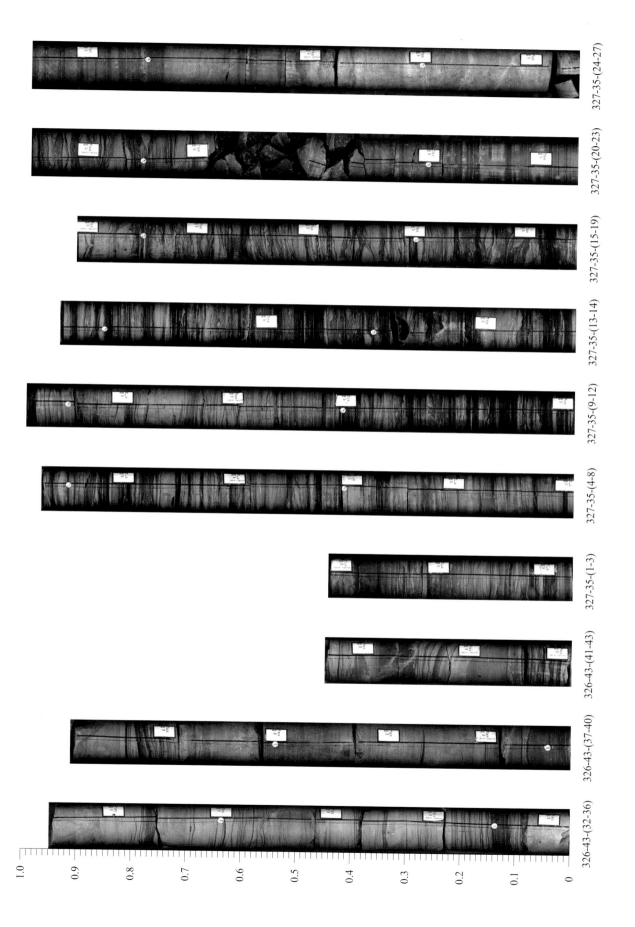

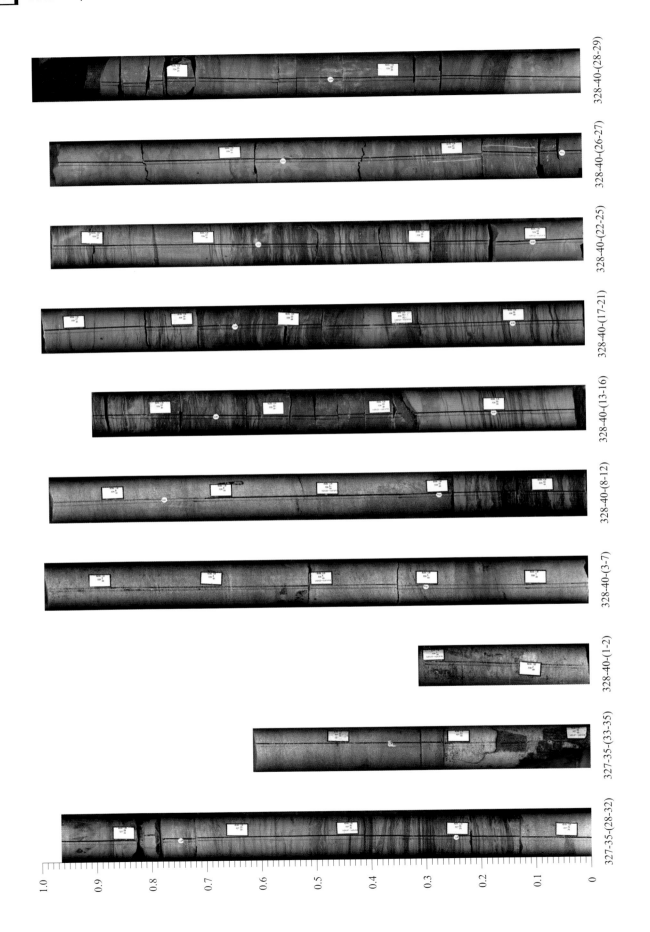

Section 5 SK-1 Core Description and Core Photographs

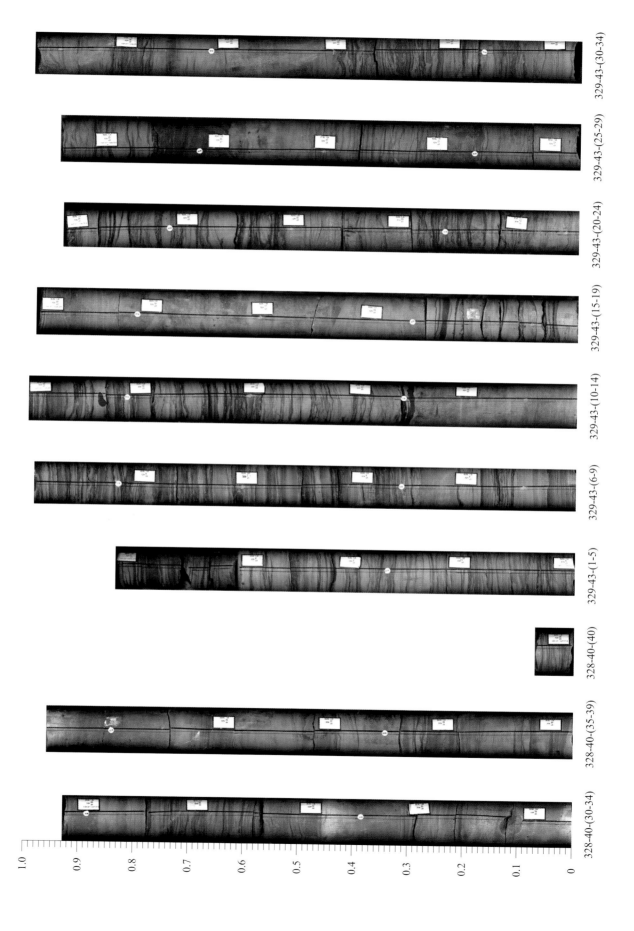

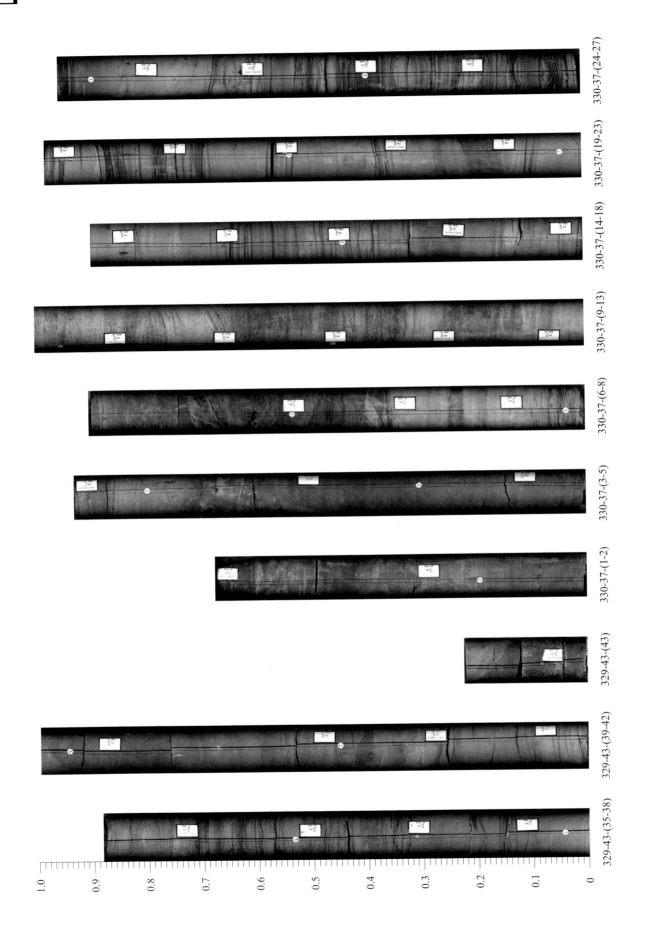

Section 5 SK-1 Core Description and Core Photographs

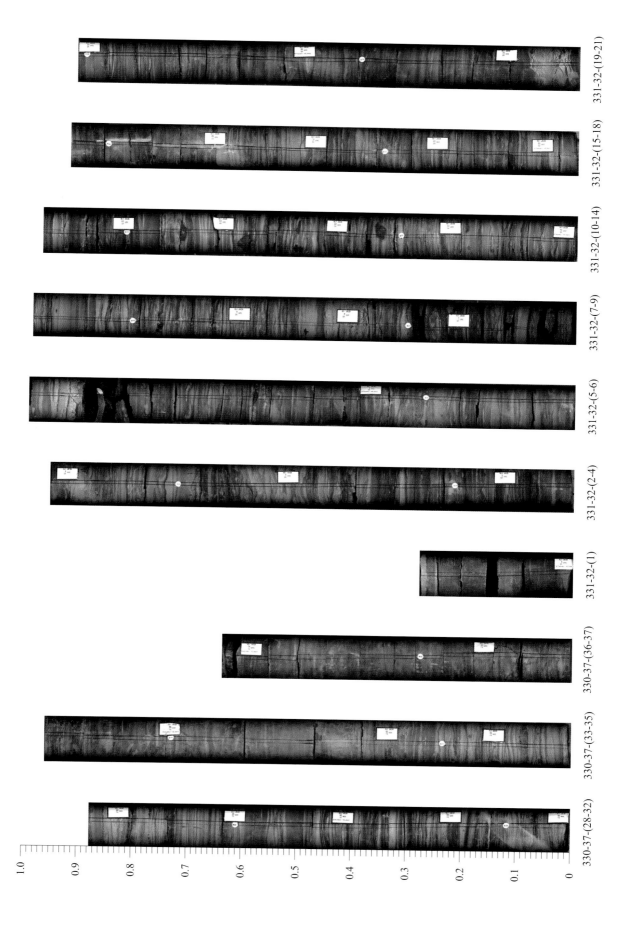

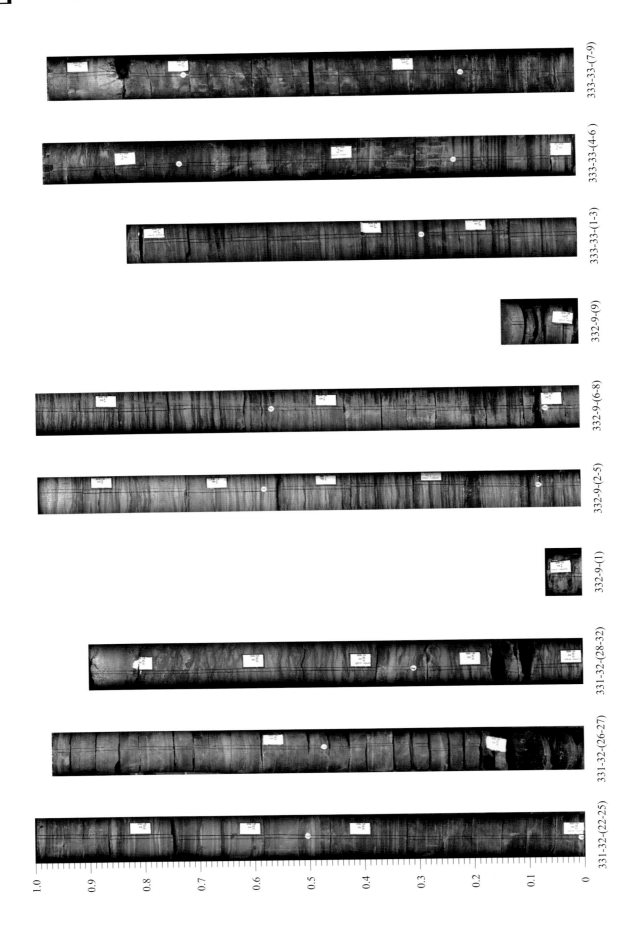

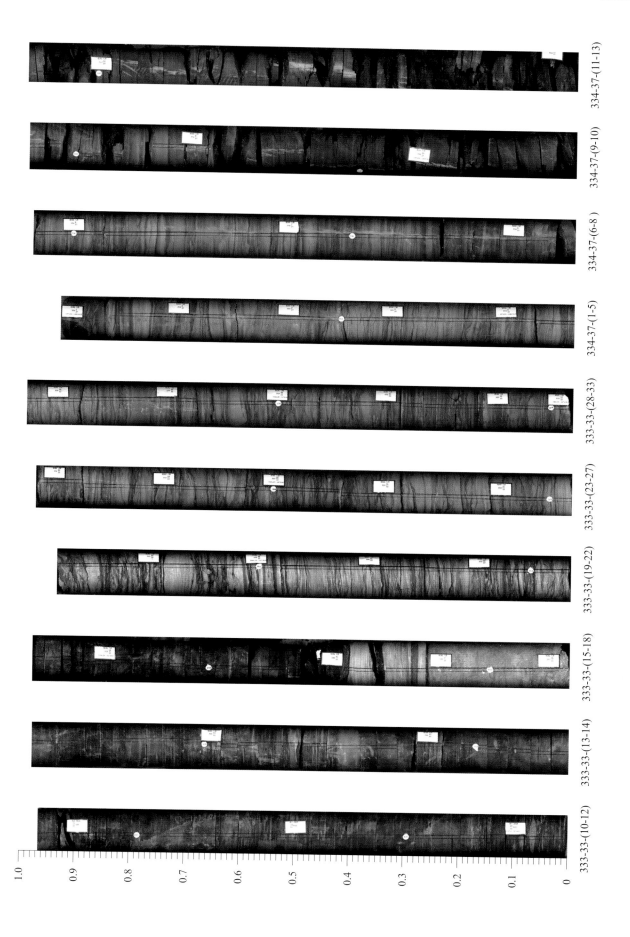

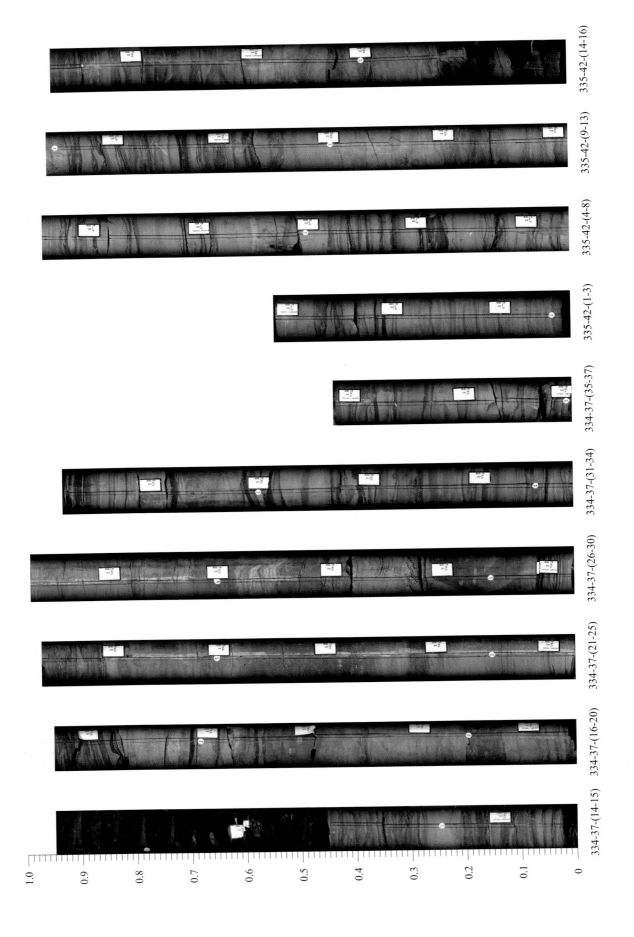

Section 5 SK-1 Core Description and Core Photographs

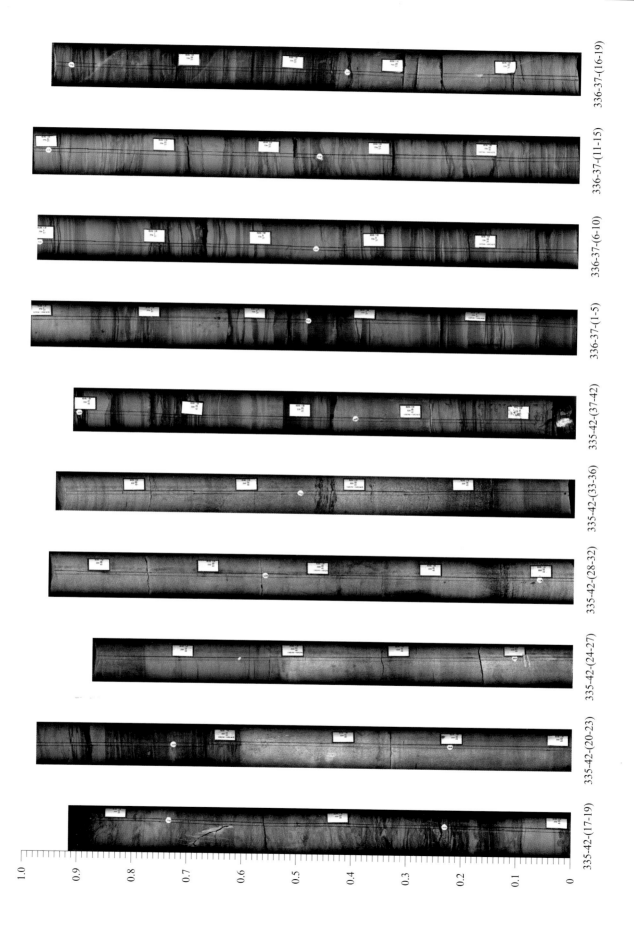

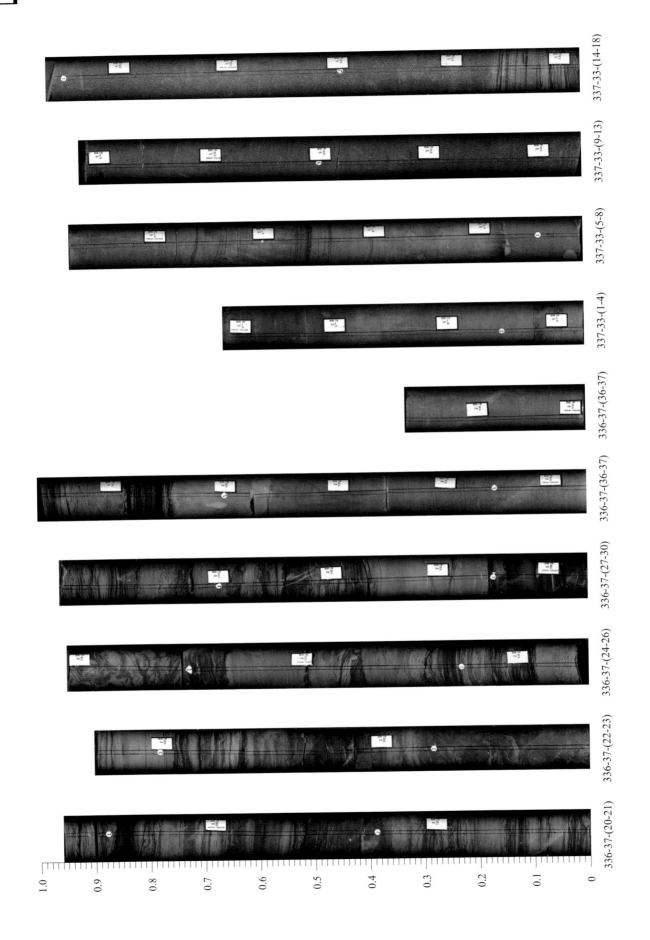

Section 5 SK–1 Core Description and Core Photographs

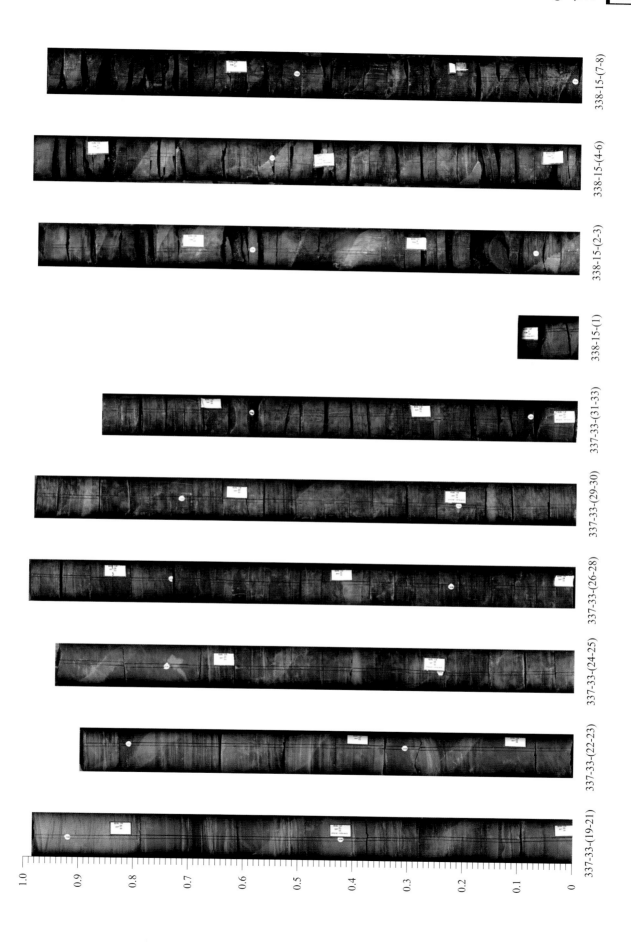

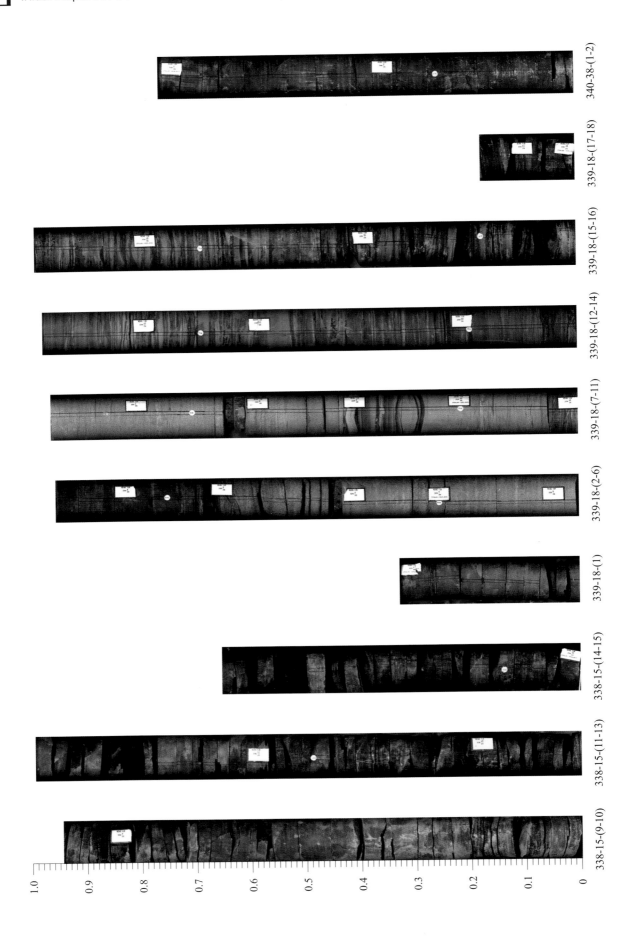

Section 5 SK–1 Core Description and Core Photographs | 625

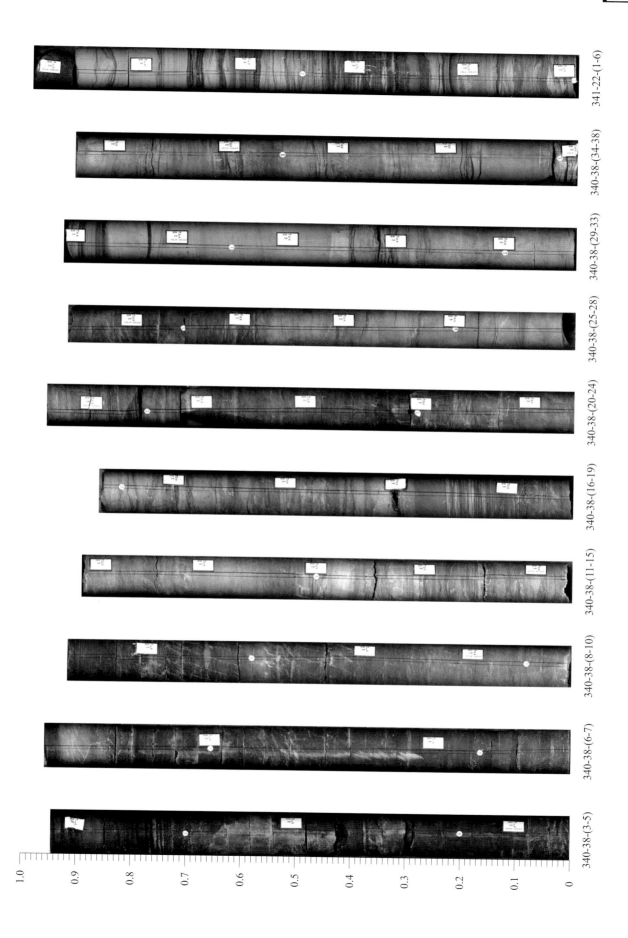

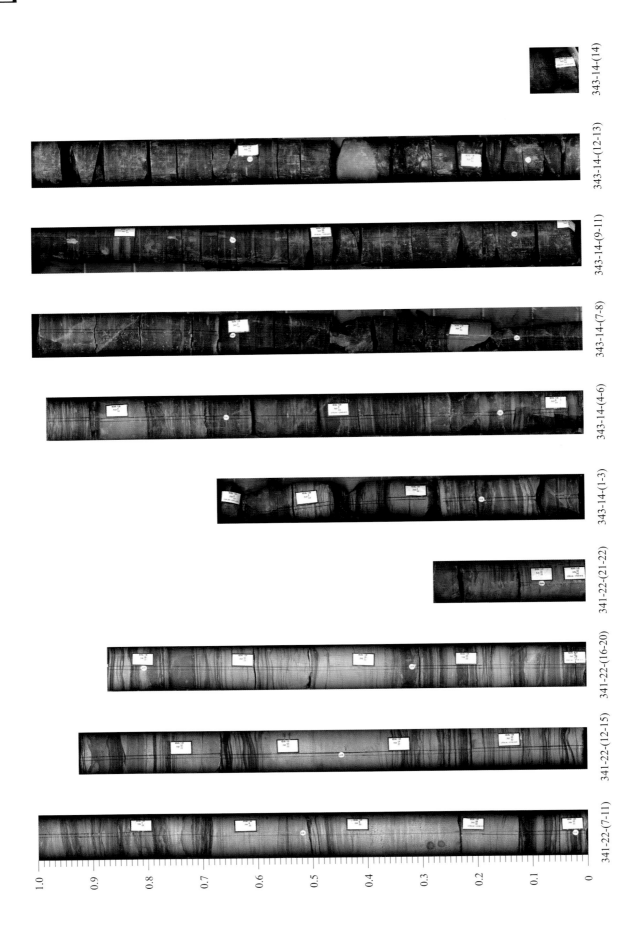

Section 5 SK-1 Core Description and Core Photographs | 627

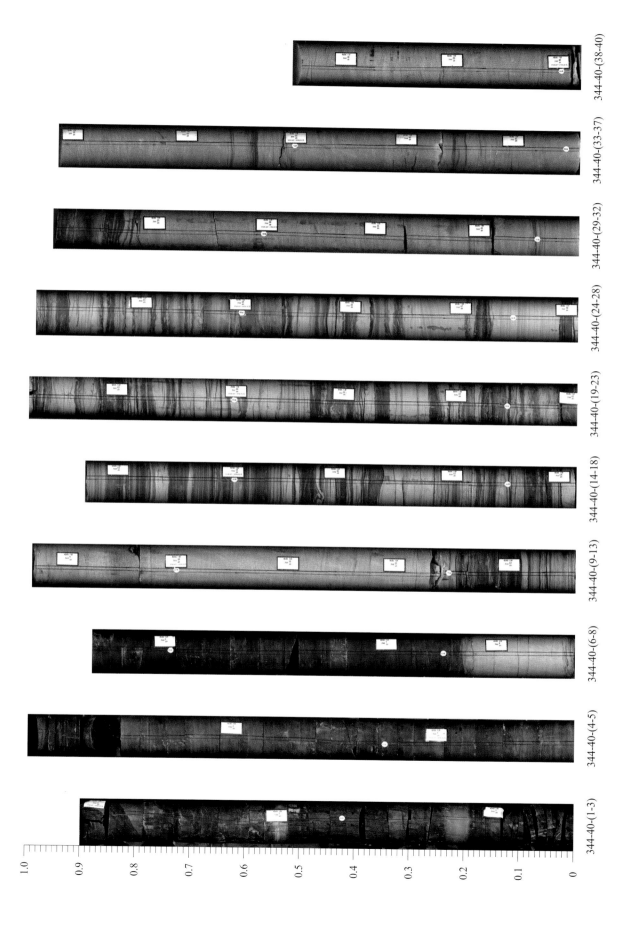

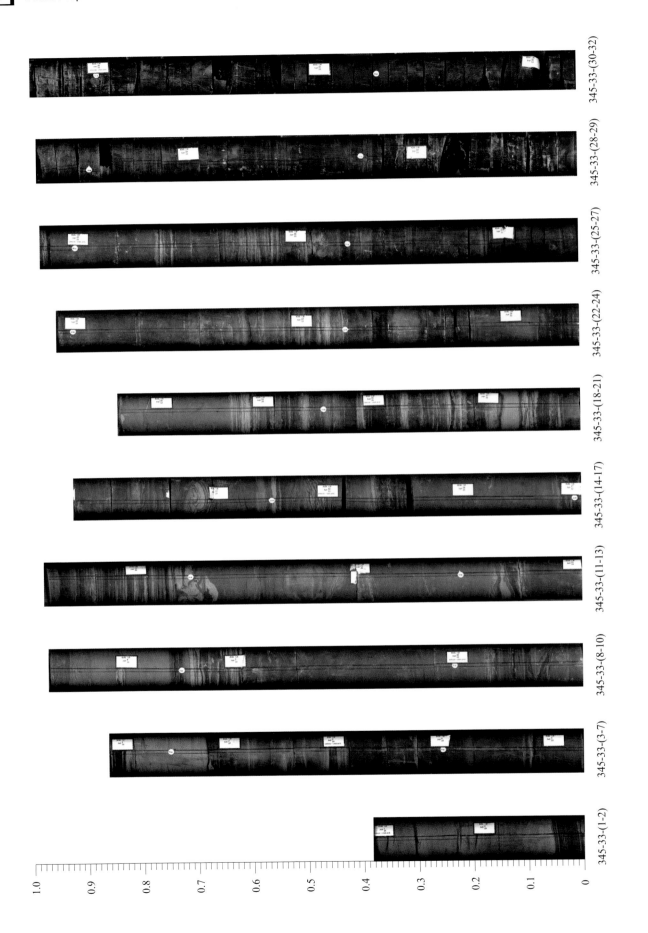

Section 5 SK-1 Core Description and Core Photographs | 629

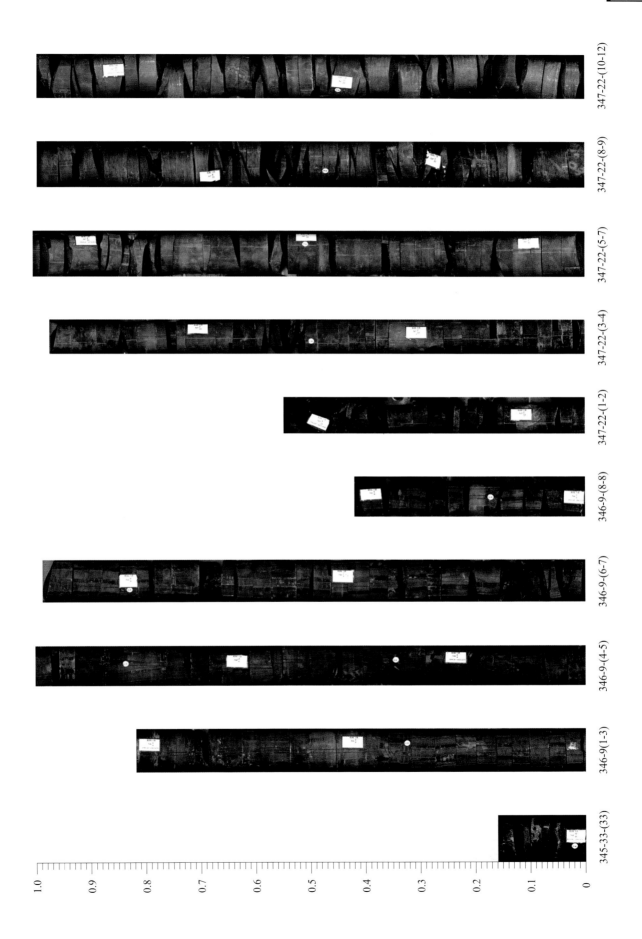

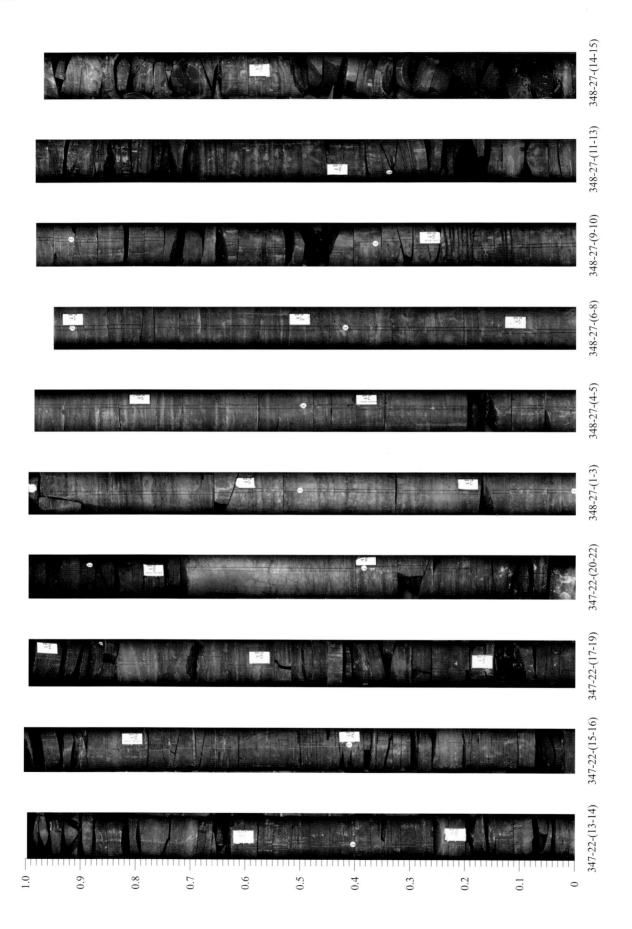

Section 5 SK-1 Core Description and Core Photographs | 631

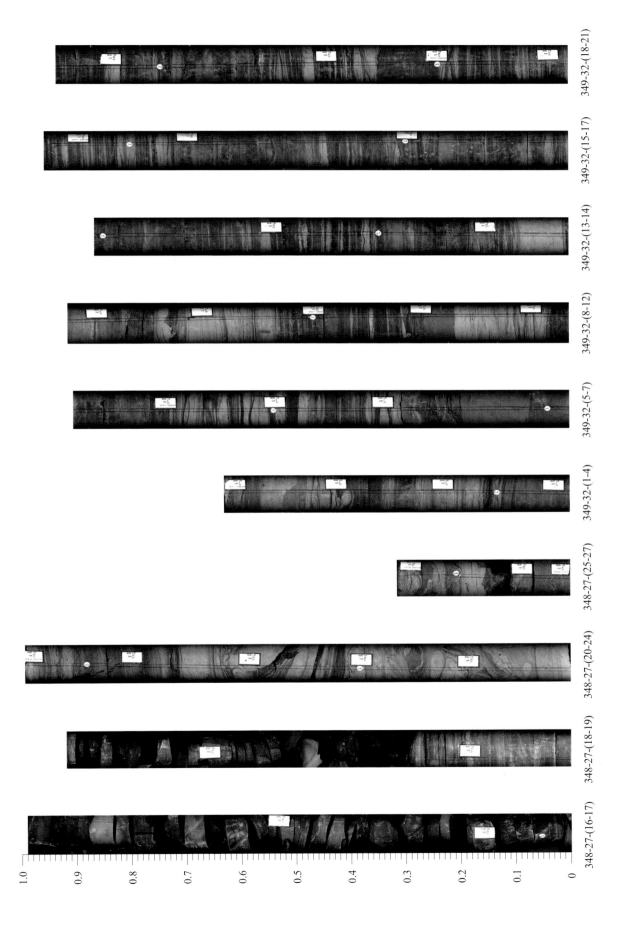

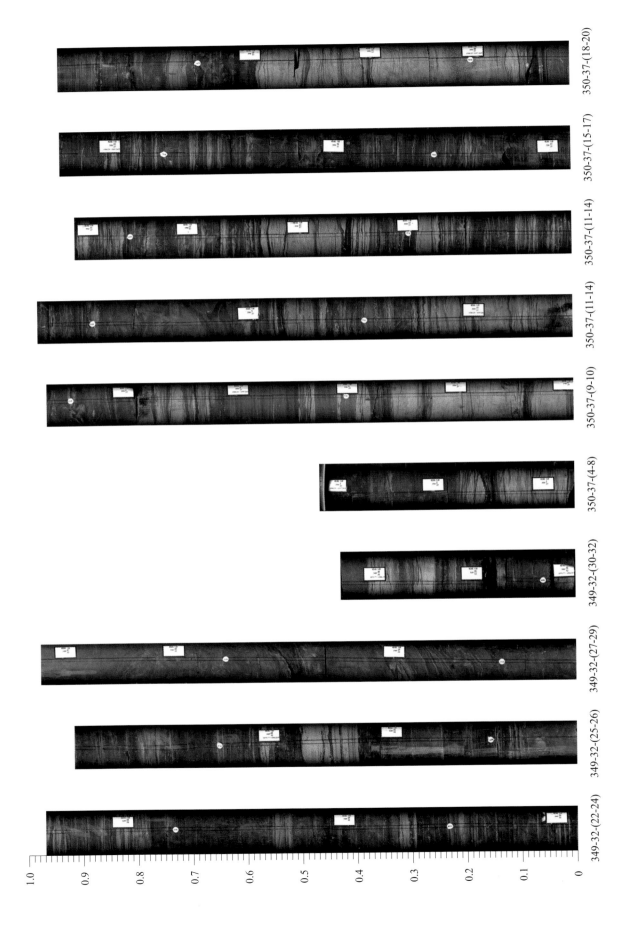

Section 5 SK-1 Core Description and Core Photographs

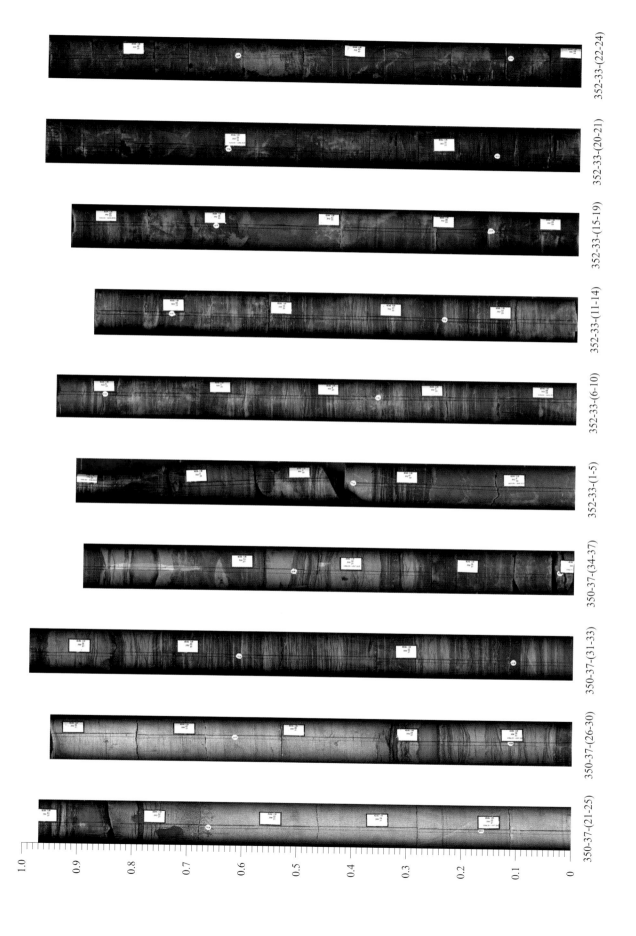

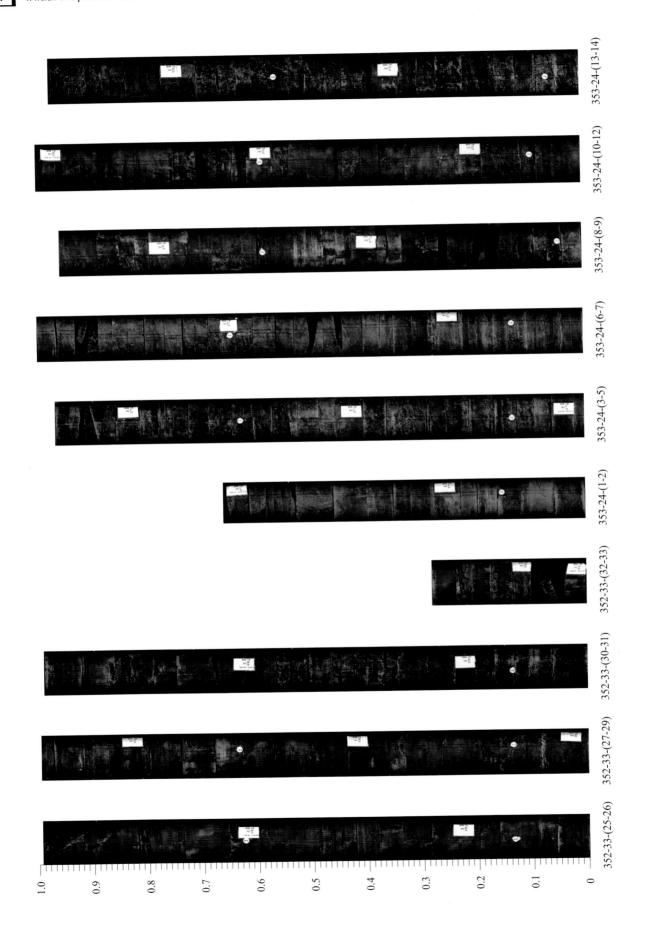

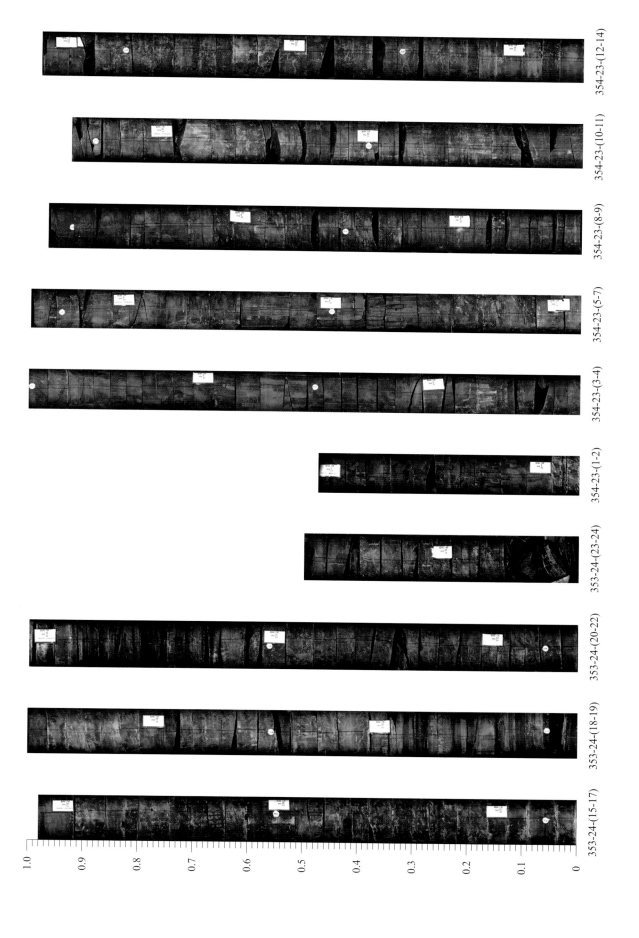

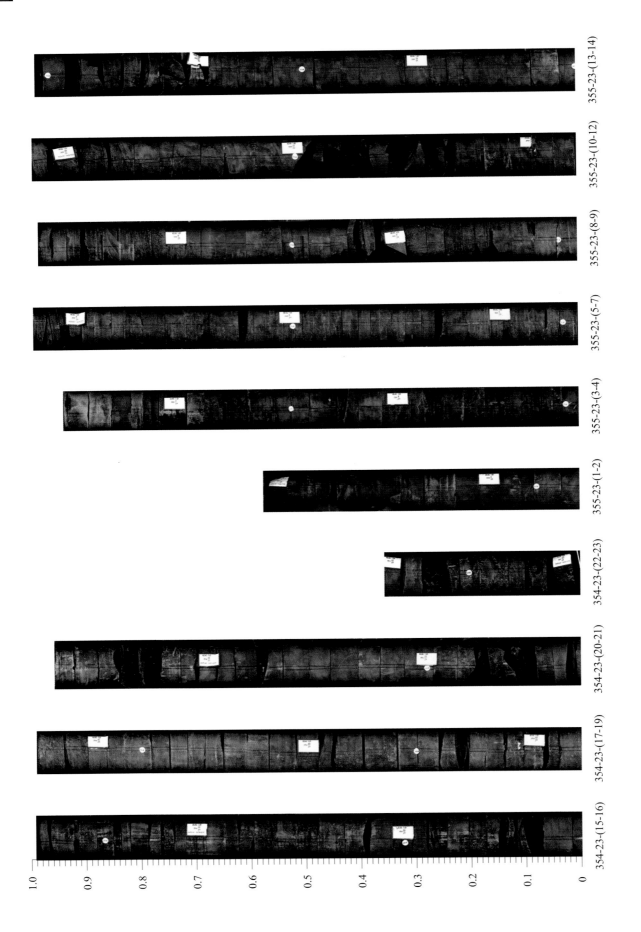

Section 5 SK–1 Core Description and Core Photographs | 637

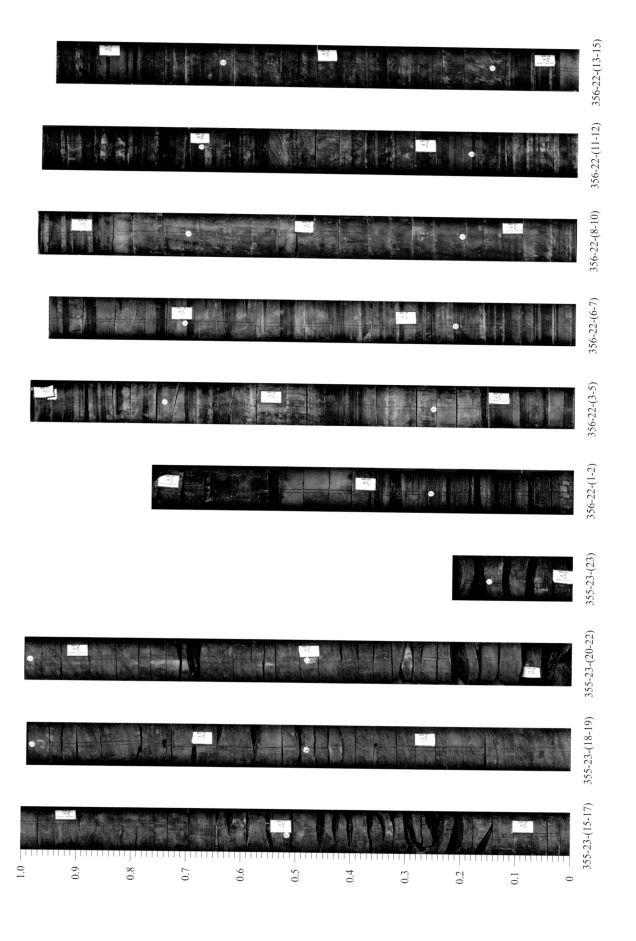

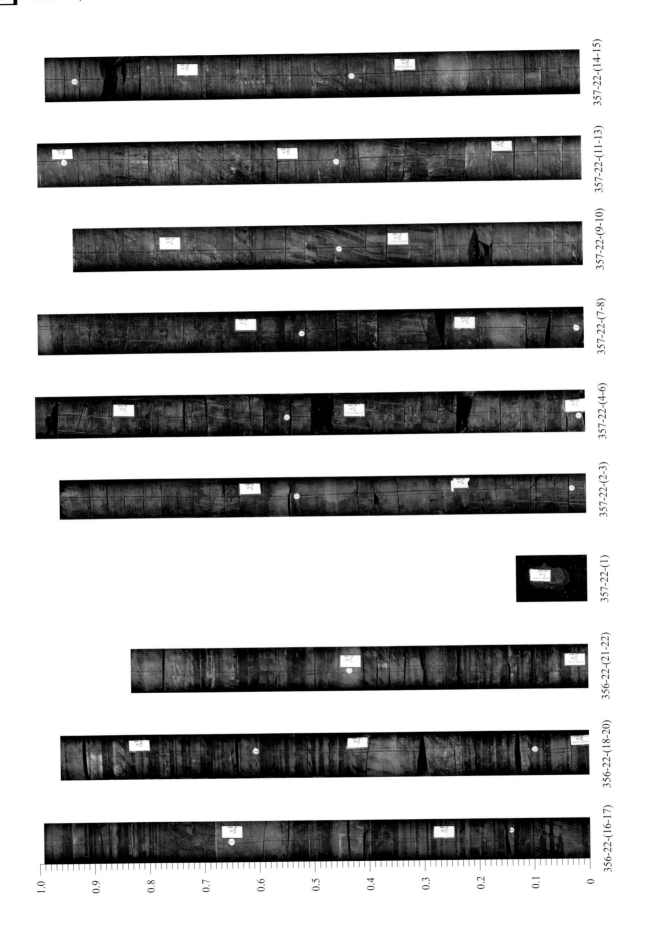

Section 5 SK-1 Core Description and Core Photographs

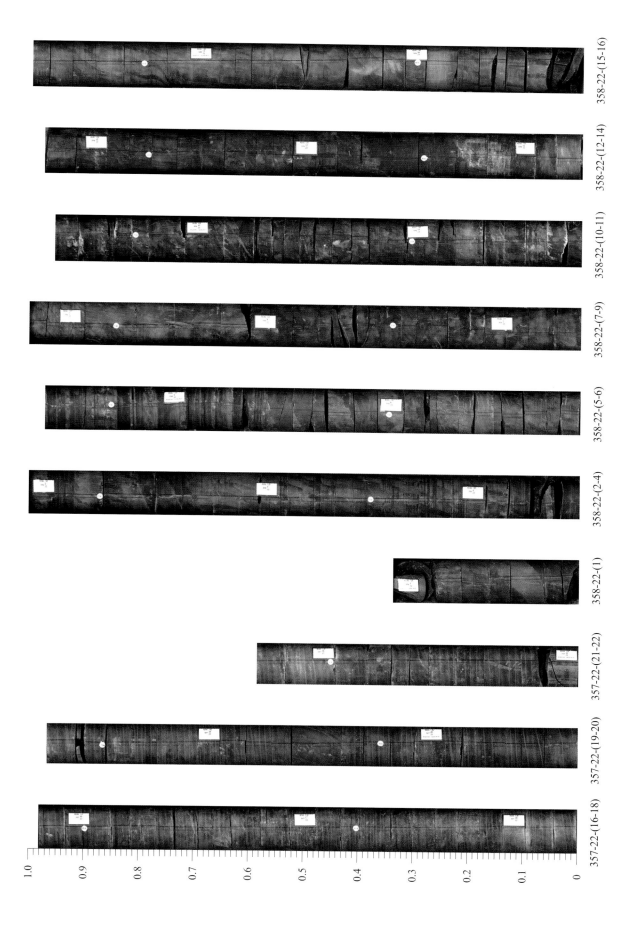

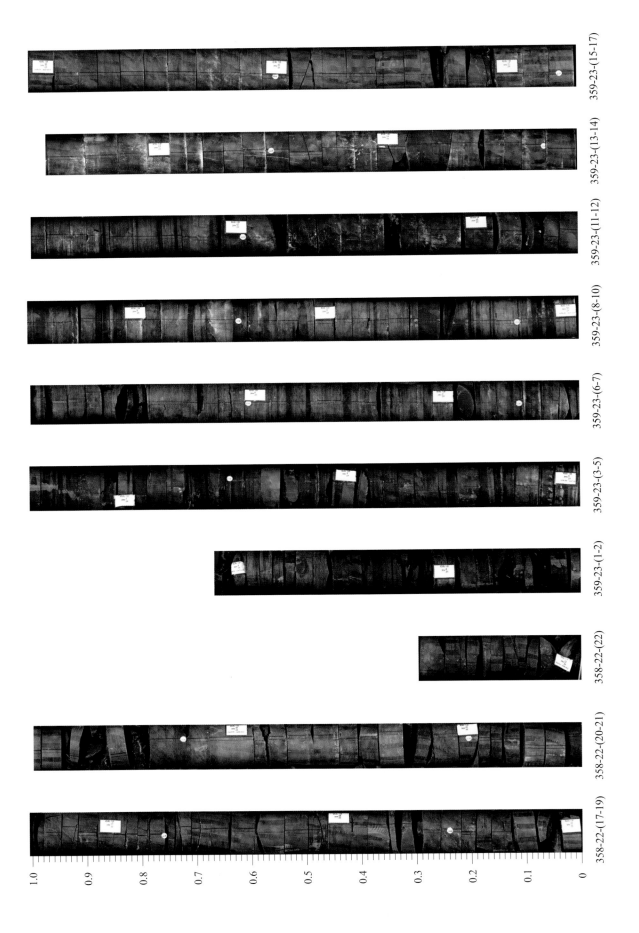

Section 5 SK-1 Core Description and Core Photographs 641

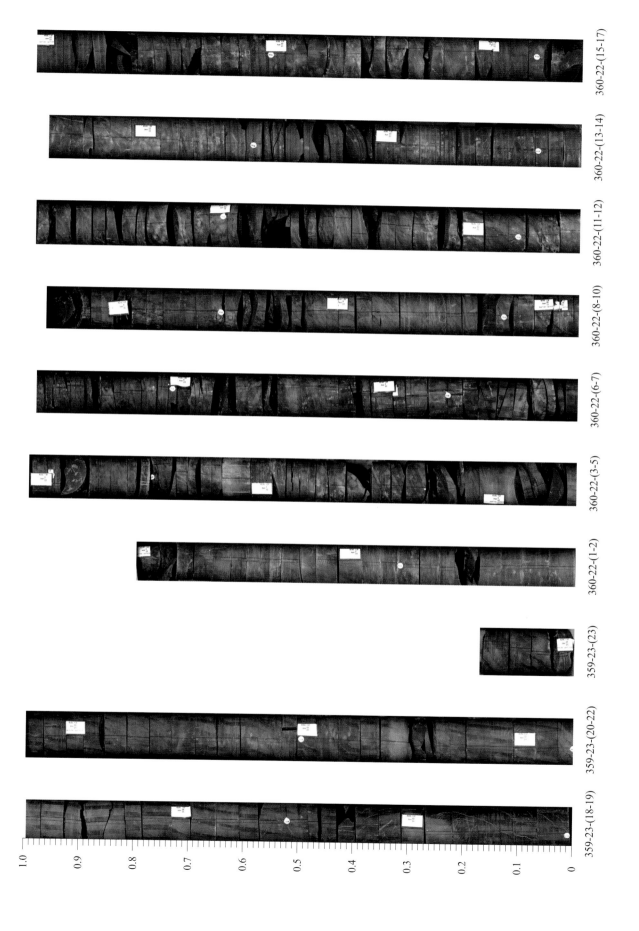

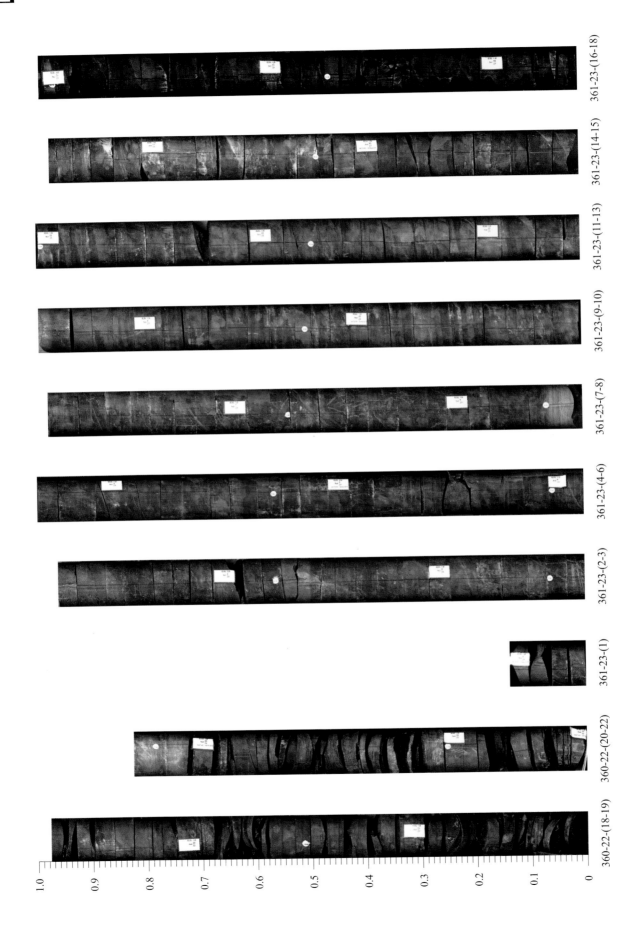

Section 5 SK-1 Core Description and Core Photographs

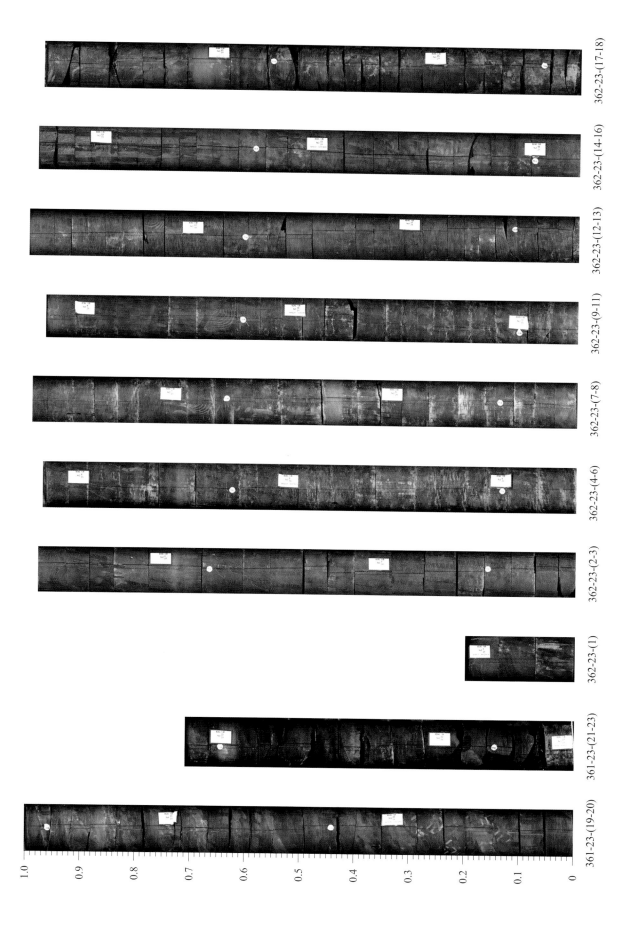

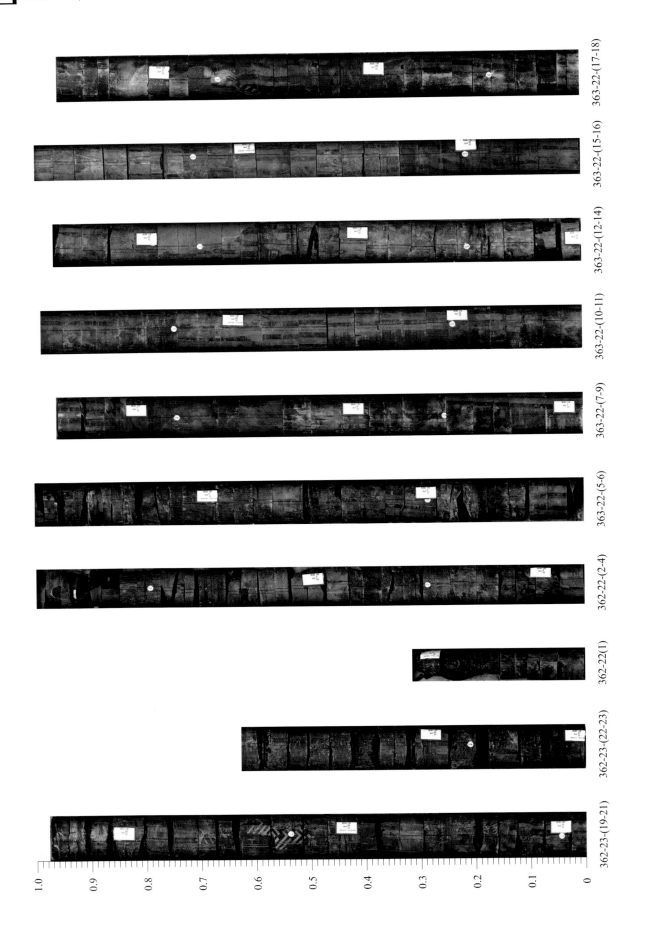

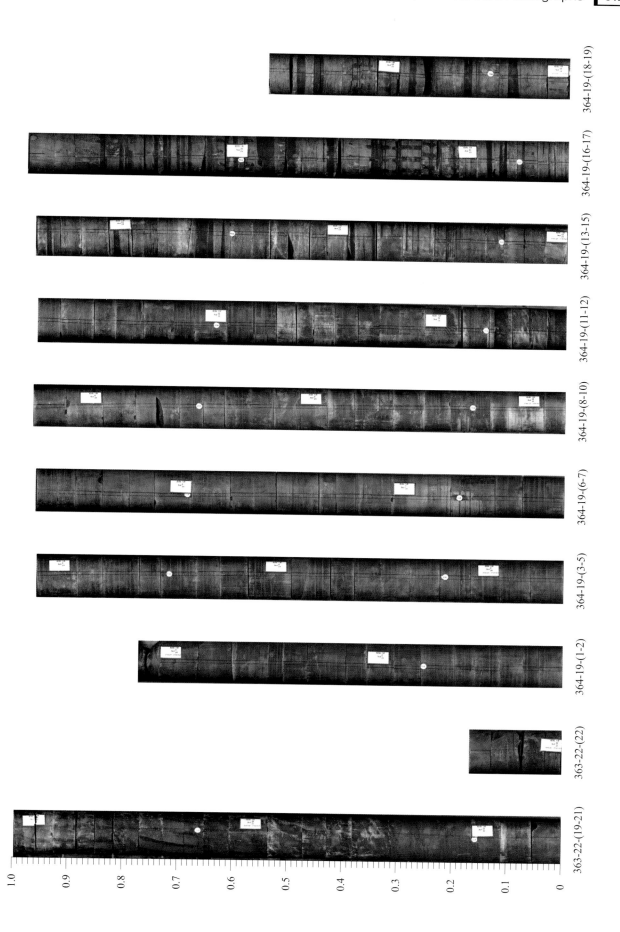

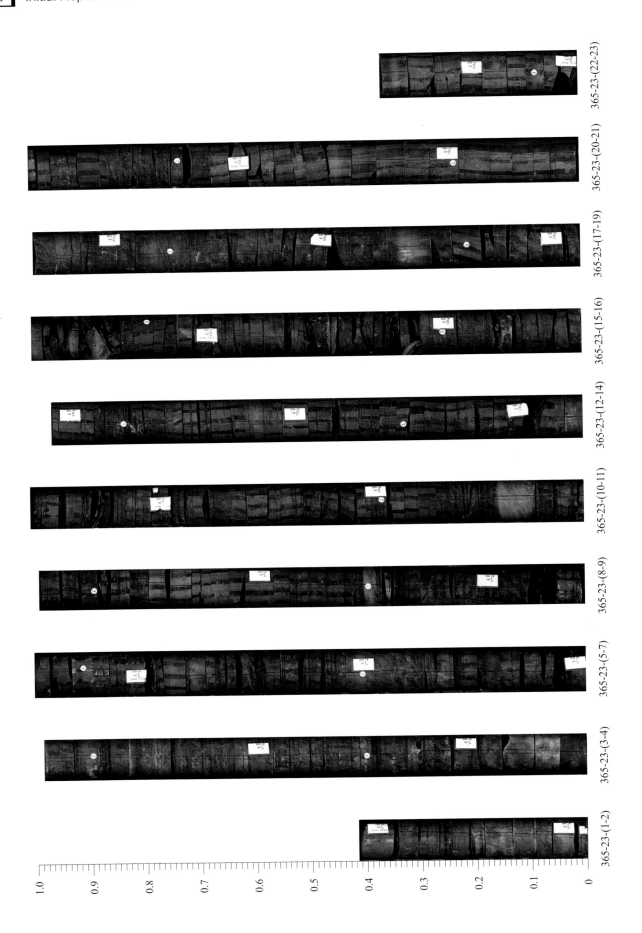

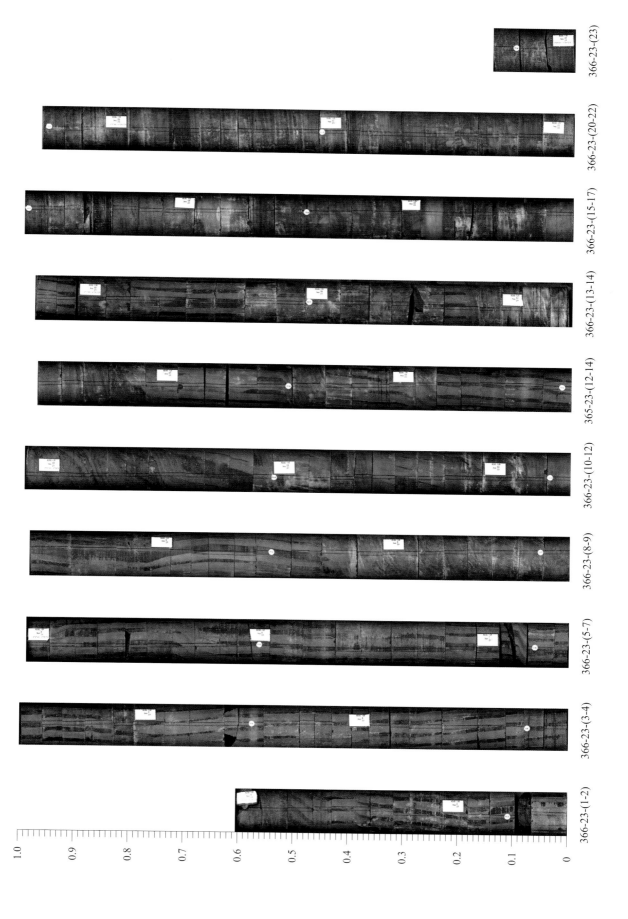

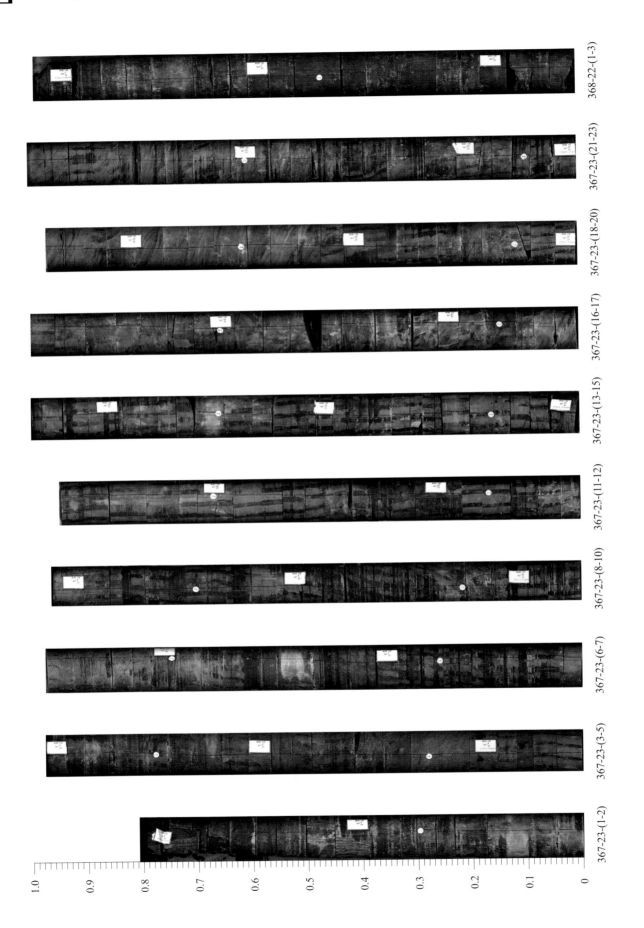

Section 5 SK-1 Core Description and Core Photographs

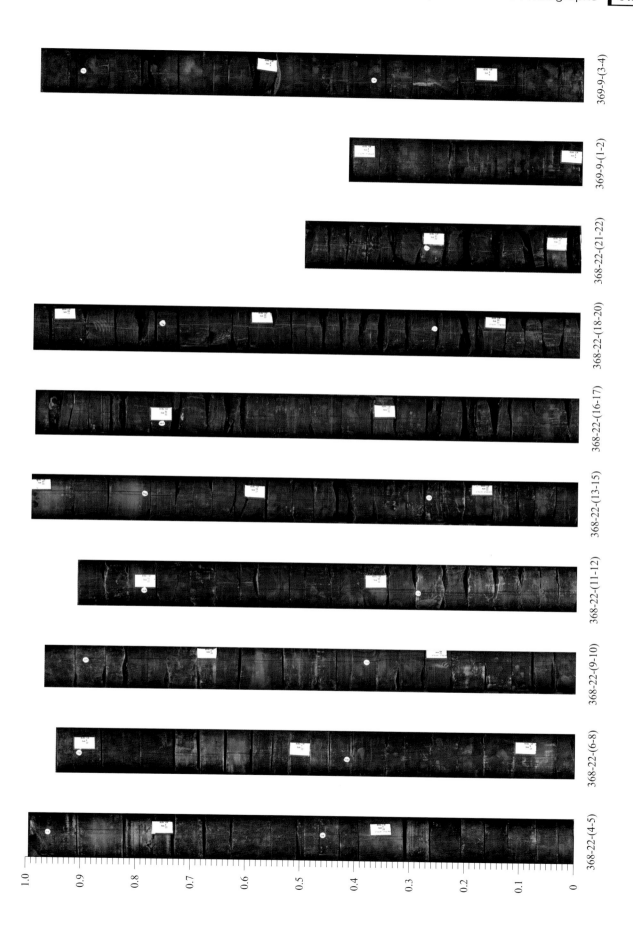

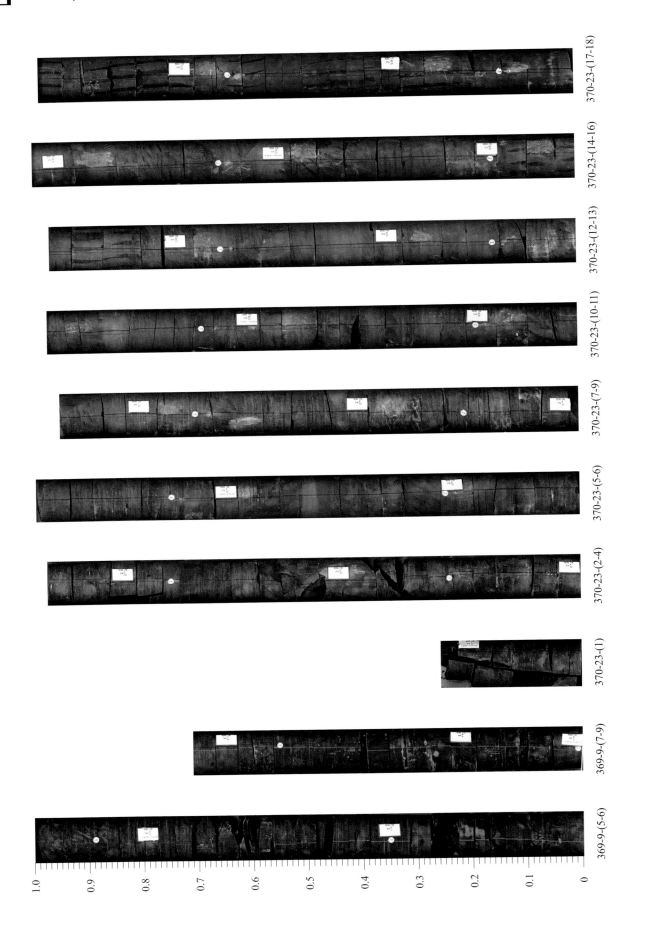

Section 5 SK-1 Core Description and Core Photographs | 651

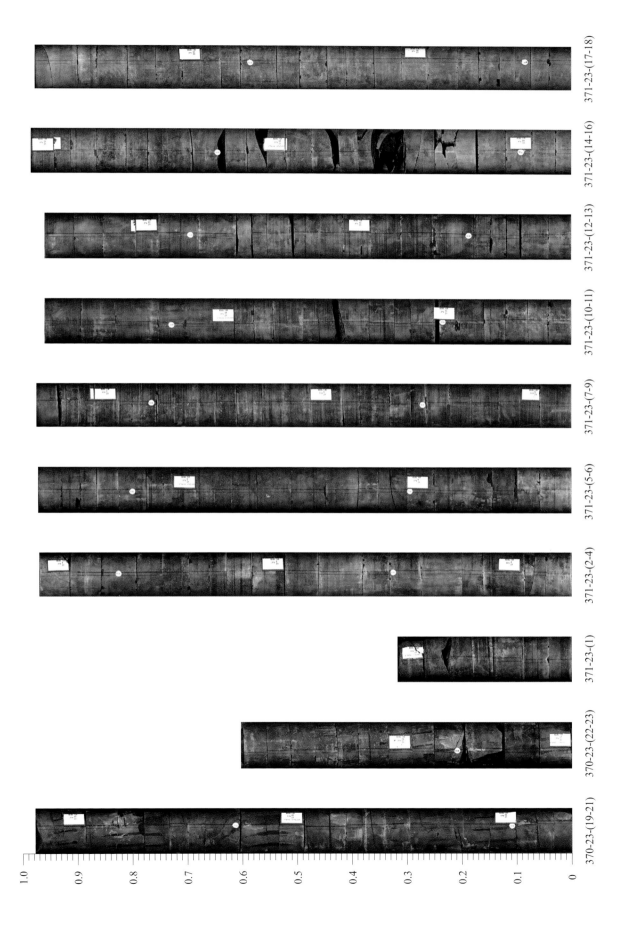

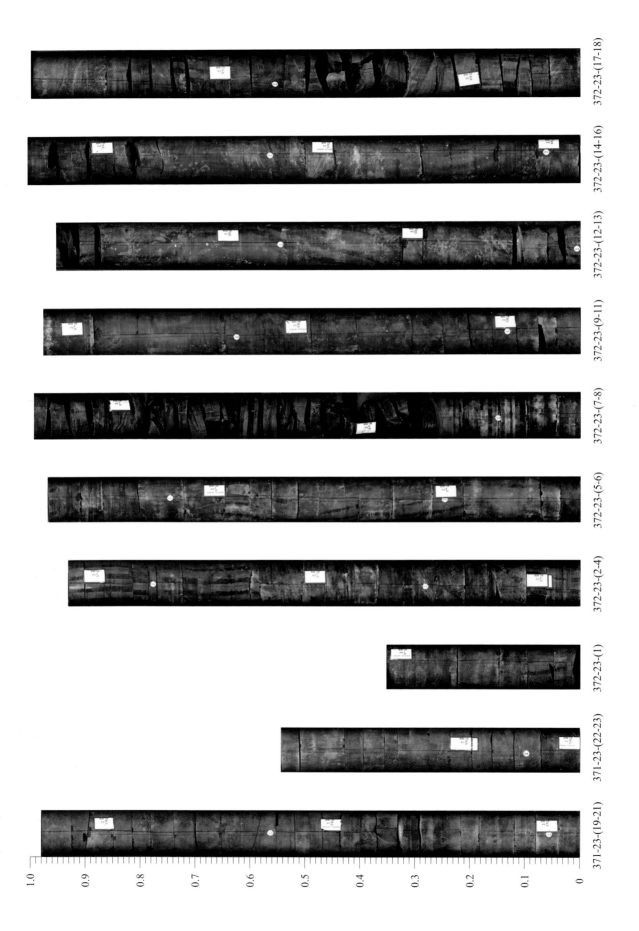

Section 5 SK-1 Core Description and Core Photographs

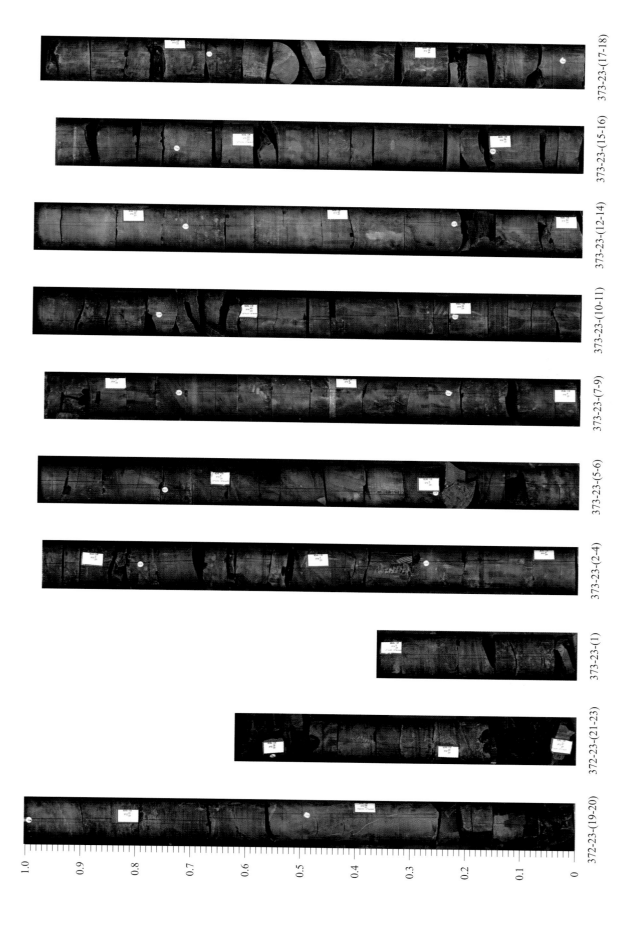

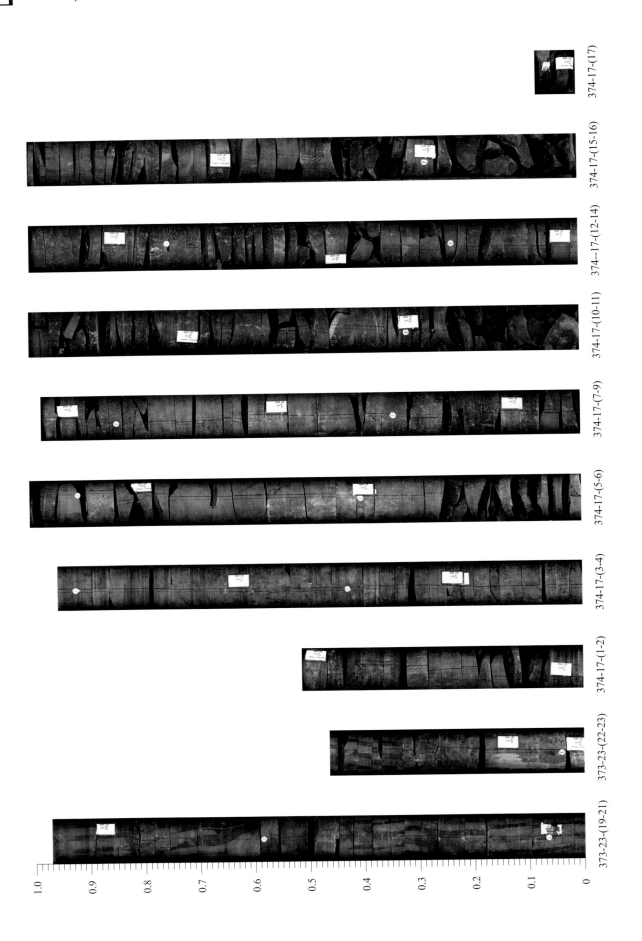

Section 5 SK-1 Core Description and Core Photographs | *655*

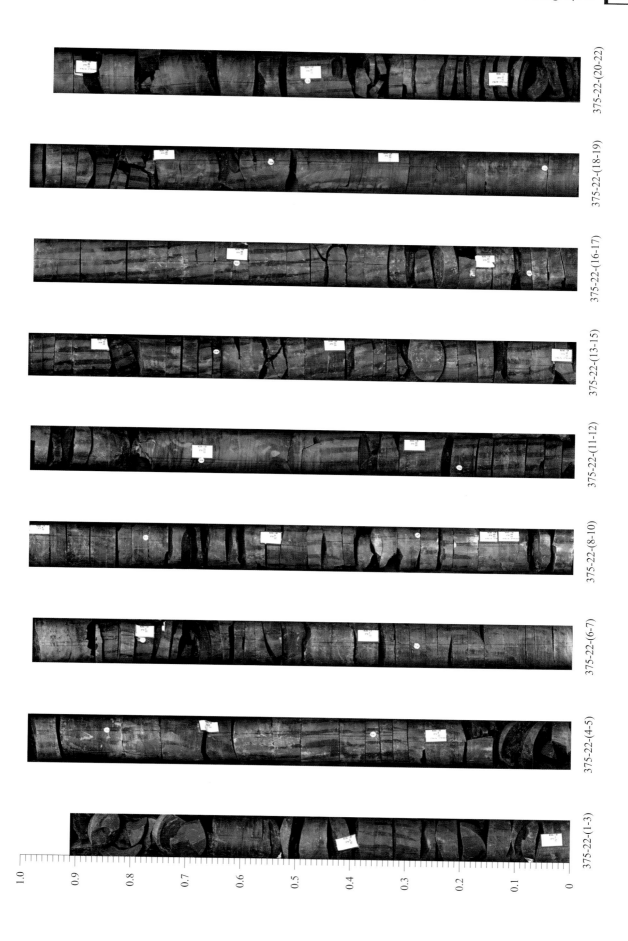

Reference

An Z S, Wang P, Shen J, et al. 2006. Geophysical survey on the tectonic and sediment distribution of Qinghai Lake Basin. Science in China Serifs D(earth Sciences). 49(8): 332—341

Argenio d'B, Fischer A G, Premoli-Silva I, et al. 2004.Approaches andcase histories. Society for Sedimentary Geology (SEPM) Special Publication, Cyclostratigraphy, 81: 1—311

Arthur M A, Dean W E, Pratt L M. 1988.Geochemical and climatic effects of increased marine organic carbon burial at the Cenomanian/Turonian boundary. Nature, 335: 714—717

Barron E, Gomez, J J, Goy A, Pieren A P. 2006. The Triassic-Jurassic boundary in Asturias (northern Spain): palynological characterisation and facies. Review of Palaeobotany and Palynology, 138:187—208

Batten D J. 1996. Upper Jurassic and Cretaceous miospores. In: Jansonius J, McGregor D C (eds). Palynology, Principles and Applications. American Association of Stratigraphic Palynologists Foundation 2. 807—830

Bice K L, Birgel D, Meyers P A, Dahl K A, Hinrichs K U, Norris R D. 2006. A multiple proxy and model study of Cretaceous upper ocean temperatures and atmospheric CO_2 concentrations. Paleoceanography, 21: PA2002

Bice K L, Huber B T, Norris R D. 2003. Extreme polar warmth during the Cretaceous greenhouse? Paradox of the late Turonian delta O-18 record at Deep Sea Drilling Project Site 511. Paleoceanography,18(2):1031—1041

Bornemann A, Pross J, Reichelt K, Herrle J O, Hemleben C, Mutterlose J. 2005. Reconstruction of short-term palaeoceanographic changes during the formation of the Late Albian "Niveau Breistroffer" black shales (Oceanic Anoxic Event 1d, SE France). Journal of the Geological Society, 162: 623—639

Boucot A J, Chen X, Scotese C R, Fan J X. 2009. Reconstruction of Global Paleoclimate in Phanerozoic. Beijing: Science Press (in Chinese)

Broecker W S. 1997. Thermohaline circulation, the achilles heel of our climate system: will man-made CO_2 upset the current balance? Science, 278(5343):1582—1588

Buonocunto F P, Sprovieri M, Bellanca A, et al. 2002. Cyclostratigraphy and high-frequency carbon isotope fluctuations in Upper Cretaceous shallow-water carbonates, southern Italy. Sedimentology, 49: 1321—1337

Cande S C, Kent D V. 1995. Revised calibration of the geomagnetic polarity timescale for the Late Cretaceous and Cenozoic. Journal of Geophysical Research, 100:6093—6095

Caron M, Robaszynski F, Amedro F, et al. 1999. Estimation de la durée de lévénement anoxique global au passage Cénomanien/Turonien: Approche cyclostratigraphique dans la formation Bahloul en Tunisie central. Bulletin de la Société Géologique de France, 170(2): 145—160

Chamberlain C P, Wan X Q, Graham S A, Carroll A R, Doebbert A C, Sageman B B, Blisniuk P, Kent-Corson M L, Wang Z, Wang C S. 2013. Stable isotopic evidence for climate and basin evolution of the Late Cretaceous Songliao Basin,

China. Palaeogeography, Palaeoclimatology, Palaeoecology, 385: 106—124

Chen P J. 1987. Cretaceous paleogeography of China. Palaeogeography, Palaeoclimatology, Palaeoecology, 59: 49—56

Chen P J, Chang Z L. 1994. Nonmarine Cretaceous stratigraphy of eastern China. Cretaceous Research 5: 245—257

Chen P J. 1997. Coastal Mountains of Se China, desertization and saliniferous lakes of Central China during the Upper Cretaceous. Journal of Stratigraphy, 21: 203—213 (in Chinese with English abstract)

Chen P J, Dong Z M, Zhen S N. 1999. An exceptionally well preserved theropod dinosaur from the Yixian Formation of China. Nature, 391: 147—152

Chen P J. 2003. Cretaceous biostratigraphy of China, In: Zhang W T, Chen P J, Palmer A R (eds). Biostratigraphy of China. Beijing: Science Press. 465—524

Cheng R H, Wang G D, Wang P J, Gao Y F.2007. Microfacies of deep-water deposits and forming models of the Chinese Continental Scientific Drilling — SKII. Acta Geologica Sinica, 81(6):1026—1032

Cheng R H, Wang G D, Wang P J. 2008. Sedimentary cycles of the cretaceous Quantou-Nenjiang Formations and milankovitch cycles of the South Hole of the Slcore- I in the Songliao Basin. Acta Geologica Sinica, 82(1):55—64

Cheng R H, Wang G G, Wang P J, et al. 2009. Description of Cretaceous sedimentary sequence of the Yaojia Formation recovered by CCSD-SK-Is borehole in Songliao Basin: lithostratigraphy, sedimentary facies and cyclic stratigraphy. Earth Science Frontiers, 16(2):140—151

Clarke L J, Jenkyns H C. 1999. New oxygen isotope evidence for longterm Cretaceous climatic change in the Southern Hemisphere.Geology, 27(8):699—702

Cohen A, Coe A, Harding L, Schwark, L.2004. Osmium isotope evidence for the regulation of atmospheric CO_2 by continental weathering. Geology, 32:157—160

Colman S M, Yu S Y, An Z, Shen J, Henderson A C G. 2007. Late Cenozoic climate changes in China's western interior: a review of research on Lake Qinghai and comparison with other records. Quaternary Science Reviews, 26(17-18):2281—2300

Coolen M J L, Overmann J. 2007. 217,000-year-old DNA sequences of green sulfur bacteria in Mediterranean sapropels and their implications for the reconstruction of the paleoenvironment. Environmental Microbiology, 9(1): 238—249

Daniel L, Albritton, D L, Allen, M R. 2001. Summary for Policymakers, IPCC Third Assessment Report 2 Climate Change: the Scientific Basis. WMO &UNEP, 201

DeConto R M, Pollard D. 2003. Rapid Cenozoic glaciation of Antarctica induced by declining atmospheric CO_2. Nature, 421(6920): 245—249

DeConto R M, Hay W W, Thompson S L, Bergengren J. 1999. Late Cretaceous climate and vegetation interactions: Cold continental interior paradox. In: Barrera E, Johnson C C (eds). Evolution of the Cretaceous Ocean-Climate System. Geological Society of America Special Paper 332, Boulder, Colorado. 391-406

Demaison G J, Moore G T. 1980. Anoxic environments and oil source bed genesis. Organic Geochemistry, 2(1): 9—31

Demaison G, Holck A J J, Jones R W, Moore G T. 1984. Predictive source bed stratigraphy; a guide to regional petroleum occurrence. Congres Mondial du Petrole, 11(2):17

Deng C L, He H Y, Pan Y X, Zhu R X. 2013. Chronology of the terrestrial Upper Cretaceous in the Songliao Basin, northeast Asia. Palaeogeography Palaeoclimatology Palaeoecology, 385: 44—54

Deng S W, Chen F. 1998. Paleobotanological assemblage and age in Early Cretaceous strata of NE China. Petroleum Exploration and Development 25(1): 35—38 (in Chinese with English Abstract)

Detlev R, Vlad G, Heinz H, Andrea F. 2004. The AIG10 drilling project (Aigion, Greece): interpretation of the litholog in the context of regional geology and tectonics. Comptes Rendus Geosciences, 336: 415—423

Ding R, Shu P, Ji X, et al. 2007. SHRIMP zircon U-Pb age and geological meaning of reservoir volcanic rocks in Qingshen Gas Field of the Songliao Basin, NE China. Journal of Jilin University (Earth Science Edition) 37(3):525—530

Du X B, Xie X N, Lu Y C, Ren J Y, Zhang S, Lang P L, Cheng T, Su M, Zhang C. 2011. Distribution of continental red paleosols and their forming mechanisms in the Late Cretaceous Yaojia Formation of the Songliao Basin, NE China. Cretaceous Research, 32: 244—257

Einsele G.2004. Sedimentary Basins. Berlin Heidelberg Neu York: Springer,

Erba E, Bartolini A, Larson R L. 2000. Valanginian Weissert oceanic anoxic event. Geology, 32: 149—152

Fang D J, Wang Z L, Jin G H, Gao R Q, Ye D Q, Xie J L. 1990. Cretaceous magnetostratigraphy in the Songliao Basin, China. Science in China (Series B), 33:246—256

Feng Z Q, Jia C Z, Xie X N, et al. 2010. Tectonostratigraphic units and stratigraphic sequences of the nonmarine Songliao basin, northeast China.Basin research, 22:79—95

Feng Z Q, Zhang S, Crossw T A, et al. 2010. Lacustrine turbidite channels and fans in the Mesozoic Songliao Basin,China. Basin research,22:96—107

Fiet N, Quidelleur X, Parize O, et al. 2006. Lower Cretaceous stage durations combining radiometric data and orbital chronology: towards a more stable relative time scale? Earth and Planetary Science Letters, 246: 407—417

Gale A S, Hardenbol J, Hathway B, et al. 2002. Global correlation of Cenomanian (Upper Cretaceous) sequences: evidence for Milankovitch control on sea level. Geology, 30(4): 291—294

Gallet Y, Hulot, G. 1997. Stationary and nonstationary behaviour within the geomagnetic polarity time scale. Geophysical Research Letters, 24:1875—1878

Gao Y F, Wang P J, Cheng R H, et al. 2009. Description of Cretaceous sedimentary sequence of the First Member of the Qingshankou Formation recovered by CCSD-SK-I s borehole in Songliao Basin: lithostratigraphy, sedimentary facies, and cyclic stratigraphy. Earth Science Frontiers,16(2):314—323

Gao R, Cai X. 1997.Hydrocarbon Formation Conditions and Distribution Rules in the Songliao Basin. Beijing: Petroleum Industry Press (in Chinese)

Gao R Q, Zhang Y, Cui T C, 1994. Cretaceous Oil and Gss Strata of Songliao Basin. Beijing: Petroleum Industry Press (in Chinese)

Gao R Q, Zhao C B, Qiao X Y, Zheng Y L, Yan F Y, Wan C B, 1999. Cretaceous Oil Strata Palynology from Songliao Basin. Beijing:Geological Publishing House (in Chinese)

Gordon W A.1973. Marine life and ocean surface currents in the Cretaceous. Journal of Geology, 81: 269—284

Gradstein F M, Ogg J G, Smith A G. 2004. A Geologic Time Scale 2004. Cambridge: Cambridge University Press. 589

Haq B U, Hardenbol J, Vail P R. 1987. Chronology of fluctuating sea levels since the Triassic. Science, 235(4793):1156—

1167

Hay W W. 2008. Evolving ideas about the Cretaceous climate and ocean circulation. Cretaceous Research, 29: 725—753

Hay W W. 2011. Can humans force a return to a 'Cretaceous' climate? Sedimentary Geology, 235:5—26

Hays J D.1976. Variations in the Earth's Orbit: pacemaker of the ice ages. Science, 194: 1121—1132

He H, Deng C, Wang P, Pan Y, Zhu R. 2012. Toward age determination of the termination of the Cretaceous Normal Superchron. Geochemistry Geophysics Geosystems, 13: Q02002. doi:10.1029/2011GC003901

Heimhofer U, Hochuli P A, Burla S, Dinis J M L, Weissert H. 2005. Timing of Early Cretaceous angiosperm diversification and possible links to major paleoenvironmental change. Geology, 33(2):141—144

Hennebert M, Dupuis C. 2003. Use of cyclostratigraphy to build a high-resolution time-scale encompassing the Cretaceous-Palaeogene boundary in the Ain Settara section (Kalaat Senan, Central Tunisia). Geobios, 36: 707—718

Herrle J O, Pross J, Friedrich O K, Wang C S, et al. 2008. Scientific drilling of the Terrestrial Cretaceous Songliao Basin. Scientific Drilling, 6:60, 61

Hu X M, Wang C S, Scott R W, Wagreich M, Jansa L. 2009. Cretaceous Oceanic Red Beds: Stratigraphy, Composition, Origins and Paleoceanographic and Paleoclimatic Significance. SEPM Special Publication No. 91 (Society for Sedimentary Geology). 13—33

Huang F T, Huang Q H. 1998. Rhythm of geological events and interaction of different geospheres in Mesozoic of Songliao Basin. Petroleum Exploration and Development, 25(5):86—89 (in Chinese with English Abstract)

Huang Q H. 2007. Upper Cretaceous Stratigraphy and Micropaelontological biotas in Songliao Basin. PhD Thesis Chinese Academy of Geosciences, Beijing (in Chinese with English abstract)

Huang Q H, Huang F T, Hou Q J. 1999. The Late Mesozoic bio-evolution and environmental changes in Songliao basin. Petroleum Exploration and Development, 26: 1—5 (in Chinese with English Abstract)

Hubbard R N L B, Boulter M C. 1983. Reconstruction of Palaeogene climate from palynological evidence. Nature, 301: 147—150

Huber B T, Hodell D A, Hamilton C P. 1995. Middle-Late Cretaceous climate of the southern high-latitudes-stable isotopic evidence for minimal equator-to-pole thermal-gradients. Geological Society of America Bulltin, 107(10): 1164—1191

Huber B T, Norris R D, MacLeod K G. 2002. Deep-sea paleotemperature record of extreme warmth during the Cretaceous. Geology, 30(2):123—126

Hullot G, Gallet Y. 2003. Do superchrons occur without any palaeomagnetic warning? Earth and Planetary Science Letters, 210: 191—201

Inagaki F, Okada H, Tsapin A I, Nealson K H. 2005. The Paleome: a sedimentary genetic record of past microbial communities. Astrobiology, 5(2): 141—153

Jarzen D M, Norris G. 1975. Evolutionary significance and botanical relationships of Cretaceous angiosperm pollen in the western Canadian Interior. Geoscience and Man, 25: 47—60

Ji Q, Li H Q, Bowe L M, et al. 2004. Early Cretaceous Archaefructus eoflora sp nov with bisexual flowers from Beipiao, Western Liaoning, China. Acta Geologica Sinica, 78(4):883—896

Ji Q, Currie P J, Norell M A, et al. 1998. Two feathered dinosaurs from northeastern China. Nature, 393:753—761

Ji Q, Norell M A, Gao K Q et al. 2001. The distribution of integumentary structures in a feathered dinosaur. Nature, 410: 1084—1088

Ji Q, Luo Z X, Ji Shu'an.1999. Chinese triconodont mammal and mosaic evolution of the mammalian skeleton. Nature, 398: 326—330

Ji Q, Luo Z X, Yuan C X, et al. 2002. The earliest known eutherian mammal. Nature, 416: 816—822

Jia C Z.2007. The Characteristics of intra-continental deformation and hydrocarbon distribution controlled by the himalayan tectonic movements in China. Earth Science Frontiers, 14(4): 96—104

Jia J, Wang P, Wan X.2008. Chronostratigraphy of the Yingcheng Formation in the Songliao Basin, Cretaceous, NE China. Geological Review, 54(4):439—448

Kalkreuth W D., McIntyre D J, Richardson R J H, 1993. The geology, petrography and palynology of Tertiary coals from the Eureka Sound Group at Strathcona Fiord and Bache Peninsula, Ellesmere Island, Arctic Canada. International Journal of Coal Geology, 24:75—111

Kiessling W, Claeys P. 2001. A geographic database approach to the KT boundary. In: Buffetaut E, Koeberl C (eds). Geological and Biological Effects of Impact Events. Berlin: Springer. 83—140

Klinger H C, Kakabadze M V. Kennedy W J. 1984. Upper Barremian (Cretaceous) heteroceratid ammonites from South Africa and the Caucasus and their palaeobiogeographic significance. Journal of Molluscan Studies, 50:43—60

Kuhnt W, Luderer F, Nederbragt S, et al. 2005. Orbital-scale record of the Late Cenomanian–Turonian oceanic anoxic event (OAE-2) in the Tarfaya Basin (Morocco). International Journal of Earth Sciences (Geol Rundsch), 94: 147—159

Kuypers M M, Pancost R D, Sinninghe D J S. 1999. A large and abrupt fall in atmospheric CO_2 concentration during Cretaceous times. Nature, 399:342—345

Larsson L M, Vajda V, Dybkjar K. 2010. Vegetation and climate in the latest Oligocene-earliest Miocene in Jylland, Denmark. Review of Palaeobotany and Palynology, 159: 166—176

Laskar J, Robutel P, Joutel F, et al. 2004. A long-term numerical solution for the insolation quantities of the Earth. Astronomy and Astrophysics. 428: 261—285

Latta D K, Anastasio D J, Hinnov L A, et al. 2006. Magnetic record of Milankovitch rhythms in lithologically noncyclic marine carbonates. Geology, 34(1): 29—32

Li D S.1996. Basic characteristics of oil and gas basins in China. Journal of Sourheart Asian Earth Sciences, 13: 299—304

Li G, Battent D J. 2004. Revision of the conchostracan genera Cratostracus and Porostracus from Cretaceous deposits in north-east China. Cretaceous Research, 25(6):919—926

Li G, Batten D J. 2005. Revision of the conchostracan genus Estherites from the Upper Cretaceous Nenjiang Formation of the Songliao Basin and its biogeographic significance in China. Cretaceous Research,26(6): 920—929

Li H, Liu Y H, Luo N, et al. 2006. Biodegradation of benzene and its derivatives by a psychrotolerant and moderately haloalkaliphilic Planococcus sp. strain ZD22. Research Microbiology, 157(7):629—636

Li W, Li J. 2005. Albian palynological assemblage from the borehole YU-302 in Yushu--with focus on the age of the denglouku formation in Songliao Basin. Acta Palaeontologica Sinica, 44(2):209—228

Li W. 2001.Palynoflora from the Quantou Formation of Songllao Basin, NE China and its bearing on the upper-lower cretaceous boundary. Acta Palaeontologica Sinica, 40(2):153—176

Lini A, Weissert H, Erba E.1992. The Valanginian carbon isotope event: a first episode of greenhouse climate conditions during the Cretaceous. Terra Nova, 4: 374—384

Liu F L, Xu Z Q, Katayama I, et al. 2001. Mineral inclusion in zircons of para- and orthogneiss from pre-pilot drillhole CCSD-PP1, Chinese Continental Scientific Drilling Project. Lithos, 59:199—215

Liu Z C, Chen Y, Yuan L W. 2001. The paleoclimate change of Qaidam Basin during the last 2.85 Ma recorded by Gamma-ray logging. Science in China, 44(2): 133—145

Liu G W, Leopold E B. 1994. Climatic comparison of Miocene pollen floras from northern East-China and south-central Alaska, USA. Palaeogeography, Palaeoclimatology, Palaeoecology 108:217—228

Liu G W, Leopold EB, Liu Y, Wang W M, Yu Z Y, Tong G B. 2002. Palynological record of Pliocene climate events in North China. Review of Palaeobotany and Palynology, 119:335—340

Locklair R E, Sageman B B.2008. Cyclostratigraphy of the Upper Cretaceous Niobrara Formation, western interior, U S A: a Coniacian–Santonian orbital timescale. Earth and Planetary Science Letters, 269: 540—553

McDonald I, Irvine G J, De Vos E, Gale A S, Reimold W U. 2006. Geochemical search for impact signatures in possible impact-generated units associated with the Jurassic-Cretaceous boundary in southern England and northern France. In: Cockell C, Koeberl C, Gilmour I (eds). Impact Studies – Biological Processes Associated with Impact Events. Heidelberg-Berlin-New York: Springer. 257—279

McFadden P L. 1991. Randomness, chaos, and the paleomagnetic record: implications for our understanding of the geodynamo. Geophysical and Astrophysical Fluid Dynamics, 60(1-4): 414, 415

McFadden P L, Merrill R T. 1984. Lower mantle convection and geomagnetism. Journal of Geophysical Research, 89: 3354—3362

McFadden P L, Merrill R T. 2000. Evolution of the geomagnetic reversal rate since 160 Ma: is the process continuous? J Geophys Res, 105: 28455—28460

Meyerhoff A A, Teichert C. 1971. Contiental drift, III : Late Paleozoic centers and Devonian-Eocene coal distribution. Journal of Geology, 79: 285—321

Milankovtich M. 1941. Kano der Erdbestrahhlung und seine Anwendung auf das Eiszeitenproblem. Academic Serbe, 133: 1—633

Miller K G, Wight J K, Fairbanks R D. 1991. Unlocking the ice house : Oligocene Miocene oxygen isotopes, eustacy, and margin erosion. J Geophys Res, 96:6829—6848

Nichols D J, Sweet A R. 1993. Biostratigraphy of Upper Cretaceous non-marine palynofloras in a northesouth transect of the Western Interior Basin. In: Caldwell W G E, Kauffman E G (eds). Evolution of the Western Interior Basin. Geological Association of Canada Special Paper, 39. 539—584

Norris G, Jarzen D M, Awai-Thorne B V. 1975. Evolution of the Cretaceous terrestrial palynoflora in western Canada. Geological Association of Canada, Special Paper 13. 333—364

Paillard D, Labeyrie L, Yiou P. 1996. Macintosh program performs time-series analysis. Eos, 77: 379

Pancost R D. 2004. The palaeoclimatic utility of terrestrial biomarkers in marine sediments. Marine Chemistry, 92: 239—261

Parrish J T. 1998. Interpreting Pre-Quaternary Climate from the Rock Record. New York :Columbia University Press.

Prokoph A, Agterberg F P.1999. Detection of sedimentary cyclicity and stratigraphic completeness by wavelet analysis: an application to late Albian cyclostratigraphy of the western Canada sedimentary basin. Journal of Sedimentary Research, 69(4): 862—875

Prokoph A, Thurow J. 2001. Orbital forcing in a Boreal Cretaceous epeiric sea: high-resolution analysis of core and logging data (Upper Albian of the Kirchrode I drill core - Lower Saxony Basin, NW Germany). Palaeogeography Palaeoclimatology Palaeoecology, 174(1-3):67—96

Prokoph A, Villeneuve M, Agterberg F P. 2001. Geochronology and calibration of global Milankovitch cyclicity at the Cenomanian-Turonian boundary. Geology, 29(6): 523—526

Puceat E, Lecuyer C, Reisberg L. 2005. Neodymium isotope evolution of NW Tethyan upper ocean waters throughout the Cretaceous. Earth and Planetary Science Letters, 236: 705—720

Ren J Y, Kensaku T, Li S T , et al. 2002. Late Mesozoic and Cenozoic rifting and its dynamic setting in Eastern China and adjacent areas. Tectonophysics, 344: 175—205

Rio D, Silva I P, Capraro L. 2003. The geological time scale and the Italian stratigraphic record. Episodes, 26(3): 259—263

Sageman B B, Rich J, Arthur M A, et al. 1997. Evidence for Milankovitch periodicities in Cenomanian–Turonian lithologic and geochemical cycles, Western Interior USA. Journal of Sedimentary Research, 67: 286—302

Schulte P, Alegret L, Arenillas I, Arz J A, Barton P J, Bown P R, Bralower T J, Christeson G L, Claeys P, Cockell C S, Collins G S, Deutsch A, Goldin T J, Goto K, Grajales-Nishimura J M, Grieve R A, Gulick S P, Johnson K R, Kiessling W, Koeberl C, Kring D A, MacLeod K G, Matsui T, Melosh J, Montanari A, Morgan J V, Neal C R, Nichols D J, Norris R D, Pierazzo E, Ravizza G, Rebolledo-Vieyra M, Reimold W U, Robin E, Salge T, Speijer R P, Sweet A R, Urrutia-Fucugauchi J, Vajda V, Whalen M T, Willumsen P S. 2010. The Chicxulub asteroid impact and mass extinction at the Cretaceous-Paleogene boundary. Science, 327:1214—1218

Schulz M, Mudelsee M. 2002. REDFIT: Estimating red-noise spectra directly from unevenly spaced paleoclimatic time series. Computers and Geosciences, 28(3): 421—426

Sewall JO, van der Wal R S W, van der Zwan K, van Oosterhout C, Dijkstra H A, Scotese CR. 2007. Climate model boundary conditions for four Cretaceous time slices. Clim Past, 3: 647—657

Sha J. 2007. Cretaceous stratigraphy of northeast China: non-marine and marine correlation. Cretaceous Research, 28: 146—170

Skelton P W, Spicer R A, Kelly S P, Gilmour I, 2003. The Cretaceous World. Cambridge (UK): Cambridge University Press. 360

Song T G.1997. Inversion styles in the Songliao Basin (northeast China) and estimation of the degree of inversion. Tectonophysics, 283: 173—188

Sprovieri M, Coccioni R, Lirer F, et al. 2006. Orbital tuning of a lower Cretaceous composite record (Maiolica Formation, central Italy). Paleoceanography, 21(4): PA4212

Sun G, Dilcher D, Zheng S L, Zhou Z K.1998. In search of the first flower: a Jurassic angiosperm, Archaeofructus, from Northeast China. Science, 282: 1692—1695

Sun G, Ji Q, Dilcher D, et al. 2002. Archaefructus, a new basal angiosperm family. Science, 296: 899—904

Tarduno J A, Brinkman D B, Renne P R, et al. 1998. Evidence for extreme climatic warmth from Late Cretaceous arctic vertebrates. Science, 282(5397): 2241—2243

Tian ZY, Han P. 1993. Structure analysis and formation mechanism of Meso-Cenozoic Basin in East China. Petroleum Exploration and Development, 20: 1—8 (in Chinese with English abstract)

Tissot B P, Welte D H.1984. Petroleum Formation and Occurrence. New York: Springer

Torrence C, Compo G P.1998. A practical guide to wavelet analysis. Bulletin of the American Meteorological Society, 79(1): 61—78

Truman P Y. 2000. Restoration ecology and conservation biology. Biological Conservation, 92: 73—83

Upchurch G R, Otto-Bliesner B L, Scotese C R. 1998. Vegetation–atmosphere interactions and their role in global warming during theLatest Cretaceous. Phil Trans R Soc Lond B, 353: 97—112

van der Zwan C J, Boulter M C, Hubbard R N L B. 1985. Climatic change during the Lower Carboniferous in Euramerica, based on multivariate statistical analysis of palynological data. Palaeogeography, Palaeoclimatology, Palaeoecology, 52: 1—20

Volkman J K, Barrett S M, Blackburn S I, Mansour M P, Sikes E L, Gelin G F. 1998. Microalgal biomarkers: a review of recent research developments. Organic Geochemistry, 29: 1163—1179

Wan X Q, Wignall P B, Zhao W J. 2003. The Cenomanian-Turonian extinction and oceanic anoxic event: evidence from South Tibet. Paleogeography, Paleoclimatology, Paleoecology, 199(3-4):283—298

Wan X Q, Scott R W, Wang P J, He H Y, Deng C L, Feng Z H, Huang Q H. 2013. Late Cretaceous stratigraphy, Songliao Basin, NE China: SK-1 cores. Palaeogeography, Palaeoclimatology, Palaeoecology, 385: 31—43

Wang C S, Feng Z Q, Zhang L M, Huang Y J, Cao K, Wang P J, Zhao B. 2013. Cretaceous paleogeography and paleoclimate and the setting of SKI borehole sites in Songliao Basin, northeast China. Palaeogeography, Palaeoclimatology, Palaeoecology, 385: 17—30

Wang C S, Hu X M, Sarti M, Scott R W, Li X H. 2005. Upper Cretaceous oceanic red beds in southern Tibet: a major change from anoxic to oxic, deep-sea environments. Cretaceous Research, 26: 21—32

Wang C S, Huang Y J, Zhao X X. 2009. Unlocking a Cretaceous geologic and geophysical puzzle: Scientific drilling of Songliao Basin in northeast China.The Leading Edge, 340—344

Wang D, Zhang W, Zhang X X, et al. 2007. Drilling Technology of China Continental Scientific Drilling Project. Beijing: Science Press (in Chinese)

Wang D P, Liu L, Zhang L P, Lv CJ. 1995. The Palaeoclimate, Depositional Cycle and Sequence Stratigraphy of Songliao Basin. Jilin:Jilin University Press (in Chinese)

Wang G D, Cheng R H, Wang P J, et al. 2009. Description of Cretaceous sedimentary sequence of the Quantou Formation recovered by CCSD-SK-Is borehole in Songliao Basin: lithostratigraphy, sedimentary facies and cyclic stratigraphy. Earth Science Frontiers, 16(2):324—338

Wang P X. 2000. Deep sea research and Earth sciences in new century. In: Lu Y X (ed). Overwiew and Perspective of Sciences and Technology in Past 100 Years. Shanghai : Shanghai Education Press

Wang P J, Chen F K, Chen S M, et al. 2006. Geochemical and Nd-Sr-Pb isotopic composition of Mesozoic volcanic rocks in the Songliao Basin, NE China. Geochemical Journal, 40: 1—11

Wang P J, Du X D, Wang J, Wang D P. 1996. Chronostratigraphy and stratigraphic classification of the Cretaceous of the Songliao Basin. Acta Geologica Sinica (English Edition), 9(2): 207—217

Wang P J, Gao Y F, Cheng R H, et al. 2009. Description of Cretaceous sedimentary sequence of the second and third Member of the Qingshankou Formation recovered by CCSD-SK-Is borehole in Songliao Basin: lithostratigraphy, sedimentary facies and cyclic stratigraphy. Earth Science Frontiers,16(2):288—313

Wang P J, Liu W Z, Shan X L, Bian W H, Ren Y G. 2001. Event Sedimentology: Introduction, Example, and Application. Changchun, China (in Chinese):Science & Technology Publishing House

Wang P J, Liu W Z, Wang D P. 1994. The application of mudstone bulk chemical composition statistics method to the Songliao basin analysis (Cretaceous, North East China). Sedimentary Facies and Palaeogeography, 14: 55—64(in Chinese with English Abstract)

Wang P J, Liu W Z, Yin X Y, et al. 2002b. Marine ingressive events recorded in epicontinental sequences: example from the Cretaceous Songliao Basin of NE China in comparison with the Triassic Central Europe Basin of SW Germany. J Geosci Res NE Asia, 5(1): 35—42

Wang P J, Liu W Z, Wang S X, Song W H. 2002a. $^{40}Ar/^{39}Ar$ and K/Ar dating on the volcanic rocks in the Songliao basin, NE China: constraints on stratigraphy and basin dynamics. International Journal of Earth Sciences, 91:331—340

Wang P J, Xie X A, Frank M, et al. 2007. The Cretaceous Songliao Basin: valcanogenic succession, sedimentary sequence and tectonic evolution, NE China. Acta Geologica Sinica, 81: 1002—1011

Wang Z, Lu H N, Zhao C B. 1985. Cretaceous Charophytes from Songliao Basin and Adjacent Areas. Heilongjiang (in Chinese): Heilongjiang Science and Technology Press

Watson M P, Hayward A B, Parkinson D N, et al. 1987. Plate tectonic history, basin development and petroleum source rock deposition onshore China, Marine and Petroleum Geology, 4:205—225

Weissert H, Lini A, Follmi K B, Kuhn O.1998. Correlation of Early Cretaceous carbon isotope stratigraphy and platform drowning events: a possible link? Palaeogeography, Palaeoclimatology, Palaeoecology, 137:189—203

White J M, Ager T A, Adam D P, Leopold E B, Liu G, Jette H, Schweger C E. 1997. An 18 million year record of vegetation and climate change in northwestern Canada and Alaska: tectonic and global climatic correlates. Palaeogeography, Palaeoclimatology, Palaeoecology, 130: 293—306

Wilson P A, Norris R D. 2001. Warm tropical ocean surface and global anoxia during the mid-Cretaceous period. Nature, 412: 425—428

Wilson P A, Norris R D, Cooper M J. 2002. Testing the Cretaceous greenhouse hypothesis using glassy foraminiferal calcite from the core of the Turonian tropics on Demerara Rise. Geology, 30(7):607—610

Wu H C, Zhang Sh H, Huang Q H. 2008. Establishment of floating astronomical time scale for the terrestrial Late Cretaceous Qingshankou Formation in the Songliao Basin of Northeast China. Earth Science Frontiers, 15(4):159—169

Wu H C, Zhang S H, Sui S W, et al. 2007. Recognition of milankovitch cycles in the natural gamma—ray logging of Upper Cretaceous terrestrial strata in the Songliao Basin. Acta Geologica Sinica, 81(6): 996—1001

Wu H C, Zhang S H, Jiang G Q, Huang Q H. 2009. The floating astronomical time scale for the terrestrial Late Cretaceous Qingshankou Formation from the Songliao Basin of Northeast China and its stratigraphic and paleoclimate implications. Earth and Planetary Science Letters, 308—323

Xie X N, Jiao J J, Tang Z H, Zheng C M. 2003. Evolution of abnormally low pressure and its implications for the hydrocarbon system in the Southeast Uplift zone of Songliao Basin, China. AAPG Bull, 87: 99—119

Xu Z Q, Yang J S, Zhang Z M, et al. 2005. Completion and achievement of the Chinese Continental Scientific Drilling (CCSD) Project. Geology in China, 32(2):177—183

Yang W L.1985. Daqing oil field, People's Republic of China; a giant field with of nomarine origin. AAPG Bulletin, 69(7): 1101—1111

Ye D Q, Zhong X C. 1990. Cretaceous Strata in Oil-bearing Provinces. Beijing:Petroleum Industry Press

Ye D Q, Huang Q H, Zhang Y, Chen C R. 2002. Cretaceous Ostracoda Biostratigraphy in Songliao Basin. Beijing (in Chinese):Petroleum Industry Press.

Zakharov Y D, Boriskina N G, Ignatyev A V, Tanabe K, Shigeta Y, Popov A M, Afanasyeva T B, Maeda H. 1999. Palaeotemperature curve for the Late Cretaceous of the northwestern circum-Pacific. Cretaceous Research, 20: 685—697

Zakharov Y D, Sha J G, Popov A M, Safronov P P, Shorochova S A, Volynets E B, Biakov A S. Burago V I, Zimina V G, Konovalova I V. 2009. Permian to earliest Cretaceous climatic oscillations in the eastern Asian continental margin (Sikhote-Alin area), as indicated by fossils and isotope data. GFF 131: 25—47

Zakharov Y D, Shigeta Y, Popov A M, Velivetskaya T A, Afanasyeva T B. 2011. Cretaceous climatic oscillations in the Bering area (Alaska and Koryak Upland): isotopic and palaeontological evidence. Sedimentary Geology, 235: 122—131

Zhang Y Y, Bao L N. 2009. Cretaceous phytoplankton assemblages from Songke Core-1, North and South (SK-1, N and S) of Songliao Basin, Northeast China. Acta petrologica sinica,83(5): 868—874

Zhang Z M, Shen K, Xiao Y L, Hoefs J, Liou J G.2006. Mineral and fluid inclusions in zircon of UHP metamorphic rocks from the CCSD-main drill hole: A record of metamorphism and fluid activity. Lithos, 92(3-4):378—398

Zhang Z M, Xiao Y L, Shen K, Gao Y J. 2005. Garnet growth compositional zonation and metamorphic P-T path of the ultrahigh-pressure eclogites from the Sulu orogenic belt, eastern Central China. Acta Petrologica Sinica, 21(3):809—818

Zhang M M, Zhou J J. 1976. Discovery of Lycoptera. Vertebratology and Anthropotology 14(3): 146—153

Zhang M M, Zhou J J, Liu Z C. 1977. Age and deposition environment of Cretaceous fish fossil-bearing strata in NE China. Vertebratology and Anthropotology, 15(3): 194—197

Zhang W, Chen P, Palmer A R. 2003. Biostratigraphy of China. Beijing:Science Press of China

Zhang Y Y. 1999. The evolutionary succession of Cretaceous angiosperm pollen in China. Acta Palaeontologica Sinica, 38: 435—453

Appendix

Appendix 1 Core Information Table of SK-1s

Barrel	Barrel depth/m	Strata	Coring technique	Coring footage /m	Core length /m	Recovery ratio /%	Total footage (test Barrels not included) /%	Total core length (test Barrels not included) /%	Total recovery ratio (test Barrels not included) /%
Test 1	955.00—959.70	K_2n^2	Regular coring	4.7	3.21	68.3			
Test 2	959.70—963.58	K_2n^2	Confined coring	3.88	5.05	130.15			
Test 3	963.58—967.96	K_2n^2	Directional	4.38	0	0			
Test 4	967.96—968.17	K_2n^2	Confined coring	0.21	4.55	2166.7			
1	968.17—968.27	K_2n^2	Directional	0.1	0	0	0.1	0	0
2	968.27—971.76	K_2n^2	Directional	3.49	0	0	3.59	0	0
3	961.76—972.26	K_2n^2	Confined coring	0.5	4.09	818	4.09	4.09	100
4	972.26—981.27	K_2n^2	Regular coring	9.01	8.62	95.67	13.1	12.71	97.02
5	981.27—991.07	K_2n^2	Regular coring	9.8	10.19	103.98	22.9	22.9	100
6	991.07—1000.62	K_2n^2	Regular coring	9.55	9.55	100	32.45	32.45	100
7	1000.62—1011.86	K_2n^2	Regular coring	11.24	11.24	100	43.69	43.69	100
8	1011.86—1022.50	K_2n^{1+2}	Regular coring	10.64	10.49	98.59	54.33	54.18	99.72
9	1022.50—1025.13	K_2n^1	Directional	2.63	2.63	100	56.96	56.81	99.74
10	1025.13—1035.96	K_2n^1	Regular coring	10.83	5.19	47.92	67.79	62	91.46
11	1035.96—1036.28	K_2n^1	Regular coring	0.32	4.09	1278.13	68.11	66.09	97.03
12	1036.28—1048.24	K_2n^1	Regular coring	11.96	12.11	101.3	80.07	78.2	97.66
13	1048.24—1060.25	K_2n^1	Regular coring	12.01	12.01	100	92.08	90.21	97.97
14	1060.25—1065.19	K_2n^1	Confined coring	4.94	4.94	100	97.02	95.15	98.07
15	1065.19—1069.21	K_2n^1	Confined coring	4.02	4.02	100	101.04	99.17	98.15
16	1069.21—1074.32	K_2n^1	Confined coring	5.11	5.11	100	106.15	104.28	98.24
17	1074.32—1078.82	K_2n^1	Confined coring	4.5	4.5	100	110.65	108.78	98.31

Continued

Barrel	Barrel depth/m	Strata	Coring technique	Coring footage/m	Core length/m	Recovery ratio/%	Total footage (test Barrels not included)/%	Total core length (test Barrels not included)/%	Total recovery ratio (test Barrels not included)/%
18	1078.82—1083.50	K_2n^1	Confined coring	4.68	4.68	100	115.33	113.46	98.38
19	1083.50—1087.96	K_2n^1	Confined coring	4.46	4.46	100	119.79	117.92	98.44
20	1087.96—1092.54	K_2n^1	Confined coring	4.58	4.58	100	124.37	122.5	98.5
21	1092.54—1097.12	K_2n^1	Confined coring	4.58	4.58	100	128.95	127.08	98.55
22	1097.12—1100.57	K_2n^1	Confined coring	3.45	3.45	100	132.4	130.53	98.69
23	1100.57—1105.14	K_2n^1	Directional	4.57	3.35	73.3	136.97	133.88	97.71
24	1105.14—1105.24	K_2n^1	Confined coring	0.1	0.67	670	137.07	134.55	98.16
25	1105.24—1116.22	K_2n^1	Regular coring	10.98	10.98	100	148.05	145.53	98.3
26	1116.22—1127.33	K_2y^{2+3}	Regular coring	11.11	11.11	100	159.64	156.64	98.42
27	1127.33—1139.54	K_2y^{2+3}	Regular coring	12.21	12.21	100	171.37	168.85	98.53
28	1139.54—1150.54	K_2y^{2+3}	Regular coring	11	11	100	182.37	179.85	98.62
29	1150.54—1153.42	K_2y^{2+3}	Directional	2.88	2.7	98.75	185.25	182.55	98.54
30	1153.42—1165.25	K_2y^{2+3}	Regular coring	11.83	11.95	101.01	197.08	194.5	98.69
31	1165.25—1169.92	K_2y^{2+3}	Regular coring	4.67	4.67	100	201.75	199.17	98.72
32	1169.92—1181.35	K_2y^{2+3}	Regular coring	11.43	11.43	100	213.18	210.6	98.79
33	1181.35—1188.13	K_2y^{2+3}	Regular coring	6.78	6.78	100	219.96	217.38	98.83
34	1188.13—1200.72	K_2y^{2+3}	Regular coring	12.59	12.4	98.49	232.55	229.78	98.81
35	1200.72—1204.76	K_2y^{2+3}	Directional	4.04	4.22	104.46	236.59	234	98.91
36	1204.76—1215.51	K_2y^{2+3}	Regular coring	10.75	10.75	100	247.34	244.75	98.95
37	1215.51—1227.04	K_2y^{2+3}	Regular coring	11.53	11.53	100	258.87	256.28	99

Continued

Barrel	Barrel depth/m	Strata	Coring technique	Coring footage/m	Core length /m	Recovery ratio /%	Total footage (test Barrels not included) /%	Total core length (test Barrels not included) /%	Total recovery ratio (test Barrels not included) /%
38	1227.04—1238.84	K_2y^{2+3}	Regular coring	11.8	11.8	100	270.67	268.08	99.04
39	1238.84—1249.95	K_2y^{2+3}	Regular coring	11.11	11.11	100	281.78	279.19	99.08
40	1249.95—1256.53	K_2n^1	Directional	6.58	6.58	100	288.36	285.77	99.1
41	1256.53—1268.51	K_2n^1	Regular coring	11.98	11.98	100	300.24	297.75	99.14
42	1268.51—1280.85	K_2n^1	Regular coring	12.34	12.34	100	312.68	310.09	99.17
43	1280.85—1292.63	K_2qn^{2+3}	Regular coring	11.78	11.78	100	324.46	321.87	99.2
44	1292.63—1304.50	K_2qn^{2+3}	Regular coring	11.87	11.87	100	336.33	333.74	99.23
45	1304.50—1310.00	K_2qn^{2+3}	Regular coring	5.5	5.5	100	341.83	339.24	99.24
46	1310.00—1315.00	K_2qn^{2+3}	Directional	5	5	100	346.83	344.24	99.25
47	1315.00—1327.02	K_2qn^{2+3}	Regular coring	12.02	12.02	100	358.85	356.26	99.28
48	1327.02—1339.03	K_2qn^{2+3}	Regular coring	12.01	12.01	100	370.86	368.27	99.3
49	1339.03—1356.21	K_2qn^{2+3}	Regular coring	17.18	16.2	94.3	388.04	384.47	99.08
50	1356.21—1361.84	K_2qn^{2+3}	Directional	5.63	6.6	117.23	393.67	391.07	99.34
51	1361.84—1373.77	K_2qn^{2+3}	Regular coring	11.93	11.93	100	405.6	403	99.36
52	1373.77—1386.14	K_2qn^{2+3}	Regular coring	12.37	12.37	100	417.97	415.37	99.38
53	1386.14—1398.16	K_2qn^{2+3}	Regular coring	12.02	12.02	100	429.99	427.39	99.4
54	1398.16—1410.42	K_2qn^{2+3}	Regular coring	12.26	12.26	100	442.25	439.65	99.41
55	1410.42—1416.93	K_2qn^{2+3}	Directional	6.51	6.51	100	448.76	446.16	99.42
56	1416.93—1429.05	K_2qn^{2+3}	Regular coring	12.12	12.12	100	460.88	458.28	99.44
57	1429.05—1441.41	K_2qn^{2+3}	Regular coring	12.36	12.36	100	473.24	470.64	99.45
58	1441.41—1453.69	K_2qn^{2+3}	Regular coring	12.28	12.28	100	485.52	482.92	99.46

Continued

Appendix | 669

Barrel	Barrel depth/m	Strata	Coring technique	Coring footage/m	Core length /m	Recovery ratio/%	Total footage (test Barrels not included)/%	Total core length (test Barrels not included)/%	Total recovery ratio (test Barrels not included)/%
59	1453.69—1464.48	K_2qn^{2+3}	Regular coring	10.79	10.79	100	496.31	493.71	99.48
60	1464.48—1471.24	K_2qn^{2+3}	Directional	6.76	6.76	100	503.07	500.47	99.48
61	1471.24—1483.55	K_2qn^{2+3}	Regular coring	12.31	12.31	100	515.38	512.78	99.5
62	1483.55—1496.09	K_2qn^{2+3}	Regular coring	12.54	12.54	100	527.92	525.32	99.51
63	1496.09—1508.09	K_2qn^{2+3}	Regular coring	12	12	100	539.92	537.32	99.52
64	1508.09—1516.39	K_2qn^{2+3}	Regular coring	8.3	8.3	100	548.22	545.62	99.53
65	1516.39—1523.23	K_2qn^{2+3}	Directional	6.84	6.84	100	555.06	552.46	99.53
66	1523.23—1535.55	K_2qn^{2+3}	Regular coring	12.32	12.32	100	567.38	564.78	99.54
67	1535.55—1547.78	K_2qn^{2+3}	Regular coring	12.23	12.23	100	579.61	577.01	99.55
68	1547.78—1560.38	K_2qn^{2+3}	Regular coring	12.6	12.6	100	592.21	589.61	99.56
69	1560.38—1569.34	K_2qn^{2+3}	Regular coring	8.96	8.96	100	601.17	598.57	99.57
70	1569.34—1576.07	K_2qn^{2+3}	Directional	6.73	6.73	100	607.9	605.3	99.57
71	1576.07—1588.23	K_2qn^{2+3}	Regular coring	12.13	12.13	100	620.06	617.46	99.58
72	1588.23—1598.23	K_2qn^{2+3}	Regular coring	12.16	12.16	100	630.06	627.46	99.59
73	1598.23—1610.49	K_2qn^{2+3}	Regular coring	12.26	12.26	100	642.32	639.73	99.6
74	1610.49—1620.97	K_2qn^{2+3}	Regular coring	10.48	10.48	100	652.8	650.2	99.6
75	1620.97—1627.69	K_2qn^{2+3}	Directional	6.72	6.72	100	659.52	656.92	99.61
76	1627.69—1639.11	K_2qn^{2+3}	Regular coring	11.42	11.42	100	670.94	668.34	99.61
77	1639.11—1646.01	K_2qn^{2+3}	Regular coring	6.9	6.9	100	677.84	675.24	99.62
78	1646.01—1647.10	K_2qn^{2+3}	Confined coring	1.09	1.09	100	678.93	676.33	99.62
79	1647.10—1659.10	K_2qn^{2+3}	Regular coring	12	10.4	100	690.93	686.733	99.62

Continued

Barrel	Barrel depth/m	Strata	Coring technique	Coring footage/m	Core length/m	Recovery ratio/%	Total footage (test Barrels not included) /%	Total core length (test Barrels not included) /%	Total recovery ratio (test Barrels not included) /%
80	1659.10—1671.87	K_2qn^{2+3}	Regular coring	12.77	14.37	112.53	703.7	701.1	99.63
81	1671.87—1678.73	K_2qn^{2+3}	Directional	6.86	6.86	100	710.56	707.96	99.63
82	1678.73—1690.77	K_2qn^{2+3}	Regular coring	12.04	12.04	100	722.6	720	99.64
83	1690.77—1702.77	K_2qn^{2+3}	Regular coring	12	12	100	734.6	732	99.65
84	1702.77—1714.47	K_2qn^{2+3}	Regular coring	11.7	11.7	100	746.3	743.7	99.65
85	1714.47—1715.27	K_2qn^{2+3}	Confined coring	0.8	0.8	100	747.1	744.5	99.65
86	1715.27—1727.03	K_2qn^{2+3}	Regular coring	11.76	11.76	100	758.86	756.26	99.66
87	1727.03—1739.54	K_2qn^{2+3}	Regular coring	12.51	12.51	100	771.37	768.77	99.66
88	1739.54—1751.63	K_2qn^{2+3}	Regular coring	12.09	12.09	100	783.46	780.86	99.67
89	1751.63—1758.46	K_2qn^{2+3}	Directional	6.83	6.83	100	790.29	787.69	99.67
90	1758.46—1771.22	K_2qn^{2+3}	Regular coring	12.76	12.76	100	803.05	800.45	99.68
91	1771.22—1782.93	K_2qn^{2+3}	Regular coring	11.71	11.71	100	814.76	812.16	99.68
92	1782.93—1794.44	K_2q^4	Regular coring	11.51	11.51	100	826.27	823.67	99.69
93	1794.44—1806.32	K_2q^4	Regular coring	11.88	11.88	100	838.15	835.55	99.69
94	1806.32—1813.20	K_2q^4	Directional	6.88	6.88	100	845.03	842.43	99.69
95	1813.20—1821.88	K_2q^4	密闭	8.68	8.68	100	853.71	851.11	99.7
96	1821.88—1834.20	K_2q^4	Regular coring	12.32	12.32	100	866.03	863.43	99.7
97	1834.20—1846.11	K_2q^4	Regular coring	11.91	11.91	100	877.94	875.34	99.7
98	1846.11—1857.61	K_2q^4	Regular coring	11.5	11.5	100	889.44	886.84	99.71
99	1857.61—1864.54	K_2q^4	Directional	6.93	6.93	100	896.37	893.77	99.71
100	1864.54—1877.08	K_2q^4	Regular coring	12.54	12.54	100	908.91	906.31	99.71

Continued

Barrel	Barrel depth/m	Strata	Coring technique	Coring footage/m	Core length/m	Recovery ratio/%	Total footage (test Barrels not included) /%	Total core length (test Barrels not included) /%	Total recovery ratio (test Barrels not included) /%
101	1877.08—1889.59	K_2q^3	Regular coring	12.51	12.51	100	921.42	918.82	99.72
102	1889.59—1902.10	K_2q^3	Regular coring	12.51	12.51	100	933.93	931.33	99.72
103	1902.10—1911.70	K_2q^3	Regular coring	9.6	9.6	100	943.53	940.93	99.72
104	1911.70—1915.00	K_2q^3	Directional	3.3	3.3	100	946.83	944.23	99.73
Coring statistics	Regual coring: 71; barrels: 793.14 m; Directional: 19; barrels 95.44 m; Confined coring: 13; barrels: 46.97 m; sealed coring: 8.68 m; total footage: 946.83 m; total core length: 944.23 m; total core recovery ratio: 99.73% Note: Test coring statistics not included								

Appendix 2 Core Information Table of SK-1n

Barrel	Barrel depth/m	Strata	Coring technique	Coring footage/m	Core length/m	Core recovery ratio/%	Coring footage/m	Core length/m	Total core recovery ratio/%
1	163.28—165.31	Nf	Confined coring	0.54	0.54	100.00	0.54	0.54	100.00
2	165.31—165.85	Nf	Confined coring	0.54	0.40	74.07	1.08	0.94	87.04
3	165.85—166.95	Nf	Confined coring	1.10	0.60	54.55	2.18	1.54	70.64
4	166.95—167.87	Nf	Regual coring	0.92	0.92	100.00	3.10	2.46	79.35
5	167.87—168.70	Nf	Regual coring	0.83	0.50	60.24	3.93	2.96	75.32
6	168.70—169.36	Nf	Regual coring	0.66	0.62	93.94	4.59	3.58	78.00
7	169.36—170.19	Nf	Regual coring	0.83	0.83	100.00	5.42	4.41	81.37
8	170.19—171.09	Nf	Regual coring	0.90	0.64	71.11	6.32	5.05	79.91
9	171.09—172.27	Nf	Confined coring	1.18	0.60	50.85	7.50	5.65	75.33
10	172.27—173.01	Nf	Confined coring	0.74	0.70	94.59	8.24	6.35	77.06
11	173.01—173.53	Nf	Confined coring	0.52	0.52	100.00	8.76	6.87	78.42
12	173.53—174.70	Nf	Confined coring	1.17	1.17	100.00	9.93	8.04	80.97
13	174.70—175.25	Nf	Confined coring	0.55	0.50	90.91	10.48	8.54	81.49
14	175.25—176.18	Nf	Confined coring	0.93	0.93	100.00	11.41	9.47	83.00
15	176.18—177.44	Nf	Confined coring	1.26	0	0	12.67	9.47	74.74
16	177.44—177.96	Nf	Confined coring	0.52	1.78	342.31	13.19	11.25	85.29
17	177.96—179.16	Nf	Confined coring	1.20	0.60	50.00	14.39	11.85	82.35
18	179.16—179.72	Nf	Confined coring	0.56	0.56	100.00	14.95	12.41	83.01
19	179.72—180.30	Nf	Confined coring	0.58	0.45	77.59	15.53	12.86	82.81
20	180.30—180.57	Nf	Confined coring	0.27	0.27	100.00	15.80	13.13	83.10
21	180.57—181.26	Nf	Confined coring	0.69	0.66	95.65	16.49	13.79	83.63

Appendix | 673

Continued

Barrel	Barrel depth/m	Strata	Coring technique	Coring footage/m	Core length/m	Core recovery ratio/%	Coring footage/m	Core length/m	Total core recovery ratio/%
22	181.26—182.32	Nf	Confined coring	1.06	1.06	100.00	17.55	14.85	84.62
23	182.32—183.62	Nf	Confined coring	1.30	1.30	100.00	18.85	16.15	85.68
24	183.62—184.71	Nf	Confined coring	1.09	1.03	94.50	19.94	17.18	86.16
25	184.71—186.08	Nf	Confined coring	1.37	0.38	27.74	21.31	17.56	82.40
26	186.08—187.35	Nf	Confined coring	1.27	1.23	96.85	22.58	18.79	83.22
27	187.35—188.14	Nf	Confined coring	0.79	0.46	58.23	23.37	19.25	82.37
28	188.14—188.80	Nf	Confined coring	0.66	0	0	24.03	19.25	80.11
29	188.80—189.46	Nf	Confined coring	0.66	0.65	98.48	24.69	19.90	80.60
30	189.46—189.88	Nf	Confined coring	0.42	0	0	25.11	19.90	79.25
31	189.88—190.22	Nf	Confined coring	0.34	0.45	132.35	25.45	20.35	79.96
32	190.22—191.22	Nf	Confined coring	1.00	0.89	89.00	26.45	21.24	80.30
33	191.22—192.20	Nf	Confined coring	0.98	0.60	61.22	27.43	21.84	79.62
34	192.2—192.99	Nf	Confined coring	0.79	0.80	101.27	28.22	22.64	80.23
35	192.99—193.54	Nf	Confined coring	0.55	0.50	90.91	28.77	23.14	80.43
36	193.54—194.41	Nf	Confined coring	0.87	0.44	50.57	29.64	23.58	79.55
37	194.41—194.69	Nf	Confined coring	0.28	0	0	29.92	23.58	78.81
38	194.69—195.52	Nf	Confined coring	0.83	1.32	159.04	30.75	24.90	80.98
39	195.52—196.55	Nf	Confined coring	1.03	0.63	61.17	31.78	25.53	80.33
40	196.55—197.52	Nf	Confined coring	0.97	0.32	32.99	32.75	25.85	78.93
41	197.52—198.05	Nf	Confined coring	0.53	0.53	100.00	33.28	26.38	79.27
42	198.05—198.69	Nf	Confined coring	0.64	0.64	100.00	33.92	27.02	79.66

Continued

Barrel	Barrel depth/m	Strata	Coring technique	Coring footage/m	Core length/m	Core recovery ratio/%	Coring footage/m	Core length/m	Total core recovery ratio/%
43	198.69—199.44	Nf	Confined coring	0.75	0.49	65.33	34.67	27.51	79.35
44	199.44—200.36	Nf	Confined coring	0.92	0.88	95.65	35.59	28.39	79.77
45	200.36—201.34	Nf	Confined coring	0.98	0.93	94.90	36.57	29.32	80.18
46	201.34—201.88	Nf	Confined coring	0.54	0.39	72.22	37.11	29.71	80.06
47	201.88—202.77	Nf	Confined coring	0.89	0.89	100.00	38.00	30.60	80.53
48	202.77—203.74	Nf	Confined coring	0.97	0.97	100.00	38.97	31.57	81.01
49	203.74—204.81	Nf	Confined coring	1.07	0.80	74.77	40.04	32.37	80.84
50	204.81—205.88	Nf	Confined coring	1.07	1.07	100.00	41.11	33.44	81.34
51	205.88—207.07	Nf	Confined coring	1.19	1.19	100.00	42.30	34.63	81.87
52	207.07—208.24	Nf	Confined coring	1.17	1.17	100.00	43.47	35.80	82.36
53	208.24—209.43	Nf	Confined coring	1.19	0.77	64.71	44.66	36.57	81.89
54	209.43—210.28	Nf	Confined coring	0.85	0.25	29.41	45.51	36.82	80.91
55	210.28—210.51	Nf	Confined coring	0.23	0	0	45.74	36.82	80.50
56	210.51—210.57	Nf	Confined coring	0.06	0.69	1150.00	45.80	37.51	81.90
57	210.57—210.93	K_2m^2	Confined coring	0.36	0	0	46.16	37.51	81.26
58	210.93—211.44	K_2m^2	Confined coring	0.51	0.78	152.94	46.67	38.29	82.04
59	211.44—212.48	K_2m^2	Confined coring	1.04	0	0	47.71	38.29	80.26
60	212.48—213.03	K_2m^2	Regual coring	0.55	0.87	158.18	48.26	39.16	81.14
61	213.03—214.33	K_2m^2	Regual coring	1.30	0	0	49.56	39.16	79.02
62	214.33—214.62	K_2m^2	Regual coring	0.29	1.59	548.28	49.85	40.75	81.75
63	214.62—216.01	K_2m^2	Confined coring	1.39	0	0	51.24	40.75	79.53

Continued

Barrel	Barrel depth/m	Strata	Coring technique	Coring footage/m	Core length/m	Core recovery ratio/%	Coring footage/m	Core length/m	Total core recovery ratio/%
64	216.01—216.48	K_2m^2	Regual coring	0.47	1.40	297.87	51.71	42.15	81.51
65	216.48—217.52	K_2m^2	Confined coring	1.04	0.70	67.31	52.75	42.85	81.23
66	217.52—218.58	K_2m^2	Confined coring	1.06	1.40	132.08	53.81	44.25	82.23
67	218.58—220.09	K_2m^2	Confined coring	1.51	1.15	76.16	55.32	45.40	82.07
68	220.09—221.29	K_2m^2	Confined coring	1.20	1.40	116.67	56.52	46.80	82.80
69	221.29—222.85	K_2m^2	Confined coring	1.56	1.56	100.00	58.08	48.36	83.26
70	222.85—224.62	K_2m^2	Confined coring	1.77	0	0	59.85	48.36	80.80
71	224.62—225.02	K_2m^2	Regual coring	0.40	1.56	390.00	60.25	49.92	82.85
72	225.02—226.36	K_2m^2	Confined coring	1.34	1.58	117.91	61.59	51.50	83.62
73	226.36—227.59	K_2m^2	Confined coring	1.23	1.23	100.00	62.82	52.73	83.94
74	227.59—229.19	K_2m^2	Confined coring	1.60	1.60	100.00	64.42	54.33	84.34
75	229.19—231.12	K_2m^2	Confined coring	1.93	1.00	51.81	66.35	55.33	83.39
76	231.12—232.09	K_2m^2	Confined coring	0.97	1.27	130.93	67.32	56.60	84.08
77	232.09—233.56	K_2m^2	Confined coring	1.47	0	0	68.79	56.60	82.28
78	233.56—233.85	K_2m^2	Regual coring	0.29	0	0	69.08	56.60	81.93
79	233.85—234.24	K_2m^2	Regual coring	0.39	0.51	130.77	69.47	57.11	82.21
80	234.24—234.79	K_2m^2	Regual coring	0.55	0.42	76.36	70.02	57.53	82.16
81	234.79—235.78	K_2m^2	Regual coring	0.99	1.12	113.13	71.01	58.65	82.59
82	235.78—236.98	K_2m^2	Regual coring	1.20	1.10	91.67	72.21	59.75	82.74
83	236.98—237.35	K_2m^2	Regual coring	0.37	0.47	127.03	72.58	60.22	82.97
84	237.35—238.15	K_2m^2	Regual coring	0.80	0.73	91.25	73.38	60.95	83.06

Continued

Barrel	Barrel depth/m	Strata	Coring technique	Coring footage/m	Core length/m	Core recovery ratio/%	Coring footage/m	Core length/m	Total core recovery ratio/%
85	238.15—239.18	K_2m^2	Regual coring	1.03	0.53	51.46	74.41	61.48	82.62
86	239.18—240.04	K_2m^2	Regual coring	0.86	0.80	93.02	75.27	62.28	82.74
87	240.04—241.23	K_2m^2	Regual coring	1.19	1.19	100.00	76.46	63.47	83.01
88	241.23—242.73	K_2m^2	Regual coring	1.50	1.23	82.00	77.96	64.70	82.99
89	242.73—244.00	K_2m^2	Regual coring	1.27	0.90	70.87	79.23	65.60	82.80
90	244.00—245.00	K_2m^2	Regual coring	1.00	1.11	111.00	80.23	66.71	83.15
91	245.00—245.82	K_2m^2	Regual coring	0.82	0	0	81.05	66.71	82.31
92	245.82—245.92	K_2m^2	Regual coring	0.10	0.90	900.00	81.15	67.61	83.31
93	245.92—247.42	K_2m^2	Regual coring	1.50	1.49	99.33	82.65	69.10	83.61
94	247.42—248.79	K_2m^2	Regual coring	1.37	1.36	99.27	84.02	70.46	83.86
95	248.79—251.04	K_2m^2	Regual coring	2.25	2.18	96.89	86.27	72.64	84.20
96	251.04—256.18	K_2m^2	Regual coring	5.14	5.04	98.05	91.41	77.68	84.98
97	256.18—258.08	K_2m^2	Confined coring	1.90	1.80	94.74	93.31	79.48	85.18
98	258.08—260.24	K_2m^2	Confined coring	2.16	2.16	100.00	95.47	81.64	85.51
99	260.24—262.43	K_2m^2	Regual coring	2.19	2.16	98.63	97.66	83.80	85.81
100	262.43—266.35	K_2m^2	Regual coring	3.92	3.80	96.94	101.58	87.60	86.24
101	266.35—271.49	K_2m^2	Regual coring	5.14	5.10	99.22	106.72	92.70	86.86
102	271.49—273.83	K_2m^2	Regual coring	2.34	2.30	98.29	109.06	95.00	87.11
103	273.83—281.61	K_2m^2	Regual coring	7.78	7.78	100.00	116.84	102.78	87.97
104	281.61—287.20	K_2m^2	Regual coring	5.59	1.20	21.47	122.43	103.98	84.93
105	287.20—287.81	K_2m^2	Regual coring	0.61	0.70	114.75	123.04	104.68	85.08

Continued

Barrel	Barrel depth/m	Strata	Coring technique	Coring footage/m	Core length/m	Core recovery ratio/%	Coring footage/m	Core length/m	Total core recovery ratio/%
106	287.81—293.49	K_2m^2	Regual coring	5.68	5.83	102.64	128.72	110.51	85.85
107	293.49—300.28	K_2m^2	Regual coring	6.79	6.70	98.67	135.51	117.21	86.50
108	300.28—303.25	K_2m^2	Regual coring	2.97	2.97	100.00	138.48	120.18	86.79
109	303.25—309.51	K_2m^2	Regual coring	6.26	6.00	95.85	144.74	126.18	87.18
110	309.51—313.53	K_2m^2	Regual coring	4.02	4.02	100.00	148.76	130.20	87.52
111	313.53—319.26	K_2m^2	Regual coring	5.73	4.77	83.25	154.49	134.97	87.36
112	319.26—325.44	K_2m^2	Regual coring	6.18	5.26	85.11	160.67	140.23	87.28
113	325.44—329.06	K_2m^2	Regual coring	3.62	2.28	62.98	164.29	142.51	86.74
114	329.06—329.42	K_2m^2	Regual coring	0.36	1.60	444.44	164.65	144.11	87.53
115	329.42—330.01	K_2m^2	Regual coring	0.59	0.68	115.25	165.24	144.79	87.62
116	330.01—333.49	K_2m^2	Regual coring	3.48	3.62	104.02	168.72	148.41	87.96
117	333.49—334.27	K_2m^2	Regual coring	0.78	0.32	41.03	169.50	148.73	87.75
118	334.27—339.95	K_2m^2	Regual coring	5.68	3.62	63.73	175.18	152.35	86.97
119	339.95—340.27	K_2m^2	Regual coring	0.32	2.27	709.38	175.50	154.62	88.10
120	340.27—348.22	K_2m^2	Regual coring	7.95	8.05	101.26	183.45	162.67	88.67
121	348.22—348.98	K_2m^2	Regual coring	0.76	0.45	59.21	184.21	163.12	88.55
122	348.98—353.36	K_2m^2	Regual coring	4.38	3.94	89.95	188.59	167.06	88.58
123	353.36—361.27	K_2m^2	Regual coring	7.91	8.24	104.17	196.50	175.30	89.21
124	361.27—363.80	K_2m^2	Regual coring	2.53	1.74	68.77	199.03	177.04	88.95
125	363.80—368.64	K_2m^2	Regual coring	4.84	2.87	59.30	203.87	179.91	88.25
126	368.64—369.93	K_2m^2	Regual coring	1.29	0.62	48.06	205.16	180.53	87.99

Continued

Barrel	Barrel depth/m	Strata	Coring technique	Coring footage/m	Core length/m	Core recovery ratio/%	Coring footage/m	Core length/m	Total core recovery ratio/%
127	369.93—372.56	K_2m^2	Regual coring	2.63	3.00	114.07	207.79	183.53	88.32
128	372.56—378.83	K_2m^2	Regual coring	6.27	4.53	72.25	214.06	188.06	87.85
129	378.83—380.96	K_2m^2	Regual coring	2.13	1.46	68.54	216.19	189.52	87.66
130	380.96—385.78	K_2m^2	Regual coring	4.82	3.41	70.75	221.01	192.93	87.29
131	385.78—385.83	K_2m^2	Regual coring	0.05	0.15	300.00	221.06	193.08	87.34
132	385.83—393.88	K_2m^2	Regual coring	8.05	7.80	96.89	229.11	200.88	87.68
133	393.88—400.05	K_2m^2	Regual coring	6.17	6.23	100.97	235.28	207.11	88.03
134	400.05—408.11	K_2m^2	Regual coring	8.06	6.90	85.61	243.34	214.01	87.95
135	408.11—414.61	K_2m^2	Regual coring	6.50	6.90	106.15	249.84	220.91	88.42
136	414.61—417.54	K_2m^2	Regual coring	2.93	1.70	58.02	252.77	222.61	88.07
137	417.54—425.44	K_2m^2	Regual coring	7.90	8.44	106.84	260.67	231.05	88.64
138	425.44—433.84	K_2m^2	Regual coring	8.40	8.49	101.07	269.07	239.54	89.03
139	433.84—441.75	K_2m^2	Regual coring	7.91	7.30	92.29	276.98	246.84	89.12
140	441.75—445.01	K_2m^2	Regual coring	3.26	3.00	92.02	280.24	249.84	89.15
141	445.01—445.38	K_2m^2	Regual coring	0.37	0.33	89.19	280.61	250.17	89.15
142	445.38—449.58	K_2m^2	Regual coring	4.20	1.50	35.71	284.81	251.67	88.36
143	449.58—455.59	K_2m^2	Regual coring	6.01	6.38	106.16	290.82	258.05	88.73
144	455.59—461.26	K_2m^2	Regual coring	5.67	5.20	91.71	296.49	263.25	88.79
145	461.26—463.38	K_2m^2	Regual coring	2.12	1.56	73.58	298.61	264.81	88.68
146	463.38—465.82	K_2m^2	Regual coring	2.44	3.09	126.64	301.05	267.90	88.99
147	465.82—468.18	K_2m^2	Confined coring	2.36	1.32	55.93	303.41	269.22	88.73

Continued

Barrel	Barrel depth/m	Strata	Coring technique	Coring footage/m	Core length/m	Core recovery ratio/%	Coring footage/m	Core length/m	Total core recovery ratio/%
148	468.18—468.32	K_2m^2	Regual coring	0.14	0	0	303.55	269.22	88.69
149	468.32—472.53	K_2m^2	Regual coring	4.21	3.96	94.06	307.76	273.18	88.76
150	472.53—479.31	K_2m^2	Regual coring	6.78	6.93	102.21	314.54	280.11	89.05
151	479.31—481.13	K_2m^2	Regual coring	1.82	1.92	105.49	316.36	282.03	89.15
152	481.13—483.71	K_2m^2	Regual coring	2.58	0	0	318.94	282.03	88.43
153	483.71—483.84	K_2m^2	Regual coring	0.13	2.46	1892.31	319.07	284.49	89.16
154	483.84—491.33	K_2m^2	Regual coring	7.49	7.74	103.34	326.56	292.23	89.49
155	491.33—499.20	K_2m^2	Regual coring	7.87	7.50	95.30	334.43	299.73	89.62
156	499.20—507.20	K_2m^2	Regual coring	8.00	8.00	100.00	342.43	307.73	89.87
157	507.20—514.71	K_2m^2	Regual coring	7.51	2.90	38.62	349.94	310.63	88.77
158	514.71—514.99	K_2m^2	Regual coring	0.28	2.20	785.71	350.22	312.83	89.32
159	514.99—516.71	K_2m^2	Regual coring	1.72	1.46	84.88	351.94	314.29	89.30
160	516.71—524.42	K_2m^2	Regual coring	7.71	7.27	94.29	359.65	321.56	89.41
161	524.42—531.61	K_2m^2	Regual coring	7.19	7.39	102.78	366.84	328.95	89.67
162	531.61—539.67	K_2m^2	Regual coring	8.06	7.50	93.05	374.90	336.45	89.74
163	539.67—546.98	K_2m^2	Regual coring	7.31	8.45	115.60	382.21	344.90	90.24
164	546.98—550.58	K_2m^2	Regual coring	3.60	2.90	80.56	385.81	347.80	90.15
165	550.58—557.96	K_2m^2	Regual coring	7.38	7.78	105.42	393.19	355.58	90.43
166	557.96—564.46	K_2m^2	Regual coring	6.50	6.61	101.69	399.69	362.19	90.62
167	564.46—572.58	K_2m^2	Regual coring	8.12	8.07	99.38	407.81	370.26	90.79
168	572.58—581.10	K_2m^2	Regual coring	8.52	8.56	100.47	416.33	378.82	90.99

Continued

Barrel	Barrel depth/m	Strata	Coring technique	Coring footage/m	Core length/m	Core recovery ratio/%	Coring footage/m	Core length/m	Total core recovery ratio/%
169	581.10—588.90	K_2m^2	Regual coring	7.80	7.80	100.00	424.13	386.62	91.16
170	588.90—596.72	K_2m^2	Regual coring	7.82	6.50	83.12	431.95	393.12	91.01
171	596.72—603.71	K_2m^2	Regual coring	6.99	6.25	89.41	438.94	399.37	90.99
172	603.71—609.09	K_2m^2	Regual coring	5.38	5.38	100.00	444.32	404.75	91.09
173	609.09—616.78	K_2m^2	Regual coring	7.69	7.29	94.80	452.01	412.04	91.16
174	616.78—625.30	K_2m^2	Regual coring	8.52	7.60	89.20	460.53	419.64	91.12
175	625.30—632.52	K_2m^2	Regual coring	7.22	7.62	105.54	467.75	427.26	91.34
176	632.52—638.52	K_2m^2	Regual coring	6.00	0	0	473.75	427.26	90.19
177	638.52—638.72	K_2m^2	Regual coring	0.20	6.20	3100.00	473.95	433.46	91.46
178	638.72—646.90	K_2m^2/K_2m^1	Regual coring	8.18	5.23	63.94	482.13	438.69	90.99
179	646.90—648.38	K_2m^1	Regual coring	1.48	4.50	304.05	483.61	443.19	91.64
180	648.38—656.10	K_2m^1	Regual coring	7.72	5.53	71.63	491.33	448.72	91.33
181	656.10—662.92	K_2m^1	Regual coring	6.82	8.51	124.78	498.15	457.23	91.79
182	662.92—670.71	K_2m^1	Regual coring	7.79	8.23	105.65	505.94	465.46	92.00
183	670.71—679.24	K_2m^1	Regual coring	8.53	6.39	74.91	514.47	471.85	91.72
184	679.24—686.39	K_2m^1	Regual coring	7.15	6.31	88.25	521.62	478.16	91.67
185	686.39—694.71	K_2m^1	Regual coring	8.32	3.37	40.50	529.94	481.53	90.87
186	694.71—695.08	K_2m^1	Regual coring	0.37	0.55	148.65	530.31	482.08	90.91
187	695.08—699.27	K_2m^1	Regual coring	4.19	2.32	55.37	534.50	484.40	90.63
188	699.27—704.69	K_2m^1	Regual coring	5.42	6.30	116.24	539.92	490.70	90.88
189	704.69—713.49	K_2m^1	Regual coring	8.80	8.23	93.52	548.72	498.93	90.93

Continued

Barrel	Barrel depth/m	Strata	Coring technique	Coring footage/m	Core length/m	Core recovery ratio/%	Coring footage/m	Core length/m	Total core recovery ratio/%
190	713.49—717.72	K_2m^1	Regual coring	4.23	4.29	101.42	552.95	503.22	91.01
191	717.72—721.66	K_2m^1	Regual coring	3.94	3.39	86.04	556.89	506.61	90.97
192	721.66—730.31	K_2m^1	Regual coring	8.65	5.38	62.20	565.54	511.99	90.53
193	730.31—734.50	K_2m^1	Regual coring	4.19	6.72	160.38	569.73	518.71	91.04
194	734.50—738.77	K_2m^1	Regual coring	4.27	1.10	25.76	574.00	519.81	90.56
195	738.77—743.59	K_2m^1	Regual coring	4.82	7.67	159.13	578.82	527.48	91.13
196	743.59—751.09	K_2m^1	Regual coring	7.50	8.76	116.80	586.32	536.24	91.46
197	751.09—759.99	K_2m^1	Regual coring	8.90	8.43	94.72	595.22	544.67	91.51
198	759.99—768.80	K_2m^1	Regual coring	8.81	8.15	92.51	604.03	552.82	91.52
199	768.80—771.13	K_2m^1	Regual coring	2.33	2.02	86.70	606.36	554.84	91.50
200	771.13—776.91	K_2m^1	Regual coring	5.78	4.15	71.80	612.14	558.99	91.32
201	776.91—783.84	K_2m^1	Regual coring	6.93	7.35	106.06	619.07	566.34	91.48
202	783.84—786.80	K_2m^1	Regual coring	2.96	1.75	59.12	622.03	568.09	91.33
203	786.80—787.89	K_2m^1	Regual coring	1.09	0.96	88.07	623.12	569.05	91.32
204	787.89—795.69	K_2m^1/K_2f	Regual coring	7.80	5.99	76.79	630.92	575.04	91.14
205	795.69—802.63	K_2f	Regual coring	6.94	8.89	128.10	637.86	583.93	91.55
206	802.63—805.46	K_2f	Regual coring	2.83	3.25	114.84	640.69	587.18	91.65
207	805.46—807.47	K_2f	Regual coring	2.01	2.01	100.00	642.70	589.19	91.67
208	807.47—815.08	K_2f	Regual coring	7.61	7.61	100.00	650.31	596.80	91.77
209	815.08—823.61	K_2f	Regual coring	8.53	8.53	100.00	658.84	605.33	91.88
210	823.61—832.23	K_2f	Regual coring	8.62	8.31	96.40	667.46	613.64	91.94

Continued

Barrel	Barrel depth/m	Strata	Coring technique	Coring footage/m	Core length/m	Core recovery ratio/%	Coring footage/m	Core length/m	Total core recovery ratio/%
211	832.23—840.19	K_2f	Regual coring	7.96	8.27	103.89	675.42	621.91	92.08
212	840.19—848.84	K_2f	Regual coring	8.65	7.62	88.09	684.07	629.53	92.03
213	848.84—856.44	K_2f	Regual coring	7.60	8.53	112.24	691.67	638.06	92.25
214	856.44—865.02	K_2f	Regual coring	8.58	5.55	64.69	700.25	643.61	91.91
215	865.02—865.55	K_2f	Regual coring	0.53	1.72	324.53	700.78	645.33	92.09
216	865.55—865.74	K_2f	Regual coring	0.19	1.51	794.74	700.97	646.84	92.28
217	865.74—872.77	K_2f	Regual coring	7.03	0	0	708.00	646.84	91.36
218	872.77—872.87	K_2f	Regual coring	0.10	4.58	4580.00	708.10	651.42	92.00
219	872.87—872.96	K_2f	Regual coring	0.09	1.30	1444.44	708.19	652.72	92.17
220	872.96—877.94	K_2f	Regual coring	4.98	4.70	94.38	713.17	657.42	92.18
221	877.94—886.68	K_2f	Regual coring	8.74	9.04	103.43	721.91	666.46	92.32
222	886.68—890.08	K_2f	Regual coring	3.40	2.46	72.35	725.31	668.92	92.23
223	890.08—890.93	K_2f	Regual coring	0.85	1.08	127.06	726.16	670.00	92.27
224	890.93—898.56	K_2f	Regual coring	7.63	7.53	98.69	733.79	677.53	92.33
225	898.56—898.60	K_2f	Regual coring	0.04	0	0	733.83	677.53	92.33
226	898.60—905.45	K_2f	Regual coring	6.85	6.85	100.00	740.68	684.38	92.40
227	905.45—914.01	K_2f	Regual coring	8.56	8.84	103.27	749.24	693.22	92.52
228	914.01—920.79	K_2f	Regual coring	6.78	6.68	98.53	756.02	699.90	92.58
229	920.79—925.80	K_2f	Regual coring	5.01	4.75	94.81	761.03	704.65	92.59
230	925.8—934.24	K_2f	Regual coring	8.44	8.68	102.84	769.47	713.33	92.70
231	934.24—937.90	K_2f	Regual coring	3.66	2.62	71.58	773.13	715.95	92.60

Continued

Barrel	Barrel depth/m	Strata	Coring technique	Coring footage/m	Core length/m	Core recovery ratio/%	Coring footage/m	Core length/m	Total core recovery ratio/%
232	937.90—945.89	K_2f	Regual coring	7.99	8.83	110.51	781.12	724.78	92.79
233	945.89—947.22	K_2f	Regual coring	1.33	1.33	100.00	782.45	726.11	92.80
234	947.22—947.70	K_2f	Regual coring	0.48	0	0	782.93	726.11	92.74
235	947.70—956.62	K_2f	Regual coring	8.92	4.92	55.16	791.85	731.03	92.32
236	956.62—962.17	K_2f	Regual coring	5.55	3.40	61.26	797.40	734.43	92.10
237	962.17—966.73	K_2f	Regual coring	4.56	9.11	199.78	801.96	743.54	92.72
238	966.73—973.87	K_2f	Regual coring	7.29	7.13	97.81	809.25	750.67	92.76
239	973.87—975.56	K_2f	Regual coring	1.54	1.47	95.45	810.79	752.14	92.77
240	975.56—977.30	K_2f	Regual coring	2.50	1.86	74.40	813.29	754.00	92.71
241	977.30—984.82	K_2f	Regual coring	6.76	6.90	102.07	820.05	760.90	92.79
242	984.82—993.78	K_2f	Regual coring	8.96	8.95	99.89	829.01	769.85	92.86
243	993.78—999.81	K_2f	Regual coring	6.03	5.64	93.53	835.04	775.49	92.87
244	999.81—1008.13	K_2f	Regual coring	8.32	8.83	106.13	843.36	784.32	93.00
245	1008.13—1016.68	K_2f	Regual coring	8.55	8.60	100.58	851.91	792.92	93.08
246	1016.68—1020.76	K_2f/K_2n^5	Regual coring	4.08	4.10	100.49	855.99	797.02	93.11
247	1020.76—1028.49	K_2n^5	Regual coring	7.73	0.56	7.24	863.72	797.58	92.34
248	1028.49—1028.54	K_2n^5	Regual coring	0.05	6.87	13740.00	863.77	804.45	93.13
249	1028.54—1029.50	K_2n^5	Regual coring	0.96	0	0	864.73	804.45	93.03
250	1029.50—1034.81	K_2n^5	Regual coring	5.31	5.31	100.00	870.04	809.76	93.07
251	1034.81—1041.77	K_2n^5	Regual coring	6.96	7.76	111.49	877.00	817.52	93.22
252	1041.77—1047.80	K_2n^5	Regual coring	6.03	5.67	94.03	883.03	823.19	93.22

Continued

Barrel	Barrel depth/m	Strata	Coring technique	Coring footage/m	Core length/m	Core recovery ratio/%	Coring footage/m	Core length/m	Total core recovery ratio/%
253	1047.80—1054.57	K_2n^5	Regual coring	6.77	2.92	43.13	889.80	826.11	92.84
254	1054.57—1058.92	K_2n^5	Regual coring	4.35	3.80	87.36	894.15	829.91	92.82
255	1058.92—1065.89	K_2n^5	Regual coring	6.97	7.85	112.63	901.12	837.76	92.97
256	1065.89—1074.63	K_2n^5	Regual coring	8.74	9.23	105.61	909.86	846.99	93.09
257	1074.63—1076.94	K_2n^5	Regual coring	2.31	2.31	100.00	912.17	849.30	93.11
258	1076.94—1081.43	K_2n^5	Regual coring	4.49	3.91	87.08	916.66	853.21	93.08
259	1081.43—1084.87	K_2n^5	Regual coring	3.44	3.92	113.95	920.10	857.13	93.16
260	1084.87—1094.44	K_2n^5	Regual coring	9.57	9.07	94.78	929.67	866.20	93.17
261	1094.44—1102.31	K_2n^5	Regual coring	7.87	7.55	95.93	937.54	873.75	93.20
262	1102.31—1102.84	K_2n^5	Regual coring	0.53	0.00	0	938.07	873.75	93.14
263	1102.84—1104.25	K_2n^5	Regual coring	1.41	2.08	147.52	939.48	875.83	93.22
264	1104.25—1107.98	K_2n^5	Regual coring	3.73	3.28	87.94	943.21	879.11	93.20
265	1107.98—1110.49	K_2n^5	Regual coring	2.51	0	0	945.72	879.11	92.96
266	1110.49—1112.69	K_2n^5	Regual coring	2.35	0	0	948.07	879.11	92.73
267	1112.69—1113.75	K_2n^5	Regual coring	0.91	5.26	578.02	948.98	884.37	93.19
268	1113.75—1122.00	K_2n^5	Regual coring	8.25	4.43	53.70	957.23	888.80	92.85
269	1122.00—1124.35	K_2n^5	Regual coring	2.35	5.59	237.87	959.58	894.39	93.21
270	1124.35—1132.98	K_2n^5	Regual coring	8.63	6.37	73.81	968.21	900.76	93.03
271	1132.98—1140.12	K_2n^5	Regual coring	7.14	7.10	99.44	975.35	907.86	93.08
272	1140.12—1149.10	K_2n^5	Regual coring	8.98	1.67	18.60	984.33	909.53	92.40
273	1149.10—1149.44	K_2n^5	Regual coring	0.34	3.10	911.76	984.67	912.63	92.68

Continued

Barrel	Barrel depth/m	Strata	Coring technique	Coring footage/m	Core length/m	Core recovery ratio/%	Coring footage/m	Core length/m	Total core recovery ratio/%
274	1149.44—1150.91	K_2n^5	Regual coring	1.47	1.34	91.16	986.14	913.97	92.68
275	1150.91—1159.30	K_2n^5	Regual coring	8.39	9.43	112.40	994.53	923.40	92.85
276	1159.30—1165.26	K_2n^5	Regual coring	5.96	4.30	72.15	1000.49	927.70	92.72
277	1165.26—1167.71	K_2n^5	Regual coring	2.45	1.03	42.04	1002.94	928.73	92.60
278	1167.71—1174.08	K_2n^5	Regual coring	6.37	6.12	96.08	1009.31	934.85	92.62
279	1174.08—1179.49	K_2n^5	Regual coring	5.41	7.34	135.67	1014.72	942.19	92.85
280	1179.49—1185.60	K_2n^5	Regual coring	6.11	1.01	16.53	1020.83	943.20	92.40
281	1185.60—1189.20	K_2n^5	Regual coring	3.60	0.00	0	1024.43	943.20	92.07
282	1189.20—1189.50	K_2n^5	Regual coring	0.30	1.75	583.33	1024.73	944.95	92.21
283	1189.50—1191.49	K_2n^5	Regual coring	1.99	0.92	46.23	1026.72	945.87	92.13
284	1191.49—1192.26	K_2n^5	Regual coring	0.77	1.50	194.81	1027.49	947.37	92.20
285	1192.26—1194.71	K_2n^5	Regual coring	2.45	2.68	109.39	1029.94	950.05	92.24
286	1194.71—1196.77	K_2n^5	Regual coring	2.06	0.49	23.79	1032.00	950.54	92.11
287	1196.77—1197.53	K_2n^5	Regual coring	0.76	1.52	200.00	1032.76	952.06	92.19
288	1197.53—1200.94	K_2n^5	Regual coring	3.41	2.08	61.00	1036.17	954.14	92.08
289	1200.94—1205.23	K_2n^5	Regual coring	4.29	5.62	131.00	1040.46	959.76	92.24
290	1205.23—1213.93	K_2n^5	Regual coring	8.70	8.70	100.00	1049.16	968.46	92.31
291	1213.93—1220.42	K_2n^5	Regual coring	6.49	6.49	100.00	1055.65	974.95	92.36
292	1220.42—1222.25	K_2n^5	Regual coring	1.83	1.51	82.51	1057.48	976.46	92.34
293	1222.25—1229.20	K_2n^5	Regual coring	6.95	6.95	100.00	1064.43	983.41	92.39
294	1229.20—1237.73	K_2n^5	Regual coring	8.53	8.50	99.65	1072.96	991.91	92.45

Continued

Barrel	Barrel depth/m	Strata	Coring technique	Coring footage/m	Core length/m	Core recovery ratio/%	Coring footage/m	Core length/m	Total core recovery ratio/%
295	1237.73—1241.50	K_2n^5	Regual coring	3.77	3.70	98.14	1076.73	995.61	92.47
296	1241.50—1246.73	K_2n^5	Regual coring	5.23	5.18	99.04	1081.96	1000.79	92.50
297	1246.73—1251.42	K_2n^5/K_2n^4	Regual coring	4.69	4.44	94.67	1086.65	1005.23	92.51
298	1251.42—1251.57	K_2n^4	Regual coring	0.15	0.15	100.00	1086.80	1005.38	92.51
299	1251.57—1258.91	K_2n^4	Regual coring	7.34	7.34	100.00	1094.14	1012.72	92.56
300	1258.91—1266.74	K_2n^4	Regual coring	7.83	8.20	104.73	1101.97	1020.92	92.64
301	1266.74—1275.88	K_2n^4	Regual coring	9.14	9.14	100.00	1111.11	1030.06	92.71
302	1275.88—1280.58	K_2n^4	Regual coring	4.70	4.36	92.77	1115.81	1034.42	92.71
303	1280.58—1287.09	K_2n^4	Regual coring	6.51	6.59	101.23	1122.32	1041.01	92.76
304	1287.09—1294.78	K_2n^4	Regual coring	7.69	7.62	99.09	1130.01	1048.63	92.80
305	1294.78—1297.22	K_2n^4	Regual coring	2.44	2.20	90.16	1132.45	1050.83	92.79
306	1297.22—1299.19	K_2n^4	Regual coring	1.97	0	0	1134.42	1050.83	92.63
307	1299.19—1306.21	K_2n^4	Regual coring	7.02	9.14	130.20	1141.44	1059.97	92.86
308	1306.21—1312.07	K_2n^4	Regual coring	5.86	5.58	95.22	1147.30	1065.55	92.87
309	1312.07—1320.94	K_2n^4	Regual coring	8.87	8.83	99.55	1156.17	1074.38	92.93
310	1320.94—1325.37	K_2n^4	Regual coring	4.43	4.20	94.81	1160.60	1078.58	92.93
311	1325.37—1331.22	K_2n^4	Regual coring	5.85	5.46	93.33	1166.45	1084.04	92.93
312	1331.22—1338.82	K_2n^4	Regual coring	7.60	7.58	99.74	1174.05	1091.62	92.98
313	1338.82—1347.84	K_2n^4	Regual coring	9.02	9.24	102.44	1183.07	1100.86	93.05
314	1347.84—1356.42	K_2n^4	Regual coring	8.58	8.47	98.72	1191.65	1109.33	93.09
315	1356.42—1361.62	K_2n^4	Regual coring	5.20	5.25	100.96	1196.85	1114.58	93.13

Continued

Barrel	Barrel depth/m	Strata	Coring technique	Coring footage/m	Core length/m	Core recovery ratio/%	Coring footage/m	Core length/m	Total core recovery ratio/%
316	1361.62—1363.71	K_2n^4	Regual coring	2.09	0	0	1198.94	1114.58	92.96
317	1363.71—1363.81	K_2n^4	Regual coring	0.10	1.17	1170.00	1199.04	1115.75	93.05
318	1363.81—1367.50	K_2n^4	Regual coring	3.69	3.30	89.43	1202.73	1119.05	93.04
319	1367.50—1373.44	K_2n^4	Regual coring	5.94	6.20	104.38	1208.67	1125.25	93.10
320	1373.44—1376.88	K_2n^4	Regual coring	3.44	2.80	81.40	1212.11	1128.05	93.06
321	1376.88—1385.37	K_2n^4	Regual coring	8.49	8.43	99.29	1220.60	1136.48	93.11
322	1385.37—1387.86	K_2n^4	Regual coring	2.49	1.73	69.48	1223.09	1138.21	93.06
323	1387.86—1393.93	K_2n^4	Regual coring	6.07	6.60	108.73	1229.16	1144.81	93.14
324	1393.93—1402.62	K_2n^4	Regual coring	8.69	8.63	99.31	1237.85	1153.44	93.18
325	1402.62—1411.57	K_2n^4	Regual coring	8.95	9.02	100.78	1246.80	1162.46	93.24
326	1411.57—1420.57	K_2n^4	Regual coring	9.00	8.98	99.78	1255.80	1171.44	93.28
327	1420.57—1429.22	K_2n^4	Regual coring	8.65	7.87	90.98	1264.45	1179.31	93.27
328	1429.22—1437.57	K_2n^4/K_2n^3	Regual coring	8.35	8.89	106.47	1272.80	1188.20	93.35
329	1437.57—1446.22	K_2n^3	Regual coring	8.65	8.76	101.27	1281.45	1196.96	93.41
330	1446.22—1455.30	K_2n^3	Regual coring	9.08	8.86	97.58	1290.53	1205.82	93.44
331	1455.30—1464.07	K_2n^3	Regual coring	8.77	8.77	100.00	1299.30	1214.59	93.48
332	1464.07—1466.35	K_2n^3	Regual coring	2.28	2.23	97.81	1301.58	1216.82	93.49
333	1466.35—1474.96	K_2n^3	Regual coring	8.61	8.53	99.07	1310.19	1225.35	93.52
334	1474.96—1483.96	K_2n^3	Regual coring	9.00	9.00	100.00	1319.19	1234.35	93.57
335	1483.96—1492.86	K_2n^3	Regual coring	8.90	8.90	100.00	1328.09	1243.25	93.61
336	1492.86—1502.02	K_2n^3	Regual coring	9.16	8.99	98.14	1337.25	1252.24	93.64

Continued

Barrel	Barrel depth/m	Strata	Coring technique	Coring footage/m	Core length/m	Core recovery ratio/%	Coring footage/m	Core length/m	Total core recovery ratio/%
337	1502.02—1511.06	K_2n^3	Regual coring	9.04	9.09	100.55	1346.29	1261.33	93.69
338	1511.06—1516.68	K_2n^3	Regual coring	5.62	5.62	100.00	1351.91	1266.95	93.72
339	1516.68—1521.35	K_2n^3/K_2n^2	Regual coring	4.67	4.32	92.51	1356.58	1271.27	93.71
340	1521.35—1530.46	K_2n^2	Regual coring	9.11	9.04	99.23	1365.69	1280.31	93.75
341	1530.46—1538.93	K_2n^2	Regual coring	8.47	4.08	48.17	1374.16	1284.39	93.47
342	1538.93—1541.00	K_2n^2	Regual coring	2.07	0.00	0	1376.23	1284.39	93.33
343	1541.00—1541.65	K_2n^2	Regual coring	0.65	4.73	727.69	1376.88	1289.12	93.63
344	1541.65—1550.51	K_2n^2	Regual coring	8.86	9.03	101.92	1385.74	1298.15	93.68
345	1550.51—1559.45	K_2n^2	Regual coring	8.94	9.02	100.89	1394.68	1307.17	93.73
346	1559.45—1562.62	K_2n^2	Regual coring	3.17	3.26	102.84	1397.85	1310.43	93.75
347	1562.62—1571.35	K_2n^2	Regual coring	8.73	8.38	95.99	1406.58	1318.81	93.76
348	1571.35—1579.77	K_2n^2	Regual coring	8.42	8.71	103.44	1415.00	1327.52	93.82
349	1579.77—1588.25	K_2n^2	Regual coring	8.48	8.48	100.00	1423.48	1336.00	93.85
350	1588.25—1597.28	K_2n^2	Regual coring	9.03	9.03	100.00	1432.51	1345.03	93.89
351	1597.28—1606.06	K_2n^2	Regual coring	8.78	8.70	99.09	1441.29	1353.73	93.92
352	1606.06—1606.52	K_2n^2	Regual coring	0.46	0.21	45.65	1441.75	1353.94	93.91
353	1606.52—1615.41	K_2n^2	Regual coring	8.89	9.00	101.24	1450.64	1362.94	93.95
354	1615.41—1624.12	K_2n^2	Regual coring	8.71	8.71	100.00	1459.35	1371.65	93.99
355	1624.12—1632.72	K_2n^2	Regual coring	8.60	8.60	100.00	1467.95	1380.25	94.03
356	1632.72—1641.14	K_2n^2	Regual coring	8.42	8.42	100.00	1476.37	1388.67	94.06
357	1641.14—1649.90	K_2n^2	Regual coring	8.76	8.45	96.46	1485.13	1397.12	94.07

Continued

Barrel	Barrel depth/m	Strata	Coring technique	Coring footage/m	Core length/m	Core recovery ratio/%	Coring footage/m	Core length/m	Total core recovery ratio/%
358	1649.90—1658.38	K_2n^2	Regual coring	8.48	8.48	100.00	1493.61	1405.60	94.11
359	1658.38—1667.02	K_2n^2	Regual coring	8.64	8.64	100.00	1502.25	1414.24	94.14
360	1667.02—1675.55	K_2n^2	Regual coring	8.53	8.26	96.83	1510.78	1422.50	94.16
361	1675.55—1684.13	K_2n^2	Regual coring	8.58	8.65	100.82	1519.36	1431.15	94.19
362	1684.13—1692.80	K_2n^2	Regual coring	8.67	8.67	100.00	1528.03	1439.82	94.23
363	1692.80—1701.15	K_2n^2	Regual coring	8.35	8.30	99.40	1536.38	1448.12	94.26
364	1701.15—1709.56	K_2n^2	Regual coring	8.41	7.16	85.14	1544.79	1455.28	94.21
365	1709.56—1717.01	K_2n^2	Regual coring	7.45	8.59	115.30	1552.24	1463.87	94.31
366	1717.01—1725.72	K_2n^2	Regual coring	8.71	8.63	99.08	1560.95	1472.50	94.33
367	1725.72—1734.39	K_2n^2	Regual coring	8.67	8.60	99.19	1569.62	1481.10	94.36
368	1734.39—1742.58	K_2n^2	Regual coring	8.19	8.19	100.00	1577.81	1489.29	94.39
369	1742.58—1745.62	K_2n^2	Regual coring	3.04	3.00	98.68	1580.85	1492.29	94.40
370	1745.62—1754.14	K_2n^2	Regual coring	8.52	8.71	102.23	1589.37	1501.00	94.44
371	1754.14—1762.95	K_2n^2	Regual coring	8.81	8.65	98.18	1598.18	1509.65	94.46
372	1762.95—1771.53	K_2n^2	Regual coring	8.58	8.74	101.86	1606.76	1518.39	94.50
373	1771.53—1780.06	K_2n^2	Regual coring	8.53	8.53	100.00	1615.29	1526.92	94.53
374	1780.06—1786.34	K_2n^2/K_2n^1	Regual coring	6.28	6.28	100.00	1621.57	1533.20	94.55
375	1786.34—1795.18	K_2n^1	Regual coring	8.84	8.45	95.59	1630.41	1541.65	94.56
Coring statistics	First spud-in 90 barrels, footage 80.23 m, core length 66.71 m, core recovery ratio 83.15%; second spud-in 285 barrels, footage 1550.18 m, core length 1474.94 m, core recovery ratio 95.15%, total footage of first and second spud-in 1630.41 m, total core length 1541.65 m, total core recovery ratio 94.56%								

Acknowledgements

The Continental Scientific Drilling Project of Cretaceous Songliao Basin (SK-1) is an important component of the National Key Basic Research Program (973) "Cretaceous Major Geological Events in Earth Surface System and Greenhouse Climate Change" (short for "Cretaceous 973 Projcet" No. 2006CB701400). We appreciate that the Ministry of Science and Technology of China and the Daqing Oildfield Company Ltd. financially supported the Cretaceous 973 Project and SK-1.

The SK-1 drilling is implemented mainly by China University of Geoscience Beijing and the Daqing Oilfield Company Ltd. Jilin University and Institute of Exploration Techniques, Chinese Academy of Geological Sciences participate in the drilling process. We surely appreciate all the hard work of everyone in the above institutions during drilling site selection, drilling implementation and core storage processes.

The Daqing Drilling Engineering Company No.3 Drilling Branch Company and the Institute of Exploration Techniques, Chinese Academy of Geological Sciences are in charge of the drilling and coring processes of SK-1. The Daqing Drilling Engineering Company No.1 Geological Logging Branch Company is in charge of the geological logging process and geological design of SK-1. The Institute of Exploration and Development of Daqing Oilfield Company Ltd. participates drilling site selection and core handling and storage. The Daqing Drilling Engineering Company Logging Branch Company and the Liaohe Oilfield Logging Company are responsible for the logging process of SK-1. China University of Geosciences Beijing and Jilin University take parts in site selection, geological and engineering design, in-site supervision and core handling. Appreciations are given to the related staff in the above institutions.

We thank supports for the Cretaceous 973 Project from China University of Geosciences Beijing, Daqing Oilfield Company, Jilin University, Institute of Exploration Techniques in Chinese Academy of Geological Sciences, Institute of Geology and Geophysics in Chinese Academy of Sciences, Nanjing Institute of Geology and Palaeontology in Chinese Academy of Sciences, Chengdu University of Technology, China University of Petroleum Beijing.

We thank supports for SK-1 scientific researches by Stanford University and Miami University.

Special appreciations are given to Prof. Wan Gang, Minister of Science and Technology of China, and Dr. Rolf Emmermann, Chair of the ICDP Executive Committee for their forewords to this book.